Surface Modeling
High Accuracy and
High Speed Methods

Applied Ecology
and Environmental Management

A SERIES

Series Editor
Sven E. Jørgensen
Copenhagen University, Denmark

Handbook of Ecological Indicators for Assessment of
Ecosystem Health, Second Edition
Sven E. Jørgensen, Fu-Liu Xu, and Robert Costanza

Surface Modeling: High Accuracy and High Speed Methods
Tian-Xiang Yue

ADDITIONAL VOLUMES IN PREPARATION

Surface Modeling
High Accuracy and
High Speed Methods

Tian-Xiang Yue

CRC Press
Taylor & Francis Group
Boca Raton London New York

CRC Press is an imprint of the
Taylor & Francis Group, an **informa** business

CRC Press
Taylor & Francis Group
6000 Broken Sound Parkway NW, Suite 300
Boca Raton, FL 33487-2742

First issued in paperback 2017

© 2011 by Taylor and Francis Group, LLC
CRC Press is an imprint of Taylor & Francis Group, an Informa business

No claim to original U.S. Government works

ISBN-13: 978-1-4398-1758-2 (hbk)
ISBN-13: 978-1-138-07567-2 (pbk)

Library of Congress Cataloging-in-Publication Data

Yue, Tian-Xiang.
   Surface modeling : high accuracy and high speed methods / author, Tian-Xiang Yue.
      p. cm. -- (Applied ecology and environmental management)
   "A CRC title."
   Includes bibliographical references and index.
   ISBN 978-1-4398-1758-2 (alk. paper)
   1. Geomorphology--Data processing. 2. Earth--Computer simulation. 3. High performance computing. I. Title. II. Series.

GB400.42.E4Y84 2010
551.4109992--dc22
                                       2010028763

Visit the Taylor & Francis Web site at
http://www.taylorandfrancis.com

and the CRC Press Web site at
http://www.crcpress.com

Contents

Section II   Surface Modeling of Environment Components of Earth Surface

Preface

Error problems and slow computational-speed problems are the two critical challenges geographical information systems (GISs) are currently facing. The error problem makes GISs useless for their special applications where high accuracy is required. The slow computational-speed problem makes real-time visualization and dynamical simulation in three dimensions difficult to be realized. In this book, we aim to find solutions to these problems that have long troubled GIS, by developing high-accuracy and high-speed methods for surface modeling.

In Chapter 1, we review the existing classical methods for surface modeling and analyze their shortcomings. In Chapter 2, a novel method of high-accuracy surface modeling (HASM) is developed in terms of the fundamental theorem of surfaces. In Chapter 3, the coefficient matrix is analyzed. It is found that HASM is a sparse and symmetric positive-definite system that has the highest computational speed when the preconditioned conjugate gradient algorithm is used to solve its equation set. However, HASM also has a huge computation cost and slow computational speed, and so has not been widely applied as yet. Thus, we develop an adaptive method of HASM (HASM-AM) in Chapter 4, a multigrid method of HASM (HASM-MG) in Chapter 5, and an adjustment method of HASM (HASM-AD) in Chapter 6. Numerical tests and real-world studies demonstrate that HASM-AM, HASM-MG, and HASM-AD have highly accelerated computational speed, especially for simulations with huge computational work. They also have greatly increased simulation accuracy and are able to provide a solution to the multiscale problem. In Chapter 7, an optimal control method of high-accuracy surface modeling (HASM-OC) is developed to meet the specific requirement of extremely high accuracy (HASM-OC). In Chapter 8, an HASM-based method for dynamic simulation (HASM-FDS) is developed on the basis of matrix factorization in order to realize dynamic simulation in three dimensions. The HASM methods are then applied to simulate terrains in Chapter 9, climate change in Chapter 16, ecosystem change in Chapter 17, land cover in Chapter 18, and soil properties in Chapter 19. HASM's potential for simulating population distribution is demonstrated in Chapter 10, human carrying capacity in Chapter 11, ecosystem services in Chapter 12, ecological diversity in Chapter 13, change detection in Chapter 14, and wind velocity in Chapter 15. In Chapter 20, we review the development of ecological modeling on a global level since the middle of the nineteenth century and examine worldwide collaborations since the middle of the 1980s. Finally, we discuss the problems that exist in surface modeling on a global level, and evaluate possible solutions to these problems.

This book would be helpful for people involved in geographical information science, ecological informatics, landscape ecology, geography, computer-aided design, global earth observation, and planetary surface modeling.

Acknowledgments

This work is supported by the China National Science Fund for Distinguished Young Scholars (40825003) and the National Key Technologies R&D Program of the Ministry of Science and Technology of the People's Republic of China (2006BAC08B). My major collaborators include Dr. Ze-Meng Fan, Dr. Zheng-Ping Du, Dr. Chuan-Fa Chen, Dr. Dun-Jiang Song, Dr. Ying-An Wang, Dr. Yong-Zhong Tian, Dr. Wei Wang, Dr. Shi-Hai Wang, Dr. Qi-Quan Li, Dr. Qing Wang, Yin-Jun Song, Dr. Wen-Jiao Shi, Professor Bing Xu, Dr. Sheng-Nan Ma, Dr. Yi-Min Lu, Dr. Hong-Sheng Huang, Dr. Xiao-Fang Sun, and Dr. Chang-Qing Yan.

Harriet Davis reviewed all the chapters of this book. I would like to express my sincere appreciation to all persons, both named and unnamed, who have helped me complete this book.

Section I

Principles and Methods

1

Introduction

1.1 Surface Modeling

Surface modeling refers to the process of simulating a surface through a scattered point-form data set, a line-form data set and/or an area-form data set. Surface modeling is aimed at formulating an object in a grid system, in which each grid cell contains an estimate of the object that is representative for that particular location. Representing data in grid form has at least four advantages (Martin and Bracken, 1991; Deichmann, 1996; Yue et al., 2008, 2009, 2010): (1) a regular grid can be easily reaggregated to any areal arrangement required; (2) producing ecological data in grid form is one way of ensuring compatibility between heterogeneous data sets; (3) data in grid form makes the fusion of multiresolution and multisource information easier, and (4) converting data into grid form can provide a way to avoid some of the problems imposed by artificial boundaries.

Various methods for surface modeling have been developed since the early 1950s. Krige (1951) proposed the basic idea of Kriging interpolation that belongs to the family of estimation algorithms of linear least-squares and interpolates the value in a random field at an unobserved location from observations of its value at nearby locations. Bengtsson and Nordbeck (1964) suggested a method for linear interpolation in a triangulated irregular network (TIN) to fit irregularly distributed points into a surface in each triangle. Zienkiewicz (1967) developed the finite element method (FEM) and Akima (1978a,b) programmed the FEM algorithm for surface fitting. Shepard (1968) proposed a surface simulation formula, that is, the inverse distance weighted (IDW) method, which is highly dependent on an appropriate weight function (Franke, 1982). Hardy (1971) formulated a multiquadric method (MQM) under the assumption that any smooth mathematical surface and smooth irregular surface might be approximated to any desired degree of exactness by the summation of mathematically defined surfaces. Harder and Desmarais (1972) discovered a method of surface splines, which is based on the small deflection equation of an infinite plate, while Maude (1973) inspired a rectangle-based blending method that generalized the idea of deficient quintic splines to several variables. Triangle-based blending methods were developed for surface estimation by improving IDW (Gold et al., 1977; McLain, 1976). Talmi and Gilat (1977) constructed a method for smooth approximation of data on the basis of the spline idea, which had many advantages compared to polynomial interpolation. A method of Delta Iteration was then proposed for producing a surface approximation (Foley and Nielson, 1980), which is based on Bernstein's polynomial and bicubic spline.

Surface modeling began to be used in the 1960s, with the general availability of computers (Lo and Yeung, 2002), but because it requires powerful software and a large amount of spatially explicit data, its development was limited before the 1990s. The major advances that enabled the use of surface modeling included trend surface analysis (TSA) (Ahlberg et al., 1967; Schroeder and Sjoquist, 1976; Legendre and Legendre, 1983), the digital terrain model (Stott, 1977), surface approximation (Long, 1980), spatial simulation of wetland habitats (Sklar et al., 1985), spatial pattern matching (Costanza, 1989), spatial prediction (Turner et al., 1989), and modeling coastal landscape dynamics (Costanza et al., 1990). Surface modeling has greatly progressed since the early 1990s, with the rapid development of remote sensing (RS) and a geographical information system (GIS), as well as the accumulation of spatially explicit data.

Surface modeling has been extensively used to analyze and understand the spatial phenomena of ecological processes since the 1980s. For example, a dynamic spatial simulation model was designed to project habitat changes as a function of marsh type, hydrology, subsidence, and sediment transport for a generalized coastal wetland area (Sklar et al., 1985). The performance of spatial simulation models was then evaluated to compare spatial patterns (Turner et al., 1989). A surface modeling workstation was also developed to implement and test spatial ecosystem models in a convenient desktop environment (Costanza and Maxwell, 1991). A method for surface modeling was developed for simulating ecosystems with spatial heterogeneity (Gao, 1996). A pairwise potential model was introduced to describe the spatial dependency of dominant grasses and forbs (Reich et al., 1997). A patch-based spatial modeling approach was developed to take into account spatial heterogeneity and its effects on ecological processes (Wu and Levin, 1997). A daily weather generator was proposed to operate at a half-degree scale of the Earth's terrestrial surface (Friend, 1998). The spatial models for habitat selection of marsh-breeding bird species were developed using artificial neural networks (Oezesmi and Oezesmi, 1999). GIS-based spatial modeling was used to simulate geographic distribution of wildlife populations (Ji and Jeske, 2000). An integrated model was developed to study the spatial interactions between soil and groundwater that may affect the nitrogen delivery to stream water in rural catchments (Beaujouan et al., 2001).

Various surface models of the global carbon cycle were developed on a global level, and these models are very sensitive to the spatial distributions of productivity, living biomass, and soil organics (Haxeltine and Prentice, 1996; Svirezhev, 2002). For example, a spatially explicit grid-based model was developed to explore how alterations to the disturbance regime (past and future) may influence the landscape structure (Perry and Enright, 2002). Similarly, regression modeling has been used to establish relationships between a response variable and a set of spatial predictors (Lehmann et al., 2003). A comprehensive review of spatial forest planning demonstrated that a hybrid-modeling technique would be very important for the development of spatial forest modeling in the future (Baskent and Keles, 2005). Principal coordinates of neighbor matrices (PCNM) was proposed to create spatial predictors that could easily be incorporated into regression or canonical analysis models (Dray et al., 2006). A numerical terrestrial scheme was produced to simulate energy, water, and carbon storage on land and exchange with the atmosphere in a spatial context on the basis of globally averaged components of zero-dimensional box models (Williamsona et al., 2006). Earth surface systems and components of the environment of the Earth's surface were spatially and explicitly simulated on a national level (Yue et al., 2005a,b,c, 2006, 2007a,b). Surface modeling of global climate change scenarios (IPCC, 2000) have been used to investigate the global impacts of climate change on ecosystems, carbon sink, food security, biomass supply, water resources, malaria distribution, and coastal flooding (Hulme et al., 1999; Parry et al., 2004; Rokityanskiy et al., 2007). A framework for an appropriate modeling scale was developed to identify proper spatial and temporal scales for modeling N flows from watersheds (Dumont et al., 2008). The spatial autocorrelation is used for analyzing landscape heterogeneity (Uuemaa et al., 2008).

The accuracy of the classic methods for surface modeling has been comparatively analyzed by many scholars. For instance, Yakowitz and Szidarovsky (1985) claimed that nonparametric regression was more robust; arguing that it predicted kriging on correlated data and was better when the data contained various types of trends. However, having compared IDW, kriging, and spline by using a carefully and specially designed survey of soil pH, Laslett and McBratney (1990) concluded that kriging was in fact the most effective method. Weber and Englund (1992) contested this, believing IDW to be far superior. Wood and Fisher (1993) assessed the simulation accuracy of IDW, contour flood filling, simultaneous over-relaxation and spline, and concluded that each of the techniques emphasized a particular aspect of the elevation model and no single technique was sufficient. Hutchinson and Gessler (1994) showed that spline was more accurate than kriging, while Myers (1994) concluded that interpolation from spatial data could be performed in various ways by invoking different assumptions and models. Mitas and Mitasova (1999) found that different methods could produce quite different spatial representations and in-depth knowledge of the phenomenon was needed to evaluate which one was the closest to reality. Laslett and McBratney (1990) calculated accuracies of kriging and spline in terms of several real data sets, and reported that kriging never performed worse than spline. Brus et al. (1996) used kriging, IDW, and spline to estimate soil properties and found that kriging was more reliable in the sense that it estimated all soil properties well. Desmet (1997) applied IDW, kriging, and spline and constructed a digital elevation model from irregularly spaced sample points and found that spline yielded the best results with regard to accuracy. Borga and Vizzaccaro (1997) compared kriging with MQM and concluded that kriging performed better at lower gauge density, while the accuracy of both estimators is similar at higher gauge density. Carrara et al. (1997) found that each method exhibited advantages and pitfalls; their respective applicability is largely dependent on the way in which source data are collected and stored. Caruso and Uarta (1998) compared IDW, kriging, MQM, and the tension finite difference method and found that each method had its own advantages and drawbacks, which strongly depended on the characteristics of the set of point data. Zimmerman et al. (1999) demonstrated that kriging was substantially superior to IDW over all levels of surface type, sampling pattern, noise, and correlation. Kravchenko (2003) found that kriging with known variogram parameters, performed significantly better than IDW for most of the case studies; however, it was very less accurate than IDW when a reliable sample variogram could not be obtained because of either an insufficient number of data points or too large a distance between the data points. Aguilar et al. (2005) comparatively analyzed IDW and radial basis functions (RBF) groups, including MQM, the inverse multiquadric method (IMQM), the multilog method (MLM), natural cubic splines (NCS), and thin plate splines (TPS); the results showed that MQM in the RBF group was the best method in terms of accuracy and numerical stability. Chaplot et al. (2006) evaluated the performance of IDW, ordinary kriging (OK), universal kriging (UK), MQM, and regularized spline with tension (RST). Their results indicated that a few differences existed between the methods when the sampling density was high; kriging had a higher accuracy if the spatial structure of altitude was strong, IDW and RST performed better when the spatial structure of height was weak, and MQM performed well in the mountainous areas.

Several scholars have also examined the role of computational speed in their comparative studies. Franke (1982) evaluated IDW, spline, TIN, FEM, and MQM and concluded that IDW performed well because it required moderate storage and computation time, while Pringle et al. (2009) argued that the primary disadvantage of kriging was its relatively slow computing speed. Katzil and Doytsher (2000) compared the computational speed of different methods, and concluded that improved cubic spline was 85 times faster than kriging when applied to a large area grid. Mitas and Mitasova (1999) pointed out that accuracy and high computational speed are seen as high priorities, and related comparative studies demonstrate that the high computational cost remains a significant problem of surface modeling (Lee et al., 1997; Aguilar et al., 2005). Cornford (2005) reviewed the comparison studies on different methods for surface modeling, and concluded that a good prior knowledge of the desired effect is crucial, and that the method of surface modeling chosen should be based on computational speed, cost of implementation, scaling with data size, and the ability to make probabilistic predictions.

1.2 Classical Methods

Surface modeling is an important subject of GIS. The dominating classical approaches include TSA, IDW, TIN, kriging, and spline.

1.2.1 Trend Surface Analysis

TSA is the simplest approach to generating a surface, which uses sparse data to fit a trend surface by least-squares regression on the spatial coordinates. The trend surface could be approximately formulated as a polynomial surface. The n orders of the polynomial surface can be formulated as

$$z(x,y) = \sum_{i=0}^{n} \sum_{j=0}^{n-i} a_{i,j} \cdot x^i \cdot y^j \tag{1.1}$$

where $z(x, y)$ is the value of the trend surface at the grid (x, y); x and y are the independent variables; $a_{i,j}$ ($i = 1, \ldots, n; j = 1, \ldots, n$) are parameters to be simulated; n is the order of the polynomial surface.

The shortcoming of this simple method is that it loses detail because powerful smoothing and variation in one part of the region affects the fit of the surface everywhere. TSA assumes that the residuals from the fitted surface are independent of one another, which is almost always violated and as a result no true estimation variance can be calculated (Oliver and Webster, 1990).

1.2.2 Inverse Distance Weighted Method

IDW can be formulated as (Shepard, 1968)

$$z_{i,j} = \left(\sum_{k=1}^{m_{i,j}} \frac{1}{(d_{i,j,k})^a} \right)^{-1} \sum_{k=1}^{m_{i,j}} \frac{z_k}{(d_{i,j,k})^a} \tag{1.2}$$

where $z_{i,j}$ indicates the estimate made at the location (i, j) where no measurements are available and an estimate is required, which is a function of the $m_{i,j}$

neighbor observations; z_k is the kth neighbor observation; $d_{i,j,k}$ is the distance from z_k to (i, j); and a is the parameter to be determined.

IDW uses an inverse distance weighting function to determine the interpolation value for any given point within the calculated area (Lee and Angelier, 1994). IDW allows an individual spatial analysis of each independent variable (Julià et al., 2004) and is a good method when all the variables have similar weight (Sinowski et al., 1997). However, IDW fails to incorporate the spatial structure and ignore information beyond the neighborhood (Zhao et al., 2005; Magnussen et al., 2007).

1.2.3 Triangulated Irregular Network

TIN is a popular model for representing surface models in GIS (Bengtsson and Nordbeck, 1964), because it has a simple data structure and can easily be rendered using common graphics hardware (Yang et al., 2000). TIN is composed of nodes, edges, triangles, topology, and a hull. It is one of the basic models for representing digital terrain—a piece of land on the Earth's surface (Tse and Gold, 2004). Suppose, points (x_1, y_1, z_1), (x_2, y_2, z_2), and (x_3, y_3, z_3) are the three nodes of a triangle; the equation, in a linear function, of the triangle surface is then given as

$$z = a \cdot x + b \cdot y + c \tag{1.3}$$

where,

$$a = \frac{\begin{vmatrix} z_1 & y_1 & 1 \\ z_2 & y_2 & 1 \\ z_3 & y_3 & 1 \\ x_1 & y_1 & 1 \\ x_2 & y_2 & 1 \\ x_3 & y_3 & 1 \end{vmatrix}}{}, \quad b = \frac{\begin{vmatrix} x_1 & z_1 & 1 \\ x_2 & z_2 & 1 \\ x_3 & z_3 & 1 \\ x_1 & y_1 & 1 \\ x_2 & y_2 & 1 \\ x_3 & y_3 & 1 \end{vmatrix}}{}, \quad c = \frac{\begin{vmatrix} x_1 & y_1 & z_1 \\ x_2 & y_2 & z_2 \\ x_3 & y_3 & z_3 \\ x_1 & y_1 & 1 \\ x_2 & y_2 & 1 \\ x_3 & y_3 & 1 \end{vmatrix}}{}$$

A TIN-based surface can then be represented by a set of linear functions. The value z of each point within a triangle can be calculated by the linear function based on its location (x, y). TIN models of the terrain surface are well known in GIS, but they ignore nonlinear information and are unable to represent cliffs, caves, or holes (Tse and Gold, 2004). TIN models cause well-known artifacts that are easily recognized as linear breaks to occur in a regular rectangular or triangular pattern (Schneider, 2001).

1.2.4 Kriging

Kriging is the term used for a family of generalized least-squares regression algorithms (Krige, 1951) and is named after the South African mining engineer, D. G. Krige (Kleijnen and van Beers, 2005). It is a fundamental tool in geostatistics. It includes OK, co-kriging, and disjunctive kriging (Kleijnen, 2009).

OK attempts to produce a set of estimates, for which the variance of errors is minimum. Random variables, V_1, V_2, ..., V_n, for available samples, are to be combined in a weighted linear combination to form the estimate, V_0:

$$V_0 = \sum_{i=1}^{n} w_i \cdot V_i \tag{1.4}$$

where w_1, w_2, ..., w_n are the weights to be estimated and $\sum_{i=1}^{n} w_i = 1$.

$$\sigma(R_0) = c_{0,0} + \sum_{i=1}^{n}\sum_{j=1}^{n} w_i \cdot w_j \cdot c_{i,j} - 2\sum_{i=1}^{n} w_i \cdot c_{i,0} \tag{1.5}$$

In Equation 1.5, R_0 is the difference between the true value and the corresponding estimate; $c_{i,j}$ is the covariance of V_i and V_j ($i = 0, 1, 2, ..., n, j = 0, 1, 2, ..., n$); when $i = j$, $c_{i,i}$ is variance of V_i.

When the Lagrange parameter, μ, is introduced into the Equation 1.5, the constrained minimization problem can be converted into the following unconstrained one, under the unbiasedness condition, $\sum_{i=1}^{n} w_i = 1$:

$$\begin{bmatrix} c_{1,1} & \cdots & c_{1,n} & 1 \\ \vdots & \ddots & \vdots & \vdots \\ c_{n,1} & \cdots & c_{n,n} & 1 \\ 1 & \cdots & 1 & 0 \end{bmatrix} \cdot \begin{bmatrix} w_1 \\ \vdots \\ w_n \\ \mu \end{bmatrix} = \begin{bmatrix} c_{1,0} \\ \vdots \\ c_{n,0} \\ 1 \end{bmatrix} \tag{1.6}$$

Let $C = \begin{bmatrix} c_{1,1} & \cdots & c_{1,n} & 1 \\ \vdots & \ddots & \vdots & \vdots \\ c_{n,1} & \cdots & c_{n,n} & 1 \\ 1 & \cdots & 1 & 0 \end{bmatrix}$, $w = \begin{bmatrix} w_1 \\ \vdots \\ w_n \\ \mu \end{bmatrix}$, and $D = \begin{bmatrix} c_{1,0} \\ \vdots \\ c_{n,0} \\ 1 \end{bmatrix}$, then the OK system can be formulated as

$$C \cdot w = D \tag{1.7}$$

The set of weights that would produce unbiased estimates with the minimum error variance is given by

$$w = C^{-1} \cdot D \tag{1.8}$$

OK is expected to be the best linear unbiased estimator. It is linear, because its estimates are weighted linear combinations of the available data; it is unbiased, because it tries to have the mean residual or error equal to zero. It is a good linear unbiased estimator, because it aims at minimizing the variance of the errors.

However, the goals of OK are ambitious and practically unattainable, since the mean error and the variance of errors are always unknown (Isaaks and Srivastava, 1989). Therefore, it could not be guaranteed that the mean error is exactly zero and that the variance of errors is minimized. The best we can do is to build a model of the available data, in which both the error and the error variance can be calculated and then choose weights for the nearby samples that ensure the average error of the model to be zero and the error variance is minimized.

Co-kriging is the logical extension of OK when two or more variables are spatially interdependent (Oliver and Webster, 1990). Disjunctive kriging provides minimum variance estimation of a property through nonlinear combinations of data and the probability that the true value equals or exceeds some defined threshold can be determined (Yates et al., 1986).

1.2.5 Spline

The specific types of spline include uniform rational basis-spline, uniform nonrational basis-spline, nonuniform nonrational basis-spline, and nonuniform rational basis-spline. A basis-spline, is a complete piecewise cubic polynomial consisting of any number of curve segments (Watt, 2000). It is a cubic segment over a certain interval, which goes from one interval to the next. Each curve segment is defined by four control points and each control point influences four and only four curve segments, which is the local control property of the basis-spline curve. The entire set of curve segments, $Q(u)$, is defined as one basis-spline curve

$$Q(u) = \sum_{i=0}^{m} P_i \cdot B_i(u) \qquad (1.9)$$

where i is a nonlocal control point number; u is a global parameter; P_i is the ith control point; m is total number of the control points; and $B_i(u)$ is the ith basis-spline.

If knots, that is the join points on the values of x between segments, are spaced at equal intervals of the parameter x, the basis-spline is termed a uniform one. A nonuniform basis-spline (NURBS) is a curve where parametric intervals between successive knot values are not necessarily equal. A rational curve is a curve defined in four-dimensional space (known as projective space), which is then projected into three-dimensional space. The rational form of the NURBS is one of the commonest forms used in practice. Rational curves enjoy all the properties of nonrational curves. The control points of a nonrational curve can only be transformed under an affine transformation. The perspective transformation used in computer graphics is not affine and the nonrational curves whose control points are transformed into image space cannot be correctly generated in this space.

A Spline surface patch can be formulated as

$$Q(u,v) = \sum_{i=0}^{n} \sum_{j=0}^{m} P_{i,j} \cdot B_i(u) \cdot B_j(v) \tag{1.10}$$

where $P_{i,j}$ is an array of control points; $B_i(u)$ and $B_j(v)$ are the univariate cubic basis-splines.

However, a few types of surfaces only fit the formulation of univariate cubic basis-splines, which leads to a considerable error of surface simulation at large.

1.3 Error and Multiscale Issues

A substantial component of research in Earth surface systems involves identifying the most important processes and developing computational models of key interactions (Billen et al., 2008), which deal with diverse data sets, ranging in spatial scales from nanometers to thousands of kilometers, and vary on timescales from seconds to billions of years. Error and multiscale problems have long troubled surface modeling.

1.3.1 Error Issues

Errors of surface modeling are caused by data acquisition, modeling errors, and design errors due to assumptions made in the design calculations (Stott, 1977). An overview of computer-intensive statistical methods concluded that error estimates had a significant impact on surface modeling (Elston and Buckland, 1993). A soil survey on different spatial scales showed large discrepancies between the values of soil water-holding capacity derived from a finer scale on county level and the U.S. Soil Conservation Service State Soil Geographic Data Base (Lathrop et al., 1995). Large differences between the estimates of global spatial models and Chinese data were found in temperate evergreen and deciduous forests, because variance in climate conditions both over a large spatial scale and a range of temporal scales had a strong influence on the accuracy of estimations by climate–vegetation models (Jiang et al., 1999). The joint effect of spatial and temporal processes makes a general interpretation of snow cover very difficult, because the actual influences of topography on turbulent air flow and actual snowfall are obviously not well represented by simple spatial models, and errors are likely to occur on steep slopes and cliffs (Tappeiner et al., 2001). The method of multivariate TSA, which is now widely applied in various fields of ecology, presents problems because of its coarseness; finer structures cannot be adequately modeled by this method (Borcard and Legendre, 2002). A number of problems may arise with the data model, including the appropriate scale of observations and locational accuracy when ecological field data are used with GIS and databases as well as errors associated with digital terrain models (Austin, 2002). The two general assumptions of ordinary least-squares (OLS) are that observations are independent and variance is homogeneous among samples (Shi et al., 2006); but the assumptions of independence and constant variance of OLS are often violated due to spatial effects on variables sampled across a landscape, which leads to a biased estimation of the standard errors of model parameters and consequently misleading significance tests (Zhang et al., 2005).

1.3.2 Multiscale Issues

In the early 1980s, the multiscale issue was recognized as a fundamental problem in GIS (Li, 2007). In 1983, a small group of leading scientists was gathered by National Aeronautics and Space Administration (NASA) to define research priorities in GIS and multiscale representation was identified as one of them. Since the early 1990s, multiscale representation has become a popular research topic in geographical information science. In 1996, the University Consortium for Geographic Information Science (UCGIS) selected multiscale representation as one of the 10 research priorities. By the late 1990s, multiscale representation gained a boost with the funding of the project of Automated Generalization New Technology (AGENT) by the European Union. In 2000, the International Society for Photogrammetry and Remote Sensing established a working group on multiscale representation. By 2003 this was adopted as one of the 14-long-term research priorities of UCGIS (Usery and McMaster, 2005). Since then, special sessions on multiscale representation have been organized at conferences, and special issues on this topic have been published in various journals.

Four scale concepts could be distinguished in the spatial–temporal domain: cartographic scale; scale of spatial extent; scale of action; and spatial resolution (Malenovsky et al., 2007). The cartographic scale refers to the ratio of a distance on a map against the corresponding distance on the ground. The scale of spatial extent refers to the area of the investigation. The scale of action represents a level at which a certain process phenomenon is best observed. Spatial resolution refers to the smallest distinguishable parts of an object, for instance, a pixel size of the raster map or RS image.

At each scale, a set of spatially explicit indicators needs to be identified to characterize pressures, conditions, trends, and scenarios of ecosystem types and underlying structural features of ecosystems. For any size patch of the Earth's surface that is chosen to be defined as an Earth surface system or an ecosystem, there are a set of factors external to the ecosystem, which influence how it functions and, in turn, flows of material and energy that extend beyond the ecosystem. The larger the scale, the more inclusive it is of these flows of material and energy. However, studies undertaken at larger scales lose the site specificity that policymakers often need. For instance, local case studies provide a refined understanding of local issues based on long-term investigation at specific locations, but the possibility of generalizing findings is limited by the geographic coverage of the studies and the locality-specific conditions, especially given the heterogeneous environment of the study area. Regional studies based on RS and census data enable wide geographic coverage, but are limited in their understanding of the underlying mechanisms of ecosystem change, which do not incorporate the decision making of ecosystem managers (Castella et al., 2007). In other words, there is no single scale at which we can obtain a full understanding of ecosystems.

In conclusion, the error and multiscale problems may mainly be caused by the various drawbacks of dominating classical methods for surface modeling (Table 1.1), and by the limited capacity of the current GIS.

TABLE 1.1

Basic Ideas and Shortcomings of the Classical Methods

Method	Description of Method	Shortcomings
TSA	Sparse data are used to fit a trend surface by least-squares regression on the spatial coordinates	The assumption of the residuals from the fitted surface being independent of one another is almost always violated
IDW	An inverse distance weighting function is used to determine the interpolation value for any given point within the calculated area	It fails to incorporate the spatial structure and ignores information beyond the neighborhood
TIN	The value of each point within a triangle is calculated by a linear function based on its location	It ignores nonlinear information and is unable to represent cliffs, caves, or holes
Kriging	It tries to have the mean residual or error equal to zero, and aims at minimizing the variance of the errors	The goals are practically unattainable, since the mean error and the variance of the errors are always unknown
Spline	Univariate cubic basis-splines are approximately used to simulate surfaces	A few types of surfaces fit the formulation of univariate cubic basis-splines

1.4 Geographical Information Systems

There are many definitions of GIS. For instance, Burrough (1986) defined GIS as a powerful set of tools for collecting, storing, retrieving at will, transforming, and displaying spatial data from the real world. Parker (1988) defined it as an information technology which stores, analyzes, and displays both spatial and nonspatial data. Lo and Yeung (2002) generally defined GIS as a set of computer-based systems for managing geographical data, which is then used to solve spatial problems, whereas Longley et al. (2005) defined GIS as a tool for performing operations on geographical data that are too tedious, expensive, or inaccurate to be performed by hand.

1.4.1 Historical Development

As a computer-dependent application, the origin of GIS could be traced back to research and development in electronic data processing in the 1940s and 1950s, which finally led to the successful implementation of computer-aided graphical data processing and a database management system. In the early 1960s, R.F. Tomlison conceptualized Canada GIS, which was the first of its kind, to address the needs of land and resource information management of the federal government of Canada, which became operational in 1971 (Lo and Yeung, 2002). In 1963, Howard H. Fisher used the computer to make simple maps by printing statistical values on a grid of plain paper, which included a set of modules for analyzing data, which could be manipulated to produce choropleth or isoline interpolations. The results were then displayed in many ways using overprinting of line-printer characters to produce suitable gray scales. Although cartographers had begun to adopt computer techniques in the 1960s, they were mostly limited to automatic draft maps. For traditional cartography, the new computer technology did not change fundamental attitudes to map-making (Burrough and McDonnell, 1998).

In the early 1970s, the Swedish Land Data Bank was developed to automate land and property registration (Adersson, 1987), while in Britain, the Local Authority Management Information System and the Joint Information System were developed for use by local governments to control and monitor land use. By the late 1970s, there had been considerable investment in the development and application of computer-assisted cartography. Hundreds of computer programs and systems were developed for various mapping applications. The history of using computers for mapping and spatial analysis shows that there were parallel developments in automated data capture, data analysis, and presentation in related fields, which resulted in the emergence of general purpose GIS. Map data processing was the primary focus of GIS; spatial analysis functionality at this time was rather limited (Lo and Yeung, 2002).

Topological principles in cartography, proposed by Corbett (1979), was a milestone of GIS development, allowing geographical data to be stored in a simple structure capable of representing what they are, where they are, and how they are spatially associated with one another. In 1982, the vector-based ArcInfo GIS software package was released by the Environmental Systems Research Institute. The software stored graphic data in the topological structure, and attribute data in the tabular structure (Morehouse, 1989). By the late 1980s, many other GIS software packages were developed using a similar data model.

By the 1990s, computer technology had provided huge amount of processing power and data storage capacity. With advances in operating systems, computer graphics, data management systems, human–computer interaction, and graphical-user interface design, GIS became a multiplatform application that ran on different kinds of computer, both as a stand-alone application and a time-sharing system (Lo and Yeung, 2002). The development of Web GIS has allowed expensive data and software to be shared. Standardization in interfaces between data programs and other programs has made it much easier to provide the functionality for handling large amount of data and GIS has considerably developed in many aspects. A great deal of guidance has been provided on how to set up computer mapping and GIS projects efficiently, and the basic functionality required for handling spatial data has been widely accepted. Although these advances have promoted a considerable development of GIS, which has now entered the age of Geographical Information Infrastructure (National Academy of Public Administration, 1998), the basic spatial models used in modern GIS are not dissimilar from those used 20 years ago. New insights into spatial and temporal modeling still need to be made if GIS is to develop beyond a mere technology (Burrough and McDonnell, 1998).

1.4.2 Error Problem

GIS is an essential platform for the construction of spatial models; however, errors are ubiquitous in the current GIS. The word "error" can be used in a variety of senses (Abler, 1987). Unwin (1995) defined it as the difference between reality and the representation of the reality. Defined in this way, error is related to accuracy.

When field data are incorporated into GIS, a common mistake is to assume that the error can be simply equated to the measurement error at the sampled points and quoted as a simple global statement, which does not address the spatial variation (Fisher, 1991). Monckton (1994) criticized the spatially uniform error assumption as untenable. He believed that a complete specification of the error should include not only the spatial field of its mean, but also its variance and spatial dependence. The problems of error in field models have proved to be very hard to address using a well-developed theory (Unwin, 1995).

The process of integrating RS data into GIS usually includes five steps. These are data acquisition, data processing, data analysis, error assessment, and final product presentation. Errors may accumulate throughout the process in an additive or a multiplicative fashion (Lunetta et al., 1991). Data acquisition errors may arise from geometric aspects, sensor systems, platforms, ground control, and/or scene considerations. Data-processing errors may be caused by geometric rectification such as resampling and data conversion from raster to vector format and from vector to raster format. Data analysis errors may arise from classification systems, data generalization, and quantitative analysis of relationships between data variables and the subsequent inferences that might develop. Error of error assessment might be mainly produced by the expression of locational accuracy, discrete multivariate, and reporting standards. Final product presentation errors include attribute errors and spatial errors that may be introduced through the use of base maps with different scales, different national horizontal datum in the source materials, and different minimum mapping units which are then resampled to a final minimum mapping unit.

These errors can be broken down into two categories inherent or operational (Vitek et al., 1984). The inherent error is the error present in source documents. The operational error is accumulated through data capture and manipulation functions of GIS (Walsh et al., 1987). Inherent errors include errors from sampling and attribute errors in source data. Operational errors include positional errors and identification errors. Positional errors stem from inaccuracies in the horizontal placement of boundaries and identification errors occur when there is a mislabeling of areas on thematic maps (Newcomer and Szajgin, 1984). Models as simulations of the real world often simplify the complexity of the real world and are therefore obviously open to error. Inherent errors can be propagated through the simulation process and become manifest in the final products (Huang and Lee, 2005). Although there are many types and sources of error and uncertainty in geographical data and their processing, the problem is not simply technical and arises from an evident inadequacy of GIS (Unwin, 1995). Integration of data from different sources and indifferent formats, at different original scales, plus inherent errors, can yield a product of questionable accuracy. The manipulation of thematic overlays within GIS to derive model variables are susceptible to inherent and operational errors, from which results may have such error margins as to be useless for specific applications (Burrough, 1986). Any decision based on such products would thus be flawed (Walsh et al., 1987).

1.4.3 Multiscale Problem

Earth surface systems are constantly changing, not only over space but also in time. Relevant processes might span over several temporal and spatial scales. The understanding of spatial and temporal processes and their interrelations is central to an understanding of the complex behavior of the earth surface systems (Pang and Shi, 2002).

The scale issue is also an inherent part of ecology (Withers and Meentemeyer, 1999). From the early 1950s to the early 1970s, many ecologists (Greig-Smith, 1952; Gould, 1966; Hutchinson, 1971) tried to incorporate scale into environmental biology. From the early 1970s to the 1980s, many ecologists focused increasing attention on the problem of spatial scale (Allen and Starr, 1982; Golley, 1989; Gosz and Sharpe, 1989; Wiens, 1989). In the 1990s, the problem of scale becomes a central issue in ecology, for unifying population biology and ecosystem science, and marrying basic ecology and applied ecology (Levin, 1992).

The explosion of interest in scale has created many methods for scale. For instance, interpolation brings multiple phenomena measured at different resolution into a common coordinate grid with a single size (Ehleringer and Field, 1993), while the multiple-variables scaling method simultaneously examines each variable at different scales (Holling, 1992). In the 1990s, spatially explicit models were developed to demonstrate scale-sensitive issues (Holt et al., 1995), and fractal geometry is now used to treat the dependence of various phenomena on scales (Mandelbrot, 1982). Resampling techniques are used to frame samples within a hierarchical framework to assess how scale affects ecosystem characteristics (Cressie, 1993) and neural models are used to test scale effects resulting from changes in grain size and spatial structure (Milne, 1992). The hierarchy theory is also employed to address issues of spatial scale; this theory implies that an ecosystem is composed of interacting components and is itself a component of a larger system (O'Neill et al., 1989). However, they are not generalized in GIS as a module.

1.4.4 Real-Time Problem

Time can be characterized as the fourth dimension of the physical space–time continuum (Asproth et al., 1995). From a human point of view, a concrete system can move in any direction on the spatial dimensions, but only forward on the temporal dimension (Miller, 1978). Static objects can be defined as objects that do not change in a short time period. GIS systems generally deal with static information. However, in many situations, the information in GIS applications changes dynamically. Quite often, it is desirable to combine static information with dynamic information. Lunetta et al.'s (1991) study shows that major impediments to the analysis of spatial data arise from a lack of well-documented methods in terms of error accumulation, and errors that occur due to the static representation of dynamic ecosystem components suggest that a real-time method must be developed. Methods of assessing the accuracy of dynamic images are also inadequate and must be further researched (Martin, 1989).

Real-time means momentary, that is, the same moment as any given event happens. In real-time systems, this implies momentary updates (Asproth et al., 1995). However, it is impossible to achieve momentary updates, as there is always some delay. The acceptable delay length for a real-time system depends on how dynamic the processes are and how time-critical the decisions are. The rapid development of computing technology in recent years has made real-time spatial analysis and real-time data visualization realizable, although current GIS software and interfaces do not encompass a set of technical and real-time functions (Valsecchi et al., 1999).

GIS provides powerful functionality for spatial analysis, data overlay, and storage. However, these spatially oriented systems lack the ability to represent temporal dynamics (Peuquet and Niu, 1995). In other words, GIS prefers a static view and generally lacks the representation of dynamics (Isenegger, 2005). The current generation of commercial GISs is unable to facilitate real-time decision-making without significant modifications or integration with external models. For instance, research results from Sun et al. (2002) showed that current GIS needs great calculation and are difficult to realize in real-time visualization. Zerger and Smith (2003) indicated that many commercial GISs have evolved from mapping tools and their functionality has only recently been extended to modeling and simulation; however, the integration of modeling and simulation with GIS does not have a real-time capability and requires events to be premodeled.

In general, GIS today remains a technology for static data, which is a major impediment to its use in spatial modeling (Goodchild, 1995). The technology of digital geographies has found the representation of change in time extremely hard to handle (Unwin and Fisher, 2005).

1.4.5 Direct Modeling Problem

Some researchers have noted that GIS has been restricted to producing cartographic products rather than spatial modeling. GIS was conventionally developed using a hybrid approach that handled graphical and descriptive geographical data separately. This geo-relational data model was the norm for GIS implementation until the late 1990s (Lo and Yeung, 2002). GIS was usually used as a means of overlaying maps (Oliver and Webster, 1990). Almost all mathematical models are too complex to be run directly from the present state-of-the-art GIS. They often run outside the GIS. In these cases, the GIS is used to supply the input data at an appropriate resolution and to display the results graphically in combination with other relevant spatial data (Heuvelink et al., 1989). Brooks and Tidwell (1993) found that physical hazard modeling must be carried out external to the GIS because the numerical modeling capabilities of GIS are limited. Newkirk (1993) stated that numerical modeling is more effectively achieved outside GIS because GIS has limited capabilities for modeling and simulation.

The integration of GIS and simulation models can be categorized into loose coupling and deep coupling (Bell et al., 2000). Most integration is in the loose coupling category that integrates GIS with simulation models through exchanging data files (Wang, 2005). This approach often requires human intervention, which can become a barrier to automating the operation process. The deep coupling approach links the GIS and simulation models with a common user interface, in which they can, in fact, remain as separate systems (Fedra, 1996).

The restrictions of current methods and tools described above are the driving force behind this book.

References

Abler, R.F. 1987. The National Science Foundation, National Center for Geographical Information and Analysis. *International Journal of Geographical Information Systems* 1: 303–326.

Adersson, S. 1987. The Swedish land data bank. *International Journal of Geographical Information Systems* 1(3): 253–263.

Aguilar, F.J., Agüera, F., Aguilar, M.A., and Carvajal, F. 2005. Effects of terrain morphology, sampling density, and interpolation methods on grid DEM accuracy. *Photogrammetric Engineering and Remote Sensing* 71: 805–816.

Ahlberg, J.H., Nilson, E.N., and Walsh, J.L. 1967. *The Theory of Splines and Their Application*. New York: Academic Press.

Akima, H. 1978a. A method of bivariate interpolation and smooth surface fitting for irregularly distributed data points. *ACM Transactions on Mathematical Software* 4(2): 148–159.

Akima, H. 1978b. An ALGORITHM 526: Bivariate interpolation and smooth surface fitting for irregularly distributed data points. *ACM Transactions on Mathematical Software* 4(2): 160–164.

Allen, T.F.H. and Starr, T.B. 1982. *Hierarchy: Perspectives for Ecological Complexity*. Chicago: University of Chicago Press.

Asproth, V., Hakansson, A., and Revay, P. 1995. Dynamic information in GIS systems. *Computers, Environment and Urban Systems* 19(2): 107–115.

Austin, M.P. 2002. Spatial prediction of species distribution: An interface between ecological theory and statistical modeling. *Ecological Modeling* 157: 101–118.

Baskent, E.Z. and Keles, S. 2005. Spatial forest planning: A review. *Ecological Modeling* 188: 145–173.

Beaujouan, V., Durand, P., and Ruiz, L. 2001. Modeling the effect of the spatial distribution of agricultural practices on nitrogen fluxes in rural catchments. *Ecological Modeling* 137: 93–105.

Bell, M., Dean, C., and Blake, M. 2000. Forecasting the pattern of urban growth with PUP: A web-based model interfaced with GIS and 3D animation. *Computers, Environment and Urban Systems* 24: 559–581.

Bengtsson, B.E. and Nordbeck, S. 1964. Construction of isarithms and isarithmic maps by computers. *BIT* 4: 87–105.

Billen, M.I., Kreylos, O., Hamann, B., Jadamec, M.A., Kellogg, L.H., Staadt, O., and Sumner, D.Y. 2008. A geoscience perspective on immersive 3D gridded data visualization. *Computers and Geosciences* 34(9): 1056–1072.

Borcard, D. and Legendre, P. 2002. All-scale spatial analysis of ecological data by means of principal coordinates of neighbor matrices. *Ecological Modelling* 153: 51–68.

Borga, M. and Vizzaccaro, A. 1997. On the interpolation of hydrologic variables: Formal equivalence of multiquadratic surface fitting and Kriging. *Journal of Hydrology* 195: 160–171.

Brooks, C. and Tidwell, J. 1993. Regional uses of geographic information systems technology. *Proceedings of the Seventeenth Annual Conference of the Association of Floodplain Managers*, Atlanta, GA, pp. 91–96.

Brus, D.J., De Gruijter, J.J., Marsman, B.A., Visschers, R., Bregt, A.K., Breeuwsma, A., and Bouma, J. 1996. The performance of spatial interpolation methods and choropleth maps to estimate properties at points: A soil survey case study. *Environmetrics* 7: 1–16.

Burrough, P.A. 1986. *Principles of Geographical Information Systems for Land Resources Assessment*. Oxford: Clarendon Press.

Burrough, P.T. and McDonnell, R.A. 1998. *Principal of Geographical Information Systems*. New York: Oxford University Press.

Carrara, A., Bitelli, G., and Carla, R. 1997. Comparison of techniques for generating digital terrain models from contour lines. *International Journal of Geographical Information Science* 11: 451–473.

Caruso, C. and Uarta, F. 1998. Interpolation methods comparison. *Computers and Mathematics with Applications* 35(12): 109–126.

Castella, J., Kam, S.P., Quang, D.D., Verburg, P.H., and Hoanh, C.T. 2007. Combining top-down and bottom-up modeling approaches of land-use/cover change to support public policies: Application to sustainable management of natural resources in northern Vietnam. *Land Use Policy* 24: 531–545.

Chaplot V., Darboux, F., Bourennane, H., Leguédois, S., Silvera, N., and Phachomphon, K. 2006. On the accuracy of interpolation techniques in digital elevation models for various landscape morphologies, surface areas and sampling densities. *Geomorphology* 77: 126–141.

Corbett, J.P. 1979. *Topological Principles in Cartography*. Washington, DC: US Government Printing Office.

Cornford, D. 2005. Are comparative studies a waste of time? SIC2004 examined. In *Spatial Interpolation Comparison 2004*, ed., G. Dubois, 61–70. Luxembourg: Office for Official Publication of the European Communities.

Costanza, R. 1989. Model goodness of fit: A multiple resolution procedure. *Ecological Modelling* 47: 199–215.

Costanza, R. and Maxwell, T. 1991. Spatial ecosystem modeling using parallel processors. *Ecological Modelling* 58 (1–4): 159–183.

Costanza, R., Sklar, F.H., and White, M.L. 1990. Modeling costal landscape dynamics. *BioScience* 40(2): 91–107.

Cressie, N. 1993. *Statistics for Spatial Data*. New York: John Wiley.

Deichmann, U. 1996. *A Review of Spatial Population Database Design and Modeling*. Technical Report 96–3, National Center for Geographic Information and Analysis, USA.

Desmet, P.J.J. 1997. Effects of interpolation errors on the analysis of DEM. *Earth Surface Processes and Landforms* 22: 563–580.

Dray, S., Legendre, P., and Peres-Neto, P.R. 2006. Spatial modeling: A comprehensive framework for principal coordinate analysis of neighbour matrices (PCNM). *Ecological Modelling* 196: 483–493.

Dumont, E., Bakkerb, E.J., Bouwmanc, L., Kroezed, C., Leemansd, R., and Steine, A. 2008. A framework to identify appropriate spatial and temporal scales for modeling *N* flows from watersheds. *Ecological Modelling* 212: 256–272.

Ehleringer, J.R. and Field, C.B. 1993. *Scaling Physiological Processes: Leaf to Globe*. San Diego, CA: Academic Press.

Elston, D.A. and Buckland, S.T. 1993. Statistical modeling of regional GIS data: An overview. *Ecological Modelling* 67(1): 81–102.

Fedra, K. 1996. Distributed models and embedded GIS. In *GIS and Environmental Modelling: Progress and Research Issues*, eds. M.F. Goodchild, L.T. Steyaert, B.O. Parks, C. Johnston, D. Maidment, M. Crane, and S. Glendinning, 413–417. Fort Collins, CO: GIS World Books.

Fisher, P.F. 1991. First experiments in viewed uncertainty: The accuracy of the viewed area. *Photogrammetric Engineering and Remote Sensing* 57: 1321–1327.

Foley, T.A. and Nielson, G.M. 1980. Multivariate interpolation to scattered data using delta iteration. In *Approximation Theory III*, ed., Cheney, E.W., 419–424. New York: Academic Press.

Franke, R. 1982. Scattered data interpolation: Tests of some methods. *Mathematics of Computation* 38: 181–200.

Friend, A.D. 1998. Parameterisation of a global daily weather generator for terrestrial ecosystem modeling. *Ecological Modelling* 109: 121–140.

Gao, Q. 1996. Dynamic modeling of ecosystems with spatial heterogeneity: A structured approach implemented in windows environment. *Ecological Modelling* 85: 241–252.

Gold, C.M., Charters, J.D., and Ramsden, J. 1977. Automated contour mapping using triangular element data structures and an interpolant over each irregular triangular domain. *ACM SIGGRAPH Computer Graphics* 11: 170–175.

Golley, F.B. 1989. Paradigm shift. *Landscape Ecology* 3(2): 65–66.

Goodchild, M.F. 1995. Geographic information systems and geographic research. In *Ground Truth: The Social Implications of Geographic Information Systems*, ed., J. Pickles, 31–50. New York: Guilford.

Gosz, J.R. and Sharpe, P.J.H. 1989. Broad-scale concepts for interaction of climate, topography, and biota at biome transitions. *Landscape Ecology* 3: 229–243.

Gould, S.J. 1966. Allometry and size in ontogeny and phylogeny. *Biological Review* 41: 587–640.

Greig-Smith, P. 1952. The use of random and contiguous quadrats in the study of the structure of plant communities. *Annals of Botany* 16: 293–316.

Harder, R.L. and Desmarais, R.N. 1972. Interpolation using surface Splines. *Journal of Aircraft* 9: 189–191.

Hardy, R.L. 1971. Multiquadric equation of topography and other irregular surfaces. *Journal of Geophysical Research* 76: 1905–1915.

Haxeltine, A. and Prentice, I.C. 1996. BIOME3: An equilibrium terrestrial biosphere model based on ecophysiological constraints, resource availability, and competition among plant functional types. *Global Biogeochemical Cycles* 10(4): 693–710.

Heuvelink, G.B.M., Burrough, P.A., and Stein, A. 1989. Propagation of errors in spatial modeling with GIS. *International Journal of Geographical Information Science* 3(4): 303–322.

Holling, C.S. 1992. Cross-scale morphology, geometry, and dynamics of ecosystems. *Ecological Monographs* 62(4): 447–502.

Holt, R.D., Pacala, S.W., Smith, T.W., and Liu, J.G. 1995. Linking contemporary vegetation models with spatially explicit animal population-models. *Ecological Applications* 5(1): 20–27.

Huang, Z. and Lees, B. 2005. Representing and reducing error in natural-resource classification using model combination. *International Journal of Geographical Information Science* 19(5): 603–621.

Hulme, M., Mitchell, J., Ingram, W., Lowe, J., Johns, T., New, M., and Viner, D. 1999. Climate change scenarios for global impacts studies. *Global Environmental Change* 9: S3–S19.

Hutchinson, G.E. 1971. Banquet address: Scale effects in ecology. In *Spatial Patterns and Statistical Distribution*, eds., G.P. Patil, E.C. Pielou, and W.E. Waters, 17–22. Pennsylvania, PA: Pennsylvania State University Press.

Hutchinson, M.F. and Gessler, F.R. 1994. Splines: More than just a smooth interpolator. *Geoderma* 62: 45–67.

IPCC. 2000. *Emissions Scenarios: A Special Report of Working Group III of the Intergovernmental Panel on Climate Change.* Cambridge: Cambridge University Press.

Isaaks, E.H. and Srivastava, R.M. 1989. *Applied Geostatistics.* New York: Oxford University Press.

Isenegger, D., Price, B., Wu, Y., Fischlin, A., Frei, U., Weibel, R., and Allgöwer, B. 2005. IPODLAS—A software architecture for coupling temporal simulation systems, VR and GIS. *ISPRS Journal of Photogrammetry and Remote Sensing* 60: 34–47.

Ji, W. and Jeske, C. 2000. Spatial modeling of the geographic distribution of wildlife populations: A case study in the lower Mississippi River region. *Ecological Modelling* 132: 95–104.

Jiang, H., Apps, M.J., Zhang, Y.l., Peng, C.H., and Woodard, P.M. 1999. Modeling the spatial pattern of net primary productivity in Chinese forests. *Ecological Modelling* 122: 275–288.

Julià, M.F., Monreal, T.E., Jiménez, A.S., and Meléndez, E.G. 2004. Constructing a saturated hydraulic conductivity map of Spain using pedotransfer functions and spatial prediction. *Geoderma* 123(3–4): 257–277.

Katzil, Y. and Doytsher, Y. 2000. Height estimation methods for filling gaps in gridded DTM. *Journal of Surveying Engineering* 126(4): 145–162.

Kleijnen, J.P.C. 2009. Kriging metamodeling in simulation: A review. *European Journal of Operational Research* 192(3): 707–716.

Kleijnen, J.P.C. and van Beers, W.C.M. 2005. Robustness of kriging when interpolating in random simulation with heterogeneous variances: Some experiments. *European Journal of Operational Research* 165(3): 826–834.

Kravchenko, A.N. 2003. Influence of spatial structure on accuracy of interpolation methods. *Soil Science Society of America Journal* 67: 1564–1571.

Krige, D.G. 1951. A statistical approach to some basic mine valuation problems on the Witwatersrand. *The Journal of the Chemical, Metallurgical and Mining Society of South Africa* 52(6): 119–139.

Laslett, G. M. and Mcbratney, A. B. 1990. Further comparison of spatial methods for predicting soil pH. *Soil Science Society of America Journal* 54: 1553–1558.

Lathrop, Jr. R.G., Aber, J.D., and Bognar, J.A. 1995. Spatial variability of digital soil maps and its impact on regional ecosystem modeling. *Ecological Modelling* 82: 1–10.

Lee, J.C. and Angelier, J. 1994. Paleostress trajectory maps based on the results of local determinations: The "lissage" program. *Computers and Geosciences* 20(2): 161–191.

Lee, S.Y., Wolberg, G., and Shin, S.Y. 1997. Scattered data interpolation with multilevel B-splines. *IEEE Transactions on Visualization and Computer Graphics* 3(3): 1–17.

Legendre, L. and Legendre, P. 1983. *Numerical Ecology*. Amsterdam: Elsevier Scientific Pul. Co.

Lehmann, A., Overton, J.M., and Leathwick, J.R. 2003. GRASP: Generalized regression analysis and spatial prediction. *Ecological Modelling* 160: 165–183.

Levin, S.A. 1992. The problem of pattern and scale in ecology. *Ecology* 73(6): 1943–1967.

Li, Z.L. 2007. *Algorithmic Foundation of Multi-Scale Spatial Representation*. London: CRC Press.

Lo, C.P. and Yeung, A.K.W. 2002. *Concepts and Techniques of Geographic Information Systems*. Upper Saddle River, NJ: Prentice-Hall.

Long, G.E. 1980. Surface approximation: A deterministic approach to modeling spatially variable systems. *Ecological Modelling* 8: 333–343.

Longley, P.A., Goodchild, M.F., Maguire, D.J., and Rhind, D.W. 2005. *Geographical Information Systems and Science*. Chichester, England: John Wiley & Sons, Ltd.

Lunetta, R.S., congalton, R.G., Fenstermaker, L.K., Jensen, J.R., McGwire, K.C., and Tinney, L.R. 1991. Remote sensing and geographic information system data integration: Error sources and research issues. *Photogrammetric Engineering and Remote Sensing* 57(6): 677–687.

Magnussen, S., Næsset, E., and Wulder, M.A. 2007. Efficient multi-resolution spatial predictions for large data arrays. *Remote Sensing of Environment* 109(4): 451–463.

Malenovsky, Z., Bartholomeus, H.M., Acerbi-Junior, F.W., Schopfer, J., Painter, T.H., Epema, G.F., and Bregt, A.K. 2007. Scaling dimensions in spectroscopy of soil and vegetation. *International Journal of Applied Earth Observation and Geoinformation* 9: 137–164.

Mandelbrot, B.B. 1982. *The Fractal Geometry of Nature*. San Francisco: W. H. Freeman.

Martin, L.R.G. 1989. Accuracy assessment of Landsat-based visual change detection methods applied to the rural–urban fringe. *Photogrammetric Engineering and Remote Sensing* 55(2): 209–215.

Martin, D. and Bracken, I. 1991. Techniques for modeling population-related raster databases. *Environment and Planning* A23: 1069–1075.

Maude, A.D. 1973. Interpolation—mainly for grapher plotters. *The Computer Journal* 16(1): 64–65.

Mclain, D.H. 1976. Two-dimensional interpolation from random data. *The Computer Journal* 19: 178–181.

Miller, J.G. 1978. *Living Systems*. New York: McGraw-Hill.

Milne, B.T. 1992. Spatial aggregation and neutral models in fractal landscapes. *American Naturalist* 139 (1): 32–57.

Mitas, L. and Mitasova, H. 1999. Spatial interpolation. In *Geographical Information Systems: Principles, Techniques, Management and Applications*, eds., P. Longley, K.F. Goodchild, D.J. Maguire and D.W. Rhind, 481–492. New York, NY: Wiley.

Monckton, C. 1994. An investigation into the spatial structure of error in digital elevation data. In *Innovations in GIS*, ed., M.F. Worboys, 201–211. London: Taylor & Francis.

Morehouse, S. 1989. The architecture of ARC/INFO. In *Auto-Carto 9 Proceedings, Falls Church*. VA: American Society of Photogrammetry and Remote Sensing, pp. 266–277.

Myers, D.E. 1994. Spatial interpolation: An overview. *Geoderma* 62: 17–28.

National Academy of Public Administration. 1998. *Geographic Information for the 21st Century: Building a Strategy for the Nation.* Washington, DC: National Academy of Public Administration.

Newcomer, J.A. and Szajgin, J. 1984. Accumulation of thematic map error in digital overlay analysis. *The American Cartographer* 11(1): 58–62.

Newkirk, R.T. 1993. Extending geographic information systems for risk analysis and management. *Journal of Contingencies and Crisis Management* 1: 203–206.

Oezesmi, S.L. and Oezesmi, U. 1999. An artificial neural network approach to spatial habitat modeling with inter-specific interaction. *Ecological Modelling* 116: 15–31.

Oliver, M.A. and Webster, R. 1990. Kringing: A method of interpolation for geographical information systems. *Journal of Geographical Information Science* 4(3): 313–332.

O'Neill, R.V., Johnson, A.R., and King, A.W. 1989. A hierarchical framework for the analysis of scale. *Landscape Ecology* 3: 193–205.

Pang, M.Y.C. and Shi, W.Z. 2002. Development of a process-based model for dynamic interaction in spatio-temporal GIS. *Geoinformatica* 6(4): 323–344.

Parker, H.D. 1988. The unique qualities of a geographic information system: A commentary. *Photogrammetric Engineering and Remote Sensing* 54: 1547–1549.

Parry, M.L., Rosenzweig, C., Iglesias, A., Livermore, M., and Fischer, G. 2004. Effects of climate change on global food production under SRES emissions and socio-economic scenarios. *Global Environmental Change* 14: 53–67.

Perry, G.L.W. and Enright, N.J. 2002. Spatial modeling of landscape composition and pattern in a maquis–forest complex, Mont Do, New Caledonia. *Ecological Modelling* 152: 279–302.

Peuquet, D.J. and Niu, D.A. 1995. An event-based spatiotemporal data model (Estdm) for temporal analysis of geographical data. *International Journal of Geographical Information Systems* 9(1): 7–24.

Pringle, M.J., Schmidt, M., and Muir, J.S. 2009. Geostatistical interpolation of SLC-off Landsat ETM+ images. *ISPRS Journal of Photogrammetry and Remote Sensing* 64(6): 654–664.

Reich, R.M., Bonham, C.D., and Metzger, K.L. 1997. Modeling small-scale spatial interaction of shortgrass prairie species. *Ecological Modelling* 101: 163–174.

Rokityanskiy, D., Benítez, P.C., Kraxner, F., McCallum, I., Obersteiner, M., Rametsteiner, E., and Yamagata, Y. 2007. Geographically explicit global modeling of land-use change, carbon sequestration, and biomass supply. *Technological Forecasting and Social Change* 74(7): 1057–1082.

Schneider, B. 2001. Phenomenon-based specification of the digital representation of terrain surfaces. *Transactions in GIS* 5(1): 39–52.

Schroeder, L.D. and Sjoquist, D.L. 1976. Investigation of population density gradients using trend surface analysis. *Land Economics* 52: 382–392.

Shepard, D. 1968. A two-dimensional interpolation function for irregularly-spaced data. In *Proceedings of the 1968 23rd ACM National Conference,* 517–524. New York: Association for Computing Machinery.

Shi, H.J., Laurent, E.J., LeBouton, J., Racevskis, L., Hall, K.R., Donovan, M., Doepker, R.V., Walters, M.B., Lupi, F., and Liu, J.G. 2006. Local spatial modeling of white-tailed deer distribution. *Ecological Modelling* 190: 171–189.

Sinowski, W., Scheinost, A.C., and Auerswald, K. 1997. Regionalization of soil water retention curves in a highly variable soilscape: II. Comparison of regionalization procedures using a pedotransfer function. *Geoderma* 78: 145–159.

Sklar, F.H., Costanza, R., and Day, Jr. J.W. 1985. Dynamic spatial simulation modeling of coastal wetland habitat succession. *Ecological Modelling* 29(1–4): 261–281.

Stott, J.P. 1977. Review of surface modeling. *Proceedings of Surface Modeling by Computer, a Conference jointly sponsored by the Royal Institution of Chartered Surveyors and the Institution of Civil Engineers*, London, 1976, pp. 1–8.

Sun, M., Xue, Y., Ma, A.N., and Mao, S.J. 2002. 3D visualization of large digital elevation model (DEM) data set. In *Proceedings of Computational Science—ICCS 2002*, 975–983. Berlin: Springer-Verlag.

Svirezhev, Y.M. 2002. Simple spatially distributed model of the global carbon cycle and its dynamic properties. *Ecological Modelling* 155: 53–69.

Talmi, A. and Gilat, G. 1977. Method for smooth approximation of data. *Journal of Computational Physics* 23: 93–123.

Tappeiner, U., Tappeiner, G., Aschenwald, J., Tasser, E., and Ostendorf, B. 2001. GIS-based modelling of spatial pattern of snow cover duration in an alpine area. *Ecological Modelling* 138: 265–275.

Tse, R.O.C. and Gold, C. 2004. TIN meets CAD—Extending the TIN concept in GIS. *Future Generation Computer Systems* 20(7): 1171–1184.

Turner, M.G., Costanza, R., and Sklar, F.H. 1989. Methods to evaluate the performance of spatial simulation models. *Ecological Modelling* 48(1–2): 1–18.

Unwin, D.J. 1995. Geographical information systems and the problems of 'error and uncertainty'. *Progress in Human Geography* 19(4): 549–558.

Unwin, D.J. and Fisher, P. 2005. Conclusion: Towards a research agenda. In *Re-presenting GIS*, eds., P. Fisher and D.J. Unwin, 277–281. New York: John Wiley & Sons.

Usery, E.L. and McMaster, R.B. 2005. Introduction to the UCGIS research agenda. In *A Research Agenda for Geographic Information Science*, eds., E.L. Usery and R.B. McMaster, 1–16. London: CRC Press.

Uuemaa, E., Roosaare, J., Kanal, A., and Mander, U. 2008. Spatial correlograms of soil cover as an indicator of landscape heterogeneity. *Ecological Indicators* 8(6): 783–794.

Valsecchi, P., Claramunt, C., and Peytchev, E. 1999. OSIRIS: An inter-operable system for the integration of real-time trac data within GIS. *Computers, Environment and Urban Systems* 23: 245–257.

Vitek, J.D., Walsh, S.J., and Gregory, M.S. 1984. Accuracy in geographic information systems: An assessment of inherent and operational errors. *Proceedings of PECORA IX Symposium*, Sioux Falls, SD, 2–4 October 1984, pp. 296–302.

Walsh, S.J., Lightfoot, D.R., and Bulter, D.R. 1987. Recognition and assessment of error in geographical information systems. *Photogrammetric Engineering and Remote Sensing* 53: 1423–1430.

Wang, X.H. 2005. Integrating GIS, simulation models, and visualization in trace impact analysis. *Computers, Environment and Urban Systems* 29: 471–496.

Watt, A. 2000. *3D Computer Graphics*. New York: Addison-Wesley.

Weber, D. and Englund, E. 1992. Evaluation and comparison of spatial interpolators. *Mathematical Geology* 24 (4): 381–391.

Wiens, J. 1989. Spatial scaling in ecology. *Functional Ecology* 3: 385–397.

Williamsona, M.S., Lentona, T.M., Shepherd, J.G., and Edwards, N.R. 2006. An efficient numerical terrestrial scheme (ENTS) for Earth system modeling. *Ecological Modelling* 198: 362–374.

Withers, M.A. and Meentemeyer, V. 1999. Concepts of scale in landscape ecology. In *Landscape Ecological Analysis: Issues and Applications*, eds., J.M. Klopatek and R.H. Gardner, 205–252. New York: Springer.

Wood, J.D. and Fisher, P.F. 1993. Assessing interpolation accuracy in elevation models. *IEEE Computer Graphics and Applications* 272: 48–56.

Wu, J.G. and Levin, S.A. 1997. A patch-based spatial modeling approach: Conceptual framework and simulation scheme. *Ecological Modelling* 101: 325–346.

Yakowitz, S.J. and Szidarovsky, F. 1985. A comparison of kriging with nonparametric regression methods. *Journal of Multivariate Analysis* 16(1): 21–53.

Yang, Q.H., Snyder, J.P., and Tobler, W.R. 2000. *Map Projection Transformation.* New York, NY: Taylor & Francis.

Yates, S.R., Warrick, A.W., and Myers, D.E. 1986. Disjunctive kriging: Overview of estimation and conditional probability. *Water Resources Research* 22: 615–627.

Yue, T.X., Fan, Z.M., and Liu, J.Y. 2005a. Changes of major terrestrial ecosystems in China since 1960. *Global and Planetary Change* 48: 287–302.

Yue, T.X., Wang, Y.A., Liu, J.Y., Chen, S.P., Qiu, D.S., Deng, X.Z., Liu, M.L., Tian, Y.Z., and Su, B. 2005b. Surface modeling of human population distribution in China. *Ecological Modelling* 181(4): 461–478.

Yue, T.X., Wang, Y.A., Liu, J.Y., Chen, S.P., Tian, Y.Z., and Su, B.P. 2005c. MSPD scenarios of spatial distribution of human population in China. *Population and Environment* 26(3): 207–228.

Yue, T.X., Fan, Z.M., Liu, J.Y., and Wei, B.X. 2006. Scenarios of major terrestrial ecosystems in China. *Ecological Modelling* 199: 363–376.

Yue, T.X., Du, Z.P., Song, D.J., and Gong, Y. 2007a. A new method of surface modeling and its application to DEM construction. *Geomorphology* 91(1–2): 161–172.

Yue, T.X., Fan, Z.M., and Liu, J.Y. 2007b. Scenarios of land cover in China. *Global and Planetary Change* 55: 317–342.

Yue, T.X., Du, Z.P., and Song, Y.J. 2008. Ecological models: Spatial models and geographic information systems. In *Encyclopedia of Ecology*, eds., S.E. Jørgensen, and B. Fath, 3315–3325. England: Elsevier Limited.

Yue, T.X., Song, D.J., Du, Z.P., and Wang, W. 2010. High accuracy surface modeling and its application to DEM generation. *International Journal of Remote Sensing* 31(8): 2205–2226.

Yue, T.X., Wang, Y.A., and Fan, Z.M. 2009. Surface modeling of population distribution. In *Handbook of Ecological Modeling and Informatics*, eds., S. E. Jørgensen, T.S. Chon, and F. Recknagel, 71–98. Ashurst Lodge, Ashurst, Southampton S040 7AA, UK: WIT Press.

Zerger, A. and Smith, D. I. 2003. Impediments to using GIS for real-time disaster decision support. *Computers, Environment and Urban Systems* 27(2): 123–141.

Zhao, C.Y., Nan, Z.G., and Cheng, G.D. 2005. Methods for modeling of temporal and spatial distribution of air temperature at landscape scale in the southern Qilian mountains, China. *Ecological Modelling* 189: 209–220.

Zienkiewicz, O. C. 1967. *The Finite Element Method in Structural and Continuum Mechanics.* London: McGraw-Hill.

Zimmerman, D., Pavlik, C., Ruggles, A., and Armstrong, M.P. 1999. An experimental comparison of ordinary universal kriging and inverse distance weighting. *Mathematical Geology* 31(4): 375–390.

2

Basic Method of HASM*

2.1 Introduction

A partial differential equation (PDE)-based approach might be an alternative way to find a solution to the error problem. The approach to representing surfaces as solutions for PDEs was initially developed in the early 1970s (Thompson et al., 1974). It demonstrated that a wide variety of surface shapes can be made accessible by varying the boundary conditions and parameter in the PDE (Bloor and Wilson, 1989, 1990; Protopopescu et al., 1989). PDE surfaces have demonstrated many modeling advantages in many fields of design such as surface blending, free-form surface modeling, and surface functional specifications (Pasadas and Rodríguez, 2009). The PDE approach has been used in a number of research fields. For instance, sculptured surfaces, ship hulls, and propeller blades were modeled using PDE (Bloor and Wilson, 1996), and PDE application to image analysis has become a research topic of increasing interest over the past few years. A direct map between two cortical surfaces was iteratively computed by solving a PDE defined on the source cortical surface, starting from a properly designed initial map (Shi et al., 2007). Elliptic PDEs have been used to model various kinds of real physical situations such as the flow of air pollutants, temperature deflection, electrostatic potential, velocity potential, and stream function (Modani et al., 2008).

PDE-based approaches, such as implicit surface modeling (Walder et al., 2006) and fast surface modeling, are effective for both surface boundary constraints and the generation of families of free-form surfaces. Compared to classical approaches, the PDE approach can model complex objects. According to Zhang and You (2004), an appropriate form of PDE needs to be properly specified and a suitable solution function constructed before the PDE-based approach can be effectively applied to complex surface modeling. The higher the order of a PDE, the more boundary conditions it can meet, and therefore, the modeled surface can satisfy more requirements. These different PDEs are treated separately. To handle different classes of geometric PDEs such as biharmonic equations, second order geometric equations and higher-order geometric equations, Xu and Zhang (2008) proposed a general framework for surface modeling.

---

* Dr. Zheng-Ping Du is a major collaborator of this chapter.

However, solving a PDE subject to arbitrary boundary conditions is a classic complex mathematical problem. In many cases, numerical measures have to be taken resulting in low and sometimes unstable processes, because many forms of PDEs are analytically unsolvable. From current literatures, little has been achieved to obtain accurate PDE solutions that would be fast enough for surface modeling applications.

2.2 Theoretical Formulation

In terms of the fundamental theorem of surfaces, a surface is uniquely defined by the first and the second fundamental coefficients (Henderson, 1998). The first fundamental coefficients are used to express how the surface inherits the natural inner product of R³, in which R³ is the set of triples (x, y, z) of real numbers (Carmo, 2006). The first fundamental coefficients of a surface yield information about some geometric properties of the surface, by which we can calculate the lengths of curves, the angles of tangent vectors, the areas of regions, and geodesics on the surface. These geometric properties and objects that can be determined only in terms of the first fundamental coefficients of a surface are called the intrinsic geometric properties. The collection of these geometric properties and objects forms the subject of intrinsic geometry of a surface; its properties do not depend on the shape of the surface, but only on measurements that we can carry out while staying on a surface itself (Toponogov, 2006). The second fundamental coefficients reflect the local warping of the surface, namely its deviation from the tangent plane at the point under consideration (Liseikin, 2004).

If a surface is a graph of a function $z = f(x, y)$ or $r = (x, y, f(x, y))$, the first fundamental coefficients E, F, and G can be formulated as

$$\begin{cases} E = 1 + f_x^2 \\ G = 1 + f_y^2 \\ F = f_x \cdot f_y \end{cases} \qquad (2.1)$$

The second fundamental coefficients L, M, and N can be formulated as

$$\begin{cases} L = \dfrac{f_{xx}}{\sqrt{1 + f_x^2 + f_y^2}} \\[2ex] N = \dfrac{f_{yy}}{\sqrt{1 + f_x^2 + f_y^2}} \\[2ex] M = \dfrac{f_{xy}}{\sqrt{1 + f_x^2 + f_y^2}} \end{cases} \qquad (2.2)$$

The Gauss equation set can be formulated as

$$\begin{cases} f_{xx} = \Gamma^1_{11} \cdot f_x + \Gamma^2_{11} \cdot f_y + L \cdot (E \cdot G - F^2)^{-1/2} \\ f_{yy} = \Gamma^1_{22} \cdot f_x + \Gamma^2_{22} \cdot f_y + N \cdot (E \cdot G - F^2)^{-1/2} \\ f_{xy} = \Gamma^1_{12} \cdot f_x + \Gamma^2_{12} \cdot f_y + M \cdot (E \cdot G - F^2)^{-1/2} \end{cases} \quad (2.3)$$

where

$$\Gamma^1_{11} = \frac{1}{2}(G \cdot E_x - 2F \cdot F_x + F \cdot E_y) \cdot (E \cdot G - F^2)^{-1}$$

$$\Gamma^1_{12} = \frac{1}{2}(G \cdot E_y - F \cdot G_x) \cdot (E \cdot G - F^2)^{-1}$$

$$\Gamma^1_{22} = \frac{1}{2}(2G \cdot F_y - G \cdot G_x - F \cdot G_y) \cdot (E \cdot G - F^2)^{-1}$$

$$\Gamma^2_{11} = \frac{1}{2}(2E \cdot F_x - E \cdot E_y - F \cdot E_x) \cdot (E \cdot G - F^2)^{-1}$$

$$\Gamma^2_{12} = \frac{1}{2}(E \cdot G_x - F \cdot E_y) \cdot (E \cdot G - F^2)^{-1}$$

$$\Gamma^2_{22} = \frac{1}{2}(E \cdot G_y - 2F \cdot F_y + F \cdot G_x) \cdot (E \cdot G - F^2)^{-1}$$

The Christoffel symbols of the second kind $\Gamma^1_{11}, \Gamma^1_{12}, \Gamma^1_{22}, \Gamma^2_{11}, \Gamma^2_{12}$, and Γ^2_{22}, depend only on the first fundamental coefficients and their derivatives.

If $\{(x_i, y_j)\}$ is an orthogonal division of computational domain Ω, $[0, Bx] \times [0, By]$ is the normalized computational domain and $\max\{Bx, By\} = 1$, $h = (Bx/(I + 1)) = (By/(J + 1))$ is computational step length and $\{(x_i, y_j) \mid 0 \le i \le I + 1, 0 \le j \le J + 1\}$ are grid cells in the normalized computational domain, then $f(x + h, y)$ and $f(x - h, y)$ could be formulated as the following Taylor expansion in the series

$$f(x + h, y) = f(x, y) + h\frac{\partial f(x, y)}{\partial x} + \frac{h^2}{2!}\frac{\partial^2 f(x, y)}{\partial x^2} + \frac{h^3}{3!}\frac{\partial^3 f(x, y)}{\partial x^3} + O(h^4) \quad (2.4)$$

$$f(x - h, y) = f(x, y) - h\frac{\partial f(x, y)}{\partial x} + \frac{h^2}{2!}\frac{\partial^2 f(x, y)}{\partial x^2} - \frac{h^3}{3!}\frac{\partial^3 f(x, y)}{\partial x^3} + O(h^4) \quad (2.5)$$

Formulation of Equation 2.4 minus formulation of Equation 2.5 results as follows:

$$f(x + h, y) - f(x - h, y) = 2h\frac{\partial f(x, y)}{\partial x} + \frac{2h^3}{3!}\frac{\partial^3 f(x, y)}{\partial x^3} + O(h^5) \quad (2.6)$$

Therefore,

$$f_x(x,y) = \frac{\partial f(x,y)}{\partial x} = \frac{f(x+h,y) - f(x-h,y)}{2h} - \frac{h^2}{3!}\frac{\partial^3 f(x,y)}{\partial x^3} + O(h^4) \qquad (2.7)$$

For sufficiently small h, the finite difference approximation of $f_x(x, y)$ could be expressed as

$$f_x(x,y) \approx \frac{f(x+h,y) - f(x-h,y)}{2h} \qquad (2.8)$$

Similarly, for sufficiently small h, the finite difference approximation of $f(x, y)$ could be expressed as

$$f_y(x,y) \approx \frac{f(x,y+h) - f(x,y-h)}{2h} \qquad (2.9)$$

Formulation of Equation 2.4 plus formulation of Equation 2.5 results as follows:

$$f(x+h,y) + f(x-h,y) = 2f(x,y) + \frac{2h^2}{2!}\frac{\partial^2 f(x,y)}{\partial x^2} + O(h^4) \qquad (2.10)$$

Therefore,

$$f_{xx}(x,y) = \frac{\partial^2 f(x,y)}{\partial x^2} = \frac{f(x+h,y) - 2f(x,y) + f(x-h,y)}{h^2} + O(h^2) \qquad (2.11)$$

For sufficiently small h, the finite difference approximation of $f_{xx}(x, y)$ could be expressed as

$$f_{xx}(x,y) \approx \frac{f(x+h,y) - 2f(x,y) + f(x-h,y)}{h^2} \qquad (2.12)$$

Similarly,

$$f_{yy}(x,y) \approx \frac{f(x,y+h) - 2f(x,y) + f(x,y-h)}{h^2} \qquad (2.13)$$

The finite difference of Gauss equation set can be formulated as

$$
\begin{cases}
f_{i+1,j} - 2f_{i,j} + f_{i-1,j} = (\Gamma_{11}^1)_{i,j} \cdot \dfrac{f_{i+1,j} - f_{i-1,j}}{2} \cdot h + (\Gamma_{11}^2)_{i,j} \\
\qquad \times \dfrac{f_{i,j+1} - f_{i,j-1}}{2} \cdot h + \dfrac{L_{i,j}}{\sqrt{E_{i,j} + G_{i,j} - 1}} \cdot h^2 \\
f_{i,j+1} - 2f_{i,j} + f_{i,j-1} = (\Gamma_{22}^1)_{i,j} \cdot \dfrac{f_{i+1,j} - f_{i-1,j}}{2} \cdot h + (\Gamma_{22}^2)_{i,j} \\
\qquad \times \dfrac{f_{i,j+1} - f_{i,j-1}}{2} \cdot h + \dfrac{N_{i,j}}{\sqrt{E_{i,j} + G_{i,j} - 1}} \cdot h^2 \\
f_{i+1,j+1} - f_{i-1,j+1} + f_{i-1,j-1} - f_{i+1,j-1} = 2 \cdot h \cdot (\Gamma_{12}^1)_{i,j} \cdot (f_{i+1,j} - f_{i-1,j}) \\
\qquad + 2 \cdot h \cdot (\Gamma_{12}^2)_{i,j} \cdot (f_{i,j+1} - f_{i,j-1}) \\
\qquad + \dfrac{4 \cdot M_{i,j} \cdot h^2}{\sqrt{E_{i,j} + G_{i,j} - 1}}
\end{cases}
\tag{2.14}
$$

In terms of numerical mathematics (Quarteroni et al., 2000), the iterative formulation of equation set (Equation 2.14) could be expressed as

$$
\begin{cases}
f_{i+1,j}^{(n+1)} - 2f_{i,j}^{(n+1)} + f_{i-1,j}^{(n+1)} = (\Gamma_{11}^1)_{i,j}^{(n)} \cdot \dfrac{f_{i+1,j}^{(n)} - f_{i-1,j}^{(n)}}{2} \cdot h + (\Gamma_{11}^2)_{i,j}^{(n)} \\
\qquad \times \dfrac{f_{i,j+1}^{(n)} - f_{i,j-1}^{(n)}}{2} \cdot h + \dfrac{L_{i,j}^{(n)}}{\sqrt{E_{i,j}^{(n)} + G_{i,j}^{(n)} - 1}} \cdot h^2 \\
f_{i,j+1}^{(n+1)} - 2f_{i,j}^{(n+1)} + f_{i,j-1}^{(n+1)} = (\Gamma_{22}^1)_{i,j}^{(n)} \cdot \dfrac{f_{i+1,j}^{(n)} - f_{i-1,j}^{(n)}}{2} \cdot h + (\Gamma_{22}^2)_{i,j}^{(n)} \\
\qquad \times \dfrac{f_{i,j+1}^{(n)} - f_{i,j-1}^{(n)}}{2} \cdot h + \dfrac{N_{i,j}^{(n)}}{\sqrt{E_{i,j}^{(n)} + G_{i,j}^{(n)} - 1}} \cdot h^2 \\
f_{i+1,j+1}^{(n+1)} - f_{i-1,j+1}^{(n+1)} + f_{i-1,j-1}^{(n+1)} - f_{i+1,j-1}^{(n+1)} = 2 \cdot h \cdot (\Gamma_{12}^1)_{i,j}^{(n)} \cdot \left(f_{i+1,j}^{(n)} - f_{i-1,j}^{(n)}\right) \\
\qquad + 2 \cdot h \cdot (\Gamma_{12}^2)_{i,j}^{(n)} \cdot \left(f_{i,j+1}^{(n)} - f_{i,j-1}^{(n)}\right) \\
\qquad + \dfrac{4 \cdot h^2 \cdot M_{i,j}^{(n)}}{\sqrt{E_{i,j}^{(n)} + G_{i,j}^{(n)} - 1}}
\end{cases}
\tag{2.15}
$$

where

$$
E_{i,j}^{(n)} = 1 + \left(\frac{f_{i+1,j}^{(n)} - f_{i-1,j}^{(n)}}{2h} \right)^2
$$

$$G_{i,j}^{(n)} = 1 + \left(\frac{f_{i,j+1}^{(n)} - f_{i,j-1}^{(n)}}{2h} \right)^2$$

$$F_{i,j}^{(n)} = \left(\frac{f_{i+1,j}^{(n)} - f_{i-1,j}^{(n)}}{2h} \right)\left(\frac{f_{i,j+1}^{(n)} - f_{i,j-1}^{(n)}}{2h} \right)$$

$$L_{i,j}^{(n)} = \frac{(f_{i+1,j}^{(n)} - 2f_{i,j}^{(n)} + f_{i-1,j}^{(n)}) / h^2}{\sqrt{1 + \left(\left(f_{i+1,j}^{(n)} - f_{i-1,j}^{(n)}\right) / 2h\right)^2 + \left(\left(f_{i,j+1}^{(n)} - f_{i,j-1}^{(n)}\right) / 2h\right)^2}}$$

$$N_{i,j}^{(n)} = \frac{\left(f_{i,j+1}^{(n)} - 2f_{i,j}^{(n)} + f_{i,j-1}^{(n)}\right) / h^2}{\sqrt{1 + \left(\left(f_{i+1,j}^{(n)} - f_{i-1,j}^{(n)}\right) / 2h\right)^2 + \left(\left(f_{i,j+1}^{(n)} - f_{i,j-1}^{(n)}\right) / 2h\right)^2}}$$

$$M_{i,j}^{(n)} = \frac{\left(f_{i+1,j+1}^{(n)} - f_{i+1,j-1}^{(n)}\right) / 4h^2 - \left(f_{i-1,j+1}^{(n)} - f_{i-1,j-1}^{(n)}\right) / 4h^2}{\sqrt{1 + \left(\frac{f_{i+1,j}^{(n)} - f_{i-1,j}^{(n)}}{2h}\right)^2 + \left(\frac{f_{i,j+1}^{(n)} - f_{i,j-1}^{(n)}}{2h}\right)^2}}$$

$$(\Gamma_{11}^1)_{i,j}^{(n)} = \frac{G_{i,j}^{(n)}(E_{i+1,j}^{(n)} - E_{i-1,j}^{(n)}) - 2F_{i,j}^{(n)}(F_{i+1,j}^{(n)} - F_{i-1,j}^{(n)}) + F_{i,j}^{(n)}(E_{i,j+1}^{(n)} - E_{i,j-1}^{(n)})}{4\left(E_{i,j}^{(n)} \cdot G_{i,j}^{(n)} - \left(F_{i,j}^{(n)}\right)^2\right)h}$$

$$(\Gamma_{11}^2)_{i,j}^{(n)} = \frac{2E_{i,j}^{(n)}(F_{i+1,j}^{(n)} - F_{i-1,j}^{(n)}) - E_{i,j}^{(n)}(E_{i,j+1}^{(n)} - E_{i,j-1}^{(n)}) - F_{i,j}^{(n)}(E_{i+1,j}^{(n)} - E_{i-1,j}^{(n)})}{4\left(E_{i,j}^{(n)} \cdot G_{i,j}^{(n)} - \left(F_{i,j}^{(n)}\right)^2\right)h}$$

$$(\Gamma_{22}^1)_{i,j}^{(n)} = \frac{2G_{i,j}^{(n)}(F_{i,j+1}^{(n)} - F_{i,j-1}^{(n)}) - G_{i,j}^{(n)}(G_{i+1,j}^{(n)} - G_{i-1,j}^{(n)}) - F_{i,j}^{(n)}(G_{i,j+1}^{(n)} - G_{i,j-1}^{(n)})}{4\left(E_{i,j}^{(n)} \cdot G_{i,j}^{(n)} - \left(F_{i,j}^{(n)}\right)^2\right)h}$$

$$(\Gamma_{22}^2)_{i,j}^{(n)} = \frac{E_{i,j}^{(n)}(G_{i,j+1}^{(n)} - G_{i,j-1}^{(n)}) - 2F_{i,j}^{(n)}(F_{i,j+1}^{(n)} - F_{i,j-1}^{(n)}) + F_{i,j}^{(n)}(G_{i+1,j}^{(n)} - G_{i-1,j}^{(n)})}{4\left(E_{i,j}^{(n)} \cdot G_{i,j}^{(n)} - \left(F_{i,j}^{(n)}\right)^2\right)h}$$

$$(\Gamma_{12}^1)_{i,j}^{(n)} = \frac{G_{i,j}^{(n)}(E_{i,j+1}^{(n)} - E_{i,j-1}^{(n)}) - F_{i,j}^{(n)}(G_{i+1,j}^{(n)} - G_{i-1,j}^{(n)})}{4\left(E_{i,j}^{(n)} \cdot G_{i,j}^{(n)} - \left(F_{i,j}^{(n)}\right)^2\right)h}$$

$$(\Gamma_{12}^2)_{i,j}^{(n)} = \frac{E_{i,j}^{(n)}(G_{i+1,j}^{(n)} - G_{i-1,j}^{(n)}) - F_{i,j}^{(n)}(F_{i,j+1}^{(n)} - F_{i,j-1}^{(n)})}{4\left(E_{i,j}^{(n)} \cdot G_{i,j}^{(n)} - \left(F_{i,j}^{(n)}\right)^2\right)h}$$

Suppose, $\mathbf{z}^{(n+1)} = \left(f_{1,1}^{(n+1)}, \ldots, f_{1,J}^{(n+1)}, \ldots, f_{I-1,1}^{(n+1)}, \ldots, f_{I-1,J}^{(n+1)}, f_{I,1}^{(n+1)}, \ldots, f_{I,J}^{(n+1)} \right)^{\mathrm{T}}$, $n \geq 0$;
$\mathbf{z}^{(0)} = (\tilde{f}_{1,1}, \ldots, \tilde{f}_{1,J}, \ldots, \tilde{f}_{I-1,1}, \ldots, \tilde{f}_{I-1,J}, \tilde{f}_{I,1}, \ldots, \tilde{f}_{I,J})^{\mathrm{T}}$ indicates interpolations based on sampling points. Then, the first equation of equation set (Equation 2.15) can be formulated as

$$A \cdot \mathbf{z}^{(n+1)} = \mathbf{d}^{(n)} \tag{2.16}$$

where

$$\mathbf{d}^{(n)} = \left[\mathbf{d}_1^{(n)}, \mathbf{d}_2^{(n)}, \ldots, \mathbf{d}_{I-1}^{(n)}, \mathbf{d}_I^{(n)} \right]^{\mathrm{T}}$$

represents the vector of the right-hand item of Equation 2.16; A represents the coefficient matrix of Equation 2.16

$$
\mathbf{d}_1^{(n)} =
\begin{bmatrix}
\dfrac{f_{2,1}^{(n)} - f_{0,1}^{(n)}}{2} \cdot (\Gamma_{11}^1)_{1,1}^{(n)} \cdot h + \dfrac{f_{1,2}^{(n)} - f_{1,0}^{(n)}}{2} \cdot \left(\Gamma_{11}^2\right)_{1,1}^{(n)} \cdot h \\[2mm]
+ \dfrac{L_{1,1}^{(n)}}{\sqrt{E_{1,1}^{(n)} + G_{1,1}^{(n)} - 1}} \cdot h^2 - f_{0,1}^{(n+1)} \\[4mm]
\dfrac{f_{2,2}^{(n)} - f_{0,2}^{(n)}}{2} \cdot (\Gamma_{11}^1)_{1,2}^{(n)} \cdot h + \dfrac{f_{1,3}^{(n)} - f_{1,1}^{(n)}}{2} \cdot \left(\Gamma_{11}^2\right)_{1,2}^{(n)} \cdot h \\[2mm]
+ \dfrac{L_{1,2}^{(n)}}{\sqrt{E_{1,2}^{(n)} + G_{1,2}^{(n)} - 1}} \cdot h^2 - f_{0,2}^{(n+1)} \\[4mm]
\vdots \\[2mm]
\dfrac{f_{2,j}^{(n)} - f_{0,j}^{(n)}}{2} \cdot (\Gamma_{11}^1)_{1,j}^{(n)} \cdot h + \dfrac{f_{1,j+1}^{(n)} - f_{1,j-1}^{(n)}}{2} \cdot \left(\Gamma_{11}^2\right)_{1,j}^{(n)} \cdot h \\[2mm]
+ \dfrac{L_{1,j}^{(n)}}{\sqrt{E_{1,j}^{(n)} + G_{1,j}^{(n)} - 1}} \cdot h^2 - f_{0,j}^{(n+1)} \\[4mm]
\vdots \\[2mm]
\dfrac{f_{2,J-1}^{(n)} - f_{0,J-1}^{(n)}}{2} \cdot (\Gamma_{11}^1)_{1,J-1}^{(n)} \cdot h + \dfrac{f_{1,J}^{(n)} - f_{1,J-2}^{(n)}}{2} \cdot \left(\Gamma_{11}^2\right)_{1,J-1}^{(n)} \cdot h \\[2mm]
+ \dfrac{L_{1,J-1}^{(n)}}{\sqrt{E_{1,J-1}^{(n)} + G_{1,J-1}^{(n)} - 1}} \cdot h^2 - f_{0,J-1}^{(n+1)} \\[4mm]
\dfrac{f_{2,J}^{(n)} - f_{0,J}^{(n)}}{2} \cdot (\Gamma_{11}^1)_{1,J}^{(n)} \cdot h + \dfrac{f_{1,J+1}^{(n)} - f_{1,J-1}^{(n)}}{2} \cdot \left(\Gamma_{11}^2\right)_{1,J}^{(n)} \cdot h \\[2mm]
+ \dfrac{L_{1,J}^{(n)}}{\sqrt{E_{1,J}^{(n)} + G_{1,J}^{(n)} - 1}} \cdot h^2 - f_{0,J}^{(n+1)}
\end{bmatrix}_{J \times 1}^{\mathrm{T}}
$$

For $i = 2, 3, \ldots, I-2, I-1$,

$$\mathbf{d}_i^{(n)} =$$

$$
\left[
\begin{array}{l}
\dfrac{f_{i+1,1}^{(n)} - f_{i-1,1}^{(n)}}{2} \cdot \left(\Gamma_{11}^1\right)_{i,1}^{(n)} \cdot h + \dfrac{f_{i,2}^{(n)} - f_{i,0}^{(n)}}{2} \cdot \left(\Gamma_{11}^2\right)_{i,1}^{(n)} \cdot h + \dfrac{L_{i,1}^{(n)}}{\sqrt{E_{i,1}^{(n)} + G_{i,1}^{(n)} - 1}} \cdot h^2 \\[4mm]
\dfrac{f_{i+1,2}^{(n)} - f_{i-1,2}^{(n)}}{2} \cdot \left(\Gamma_{11}^1\right)_{i,2}^{(n)} \cdot h + \dfrac{f_{i,3}^{(n)} - f_{i,1}^{(n)}}{2} \cdot \left(\Gamma_{11}^2\right)_{i,2}^{(n)} \cdot h + \dfrac{L_{i,2}^{(n)}}{\sqrt{E_{i,2}^{(n)} + G_{i,2}^{(n)} - 1}} \cdot h^2 \\[4mm]
\vdots \\[2mm]
\dfrac{f_{i+1,j}^{(n)} - f_{i-1,j}^{(n)}}{2} \cdot \left(\Gamma_{11}^1\right)_{i,j}^{(n)} \cdot h + \dfrac{f_{i,j+1}^{(n)} - f_{i,j-1}^{(n)}}{2} \cdot \left(\Gamma_{11}^2\right)_{i,j}^{(n)} \cdot h + \dfrac{L_{i,j}^{(n)}}{\sqrt{E_{i,j}^{(n)} + G_{i,j}^{(n)} - 1}} \cdot h^2 \\[4mm]
\vdots \\[2mm]
\dfrac{f_{i+1,J-1}^{(n)} - f_{i-1,J-1}^{(n)}}{2} \cdot \left(\Gamma_{11}^1\right)_{i,J-1}^{(n)} \cdot h + \dfrac{f_{i,J}^{(n)} - f_{i,J-2}^{(n)}}{2} \cdot \left(\Gamma_{11}^2\right)_{i,J-1}^{(n)} \cdot h + \dfrac{L_{i,J-1}^{(n)}}{\sqrt{E_{i,J-1}^{(n)} + G_{i,J-1}^{(n)} - 1}} \cdot h^2 \\[4mm]
\dfrac{f_{i+1,J}^{(n)} - f_{i-1,J}^{(n)}}{2} \cdot \left(\Gamma_{11}^1\right)_{i,J}^{(n)} \cdot h + \dfrac{f_{i,J+1}^{(n)} - f_{i,J-1}^{(n)}}{2} \cdot \left(\Gamma_{11}^2\right)_{i,J}^{(n)} \cdot h + \dfrac{L_{i,J}^{(n)}}{\sqrt{E_{i,J}^{(n)} + G_{i,J}^{(n)} - 1}} \cdot h^2
\end{array}
\right]_{J \times 1}^{T}
$$

$$\mathbf{d}_I^{(n)} =$$

$$
\left[
\begin{array}{l}
\dfrac{f_{I+1,1}^{(n)} - f_{I-1,1}^{(n)}}{2} \cdot \left(\Gamma_{11}^1\right)_{I,1}^{(n)} \cdot h + \dfrac{f_{I,2}^{(n)} - f_{I,0}^{(n)}}{2} \cdot \left(\Gamma_{11}^2\right)_{I,1}^{(n)} \cdot h + \dfrac{h^2 \cdot L_{I,1}^{(n)}}{\sqrt{E_{I,1}^{(n)} + G_{I,1}^{(n)} - 1}} - f_{I+1,1}^{(n+1)} \\[4mm]
\dfrac{f_{I+1,2}^{(n)} - f_{I-1,2}^{(n)}}{2} \cdot \left(\Gamma_{11}^1\right)_{I,2}^{(n)} \cdot h + \dfrac{f_{I,3}^{(n)} - f_{I,1}^{(n)}}{2} \cdot \left(\Gamma_{11}^2\right)_{I,2}^{(n)} \cdot h + \dfrac{h^2 \cdot L_{I,2}^{(n)}}{\sqrt{E_{I,2}^{(n)} + G_{I,2}^{(n)} - 1}} - f_{I+1,2}^{(n+1)} \\[4mm]
\vdots \\[2mm]
\dfrac{f_{I+1,j}^{(n)} - f_{I-1,j}^{(n)}}{2} \cdot \left(\Gamma_{11}^1\right)_{I,j}^{(n)} \cdot h + \dfrac{f_{I,j+1}^{(n)} - f_{I,j-1}^{(n)}}{2} \cdot \left(\Gamma_{11}^2\right)_{I,j}^{(n)} \cdot h + \dfrac{h^2 \cdot L_{I,j}^{(n)}}{\sqrt{E_{I,j}^{(n)} + G_{I,j}^{(n)} - 1}} - f_{I+1,j}^{(n+1)} \\[4mm]
\vdots \\[2mm]
\dfrac{f_{I+1,J-1}^{(n)} - f_{I-1,J-1}^{(n)}}{2} \cdot \left(\Gamma_{11}^1\right)_{I,J-1}^{(n)} \cdot h + \dfrac{f_{I,J}^{(n)} - f_{I,J-2}^{(n)}}{2} \cdot \left(\Gamma_{11}^2\right)_{I,J-1}^{(n)} \cdot h \\[3mm]
\quad + \dfrac{h^2 \cdot L_{I,J-1}^{(n)}}{\sqrt{E_{I,J-1}^{(n)} + G_{I,J-1}^{(n)} - 1}} - f_{I+1,J-1}^{(n+1)} \\[4mm]
\dfrac{f_{I+1,J}^{(n)} - f_{I-1,J}^{(n)}}{2} \cdot \left(\Gamma_{11}^1\right)_{I,J}^{(n)} \cdot h + \dfrac{f_{I,J+1}^{(n)} - f_{I,J-1}^{(n)}}{2} \cdot \left(\Gamma_{11}^2\right)_{I,J}^{(n)} \cdot h + \dfrac{h^2 \cdot L_{I,J}^{(n)}}{\sqrt{E_{I,J}^{(n)} + G_{I,J}^{(n)} - 1}} - f_{I+1,J}^{(n+1)}
\end{array}
\right]_{J \times 1}^{T}
$$

$$
\mathbf{A} = \begin{bmatrix} -2\mathbf{I}_J & \mathbf{I}_J & & & & \\ \mathbf{I}_J & -2\mathbf{I}_J & \mathbf{I}_J & & & \\ & \ddots & \ddots & \ddots & & \\ & & \mathbf{I}_J & -2\mathbf{I}_J & \mathbf{I}_J & \\ & & & \mathbf{I}_J & -2\mathbf{I}_J \end{bmatrix}_{(I \cdot J) \times (I \cdot J)}
$$

where $\mathbf{I}_J$ is the $J \times J$ unit matrix.

The second equation of equation set (Equation 2.15) can be formulated as

$$
\mathbf{B} \cdot \mathbf{z}^{(n+1)} = \mathbf{q}^{(n)} \tag{2.17}
$$

where

$$
\mathbf{q}^{(n)} = \left[\mathbf{q}_1^{(n)}, \mathbf{q}_2^{(n)}, \dots, \mathbf{q}_{I-1}^{(n)}, \mathbf{q}_I^{(n)} \right]^{\mathrm{T}}
$$

is vector of the right-hand item of the Equation 2.17; $\mathbf{B}$ is the coefficient matrix of Equation 2.17,

$$
\mathbf{B} = \begin{bmatrix} \mathbf{B}_J & & \\ & \ddots & \\ & & \mathbf{B}_J \end{bmatrix}_{(I \times J) \times (I \times J)}
$$

and

$$
\mathbf{B}_J = \begin{bmatrix} -2 & 1 & & & & & \\ 1 & -2 & 1 & & & & \\ & \ddots & \ddots & \ddots & & & \\ & & 1 & -2 & 1 & & \\ & & & \ddots & \ddots & \ddots & \\ & & & & 1 & -2 & 1 \\ & & & & & 1 & -2 \end{bmatrix}_{J \times J}
$$

For $i = 1, 2, \ldots, I$,

$$\mathbf{q}_i^{(n)} =$$

$$
\begin{bmatrix}
\dfrac{f_{i+1,1}^{(n)} - f_{i-1,1}^{(n)}}{2} \cdot (\Gamma_{22}^1)_{i,1}^{(n)} \cdot h + \dfrac{f_{i,2}^{(n)} - f_{i,0}^{(n)}}{2} \cdot (\Gamma_{22}^2)_{i,1}^{(n)} \cdot h + \dfrac{N_{i,1}^{(n)}}{\sqrt{E_{i,1}^{(n)} + G_{i,1}^{(n)} - 1}} \cdot h^2 - f_{i,0}^{(n+1)} \\[3mm]
\dfrac{f_{i+1,2}^{(n)} - f_{i-1,2}^{(n)}}{2} \cdot (\Gamma_{22}^1)_{i,2}^{(n)} \cdot h + \dfrac{f_{i,3}^{(n)} - f_{i,1}^{(n)}}{2} \cdot (\Gamma_{22}^2)_{i,2}^{(n)} \cdot h + \dfrac{N_{i,2}^{(n)}}{\sqrt{E_{i,2}^{(n)} + G_{i,2}^{(n)} - 1}} \cdot h^2 \\[3mm]
\vdots \\[3mm]
\dfrac{f_{i+1,j}^{(n)} - f_{i-1,j}^{(n)}}{2} \cdot (\Gamma_{22}^1)_{i,j}^{(n)} \cdot h + \dfrac{f_{i,j+1}^{(n)} - f_{i,j-1}^{(n)}}{2} \cdot (\Gamma_{22}^2)_{i,j}^{(n)} \cdot h + \dfrac{N_{i,j}^{(n)}}{\sqrt{E_{i,j}^{(n)} + G_{i,j}^{(n)} - 1}} \cdot h^2 \\[3mm]
\vdots \\[3mm]
\dfrac{f_{i+1,J-1}^{(n)} - f_{i-1,J-1}^{(n)}}{2} \cdot (\Gamma_{22}^1)_{i,J-1}^{(n)} \cdot h + \dfrac{f_{i,J}^{(n)} - f_{i,J-2}^{(n)}}{2} \cdot (\Gamma_{22}^2)_{i,J-1}^{(n)} \cdot h + \dfrac{N_{i,J-1}^{(n)}}{\sqrt{E_{i,J-1}^{(n)} + G_{i,J-1}^{(n)} - 1}} \cdot h^2 \\[3mm]
\dfrac{f_{i+1,J}^{(n)} - f_{i-1,J}^{(n)}}{2} \cdot (\Gamma_{22}^1)_{i,J}^{(n)} \cdot h + \dfrac{f_{i,J+1}^{(n)} - f_{i,J-1}^{(n)}}{2} \cdot (\Gamma_{22}^2)_{i,J}^{(n)} \cdot h + \dfrac{N_{i,J}^{(n)}}{\sqrt{E_{i,J}^{(n)} + G_{i,J}^{(n)} - 1}} \cdot h^2 - f_{i,J+1}^{(n+1)}
\end{bmatrix}_{J \times 1}^T
$$

The third equation of equation set (Equation 2.15) can be expressed as

$$\mathbf{C} \cdot \mathbf{z}^{(n+1)} = \mathbf{h}^{(n)} \tag{2.18}$$

where

$$\mathbf{h}^{(n)} = \left[\mathbf{h}_1^{(n)}, \, \mathbf{h}_2^{(n)}, \, \ldots, \, \mathbf{h}_{I-1}^{(n)}, \, \mathbf{h}_I^{(n)} \right]^T$$

is the vector of the right-hand item of Equation 2.18; $\mathbf{C}$ is the coefficient matrix of Equation 2.18,

$$
\mathbf{C} =
\begin{bmatrix}
0 & \mathbf{C}_J & & & \\
\mathbf{C}_J & 0 & \mathbf{C}_J & & \\
& \ddots & \ddots & \ddots & \\
& & \mathbf{C}_J & 0 & \mathbf{C}_J \\
& & & \mathbf{C}_J & 0
\end{bmatrix}_{(I \cdot J) \times (I \cdot J)}
$$

and

$$\mathbf{C}_J = \begin{bmatrix} 0 & -1 & & & \\ 1 & 0 & -1 & & \\ & \ddots & \ddots & \ddots & \\ & & 1 & 0 & -1 \\ & & & 1 & 0 \end{bmatrix}_{J \times J}$$

$$\mathbf{h}_1^{(n)} = \begin{bmatrix} 2h\left((\Gamma_{12}^1)_{1,1}^{(n)}(f_{2,1}^{(n)} - f_{0,1}^{(n)}) + \left(\Gamma_{12}^2\right)_{1,1}^{(n)}(f_{1,2}^{(n)} - f_{1,0}^{(n)})\right) \\ + \dfrac{4M_{1,1}^{(n)} \cdot h^2}{\sqrt{E_{1,1}^{(n)} + G_{1,1}^{(n)} - 1}} + f_{0,2}^{(n+1)} - f_{0,0}^{(n+1)} + f_{2,0}^{(n+1)} \\ \\ 2h\left((\Gamma_{12}^1)_{1,2}^{(n)}(f_{2,2}^{(n)} - f_{0,2}^{(n)}) + \left(\Gamma_{12}^2\right)_{1,2}^{(n)}(f_{1,3}^{(n)} - f_{1,1}^{(n)})\right) \\ + \dfrac{4M_{1,2}^{(n)} \cdot h^2}{\sqrt{E_{1,2}^{(n)} + G_{1,2}^{(n)} - 1}} + f_{0,3}^{(n+1)} - f_{0,1}^{(n+1)} \\ \vdots \\ 2h\left((\Gamma_{12}^1)_{1,j}^{(n)}(f_{2,j}^{(n)} - f_{0,j}^{(n)}) + \left(\Gamma_{12}^2\right)_{1,j}^{(n)}(f_{1,j+1}^{(n)} - f_{1,j-1}^{(n)})\right) \\ + \dfrac{4M_{1,j}^{(n)} \cdot h^2}{\sqrt{E_{1,j}^{(n)} + G_{1,j}^{(n)} - 1}} + f_{0,j+1}^{(n+1)} - f_{0,j-1}^{(n+1)} \\ \vdots \\ 2h\left((\Gamma_{12}^1)_{1,J-1}^{(n)}(f_{2,J-1}^{(n)} - f_{0,J-1}^{(n)}) + \left(\Gamma_{12}^2\right)_{1,J-1}^{(n)}(f_{1,J}^{(n)} - f_{1,J-2}^{(n)})\right) \\ + \dfrac{4M_{1,J-1}^{(n)} \cdot h^2}{\sqrt{E_{1,J-1}^{(n)} + G_{1,J-1}^{(n)} - 1}} + f_{0,J}^{(n+1)} - f_{0,J-2}^{(n+1)} \\ \\ 2h\left((\Gamma_{12}^1)_{1,J}^{(n)}(f_{2,J}^{(n)} - f_{0,J}^{(n)}) + \left(\Gamma_{12}^2\right)_{1,J}^{(n)}(f_{1,J+1}^{(n)} - f_{1,J-1}^{(n)})\right) \\ + \dfrac{4M_{1,J}^{(n)} \cdot h^2}{\sqrt{E_{1,J}^{(n)} + G_{1,J}^{(n)} - 1}} + f_{0,J+1}^{(n+1)} - f_{0,J-1}^{(n+1)} - f_{2,J+1}^{(n+1)} \end{bmatrix}_{J \times 1}^{T}$$

For $i = 2, 3, \ldots, I - 2, I - 1,$

$\mathbf{h}_i^{(n)} =$

$$
\begin{bmatrix}
2h\left((\Gamma_{12}^1)_{i,1}^{(n)}(f_{i+1,1}^{(n)} - f_{i-1,1}^{(n)}) + \left(\Gamma_{12}^2\right)_{i,1}^{(n)}(f_{i,2}^{(n)} - f_{i,0}^{(n)})\right) + \dfrac{4M_{i,1}^{(n)} \cdot h^2}{\sqrt{E_{i,1}^{(n)} + G_{i,1}^{(n)} - 1}} + f_{i+1,0}^{(n+1)} \\[4mm]
2h\left((\Gamma_{12}^1)_{i,2}^{(n)}(f_{i+1,2}^{(n)} - f_{i-1,2}^{(n)}) + \left(\Gamma_{12}^2\right)_{i,2}^{(n)}(f_{i,3}^{(n)} - f_{i,1}^{(n)})\right) + \dfrac{4M_{i,2}^{(n)} \cdot h^2}{\sqrt{E_{i,2}^{(n)} + G_{i,2}^{(n)} - 1}} \\[4mm]
\vdots \\[2mm]
2h\left((\Gamma_{12}^1)_{i,j}^{(n)}(f_{i+1,j}^{(n)} - f_{i-1,j}^{(n)}) + \left(\Gamma_{12}^2\right)_{i,j}^{(n)}(f_{i,j+1}^{(n)} - f_{i,j-1}^{(n)})\right) + \dfrac{4M_{i,j}^{(n)} \cdot h^2}{\sqrt{E_{i,j}^{(n)} + G_{i,j}^{(n)} - 1}} \\[4mm]
\vdots \\[2mm]
2h\left((\Gamma_{12}^1)_{i,J-1}^{(n)}(f_{i+1,J-1}^{(n)} - f_{i-1,J-1}^{(n)}) + \left(\Gamma_{12}^2\right)_{i,J-1}^{(n)}(f_{i,J}^{(n)} - f_{i,J}^{(n)})\right) + \dfrac{4M_{i,J-1}^{(n)} \cdot h^2}{\sqrt{E_{i,J-1}^{(n)} + G_{i,J-1}^{(n)} - 1}} \\[4mm]
2h\left((\Gamma_{12}^1)_{i,J}^{(n)}(f_{i+1,J}^{(n)} - f_{i-1,J}^{(n)}) + \left(\Gamma_{12}^2\right)_{i,J}^{(n)}(f_{i,J+1}^{(n)} - f_{i,J-1}^{(n)})\right) + \dfrac{4M_{i,J}^{(n)} \cdot h^2}{\sqrt{E_{i,J}^{(n)} + G_{i,J}^{(n)} - 1}} - f_{i+1,J+1}^{(n+1)}
\end{bmatrix}_{J \times 1}^{T}
$$

$$
\mathbf{h}_I^{(n)} =
\begin{bmatrix}
\begin{aligned}
&2h\left((\Gamma_{12}^1)_{I,1}^{(n)}(f_{I+1,1}^{(n)} - f_{I-1,1}^{(n)}) + \left(\Gamma_{12}^2\right)_{I,1}^{(n)}(f_{I,2}^{(n)} - f_{I,0}^{(n)})\right)\\
&\quad + \frac{4h^2 \cdot M_{I,1}^{(n)}}{\sqrt{E_{I,1}^{(n)} + G_{I,1}^{(n)} - 1}} - f_{I+1,2}^{(n+1)} + f_{I+1,0}^{(n+1)} - f_{I-1,0}^{(n+1)}
\end{aligned} \\[6mm]
\begin{aligned}
&2h\left((\Gamma_{12}^1)_{I,2}^{(n)}(f_{I+1,2}^{(n)} - f_{I-1,2}^{(n)}) + \left(\Gamma_{12}^2\right)_{I,2}^{(n)}(f_{I,3}^{(n)} - f_{I,1}^{(n)})\right)\\
&\quad + \frac{4h^2 \cdot M_{I,2}^{(n)}}{\sqrt{E_{I,2}^{(n)} + G_{I,2}^{(n)} - 1}} - f_{I+1,3}^{(n+1)} + f_{I+1,1}^{(n+1)}
\end{aligned} \\[6mm]
\vdots \\[2mm]
\begin{aligned}
&2h\left((\Gamma_{12}^1)_{I,j}^{(n)}(f_{I+1,j}^{(n)} - f_{I-1,j}^{(n)}) + \left(\Gamma_{12}^2\right)_{I,j}^{(n)}(f_{I,j+1}^{(n)} - f_{I,j-1}^{(n)})\right)\\
&\quad + \frac{4h^2 \cdot M_{I,j}^{(n)}}{\sqrt{E_{I,j}^{(n)} + G_{I,j}^{(n)} - 1}} - f_{I+1,j+1}^{(n+1)} + f_{I+1,j-1}^{(n+1)}
\end{aligned} \\[6mm]
\vdots \\[2mm]
\begin{aligned}
&2h\left((\Gamma_{12}^1)_{I,J-1}^{(n)}(f_{I+1,J-1}^{(n)} - f_{I-1,J-1}^{(n)}) + \left(\Gamma_{12}^2\right)_{I,J-1}^{(n)}(f_{I,J}^{(n)} - f_{I,J-2}^{(n)})\right)\\
&\quad + \frac{4h^2 \cdot M_{I,J-1}^{(n)}}{\sqrt{E_{I,J-1}^{(n)} + G_{I,J-1}^{(n)} - 1}} - f_{I+1,J}^{(n+1)} + f_{I+1,J-2}^{(n+1)}
\end{aligned} \\[6mm]
\begin{aligned}
&2h\left((\Gamma_{12}^1)_{I,J}^{(n)}(f_{I+1,J}^{(n)} - f_{I-1,J}^{(n)}) + \left(\Gamma_{12}^2\right)_{I,J}^{(n)}(f_{I,J+1}^{(n)} - f_{I,J-1}^{(n)})\right)\\
&\quad + \frac{4h^2 \cdot M_{I,J}^{(n)}}{\sqrt{E_{I,J}^{(n)} + G_{I,J}^{(n)} - 1}} - f_{I+1,J+1}^{(n+1)} + f_{I+1,J-1}^{(n+1)} + f_{I-1,J+1}^{(n+1)}
\end{aligned}
\end{bmatrix}_{J \times 1}^{T}
$$

2.3 Optimum Formulation

The first, second, and third equation of the Gauss equation set are successively marked as a, b, and c. The surface model based on Equation 2.16, HASM1a, can be expressed as a constraint least-squares approximation

$$\begin{cases} \min \left\| \mathbf{A} \cdot \mathbf{z}^{(n+1)} - \mathbf{d}^{(n)} \right\|_2 \\ \text{s.t.} \quad \mathbf{S} \cdot \mathbf{z}^{(n+1)} = \mathbf{k} \end{cases} \tag{2.19}$$

where $\mathbf{S}$ and $\mathbf{k}$ are respectively the sampling points and corresponding values of $z = f(x, y)$ at the sampling points; the "min" means "minimize"; the "s.t." is the abbreviation of "subject to," which means that $\min \left\| \mathbf{A} \cdot \mathbf{z}^{(n+1)} - \mathbf{d}^{(n)} \right\|_2$ is subject to $\mathbf{S} \cdot \mathbf{z}^{(n+1)} = \mathbf{k}$. If $f_{i,j}$ is value of $z = f(x, y)$ at the pth sampling point (x_i, y_j), then $s_{p,(i-1)\times J+j} = 1$ and $k_p = \bar{f}_{i,j}$.

For sufficiently large λ, HASM1a can be formulated as an unconstrained least-squares approximation (Golub and van Loan, 1996)

$$\mathbf{z}^{(n+1)} = (\mathbf{A}^T \cdot \mathbf{A} + \lambda^2 \cdot \mathbf{S}^T \cdot \mathbf{S})^{-1}(\mathbf{A}^T \cdot \mathbf{d}^{(n)} + \lambda^2 \cdot \mathbf{S}^T \cdot \mathbf{k}) \tag{2.20}$$

Similarly, the surface model based on the Equation 2.17, HASM1b, can be expressed as

$$\mathbf{z}^{(n+1)} = (\mathbf{B}^T \cdot \mathbf{B} + \lambda^2 \cdot \mathbf{S}^T \cdot \mathbf{S})^{-1}(\mathbf{B}^T \cdot \mathbf{q}^{(n)} + \lambda^2 \cdot \mathbf{S}^T \cdot \mathbf{k}) \tag{2.21}$$

The surface model based on Equation 2.18, HASM1c, can be formulated as

$$\mathbf{z}^{(n+1)} = (\mathbf{C}^T \cdot \mathbf{C} + \lambda^2 \cdot \mathbf{S}^T \cdot \mathbf{S})^{-1}(\mathbf{C}^T \cdot \mathbf{h}^{(n)} + \lambda^2 \cdot \mathbf{S}^T \cdot \mathbf{k}) \tag{2.22}$$

The surface model based on Equations 2.16 and 2.17, HASM2ab, can be expressed as

$$\mathbf{z}^{(n+1)} = (\mathbf{A}^T \cdot \mathbf{A} + \mathbf{B}^T \cdot \mathbf{B} + \lambda^2 \cdot \mathbf{S}^T \cdot \mathbf{S})^{-1}$$
$$\times (\mathbf{A}^T \cdot \mathbf{d}^{(n)} + \mathbf{B}^T \cdot \mathbf{q}^{(n)} + \lambda^2 \cdot \mathbf{S}^T \cdot \mathbf{k}) \tag{2.23}$$

The surface model based on Equations 2.16 and 2.18, HASM2ac, can be expressed as

$$\mathbf{z}^{(n+1)} = (\mathbf{A}^T \cdot \mathbf{A} + \mathbf{C}^T \cdot \mathbf{C} + \lambda^2 \cdot \mathbf{S}^T \cdot \mathbf{S})^{-1}(\mathbf{A}^T \cdot \mathbf{d}^{(n)} + \mathbf{C}^T \cdot \mathbf{h}^{(n)} + \lambda^2 \cdot \mathbf{S}^T \cdot \mathbf{k}) \tag{2.24}$$

The one based on Equations 2.17 and 2.18, HASM2bc, is formulated as

$$\mathbf{z}^{(n+1)} = (\mathbf{B}^T \cdot \mathbf{B} + \mathbf{C}^T \cdot \mathbf{C} + \lambda^2 \cdot \mathbf{S}^T \cdot \mathbf{S})^{-1}(\mathbf{B}^T \cdot \mathbf{q}^{(n)} + \mathbf{C}^T \cdot \mathbf{h}^{(n)} + \lambda^2 \cdot \mathbf{S}^T \cdot \mathbf{k}) \tag{2.25}$$

The one based on the three equations, HASM3abc, is formulated as

$$\mathbf{z}^{(n+1)} = (\mathbf{A}^T \cdot \mathbf{A} + \mathbf{B}^T \cdot \mathbf{B} + \mathbf{C}^T \cdot \mathbf{C} + \lambda^2 \cdot \mathbf{S}^T \cdot \mathbf{S})^{-1}$$
$$\times (\mathbf{A}^T \cdot \mathbf{d}^{(n)} + \mathbf{B}^T \cdot \mathbf{q}^{(n)} + \mathbf{C}^T \cdot \mathbf{h}^{(n)} + \lambda^2 \cdot \mathbf{S}^T \cdot \mathbf{k}) \quad (2.26)$$

In order to analyze errors and compute the time of the seven HASM expressions, a test surface $f(x,y) = 3 + 2\sin(2\pi \cdot x) \cdot \sin(2\pi \cdot y) + 13$ is selected (Figure 2.1), of which the computational domain is $[0\ 1] \times [0\ 1]$. The mean absolute error MAE and mean relative error MRE are respectively formulated as

$$MAE = \frac{1}{(I+2) \cdot (J+2)} \sum_i \sum_j |f_{i,j} - Sf_{i,j}| \quad (2.27)$$

$$MRE = \frac{1}{(I+2) \cdot (J+2)} \sum_i \sum_j \left| \frac{f_{i,j} - Sf_{i,j}}{f_{i,j}} \right| \cdot 100\% \quad (2.28)$$

where $f_{i,j}$ is the true value of $f(x, y)$ and $Sf_{i,j}$ is the simulated value of $f(x, y)$ at (x_i, y_j); I and J are respectively the number of grid cells in the directions x and y.

The test results (Table 2.1, Figures 2.2 and 2.3) show that both HASM1a and HASM1b have almost the same absolute and relative errors and computing times. HASM1a and HASM1b coefficient matrix structures are quite good, which reduce the computing time. However, the errors of HASM1a

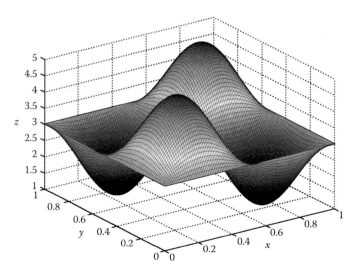

FIGURE 2.1
Test surface.

TABLE 2.1

Calculated Errors of Individual Equation and Different Combinations of Gauss Equations at 20th HASM Iteration

Grid Number in the Computational Domain		256	576	1024	1600	2304	3136
HASM1a	MAE	0.0532	0.0281	0.017	0.0112	0.008	0.0059
	MRE	0.0225	0.0118	0.0071	0.0047	0.0034	0.0025
HASM1b	MAE	0.0532	0.0282	0.0169	0.0112	0.0079	0.0059
	MRE	0.0225	0.0118	0.0071	0.0047	0.0033	0.0025
HASM1c	MAE	Defect	Defect	Defect	Defect	Defect	Defect
	MRE	Defect	Defect	Defect	Defect	Defect	Defect
HASM2ab	MAE	5.79×10^{-5}	7.22×10^{-6}	5.32×10^{-6}	4.36×10^{-6}	3.69×10^{-6}	3.24×10^{-6}
	MRE	2.93×10^{-5}	4.10×10^{-6}	4.02×10^{-6}	3.68×10^{-6}	3.28×10^{-6}	2.85×10^{-6}
HASM2ac	MAE	Overflow	Overflow	Overflow	Overflow	Overflow	Overflow
	MRE	Overflow	Overflow	Overflow	Overflow	Overflow	Overflow
HASM2bc	MAE	Overflow	Overflow	Overflow	Overflow	Overflow	Overflow
	MRE	Overflow	Overflow	Overflow	Overflow	Overflow	Overflow
HASM3abc	MAE	0.0229	0.0098	0.0055	0.0034	0.0023	0.0017
	MRE	0.0169	0.0082	0.0049	0.0031	0.0022	0.0016

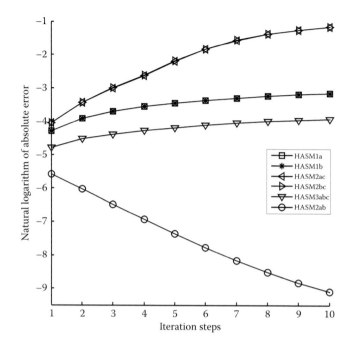

FIGURE 2.2
Absolute errors of all possible surface model formulations in the iterative process.

and HASM1b slowly become bigger with the increase of iteration steps and cannot reach convergence.

The coefficient matrix of HASM1c is singular, which leads to the defect in HASMc simulation. HASM2ac and HASM2bc spend the longest computing time and create overflow results, because equation c has a very bad structure of coefficient matrix and both equations a and b are unable to counter the impact of equation c. The errors of HASM3abc are smaller than those of HASM1a and HASM1b. HASM3abc has much bigger absolute and relative errors and a longer CPU time cost compared to HASM2ab, because HASM3abc includes equation c as well. In other words, equation c not only destroys the structure of HASM3abc coefficient matrix, but also increases both simulation errors and the computing time.

HASM2ab has a favorable structure of coefficient matrix and simulates a surface in both x and y directions simultaneously. Although HASM2ab's computing time is a little longer than HASM1a and HASM1b, which only simulate a surface in one direction (either x or y). HASM2ab's errors are three orders of magnitude lower compared to those of HASM1a and HASM1b. HASM2ab is the optimum formulation of high-accuracy surface modeling (HASM).

Numerical tests show that seven different surface models could be developed in terms of different combinations of the equations in equation set

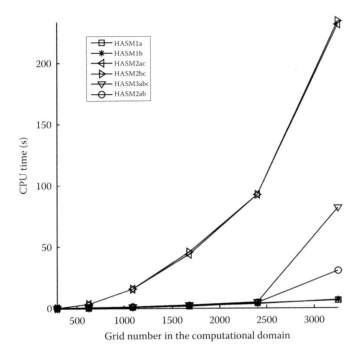

FIGURE 2.3
CPU time of all possible surface model formulations in the iterative process.

(Equation 2.3). Errors and the computing time of these surface models differ from each other. The one with the least error and less computing time is named HASM-B and its master equation set could be formulated as

$$\begin{cases} f_{xx} = \Gamma_{11}^1 \cdot f_x + \Gamma_{11}^2 \cdot f_y + L \cdot (E \cdot G - F^2)^{-1/2} \\ f_{yy} = \Gamma_{22}^1 \cdot f_x + \Gamma_{22}^2 \cdot f_y + N \cdot (E \cdot G - F^2)^{-1/2} \end{cases} \tag{2.29}$$

2.4 Iterative Formulation of HASM

If the maximum lengths of the computational domain Λ in the x and y directions are respectively m_x and m_y, Λ can be included in the rectangular domain $[0, m_x] \times [0, m_y]$. If the spatial resolution of DEM is $SR \times SR$, lattice number in direction x would be $I + 2$ and lattice number in direction y would be $J + 2$, in which $I = round(m_x / h) - 2$ and $J = round(m_y / h) - 2$ where $b = round(a)$ means that b is the integer closest to the real number a. If the computational domain $[\{0, m_x] \times [\{0, m_y]$ is normalized to $[0, m_x/(\max(m_x,$

$m_y))] \times [0, m_y/(\max(m_x, m_y))]$, where $\max(m_x, m_y)$ means the maximum one of m_x and m_y, and the normalized Λ is expressed as Ω, it would be found that $\Omega \subseteq [0, m_x/(\max(m_x, m_y))] \times [0, m_y/(\max(m_x, m_y))]$.

If h represents simulation step length, the relationship between h and spatial resolution $SR \times SR$ could be formulated as $h = SR/(\max(m_x, m_y))$. The central point of lattice $(0.5h + (i-1)h, 0.5h + (j-1)h)$ could be expressed as (x_i, y_j), in which $i = 0, 1, 2, \ldots, I, I+1$ and $j = 0, 1, 2, \ldots, J, J+1$.

If $f_{i,j}^{(n)}$ ($n \geq 0$) represents the iterants of $f(x, y)$ at (x_i, y_j) in the nth iterative step, in which $f_{i,j}^{(0)} = \tilde{f}_{i,j}$ and $\{\tilde{f}_{i,j}\}$ is interpolations based on sampling values $\{\bar{f}_{i,j}\}$, in terms of numerical mathematics (Quarteroni et al., 2000), the iterative formulation of finite difference of equation set (Equation 2.29) can be expressed as

$$\begin{cases} \dfrac{f_{i+1,j}^{(n+1)} - 2f_{i,j}^{(n+1)} + f_{i-1,j}^{(n+1)}}{h^2} = \left(\Gamma_{11}^1\right)_{i,j}^{(n)} \dfrac{f_{i+1,j}^{(n)} - f_{i-1,j}^{(n)}}{2h} \\[2ex] \qquad\qquad + \left(\Gamma_{11}^2\right)_{i,j}^{(n)} \dfrac{f_{i,j+1}^{(n)} - f_{i,j-1}^{(n)}}{2h} + \dfrac{L_{i,j}^{(n)}}{\sqrt{E_{i,j}^{(n)} + G_{i,j}^{(n)} - 1}} \\[3ex] \dfrac{f_{i,j+1}^{(n+1)} - 2f_{i,j}^{(n+1)} + f_{i,j-1}^{(n+1)}}{h^2} = \left(\Gamma_{22}^1\right)_{i,j}^{(n)} \dfrac{f_{i+1,j}^{(n)} - f_{i-1,j}^{(n)}}{2h} \\[2ex] \qquad\qquad + \left(\Gamma_{22}^2\right)_{i,j}^{(n)} \dfrac{f_{i,j+1}^{(n)} - f_{i,j-1}^{(n)}}{2h} + \dfrac{N_{i,j}^{(n)}}{\sqrt{E_{i,j}^{(n)} + G_{i,j}^{(n)} - 1}} \end{cases} \qquad (2.30)$$

where $E_{i,j}^{(n)}$, $F_{i,j}^{(n)}$, $G_{i,j}^{(n)}$, $L_{i,j}^{(n)}$, $N_{i,j}^{(n)}$, $(\Gamma_{11}^1)_{i,j}^{(n)}$, $(\Gamma_{11}^2)_{i,j}^{(n)}$, $(\Gamma_{22}^1)_{i,j}^{(n)}$, and $(\Gamma_{22}^2)_{i,j}^{(n)}$ have the same formulations as appeared in equation sets (Equation 2.15).

Equation 2.30 is a formulation of HASM-B when all grid cells have an equal size. But if the grid cell has a changeable size and its edge lengths in x and y directions are respectively $hx_i = x_i - x_{i-1}$ and $hy_j = y_j - y_{j-1}$, then HASM-B equation set with a changing grid cell size can be formulated as

$$\begin{cases} \dfrac{\left(\left(f_{i+1,j}^{(n+1)} - f_{i,j}^{(n+1)}\right)/hx_{i+1}\right) - \left(\left(f_{i,j}^{(n+1)} - f_{i-1,j}^{(n+1)}\right)/hx_i\right)}{0.5 \cdot (hx_i + hx_{i+1})} \\[3ex] = \left(\Gamma_{11}^1\right)_{i,j}^{(n)} \cdot \dfrac{f_{i+1,j}^{(n)} - f_{i-1,j}^{(n)}}{hx_i + hx_{i+1}} + \left(\Gamma_{11}^2\right)_{i,j}^{(n)} \cdot \dfrac{f_{i,j+1}^{(n)} - f_{i,j-1}^{(n)}}{hy_j + hy_{j+1}} + \dfrac{L_{i,j}^{(n)}}{\sqrt{E_{i,j}^{(n)} + G_{i,j}^{(n)} - 1}} \\[3ex] \dfrac{\left(\left(f_{i,j+1}^{(n+1)} - f_{i,j}^{(n+1)}\right)/hy_{j+1}\right) - \left(\left(f_{i,j}^{(n+1)} - f_{i,j-1}^{(n+1)}\right)/hy_j\right)}{0.5 \cdot (hy_j + hy_{j+1})} \\[3ex] = \left(\Gamma_{22}^1\right)_{i,j}^{(n)} \cdot \dfrac{f_{i+1,j}^{(n)} - f_{i-1,j}^{(n)}}{hx_i + hx_{i+1}} + \left(\Gamma_{22}^2\right)_{i,j}^{(n)} \cdot \dfrac{f_{i,j+1}^{(n)} - f_{i,j-1}^{(n)}}{hy_i + hy_{i+1}} + \dfrac{N_{i,j}^{(n)}}{\sqrt{E_{i,j}^{(n)} + G_{i,j}^{(n)} - 1}} \end{cases} \qquad (2.31)$$

where

$$E_{i,j}^{(n)} = 1 + \left(\frac{f_{i+1,j}^{(n)} - f_{i-1,j}^{(n)}}{hx_i + hx_{i+1}} \right)^2 ;$$

$$F_{i,j}^{(n)} = \left(\frac{f_{i+1,j}^{(n)} - f_{i-1,j}^{(n)}}{hx_i + hx_{i+1}} \right) \cdot \left(\frac{f_{i,j+1}^{(n)} - f_{i,j-1}^{(n)}}{hy_j + hy_{j+1}} \right) ;$$

$$G_{i,j}^{(n)} = 1 + \left(\frac{f_{i,j+1}^{(n)} - f_{i,j-1}^{(n)}}{hy_j + hy_{j+1}} \right)^2 ;$$

$$L_{i,j}^{(n)} = \frac{\left(\frac{f_{i+1,j}^{(n)}}{hx_{i+1}} \right) - \left(\left(\frac{1}{hx_{i+1}} \right) + \left(\frac{1}{hx_i} \right) \right) \cdot f_{i,j}^{(n)} + \left(\frac{f_{i-1,j}^{(n)}}{hx_i} \right)}{(hx_i + hx_{i+1})/2 \sqrt{1 + \left(\frac{f_{i+1,j}^{(n)} - f_{i-1,j}^{(n)}}{(hx_i + hx_{i+1})} \right)^2 + \left(\frac{f_{i,j+1}^{(n)} - f_{i,j-1}^{(n)}}{(hy_j + hy_{j+1})} \right)^2}} ;$$

$$N_{i,j}^{(n)} = \frac{\left(\frac{f_{i,j+1}^{(n)}}{hy_{j+1}} \right) - \left(\left(\frac{1}{hy_{j+1}} \right) + \left(\frac{1}{hy_j} \right) \right) \cdot f_{i,j}^{(n)} + \left(\frac{f_{i,j-1}^{(n)}}{hy_j} \right)}{(hy_j + hy_{j+1})/2 \sqrt{1 + \left(\frac{f_{i+1,j}^{(n)} - f_{i-1,j}^{(n)}}{(hx_i + hx_{i+1})} \right)^2 + \left(\frac{f_{i,j+1}^{(n)} - f_{i,j-1}^{(n)}}{hy_j + hy_{j+1}} \right)^2}} ;$$

$$(\Gamma_{11}^1)_{i,j}^{(n)} =$$

$$\frac{G_{i,j}^{(n)} \cdot (E_{i+1,j}^{(n)} - E_{i-1,j}^{(n)}) - 2F_{i,j}^{(n)} \cdot (F_{i+1,j}^{(n)} - F_{i-1,j}^{(n)}) + F_{i,j}^{(n)} \cdot (E_{i,j+1}^{(n)} - E_{i,j-1}^{(n)}) \cdot \left(\frac{(hx_i + hx_{i+1})}{(hy_j + hy_{j+1})} \right)}{\left(E_{i,j}^{(n)} \cdot G_{i,j}^{(n)} - \left(F_{i,j}^{(n)} \right)^2 \right) \cdot (hx_i + hx_{i+1})} ;$$

$$(\Gamma_{22}^1)_{i,j}^{(n)} =$$

$$\frac{2G_{i,j}^{(n)} \cdot (F_{i,j+1}^{(n)} - F_{i,j-1}^{(n)}) - G_{i,j}^{(n)} \cdot (G_{i+1,j}^{(n)} - G_{i-1,j}^{(n)}) \cdot \left(\frac{(hy_j + hy_{j+1})}{(hx_i + hx_{i+1})} \right) - F_{i,j}^{(n)} \cdot (G_{i,j+1}^{(n)} - G_{i,j-1}^{(n)})}{\left(E_{i,j}^{(n)} \cdot G_{i,j}^{(n)} - \left(F_{i,j}^{(n)} \right)^2 \right) \cdot (hy_j + hy_{j+1})} ;$$

$(\Gamma_{11}^2)_{i,j}^{(n)} =$

$$\frac{2E_{i,j}^{(n)}(F_{i+1,j}^{(n)} - F_{i-1,j}^{(n)}) \cdot \left(\dfrac{(hy_j + hy_{j+1})}{(hx_i + hx_{i+1})}\right) - E_{i,j}^{(n)} \cdot (E_{i,j+1}^{(n)} - E_{i,j-1}^{(n)}) - F_{i,j}^{(n)} \cdot (E_{i,j+1}^{(n)} - E_{i,j-1}^{(n)})}{\left(E_{i,j}^{(n)} \cdot G_{i,j}^{(n)} - \left(F_{i,j}^{(n)}\right)^2\right) \cdot (hy_j + hy_{j+1})};$$

$(\Gamma_{22}^2)_{i,j}^{(n)} =$

$$\frac{E_{i,j}^{(n)} \cdot (G_{i,j+1}^{(n)} - G_{i,j-1}^{(n)}) - 2 \cdot F_{i,j}^{(n)} \cdot (F_{i,j+1}^{(n)} - F_{i,j-1}^{(n)}) + F_{i,j}^{(n)} \cdot (G_{i+1,j}^{(n)} - G_{i-1,j}^{(n)}) \cdot \left(\dfrac{(hy_j + hy_{j+1})}{(hx_i + hx_{i+1})}\right)}{\left(E_{i,j}^{(n)} \cdot G_{i,j}^{(n)} - \left(F_{i,j}^{(n)}\right)^2\right) \cdot (hy_j + hy_{j+1})}.$$

The formulation based on nonuniform grid-cell size is useful for simulating the heterogeneous surfaces. Smaller grid cell size could be designed in the area where surface is highly variable and larger grid cell size could be adopted in gently changing areas, by which simulation accuracy could be improved and the computing time shortened.

The matrix formulation of equation set (Equation 2.30 or 2.31) can be expressed as

$$\begin{cases} \mathbf{A} \cdot \mathbf{z}^{(n+1)} = \mathbf{d}^{(n)} \\ \mathbf{B} \cdot \mathbf{z}^{(n+1)} = \mathbf{q}^{(n)} \end{cases} \tag{2.32}$$

where $\mathbf{z}^{(n+1)} = (f_{1,1}^{(n+1)}, \ldots, f_{1,J}^{(n+1)}, \ldots, f_{I,1}^{(n+1)}, \ldots, f_{I,J}^{(n+1)})^{\mathrm{T}}$; $\mathbf{A}$ and $\mathbf{B}$, respectively, represent coefficient matrixes of the first equation and the second equation in equation set (Equation 2.30 or 2.31); $\mathbf{d}^{(n)}$ and $\mathbf{q}^{(n)}$ are, respectively, the right-hand side vectors of equation set (Equation 2.30 or 2.31).

If $\bar{f}_{i,j}$ is value of $z = f(x, y)$ at the pth sampled point (x_i, y_j) in the computational domain, the simulation value should be equal or approximate to the sampling value at this lattice so that a constraint equation set is added to equation set (Equation 2.30). HASM-B could be formulated as

$$\begin{cases} \min \left\| \begin{bmatrix} \mathbf{A} \\ \mathbf{B} \end{bmatrix} \cdot \mathbf{z}^{(n+1)} - \begin{bmatrix} \mathbf{d}^{(n)} \\ \mathbf{q}^{(n)} \end{bmatrix} \right\| \\ \text{s.t.} \quad \mathbf{S} \cdot \mathbf{z}^{(n+1)} = \mathbf{k} \end{cases} \tag{2.33}$$

where the nonzero element of the sample matrix can be expressed as $s_{p,(i-1)\times I+j} = 1$ and the nonzero element of the sample vector as $k_p = \bar{f}_{i,j}$.

For sufficiently large λ, HASM-B could be transferred into unconstrained least-squares approximation

$$\min_{f} \left\| \begin{bmatrix} \mathbf{A} \\ \mathbf{B} \\ \lambda \cdot \mathbf{S} \end{bmatrix} \mathbf{z}^{(n+1)} - \begin{bmatrix} \mathbf{d}^{(n)} \\ \mathbf{q}^{(n)} \\ \lambda \cdot \mathbf{k} \end{bmatrix} \right\| \tag{2.34}$$

or

$$\begin{bmatrix} \mathbf{A}^T & \mathbf{B}^T & \lambda \cdot \mathbf{S}^T \end{bmatrix} \begin{bmatrix} \mathbf{A} \\ \mathbf{B} \\ \lambda \cdot \mathbf{S} \end{bmatrix} \mathbf{z}^{(n+1)} = \begin{bmatrix} \mathbf{A}^T & \mathbf{B}^T & \lambda \cdot \mathbf{S}^T \end{bmatrix} \begin{bmatrix} \mathbf{d}^{(n)} \\ \mathbf{q}^{(n)} \\ \lambda \cdot \mathbf{k} \end{bmatrix} \tag{2.35}$$

The parameter λ is the weight of the sampling points and determines the contribution of the sampling points to the simulated surface. λ could be a real number, which means all sampling points have the same weight, or a vector, which means every sampling point has its own weight. An area affected by a sampling point in a complex region is smaller than that in a flat region. Therefore, a smaller value of λ is selected in a complex region and a bigger value of λ is selected in a flat region.

According to the theory of differential geometry, a necessary and sufficient condition for a surface to be plane is that all the second fundamental coefficients equals zero at all points of the surface (Somasundaram, 2005). Therefore, we suppose that if L, M, and N do not vanish at all points of a piece of a surface, the piece of the surface could be simulated by HASM-B. If $L = M = N \equiv 0$ on a piece of a surface, this surface could be expressed by triangles. If the three nodes of a triangle are (x_1, y_1, z_1), (x_2, y_2, z_2), and (x_3, y_3, z_3), the triangle could be formulated as

$$z = a \cdot x + b \cdot y + c \tag{2.36}$$

where

$$a = \frac{\begin{vmatrix} z_1 & y_1 & 1 \\ z_2 & y_2 & 1 \\ z_3 & y_3 & 1 \end{vmatrix}}{\begin{vmatrix} x_1 & y_1 & 1 \\ x_2 & y_2 & 1 \\ x_3 & y_3 & 1 \end{vmatrix}}$$

$$b = \begin{vmatrix} x_1 & z_1 & 1 \\ x_2 & z_2 & 1 \\ x_3 & z_3 & 1 \\ x_1 & y_1 & 1 \\ x_2 & y_2 & 1 \\ x_3 & y_3 & 1 \end{vmatrix}$$

$$c = \begin{vmatrix} x_1 & y_1 & z_1 \\ x_2 & y_2 & z_2 \\ x_3 & y_3 & z_3 \\ x_1 & y_1 & 1 \\ x_2 & y_2 & 1 \\ x_3 & y_3 & 1 \end{vmatrix}$$

2.5 Numerical Tests

2.5.1 Comparative Error Analyses

In addition to the newly developed HASM-B, there are many other methods of surface modeling which have been widely used in various GIS applications in recent years, such as IDW, TIN, kriging, and spline. These classical surface models are used to comparatively analyze HASM-B errors.

HASM-B uses existing data, such as points or contours, to globally fit a surface through several iterative simulation steps. This surface is then used to interpolate a value at an unknown point. The unknown points are located on a regular space lattice. The iterative simulation steps are summarized as follows: (1) conducting interpolation on the computational domain Ω in terms of sampling data $(x_i, y_j, \bar{f}_{i,j})$, from which we can get interpolated approximate values $\{\tilde{f}_{i,j}\}$ at point (x_i, y_j); (2) letting $f_{i,j}^{(0)} = \tilde{f}_{i,j}$ and calculating the first fundamental coefficients $E_{i,j}^{(n)}$, $F_{i,j}^{(n)}$ and $G_{i,j}^{(n)}$ and the second fundamental coefficients $L_{i,j}^{(n)}$ and $N_{i,j}^{(n)}$ as well as the Christoffel symbols of the second kind of HASM-B equations in terms of $\{f_{i,j}^{(n)}\}$; (3) for $n \geq 0$, we can get $\{f_{i,j}^{(n+1)}\}$ by solving the HASM-B equations; (4) the iterative process is repeated until simulation accuracy is satisfied.

$$f(x,y) = 3(1-x)^2 e^{(-x^2-(y+1)^2)} - 10\left(\frac{x}{5} - x^2 - y^5\right) e^{(-x^2-y^2)} - \frac{1}{3} e^{(-(x+1)^2-y^2)},$$

Gaussian synthetic surface, is taken as the test surface (Figure 2.4a), so that the "true" value can be predetermined to avoid uncertainty caused by uncontrollable data errors. Its computational domain is $[-3,3] \times [-3,3]$ and $-6.6 < f(x, y) < 8.1$. In ArcMap's Visual Basic for Application (VBA) environment, ArcObjects' functions are used to create data in grid format with spatial resolution on 0.08×0.08 in terms of the formulation of Gaussian synthetic surface. The sampling contour data are produced by employing the Contour command of the Spatial Analyst module.

HASM-B and the classical methods are used to simulate the test surface, in which grid spacing is selected as $H = 0.08$ and sampling interval as $5 \cdot H$. The shaded relief maps are created by the HillShade command in the surface analysis menu of the Spatial Analyst module. The simulation error is calculated by *RMSE*, which is expressed as

$$RMSE = \sqrt{\frac{1}{76 \cdot 76} \sum_{i=0}^{75} \sum_{j=0}^{75} (f_{i,j} - Sf_{i,j})^2} \tag{2.37}$$

where $f_{i,j}$ is the true value of $f(x, y)$ and $Sf_{i,j}$ is the simulated value of $f(x, y)$ at (x_i, y_j).

The simulation results show that surfaces simulated by the classical methods have been considerably deformed, while HASM-B has a perfect simulation result (Figure 2.4b). The contour map simulated by HASM-B (Figure 2.5b) is almost same as the original Gaussian synthetic surface (Figure 2.5a). Spline produced three more contour lines compared to the original (Figure 2.5c). TIN made one contour line become two crossing contour lines and the other contour line disappears (Figure 2.5d), while kriging and IDW made contour lines out of shape (Figure 2.5e and f). *RMSE* of HASM-B in the fourth iterative step is 1.4802 times less than spline, 1.7717 times less than TIN, 5.0675 times less than kriging, and 7.2143 times less than IDW (Table 2.2).

2.5.2 Comparative Analysis of Simulating Different Surfaces

We select eight mathematical surfaces which represent different types of shapes (Figure 2.6) with the following formulas:

$$f_1(x,y) = \cos(10y) + \sin(10(x - y)) \tag{2.38}$$

$$f_2(x,y) = e^{(-(5-10x)^2/2)} + 0.75e^{(-(5-10y)^2/2)} + 0.75e^{(-(5-10x)^2/2)} e^{(-(5-10y)^2/2)} \tag{2.39}$$

$$f_3(x,y) = \sin(2\pi \cdot y) \cdot \sin(\pi \cdot x) \tag{2.40}$$

$$f_4(x,y) = 0.75e^{(-((9x-2)^2+(9y-2)^2)/4)} + 0.75e^{(-(9x+1)^2/49-(9y+1)/10)}$$
$$+0.5e^{(-((9x-7)^2+(9y-3)^2)/4)} - 0.2e^{(-(9x-4)^2-(9y-7)^2)} \tag{2.41}$$

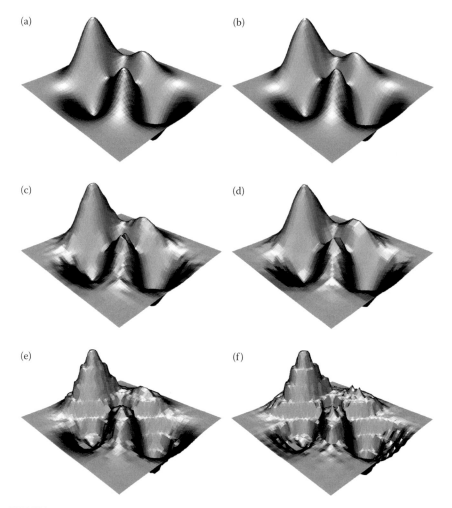

FIGURE 2.4
Comparison between HASM-B simulation surface and surfaces simulated by classical methods: (a) the original Gaussian synthetic surface; (b) HASM-B surface; (c) Spline surface; (d) TIN surface; (e) Kriging surface; and (f) IDW surface.

$$f_5(x,y) = \frac{1}{9}(\tanh(9y - 9x) + 1) \tag{2.42}$$

$$f_6(x,y) = \frac{1.25 + \cos(5.4y)}{6(1 + (3x - 1)^2)} \tag{2.43}$$

$$f_7(x,y) = \frac{1}{3}e^{-(81/16)((x-0.5)^2 + (y-0.5)^2)} \tag{2.44}$$

$$f_8(x,y) = \frac{1}{3}e^{-(81/4)((x-0.5)^2 + (y-0.5)^2)} \tag{2.45}$$

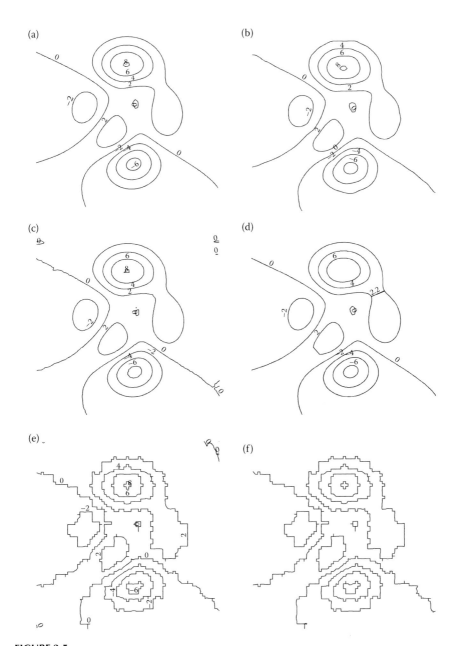

FIGURE 2.5
Comparison between contour lines of the surface simulated by HASM-B and those simulated by classical methods: (a) the original contour lines of Gaussian synthetic surface; (b) HASM-B contour lines; (c) spline contour lines; (d) TIN contour lines; (e) kriging contour lines; and (f) IDW contour lines.

TABLE 2.2

Comparative Analyses of HASM-B Error

Method	IDW	Kriging	TIN	Spline	HASM-B
RMSE	0.3636	0.2554	0.0893	0.0746	0.0504
The ratio of HASM-B error to ones of the classic methods	7.2143	5.0675	1.7718	1.4802	

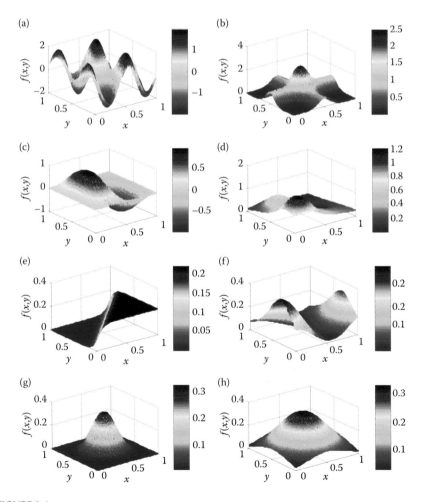

FIGURE 2.6

The eight test surfaces used for model evaluation: (a) $f_1(x, y)$, (b) $f_2(x, y)$, (c) $f_3(x, y)$, (d) $f_4(x, y)$, (e) $f_5(x, y)$, (f) $f_6(x, y)$, (g) $f_7(x, y)$, and (h) $f_8(x, y)$.

In the normalized computational domain, 14641 (121 × 121) lattices were created by an orthogonal division of the computational domain. Therefore, in simulation processes using HASM-B and the classical interpolation methods, the grid spacing (or simulation step length) was set as $h_{x_i} = h_{y_j} = h = 1/120$. Two sampling methods, uniform sampling and random sampling, were adopted to comparatively analyze errors of HASM-B and the classical interpolation methods. For the uniform sampling, $5h$ was selected as the sampling interval to have 625 sampling points. For random sampling, 1296 points were randomly sampled.

The *RMSE* in this case is expressed as

$$RMSE = \sqrt{\frac{1}{121 \cdot 121} \sum_{i=0}^{120} \sum_{j=0}^{120} (f_{i,j} - Sf_{i,j})^2} \qquad (2.46)$$

where $Sf_{i,j}$ is the numerical solution of $f(x, y)$ at point (x_i, y_j); $f_{i,j}$ is true value of f at the lattice (i, j).

Under uniform sampling (Tables 2.3 and 2.4), the accuracy of HASM-B compared with the other sampling methods is the smallest for the test surface $f_5(x, y)$. However, the HASM-B accuracy is still 3.85, 17.7, 35, and 171 times higher than the accuracy of spline, kriging, TIN, and IDW, respectively. Test surface $f_8(x, y)$ gives HASM-B the highest accuracy; 11.4, 40.8, 70.3, and 647 times higher than that of spline, kriging, TIN, and IDW, respectively. On an average, the accuracy of HASM-B simulation is 9.15 times higher than that of spline, 41.8 times higher than kriging, 53.9 times higher than TIN, and 276 times higher than IDW.

Under random sampling (Tables 2.5 and 2.6), HASM-B simulation accuracies of the test surfaces $f_1(x, y), f_2(x, y), f_3(x, y), f_4(x, y), f_5(x, y), f_6(x, y), f_7(x, y)$, and $f_8(x, y)$ are 47.1, 62.3, 77.7, 108, 56.8, 108, 206, and 198 times as much as the accuracies of the classical interpolation methods, on an average. The same method has different levels of power to simulate different test surfaces. For instance, when IDW is used to simulate the surface $f_8(x, y)$ under random sampling, its

TABLE 2.3

RMSE of Simulation Methods under Uniform Sampling

Test Surface	HASM-B	Spline	Kriging	TIN	IDW
$f_1(x, y)$	1.2463×10^{-3}	3.9862×10^{-3}	1.4295×10^{-1}	2.8509×10^{-2}	1.4030×10^{-1}
$f_2(x, y)$	2.3406×10^{-4}	1.9878×10^{-3}	4.7567×10^{-3}	9.0201×10^{-3}	4.6162×10^{-2}
$f_3(x, y)$	4.3798×10^{-5}	6.5524×10^{-4}	2.4732×10^{-3}	4.1618×10^{-3}	3.9796×10^{-3}
$f_4(x, y)$	5.8282×10^{-5}	4.3721×10^{-4}	1.2908×10^{-3}	2.4557×10^{-3}	1.3953×10^{-2}
$f_5(x, y)$	2.8501×10^{-5}	1.0962×10^{-4}	5.0507×10^{-4}	9.9612×10^{-4}	4.8667×10^{-3}
$f_6(x, y)$	1.8963×10^{-5}	1.0445×10^{-4}	2.5823×10^{-4}	4.9668×10^{-4}	3.9796×10^{-3}
$f_7(x, y)$	8.6858×10^{-6}	1.5837×10^{-4}	4.2607×10^{-4}	8.7978×10^{-4}	4.6944×10^{-3}
$f_8(x, y)$	5.7448×10^{-6}	6.5652×10^{-5}	2.3417×10^{-4}	4.0412×10^{-4}	3.7144×10^{-3}

TABLE 2.4

Ratio of *RMSE* of a Classical Method to that of HASM-B under Uniform Sampling

Test Surface	Spline	Kriging	TIN	IDW	Average
$f_1(x, y)$	3.20	115	22.9	113	63.3
$f_2(x, y)$	8.49	20.3	38.5	197	66.1
$f_3(x, y)$	15	56.5	95	90.9	64.3
$f_4(x, y)$	7.5	22.1	42.1	239	77.8
$f_5(x, y)$	3.85	17.7	35	171	56.8
$f_6(x, y)$	5.51	13.6	26.2	210	63.8
$f_7(x, y)$	18.2	49.1	101	540	177
$f_8(x, y)$	11.4	40.8	70.3	647	192
On an average	9.15	41.8	53.9	276	

TABLE 2.5

RMSE of Simulation Methods under Random Sampling

Test Surface	HASM-B	Spline	Kriging	TIN	IDW
$f_1(x, y)$	7.7636×10^{-4}	4.8223×10^{-3}	1.1371×10^{-1}	2.6671×10^{-2}	1.0430×10^{-3}
$f_2(x, y)$	2.0898×10^{-4}	1.6529×10^{-3}	8.3032×10^{-3}	9.8964×10^{-3}	3.2262×10^{-2}
$f_3(x, y)$	1.0532×10^{-4}	5.9686×10^{-4}	3.7706×10^{-3}	3.7984×10^{-3}	2.4583×10^{-2}
$f_4(x, y)$	4.0324×10^{-5}	3.8536×10^{-4}	2.2148×10^{-3}	2.5441×10^{-3}	1.2291×10^{-2}
$f_5(x, y)$	2.8466×10^{-5}	2.1296×10^{-4}	8.8910×10^{-4}	1.0277×10^{-3}	4.3340×10^{-3}
$f_6(x, y)$	1.0014×10^{-5}	8.0028×10^{-5}	5.0024×10^{-4}	4.7327×10^{-4}	3.2566×10^{-3}
$f_7(x, y)$	7.0024×10^{-6}	1.4075×10^{-4}	7.5926×10^{-4}	8.5749×10^{-4}	4.0161×10^{-3}
$f_8(x, y)$	5.2814×10^{-6}	5.3117×10^{-5}	4.7720×10^{-4}	3.8095×10^{-4}	3.2805×10^{-3}

TABLE 2.6

Ratio of *RMSE* of a Classical Method to that of HASM-B under Random Sampling

Test Surface	Spline	Kriging	TIN	IDW	On Average
$f_1(x, y)$	6.21	146	34.4	1.34	47.1
$f_2(x, y)$	7.91	39.7	47.4	154	62.3
$f_3(x, y)$	5.67	35.8	36.1	233	77.7
$f_4(x, y)$	9.56	54.9	63.1	305	108
$f_5(x, y)$	7.48	31.2	36.1	152	56.8
$f_6(x, y)$	7.99	50	47.3	325	108
$f_7(x, y)$	20.1	108	122	574	206
$f_8(x, y)$	10.1	90.4	72.1	621	198
On an average	9.37	69.6	57.4	296	

RMSE is 621 times as much as that of HASM-B, but *RMSE* of IDW is only 1.34 times of that of HASM-B for $f_1(x, y)$.

The sampling methods greatly affect the accuracy of different interpolation methods simulating the particular test surface. For example, *RMSE* of IDW simulating the test surface $f_1(x, y)$ is only 1.34 times as much as that of HASM-B under random sampling, but 113 times under uniform sampling. For the simulation of test surface $f_3(x, y)$, the accuracy of HASM-B is 5.67, 35.8, 36.1, and 233 times higher than that of Spline, Kriging, TIN, and IDW, respectively under random sampling, but 15, 56.5, 95, and 90.9 times higher under uniform sampling.

Many error assessments simply reported a single number to express the accuracy without regard for the location. If all errors balance out, nonsite

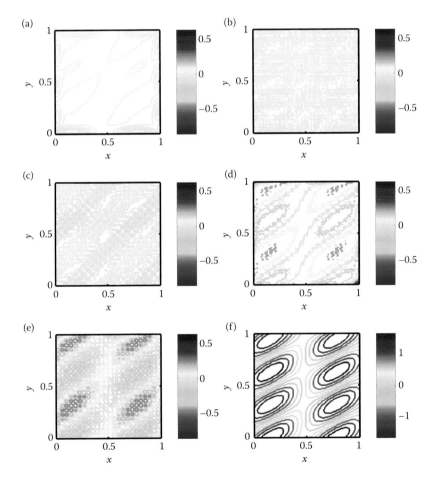

FIGURE 2.7
Error distribution contours of simulations applied to $f_1(x, y)$. (a) HASM-B, (b) spline, (c) TIN, (d) kriging, (e) IDW, and (f) contour chart of $f_1(x, y)$.

specific accuracy assessments give misleading results (Lunetta et al., 1991). Therefore, we introduce an error matrix for every method of surface modeling for each test surface, $\{er_{i,j}\}$. The error $er_{i,j}$ at the lattice point (or grid cell) (i,j) is formulated as

$$er_{i,j} = Sf_{i,j} - f_{i,j} \qquad (2.47)$$

where $f_{i,j}$ is the true value of f and $Sf_{i,j}$ is the simulated value of f at the lattice point (i,j).

Figures 2.7 through 2.14 indicate that the errors of HASM-B are bigger along the boundaries of the resultant surfaces or in the areas near the peaks.

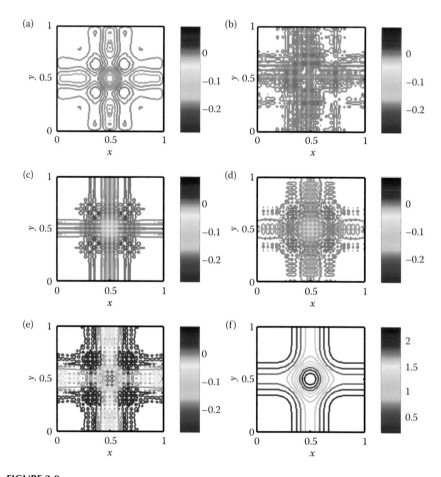

FIGURE 2.8
Error distribution contours of simulations applied to $f_2(x, y)$. (a) HASM-B, (b) spline, (c) TIN, (d) kriging, (e) IDW, and (f) contour chart of $f_2(x, y)$.

If the test surfaces are symmetric, the error distributions of HASM-B, IDW, TIN, and kriging simulations are also symmetric. However, the error distribution of the spline simulation for the symmetric test surface is asymmetric, and the errors tend to be uniformly distributed.

2.6 Discussion and Conclusions

The simulation process of HASM-B include five steps: (1) to normalize the computational domain as $[0, Bx] \times [0, By]$, in which $\max(Bx, By) = 1$, and

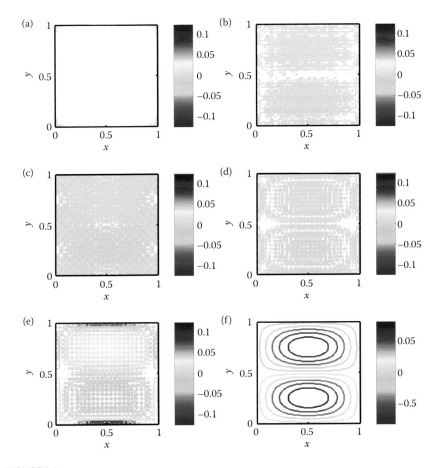

FIGURE 2.9
Error distribution contours of simulations applied to $f_3(x, y)$. (a) HASM-B, (b) spline, (c) TIN, (d) kriging, (e) IDW, and (f) contour chart of $f_3(x, y)$.

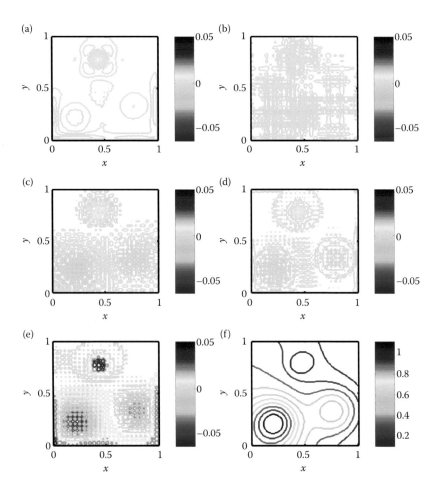

FIGURE 2.10
Error distribution contours of simulations applied to $f_4(x, y)$. (a) HASM-B, (b) spline, (c) TIN, (d) kriging, (e) IDW, and (f) contour chart of $f_4(x, y)$.

to discretize the normalized domain, (2) to interpolate in terms of the sampling data $(x_i, y_j, \bar{f}_{i,j})$ to calculate $\{\tilde{f}_{i,j}\}$ at each grid cell (x_i, y_j) and let $\{f_{i,j}^{(0)}\} = \{\tilde{f}_{i,j}\}$, (3) to calculate the first and second fundamental coefficients in terms of $\{f_{i,j}^{(n)}\}$ ($n \geq 0$) and then calculate the right-hand items of equation set (Equation 2.34); (4) for $n \geq 0$, to solve equation set (Equation 2.34) and obtain $\{f_{i,j}^{(n+1)}\}$; (5) to repeat the iterative process until the error would have been less than the convergence criterion ε. The error could be calculated by whatever $\max_{i,j}(|f_{i,j}^{(n)} - f_{i,j}^{(n+1)}|)/(|f_{i,j}^{(n+1)}|)$, $\max_{i,j}(|f_{i,j}^{(n)} - f_{i,j}^{(n+1)}|)$, $\Sigma_{i,j}(|f_{i,j}^{(n)} - f_{i,j}^{(n+1)}|)/(|f_{i,j}^{(n+1)}| \cdot I \cdot \tilde{J})$, $\Sigma_{i,j}(|f_{i,j}^{(n)} - f_{i,j}^{(n+1)}|)/I \cdot J$, or *RMSE*, according to the application requirement.

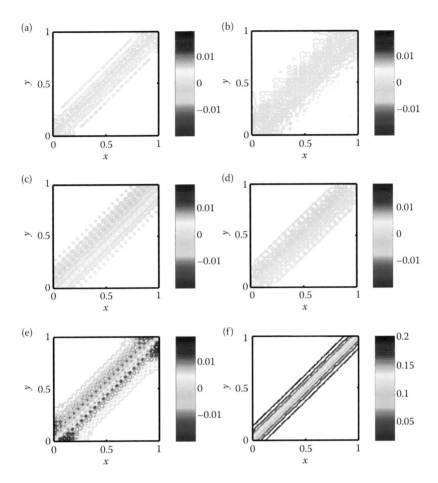

FIGURE 2.11
Error distribution contours of simulations applied to $f_5(x, y)$. (a) HASM-B, (b) spline, (c) TIN, (d) kriging, (e) IDW, and (f) contour chart of $f_5(x, y)$.

In the numerical tests, various test surfaces including symmetric, asymmetric, single-peak, multipeak, gently changing, and sharply changing surfaces, were simulated by HASM-B and the classical methods such as TIN, spline, IDW, and kriging. The comparative error analyzes indicate that HASM-B errors are much smaller than those of the classical methods (Yue et al., 2010).

TIN-surfaces are constructed by triangulating a set of vertices (points), for which the vertices are connected with a series of edges to form a network of triangles. Each triangular facet describes the behavior of a portion of a TIN-surface and two triangular facets become one rectangle facet when the vertices are evenly distributed, which would cause the so-called peak truncation and pit-fill problem (Yue et al., 2007).

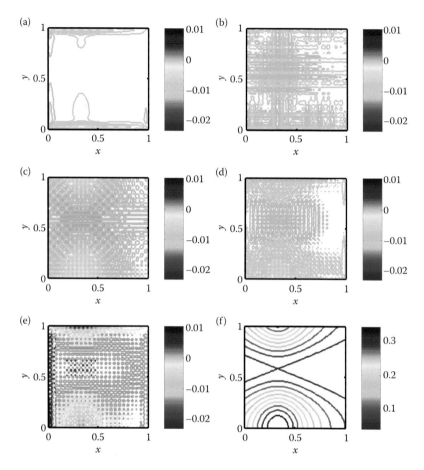

FIGURE 2.12
Error distribution contours of simulations applied to $f_6(x, y)$. (a) HASM-B, (b) spline, (c) TIN, (d) kriging, (e) IDW, and (f) contour chart of $f_6(x, y)$.

Each facet of spline-surface is estimated by univariate cubic basis-splines. However, a few facets of surface could be formulated into this kind of mathematical function. Spline-surface might mostly have an oscillation problem.

Cell values of IDW-surfaces are estimated by averaging the values of sample data points in the neighborhood of each processing cell. The closer a point is to the center of the cell being estimated, the bigger weight it has in the average calculation process, which sometimes causes "bull's eyes" problems.

Kriging is an appropriate method for issues that have a spatially correlated distance or directional bias in the data. Kriging fits a mathematical function to a specified number of points, or all points within a specified radius, to determine the output value for each location, which might make the kriging-surface fragmental.

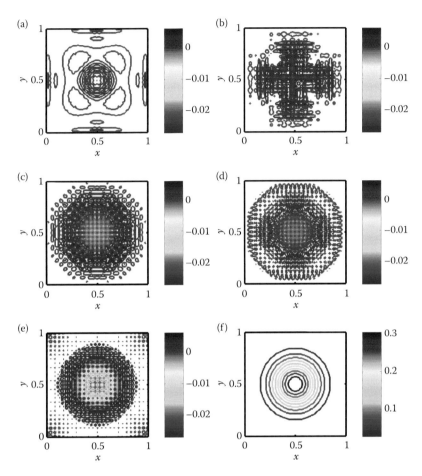

FIGURE 2.13
Error distribution contours of simulations applied to $f_7(x, y)$. (a) HASM-B, (b) spline, (c) TIN, (d) kriging, (e) IDW, and (f) contour chart of $f_7(x, y)$.

HASM-B has theoretically provided the solution to error problems that have long troubled surface modeling. Earth observations from satellites have been recognized as an excellent tool for providing both global and local views, and HASM-B could be used to support earth observation image processing. A great variety of satellites and sensors are commercially or experimentally available. Recent developments show that synthetic aperture radars (SAR) can provide high-resolution images, independent from cloud coverage. Conventionally, these images consist of a large amount of data, but without combining the derived space-borne information with suitable ancillary data, especially ground references, present in different kinds of information systems, the resulting information does not accurately reflect complex interactions on the earth's surface. However, when the global positioning system

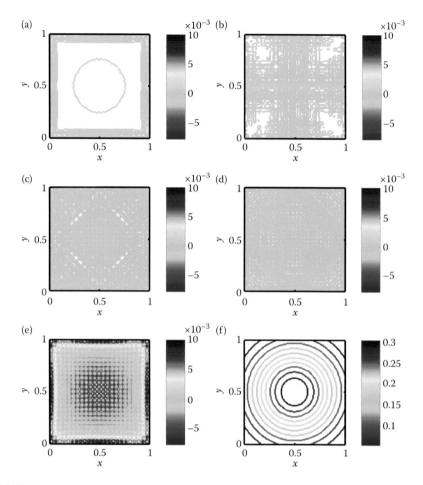

FIGURE 2.14
Error distribution contours of simulations applied to $f_8(x, y)$. (a) HASM-B, (b) spline, (c) TIN, (d) kriging, (e) IDW, and (f) contour chart of $f_8(x, y)$.

is connected with HASM-B, these satellite images can be used in the field (Zingler et al., 1999).

Mercator's projection can be used for the transformation between the geographical coordinate system and the plane rectangular Cartesian coordinate system, which is formulated as (Maling, 1992)

$$\begin{cases} x = a \cdot \lambda \\ y = a \cdot \ln \tan\left(\dfrac{\pi}{4} + \dfrac{\phi}{2}\right)\left(\dfrac{1 - e \cdot \sin\phi}{1 + e \cdot \sin\phi}\right)^{e/2} + C \end{cases} \tag{2.48}$$

where x is the abscissa of the plane Cartesian coordinate system, of which the origin is located on equator at its intersection with Greenwich Meridian; y is ordinate, which coincides with the Greenwich Meridian; ϕ is latitude; λ is longitude; a is the major semi-axis of earth; $e = ((a^2 - b^2) / a^2)^{1/2}$ is the first eccentricity of earth defined by a and b, in which b is minor semi-axis of earth; C is a constant of map projection.

HASM-B has a huge computation cost, because it must use an equation set for simulating each lattice of a surface. If a surface consists of 2.25 million grid cells and HASM-B is operated on a typical personal computer available today, the simulation process would take about 10 h—the computation time being approximately proportional to the third power of the total number of grid cells. A future research focus (Yue and Song, 2008) is to shorten the computing time by introducing the multigrid method, adaptive algorithm, and parallel computing technology.

References

Bloor, M.I.G. and Wilson, M.J. 1989. Generating blend surfaces using partial differential equations. *Computer-Aided Design* 21(3): 165–171.

Bloor, M.I.G. and Wilson, M.J. 1990. Using partial differential equations to generate free-form surfaces. *Computer-Aided Design* 22(4): 202–212.

Bloor, M.I.G. and Wilson, M.J. 1996. Spectral approximations to PDE surfaces. *Computer-Aided Design* 28(2): 145–152.

Carmo, M.P. 2006. *Differential Geometry of Curves and Surfaces*. Beijing: China Machine Press.

Golub, G.H. and van Loan, C.F. 1996. *Matrix Computation*. Baltimore: The Johns Hopkins University Press.

Henderson, D.W. 1998. *Differential Geometry*. London: Prentice-Hall, Inc.

Liseikin, V.D. 2004. *A Computational Differential Geometry Approach to Grid Generation*. Berlin: Springer.

Lunetta, R.S., Congalton, R.G., Fenstermaker, L.K., Jensen, J.R., Mcgwire, K.C., and Tinney, L.R. 1991. Remote sensing and geographic information system data integration: Error sources and research issues. *Photogrammetric Engineering and Remote Sensing* 57: 677–687.

Maling, D.H. 1992. *Coordinate Systems and Map Projections*. New York, NY: Pergamon Press.

Modani, M., Sharan, M., and Rao, S.C.S. 2008. Numerical solution of elliptic partial differential equations on parallel systems. *Applied Mathematics and Computation* 195: 162–182.

Pasadas, M. and Rodríguez, M.L. 2009. Construction and approximation of surfaces by discrete PDE splines on a polygonal domain. *Journal of Computational and Applied Mathematics* 59(1): 205–218.

Protopopescu, V., Santoro, R.T., and Dockery, J. 1989. Combat modeling with partial differential equations. *European Journal of Operational Research* 3(2): 178–183.

Quarteroni, A., Sacco, R., and Saleri, F. 2000. *Numerical Mathematics*. New York, NY: Springer.

Shi, Y.G., Thompson, P. M., Dinov, I., Osher, St., and Tog, A.W. 2007. Direct cortical mapping via solving partial differential equations on implicit surfaces. *Medical Image Analysis* 11: 207–223.

Somasundaram, D. 2005. *Differential Geometry*. Harrow, UK: Alpha Science International Ltd.

Thompson, J.F., Thames, F.C., and Mastin, C.W. 1974. Automatic numerical generation of body-fitted curvilinear coordinate system for field containing any number of arbitrary two-dimensional bodies. *Journal of Computational Physics* 15(3): 299–319.

Toponogov, V.A. 2006. *Differential Geometry of Curves and Surfaces*. New York, NY: Birkhauser Boston.

Walder, C., Schoelkopf, B., and Chapelle, O. 2006. Implicit surface modeling with a globally regularized basis of compact support. *EUROGRAPHICS* 25: 635–644.

Xu, G.L. and Zhang, Q. 2008. A general framework for surface modeling using geometric partial differential equations. *Computer Aided Geometric Design* 25(3): 181–202.

Yue, T.X., Du Z.P., and Song, Y.J. 2008. Ecological models: Spatial models and Geographic Information Systems. In *Encyclopedia of Ecology*, eds. S.E. Jørgensen, and B. Fath, 3315–3325. England: Elsevier Limited.

Yue, T.X., Du Z.P., Song, D.J., and Gong, Y. 2007. A new method of surface modeling and its application to DEM construction. *Geomorphology* 91(1–2): 161–172.

Yue, T.X., Song, D.J., Du Z.P., and Wang, W. 2010. High accuracy surface modeling and its application to DEM generation. *International Journal of Remote Sensing* 31(8): 2205–2226.

Zhang, J.J. and You, L.H. 2004. Fast surface modeling using a 6th order PDE. *Eurographics* 23: 311–320.

Zingler, M., Fischer, P., and Lichtenegger, J. 1999. Wireless field data collection and EO-GIS-GPS integration. *Computers, Environment and Urban Systems* 23: 305–313.

3

Comparative Analyses of Various HASM Algorithms*

3.1 Introduction

3.1.1 Algorithms

The approaches for solving the large sparse linear systems can be distinguished into direct and iterative algorithms (Fox et al., 1948; Dietl, 2007). Direct algorithms, which are based on factorization of the coefficient matrix into easily invertible matrices, do not provide a solution until all steps of the algorithm are processed. By contrast, iterative algorithms produce an approximate solution at each iteration step, which is improved step by step.

Direct algorithms are very robust and require a predictable amount of resources in terms of time and storage (George and Liu, 1981; Duff et al., 1986). It is possible to efficiently solve linear systems of fairly large size in a reasonable amount of time, particularly when the underlying problem is two-dimensional. Direct solvers are also the method of choice in certain areas not governed by PDEs (Benz, 2002). However, for extremely large-scale problems, iterative methods are the only option available because they require smaller storage and generally fewer operations than direct methods. Preconditioning is necessary when iterative methods fail in some applications.

A matrix is expressed as the product of two essentially triangular matrices, one of them a permutation of a lower triangular matrix and the other an upper triangular matrix, which is often called LU factorization. Gaussian elimination algorithm is the most famous direct algorithm, which is based on LU factorization of the system matrix in a lower and an upper triangular matrix. If the system matrix is conjugate and positive definite, the LU factorization can be reformulated as Choleski factorization. Fox et al. (1948) comparatively analyzed Gaussian elimination algorithm, a method of orthogonal vectors, and the Choleski method in a numerical test. They found that the Choleski method was the quicker and the more accurate of the two for symmetric matrices. Simon (1989) claimed that direct sparse solvers would remain the method of choice for irregularly structured problems. However, for many problems related to PDEs, realistic three-dimensional modeling

---

* Dr. Chuan-Fa Chen and Dr. Zheng-Ping Du are major collaborators of this chapter.

was not feasible, because the elimination process became too complex. Direct algorithms have high memory requirement, and may have to be replaced by iterative methods (Saad and van der Vorst, 2000).

If a solution is not adequate, a self-correcting method of successive approximation might provide a sequence that converges to the true solution. Methods of successive approximation are often called iterative. The term, relaxation, is also used with varying connotations in connection with methods of successive approximation to designate a method in which each step to be taken is selected according to a certain criterion from a finite number of allowable alternatives (Householder, 1964). The relaxation method was originally used as a means of calculating stresses in highly redundant pin-jointed frameworks (Southwell, 1940). This method was especially effective when human insight guided the entire course of computations (Varga, 1962).

3.1.2 Preconditioning

Preconditioning was crucial in the development of efficient solvers for challenging problems in scientific computation (Benzi, 2002). Preconditioning refers to transforming the system, $\mathbf{W} \cdot \mathbf{z}^{(n+1)} = \mathbf{v}^{(n)}$, into another system with more favorable properties for iterative solution by improving the spectral properties of the coefficient matrix. A preconditioner is a matrix that effects such a transformation of the original system that it accelerates the convergence of iterative methods. If $\mathbf{P}$ is a nonsingular matrix that approximates $\mathbf{W}$, that is, $\mathbf{P}^{-1} \cdot \mathbf{W}$ is close to a unit matrix (Faddeev and Faddeeva, 1963), then the linear system $\mathbf{P}^{-1} \cdot \mathbf{W} \cdot \mathbf{Z} = \mathbf{P}^{-1} \cdot \mathbf{V}$ has the same solution as $\mathbf{W} \cdot \mathbf{z}^{(n+1)} = \mathbf{v}^{(n)}$ but may be easier to be solved, in which case $\mathbf{P}$ is the preconditioner.

Various preconditioning procedures have been constructed for different applications. For instance, semianalytical preconditioning was proposed for solving large-scale boundary-value practical problems (Bulgakov, 1993). A factorized preconditioning procedure of successive approximations was developed for the computation of the stable approximate solutions of a large system of linear equations with dense, noncontractive and ill-conditioned matrices (Timonov, 2001). A preconditioning scheme was developed for radial basis functions with domain decomposition methods on the basis of constructing least-squares approximation cardinal basis functions (Ling and Kansa, 2004). The approximate cardinal basis function preconditioning technique was used to solve PDEs with radial basis functions (Brown et al., 2005), and a high-order Sobolev preconditioning strategy was described for the linear part of a nonlinear differential equation (Richardson, 2005). An aggregation-based algebraic multilevel preconditioning was proposed for linear systems arising from second-order scalar elliptic PDEs discretized by finite differences or finite elements with nodal basis functions (Notay, 2006). Darvishi's preconditioning was used to solve system of time-dependent PDEs (Javidi, 2006), while multilevel functional preconditioning was presented for optimal shape design (Courty and Dervieux, 2006). Additive and multiplicative two-level spectral preconditioning was introduced for the solution of general symmetric and unsymmetric linear systems (Carpentier et al., 2007). Block preconditioners for real equivalent formulations were developed for the original complex formulation (Benzi and Bertaccini, 2008). A preconditioner based on a coarse scale correction with an incomplete LU factorization with zero fill-ins as a smoother was proposed for discontinuous Galerkin discretizations of the Navier–Stokes equations (Persson and Peraire, 2008). A piecewise linear finite element preconditioner was analyzed in terms of condition numbers for the high-order element discretizations applied to a model elliptic operator (Kim and Kim, 2009). Schur complement based preconditioners were also proposed, which take into account the active and inactive set structure of the problem (Hintermueller et al., 2009), and equivalent operator preconditioning was developed for elliptic problems (Axelsson and Karátson, 2009). In short, preconditioning has always been an essential technique for improving the convergence of iterative methods.

The coefficient matrix of HASM, **W**, is a symmetric positive definite matrix and a large percentage of the elements of the coefficient matrix is zero. For this special large linear system, we will comparatively analyze the computational efficiencies of Gaussian elimination algorithm (HASM-GE), the square-root method (HASM-SR), Gauss–Seidel iterative algorithm (HASM-GS), the preconditioned Gauss–Seidel method (HASM-PGS), the conjugate gradient method (HASM-CG), and the preconditioned conjugate gradient method (HASM-PCG).

3.2 Direct Algorithms of HASM

If $\{(x_i, y_j) | 0 \leq i \leq I + 1, 0 \leq j \leq J + 1\}$ are the grid cells in the normalized computational domain and h is the computational step length, then the finite difference of the master equation set of HASM can be formulated as (Yue et al., 2007, 2008, 2010; Yue and Song, 2008)

$$
\begin{cases}
f_{i+1,j} - 2f_{i,j} + f_{i-1,j} = (\Gamma^1_{11})_{i,j} \cdot \dfrac{f_{i+1,j} - f_{i-1,j}}{2} \cdot h + (\Gamma^2_{11})_{i,j} \cdot \dfrac{f_{i,j+1} - f_{i,j-1}}{2} \cdot h + \dfrac{L_{i,j}}{\sqrt{E_{i,j} + G_{i,j} - 1}} \cdot h^2 \\[4mm]
f_{i,j+1} - 2f_{i,j} + f_{i,j-1} = (\Gamma^1_{22})_{i,j} \cdot \dfrac{f_{i+1,j} - f_{i-1,j}}{2} \cdot h + (\Gamma^2_{22})_{i,j} \cdot \dfrac{f_{i,j+1} - f_{i,j-1}}{2} \cdot h + \dfrac{N_{i,j}}{\sqrt{E_{i,j} + G_{i,j} - 1}} \cdot h^2
\end{cases}
$$

$$(3.1)$$

where

$$
E_{i,j} = 1 + \left(\frac{f_{i+1,j} - f_{i-1,j}}{2h} \right)^2
$$

$$G_{i,j} = 1 + \left(\frac{f_{i,j+1} - f_{i,j-1}}{2h} \right)^2$$

$$F_{i,j} = \left(\frac{f_{i+1,j} - f_{i-1,j}}{2h} \right) \left(\frac{f_{i,j+1} - f_{i,j-1}}{2h} \right)$$

$$L_{i,j} = \frac{(f_{i+1,j} - 2f_{i,j} + f_{i-1,j}) / h^2}{\sqrt{1 + \left((f_{i+1,j} - f_{i-1,j}) / 2h \right)^2 + \left((f_{i,j+1} - f_{i,j-1}) / 2h \right)^2}}$$

$$N_{i,j} = \frac{(f_{i,j+1} - 2f_{i,j} + f_{i,j-1}) / h^2}{\sqrt{1 + \left((f_{i+1,j} - f_{i-1,j}) / 2h \right)^2 + \left((f_{i,j+1} - f_{i,j-1}) / 2h \right)^2}}$$

$$(\Gamma_{11}^1)_{i,j} = \frac{G_{i,j}(E_{i+1,j} - E_{i-1,j}) - 2F_{i,j}(F_{i+1,j} - F_{i-1,j}) + F_{i,j}(E_{i,j+1} - E_{i,j-1})}{4(E_{i,j} \cdot G_{i,j} - (F_{i,j})^2)h}$$

$$(\Gamma_{11}^2)_{i,j} = \frac{2E_{i,j}(F_{i+1,j} - F_{i-1,j}) - E_{i,j}(E_{i,j+1} - E_{i,j-1}) - F_{i,j}(E_{i+1,j} - E_{i-1,j})}{4(E_{i,j} \cdot G_{i,j} - (F_{i,j})^2)h}$$

$$(\Gamma_{22}^1)_{i,j} = \frac{2G_{i,j}(F_{i,j+1} - F_{i,j-1}) - G_{i,j}(G_{i+1,j} - G_{i-1,j}) - F_{i,j}(G_{i,j+1} - G_{i,j-1})}{4(E_{i,j} \cdot G_{i,j} - (F_{i,j})^2)h}$$

$$(\Gamma_{22}^2)_{i,j} = \frac{E_{i,j}(G_{i,j+1} - G_{i,j-1}) - 2F_{i,j}(F_{i,j+1} - F_{i,j-1}) + F_{i,j}(G_{i+1,j} - G_{i-1,j})}{4(E_{i,j} \cdot G_{i,j} - (F_{i,j})^2)h}.$$

Suppose, $\mathbf{z} = (f_{1,1}, \ldots, f_{1,J}, \ldots, \ldots, f_{I-1,1}, \ldots, f_{I-1,J}, f_{I,1}, \ldots, f_{I,J})^{\mathrm{T}}$, $\bar{f}_{i,j}$ is the sampled value of $z = f(x, y)$ at the pth sampled point (x_i, y_j), $\tilde{\mathbf{z}} = (\tilde{f}_{1,1}, \ldots, \tilde{f}_{1,J}, \ldots, \tilde{f}_{I-1,1}, \ldots, \tilde{f}_{I-1,J}, \tilde{f}_{I,1}, \ldots, \tilde{f}_{I,J})^{\mathrm{T}}$ indicates interpolations based on sampling points $\{f_{i,j}\}$. Then, HASM could be formulated as

$$\begin{bmatrix} \mathbf{A}^T & \mathbf{B}^T & \lambda \cdot \mathbf{S}^T \end{bmatrix} \begin{bmatrix} \mathbf{A} \\ \mathbf{B} \\ \lambda \cdot \mathbf{S} \end{bmatrix} \mathbf{z} = \begin{bmatrix} \mathbf{A}^T & \mathbf{B}^T & \lambda \cdot \mathbf{S}^T \end{bmatrix} \begin{bmatrix} \tilde{\mathbf{d}} \\ \tilde{\mathbf{q}} \\ \lambda \cdot \mathbf{k} \end{bmatrix} \qquad (3.2)$$

where the nonzero element of the sample matrix $\mathbf{S}$ can be expressed as $S_{p,(i-1)\times J+j} = 1$; the nonzero element of the sample vector $\mathbf{k}$ is $k_p = \bar{f}_{i,j}$; $\tilde{\mathbf{d}} = [\tilde{\mathbf{d}}_1, \tilde{\mathbf{d}}_2, \ldots, \tilde{\mathbf{d}}_{I-1}, \tilde{\mathbf{d}}_I]^T$; $\tilde{\mathbf{q}} = [\tilde{\mathbf{q}}_1, \tilde{\mathbf{q}}_2, \ldots, \tilde{\mathbf{q}}_{I-1}, \tilde{\mathbf{q}}_I]^T$;

$$
\tilde{\mathbf{d}}_1 = \begin{bmatrix}
\dfrac{\tilde{f}_{2,1} - \tilde{f}_{0,1}}{2} \cdot (\Gamma^1_{11})_{1,1} \cdot h + \dfrac{\tilde{f}_{1,2} - \tilde{f}_{1,0}}{2} \cdot \left(\tilde{\Gamma}^2_{11}\right)_{1,1} \cdot h + \dfrac{\tilde{L}_{1,1}}{\sqrt{\tilde{E}_{1,1} + \tilde{G}_{1,1} - 1}} \cdot h^2 - \tilde{f}_{0,1} \\[3ex]
\dfrac{\tilde{f}_{2,2} - \tilde{f}_{0,2}}{2} \cdot (\tilde{\Gamma}^1_{11})_{1,2} \cdot h + \dfrac{\tilde{f}_{1,3} - \tilde{f}_{1,1}}{2} \cdot \left(\tilde{\Gamma}^2_{11}\right)_{1,2} \cdot h + \dfrac{\tilde{L}_{1,2}}{\sqrt{\tilde{E}_{1,2} + \tilde{G}_{1,2} - 1}} \cdot h^2 - \tilde{f}_{0,2} \\[2ex]
\vdots \\[2ex]
\dfrac{\tilde{f}_{2,j} - \tilde{f}_{0,j}}{2} \cdot (\tilde{\Gamma}^1_{11})_{1,j} \cdot h + \dfrac{\tilde{f}_{1,j+1} - \tilde{f}_{1,j-1}}{2} \cdot \left(\tilde{\Gamma}^2_{11}\right)_{1,j} \cdot h + \dfrac{\tilde{L}_{1,j}}{\sqrt{\tilde{E}_{1,j} + \tilde{G}_{1,j} - 1}} \cdot h^2 - \tilde{f}_{0,j} \\[2ex]
\vdots \\[2ex]
\dfrac{\tilde{f}_{2,J-1} - \tilde{f}_{0,J-1}}{2} \cdot (\tilde{\Gamma}^1_{11})_{1,J-1} \cdot h + \dfrac{\tilde{f}_{1,J} - \tilde{f}_{1,J-2}}{2} \cdot \left(\tilde{\Gamma}^2_{11}\right)^n_{1,J-1} \cdot h + \dfrac{\tilde{L}_{1,J-1}}{\sqrt{\tilde{E}_{1,J-1} + \tilde{G}_{1,J-1} - 1}} \cdot h^2 - \tilde{f}_{0,J-1} \\[3ex]
\dfrac{\tilde{f}_{2,J} - \tilde{f}_{0,J}}{2} \cdot (\tilde{\Gamma}^1_{11})_{1,J} \cdot h + \dfrac{\tilde{f}_{1,J+1} - \tilde{f}_{1,J-1}}{2} \cdot \left(\tilde{\Gamma}^2_{11}\right)_{1,J} \cdot h + \dfrac{\tilde{L}_{1,J}}{\sqrt{\tilde{E}_{1,J} + \tilde{G}_{1,J} - 1}} \cdot h^2 - \tilde{f}_{0,J}
\end{bmatrix}^T_{J \times 1} ;
$$

for $i = 2, 3, \ldots, I-2, I-1$,

$$
\tilde{\mathbf{d}}_i = \begin{bmatrix}
\dfrac{\tilde{f}_{i+1,1} - \tilde{f}_{i-1,1}}{2} \cdot (\tilde{\Gamma}^1_{11})_{i,1} \cdot h + \dfrac{\tilde{f}_{i,2} - \tilde{f}_{i,0}}{2} \cdot \left(\tilde{\Gamma}^2_{11}\right)_{i,1} \cdot h + \dfrac{\tilde{L}_{i,1}}{\sqrt{\tilde{E}_{i,1} + \tilde{G}_{i,1} - 1}} \cdot h^2 \\[3ex]
\dfrac{\tilde{f}_{i+1,2} - \tilde{f}_{i-1,2}}{2} \cdot (\tilde{\Gamma}^1_{11})_{i,2} \cdot h + \dfrac{\tilde{f}_{i,3} - \tilde{f}_{i,1}}{2} \cdot \left(\tilde{\Gamma}^2_{11}\right)_{i,2} \cdot h + \dfrac{\tilde{L}_{i,2}}{\sqrt{\tilde{E}_{i,2} + \tilde{G}_{i,2} - 1}} \cdot h^2 \\[2ex]
\vdots \\[2ex]
\dfrac{\tilde{f}_{i+1,j} - \tilde{f}_{i-1,j}}{2} \cdot (\tilde{\Gamma}^1_{11})_{i,j} \cdot h + \dfrac{\tilde{f}_{i,j+1} - \tilde{f}_{i,j-1}}{2} \cdot \left(\tilde{\Gamma}^2_{11}\right)_{i,j} \cdot h + \dfrac{\tilde{L}_{i,j}}{\sqrt{\tilde{E}_{i,j} + \tilde{G}_{i,j} - 1}} \cdot h^2 \\[2ex]
\vdots \\[2ex]
\dfrac{\tilde{f}_{i+1,J-1} - \tilde{f}_{i-1,J-1}}{2} \cdot (\tilde{\Gamma}^1_{11})_{i,J-1} \cdot h + \dfrac{\tilde{f}_{i,J} - \tilde{f}_{i,J-2}}{2} \cdot \left(\tilde{\Gamma}^2_{11}\right)_{i,J-1} \cdot h + \dfrac{\tilde{L}_{i,J-1}}{\sqrt{\tilde{E}_{i,J-1} + \tilde{G}_{i,J-1} - 1}} \cdot h^2 \\[3ex]
\dfrac{\tilde{f}_{i+1,J} - \tilde{f}_{i-1,J}}{2} \cdot (\tilde{\Gamma}^1_{11})_{i,J} \cdot h + \dfrac{\tilde{f}_{i,J+1} - \tilde{f}_{i,J-1}}{2} \cdot \left(\tilde{\Gamma}^2_{11}\right)_{i,J} \cdot h + \dfrac{\tilde{L}_{i,J}}{\sqrt{\tilde{E}_{i,J} + \tilde{G}_{i,J} - 1}} \cdot h^2
\end{bmatrix}^T_{J \times 1} ;
$$

$$
\tilde{\mathbf{d}}_I =
\begin{bmatrix}
\dfrac{\tilde{f}_{I+1,1} - \tilde{f}_{I-1,1}}{2} \cdot (\tilde{\Gamma}_{11}^1)_{I,1} \cdot h + \dfrac{\tilde{f}_{I,2} - \tilde{f}_{I,0}}{2} \cdot \left(\tilde{\Gamma}_{11}^2\right)_{I,1} \cdot h + \dfrac{h^2 \cdot \tilde{L}_{I,1}}{\sqrt{\tilde{E}_{I,1} + \tilde{G}_{I,1}^n - 1}} - \tilde{f}_{I+1,1} \\[2ex]
\dfrac{\tilde{f}_{I+1,2} - \tilde{f}_{I-1,2}}{2} \cdot (\tilde{\Gamma}_{11}^1)_{I,2} \cdot h + \dfrac{\tilde{f}_{I,3} - \tilde{f}_{I,1}}{2} \cdot \left(\tilde{\Gamma}_{11}^2\right)_{I,2} \cdot h + \dfrac{h^2 \cdot \tilde{L}_{I,2}}{\sqrt{\tilde{E}_{I,2} + \tilde{G}_{I,2} - 1}} - \tilde{f}_{I+1,2} \\[2ex]
\vdots \\[1ex]
\dfrac{\tilde{f}_{I+1,j} - \tilde{f}_{I-1,j}}{2} \cdot (\tilde{\Gamma}_{11}^1)_{I,j} \cdot h + \dfrac{\tilde{f}_{I,j+1} - \tilde{f}_{I,j-1}}{2} \cdot \left(\tilde{\Gamma}_{11}^2\right)_{I,j} \cdot h + \dfrac{h^2 \cdot \tilde{L}_{I,j}}{\sqrt{\tilde{E}_{I,j} + \tilde{G}_{I,j} - 1}} - \tilde{f}_{I+1,j} \\[2ex]
\vdots \\[1ex]
\dfrac{\tilde{f}_{I+1,J-1} - \tilde{f}_{I-1,J-1}}{2} \cdot (\tilde{\Gamma}_{11}^1)_{I,J-1} \cdot h + \dfrac{\tilde{f}_{I,J} - \tilde{f}_{I,J-2}}{2} \cdot \left(\tilde{\Gamma}_{11}^2\right)_{I,J-1} \cdot h + \dfrac{h^2 \cdot \tilde{L}_{I,J-1}}{\sqrt{\tilde{E}_{I,J-1} + \tilde{G}_{I,J-1} - 1}} - \tilde{f}_{I+1,J-1} \\[2ex]
\dfrac{\tilde{f}_{I+1,J} - \tilde{f}_{I-1,J}}{2} \cdot (\tilde{\Gamma}_{11}^1)_{I,J} \cdot h + \dfrac{\tilde{f}_{I,J+1} - \tilde{f}_{I,J-1}}{2} \cdot \left(\tilde{\Gamma}_{11}^2\right)_{I,J} \cdot h + \dfrac{h^2 \cdot \tilde{L}_{I,J}}{\sqrt{\tilde{E}_{I,J} + \tilde{G}_{I,J} - 1}} - \tilde{f}_{I+1,J}
\end{bmatrix}^T_{J \times 1} ;
$$

for $i = 1, 2, \ldots, I,$

$$
\tilde{\mathbf{q}}_i =
\begin{bmatrix}
\dfrac{\tilde{f}_{i+1,1} - \tilde{f}_{i-1,1}}{2} \cdot (\tilde{\Gamma}_{22}^1)_{i,1} \cdot h + \dfrac{\tilde{f}_{i,2} - \tilde{f}_{i,0}}{2} \cdot \left(\tilde{\Gamma}_{22}^2\right)_{i,1} \cdot h + \dfrac{\tilde{N}_{i,1}}{\sqrt{\tilde{E}_{i,1} + \tilde{G}_{i,1} - 1}} \cdot h^2 - \tilde{f}_{i,0} \\[2ex]
\dfrac{\tilde{f}_{i+1,2} - \tilde{f}_{i-1,2}}{2} \cdot (\tilde{\Gamma}_{22}^1)_{i,2} \cdot h + \dfrac{\tilde{f}_{i,3} - \tilde{f}_{i,1}}{2} \cdot \left(\tilde{\Gamma}_{22}^2\right)_{i,2} \cdot h + \dfrac{\tilde{N}_{i,2}}{\sqrt{\tilde{E}_{i,2} + \tilde{G}_{i,2} - 1}} \cdot h^2 \\[2ex]
\vdots \\[1ex]
\dfrac{\tilde{f}_{i+1,j} - \tilde{f}_{i-1,j}}{2} \cdot (\tilde{\Gamma}_{22}^1)_{i,j} \cdot h + \dfrac{\tilde{f}_{i,j+1} - \tilde{f}_{i,j-1}}{2} \cdot \left(\tilde{\Gamma}_{22}^2\right)_{i,j} \cdot h + \dfrac{\tilde{N}_{i,j}}{\sqrt{\tilde{E}_{i,j} + \tilde{G}_{i,j} - 1}} \cdot h^2 \\[2ex]
\vdots \\[1ex]
\dfrac{\tilde{f}_{i+1,J-1} - \tilde{f}_{i-1,J-1}}{2} \cdot (\tilde{\Gamma}_{22}^1)_{i,J-1} \cdot h + \dfrac{\tilde{f}_{i,J} - \tilde{f}_{i,J-2}}{2} \cdot \left(\tilde{\Gamma}_{22}^2\right)_{i,J-1} \cdot h + \dfrac{\tilde{N}_{i,J-1}}{\sqrt{\tilde{E}_{i,J-1} + \tilde{G}_{i,J-1} - 1}} \cdot h^2 \\[2ex]
\dfrac{\tilde{f}_{i+1,J} - \tilde{f}_{i-1,J}}{2} \cdot (\tilde{\Gamma}_{22}^1)_{i,J} \cdot h + \dfrac{\tilde{f}_{i,J+1} - \tilde{f}_{i,J-1}}{2} \cdot \left(\tilde{\Gamma}_{22}^2\right)_{i,J} \cdot h + \dfrac{\tilde{N}_{i,J}}{\sqrt{\tilde{E}_{i,J} + \tilde{G}_{i,J} - 1}} \cdot h^2 - \tilde{f}_{i,J+1}
\end{bmatrix}^T_{J \times 1} ;
$$

$$
\mathbf{A} =
\begin{bmatrix}
-2\mathbf{I}_{J \times J} & \mathbf{I}_{J \times J} & & & \\
\mathbf{I}_{J \times J} & -2\mathbf{I}_{J \times J} & \mathbf{I}_{J \times J} & & \\
& \ddots & \ddots & \ddots & \\
& & \mathbf{I}_{J \times J} & -2\mathbf{I}_{J \times J} & \mathbf{I}_{J \times J} \\
& & & \mathbf{I}_{J \times J} & -2\mathbf{I}_{J \times J}
\end{bmatrix}_{(I \cdot J) \times (I \cdot J)} ;
$$

$$\mathbf{I}_{J \times J} = \begin{bmatrix} 1 & 0 & & & & \\ 0 & 1 & 0 & & & \\ & \ddots & \ddots & \ddots & & \\ & & 0 & 1 & 0 \\ & & & 0 & 1 \end{bmatrix}_{J \times J} ;$$

$$\mathbf{B} = \begin{bmatrix} \mathbf{B}_{J \times J} & & \\ & \ddots & \\ & & \mathbf{B}_{J \times J} \end{bmatrix}_{(I \cdot J) \times (I \cdot J)} ;$$

$$\mathbf{B}_{J \times J} = \begin{bmatrix} -2 & 1 & & & & & \\ 1 & -2 & 1 & & & & \\ & \ddots & \ddots & \ddots & & & \\ & & 1 & -2 & 1 & & \\ & & & \ddots & \ddots & \ddots & \\ & & & & 1 & -2 & 1 \\ & & & & & 1 & -2 \end{bmatrix}_{J \times J} ;$$

$$\tilde{E}_{i,j} = 1 + \left(\frac{\tilde{f}_{i+1,j} - \tilde{f}_{i-1,j}}{2h} \right)^2$$

$$\tilde{G}_{i,j} = 1 + \left(\frac{\tilde{f}_{i,j+1} - \tilde{f}_{i,j-1}}{2h} \right)^2 ;$$

$$\tilde{F}_{i,j} = \left(\frac{\tilde{f}_{i+1,j} - \tilde{f}_{i-1,j}}{2h} \right) \left(\frac{\tilde{f}_{i,j+1} - \tilde{f}_{i,j-1}}{2h} \right) ;$$

$$\tilde{L}_{i,j} = \frac{\left(\tilde{f}_{i+1,j} - 2\tilde{f}_{i,j} + \tilde{f}_{i-1,j} \right) / h^2}{\sqrt{1 + \left(\left(\tilde{f}_{i+1,j} - \tilde{f}_{i-1,j} \right) / 2h \right)^2 + \left(\left(\tilde{f}_{i,j+1} - \tilde{f}_{i,j-1} \right) / 2h \right)^2}} ;$$

$$\tilde{N}_{i,j} = \frac{\left(\tilde{f}_{i,j+1} - 2\tilde{f}_{i,j} + \tilde{f}_{i,j-1} \right) / h^2}{\sqrt{1 + \left(\left(\tilde{f}_{i+1,j} - \tilde{f}_{i-1,j} \right) / 2h \right)^2 + \left(\left(\tilde{f}_{i,j+1} - \tilde{f}_{i,j-1} \right) / 2h \right)^2}} ;$$

$$(\tilde{\Gamma}_{11}^1)_{i,j} = \frac{\tilde{G}_{i,j}(\tilde{E}_{i+1,j} - \tilde{E}_{i-1,j}) - 2\tilde{F}_{i,j}(\tilde{F}_{i+1,j} - \tilde{F}_{i-1,j}) + \tilde{F}_{i,j}(\tilde{E}_{i,j+1} - \tilde{E}_{i,j-1})}{4\left(\tilde{E}_{i,j} \cdot \tilde{G}_{i,j} - \left(\tilde{F}_{i,j}\right)^2\right)h};$$

$$(\tilde{\Gamma}_{11}^2)_{i,j} = \frac{2\tilde{E}_{i,j}(\tilde{F}_{i+1,j} - \tilde{F}_{i-1,j}) - \tilde{E}_{i,j}(\tilde{E}_{i,j+1} - \tilde{E}_{i,j-1}) - \tilde{F}_{i,j}(\tilde{E}_{i+1,j} - \tilde{E}_{i-1,j})}{4\left(\tilde{E}_{i,j} \cdot \tilde{G}_{i,j} - \left(\tilde{F}_{i,j}\right)^2\right)h};$$

$$(\tilde{\Gamma}_{22}^1)_{i,j} = \frac{2\tilde{G}_{i,j}(\tilde{F}_{i,j+1} - \tilde{F}_{i,j-1}) - \tilde{G}_{i,j}(\tilde{G}_{i+1,j} - \tilde{G}_{i-1,j}) - \tilde{F}_{i,j}(\tilde{G}_{i,j+1} - \tilde{G}_{i,j-1})}{4\left(\tilde{E}_{i,j} \cdot \tilde{G}_{i,j} - \left(\tilde{F}_{i,j}\right)^2\right)h};$$

$$(\tilde{\Gamma}_{22}^2)_{i,j} = \frac{\tilde{E}_{i,j}(\tilde{G}_{i,j+1} - \tilde{G}_{i,j-1}) - 2\tilde{F}_{i,j}(\tilde{F}_{i,j+1} - \tilde{F}_{i,j-1}) + \tilde{F}_{i,j}(\tilde{G}_{i+1,j} - \tilde{G}_{i-1,j})}{4\left(\tilde{E}_{i,j} \cdot \tilde{G}_{i,j} - \left(\tilde{F}_{i,j}\right)^2\right)h}.$$

Let $\mathbf{W} = \begin{bmatrix} \mathbf{A}^T & \mathbf{B}^T & \lambda \cdot \mathbf{S}^T \end{bmatrix} \begin{bmatrix} \mathbf{A} \\ \mathbf{B} \\ \lambda \cdot \mathbf{S} \end{bmatrix}$, $\tilde{\mathbf{v}} = \begin{bmatrix} \mathbf{A}^T & \mathbf{B}^T & \lambda \cdot \mathbf{S}^T \end{bmatrix} \begin{bmatrix} \tilde{\mathbf{d}} \\ \tilde{\mathbf{q}} \\ \lambda \cdot \mathbf{k} \end{bmatrix}$, then

the equation set of HASM for direct algorithms can be formulated as

$$\mathbf{W} \cdot \mathbf{z} = \tilde{\mathbf{v}} \tag{3.3}$$

3.2.1 Gaussian Elimination Algorithm of HASM (HASM-GE)

$\mathbf{W} \cdot \mathbf{z} = \tilde{\mathbf{v}}$ can be expressed as,

$$\begin{cases} w_{1,1} \cdot z_1 + w_{1,2} \cdot z_2 + \cdots + w_{1,I \cdot J} \cdot z_{I \cdot J} = \tilde{v}_1 \\ w_{2,1} \cdot z_1 + w_{2,2} \cdot z_2 + \cdots + w_{2,I \cdot J} \cdot z_{I \cdot J} = \tilde{v}_2 \\ w_{I \cdot J,1} \cdot z_1 + w_{I \cdot J,2} \cdot z_2 + \cdots + w_{I \cdot J,I \cdot J} \cdot z_{I \cdot J} = \tilde{v}_{I \cdot J} \end{cases} \tag{3.4}$$

The $w_{1,1}$ and the first row is respectively selected as the first pivot element and the pivot row (Axelsson, 1994). The coefficients in the matrix are denoted by $w_{i,j}^{(1)} = w_{i,j}$ and the right-hand side by $v_i^{(1)} = \tilde{v}_i$. The first variable is eliminated from the remaining equations by multiplying the first row by $-(w_{i,1}^{(1)} / w_{1,1}^{(1)})$ and adding it to the ith row, $i = 2, 3, \ldots, I \cdot J$. Then we have

$$\begin{cases} w_{1,1}^{(1)} \cdot z_1 + w_{1,2}^{(1)} \cdot z_2 + w_{1,3}^{(1)} \cdot z_3 + \cdots + w_{1,I \cdot J}^{(1)} \cdot z_{I \cdot J} = v_1^{(1)} \\ \qquad\qquad w_{2,2}^{(2)} \cdot z_2 + w_{2,3}^{(2)} \cdot z_3 + \cdots + w_{2,I \cdot J}^{(2)} \cdot z_{I \cdot J} = v_2^{(2)} \\ \qquad\qquad\qquad\qquad\qquad \vdots \\ \qquad\qquad w_{I \cdot J,2}^{(2)} \cdot z_2 + w_{I \cdot J,3}^{(2)} \cdot z_3 + \cdots + w_{I \cdot J,I \cdot J}^{(2)} \cdot z_{I \cdot J} = v_{I \cdot J}^{(2)} \end{cases} \tag{3.5}$$

where $w_{i,j}^{(2)} = w_{i,j}^{(1)} - w_{i,1}^{(1)}/w_{1,1}^{(1)} \cdot w_{1,j}^{(1)}$ and $v_i^{(2)} = v_i^{(1)} - w_{i,1}^{(1)}/w_{1,1}^{(1)} \cdot v_1^{(1)}, 2 \le i \le I \cdot J$, and $2 \le j \le I \cdot J$. $w_{2,2}^{(2)}$ is selected as the second pivot element and the above elimination is repeated, then we get

$$
\begin{cases}
w_{1,1}^{(1)} \cdot z_1 + w_{1,2}^{(1)} \cdot z_2 + w_{1,3}^{(1)} \cdot z_3 + \cdots + w_{1,I\cdot J}^{(1)} \cdot z_{I\cdot J} = v_1^{(1)} \\
\qquad w_{2,2}^{(2)} \cdot z_2 + w_{2,3}^{(2)} \cdot z_3 + \cdots + w_{2,I\cdot J}^{(2)} \cdot z_{I\cdot J} = v_2^{(2)} \\
\qquad\qquad w_{3,3}^{(3)} \cdot z_3 + \cdots + w_{3,I\cdot J}^{(3)} \cdot z_{I\cdot J} = v_3^{(3)} \\
\qquad\qquad\qquad \vdots \\
\qquad\qquad w_{I\cdot J,3}^{(3)} \cdot z_3 + \cdots + w_{I\cdot J,I\cdot J}^{(3)} \cdot z_{I\cdot J} = v_{I\cdot J}^{(3)}
\end{cases} \tag{3.6}
$$

We can continue to repeat this elimination until we get the following equation set,

$$
\begin{cases}
w_{1,1}^{(1)} \cdot z_1 + w_{1,2}^{(1)} \cdot z_2 + w_{1,3}^{(1)} \cdot z_3 + \cdots + w_{1,I\cdot J}^{(1)} \cdot z_{I\cdot J} = v_1^{(1)} \\
\qquad w_{2,2}^{(2)} \cdot z_2 + w_{2,3}^{(1)} \cdot z_3 + \cdots + w_{2,I\cdot J}^{(2)} \cdot z_{I\cdot J} = v_2^{(2)} \\
\qquad\qquad w_{3,3}^{(3)} \cdot z_3 + \cdots + w_{3,I\cdot J}^{(3)} \cdot z_{I\cdot J} = v_3^{(3)} \\
\qquad\qquad\qquad \vdots \\
\qquad\qquad\qquad\qquad w_{I\cdot J,I\cdot J}^{(I\cdot J)} \cdot z_{I\cdot J} = v_{I\cdot J}^{(I\cdot J)}
\end{cases} \tag{3.7}
$$

In general, $w_{i,j}^{(n+1)} = w_{i,j}^{(n)} - (w_{i,n}^{(n)} / w_{i,i}^{(n)}) \cdot w_{n,j}^{(n)}$ and $v_i^{(n+1)} = v_i^{(n)} - (w_{i,n}^{(n)} / w_{i,i}^{(n)}) \cdot v_n^{(n)}$ where $n + 1 \le i \le I \cdot J$ and $n + 1 \le j \le I \cdot J$.

Then, we solve Equation 3.7 by back substitution. The solution is formulated as $z_{I\cdot J} = (v_{I\cdot J}^{(I\cdot J)}/w_{I\cdot J,I\cdot J}^{(I\cdot J)})$ and $z_i = (1/w_{i,i}^{(i)}) \cdot (v_i^{(i)} - \sum_{k=i+1}^{I\cdot J} w_{i,k}^{(i)} \cdot z_k)$ where $1 \le i \le I \cdot J - 1$.

3.2.2 Square-Root Algorithm of HASM (HASM-SR)

The coefficient matrix of HASM, **W**, can be expressed as the product of a triangular and its transposed matrix because **W** is a symmetric positive definite matrix (Faddeev and Faddeeva, 1963). Thus, let

$$
\mathbf{W} = \mathbf{U}^T \cdot \mathbf{U} \tag{3.8}
$$

$$
\mathbf{U} = \begin{bmatrix}
u_{1,1} & u_{1,2} & \cdots & u_{1,I\cdot J} \\
0 & u_{2,2} & \cdots & u_{2,I\cdot J} \\
\cdots & \cdots & \cdots & \cdots \\
0 & 0 & \cdots & u_{I\cdot J,I\cdot J}
\end{bmatrix} \tag{3.9}
$$

where $w_{i,j} = \sum_{k=1}^{i} u_{k,i} \cdot u_{k,j}$ when $i < j$, $w_{i,i} = \sum_{k=1}^{i} u_{k,i}^2$ when $i = j$.

We have

$$u_{1,1} = \sqrt{w_{1,1}} \tag{3.10}$$

$$u_{1,j} = \frac{w_{1,j}}{u_{1,1}} \tag{3.11}$$

when $i > 1$,

$$u_{i,i} = \sqrt{w_{i,i} - \sum_{k=1}^{i-1} u_{k,i}^2} \tag{3.12}$$

when $j > i$,

$$u_{i,j} = \frac{w_{i,j} - \sum_{k=1}^{i-1} u_{k,i} \cdot u_{k,j}}{u_{i,i}} \tag{3.13}$$

when $j < i$,

$$u_{i,j} = 0 \tag{3.14}$$

The equality, $\mathbf{W} \cdot \mathbf{z} = \mathbf{v}$, is equivalent to the two equalities

$$\mathbf{U}^T \cdot \boldsymbol{\rho} = \mathbf{v} \tag{3.15}$$

and

$$\mathbf{U} \cdot \mathbf{z} = \boldsymbol{\rho} \tag{3.16}$$

The elements of the vector $\boldsymbol{\rho}$ are calculated by forward substitution of equality (Equation 3.16), that is,

$$\rho_1 = \frac{v_1}{u_{1,1}} \tag{3.17}$$

when $i > 1$,

$$\rho_i = \frac{v_i - \sum_{k=1}^{i-1} u_{k,i} \cdot \rho_k}{u_{i,i}} \tag{3.18}$$

The final solution is calculated by back substitution of equality (Equation 3.17),

$$z_{I \cdot J} = \frac{\rho_{I \cdot J}}{u_{I \cdot J, I \cdot J}} \tag{3.19}$$

when $i \leq I \cdot J - 1$,

$$z_i = \frac{\rho_i - \sum_{k=i+1}^{I \cdot J} u_{i,k} \cdot z_k}{u_{i,i}} \tag{3.20}$$

3.2.3 Numerical Test

The Gaussian synthetic surface (Figure 2.4) is employed to test the computational efficiencies of HASM-GE and HASM-SR. The numerical test shows that computational efficiency of HASM-GE is much higher than that of HASM-SR (Table 3.1). The relationship between the total number of grid cells and the computing time of HASM-GE and HASM-SR can be respectively formulated as the following linear regression equations

$$t_{GE} = 0.012\, gn - 1854.7, \quad R^2 = 0.9669 \tag{3.21}$$

$$t_{SR} = 0.0387\, gn - 5896.6, \quad R^2 = 0.9682 \tag{3.22}$$

where t_{GE} is the computing time of HASM-GE, t_{SR} the computing time of HASM-SR, and gn total number of grid cells in the computational domain.

The computing time difference can be expressed as

$$\Delta t = t_{SR} - t_{GE} = 0.0268 gn - 4041.9, \quad R^2 = 0.9687 \tag{3.23}$$

In other words, the greater the total number of grid cells in the computational domain, the more efficient HASM-GE is compared with HASM-SR. Thus, HASM-GE might be the best selection if direct algorithm of HASM was used.

TABLE 3.1

Comparison of Computing Time between HASM-GE and HASM-SR

Total Number of Grid Cells	101 × 101	201 × 201	401 × 401	801 × 801	1601 × 1601
HASM-GE (second)	2.8	10.6	72.3	1706.3	29838.4
HASM-SR (second)	8	43.4	339.4	5885.5	96576.6
Δt	5.2	32.8	267.1	4179.3	66738.2

3.3 Iterative Formulation of HASM and Its Coefficient Matrix Structure

3.3.1 Iterative Formulation of HASM

The iterative formulation of the master equation set of HASM, as seen in Chapter 2, has been expressed as

$$\begin{cases} \mathbf{A} \cdot \mathbf{z}^{(n+1)} = \mathbf{d}^{(n)} \\ \mathbf{B} \cdot \mathbf{z}^{(n+1)} = \mathbf{q}^{(n)} \end{cases} \tag{3.24}$$

where $\mathbf{z}^{(n+1)} = \left(f_{1,1}^{(n+1)}, \ldots, f_{1,J}^{(n+1)}, \ldots, f_{I,1}^{(n+1)}, \ldots, f_{I,J}^{(n+1)} \right)^{T} = \left(z_{1}^{(n+1)}, \ldots, z_{J}^{(n+1)}, \ldots, \right.$

$\left. z_{(I-1)\cdot J+1}^{(n+1)}, \ldots, z_{I\cdot J}^{(n+1)} \right)^{T}$, $z_{(i-1)\cdot J+j}^{(n+1)} = f_{i,j}^{(n+1)}$ for $1 \le i \le I, 1 \le j \le J$; $\mathbf{A}$ and $\mathbf{B}$ respectively represent coefficient matrices of the first and the second equations; $\mathbf{d}^{(n)}$ and $\mathbf{q}^{(n)}$ are respectively the right-hand side vectors of the equation set, formulations of which can be found in Chapter 2; $\{(x_i, y_i) \mid x_i = i \times h, y_j = j \times h, 0 \le i \le I+1, 0 \le j \le J+1\}$ is the orthogonal division of computational domain Ω; $f_{i,j}^{(n)}$ is the value of the nth iteration of $f(x, y)$ at grid cell (x_i, y_i).

The matrices of the first equation and the second equation have the following formulations (see Figures 3.1 and 3.2), respectively.

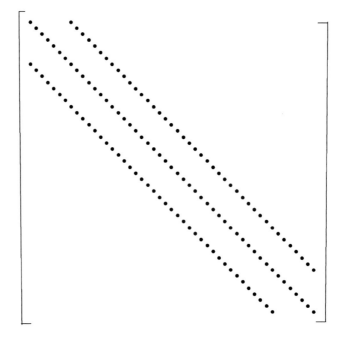

FIGURE 3.1
Distribution sketch of nonzero elements of matrix **A**.

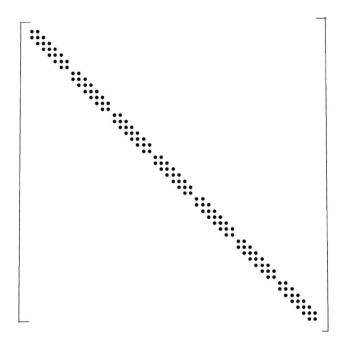

FIGURE 3.2
Distribution sketch of nonzero elements of matrix **B**.

$$
\mathbf{A} =
\begin{bmatrix}
-2\mathbf{I}_{J\times J} & \mathbf{I}_{J\times J} & & & & \\
\mathbf{I}_{J\times J} & -2\mathbf{I}_{J\times J} & \mathbf{I}_{J\times J} & & & \\
& \ddots & \ddots & \ddots & & \\
& & \mathbf{I}_{J\times J} & -2\mathbf{I}_{J\times J} & \mathbf{I}_{J\times J} \\
& & & \mathbf{I}_{J\times J} & -2\mathbf{I}_{J\times J}
\end{bmatrix}_{(I\cdot J)\times(I\cdot J)}
\tag{3.25}
$$

$$
\mathbf{B} =
\begin{bmatrix}
\mathbf{B}_{J\times J} & & \\
& \ddots & \\
& & \mathbf{B}_{J\times J}
\end{bmatrix}_{(I\cdot J)\times(I\cdot J)}
\tag{3.26}
$$

where

$$
\mathbf{I}_{J\times J} =
\begin{bmatrix}
1 & 0 & \cdots & 0 \\
0 & 1 & \cdots & 0 \\
0 & 0 & \cdots & 0 \\
0 & 0 & \cdots & 1
\end{bmatrix}_{J\times J}
$$

and

$$
\mathbf{B}_{J\times J} =
\begin{bmatrix}
-2 & 1 \\
1 & -2 & 1 \\
 & \ddots & \ddots & \ddots \\
 & & 1 & -2 & 1 \\
 & & & \ddots & \ddots & \ddots \\
 & & & & 1 & -2 & 1 \\
 & & & & & 1 & -2
\end{bmatrix}_{J\times J}.
$$

The sample equations can be formulated as

$$
\mathbf{S}\cdot \mathbf{z}^{(n+1)} = \mathbf{k} \tag{3.27}
$$

where $\mathbf{S}$ and $\mathbf{k}$, respectively, represent coefficient matrix and right-hand-side vector, which are established in terms of the sampled points.

Suppose that there are K sampled points,

$$
\mathbf{S} =
\begin{bmatrix}
0 & \cdots & 0 & 1 & 0 & \cdots & 0 & 0 & 0 & \cdots & 0 & 0 & 0 & \cdots & 0 \\
0 & \cdots & 0 & 0 & 0 & \cdots & 0 & 1 & 0 & \cdots & 0 & 0 & 0 & \cdots & 0 \\
\vdots & \vdots & \vdots & \vdots & \vdots & \vdots & \vdots & \vdots & \vdots & \vdots & \vdots & \vdots & \vdots & \vdots & \vdots \\
0 & \cdots & 0 & 0 & 0 & \cdots & 0 & 0 & 0 & \cdots & 0 & 1 & 0 & \cdots & 0
\end{bmatrix}_{K\times(I\cdot J)} \tag{3.28}
$$

If $\bar{f}_{i,j}$ is value of $z = f(x, y)$ at the pth sampled point (x_i, y_j), $s_{p,(i-1)\times J+j} = 1$, $k_p = \bar{f}_{i,j}$. There is only one nonzero element, 1, in every row of the coefficient matrix $\mathbf{S}$, making it a sparse matrix.

The solution procedure of HASM, taking the sampled points as its constraints, can be transformed into solving the following linear equation set in terms of least-squares principle

$$
\begin{bmatrix} \mathbf{A}^T & \mathbf{B}^T & \lambda \cdot \mathbf{S}^T \end{bmatrix}
\begin{bmatrix} \mathbf{A} \\ \mathbf{B} \\ \lambda \cdot \mathbf{S} \end{bmatrix}
\mathbf{z}^{(n+1)} =
\begin{bmatrix} \mathbf{A}^T & \mathbf{B}^T & \lambda \cdot \mathbf{S}^T \end{bmatrix}
\begin{bmatrix} \mathbf{d}^{(n)} \\ \mathbf{q}^{(n)} \\ \lambda \cdot \mathbf{k} \end{bmatrix} \tag{3.29}
$$

Let

$$
\mathbf{W} = \begin{bmatrix} \mathbf{A}^T & \mathbf{B}^T & \lambda \cdot \mathbf{S}^T \end{bmatrix}
\begin{bmatrix} \mathbf{A} \\ \mathbf{B} \\ \lambda \cdot \mathbf{S} \end{bmatrix}
$$

and

$$v^{(n)} = \begin{bmatrix} A^T & B^T & \lambda \cdot S^T \end{bmatrix} \begin{bmatrix} d^{(n)} \\ q^{(n)} \\ \lambda \cdot k \end{bmatrix},$$

then equation set (Equation 3.29) is formulated as

$$W \cdot z^{(n+1)} = v^{(n)} \tag{3.30}$$

3.3.2 Coefficient Matrix Structure

Expressions (Equations 3.3 and 3.30) indicate that the coefficient matrices of the formulation of HASM for direct algorithms and those for iterative algorithms are the same. They can be formulated as

$$W = A^T \cdot A + B^T \cdot B + \lambda^2 \cdot S^T \cdot S \tag{3.31}$$

or

$$W = \begin{bmatrix}
W_1 & -4I_J & I_J & 0 & \cdots & 0 & 0 & 0 & 0 \\
-4I_J & W_2 & -4I_J & I_J & \cdots & 0 & 0 & 0 & 0 \\
I_J & -4I_J & W_3 & -4I_J & \cdots & 0 & 0 & 0 & 0 \\
0 & I_J & -4I_J & W_4 & \cdots & 0 & 0 & 0 & 0 \\
\vdots & \vdots & \vdots & \vdots & \vdots & \vdots & \vdots & \vdots & \vdots \\
0 & 0 & 0 & 0 & \cdots & W_{I-3} & -4I_J & I_J & 0 \\
0 & 0 & 0 & 0 & \cdots & -4I_J & W_{I-2} & -4I_J & I_J \\
0 & 0 & 0 & 0 & \cdots & I_J & -4I_J & W_{I-1} & -4I_J \\
0 & 0 & 0 & 0 & \cdots & 0 & I_J & -4I_J & W_I
\end{bmatrix}_{I \cdot J \times I \cdot J} \tag{3.32}$$

The submatrix, W_i, has the following expression:

$$W_i = \begin{bmatrix}
w_{i,1} & -4 & 1 & 0 & \cdots & 0 & 0 & 0 & 0 \\
-4 & w_{i,2} & -4 & 1 & \cdots & 0 & 0 & 0 & 0 \\
1 & -4 & w_{i,3} & -4 & \cdots & 0 & 0 & 0 & 0 \\
0 & 1 & -4 & w_{i,4} & \cdots & 0 & 0 & 0 & 0 \\
\vdots & \vdots & \vdots & \vdots & \vdots & \vdots & \vdots & \vdots & \vdots \\
0 & 0 & 0 & 0 & \cdots & w_{i,J-3} & -4 & 1 & 0 \\
0 & 0 & 0 & 0 & \cdots & -4 & w_{i,J-2} & -4 & 1 \\
0 & 0 & 0 & 0 & \cdots & 1 & -4 & w_{i,J-1} & -4 \\
0 & 0 & 0 & 0 & \cdots & 0 & 1 & -4 & w_{i,J}
\end{bmatrix}_{J \times J} \tag{3.33}$$

The element $w_{i,j}$ of $\mathbf{W}_i$ can be calculated by the following pseudo-codes

```
If J+1≤i≤(I-1)·J% of the simulated grid cell is not
located in the first row and the last row then
    a = 6;
else,
    a = 5;
end.
If (mod(i,J)==1) || mod(i,J)==0)% start and end of
every submatrix will be
    b = 5;
else
    b = 6;
end
        w = a + b;
If this grid cell is a sampled point
        w = w + λ²;
end
```

There are at most nine nonzero elements in each row of the coefficient matrix, $\mathbf{W}$, according to formulations (Equations 3.32 and 3.33). If we divide all the elements of this coefficient matrix into I groups, meaning that every group consists of J rows, in the first J rows and the last J rows the number of nonzero elements are $5,6,\overbrace{7,\ldots,7}^{J-4},6,5$ respectively. From the $(J+1)$th row to $(2J)th$ row and from the $((I-2)\cdot J)$th row to $((I-1)\cdot J)$th row, the number of nonzero elements are $6,7,\overbrace{8,\ldots,8}^{J-4},7,6$ respectively. In other $(I-4)$ groups, the number of nonzero elements in each group with J rows is $7,8,\overbrace{9,\ldots,9}^{J-4},8,7$ respectively. The total number of nonzero elements can be formulated as $n_z = 9(I \cdot J) - 6I - 6J$. If the nonzero elements are only stored, a huge amount of storage space will be saved.

In short, the coefficient matrix of HASM, $\mathbf{W}$, is a symmetric positive-definite matrix. Both $\mathbf{W} \cdot \mathbf{z}^{(n+1)} = \mathbf{v}^{(n)}$ and $\mathbf{W} \cdot \mathbf{z} = \tilde{\mathbf{v}}$ are large sparse linear systems (Figure 3.3). They are characterized by the property that a large percentage of the elements of the coefficient matrix are zero.

3.4 Gauss–Seidel Algorithm and Preconditioned Gauss–Seidel Algorithm

3.4.1 Gauss–Seidel Algorithm

Solving a large system of linear equations by iterative methods can be dated back to 1823 (Gauss, 1823). The iterative method was then developed by Seidel

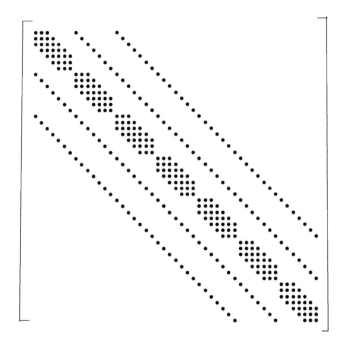

FIGURE 3.3
Distribution sketch of nonzero element of coefficient matrix **W**.

(1874). The Gauss–Seidel iteration was the starting point for succession of over-relaxation methods (Householder, 1964). These iteratively update the single elements in the solution assuming that the remaining elements are the values from previous iteration steps, which need an infinite number of iterations to converge to the optimum solution. Gauss–Seidel algorithm is used for the solution of large linear systems with diagonally dominant sparse matrices, mainly because it requires a limited amount of memory.

3.4.1.1 Matrix Expression

For the large sparse linear system $\mathbf{W} \cdot \mathbf{z}^{(n+1)} = \mathbf{v}^{(n)}$, we can express the coefficient matrix **W** as a sum of three matrices,

$$\mathbf{W} = \underset{diag}{\mathbf{W}} - \underset{lower}{\mathbf{W}} - \underset{upper}{\mathbf{W}} \tag{3.34}$$

where $\underset{diag}{\mathbf{W}}$ is the diagonal matrix, $-\underset{lower}{\mathbf{W}}$ is the strictly lower triangular matrix and $-\underset{upper}{\mathbf{W}}$ is the strictly upper triangular matrix. Then,

$$\underset{diag}{\mathbf{W}} \cdot \mathbf{z}^{(n+1)} = \mathbf{v}^{(n)} + \underset{lower}{\mathbf{W}} \cdot \mathbf{z}^{(n+1)} + \underset{upper}{\mathbf{W}} \cdot \mathbf{z}^{(n+1)} \tag{3.35}$$

The elements of $\left(\mathbf{z}^{(n+1)}\right)^{(k+1)}$ can be computed sequentially using the forward substitution:

$$
\left(z_i^{(n+1)}\right)^{(k+1)} = \frac{1}{w_{i,i}} \left(v_i^{(n)} - \sum_{j=1}^{i-1} w_{i,j} \cdot \left(z_j^{(n+1)}\right)^{(k+1)} - \sum_{j=i+1}^{I \cdot J} w_{i,j} \cdot \left(z_j^{(n+1)}\right)^{(k)} \right),
$$
$$
(3.36)
$$

$$
i = 1, 2, \ldots, I \cdot J,
$$

where $(n + 1)$ represents the $(n + 1)$th outer iteration step; $(k + 1)$ represents the $(k + 1)$th inner iteration step; $k > 0$ and $\left(\mathbf{z}^{(n+1)}\right)^{(0)} = \mathbf{z}^{(n)}$.

The Gauss–Seidel algorithm can be formulated as (Varga, 1962)

$$
\left(\mathbf{z}^{(n+1)}\right)^{(k+1)} = (\underset{diag}{\mathbf{W}} - \underset{lower}{\mathbf{W}})^{-1} \cdot \underset{upper}{\mathbf{W}} \cdot \left(\mathbf{z}^{(n+1)}\right)^{(k)} + (\underset{diag}{\mathbf{W}} - \underset{lower}{\mathbf{W}})^{-1} \cdot \mathbf{v}^{(n)} \qquad (3.37)
$$

3.4.1.2 Vector Expression

Since $\mathbf{W}$ is a symmetric and positive-definite matrix,

$$
\frac{1}{2} \left(\mathbf{W} \cdot \mathbf{z}^{(n+1)} - \mathbf{v}^{(n)} \right)^{\mathrm{T}} \cdot \mathbf{W}^{-1} \cdot \left(\mathbf{W} \cdot \mathbf{z}^{(n+1)} - \mathbf{v}^{(n)} \right)
$$
$$
= \frac{1}{2} \left(\mathbf{z}^{(n+1)}\right)^{\mathrm{T}} \cdot \mathbf{W} \cdot \mathbf{z}^{(n+1)} - \mathbf{v}^{(n)} \cdot \mathbf{z}^{(n+1)} + \frac{1}{2} \left(\mathbf{v}^{(n)}\right)^{\mathrm{T}} \cdot \mathbf{W}^{-1} \cdot \mathbf{v}^{(n)} \qquad (3.38)
$$

If we define the norm

$$
\|\mathbf{u}\|_{\mathbf{W}^{-\frac{1}{2}}} = \left\| \mathbf{W}^{-\frac{1}{2}} \cdot \mathbf{u} \right\| = \left(\mathbf{u}^{\mathrm{T}} \cdot \mathbf{W}^{-1} \cdot \mathbf{u} \right)^{\frac{1}{2}} \qquad (3.39)
$$

Equation 3.38 is equivalent to minimizing the function $\left\| \mathbf{W} \cdot \mathbf{z}^{(n+1)} - \mathbf{v}^{(n)} \right\|_{\mathbf{W}^{-\frac{1}{2}}}$.

$\mathbf{W}$ is a positive definite so that Equation 3.38 has a unique minimum. $(1/2)(\mathbf{v}^{(n)})^{\mathrm{T}} \cdot \mathbf{W}^{-1} \cdot \mathbf{v}^{(n)}$ is a constant and minimizing Equation 3.38 is equivalent to minimizing

$$
\Theta\left(\mathbf{z}^{(n+1)}\right) = \frac{1}{2} \left(\mathbf{z}^{(n+1)}\right)^{\mathrm{T}} \cdot \mathbf{W} \cdot \mathbf{z}^{(n+1)} - \mathbf{v}^{(n)} \cdot \mathbf{z}^{(n+1)} \qquad (3.40)
$$

Since $\Theta'(\mathbf{z}^{(n+1)}) = \mathbf{W} \cdot \mathbf{z}^{(n+1)} - \mathbf{v}^{(n)}$ and $\Theta''(\mathbf{z}^{(n+1)}) = \mathbf{W}$, $\mathbf{W} \cdot \mathbf{z}^{(n+1)} = \mathbf{v}^{(n)}$ and Equation 3.38 have the same solution, then solving the equation set $\mathbf{W} \cdot \mathbf{z}^{(n+1)} = \mathbf{v}^{(n)}$ is transformed into calculating the minimal value of function Θ.

Given a starting vector $\left(\mathbf{z}^{(n+1)}\right)^{(0)} \in \mathbb{R}^{(I \cdot J) \times 1}$ a way $\mathbf{e}_1 \in \mathbb{R}^{(I \cdot J) \times 1}$, $\boldsymbol{\beta}_2 = \left(\mathbf{z}^{(n+1)}\right)^{(0)} + \alpha_1 \cdot \mathbf{e}_1$ is chosen on $\boldsymbol{\beta}_1 = \left(\mathbf{z}^{(n+1)}\right)^{(0)} + \alpha \cdot \mathbf{e}_1$, so that for all real numbers α, $\Theta\left(\left(\mathbf{z}^{(n+1)}\right)^{(0)} + \alpha_1 \cdot \mathbf{e}_1\right) < \Theta\left(\left(\mathbf{z}^{(n+1)}\right)^{(0)} + \alpha \cdot \mathbf{e}_1\right)$. In other words, $\Theta(\mathbf{z})$ has its minimal value at $\boldsymbol{\beta}_2$. Then starting from $\boldsymbol{\beta}_2$, we determine a direction of search, $\mathbf{e}_2$, and choose a real number of α_2, by which $\Theta(\boldsymbol{\beta}_2 + \alpha_2 \cdot \mathbf{e}_2) < \Theta(\boldsymbol{\beta}_2 + \alpha \cdot \mathbf{e}_2)$. This way, we can get a sequence of $\alpha_1, \alpha_2, \ldots, \alpha_J \cdot J$ and $\mathbf{e}_1, \mathbf{e}_2, \ldots, \mathbf{e}_J \cdot J$. The pseudocodes can be described as (Ujevic, 2006)

```
For n = 0,1,2, …
    β₁ = z (n)

    for i = 0,1,2, …, m
    βᵢ₊₁ = βᵢ + αᵢ·eᵢ
    end for i

    z (n+1) = βₘ₊₁

Stopping criteria
end for n
```

where $\mathbf{z}^{(0)}$ is the initial vector, n represents the nth outer iteration step, m is the total inner iteration steps, $\mathbf{e}_1 = (1, 0, 0, \ldots, 0, 0)$, $\mathbf{e}_2 = (0, 1, 0, \ldots, 0, 0)$, $\ldots$, $\mathbf{e}_{J \cdot J-1} = (0, 0, 0, \ldots, 1, 0)$, $\mathbf{e}_{J \cdot J} = (0, 0, 0, \ldots, 1, 0)$, $\alpha_i = -(\tau_i/w_{i,i})$, and $\tau_i = (\mathbf{w}_i, \boldsymbol{\beta}_i) - v_i^{(n)}$, $\mathbf{w}_i$ is the ith row vector of the matrix $\mathbf{W}$, $w_{i,i}$ is the element at the ith row and ith column of the matrix $\mathbf{W}$, and $v_i^{(n)}$ is the ith element of the right-hand side vector $\mathbf{v}^{(n)}$. Taylor expansion of $\Theta(\boldsymbol{\beta}_{i+1})$ at $\boldsymbol{\beta}_i$ can be formulated as

$$\Theta(\boldsymbol{\beta}_{i+1}) = \Theta(\boldsymbol{\beta}_i + \alpha_i \cdot \mathbf{e}_i)$$

$$= \Theta(\boldsymbol{\beta}_i) + \alpha_i \cdot (\mathbf{W} \cdot \boldsymbol{\beta}_i - \mathbf{v}, \mathbf{e}_i) + \frac{1}{2}\alpha_i^2 \cdot (\mathbf{W} \cdot \mathbf{e}_i, \mathbf{e}_i)$$

$$= \Theta(\boldsymbol{\beta}_i) + \alpha_i \cdot \tau_i + \frac{1}{2}\alpha_i^2 \cdot w_{i,i} \tag{3.41}$$

Substituting $\alpha_i = -(\tau_i/w_{i,i})$ into Equation 3.41, we can get

$$\Theta(\boldsymbol{\beta}_{i+1}) - \Theta(\boldsymbol{\beta}_i) = \alpha_i \cdot \tau_i + \frac{1}{2}\alpha_i^2 \cdot w_{i,i} = -\frac{\tau_i^2}{2w_{i,i}} \leq 0 \tag{3.42}$$

If $\hat{\mathbf{z}}$ is a solution of $\mathbf{W} \cdot \mathbf{z}^{(n+1)} = \mathbf{v}^{(n)}$ and W-norm is expressed as $\left\|\mathbf{z}^{(n+1)}\right\|_{\mathbf{W}}^2 = \left(\mathbf{W} \cdot \mathbf{z}^{(n+1)}, \mathbf{z}^{(n+1)}\right)$, then

$$\left\|\boldsymbol{\beta}_{i+1} - \hat{\mathbf{z}}\right\|_{\mathbf{W}}^2 - \left\|\boldsymbol{\beta}_i - \hat{\mathbf{z}}\right\|_{\mathbf{W}}^2 = \left(\mathbf{W} \cdot \boldsymbol{\beta}_{i+1} - \mathbf{W} \cdot \hat{\mathbf{z}}, \boldsymbol{\beta}_{i+1} - \hat{\mathbf{z}}\right) - \left(\mathbf{W} \cdot \boldsymbol{\beta}_i - \mathbf{W} \cdot \hat{\mathbf{z}}, \boldsymbol{\beta}_i - \hat{\mathbf{z}}\right)$$

$$= \left(\mathbf{W} \cdot \boldsymbol{\beta}_{i+1}, \boldsymbol{\beta}_{i+1}\right) - 2\left(\mathbf{v}^{(n)}, \boldsymbol{\beta}_{i+1}\right) - \left[\left(\mathbf{W} \cdot \boldsymbol{\beta}_i, \boldsymbol{\beta}_i\right) - 2\left(\mathbf{v}^{(n)}, \boldsymbol{\beta}_i\right)\right]$$

$$= 2\Theta\left(\boldsymbol{\beta}_{i+1}\right) - 2\Theta\left(\boldsymbol{\beta}_i\right) = -\frac{\tau_i^2}{w_{i,i}} \leq 0 \tag{3.43}$$

In other words, the reduction of the function Θ is equivalent to reduction of errors in the **W**-norm. Gauss–Seidel algorithm of HASM is convergent.

3.4.2 Preconditioned Gauss–Seidel Algorithm

It was found that maximizing the convergence rate was made computationally feasible by the introduction of a new preconditioner (Evans, 1968). Preconditioning was first considered as a means of reducing the condition number in order to improve convergence of an iterative process by Cesari (1937). However, after the introduction of Turing (1948), it became a standard terminology for problem transformation in order to make solutions easier.

3.4.2.1 Matrix Expression

When the coefficient matrix, **W**, is transformed into a nonsingular M-matrix, **V**, the convergence rate of the Gauss–Seidel iterative algorithm can be considerably improved by introducing a preconditioner (Gunawardena et al., 1991),

$$\mathbf{P} = I + \mathbf{\Psi} \tag{3.44}$$

where **I** is unit matrix and

$$
\mathbf{\Psi} = \begin{bmatrix}
0 & -v_{12} & 0 & \cdots & 0 \\
0 & 0 & -v_{23} & \cdots & 0 \\
\vdots & \vdots & \vdots & \vdots & \vdots \\
0 & 0 & 0 & 0 & -v_{I \cdot J-1, I \cdot J} \\
0 & 0 & 0 & 0 & 0
\end{bmatrix}.
$$

V can be formulated as a sum of three matrices,

$$\mathbf{V} = \mathbf{V}_{diag} - \mathbf{V}_{lower} - \mathbf{V}_{upper} \tag{3.45}$$

Then,

$$\mathbf{V}_{diag}^{-1} \cdot \mathbf{V} = \mathbf{I} - \mathbf{V}_{diag}^{-1} \cdot \mathbf{V}_{lower} - \mathbf{V}_{diag}^{-1} \cdot \mathbf{V}_{upper} \tag{3.46}$$

$$\bar{\mathbf{V}} = (\mathbf{I} + \mathbf{\Psi}) \cdot \left(\mathbf{V}_{diag}^{-1} \cdot \mathbf{V} \right)$$

$$= \mathbf{I} - \mathbf{V}_{diag}^{-1} \cdot \mathbf{V}_{lower} - \mathbf{\Psi} \cdot \mathbf{V}_{diag}^{-1} \cdot \mathbf{V}_{lower} - \left(\mathbf{V}_{diag}^{-1} \cdot \mathbf{V}_{upper} - \mathbf{\Psi} + \mathbf{\Psi} \cdot \mathbf{V}_{diag}^{-1} \cdot \mathbf{V}_{upper} \right) \tag{3.47}$$

Then the modified Gauss–Seidel iteration matrix is expressed as

$$\bar{\mathbf{P}} = \left(\mathbf{I} - \mathbf{V}_{diag}^{-1} \cdot \mathbf{V}_{lower} - \mathbf{\Psi} \cdot \mathbf{V}_{diag}^{-1} \cdot \mathbf{V}_{lower} \right)^{-1} \cdot \left(\mathbf{V}_{diag}^{-1} \cdot \mathbf{V}_{upper} - \mathbf{\Psi} + \mathbf{\Psi} \cdot \mathbf{V}_{diag}^{-1} \cdot \mathbf{V}_{upper} \right) \tag{3.48}$$

It was proven that the preconditioned Gauss–Seidel algorithm (PGS) converges quite fast if some parameters are properly chosen even though the standard Gauss–Seidel algorithm diverges (Li, 2003). Niki et al. (2004) analyzed the contributions of three preconditioners, proposed respectively by Milaszewicz (1987), Gunawardena et al. (1991), and Kohno et al. (1997), as a means of accelerating the rate of convergence in the Gauss–Seidel algorithm. They found that preconditioners are effective in accelerating the convergence of the Gauss–Sidel algorithm. However, in recent years, arguments on new preconditioners have been published (Li, 2006; Noutsos and Tzoumas, 2006; Niki et al., 2008; Kohno and Niki, 2010).

3.4.2.2 Vector Expression

According to the Ujevic (2006) method, let $\boldsymbol{\beta}_{i+1} = \boldsymbol{\beta}_i + \boldsymbol{\zeta}_i, \boldsymbol{\theta}_{i+1} = \boldsymbol{\beta}_i + \boldsymbol{\zeta}_i + \gamma_i \cdot \boldsymbol{\xi}_i, \boldsymbol{\zeta}_i = -(\tau_i \cdot \mathbf{e}_i / w_{i,i})$ where $\mathbf{e}_1 = (1, 0, 0, \ldots, 0, 0)$, $\mathbf{e}_2 = (0, 1, 0, \ldots, 0, 0)$, $\ldots$, $\mathbf{e}_{I \cdot J - 1} = (0, 0, 0, \ldots, 1, 0)$, $\mathbf{e}_{I \cdot J} = (0, 0, 0, \ldots, 0, 1)$, $\tau_i = (\mathbf{w}_i, \boldsymbol{\beta}_i) - v_i^{(n)}$, $\gamma_i \in \mathbb{R}$, and $\boldsymbol{\xi}_i$ is any element of $\mathbb{R}^{(I \cdot J) \times 1}$, then

$$\Theta\left(\boldsymbol{\beta}_i + \boldsymbol{\zeta}_i + \gamma_i \cdot \boldsymbol{\xi}_i\right) = \Theta\left(\boldsymbol{\beta}_i\right) + \left(\Theta'\left(\boldsymbol{\beta}_i\right), \boldsymbol{\zeta}_i + \gamma_i \cdot \boldsymbol{\xi}_i\right) + \frac{1}{2}\left(\mathbf{W} \cdot \left(\boldsymbol{\zeta}_i + \gamma_i \cdot \boldsymbol{\xi}_i\right), \boldsymbol{\zeta}_i + \gamma_i \cdot \boldsymbol{\xi}_i\right)$$

$$= \Theta\left(\boldsymbol{\beta}_i\right) + \left(\mathbf{W} \cdot \boldsymbol{\beta}_i - \mathbf{v}^{(n)}, \boldsymbol{\zeta}_i\right) + \gamma_i \cdot \left(\mathbf{W} \cdot \boldsymbol{\beta}_i - \mathbf{v}^n, \boldsymbol{\xi}_i\right) + \frac{1}{2}\left(\mathbf{W} \cdot \boldsymbol{\zeta}_i, \boldsymbol{\zeta}_i\right)$$

$$+ \gamma_i \cdot \left(\mathbf{W} \cdot \boldsymbol{\xi}_i, \boldsymbol{\zeta}_i\right) + \frac{\gamma_i^2}{2}\left(\mathbf{W} \cdot \boldsymbol{\xi}_i, \boldsymbol{\xi}_i\right)$$

$$= \Theta\left(\boldsymbol{\beta}_{i+1}\right) + \gamma_i \cdot \left(\left(\mathbf{W} \cdot \boldsymbol{\beta}_i - \mathbf{v}^{(n)}, \boldsymbol{\xi}_i\right) + \left(\mathbf{W} \cdot \boldsymbol{\xi}_i, \boldsymbol{\zeta}_i\right)\right) + \frac{\gamma_i^2}{2}\left(\mathbf{W} \cdot \boldsymbol{\xi}_i, \boldsymbol{\xi}_i\right)$$

$$\tag{3.49}$$

If

$$\Phi(\lambda) = \gamma \cdot \left(\left(\mathbf{W} \cdot \boldsymbol{\beta}_i - \mathbf{v}^{(n)}, \boldsymbol{\xi}_i\right) + \left(\mathbf{W} \cdot \mathbf{q}_i, \boldsymbol{\zeta}_i\right)\right) + \frac{\gamma^2}{2}\left(\mathbf{W} \cdot \boldsymbol{\xi}_i, \boldsymbol{\xi}_i\right),$$

then,

$$\Phi'(\lambda) = \left(\left(\mathbf{W} \cdot \boldsymbol{\beta}_i - \mathbf{v}^{(n)}, \boldsymbol{\xi}_i\right) + \left(\mathbf{W} \cdot \boldsymbol{\xi}_i, \boldsymbol{\zeta}_i\right)\right) + \gamma \cdot \left(\mathbf{W} \cdot \boldsymbol{\xi}_i, \boldsymbol{\xi}_i\right) \tag{3.50}$$

$$\Phi''(\lambda) = (\mathbf{W} \cdot \boldsymbol{\xi}_i, \boldsymbol{\xi}_i) > 0 \tag{3.51}$$

When $\Phi'(\lambda) = 0$,

$$\gamma_i = -\frac{\left(\mathbf{W} \cdot \boldsymbol{\beta}_i - \mathbf{v}^{(n)}, \boldsymbol{\xi}_i\right) + \left(\mathbf{W} \cdot \boldsymbol{\xi}_i, \boldsymbol{\zeta}_i\right)}{\left(\mathbf{W} \cdot \boldsymbol{\xi}_i, \boldsymbol{\xi}_i\right)}.$$

Then,

$$
\Phi(\lambda) = -\frac{\left[\left(\mathbf{W}\cdot\boldsymbol{\beta}_i - \mathbf{v}^{(n)},\xi_i\right) + \left(\mathbf{W}\cdot\xi_i,\zeta_i\right)\right]^2}{\left(\mathbf{W}\cdot\xi_i,\xi_i\right)} + \frac{\left[\left(\mathbf{W}\cdot\boldsymbol{\beta}_i - \mathbf{v}^{(n)},\xi_i\right) + \left(\mathbf{W}\cdot\xi_i,\zeta_i\right)\right]^2}{2\left(\mathbf{W}\cdot\xi_i,\xi_i\right)}
$$

$$
= -\frac{\left[\left(\mathbf{W}\cdot\boldsymbol{\beta}_i - \mathbf{v}^{(n)},\xi_i\right) + \left(\mathbf{W}\cdot\xi_i,\zeta_i\right)\right]^2}{2\left(\mathbf{W}\cdot\xi_i,\xi_i\right)} \le 0 \tag{3.52}
$$

$$
\Theta\left(\boldsymbol{\theta}_{i+1}\right) - \Theta\left(\boldsymbol{\beta}_i\right) = \Theta\left(\boldsymbol{\beta}_{i+1}\right) - \Theta\left(\boldsymbol{\beta}_i\right) + \gamma_i\cdot\left(\left(\mathbf{W}\cdot\boldsymbol{\beta}_i - \mathbf{v}^{(n)},\xi_i\right) + \left(\mathbf{W}\cdot\xi_i,\zeta_i\right)\right) + \frac{\gamma_i^2}{2}\left(\mathbf{W}\cdot\xi_i,\xi_i\right)
$$

$$
= \Theta\left(\boldsymbol{\beta}_{i+1}\right) - \Theta\left(\boldsymbol{\beta}_i\right) - \frac{\left[\left(\mathbf{W}\cdot\boldsymbol{\beta}_i - \mathbf{v}^{(n)},\xi_i\right) + \left(\mathbf{W}\cdot\xi_i,\zeta_i\right)\right]^2}{2\left(\mathbf{W}\cdot\xi_i,\xi_i\right)} \le \Theta\left(\boldsymbol{\beta}_{i+1}\right) - \Theta\left(\boldsymbol{\beta}_i\right)
$$

$$\tag{3.53}$$

In short, the procedure $\boldsymbol{\theta}_{i+1}$ has a better reduction of errors than the procedure $\boldsymbol{\beta}_{i+1}$. The pseudocodes of HASM-PGS is expressed as follows:

```
for n = 0,1,2, …
    β₁ = z (n)
    for i = 0,1,2, …, m
        θᵢ₊₁ = θᵢ + ζᵢ + γᵢ · ξᵢ
    end for i
        z (n+1) = θₘ₊₁
Stopping criteria
end for n
```

where $\mathbf{z}^{(0)}$ is the initial values from interpolation; n indicates the nth outer iteration step, m is the total steps of inner iteration, $\xi_i = -(\eta_i\cdot\mathbf{e}_i/w_{i,i})$,
$\eta_i = \left(\mathbf{W}_i,\boldsymbol{\theta}_i\right) - v_i^{(n)}$, $\gamma_i = -\dfrac{\left(\mathbf{W}\cdot\boldsymbol{\theta}_i - \mathbf{v}^{(n)},\xi_i\right) + \left(\mathbf{W}\cdot\xi_i,\zeta_i\right)}{\left(\mathbf{W}\cdot\xi_i,\xi_i\right)}$, and ξ_i is any element of $\mathbb{R}^{(I\cdot J)\times 1}$.

If $\xi_i = \mathbf{e}_j$ and $i = j+1$ when $j < I\cdot J$ and $i = 1$ when $j = I\cdot J$, then,

$$
\boldsymbol{\theta}_{i+1} = \boldsymbol{\theta}_i + \zeta_i + \gamma_i\cdot\mathbf{e}_{i-1} \tag{3.54}
$$

$$
\zeta_i = -\frac{\eta_i\cdot\mathbf{e}_i}{w_{i,i}} \tag{3.55}
$$

$$\eta_i = \left(\mathbf{w}_i, \boldsymbol{\theta}_i\right) - v_i^{(n)} \tag{3.56}$$

$$\gamma_i = -\frac{\left(\mathbf{W} \cdot \boldsymbol{\theta}_i - \mathbf{v}^{(n)}, \mathbf{e}_{i-1}\right) + \left(\mathbf{W} \cdot \mathbf{e}_{i-1}, \boldsymbol{\zeta}_i\right)}{\left(\mathbf{W} \cdot \mathbf{e}_{i-1}, \mathbf{e}_{i-1}\right)} = -\frac{\eta_{i-1} - \dfrac{\eta_i}{w_{i,i}} \cdot w_{i,i-1}}{w_{i-1,i-1}}$$

$$= -\frac{\eta_{i-1}}{w_{i-1,i-1}} + \frac{w_{i,i-1}}{w_{i,i} \cdot w_{i-1,i-1}} \cdot \eta_i \tag{3.57}$$

Thus,

$$\eta_{i-1} = \left(\mathbf{w}_{i-1}, \boldsymbol{\theta}_i\right) - v_{i-1}^{(n)}$$

$$= \left(\mathbf{w}_{i-1}, \boldsymbol{\theta}_{i-1} + \boldsymbol{\zeta}_{i-1} + \gamma_{i-1} \cdot \mathbf{e}_{i-2}\right) - v_{i-1}^{(n)}$$

$$= \left(\mathbf{w}_{i-1}, \boldsymbol{\theta}_{i-1}\right) - v_{i-1}^{(n)} + \left(\mathbf{w}_{i-1}, -\frac{\eta_{i-1}}{w_{i-1,i-1}} \cdot \mathbf{e}_{i-1}\right) + \gamma_{i-1} \cdot \left(\mathbf{w}_{i-1}, \mathbf{e}_{i-2}\right)$$

$$= \eta_{i-1} - \frac{\eta_{i-1}}{w_{i-1,i-1}} \cdot w_{i-1,i-1} + \gamma_{i-1} \cdot w_{i-1,i-2} = \gamma_{i-1} \cdot w_{i-1,i-2} \tag{3.58}$$

$$\gamma_i = -\frac{\eta_{i-1}}{w_{i-1,i-1}} + \frac{w_{i,i-1}}{w_{i,i} \cdot w_{i-1,i-1}} \cdot \eta_i = -\frac{w_{i-1,i-2}}{w_{i-1,i-1}} \cdot \gamma_{i-1} + \frac{w_{i,i-1}}{w_{i,i} \cdot w_{i-1,i-1}} \cdot \eta_i \tag{3.59}$$

where $\gamma_0 = \gamma_n$; $\mathbf{e}_0 = \mathbf{e}_{IJ}$; $\mathbf{e}_{-1} = \mathbf{e}_{IJ-1}$; $\eta_0 = \eta_n$; $\zeta_0 = \zeta_n$; $w_{0,0} = w_{IJ,IJ}$; $w_{0,-1} = w_{IJ,IJ-1}$; $w_{1,0} = w_{1,IJ}$.
Let

$$\omega_i = \frac{w_{i-1,i-2}}{w_{i-1,i-1}} \tag{3.60}$$

$$\mu_i = \frac{w_{i,i-1}}{w_{i,i} \cdot w_{i-1,i-1}} \tag{3.61}$$

Then, in terms of expressions of Equations 3.60 and 3.61 as well as structure analysis of $\mathbf{W}$, $\mu_{i \cdot J+1,i \cdot J} = 0$, $\omega_{i \cdot J+2,i \cdot J+1} = 0$, $w_{i,i-1} = w_{i-1,i-2} = -4$. There are at least nine nonzero elements in every row of the coefficient matrix. In HASM-PGS computation, nonzero elements are directly used to perform the mathematical operations of addition, subtraction, multiplication and division in the Gauss–Seidel iterative process. Multiplication between nonzero elements and zero elements is avoided to reduce computational complexity.

3.4.3 Numerical Tests

The Gaussian synthetic surface (Figure 2.4) is selected as the test surface once again. We design three numerical tests. The first compares the computational efficiency of HASM-PGS with the efficiency of all Krylov subspace methods supplied by MATLAB. The second compares the computational efficiency of HASM-PGS with that of HASM-GS; the third analyzes the computational efficiency of HASM-PGS, evaluating the effect of initial values on the convergence of HASM-PGS.

3.4.3.1 Comparison of HASM-PGS with Krylov Subspace Algorithms

The Krylov subspace algorithms include BiConjugate Gradients Algorithm (BICG), BiConjugate Gradients Stabilized Algorithm (BICGSTAB), Conjugate Gradients Squared Algorithm (CGS), Least-Squares QR algorithm (LSQR) for implementation of Conjugate Gradients on the Normal Equations, Minimum Residual Algorithm (MINRES), Preconditioned Conjugate Gradients Algorithm (PCG), Quasi-Minimal Residual Algorithm (QMR), Generalized Minimum Residual Algorithm (GMRES), and Symmetric LQ Algorithm (SYMMLQ). The test is conducted with maximum iteration steps of max$it = 1000$ and a shutdown accuracy of $tol = 10^{-15}$. All Krylov subspace algorithms and HASM-PGS start with the same initial values. The computational domain has 1001×1001 grid cells with a spatial resolution of 0.006. Uniform sampling interval 4 is designed for the numerical test. The simulation error is calculated by

$$RMSE = \sqrt{\frac{1}{(I+2) \cdot (J+2)} \sum_{i=0}^{I+1} \sum_{j=0}^{J+1} (f_{i,j} - Sf_{i,j})^2}$$

where $f_{i,j}$ is true value of $f(x, y)$, $Sf_{i,j}$ is simulated value of $f(x, y)$ at (x_i, y_j), $I = J = 999$.

The test result (see Table 3.2) shows that HASM-PGS took much less time as compared to all Krylov subspace algorithms. The computational speed of HASM-PGS is 12.409 times faster than CGS, the fastest Krylov subspace algorithm, and 470.912 times faster than LSQR, the slowest Krylov subspace algorithm.

3.4.3.2 Comparison of HASM-PGS with HASM-GS

In this test, the sampling interval selected is four grid cells, the shutdown accuracy of inner iteration is 10^{-7}, and outer iteration step is 1. The computing time of HASM-PGS is compared with that of HASM-GS by changing the total grid-cell number in the computational domain.

The result (Table 3.3) indicates that HASM-PGS took much less computing time than HASM-GS did. The time difference between HASM-PGS and

TABLE 3.2

Comparison of Computational Efficiencies among Different Algorithms under Uniform Sampling

Algorithms	RMSE	Iteration Steps	Computing Time (s)	Ratio of Time Spent by HASM-PGS to Ones by Krylov Subspace Algorithms
HASM-PGS	2.65×10^{-6}	50	6.521	1
BICG	2.65×10^{-6}	96	130.002	19.936
BICGSTAB	2.65×10^{-6}	58	108.618	16.657
CGS	2.65×10^{-6}	55	80.920	12.409
LSQR	2.65×10^{-6}	907	3070.817	470.912
MINRES	2.65×10^{-6}	198	243.246	37.302
PCG	2.65×10^{-6}	96	91.860	14.087
QMR	2.65×10^{-6}	92	143.682	22.034
SYMMLQ	2.65×10^{-6}	225	358.075	54.911
GMRES	2.65×10^{-6}	77	270.904	41.543

HASM-GS is closely correlated with the total number of grid cells in the computational domain. In other words, the greater the total number of grid cells, the more efficient HASM-PGS is, compared with HASM-GS. The regression equation can be expressed as

$$\Delta t = 9 \cdot 10^{-6} \cdot gn - 7.2647, \quad R^2 = 0.9172 \tag{3.62}$$

where Δt is the difference of computing time between HASM-PGS and HASM-GS, and gn is the total number of grid cells in the computational domain.

When the sampling interval is four grid cells, the total number of grid cells is 1001×1001, and the outer iteration steps are five, the effect of inner

TABLE 3.3

Comparison of Computing Time between HASM-PGS and HASM-GS

Total Number of Grid Cells	Time Spent by HASM-GS (s)	Time Spent by HASM-PGS (s)	Time Difference (s)
101×101	0.3997	0.3952	0.0045
301×301	3.4381	3.1122	0.3259
501×501	9.5758	8.5352	1.0406
1001×1001	38.5613	34.0659	4.4955
2001×2001	157.3256	136.6111	20.7145
3001×3001	355.0508	310.8424	44.2083
4001×4001	984.5033	820.0605	164.4428

TABLE 3.4

Comparison between RMSEs of PGS and GS under Different
Inner Iterations

Inner Iteration Steps	RMSE of HAMS-GS	RMSE of HASM-PGS	RMSE Difference between HASM-PGS and HASM-GS
5	6.4517×10^{-3}	3.9587×10^{-3}	2.4390×10^{-3}
10	3.4696×10^{-3}	2.0536×10^{-3}	1.4160×10^{-3}
20	1.8455×10^{-3}	1.3144×10^{-3}	0.5311×10^{-3}
40	1.2548×10^{-3}	1.1298×10^{-3}	0.1250×10^{-3}
80	1.1217×10^{-3}	1.1122×10^{-3}	0.0090×10^{-3}
160	1.1120×10^{-3}	1.1119×10^{-3}	0.0001×10^{-3}
320	1.1119×10^{-3}	1.1119×10^{-3}	0

iteration steps on simulation accuracy is analyzed by comparing *RMSE* of HASM-PGS with that of HASM-GS.

Table 3.4 demonstrates that HASM-PGS has a higher accuracy when less inner iteration steps are conducted. HASM-PGS has a faster convergence rate compared to HASM-GS. When 320 inner iteration steps are conducted, both HASM-GS and HASM-PGS reach their stable solution, but HASM-GS needs much more computing time.

When 50 inner iteration steps are given, the outer iteration steps change. The lesser the outer iteration steps, the more the efficient HASM-PGS is (Table 3.5). HASM-PGS has a much faster convergence rate.

3.4.4.3 Effects of Initial Values on Convergence of HASM-PGS

Simulation accuracy is closely related to initial values when the computing time is fixed. Here we will fix a sampling interval of four grid cells, a total

TABLE 3.5

Comparison between PGS and GS under Different Outer Iteration Steps

Outer Iteration Steps	RMSE of HASM-GS	RMSE of HASM-PGS	RMSE Difference between HASM-GS and HASM-PGS
2	15.524×10^{-5}	15.178×10^{-5}	0.346×10^{-5}
4	6.6014×10^{-5}	6.4578×10^{-5}	0.1436×10^{-5}
8	3.4646×10^{-5}	3.4426×10^{-5}	0.022×10^{-5}
10	3.0331×10^{-5}	3.0203×10^{-5}	0.0128×10^{-5}
20	2.1085×10^{-5}	2.1065×10^{-5}	0.002×10^{-5}
25	1.8870×10^{-5}	1.8865×10^{-5}	0.0005×10^{-5}
28	1.7848×10^{-5}	1.7848×10^{-5}	0

TABLE 3.6

Simulation Results of HASM-PGS under Different PreInterpolation Methods

Interpolation Method	Spline	Linear	Nearest
RMSE	6.3×10^{-7}	4.2×10^{-4}	9.7×10^{-3}

grid-cell number of 801×801, one outer iteration step, and 50 inner iteration steps to analyze the simulation accuracies of HASM-PGS when its initial values are respectively supplied by different classical methods such as linear interpolation, nearest method, and spline.

The results indicate that RMSE of HASM-PGS using spline interpolation as its initial values is four orders of magnitude lower than RMSE of HASM-PGS using initial values, interpolated by nearest method (Table 3.6) because spline has a much higher interpolation accuracy compared to the nearest method. Therefore, initial values of HASM-PGS should be as accurate as possible.

3.5 Conjugate Gradient Algorithm and Preconditioned Conjugate Gradient Algorithm

3.5.1 Conjugate Gradient Algorithm

Krylov method started in the early 1950s with the introduction of the conjugate gradients methods (CGs). These methods were designed to construct approximate solutions in the Krylov subspace (van der Vorst, 2002). CG requires less storage space and is especially suited to handling linear systems arising from different equations approximating boundary value problems (Hestenes and Stiefel, 1952). CG was originally viewed as an acceleration technique for the effective solution of large linear systems by a succession of well-convergent approximations (Lanczos, 1952), but in 1952, two different versions of the conjugate gradient method were separately proposed (Hestenes and Stiefel, 1952; Lanczos, 1952). The method proposed by Lanczos was for symmetric positive-definite matrices, mathematically equivalent to CG, but described for the general case of nonsymmetric matrices. CG saw relatively little use for almost 20 years, during which Gaussian elimination for dense matrices and Chebyschev iteration for sparse matrices were generally the preferred solution methods. However, it was proven that CG yielded good approximate solutions than $I \cdot J$ steps for large sparse matrices (Reid, 1971) in which $I \cdot J$ is the dimension of the problem. CG came into wide use in the middle of the 1970s when vector computers and massive computer memories made it possible to use CG to solve problems (Golub and O'Leary, 1989).

In all, it took about 25 years for CG to become the method of choice for symmetric positive-definite matrices.

The HASM-CG can be regarded as a generalized least-squares method, where the minimization takes place on the Krylov subspace. For the symmetric and positive-definite matrix $\mathbf{W}$, the standard form of HASM-CG can be formulated as (Axelsson, 1994),

$$\Lambda\left(\mathbf{z}^{(n+1)}\right) = \frac{1}{2}\left(\mathbf{r}, \mathbf{W}^{-1}\cdot\mathbf{r}\right) \tag{3.63}$$

where $\mathbf{r} = \mathbf{W}\cdot\mathbf{z}^{(n+1)} - \mathbf{v}^{(n)}$ and $(\cdot,\cdot)$ is an inner product.

As seen in Equation 3.38,

$$\Lambda\left(\mathbf{z}^{(n+1)}\right) = \left\|\mathbf{W}\cdot\mathbf{z}^{(n+1)} - \mathbf{v}^{(n)}\right\|_{\mathbf{W}^{-\frac{1}{2}}} = \frac{1}{2}\left(\mathbf{z}^{(n+1)}\right)^{T}\cdot\mathbf{W}\cdot\mathbf{z}^{(n+1)} - \mathbf{v}^{(n)}\cdot\mathbf{z}^{(n+1)}$$

$$+ \frac{1}{2}\left(\mathbf{v}^{n}\right)^{T}\cdot\mathbf{W}^{-1}\cdot\mathbf{v}^{(n)} \tag{3.64}$$

The minimizer of $\Lambda(\mathbf{z}^{(n+1)})$ is the solution $\mathbf{z}^{(n+1)} = \mathbf{W}^{-1}\cdot\mathbf{v}^{(n)}$ of $\mathbf{W}\cdot\mathbf{z}^{(n+1)} = \mathbf{v}^{(n)}$. The gradient of Λ at $\mathbf{z}^{(n+1)}$ is the vector

$$\mathbf{g}\left(\mathbf{z}^{(n+1)}\right) = \left(\frac{\partial\Lambda}{\partial z_{1}^{(n+1)}}, \cdots, \frac{\partial\Lambda}{\partial z_{I\cdot J}^{(n+1)}}\right) \tag{3.65}$$

Let $\boldsymbol{\omega}$ be any nonzero vector in $\mathbb{R}^{(I\cdot J)\times 1}$ and $(\mathbf{g},\boldsymbol{\omega}) = \lim_{\tau\to 0}\frac{1}{\tau}\left(\Lambda\left(\mathbf{z}^{(n+1)} + \tau\cdot\boldsymbol{\omega}\right) - \Lambda(\mathbf{z}^{(n+1)})\right)$ We have,

$$\Lambda\left(\mathbf{z}^{(n+1)} + \tau\cdot\boldsymbol{\omega}\right) - \Lambda\left(\mathbf{z}^{(n+1)}\right) = \frac{1}{2}\cdot\left(\mathbf{r} + \tau\cdot\mathbf{W}\cdot\boldsymbol{\omega}, \mathbf{W}^{-1}(\mathbf{r} + \tau\cdot\mathbf{W}\cdot\boldsymbol{\omega})\right) - \frac{1}{2}\cdot\left(\mathbf{r}, \mathbf{W}^{-1}\cdot\mathbf{r}\right)$$

$$= \tau\cdot(\mathbf{r},\boldsymbol{\omega}) + \frac{1}{2}\cdot\tau^{2}\cdot(\boldsymbol{\omega}, \mathbf{W}\cdot\boldsymbol{\omega}) \tag{3.66}$$

Thus,

$$(\mathbf{g},\boldsymbol{\omega}) = (\mathbf{r},\boldsymbol{\omega}) \tag{3.67}$$

Equation 3.68 is valid for any ω so that $\mathbf{g} = \mathbf{r}$. In other words, the gradient of Λ equals the residual.

We will now construct a new search direction $\boldsymbol{\omega}^{(k)}$, which will be conjugately orthogonal to the previous search directions to find the minimal of Λ.

We compute $\tau = \tau_k$ such that $\Lambda\left(\left(z^{(n+1)}\right)^{(k)} + \tau \cdot \omega^{(k)}\right), -\infty < \tau < +\infty$, is minimized by τ_k. Then let the new approximation be

$$\left(z^{(n+1)}\right)^{(k+1)} = \left(z^{(n+1)}\right)^{(k)} + \tau_k \cdot \omega^{(k)} \tag{3.68}$$

Let $r^{(k)} = W \cdot \left(z^{(n+1)}\right)^{(k)} - v^{(n)}$ and $\varepsilon(\tau) = \Lambda\left(\left(z^{(n+1)}\right)^{(k)} + \tau \cdot \omega^{(k)}\right) - \Lambda\left(\left(z^{(n+1)}\right)^{(k)}\right)$. In terms of Equation 3.66,

$$\varepsilon(\tau) = \tau \cdot \left(r^{(k)}, \omega^{(k)}\right) + \frac{1}{2} \cdot \tau^2 \cdot \left(\omega^{(k)}, W \cdot \omega^{(k)}\right) \tag{3.69}$$

$\varepsilon(\tau)$ has its smallest value when $(r^{(k)}, \omega^{(k)}) + \tau \cdot (\omega^{(k)}, W \cdot \omega^{(k)}) = 0$, that is,

$$\tau = \tau_k = -\frac{\left(r^{(k)}, \omega^{(k)}\right)}{\left(\omega^{(k)}, W \cdot \omega^{(k)}\right)} \tag{3.70}$$

In terms of Equation 3.68,

$$r^{(k+1)} = r^{(k)} + \tau_k \cdot W \cdot \omega^{(k)} \tag{3.71}$$

Then

$$\left(r^{(k+1)}, \omega^{(k)}\right) = \left(r^{(k)} - \frac{\left(r^{(k)}, \omega^{(k)}\right)}{\left(\omega^{(k)}, W \cdot \omega^{(k)}\right)} \cdot W \cdot \omega^{(k)}, \omega^{(k)}\right) = 0 \tag{3.72}$$

This means that the gradient becomes orthogonal to the search direction. A new search direction, $\omega^{(k+1)}$, for the next iteration step can be constructed as follows (Hestenes and Stiefel, 1952):

$$\omega^{(k+1)} = -r^{(k+1)} + \beta_k \cdot \omega^{(k)}, \quad k = 0, 1, 2, \ldots \tag{3.73}$$

$$\beta_k = \frac{\left(r^{(k+1)}, W \cdot \omega^{(k)}\right)}{\left(\omega^{(k)}, W \cdot \omega^{(k)}\right)} \tag{3.74}$$

Then, it can be proven that

$$\left(\omega^{(k)}, W \cdot \omega^{(m)}\right) = 0, \quad m \neq k \tag{3.75}$$

$$\left(\mathbf{r}^{(k)},\mathbf{r}^{(m)}\right) = 0 \quad m \neq k \tag{3.76}$$

$$\left(\mathbf{r}^{(k)},\boldsymbol{\omega}^{(m)}\right) = 0, \quad 0 \leq m \leq k - 1, \ k \geq 1 \tag{3.77}$$

where $\mathbf{r}^{(0)} = \mathbf{W} \cdot (\mathbf{z}^{(n+1)})^{(0)} - \mathbf{v}^{(n)}$.

This means that the search directions $\boldsymbol{\omega}^{(0)}$, $\boldsymbol{\omega}^{(1)}$, ..., $\boldsymbol{\omega}^{(k)}$ are mutually conjugate and the residuals $\mathbf{r}^{(0)}$, $\mathbf{r}^{(1)}$, ..., $\mathbf{r}^{(k)}$ are mutually orthogonal. HASM-CG can be formulated as

$$\begin{cases} \tau_k = -\dfrac{\left(\mathbf{r}^{(k)},\mathbf{r}^{(k)}\right)}{\left(\boldsymbol{\omega}^{(k)},\mathbf{W} \cdot \boldsymbol{\omega}^{(k)}\right)} \\[3mm] \left(\mathbf{z}^{(n+1)}\right)^{(k+1)} = \left(\mathbf{z}^{(n+1)}\right)^{(k)} + \tau_k \cdot \boldsymbol{\omega}^{(k)} \\[3mm] \mathbf{r}^{(k+1)} = \mathbf{r}^{(k)} + \tau_k \cdot \mathbf{W} \cdot \boldsymbol{\omega}^{(k)} \\[3mm] \beta_k = \dfrac{\left(\mathbf{r}^{(k+1)},\mathbf{r}^{(k+1)}\right)}{\left(\mathbf{r}^{(k)},\mathbf{r}^{(k)}\right)} \\[3mm] \boldsymbol{\omega}^{(k+1)} = -\mathbf{r}^{(k+1)} + \beta_k \cdot \boldsymbol{\omega}^{(k)} \end{cases} \tag{3.78}$$

HASM-CG includes an initial step and an inner iteration process. The initial step is to select the initial vector $\boldsymbol{\omega}^{(0)} = -\mathbf{r}^{(0)}$ from $\mathbf{r}^{(0)} = \mathbf{W} \cdot (\mathbf{z}^{(n+1)})^{(0)} - \mathbf{v}^{(n)}$, in which $(\mathbf{z}^{(n+1)})^{(0)}$ is an estimate from the nth outer iteration step of $\mathbf{W} \cdot \mathbf{z}^{(n+1)} = \mathbf{v}^{(n)}$ where $n = 0, 1, 2, ..., n_{\max}$ and $n_{\max}$ is the maximum number of outer iteration steps. The inner iteration process is to find the minimizer of $\Lambda(\mathbf{z}^{(n+1)})$ by determining τ_k, $(\mathbf{z}^{(n+1)})^{(k+1)}$ and $\mathbf{r}^{(k+1)}$. If the inner product $(\mathbf{r}^{(k+1)},\mathbf{r}^{(k+1)})$ is not small enough to stop the inner iteration β_k, the next new search direction $\boldsymbol{\omega}^{(k+1)}$ would be calculated.

3.5.2 Preconditioned Conjugate Gradient Algorithm

A preconditioner is used to increase the convergence rate of an iterative method. An inner product can be defined by a symmetric and positive-definite matrix $\mathbf{C}$ such that

$$(\mathbf{x},\mathbf{y}) = \mathbf{x}^{\mathrm{T}} \cdot \mathbf{C} \cdot \mathbf{y} \tag{3.79}$$

Then, the pseudoresiduals $\boldsymbol{\sigma}^{(k)} = \mathbf{C}^{-1}(\mathbf{W} \cdot (\mathbf{z}^{(n+1)})^{(k)} - \mathbf{v}^{(n)}) \cdot \mathbf{C}^{-1} \cdot \mathbf{W}$ is a symmetric and positive-definite matrix. The preconditioned conjugate gradient algorithm of HASM (HASM-PCG) can be expressed as

$$
\begin{cases}
\tau_k = -\dfrac{\left(\boldsymbol{\sigma}^{(k)}, \mathbf{C} \cdot \boldsymbol{\sigma}^{(k)}\right)}{\left(\boldsymbol{\omega}^{(k)}, \mathbf{W} \cdot \boldsymbol{\omega}^{(k)}\right)} \\[3mm]
\left(\mathbf{z}^{(n+1)}\right)^{(k+1)} = \left(\mathbf{z}^{(n+1)}\right)^{(k)} + \tau_k \cdot \boldsymbol{\omega}^{(k)} \\[3mm]
\boldsymbol{\sigma}^{(k+1)} = \boldsymbol{\sigma}^{(k)} + \tau_k \cdot \mathbf{C}^{-1} \cdot \mathbf{W} \cdot \boldsymbol{\sigma}^{(k)} \\[3mm]
\beta_k = \dfrac{\left(\boldsymbol{\sigma}^{(k+1)}, \mathbf{C} \cdot \boldsymbol{\sigma}^{(k+1)}\right)}{\left(\boldsymbol{\sigma}^{(k)}, \mathbf{C} \cdot \boldsymbol{\sigma}^{(k)}\right)} \\[3mm]
\boldsymbol{\omega}^{(k+1)} = -\boldsymbol{\sigma}^{(k+1)} + \beta_k \cdot \boldsymbol{\omega}^{(k)}
\end{cases}
\tag{3.80}
$$

The formulation (Equation 3.47) now takes the form

$$
\begin{aligned}
\Lambda\left(\mathbf{z}^{(n+1)}\right) &= \frac{1}{2}\left(\boldsymbol{\sigma}, \left(\mathbf{C}^{-1} \cdot \mathbf{W}\right)^{-1} \cdot \boldsymbol{\sigma}\right) \\
&= \frac{1}{2}\left(\mathbf{C}^{-1} \cdot \mathbf{r}\right)^{\mathrm{T}} \cdot \mathbf{C} \cdot \left(\mathbf{C}^{-1} \cdot \mathbf{W}\right)^{-1} \cdot \mathbf{C}^{-1} \cdot \mathbf{r} \\
&= \frac{1}{2}\mathbf{r}^{\mathrm{T}} \cdot \mathbf{W}^{-1} \cdot \mathbf{r}
\end{aligned}
\tag{3.81}
$$

In short, HASM-PCG minimizes the same function $\mathbf{r}^{\mathrm{T}} \cdot \mathbf{W}^{-1} \cdot \mathbf{r}$, as HASM-CG does, but on the Krylov subspace $\{\mathbf{r}^{(0)}, \mathbf{W} \cdot \mathbf{C}^{-1}\mathbf{r}^{(0)}, \ldots, (\mathbf{W} \cdot \mathbf{C}^{-1})^k \mathbf{r}^{(0)}\}$. With an appropriate preconditioner, this Krylov subspace can generate vectors to make HASM-PCG minimize $\Lambda(\mathbf{z}^{(n+1)})$ much faster than HASM-CG.

3.5.3 Numerical Test

We take Gaussian synthetic surface (Figure 2.4) as a test surface. The numerical tests in Section 3.4 demonstrate that HASM-PGS is faster than HASM-GS and at least 12 times faster than all Krylov subspace algorithms. In this section, we focus on comparing the computing time of HASM-PCG with HASM-CG, HASM-PGS and HASM-GS. In this numerical test, the sampling interval is selected as 4. Initial values are interpolated by a linear method. There is one outer iteration step.

The test results (Table 3.7) indicate that HASM-PCG has the fastest computational speed compared with HASM-CG, HASM-PGS, and HASM-GS when

TABLE 3.7

Comparative Analysis of the Computing Time of HASM-PCG

Algorithm		HASM-PCG		HASM-CG		HASM-PGS		HASM-GS	
Total Number of Grid Cells	RMSE	Computing Time (s)	Inner Iteration Step	Computing Time (s)	Inner Iteration Step	Computing Time (s)	Inner Iteration Step	Computing Time (s)	Inner Iteration Step
101 × 101	0.0275	0.1876	21	0.2105	30	0.6866	200	0.8901	415
201 × 201	0.0067	0.3794	13	0.4181	18	2.6337	200	3.3227	394
301 × 301	0.0030	0.8086	12	1.0044	17	6.2149	200	7.7854	376
401 × 401	0.0017	1.4807	12	1.7117	17	11.2155	200	14.3392	374
501 × 501	0.0011	2.2711	12	2.6364	17	17.5522	200	21.7859	363
1001 × 1001	2.7×10^{-4}	8.4646	12	10.0418	17	70.4669	200	85.8354	361

the simulation processes of all the algorithms have the same shutdown error. HASM-GS takes the longest computing time. When the shutdown error is 2.7×10^{-4}, HASM-PCG needs 8.4646 seconds of computing time and 12 inner iteration steps to accomplish the simulation process, while HASM-CG, HASM-PGS, and HASM-GS need 10.0418, 70.7669, and 85.8354 s and 17, 200, and 361 inner iteration steps, respectively. In other words, HASM-PCG has the fastest convergence rate. The relationships between the computing time and the total number of grid cells of HASM-PCG, HASM-CG, HASM-PGS, and HASM-GS can be respectively formulated as

$$t_{PCG} = 8 \cdot 10^{-6} gn + 0.096, \quad R^2 = 0.9997 \tag{3.82}$$

$$t_{CG} = 10^{-5} gn + 0.0943, \quad R^2 = 0.9999 \tag{3.83}$$

$$t_{PGS} = 7 \cdot 10^{-5} gn - 0.1311, \quad R^2 = 1 \tag{3.84}$$

$$t_{GS} = 9 \cdot 10^{-5} gn + 0.1383, \quad R^2 = 0.9999 \tag{3.85}$$

where t_{PCG}, t_{CG}, t_{PGS}, and t_{GS} represent the computing time of HASM-PCG, HASM-CG, HASM-PGS, and HASM-GS respectively; gn is the total number of grid cells in the computational domain; R is the correlation coefficient between computing time and the total number of grid cells.

3.6 Conclusions and Discussion

In general, iterative algorithms have a much higher computational speed than direct algorithms. Although Gaussian algorithm is very robust, it destroys the sparsity of the HASM coefficient matrix considerably and requires a large amount of storage. Traditionally, Choleski factorization could improve the computational speed and simulation accuracy because the coefficient matrix of HASM is symmetric and positive definite. But our tests demonstrate that Choleski factorization makes Gaussian algorithm less efficient. Thus, if a direct algorithm was required, HASM-GE would be a good choice.

Gauss–Seidel iterative algorithm was widely used in the past because it had an advantage of occupying less storage space, and could maintain the sparsity of the coefficient matrix. But its slow convergence rate meant that a preconditioned Gauss–Seidel algorithm had to be developed. Three numerical tests are designed to analyze the efficiency of HASM-PGS by comparing it with the Krylov subspace algorithms BICG, BICGSTAB, CGS, LSQR, MINRES, PCG, QMR, GMRES, and SYMMLQ. The tests indicate

TABLE 3.8

Comparison of Computational Speeds of all HASM Algorithms

Algorithm	HASM-PCG	HASM-CG	HASM-PGS	HASM-GS	HASM-GE	HASM-SR
RCTTN	8×10^{-6}	10^{-5}	7×10^{-5}	9×10^{-5}	1.2×10^{-2}	3.9×10^{-2}
Times	1	1.25	8.75	11.25	1500	4875

that HASM-PGS is at least 10 times faster than all Krylov subspace algorithms, including PCG.

Conjugate gradient algorithm is widely accepted as an efficient iterative approach to large sparse linear systems. Comparative analyses show that HASM-PCG has a much faster convergence rate than HASM-PGS, although HASM-PCG requires more storage space. We note that the combination of PCG with HASM improves computational efficiency of PCG.

For sufficiently large gn, ratios of the computing time to the total number of grid cells (RCTTN) of HASM-PCG, HASM-CG, HASM-PGS, HASM-GS HASM-GE, and HASM-SR are about 8×10^{-6}, 10^{-5}, 7×10^{-5}, 9×10^{-5}, 1.2×10^{-2}, and 3.9×10^{-2} respectively in terms of Equations 3.83 through 3.86, 3.21 and 3.22. This means that the computational speed of HASM-PCG is 1.25, 8.75, 11.25, 1500, and 4875 times higher than HASM-CG, HASM-PGS, HASM-GS HASM-GE, and HASM-SR, respectively (Table 3.8). We can therefore conclude that HASM-PCG is the most efficient algorithm of HASM in terms of simulation accuracy and computational speed.

References

Axelsson, O. 1994. *Iterative Solution Methods*. New York: Cambridge University Press.

Axelsson, O. and Karátson, J. 2009. Equivalent operator preconditioning for elliptic problems. *Numerical Algorithms* 50: 297–380.

Benzi, M. 2002. Preconditioning techniques for large linear systems: A survey. *Journal of Computational Physics* 182: 418–477.

Benzi, M. and Bertaccini, D. 2008. Block preconditioning of real-valued iterative algorithms for complex linear systems. *IMA Journal of Numerical Analysis* 28: 598–618.

Brown, D., Ling, L., Kansa, E., and Levesley, J. 2005. On approximate cardinal preconditioning methods for solving PDEs with radial basis functions. *Engineering Analysis with Boundary Elements* 29: 343–353.

Bulgakov, V.E. 1993. Multi-level iterative technique and aggregation concept with semi-analytical preconditioning for solving boundary-value problems. *Communications in Numerical Methods in Engineering* 9: 649–657.

Carpentier, B., Giraud, L., and Gratton, S. 2007. Additive and multiplicative two-level spectral preconditioning for general linear systems. *SIAM Journal on Scientific Computing* 29(4): 1593–1612.

Cesari, L. 1937. Sulla risoluzione dei sistemi di equazioni lineari per approssimazioni successive. *Atti Accad. Nazionale Lincei R. Classe Sci. Fis. Mat. Nat.* 25: 422.

Courty, F. and Dervieux, A. 2006. Multi-level functional preconditioning for shape optimization. *International Journal of Computational Fluid Dynamics* 20(7): 481–490.

Dietl, G.K.E. 2007. *Linear Estimation and Detection in Krylov Subspaces.* Berlin: Springer.

Duff, I.S., Erisman, A.M., and Reid, J.K. 1986. *Direct Methods for Sparse Matrices.* Oxford: Clarendon.

Evans, D.J. 1968. The use of pre-conditioning in iterative methods for solving linear equations with symmetric positive definite matrices. *Journal of Applied Mathematics* 4(3): 295–314.

Faddeev, D.K. and Faddeeva, V.N. 1963. *Computational Methods of Linear Algebra.* San Francisco: Freeman and Company.

Fox, L., Huskey, H.D., and Wilkinson, J.H. 1948. Notes on the solution of algebraic linear simultaneous equations. *The Quarterly Journal of Mechanics and Applied Mathematics* 1: 149–173.

Gauss, C.F. 1823. Brief und Geling. *Werke* 9: 278–281.

George, A. and Liu, J.W. 1981. *Computer Solution of Large Sparse Positive Definite Systems.* Englewood Cliffs, NJ: Prentice-Hall.

Golub, G.H. and O'Leary, D.P. 1989. Some history of the conjugate gradient and Lanczos algorithms: 1948–1976. *SIAM Review* 31: 50–102.

Gunawardena, A.D., Jain, S.K., and Snyder, L. 1991. Modified iterative methods for consistent linear systems. *Linear Algebra and its Applications* 154–156: 123–143.

Hestenes, M.R. and Stiefel, E.L. 1952. Methods of conjugate gradients for solving linear systems. *Journal of Research of the National Bureau of Standards B* 49: 409–436.

Hintermueller, M., Kopacka, I., and Volkwein, S. 2009. Mesh-independence and preconditioning for solving parabolic control problems with mixed control-state constraints. *ESAIM: Control, Optimisation and Calculus of Variations* 15: 626–652.

Householder, A.S. 1964. *Theory of Matrices in Numerical Analysis.* Johnson, CO: Blaisdell Publishing Company.

Kim, S. and Kim, S.D. 2009. Preconditioning on high-order element methods using Chebyshev–Gauss–Lobatto nodes. *Applied Numerical Mathematics* 59: 316–333.

Kohno, T., Kotakemori, H., and Niki, H. 1997. Improving the modified Gauss–Seidel method for Z-matrices. *Linear Algebra and its Applications* 267: 113–123.

Kohno, T. and Niki, H. 2010. Letter to the Editor: A note on the preconditioned Gauss–Seidel (GS) method for linear systems. *Journal of Computational and Applied Mathematics* 233(9): 2413–2421.

Javidi, M. 2006. Pseudospectral method and Darvishi's preconditioning for solving system of time dependent partial differential equations. *Applied Mathematics and Computation* 176: 334–340.

Lanczos, C. 1952. Solution of systems of linear equations by minimized iterations. *Journal of Research of the National Bureau of Standards* 49(1): 33–53.

Li, W. 2003. The convergence of the modified Gauss–Seidel methods for consistent linear systems. *Journal of Computational and Applied Mathematics* 154: 97–105.

Li, W. 2006. A note on the preconditioned Gauss–Seidel (GS) method for linear systems. *Journal of Computational and Applied Mathematics* 182: 81–90.

Ling, L. and Kansa, E.J. 2004. Preconditioning for radial basis functions with domain decomposition methods. *Mathematical and Computer Modelling* 40: 1413–1427.

Milaszewicz, J.P. 1987. Improving Jacobi and Gauss–Seidel iterations. *Linear Algebra and its Applications* 93: 161–170.

Niki, H., Harada, K., Morimoto, M., and Sakakihara, M. 2004. The survey of preconditioners used for accelerating the rate of convergence in the Gauss–Seidel method. *Journal of Computational and Applied Mathematics* 164–165: 587–600.

Niki, H., Kohno, T., and Morimoto, M. 2008. The preconditioned Gauss–Seidel method faster than the SOR method. *Journal of Computational and Applied Mathematics* 219: 59–71.

Notay, Y. 2006. Aggregation-based algebraic multilevel preconditioning. *SIAM Journal on Matrix Analysis and Applications* 27(4): 998–1018.

Noutsos, D. and Tzoumas, M. 2006. On optimal improvements of classical iterative schemes for Z-matrices. *Journal of Computational and Applied Mathematics* 188: 89–106.

Persson, P.O. and Peraire, J. 2008. Newton-GMRES preconditioning for discontinuous Galerkin discretizations of the Navier–Stokes equations. *SIAM Journal on Scientific Computing* 30(6): 2709–2733.

Reid, J. K. 1971. On the method of conjugate gradients for the solution of large sparse systems of linear equations. In *Large Sparse Sets of Linear Equations: Proceedings of the Oxford Conference of the Institute of Mathematics and its Applications*, 1970, ed. J. K. Reid, 231–254. London: Academic Press.

Richardson Jr., W.B. 2005. High-order Sobolev preconditioning. *Nonlinear Analysis* 63: e1779–e1787.

Saad, Y. and van der Vorst, H. A. 2000. Iterative solution of linear systems in the 20th century. *Journal of Computational and Applied Mathematics* 123(1–2): 1–33.

Seidel, L. 1874. Ueber ein Verfahren die Gleichungen, auf welche die Methode der kleinsten Quadrate fuehrt, sowie Lineare Gleichungen ueberhaupt, durch successive Annaeherung aufzuloesen. *Akademie der Wissenschaften, Mathematisch-Naturwissenschaftliche Klasse, Abhandlungen* 11: 81–108.

Simon, H. D. 1989. Direct sparse matrix methods. In *Modern Numerical Algorithms for Supercomputers*, eds., J.C. Almond, D.M. Young, 325–344. Austin: The University of Texas.

Southwell, R.V. 1940. *Relaxation Methods in Engineering Science*. London: Oxford University Press.

Timonov, A. 2001. Factorized preconditionings of successive approximations in finite precision. *BIT* 41(3): 582–598.

Turing, A.M. 1948. Rounding-off errors in matrix processes. *The Quarterly Journal of Mechanics and Applied Mathematics* 1: 287–308.

Ujevic, N. 2006. A new iterative method for solving linear systems. *Applied Mathematics and Computation* 179: 725–730.

van der Vorst, H.A. 2002. Efficient and reliable iterative methods for linear systems. *Journal of Computational and Applied Mathematics* 149: 251–265.

Varga, R.S. 1962. *Matrix Iterative Analysis*. Englewood Cliffs, NJ: Prentice-Hall.

Yue, T.X., Du, Z.P., Song, D.J., and Gong, Y. 2007. A new method of surface modeling and its application to DEM construction. *Geomorphology* 91(1–2): 161–172.

Yue, T.X., Du, Z.P., and Song, Y.J. 2008. Ecological models: Spatial models and Geographic Information Systems. In *Encyclopedia of Ecology*, eds., S.E. Jørgensen and B. Fath, 3315–3325. England: Elsevier Limited.

Yue, T.X. and Song, Y.J. 2008. The YUE-HASM Method. In *Accuracy in Geomatics, Proceedings of the 8th International Symposium on Spatial Accuracy Assessment in Natural Resources and Environmental Sciences*, Shanghai, June 25–27, 2008, eds., D. Li, Y. Ge, and G. M. Foody, 148–153. Liverpool: World Academic Union Ltd.

Yue, T. X., Song, D. J., Du, Z. P., and Wang, W. 2010. High accuracy surface modeling and its application to DEM generation. *International Journal of Remote Sensing* 31(8): 2205–2226.

4

*Adaptive Method of HASM**

4.1 Introduction

In most numerical procedures for solving PDEs, the problem is first discretized by choosing algebraic equations on a finite-dimension approximation space. A numerical process is then devised to solve this huge system of discrete equations. The discretization process, which is unable to predict the proper resolution and the proper order of approximation at each location, produces a grid that is too fine. The algebraic system thus becomes unnecessarily large in size, while accuracy usually remains rather low (Brandt, 1977). The aim of adaptive methods is the generation of a grid that is adapted to the problem so that a given error criterion is fulfilled by the solution on this grid. An optimal grid should be as coarse as possible while meeting the criterion in order to save on computing time and memory requirements (Schmidt and Siebert, 2005). For stationary issues, a grid is almost optimal when the local errors are approximately equal for all elements. Therefore, elements where the error is large will be marked for refinement, while elements with small estimated errors are left unchanged or marked for coarsening.

Adaptive procedures for the numerical solution of differential equations started in the late 1950s. Birchfield (1960) found that the truncation error could be reduced by using smaller grid results in improved forecasts of hurricane movement trajectories, in which 150 km × 150 km grid cells only covered the neighborhood of the vortex and the remaining part of the region was covered by 300 km × 300 km grid cells. Morrison (1962) stated that in the integration of a system of ordinary differential equations, the simplest approach was to use a fixed step size; however, over some parts of the range of integration it was generally possible to take a larger step size without seriously affecting the local truncation error. Harrison (1973) suggested that the domain to be resolved with the highest resolution should be kept to a minimum in order to greatly reduce the computer space and time requirements; utilizing a gradual reduction in grid scale would essentially allow one to focus on a region of interest with a very fine grid, while still maintaining a relatively coarse resolution in the surrounding area. Ley and Elsberry (1976)

---

* Dr. Chuan-Fa Chen is a major collaborator of this chapter.

pointed out that a multiply nested grid arrangement may be more effective than maintaining the same grid size throughout the domain if a very fine resolution is only required near the center. Kurihara et al. (1979) and Kurihara and Bender (1980) proposed a two-way system in which the time integration proceeded simultaneously for a fine and a coarse resolution grid area so that the two grid areas interacted dynamically with one another. The two-way system differs from the one-way approach, in which time integration is performed first for the total domain by using a coarse resolution, and then is redone for an inner limited area by utilizing a fine resolution. Methods for the adaptive refinement of geometric grids have been increasingly used in a number of areas of numerical analysis since the late 1970s (Rheinboldt, 1980; Brackbill and Saltzman, 1982; Bastian et al., 1997; Haefner and Boy, 2003).

4.2 Test Area and Data Acquisition

Dongzhi Yuan (tableland) of the Loess Plateau, China, with its elevation of 1350 m on an average, is located in 35°28′–35°40′N and 107°39′–108°05′E (Figure 4.1). Its total area is 2778 km², of which 910 km² is relatively flat land. Dongzhi tableland was formed 2 million years ago. Most scholars believe that the Loess Plateau surface was created by wind, and came into being as a result of several million years of synthetic geological effects. Dongzhi tableland is recognized as "the first loess tableland," because it is the largest expanse of flat land with the deepest soil in the whole of China's Loess Plateau. The original Dongzhi tableland, which for a long time retained its integrity, has become a fragmented crisscross of gullies, largely as a result of

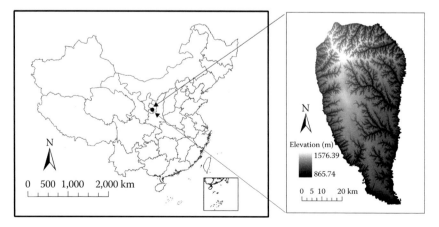

FIGURE 4.1
Location of Dongzhi tableland.

FIGURE 4.2
Distribution of gullies on Dongzhi tableland.

vegetation reduction and soil erosion. The density of gullies is 2.17 km/km^2 on an average (Figure 4.2 and Table 4.1). The eroded area is 2724 km^2, accounting for 98.1% of the total area. The serious soil erosion has led to a rapid shrinkage of the tableland surface. According to historic records, Dongzhi tableland was 110 km in length and 32 km in width between the years 618 and 917. However, the widest area of Dongzhi tableland is now 18 km, and the narrowest area is only 0.05 km, although its length remains unchanged.

There are 197 pieces of relief map that together cover all of Dongzhi tableland. These pieces have been scanned and digitized; 60 on a scale of

TABLE 4.1

Terrain Parameters of Dongzhi Tableland

Terrain Parameters	Relatively Flat Area	Eroded Tableland	Terrace	Bridge	Hill	Gully
Standard deviation of elevation (m)	20.9	43.0	64.4	74.3	60.4	69.0
Elevation on an average (m)	1371	1380	1233	1383	1326	1256
Gully density (km/km²)	0.36	0.3	0.89	1.94	2.14	2.35
Slope on an average (degree)	1.97	6.67	14.1	22.3	24.5	25.7

1/5000 and 137 on a scale of 1/10,000. The relief maps include contour lines, as well as 6692 sampled heights on high accuracy scattered across Dongzhi tableland. Errors created in the scanning process are corrected by comparing every piece of the original contour map with the contour map derived from its digital elevation model (DEM), constructed on a resolution of 1 m × 1 m in relation to data on the contour lines and sampled heights. All the scanned pieces of the contour maps are then combined by joining the ends of the contour lines in the different pieces. Gauss–Krueger projection, in which each zone is 3° of longitude in width, is adopted to transform the Beijing geographical coordinates established in 1954 into rectangular Cartesian coordinates for easier calculation, and Dongzhi tableland is projected to the 36th zone. The Huang–Hai Elevation System, established in 1956, is used as elevation datum. The sampled heights on high accuracy are then used to test the constructed DEM and develop error surfaces.

4.3 Adaptive Method of HASM

4.3.1 Basic Formulation of HASM-AM

For sufficiently large λ, as seen in Chapter 3, the equation set of HASM could be transferred into unconstrained least-squares approximation (Yue et al., 2007, 2008, 2010; Yue and Song, 2008)

$$\begin{bmatrix} \mathbf{A}_h^T & \mathbf{B}_h^T & \lambda \cdot \mathbf{S}^T \end{bmatrix} \begin{bmatrix} \mathbf{A}_h \\ \mathbf{B}_h \\ \lambda \cdot \mathbf{S} \end{bmatrix} \mathbf{z}^{(n+1)} = \begin{bmatrix} \mathbf{A}_h^T & \mathbf{B}_h^T & \lambda \cdot \mathbf{S}^T \end{bmatrix} \begin{bmatrix} \mathbf{d}_h^{(n)} \\ \mathbf{q}_h^{(n)} \\ \lambda \cdot \mathbf{k} \end{bmatrix} \quad (4.1)$$

Let

$$\mathbf{W}_h = \begin{bmatrix} \mathbf{A}_h^T & \mathbf{B}_h^T & \lambda \cdot \mathbf{S}^T \end{bmatrix} \begin{bmatrix} \mathbf{A}_h \\ \mathbf{B}_h \\ \lambda \cdot \mathbf{S} \end{bmatrix} \quad (4.2)$$

$$v_h^{(n)} = \begin{bmatrix} A_h^T & B_h^T & \lambda \cdot S^T \end{bmatrix} \begin{bmatrix} d_h^{(n)} \\ q_h^{(n)} \\ \lambda \cdot k \end{bmatrix} \tag{4.3}$$

Then equation set (Equation 4.1) is transformed into the basic formulation of adaptive method of HASM (HASM-AM):

$$W_h \cdot z^{(n+1)} = v_h^{(n)} \tag{4.4}$$

4.3.2 Error Estimators and Error Indicators

The true error is not generally available during computation, and so an error estimator is needed. Error estimators not only provide means for adaptive optimization of the grid, but also allow an assessment of the reliability of results. The overall aim is to provide accurate and computationally inexpensive error estimates. Such estimates need to be available for various physically important norms in order to be used widely. Miel (1977) proposed a stopping inequality with constant α for new applications of *a posteriori* error estimates. Babuska and Rheinboldt (1978, 1981) developed a mathematical theory for a class of *a posteriori* error estimates. Kelly et al. (1983) found that one of the main features of these *a posteriori* error estimates is that they involved local, rather than global, computations. They were given in an asymptotic form which guaranteed accuracy when linked to adaptive refinement algorithms. However, the fact that these error estimates are historically related to adaptive grid schemes, so that the asymptotic character of the error estimates is accounted for, presents a problem. De et al. (1983) proposed an error indication and an error estimation, which would provide information about where to refine a given grid and when to stop the adaptive process respectively. A value that indicates which grid cells have the larger error, without necessarily revealing what the error is, is referred to as an error indicator (Mitchell, 1989). Loehner (1987) suggested that the error indicator, (1) should be dimensionless so that the numerous key variables could be monitored at the same time, (2) should be bounded so that no further user intervention would be necessary as the solution is evolved, (3) should not only mark the regions with strong shocks to be refined, but also mark weak shocks, contact discontinuities and other weak features in the flow, and (4) should be fast.

For HASM-AM, the error estimator is formulated as

$$RMSE = \sqrt{\frac{1}{6692} \sum_{l=1}^{6692} (f_l - Sf_l)^2} \tag{4.5}$$

where *RMSE* is root mean-square error; f_l is the sampled height on high accuracy at the *l*th sampled point, $l = 1, 2, \ldots, 6692$; Sf_l is the simulated value of $f(x, y)$ in terms of data transformed from contour lines at the lth sampled point.

Simulation results of Dongzhi tableland demonstrate that *RMSE* in all landform types of Leoss Plateau become larger as the grid cell size increases, and that *RMSE* has a close linear relation with grid cell size.

For a relatively flat area

$$RMSE = 2.622 + 0.099h, \quad R^2 = 0.947 \qquad (4.6)$$

where h is the grid cell size, R is the correlation coefficient between h and *RMSE*.

For the eroded tableland,

$$RMSE = 5.825 + 0.142h, \quad R^2 = 0.963 \qquad (4.7)$$

For the loess terrace,

$$RMSE = 3.085 + 0.34h, \quad R^2 = 0.994 \qquad (4.8)$$

For the loess ridge,

$$RMSE = 6.601 + 0.342h, \quad R^2 = 0.990 \qquad (4.9)$$

For the loess hill,

$$RMSE = 7.785 + 0.375h, \quad R^2 = 0.977 \qquad (4.10)$$

For the loess gully,

$$RMSE = 9.157 + 0.397h, \quad R^2 = 0.975 \qquad (4.11)$$

In general,

$$RSME = a + b \cdot h \qquad (4.12)$$

where a and b are determined by gully density, *GD*.

According to the statistically analyzed data of Dongzhi tableland, a and b are expressed as the following linear regression equations:

$$a = 3.068 + 0.472 \cdot GD + 0.830 \cdot GD^2, \quad R^2 = 0.807 \qquad (4.13)$$

$$b = -0.038 + 0.416 \cdot GD - 0.103 \cdot GD^2, \quad R^2 = 0.818 \qquad (4.14)$$

In terms of careful analyses of the requested topographic information, the optimal-grid criterion can be formulated as

$$RMSE = 3.068 + 0.472 \cdot GD + 0.830 \cdot GD^2$$

$$+ (- 0.038 + 0.416 \cdot GD - 0.103 \cdot GD^2)h \qquad (4.15)$$

The gully density in Dongzhi tableland is 0.79 km/km². Absolute error should not exceed 40 m. Then, the spatial resolution of DEM should not be coarser than 160×160 m in terms of Equation 4.15. Thus, the starting grid-cell size (or spatial resolution) is selected as 160×160 m for the whole region of Dongzhi tableland.

The absolute error is formulated as

$$AE_l = \left| f_l - Sf_l \right| \qquad (4.16)$$

where f_l is the sampled height at the kth sampled point, Sf_l is the simulated value of $f(x,y)$ at the lth sampled point; $l = 1, 2, ..., 6692$.

Then, an error surface can be interpolated in terms of the absolute error calculated at every sampled point in combination with Equations 4.6 through 4.11. The location of the sampled point determines which equation is used to interpolate the error surface around the sampled point.

The error indicator is defined as

$$EI_{i,j} = \frac{AE_{i,j}}{40} \qquad (4.17)$$

where $EI_{i,j}$ is the error indicator at lattice (i, j); $AE_{i,j}$ is the interpolated absolute error at lattice (i, j), $i = 1, 2, ..., I, j = 1, 2, ..., J$.

4.3.3 Adaptive Refinement

The adaptive approach is intended to develop the combined processes of adaptive refinement and multigrid solution. Grid adaptation can be divided into two steps; grid marking and grid refinement. The adaptive refinement of discretization grids has proven to be a very successful tool for decreasing the size of linear systems that arise when solving PDEs (Mitchell, 1989). Local grid refinement of computational grids has successfully been applied in two and three dimensions in order to locally improve the approximate solution of systems of PDEs (Kossaczky, 1994). In the case of two-dimensional grid refinement, refinement methods can be distinguished into regular refinement, bisection, and newest-node algorithm. In bisection, a vertex of the triangle is connected to the midpoint of the opposite side of the triangle, which forms two triangles of equal area. In regular refinement, the mid-points of the sides of the triangle are connected to form similar triangles

(Bank and Welfert, 1991). In the newest-node algorithm, a triangle is always bisected by using its newest node, but without restriction on bisecting the longest edge as the bisection method (Jones and Plassmann, 1997). Mitchell (1989) compared these three methods and found that it was difficult to choose a consistently superior algorithm, but all three methods were preferable to using uniform refinement except on smooth problems. However, where local grid refinement is applied in more dimensions, many interesting issues of a physical nature are formulated. Therefore, Maubach (1995) presented a local refinement method of the bisection type in n dimensions.

Discretization of PDEs requires the subdivision of the computational domain. One special kind of subdivision is conforming triangulation (Baensch, 1991). Global refinement of a grid and local refinement are two general algorithms for refining triangular grids. In both cases, the intersection of two nondisjoint, nonidentical triangles consists either of a common vertex or a common side (Rivara, 1984). Regular division and bisection division are two basic methods used for dividing triangles in practice (Mitchell, 1989). For many important applications, the numerical solution of PDEs is the most computationally intensive part of solving a mathematical model, so a great deal of research has been conducted to find faster methods of solving PDEs at a higher resolution (Vey and Voigt, 2007). On sequential computers, methods combining adaptive grid refinement and full multigrid have been shown to have optimal efficiency for many classes of PDEs (Bern et al., 1999).

Grid refinement is a highly important step in the grid adaption process and many methods have been developed for the same. Kim and Thompson (1990) divided adaptive grid strategies into grid-point redistribution and local grid refinement. Grid-point redistribution allows a fixed number of grid points to move continuously during the simulation to increase the grid-point density. Local grid refinement schemes insert and extract grid points into a static grid to increase the grid density in some regions and reduce it in others, in order to achieve a reduction of truncation errors. For instance, Dietachmayer and Droegemeier (1992) distinguish grid-refinement methods including grid transformation and nesting strategies, into two basic categories. The first includes methods in which grid points are either added or subdivided locally to the computational domain as the calculation proceeds to provide an increased spatial resolution based on predetermined physical criteria. The second redistributes a fixed number of grid points so as to provide locally increased resolution, for which the criteria determining how the grid points are redistributed with time are critical. Behrens (2006) stated that there are two major adaptation principles. One does not change the number of grid points and interconnectivity, but instead changes the spacing between grid points by transformation functions (Brackbill and Saltzman, 1982; Dietachmayer and Droegemeier, 1992; Iselin et al., 2002); the other refines or coarsens the mesh by inserting or deleting grid points and re-meshing locally (Behrens, 2006). Any adaptive refinement scheme includes three ingredients; an optimal-grid criterion, an error indicator, and a method to refine the grid.

For HASM-AM, grid cells on which $EI_{i,j} > 1$ enter the refinement process. The refinement process is stopped for grid cells where $EI_{i,j} \leq 1$. The method to refine the grid includes seven steps

1. The whole domain is simulated on the optimal-grid cell size (or spatial resolution) of $h \times h$ for starting by operating HASM, $A_h U^{n+1} = B_h^n$, and the simulated values, U_0, are obtained on the spatial resolution of $h \times h$, in which $h = 160\text{m}$.

2. $AE_{i,j}$ ($i = 1, 2, \ldots, I; j = 1, 2, \ldots, J$) on every grid cell (i, j) is interpolated by combining the calculated absolute errors at the sampled points with equations from (6) to (11) under consideration of landforms. If $EI_{i,j} > 1$, the grid cell (i, j) is flagged for refinement.

3. The flagged grid cells are clustered into different refinement subdomains, $SD_{1,1}, SD_{1,2}, \ldots, SD_{1,K}$, where K is the number of the subdomains to be refined.

4. Every grid cell in the subdomains, $SD_{1,1}, SD_{1,2}, \ldots, SD_{1,K}$, is bisected by connecting the midpoints of two sides with their opposite sides, which forms four smaller grid cells of equal areas with a spatial resolution of $(h/2) \times (h/2)$.

5. The information on coarser grid cells is transferred to the finer grid cells.

6. Equations of HASM, $A_{h/2} U^{n+1} = B_{h/2}^n$, are solved, respectively, in the subdomains, $SD_{1,1}, SD_{1,2}, \ldots, SD_{1,K}$.

7. The process from step (2) to step (6) is repeated until all grid cells meet the requirement for accuracy.

4.4 Validation of HASM-AM

4.4.1 DEM Construction of Dongzhi Tableland

The first step is to simulate the whole of Dongzhi tableland on an optimal-grid cell size (or spatial resolution) of 160 m × 160 m. The error distribution map (Figure 4.3) shows that simulation errors closely relate to gully density. Higher the gully density, bigger is the error. All grid cells in the flat area satisfy $EI_{i,j} \leq 1$. Error analysis demonstrates that the first simulation has a maximum error of 224.7 m, a mean error of 43.2 m, and a standard deviation of 33.0 m. The number of grid cells where $EI_{i,j} > 1$ accounts for 30%.

Next, the grid cells where $EI_{i,j} > 1$ are bisected and then simulated on a spatial resolution of 80 m × 80 m. Error calculation (Figure 4.4) demonstrates that grid cells of $EI_{i,j} > 1$ account for 24% of grid cells after the second simulation. The maximum error decreases to 149.4 m, the mean error to 27.3 m, and the standard deviation to 20.1 m.

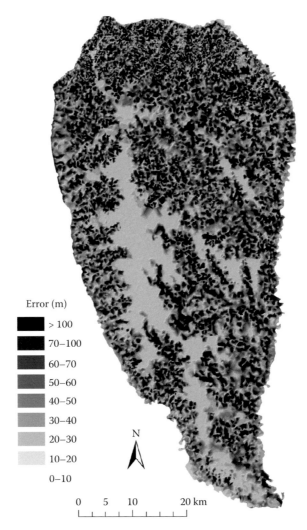

FIGURE 4.3
Error distribution map of the first simulated Dongzhi tableland on a spatial resolution of
160×160 m.

For 24% of the grid cells where $EI_{i,j} > 1$, refinement is conducted again and
then simulation implemented on a spatial resolution of 40 m × 40 m. After
the third simulation, only 3% of grid cells on which $EI_{i,j} > 1$ are left. The maximum error is 99.4 m; the mean error is 15.1 m; and the standard deviation is
10.8 m (see Figure 4.5).

The 3% of grid cells where $EI_{i,j} > 1$ are then bisected for a third time into
finer grid cells of 20 m × 20 m. According to the error distribution map
(see Figure 4.6), grid cells of $EI_{i,j} > 1$ only account for 0.1% grid cells. The

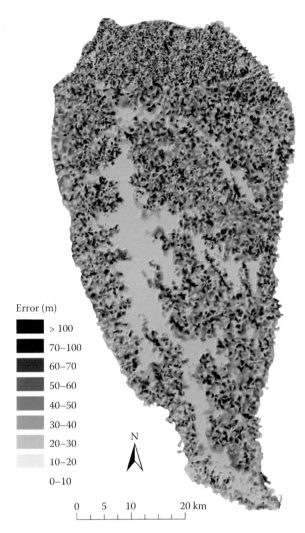

Error (m)

- > 100
- 70–100
- 60–70
- 50–60
- 40–50
- 30–40
- 20–30
- 10–20
- 0–10

N

0　5　10　　　　20 km

FIGURE 4.4
Error distribution map of simulated Dongzhi tableland after the first local refinement on a spatial resolution of 80×80 m.

maximum error decreases to 64.3 m, the mean error to 10.7 m, and the standard deviation to 6.7 m.

The remaining grid cells where $EI_{i,j} > 1$ are then refined for the fourth time. According to the error distribution map of the fifth simulation on a spatial resolution of 10×10 m (Figure 4.7), all grid cells now satisfy $EI_{i,j} \leq 1$. The final maximum error is 40 m, the mean error 8.5 m, and the standard deviation 5.3 m. The final elevation surface of Dongzhi tableland is obtained after the refinement process is completed (Figure 4.8 and Table 4.2).

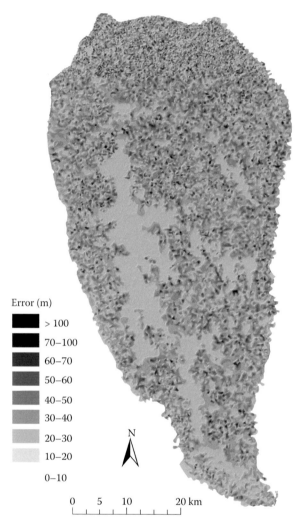

Error (m)

■ > 100

■ 70–100

■ 60–70

■ 50–60

■ 40–50

■ 30–40

■ 20–30

□ 10–20

  0–10

N

0   5   10   20 km

FIGURE 4.5
Error distribution map of simulated Dongzhi tableland after the second local refinement on a spatial resolution of 40×40 m.

4.4.2 Testing Computational Efficiency

Currently, HSAM-AM only deals with about 0.87 million of Dongzhi table-land's grid cells. However, if the whole of Dongzhi tableland's elevation surface was simulated by HASM, 27.24 million grid cells would be calculated on a spatial resolution of 10×10 m. The computational efficiency of HASM has been greatly improved.

In fact, Dongzhi tableland in its entirety is too big to be simulated on a spatial resolution of 10×10 m by classic methods such as IDW (Shepard,

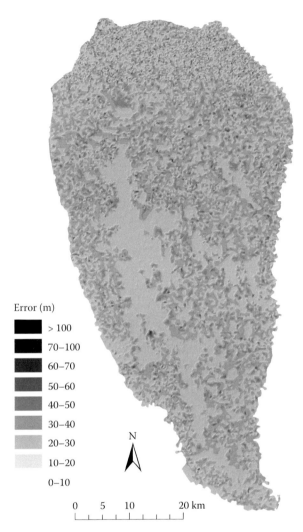

Error (m)

> 100

70–100

60–70

50–60

40–50

30–40

20–30

10–20

0–10

N

0   5   10   20 km

FIGURE 4.6
Error distribution map of simulated Dongzhi tableland after the third local refinement on a spatial resolution of 20 × 20 m.

1968), kriging (Krige, 1951), and spline (Watt, 2000). Instead, a region with an area of 26.5 km² is selected to comparatively analyze the computation time of HASM-AM. In this region, the standard deviation of elevation is 53.5 m. The region is located in 35°49′59″–35°52′30″N and 107°33′45″–107°37′30″E, where the maximum and minimum elevation are 1450 and 1225 m, respectively (Figure 4.9).

The simulated elevation surface consists of 0.265 million grid cells on a spatial resolution of 10 × 10 m. The test results (Table 4.3) show that HASM-AM

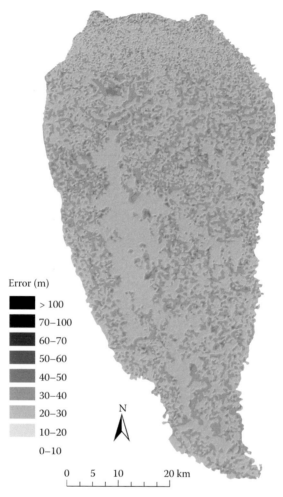

FIGURE 4.7
Error distribution map of simulated Dongzhi tableland after the fourth local refinement on a spatial resolution of 10×10 m.

has the highest accuracy and the fastest computation speed, compared to HASM, IDW, spline, and kriging. HASM-AM created 11.1 m of *RMSE* and accomplished the simulation process in 10.9 s. The computing time of *RMSE* and HASM are 11.5 m and 19.2 s, respectively. This means that the introduction of an adaptive method improved both the computational speed and simulation accuracy. Kriging, IDW, and spline took 2807, 449, and 438 s to accomplish their simulation processes and created 12.3, 12.2, and 31.5 m of *RMSE*, respectively. Both HASM-AM and HASM have much higher computational speeds and accuracy than the classic methods.

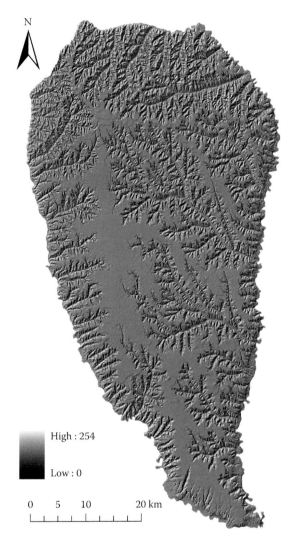

FIGURE 4.8
Shaded relief map in 3D of Dongzhi tableland simulated after the refinement process is completed. (Light from 135° with a dip angle of 45°.)

4.5 Discussion and Conclusions

Defining a global grid for the discretization of a given problem independent of the solution process is often insufficient. The adaptivity of grids is one of the major trends in numerical simulation and scientific computing. In the adaptive multigrid process, finer grids are not constructed globally. They are

TABLE 4.2

Improving Process of Accuracy by Adaptive Refinement

Simulation Process	Maximum Error (m)	Mean Error (m)	Standard Deviation (m)	Ratio of Grid Cells of $EI > 1$ (%)
First simulation of the whole domain	224.7	43.2	33.0	30
Second simulation after the first refinement	149.4	27.3	20.1	24
Third simulation after the second refinement	99.4	15.1	10.8	3
Fourth simulation after the third refinement	64.3	10.7	6.7	0.1
Fifth simulation after the fourth refinement	40.0	8.5	5.3	0

only constructed in those parts of the domain where the current discretization error is significantly large. An adaptive simulation approach with grid selection strategies is highly effective for HASM, and a desirable feature for accurate analysis and efficient simulation. The adaptive approach can be distinguished into predefined refinement and self-adaptive refinement. In predefined refinement, the refinement is determined before the solution process has started; in self-adaptive approaches, the grid refinements are carried out

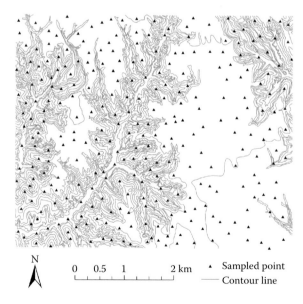

N

| 0   0.5   1   2 km | ▲ Sampled point |
| | — Contour line |

FIGURE 4.9

Sector map on a spatial scale of 1/10,000 of the test region, in which curves are contour lines and black triangles are sampled heights in high accuracy.

TABLE 4.3

Comparisons of Computing Time and Simulation Accuracy

Method	Time (s)	RMSE (m)
Kriging	2807	12.3
IDW	449	12.2
Spline	438	31.5
HASM	19.2	11.5
HASM-AM	10.9	11.1

dynamically during the solution process (Trottenberg et al., 2001). In the process of constructing the DEM of Dongzhi tableland, the predefined refinement and self-adaptive refinement are combined.

Adaptivity can help to resolve local scales that interact with global scales in a consistent way (Behrens, 2006). The most demanding issue for adaptive methods is finding a suitable refinement criterion. An adaptive method is as good as the refinement criterion that controls it. Thus, a good understanding of what accuracy really means in the context of the problem and full insight into the reason for error is required if one aims for substantial improvement in accuracy of the solution. An effective refinement criterion needs to detect those areas that cause the highest error or that are of the highest interest concerning the relevant physical features. In the case of Dongzhi tableland, it is evident that gully density and grid cell size define the most effective refinement criterion.

HASM-AM helps to speed up the computation of HASM by avoiding unnecessary calculations and saving memory. The test result shows that the adaptive method shortens the computation time of HASM from 21003.1 to 3788.4 s, and increases the computational speed by 454.4%.

The adaptive refinement technique is very successful in reducing the computational and storage requirements for solving PDEs of HASM. The adaptive refinement technique places more grid cells in areas where the local error in the solution is large instead of using a uniform mesh with grid cells evenly spaced in a domain. If a nonuniform grid is generated by an adaptive refinement algorithm and portioned into a number of sets equal to the number of processors, a parallel algorithm for the adaptive refinement can be developed, speeding up the computation of HASM-AM. The effect is HASM-AM parallelization.

References

Babuska, I. and Rheinboldt, W. 1978. Error estimates for adaptive finite element computations. *SIAM Journal on Numerical Analysis* 15: 736–754.

Babuska, I. and Rheinboldt, W. 1981. *A posteriori* error analysis of finite element solutions for one-dimensional problems. *SIAM Journal on Numerical Analysis* 18: 565–589.

Baensch, E. 1991. Local mesh refinement in two and three dimensions. *IMPACT Computing in Science and Engineering* 3: 181–191.

Bank, R.E. and Welfert, B.D. 1991. *A posteriori* error estimates for the Stokes problem. *SIAM Journal on Numerical Analysis* 28: 591–623.

Bastian, P., Birken, K., Eckstein, K., Johannsen, K., Lang, S., Neuss, N., and Rentz-Reichert, H. 1997. UG: A flexible software toolbox for solving partial differential equations. *Computing and Visualization in Sciences* 1 (1): 27–40.

Behrens, J. 2006. *Adaptive Atmospheric Modelling*. Berlin: Springer-Verlag.

Bern, M.W., Flaherty, J.E., and Luskin, M. 1999. *Grid Generation and Adaptive Algorithms*. Berlin: Springer.

Birchfield, G.E. 1960. Numerical prediction of hurricane movement with the use of a fine grid. *Journal of Meteorology* 17: 404–414.

Brackbill, J.U. and Saltzman, J.S. 1982. Adaptive zoning for singular problems in two dimensions. *Journal of Computational Physics* 46: 342–368.

Brandt, A. 1977. Multilevel adaptive solutions to boundary value problems. *Mathematics of Computation* 31: 333–390.

De, J.P., Gago, S.R., Kelly, D.W., Zienkiewicz, O.C., and Babuska, I. 1983. *A posteriori* error analysis and adaptive processes in the finite element method: Part II—Adaptive mesh refinement. *International Journal for Numerical Methods in Engineering* 19: 1621–1656.

Dietachmayer, G.S. and Droegemeier, K.K. 1992. Application of continuous dynamic grid adaptation techniques to meteorological modeling, Part I: Basic formulation and accuracy. *Monthly Weather Review* 120: 1675–1706.

Haefner, F. and Boy, S. 2003. Fast transport simulation with an adaptive grid refinement. *Ground Water* 41(2): 273–279.

Harrison, E.J. Jr. 1973. Three-dimensional numerical simulations of tropical systems utilizing nested finite grids. *Journal of the Atmospheric Sciences* 30: 1528–1543.

Iselin, J.P., Prusa, J.M., and Gutowski, W.J. 2002. Dynamic grid adaptation using the MPDATA scheme. *Monthly Weather Review* 130: 1026–1039.

Jones, M.T. and Plassmann, P.E. 1997. Adaptive refinement of unstructured finite-element meshes. *Finite Elements in Analysis and Design* 25(1–2): 41–60.

Kelly, D.W., De, J.P., Gago, S.R., Zienkiewicz, O.C., and Babuska, I. 1983. *A posteriori* error analysis and adaptive processes in the finite element method: Part I—Error analysis. *International Journal for Numerical Methods in Engineering* 19: 1621–1656.

Kim, J.K. and Thompson, J.R. 1990. Three-dimensional adaptive grid generation on a composite-block grid. *AIAA Journal* 38, 470–477.

Kossaczky, I. 1994. A recursive approach to local mesh refinement in two and three dimensions. *Journal of Computational and Applied Mathematics* 55: 275–288.

Krige, D.G. 1951. A statistical approach to some basic mine valuation problems on the Witwatersrand. *The Journal of the Chemical, Metallurgical & Mining Society of South Africa* 52(6): 119–139.

Kurihara, Y. and Bender, W.A. 1980. Design of a movable nested-mesh primitive equation model for tracking a small vortex. *Monthly Weather Review* 108: 1792–1809.

Kurihara, Y., Tripoli, G.J., and Bender, M.A. 1979. Design of a movable nested-mesh primitive equation model. *Monthly Weather Review* 107: 239–249.

Ley, G.W. and Elsberry, R.L. 1976. Forecasts of typhoon Irma using a nested-grid model. *Monthly Weather Review* 104: 1154–1161.

Loehner, R. 1987. An adaptive finite element scheme for transient problems in CFD. *Computer Methods in Applied Mechanics and Engineering* 61: 323–338.

Maubach, J.M. 1995. Local bisection refinement for n-simplicial grids generated by reflection. *SIAM Journal of Scientific Computing* 16: 210–227.

Miel, G. 1977. On *a posteriori* error estimates. *Mathematics of Computation* 31 (137): 204–213.

Mitchell, W.F. 1989. A comparison of adaptive refinement techniques for elliptic problems. *ACM Transactions on Mathematical Software* 15: 326–347.

Morrison, D. 1962. Optimal mesh size in the numerical integration of an ordinary different equation. *Journal of the Association for Computing Machinery* 9: 98–103.

Rheinboldt, W. 1980. On a theory of mesh-refinement processes. *SIAM Journal on Numerical Analysis* 17: 766–778.

Rivara, M.C. 1984. Algorithms for refining triangular grids suitable for adaptive and multigrid techniques. *International Journal for Numerical Methods in Engineering* 20: 745–756.

Schmidt, A. and Siebert, K.G. 2005. *Design of Adaptive Finite Element Software*. Berlin: Springer-Verlag.

Shepard, D. 1968. A two-dimensional interpolation function for irregularly-spaced data. In *Proceedings of the 1968 23rd ACM National Conference*, 517–524. New York: Association for Computing Machinery.

Trottenberg, U., Oosterlee, C.W., and Schueller, A. 2001. *Multigrid*. London: Academic Press.

Vey, S. and Voigt, A. 2007. AMDiS: Adaptive multidimensional simulations. *Computing and Visualization in Science* 10: 57–67.

Watt, A. 2000. *3D Computer Graphics*. New York: Addison-Wesley.

Yue, T.X., Du, Z.P., and Song, Y.J. 2008. Ecological models: Spatial models and Geographic Information Systems. *Encyclopedia of Ecology*, eds., S.E. Jørgensen and B. Fath, 3315–3325. England: Elsevier Limited.

Yue, T.X., Du, Z.P., Song, D.J., and Gong, Y. 2007. A new method of surface modeling and its application to DEM construction. *Geomorphology* 91(1–2): 161–172.

Yue, T.X. and Song, Y.J. 2008. The YUE-HASM method. In *Accuracy in Geomatics*: *Proceedings of the 8th International Symposium on Spatial Accuracy Assessment in Natural Resources and Environmental Sciences*, Shanghai, June 25–27, 2008, pp. 148–153.

Yue, T.X., Song, D.J., Du, Z.P., and Wang, W. 2010. High accuracy surface modeling and its application to DEM generation. *International Journal of Remote Sensing* 31(8): 2205–2226.

5

Multigrid Method of HASM*

5.1 Introduction

The primitive ideas of multigrid can be found from error smoothing by relaxation, total reduction, and nested iteration. Relaxation processes have been long known for error-smoothing properties (Southwell, 1935, 1946). Reduction methods were originally linked to calculations on coarser grids and recursive applications in the early 1950s (Schroeder, 1954). Nested iteration has been used for a number of years to obtain first approximations on finer grids from coarser grids. In 1962, Fedorenko described the first correct two-grid iteration and emphasized on the complementary roles of Jacobi iteration and coarse-grid correction. In 1964, he formulated the first multigrid algorithm and proved its typical convergence behavior. In 1966, Bakhvalov suggested a much more complex possibility of combining multigrid with nested iteration. In 1971, Astrakhantsev generalized Bakhvalov's convergence result in order to apply it to the general boundary conditions.

Brandt (1973) was the first to recognize the actual efficiency of multigrid, clearly outlining its main principles and practical uses. He viewed the multigrid method in two complementary ways: the first sees the coarser grids as correction grids, accelerating the convergence of a relaxation scheme on the finest grid by efficiently liquidating its smooth error components; the second regards the finer grids as correction grids, improving the accuracy on coarser grids by correcting their forcing terms (Brandt, 1977). His essential contributions to early studies included the introduction of nonlinear multigrid and adaptive techniques, the discussion of general domain and local grid refinements, a systematic application to the nested iteration idea, and a provision of the local Fourier analysis tool for theoretical investigation (Stueben and Trottenberg, 1982). Nicolaides (1975) discussed multigrid ideas in connection with finite-element discretization after his first study of multigrid for Poisson's equation in a square, while Hackbusch (1980) presented a general convergence theory of multigrid. Since the early 1980s, the number of yearly publications on mutigrid methods has considerably increased (Wesseling, 1992). However, fairly comprehensive coverage is given by two monographs written by Hackbusch (1985) and Wesseling (1992), respectively.

---

* Yin-Jun Song is a major collaborator of this chapter.

The multigrid idea is based on two principles: error smoothing and coarse grid correction (Trottenberg et al., 2001). If interpreted appropriately, multigrid is typically optimal in the sense that the number of arithmetic operations needed to solve a discrete problem is proportional to the number of unknowns in the problem considered. If designed well, the mesh-size-independent convergence factors can be made very small and the operation count per unknown per iteration step is also small. If the multigrid idea is generalized to structures other than grids, one obtains multilevel, multiscale, or multiresolution methods.

A characteristic feature of the iterative multigrid approach is that the multi-grid convergence speed is independent of grid cell size and that the number of arithmetic operations per iteration step is proportional to the number of grid points. The multigrid principle allows us to construct very efficient linear solvers for which the iterative approach is fundamental. Multigrid can also be regarded as a solution method for the (continuous) differential problem. The more efficiently the solution process is performed, the more efficiently the continuous differential background can be exploited. The method is oriented to minimize the differential error, as opposed to minimizing the algebraic error.

Although HASM-B has a much higher accuracy than the classical methods (Yue et al., 2007, 2010), it also has a much longer computing time, because it must use two PDEs to simulate each lattice of a surface. For this reason, HASM-B is not widely applied in practice. The multigrid approach is introduced into high-accuracy surface modeling in order to shorten the computing time.

5.2 Multigrid Method of HASM

5.2.1 Multigrid

If grid cell size is h, as seen in Chapter 4, HASM-B can be formulated as

$$\mathbf{W}_h \cdot \mathbf{z}^{(n+1)} = \mathbf{v}_h^{(n)} \tag{5.1}$$

If an iteration method is used to solve equation set (Equation 5.1), the iteration cycle for solving the equation set is named as the inner iteration and the procedure of updating the right-hand item $\mathbf{v}_h^{(n)}$ is named as the outer iteration.

If a coarser grid $H = 2h$ is introduced, a two-grid method can be developed. Each iteration step of the two-grid method consists of a pre-smoothing, a coarse grid correction, and a post-smoothing part (Trottenberg et al., 2001). Multigrid is the extension of two-grid levels (Φ_h, Φ_{2h}) to a sequence of levels, $(\Phi_h, \Phi_{2h}, \Phi_{4h}, \ldots, \Phi_{hmax})$, which ends with the coarsest grid Φ_{hmax}.

Let $\mathbf{z}^e = [1,1,\ldots, 1]^T_{1 \times I \cdot J}$ be the exact solution at the fine grid, then $\mathbf{v}_h^{(0)} = \mathbf{W}_h \cdot \mathbf{z}^e$. HASM-MG can be formulated as

$$\mathbf{W}_h \cdot \mathbf{z}^{(1)} = \mathbf{v}_h^{(0)} \qquad (5.2)$$

Every component of $\mathbf{z}^e$ is randomly perturbed by an error of 10^{-3} and the perturbed vector is expressed as $\mathbf{z}^1$, which is employed as the initial value of the first two-grid iteration at the finest grid on multigrid level. The Gauss–Seidel method is used for smoothing iteration, in which the pre-smoothing is $v_1 = 2$ and the post-smoothing is $v_2 = 1$.

As seen in Chapter 3, $\mathbf{W}_h \cdot \mathbf{z}^{(k+1)} = \mathbf{v}_h^{(k)}$ could be formulated as follows by Gauss–Seidel iteration

$$\underset{\text{diag}}{\mathbf{W}_h} \cdot \mathbf{z}^{(n+1)} = \mathbf{v}^{(n)} + \underset{\text{lower}}{\mathbf{W}_h} \cdot \mathbf{z}^{(n+1)} + \underset{\text{upper}}{\mathbf{W}_h} \cdot \mathbf{z}^{(n+1)} \qquad (5.3)$$

where $\underset{\text{diag}}{\mathbf{W}}$ is a diagonal matrix, $-\underset{\text{lower}}{\mathbf{W}}$ is strictly a lower triangular matrix and $-\underset{\text{upper}}{\mathbf{W}}$ is strictly an upper triangular matrix.

The elements of $(\mathbf{z}^{(n+1)})^{(k+1)}$ can be computed sequentially using forward substitution:

$$\left(z_i^{(n+1)}\right)^{(k+1)} = \frac{1}{w_{i,i}}\left(v_i^{(n)} - \sum_{j=1}^{i-1} w_{i,j} \cdot \left(z_j^{(n+1)}\right)^{(k+1)} - \sum_{j=i+1}^{I \cdot J} w_{i,j} \cdot \left(z_j^{(n+1)}\right)^{(k)}\right)$$

$$i = 1,2,\ldots, I \cdot J \qquad (5.4)$$

where $(n + 1)$ represents the $(n + 1)$th outer iteration step and $(k + 1)$ represents the $(k + 1)$th inner iteration step.

5.2.2 Numerical Tests

The numerical test is conducted by nested grids, nine-point restriction, nine-point prolongation (Figure 5.1), and W-cycle (Figure 5.2). *RMSE* is used for error analysis, that is,

$$RMSE(\mathbf{z}^{(m)}, \mathbf{z}^e) = \left(\sum_{i=1}^{I \cdot J} \left(z^{(m)}(i) - z^e(i)\right)^2 / (J \cdot I)\right)^{\frac{1}{2}} \qquad (5.5)$$

where $\mathbf{z}^{(m)}$ is the numerical solution and $\mathbf{z}^e$ is the exact solution.

$$f(x,y) = 3(1 - x)^2 e^{-x^2 - (y+1)^2} - 10\left(\frac{x}{5} - x^3 - y^5\right)e^{-x^2 - y^2} - \frac{e^{-(x+1)^2 - y^2}}{3},$$

Gaussian synthetic surface (Figure 5.3) is selected as a test surface. Its computational domain is $[-3, 3] \times [-3, 3]$ and $-6.5510 < f(x, y) < 8.1062$. The

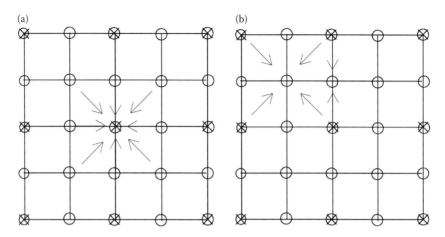

FIGURE 5.1
Information fusion on multigrids, (a) nine-point restriction, and (b) nine-point prolongation
($\bigcirc$ represents fine grid and $\otimes$ coarse grid).

computational domain is divided into 129×129 grid cells and the simulation
step is $h = 6/128 = 0.0469$. The sampling interval is selected as 0.375. The error
analysis of this simulation result shows that *RMSE* of HASM-MG is 0.007,
while *RMSE* of HASM-B is 0.0504.

Numerical tests demonstrate that HASM-MG efficiently eliminates low-
frequency errors and speeds up the inner iteration speed. Different simulation step lengths have almost no effect on the convergence property of the
inner iteration of HASM-MG. The convergence speed of HASM-MG does not
deteriorate when the discretization is refined, whereas classical iterative

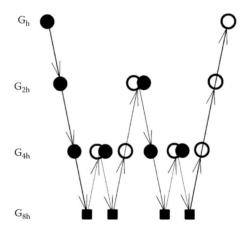

FIGURE 5.2
The structure of W-cycle: $\bullet$ represents v_1 presmoothing steps, $\bigcirc$ v_2 postsmoothing steps,
$\searrow$ restriction, $\nearrow$ prolongation, $\blacksquare$ direct solver.

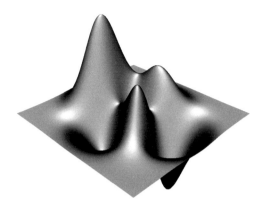

FIGURE 5.3
The Gaussian synthetic surface.

methods slow down for decreasing grid cell size (Hackbusch, 1985). The operation process of HASM-MG shows that the computing time of both inner and outer iteration is directly proportional to the computational work, for example, the total number of grid cells in the computation domain. The linear regression equation between the computing time and computational work (Table 5.1) can be formulated as

$$t = 0.0001gn - 12.118 \qquad (5.6)$$

where correlation coefficient $R^2 = 0.9997$; t represents the computing time; and gn is the total number of grid cells in the computation domain.

When a direct algorithm was used to solve HASM-B for simulating a surface with 2.25 million grids, the process took 10 h (Yue et al., 2007). However, when HASM-MG simulated a surface with 4 million grids using the same computer, the process took only 20 min. The comparison between HASM-B and HASM-MG shows that the multigrid method greatly improved both the computational speed and simulation accuracy.

The numerical test indicates that there is not a close relationship between HASM-MG's simulation accuracy and its computational time. In the process of increasing iteration steps, the number of errors first decreased, and then increased when too many iterative steps were conducted because of rounding error. Therefore, the computational time must depend on shutdown accuracy.

TABLE 5.1

Computing Time and the Total Number of Grid Cells in the Computational Domain

Total number of grid cells	129×129	257×257	513×513	1025×1025	2049×2049	4097×4097
Computing time	1.828	7.391	29.344	120.69	516.77	2228.6

5.3 Comparative Analysis of HASM-MG with OK

HASM-B has been successfully applied to constructing DEMs (Yue and Wang, 2010) and simulating ecosystem change (Yue and Song, 2008), and soil properties (Shi et al., 2009). Here we select the climatic change in the Jiangxi province of China, as a real-world example to demonstrate the applicability of HASM-MG. Ordinary Kriging (OK) has been the most widely used method for spatially simulating climate change in recent years. Thus, the simulation results of HASM-MG are compared with those of OK in this paper.

5.3.1 OK Method

We take the Gaussian synthetic surface (Figure 5.3) once again as a numeric test surface to compare the simulation accuracy of OK with that of HASM-MG. The computational domain is divided into 1025×1025 grid cells. The sampling interval is selected as 0.375, which means we have 289 sampled points.

OK is performed by the module of 3D Analyst in ArcGIS 9.2. Default values of their input parameters are used under the assumption that these default values provide reasonable results under most conditions, because they are all well-known and long-standing implementations (Reuter et al., 2007). The model of semivariogram is spherical, the search radius is variable, and the maximum number of search points is 12.

The error histogram indicates the general pattern of absolute errors from HASM-MG and OK (Figure 5.4). The vertical coordinate represents frequency, that is, the count of grids at which there is an equal absolute error. Abscissa axis represents absolute error. The absolute-error range of OK is [−0.3857 0.4249], while the absolute-error range of HASM-MG is [−0.0995 0.0920]. *RMSEs* of OK and HASM-MG are, respectively, 0.0800 and 0.0148. When the sampling interval is shortened to 0.1875, the absolute-error range of OK and HASM-MG becomes [−0.1112 0.1122] and [−0.0063 0.0096], respectively, while *RMSEs* decrease to 0.0190 and 0.0008. When the sampling interval is further shortened to 0.09375, the absolute-error range of OK and HASM-MG becomes [−0.0295 0.0305] and [−0.0016 0.0041], respectively, and *RMSEs* are reduced to 0.0049 and 0.0002. In other words, the increase of sample points makes both methods more accurate. The error surfaces (Figure 5.5) indicate that HASM-MG has a much smaller magnitude of error compared to that of OK.

5.3.2 Real-World Test

Jiangxi province is located in China, between 24°29′–30°05′N and 113° 34′–118°29′E. It borders the Yangtze river to the north, the Jiu-Lian-Shan and Da-Yu-Ling mountains to the south, the Wu-Yi and Huai-Yu mountains to the east, the Luo-Xiao mountains to the west, and the Jiu-Ling and Mu-Fu mountains to the northwest (Figure 5.6). Its area is 166,600 km², of which

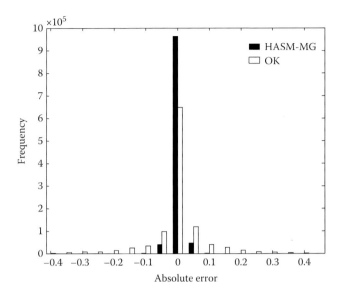

FIGURE 5.4
A comparison between the absolute errors of HASM-MG and OK.

mountainous areas account for 35.9%, hilly areas for 42.3% and plain areas for 21.8%. Jiangxi province is 620 km from south to north in length and 490 km from east to west in width. Most of Poyang lake's plain is lower than 50 m. Its hilly area has an elevation of roughly 100–500 m. The peak of the highest mountain of Jiangxi province, Huanggang mountain in the Wuyi mountains, has an elevation of 2158 m.

The daily temperature and precipitation data from 100 weather observation stations scattered over and around Jiangxi province (Figure 5.6) were selected to simulate the spatial regularity of climate change between 1960 and 1999 in terms of DEM. Simulated results show that the annual mean

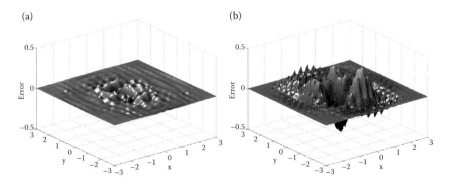

FIGURE 5.5
Error surfaces: (a) from HASM-MG, and (b) from OK.

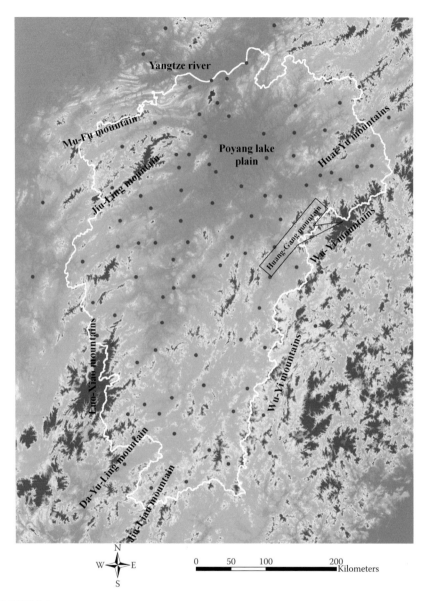

FIGURE 5.6
Mountain system and weather observation stations scattered over and around Jiangxi province.

temperature and the annual mean precipitation have very close relationships with elevation when the elevation is higher than 300 m. The spatial regularity can be formulated as

$$T = -0.0065h_e + 19.564 \tag{5.7}$$

$$P = 0.4021h_e + 1558 \qquad (5.8)$$

where T represents temperature, P precipitation, and h_e elevation (higher than 300 m); the correlation coefficient between temperature and elevation is $R^2 = 0.9917$ and the correlation between precipitation and elevation is $R^2 = 0.9239$. This DEM-based spatial regularity is used to simulate climatic change patterns in Jiangxi province.

HASM-MG is cross validated by simulating the annual mean temperature and annual mean precipitation of Jiangxi province between 1960 and 1999 in terms of the DEM-based spatial regularity by comparing the HASM-MG results with those produced by OK (Figure 5.7). Cross-validation comprises three steps: (1) several data points are removed; (2) their values are simulated using the remaining data, and (3) the data points are returned. This process continues until all the data points have been simulated in this manner and simulation error statistics can be obtained (Lloyd, 2007).

When 5% of the 100 weather observation stations are removed, *RMSE*s of annual mean temperatures and annual mean precipitations simulated by HASM-MG are 0.0182°C and 2.6149 mm, respectively, while *RMSE*s simulated by OK are 0.6007°C and 72.1292 mm, respectively (Table 5.2). When 10% of the 100 weather observation stations are removed, *RMSE* ratios of annual mean temperatures and annual mean precipitations simulated by OK to those simulated by HASM-MG are 74 and 58, respectively. When 20% of the weather observation stations are removed, HASM-MG *RMSE*s of annual mean temperatures and annual mean precipitations are 0.0237°C and 3.2484 mm, respectively, while OK *RMSE*s are 0.7818°C and 82.78 mm, respectively.

In brief, the accuracy of HASM-MG is much higher than OK. The cross-validation shows that *RMSE*s of OK tend to increase when more and more weather observation stations are removed, while HASM-MG has the highest accuracy when 10% of weather observation stations are removed.

The unbiasedness and the minimum variance conditions of OK require that the surface is spatially homogeneous with a constant variance (Olea, 1999; Wang et al., 2009). The uneven spatial distribution of weather observation stations fails to match the requirements of OK, which violates OK's conditions of unbiasedness and minimum error variance. This is likely to be the main reason why OK's errors are so huge compared to HASM-MG.

5.4 Discussion and Conclusions

The numerical and real-world tests show that the computational speed of HASM-MG is much quicker than HASM-B. HASM-MG is also more accurate than HASM-B, and the widely used classic method, OK. In addition, cross-validation indicates that the ratio of *RMSE* of OK to *RMSE* of HASM-MG is not

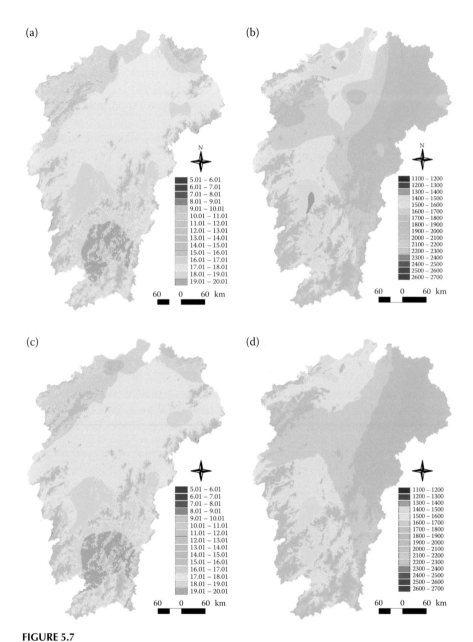

FIGURE 5.7
Comparison between OK and HASM-MG simulation results of annual mean temperature and annual mean precipitation between 1960 and 1999 in Jiangxi province: (a) and (b) are HASM-MG simulation results of the annual mean temperature and annual mean precipitation, respectively; (c) and (d) are OK simulation results of the annual mean temperature and annual mean precipitation respectively.

TABLE 5.2

Cross-Validation of Annual Mean Temperature and Annual Mean Precipitation between 1960 and 1999 in Jiangxi Province

	Annual Mean Temperature			Annual Mean Precipitation		
Ratio of removed data points	5%	10%	20%	5%	10%	20%
RMSE of OK	0.6007	0.7133	0.7818	72.1292	81.4551	82.7800
RMSE of HASM-MG	0.0182	0.0096	0.0237	2.6149	1.4079	3.2484
RMSE ratio of OK to HASM-MG	33	74	33	28	58	26

less than 26. Fast and highly accurate HASM-MG lays the foundation for three-dimensional (3D) dynamic representations and real-time visualizations.

Current GISs are mostly limited to providing representations of data in two dimensions, an issue which has been widely discussed in literature (Jones, 1989; Houlding, 1994). GIS research and development still lies largely in this traditional map-based approach (Brooks and Whalley, 2008). Until recently, 3D GIS were not practical, due to delayed responses to user interaction. However, 3D GIS is a requirement of many applications, and 3D dynamic representations may soon become practical due to advances in the efficiency of HASM-MG. It might be necessary to apply finite elements to HASM-MG to develop this efficiency.

Earth surface systems have ever been changing. The changes evolve from subsecond rates to geological timescales and involve interactions between many environmental components of Earth's surface systems. It is becoming increasingly important to understand the processes behind these changes in order for anthropogenic effects on the environment to be better managed (UNEP, 1995). Although GIS provides powerful functionality for spatial analysis, data overlay, and storage (Nyerges, 1993), it lacks the ability to represent temporal dynamics (Peuquet and Niu, 1995). In general, GIS today remains a technology for static data, which is a major impediment to its use in spatial modeling (Goodchild, 1995). The technology of digital geographies has found that the representation of change in time is extremely hard to handle because of the limitation of slow computational speeds (Unwin and Fisher, 2005). However, further development of HASM-MG might make real-time spatial analysis and real-time data visualization become a reality. There is still a great deal of potential for considerably improving the computational speed of HASM-MG by developing parallel algorithm and adaptive methods.

References

Astrakhantsev, G.P. 1971. An iterative method of solving elliptic net problems. *USSR Computational Mathematics and Mathematical Physics* 11: 171–182.

Bakhvalov, N.S. 1966. On the convergence of a relaxation method with natural constraints on the elliptic operator. *USSR Computational Mathematics and Mathematical Physics* 6: 101–135.

Brandt, A. 1973. Multi-level adaptive technique (MLAT) for fast numerical solution to boundary value problems. In *Proceedings of the 3rd International Conference on Numerical Methods in Fluid Mechanics*, eds., H. Cabannes, and R. Temam, 82–89. Berlin: Springer.

Brandt, A. 1977. Multi-level adaptive solutions to boundary-value problems. *Mathematics of Computation* 31: 333–390.

Brooks, S. and Whalley, J.L. 2008. Multilayer hybrid visualizations to support 3D GIS. *Computers, Environment and Urban Systems* 32(4): 278–292.

Fedorenko, R.P. 1962. A relaxation method for solving elliptic difference equations. *USSR Computational Mathematics and Mathematical Physics* 1: 1092–1096.

Fedorenko, R.P. 1964. The speed of convergence of one iterative process. *USSR Computational Mathematics and Mathematical Physics* 4: 227–235.

Goodchild, M.F. 1995. Geographic information systems and geographic research. In *Ground Truth: The Social Implications of Geographic Information Systems*, ed., J. Pickles, 31–50. New York: Guilford.

Hackbusch, W. 1980. Convergence of multigrid iterations applied to different equations. *Mathematics of Computation* 34: 425–440.

Hackbusch, W. 1985. *Multigrid Methods and Applications*. Berlin: Springer-Verlag.

Houlding, S.W. 1994. 3D *Geoscience Modeling: Computer Techniques for Geological Characterization*. Berlin: Springer-Verlag.

Jones, C.B. 1989. Data structures for three-dimensional spatial information systems in geology. *International Journal of Geographical Information Systems* 3(1): 15–31.

Lloyd, C.D. 2007. *Local Models for Spatial Analysis*. London: CRC Press.

Nicolaides, R.A. 1975. On multiple grid and related techniques for solving discrete elliptic systems. *Journal of Computational Physics* 19: 418–431.

Nyerges, T. 1993. Understanding the scope of GIS: Its relationship to environmental modeling. In *Environmental Modeling with GIS*, eds., M.F. Goodchild, B.O. Parks and L.T. Steyaert, 75–107. New York: Oxford University Press.

Olea, R.A. 1999. *Geostatistics for Engineers and Earth Scientists*. Kluwer, Boston: MA.

Peuquet, D.J. and Niu, D.A. 1995. An event-based spatiotemporal data model (ESTDM) for temporal analysis of geographical data. *International Journal of Geographical Information Systems* 9(1): 7–24.

Reuter, H.I., Nelson, A., and Jarvis, A. 2007. An evaluation of void-filling interpolation methods for SRTM data. *International Journal of Geographical Information Science* 21(9): 983–1008.

Schroeder, J. 1954. Zur Loesung von Potentialaufgaben mit Hilfe des Differenzenverfahrens (For the solution of potential problems with the help of the difference method). *Zeitschrift für Angewandte Mathematik und Mechanik* 34: 241–253.

Shi, W.J., Liu, J.Y., Song, Y.J., Du, Z.P., Chen, C.F., and Yue, T.X. 2009. Surface modelling of soil pH. *Geoderma* 150(1–2): 113–119.

Southwell, R.V. 1935. Stress calculation in frameworks by the method of systematic relaxation of constraints I, II. *Proceedings of the Royal Society of London Series A* 151: 56–95.

Southwell, R.V. 1946. *Relaxation Methods in Theoretical Physics*. London: Clarendon Press.

Stueben, K. and Trottenberg, U. 1982. Multigrid methods: Fundamental algorithms, model problem analysis and applications. *Multigrid Methods: Proceedings of the Conference*, Koeln-Porz, November 23–27, 1981, eds., W. Hackbusch, and U. Trottenberg, pp. 1–176. Berlin: Springer-Verlag.

Trottenberg, U., Oosterlee, C.W., and Schueller, A. 2001. *Multigrid*. London: Academic Press.

United Nations Environment Programme (UNEP). 1995. *Guidelines for Integrated Management of Coastal and Marine Areas, No. 161.* UNEP Regional Seas Reports and Studies.

Unwin, D.J. and Fisher, P. 2005. Conclusion: Towards a research agenda. In *Representing GIS*, eds., P. Fisher and D.J. Unwin, pp. 277–281. New York: John Wiley & Sons.

Wang, J.F., Christakos, G., and Hu, M.G. 2009. Modeling spatial means of surfaces with stratified nonhomogeneity. *IEEE Transactions on Geoscience and Remote Sensing* 47(12): 4167–4174.

Wesseling, P. 1992. *An Introduction to Multigrid Methods.* New York: John Wiley & Sons.

Yue, T.X., Du, Z.P., Song, D.J., and Gong, Y. 2007. A new method of surface modeling and its application to DEM construction. *Geomorphology* 91(1–2): 161–172.

Yue, T.X. and Song, Y.J. 2008. The YUE-HASM method. In *Accuracy in Geomatics: Proceedings of the 8th International Symposium on Spatial Accuracy Assessment in Natural Resources and Environmental Sciences*, Shanghai, June 25–27, 2008, pp. 148–153.

Yue, T.X., Song, D.J., Du, Z.P., and Wang, W. 2010. High accuracy surface modeling and its application to DEM generation. *International Journal of Remote Sensing* 31(8): 2205–2226.

Yue, T. X. and Wang, S. H. 2010. Adjustment computation of HASM: A high accuracy and speed method. *International Journal of Geographical Information Sciences* (in press).

6

Adjustment Computation of HASM*

6.1 Adjustment Computation

The primary purpose of an adjustment computation is to ensure that all observations are used to find the most probable values for the unknowns in the model. A combination of stochastic and functional models results in a mathematical model for the adjustment. The determination of variances and subsequently the weights of the observations are known as the stochastic model. A functional model in adjustment computations is an equation or set of equations that represents or defines an adjustment condition (Ghilani and Wolf, 2006).

An adjustment computation is rigorously based on the probability theory. The sum of the squares of the errors, times their respective weights is minimized in the adjustment computation. By enforcing this condition in any adjustment, the set of errors that is computed has the highest probability of occurrence. The adjustment computation permits all observations, regardless of their number or type to be entered into the adjustment and used simultaneously in the computations by means of least squares. Least squares have the advantage that after an adjustment is completed, a complete statistical analysis can be made out of the results. Various tests can be conducted to determine if a survey meets acceptable tolerances or if the observations must be repeated.

Parametric and conditional adjustments are two basic forms of functional models. When performing a parametric adjustment, observations are expressed in terms of unknown parameters that are never directly observed. In a conditional adjustment, the geometric conditions are enforced on the observations and corrections.

6.1.1 Parametric Adjustment

A system of linear observation equations can be represented by the matrix notation

$$\mathbf{W} \cdot \mathbf{x} = \mathbf{v} + \mathbf{r} \tag{6.1}$$

* Dr. Shi-Hai Wang is a major collaborator of this chapter.

where

$$\mathbf{x} = \begin{bmatrix} x_1 \\ x_2 \\ \vdots \\ x_n \end{bmatrix}$$

are unknowns;

$$\mathbf{W} = \begin{bmatrix} w_{11} & w_{12} & \cdots & w_{1n} \\ w_{21} & w_{22} & \cdots & w_{2n} \\ \vdots & \vdots & \vdots & \vdots \\ w_{m1} & w_{m2} & \cdots & w_{mn} \end{bmatrix}$$

are coefficients of the unknowns;

$$\mathbf{v} = \begin{bmatrix} v_1 \\ v_2 \\ \vdots \\ v_m \end{bmatrix}$$

are observed values; and

$$\mathbf{r} = \begin{bmatrix} r_1 \\ r_2 \\ \vdots \\ r_m \end{bmatrix}$$

are residuals.

The solution of Equation 6.1 should make $\mathbf{r}^T \cdot \mathbf{P} \cdot \mathbf{r} = \min$, that is,

$$\frac{\partial \left(\mathbf{r}^T \cdot \mathbf{P} \cdot \mathbf{r} \right)}{\partial \mathbf{x}} = 2\mathbf{r}^T \cdot \mathbf{P} \cdot \frac{\partial \mathbf{r}}{\partial \mathbf{x}} = 2\mathbf{r}^T \cdot \mathbf{P} \cdot \mathbf{W} = 0 \qquad (6.2)$$

where $\mathbf{P}$ is the weight matrix of $\mathbf{v}$.

Substituting Equation 6.1 into Equation 6.2 yields

$$\mathbf{W}^T \cdot \mathbf{P} \cdot \left(\mathbf{W} \cdot \mathbf{x} - \mathbf{v} \right) = 0 \qquad (6.3)$$

Then, the normal equations are formulated as

$$\mathbf{W}^T \cdot \mathbf{P} \cdot \mathbf{W} \cdot \mathbf{x} = \mathbf{W}^T \cdot \mathbf{P} \cdot \mathbf{v} \tag{6.4}$$

and

$$\mathbf{x} = \left(\mathbf{W}^T \cdot \mathbf{P} \cdot \mathbf{W}\right)^{-1} \mathbf{W}^T \cdot \mathbf{P} \cdot \mathbf{v} \tag{6.5}$$

When $\mathbf{P} = \mathbf{I}$ is an identity matrix, $\mathbf{W}^T \cdot \mathbf{W} \cdot \mathbf{x} = \mathbf{W}^T \cdot \mathbf{v}$ and $\mathbf{x} = (\mathbf{W}^T \cdot \mathbf{W})^{-1} \mathbf{W}^T \cdot \mathbf{v}$.

A system of nonlinear equations that are linearized by a Taylor-series approximation can be written as

$$\mathbf{J} \cdot \mathbf{x} = \mathbf{v} + \mathbf{r} \tag{6.6}$$

where

$$\mathbf{x} = \begin{bmatrix} dx_1 \\ dx_2 \\ \vdots \\ dx_n \end{bmatrix}$$

are unknowns;

$$\mathbf{J} = \begin{bmatrix} \dfrac{\partial f_1}{\partial x_1} & \dfrac{\partial f_1}{\partial x_2} & \cdots & \dfrac{\partial f_1}{\partial x_n} \\[2ex] \dfrac{\partial f_2}{\partial x_1} & \dfrac{\partial f_2}{\partial x_2} & \cdots & \dfrac{\partial f_2}{\partial x_n} \\[2ex] \vdots & \vdots & \vdots & \vdots \\[2ex] \dfrac{\partial f_m}{\partial x_1} & \dfrac{\partial f_m}{\partial x_2} & \cdots & \dfrac{\partial f_m}{\partial x_n} \end{bmatrix}$$

is the Jacobian matrix;

$$\mathbf{k} = \begin{bmatrix} v_1 - f_1(x_1, x_2, \ldots, x_n) \\ v_2 - f_2(x_1, x_2, \ldots, x_n) \\ \vdots \\ v_m - f_m(x_1, x_2, \ldots, x_n) \end{bmatrix}$$

are constants; and

$$\mathbf{r} = \begin{bmatrix} r_1 \\ r_2 \\ \vdots \\ r_m \end{bmatrix}$$

are residuals.

The vector of least-squares corrections in the equally weighted system is given by

$$\mathbf{x} = (\mathbf{J}^T \cdot \mathbf{J})^{-1} \mathbf{J}^T \cdot \mathbf{k} \tag{6.7}$$

The least-squares solution for a system of nonlinear equations can be found using the following five steps, (1) write the first-order Taylor-series approximation for each equation, (2) determine initial approximations for the unknowns in the equations of step 1, (3) use matrix methods to find the least-squares solution for the equations of step 1, (4) apply the corrections to the initial approximations, and (5) repeat steps 1 through 4 until the corrections become sufficiently small.

6.1.2 Conditional Adjustment

Suppose that

$$\mathbf{V}_{n \times 1} = \begin{pmatrix} v_1 \\ v_2 \\ \vdots \\ v_n \end{pmatrix}$$

are observed values with n independent stochastic errors following normal distribution,

$$\mathbf{P}_{n \times n} = \begin{pmatrix} p_1 & 0 & \cdots & 0 \\ 0 & p_2 & \cdots & 0 \\ \vdots & \vdots & \vdots & \vdots \\ 0 & 0 & \cdots & p_n \end{pmatrix}$$

is the weight matrix,

$$\mathbf{r}_{n\times1} = \begin{bmatrix} r_1 \\ r_2 \\ \vdots \\ r_n \end{bmatrix}$$

are correction values, and

$$\hat{\mathbf{v}}_{n\times1} = \begin{pmatrix} \hat{v}_1 \\ \hat{v}_2 \\ \vdots \\ \hat{v}_n \end{pmatrix}$$

are adjusted values. If m additional observations are conducted, $\hat{\mathbf{v}}_{n\times1}$ would satisfy m conditional equations

$$\mathbf{W} \cdot \hat{\mathbf{v}} + \mathbf{w_0} = 0 \tag{6.8}$$

where

$$\mathbf{W}_{m\times n} = \begin{pmatrix} a_1 & a_2 & \cdots & a_n \\ b_1 & b_2 & \cdots & b_n \\ \vdots & \vdots & \vdots & \vdots \\ q_1 & q_2 & \cdots & q_n \end{pmatrix}$$

are coefficients of the conditional equations;

$$\mathbf{w_0}_{m\times1} = \begin{pmatrix} a_0 \\ b_0 \\ \vdots \\ q_0 \end{pmatrix}$$

are constants.

As $\hat{v}_i = v_i + r_i$ ($i = 1, 2, \ldots, n$), conditional equations can be expressed as

$$\mathbf{W} \cdot \mathbf{r} + \mathbf{w} = 0 \atop {m\times n \ n\times1 \ m\times1} \tag{6.9}$$

where $\mathbf{w}_{m\times1} = \mathbf{W}_{m\times n} \cdot \mathbf{v}_{n\times1} + \mathbf{w_0}_{m\times1}$.

The unique solution of $\mathbf{r}^T \cdot \mathbf{P} \cdot \mathbf{r} = \text{minimum}$ can be obtained by means of the method of Lagrange multipliers (Kolman and Trench, 1971),

$$\Phi = \mathbf{r}^T \cdot \mathbf{P} \cdot \mathbf{r} - 2\mathbf{k}^T (\mathbf{A} \cdot \mathbf{r} + \mathbf{w}) \tag{6.10}$$

$$\frac{d\Phi}{d\mathbf{r}} = 2\mathbf{r}^T \cdot \mathbf{P} - 2\mathbf{k}^T \cdot \mathbf{W} = 0 \tag{6.11}$$

$$\underset{n \times 1}{\mathbf{r}} = \underset{n \times n}{\mathbf{P}^{-1}} \cdot \underset{n \times m}{\mathbf{W}^T} \cdot \underset{m \times 1}{\mathbf{k}} \tag{6.12}$$

where $\mathbf{k}^T = (k_a, k_b, \ldots, k_q)$ are connection numbers.

Substituting Equation 6.12 into Equation 6.9 yields

$$\underset{m \times n}{\mathbf{W}} \cdot \underset{n \times n}{\mathbf{P}^{-1}} \cdot \underset{n \times m}{\mathbf{W}^T} \cdot \underset{m \times 1}{\mathbf{k}} + \underset{m \times 1}{\mathbf{w}} = 0 \tag{6.13}$$

Let $\underset{m \times m}{\mathbf{S}} = \underset{m \times n}{\mathbf{W}} \cdot \underset{n \times n}{\mathbf{P}^{-1}} \cdot \underset{n \times m}{\mathbf{W}^T}$, then $\mathbf{S}^T = \mathbf{S}$ and normal equations of connection numbers can be expressed as the following linear symmetric equation set:

$$\underset{m \times m}{\mathbf{S}} \cdot \underset{m \times 1}{\mathbf{k}} + \underset{m \times 1}{\mathbf{w}} = 0 \tag{6.14}$$

As the rank of $\mathbf{S}$ is m, we can get a set of unique solutions of $\mathbf{k}$ from Equation 6.14. Then, the values of $\mathbf{r}$ can be obtained by substituting $\mathbf{k}$ into Equation 6.12. Finally, $\hat{\mathbf{v}} = \mathbf{v} + \mathbf{r}$ is completed.

6.1.3 Successive Conditional Adjustment

When $\mathbf{P} = \mathbf{I}$ is an identity matrix, the normal equations of connection numbers can be simplified as

$$\underset{m \times n}{\mathbf{W}} \cdot \underset{n \times m}{\mathbf{W}^T} \cdot \underset{m \times 1}{\mathbf{k}} + \underset{m \times 1}{\mathbf{w}} = 0 \tag{6.15}$$

The correction equations can be formulated as

$$\underset{n \times 1}{\mathbf{r}} = \underset{n \times m}{\mathbf{W}^T} \cdot \underset{m \times 1}{\mathbf{k}} \tag{6.16}$$

In terms of the Krueger method of successive conditional adjustment, all conditional equations can be divided into two sets,

$$\underset{m_1 \times n}{\mathbf{W}_1} \cdot \underset{n \times 1}{\mathbf{r}} + \underset{m_1 \times 1}{\mathbf{w}_1} = 0 \tag{6.17}$$

$$\underset{(m - m_1) \times n}{\mathbf{W}_2} \cdot \underset{n \times 1}{\mathbf{r}} + \underset{(m - m_1) \times 1}{\mathbf{w}_2} = 0 \tag{6.18}$$

Then, the conditional Equation 6.9 can be rewritten as

$$\begin{bmatrix} \mathbf{W}_1 \\ \mathbf{W}_2 \end{bmatrix} \mathbf{r} + \begin{bmatrix} \mathbf{w}_1 \\ \mathbf{w}_2 \end{bmatrix} = 0 \tag{6.19}$$

Let $\underset{(m-m_1)\times n}{\overline{\mathbf{W}}_2} = \underset{(m-m_1)\times n}{\mathbf{W}_2} + \underset{(m-m_1)\times m_1}{\boldsymbol{\rho}^T} \cdot \underset{m_1 \times n}{\mathbf{W}_1}$, then Equation 6.19 is transformed into

$$\begin{bmatrix} \mathbf{W}_1 \\ \overline{\mathbf{W}}_2 \end{bmatrix} \mathbf{r} + \begin{bmatrix} \mathbf{w}_1 \\ \overline{\mathbf{w}}_2 \end{bmatrix} = 0 \tag{6.20}$$

where $\boldsymbol{\rho}$ is the between matrix.

The normal equations can be expressed as

$$\begin{bmatrix} \mathbf{W}_1 \\ \overline{\mathbf{W}}_2 \end{bmatrix} \begin{bmatrix} \mathbf{W}_1^T & \overline{\mathbf{W}}_2^T \end{bmatrix} \begin{bmatrix} \mathbf{k}_1 \\ \mathbf{k}_2 \end{bmatrix} + \begin{bmatrix} \mathbf{w}_1 \\ \overline{\mathbf{w}}_2 \end{bmatrix} = 0 \tag{6.21}$$

or

$$\begin{bmatrix} \mathbf{W}_1 \cdot \mathbf{W}_1^T & \mathbf{W}_1 \cdot \overline{\mathbf{W}}_2^T \\ \overline{\mathbf{W}}_2 \cdot \mathbf{W}_1^T & \overline{\mathbf{W}}_2 \cdot \overline{\mathbf{W}}_2^T \end{bmatrix} \begin{bmatrix} \mathbf{k}_1 \\ \mathbf{k}_2 \end{bmatrix} + \begin{bmatrix} \mathbf{w}_1 \\ \overline{\mathbf{w}}_2 \end{bmatrix} = 0 \tag{6.22}$$

The transformation is to divide Equation 6.22 into two independent sets, that is,

$$\mathbf{W}_1 \cdot \overline{\mathbf{W}}_2^T = \mathbf{W}_1 \left(\mathbf{W}_2 + \boldsymbol{\rho}^T \cdot \mathbf{W}_1 \right)^T = \mathbf{W}_1 \cdot \mathbf{W}_1^T \cdot \boldsymbol{\rho} + \mathbf{W}_1 \cdot \mathbf{W}_2^T = 0 \tag{6.23}$$

In other words, $\boldsymbol{\rho}$ as the solution of Equation 6.23 guarantees that

$$\begin{bmatrix} \mathbf{W}_1 \cdot \mathbf{W}_1^T & 0 \\ 0 & \overline{\mathbf{W}}_2 \cdot \overline{\mathbf{W}}_2^T \end{bmatrix} \begin{bmatrix} \mathbf{k}_1 \\ \mathbf{k}_2 \end{bmatrix} + \begin{bmatrix} \mathbf{w}_1 \\ \overline{\mathbf{w}}_2 \end{bmatrix} = 0 \tag{6.24}$$

Then, two independent normal equation sets are obtained. They are respectively formulated as

$$\mathbf{W}_1 \cdot \mathbf{W}_1^T \cdot \mathbf{k}_1 + \mathbf{w}_1 = 0 \tag{6.25}$$

$$\overline{\mathbf{W}}_2 \cdot \overline{\mathbf{W}}_2^T \cdot \mathbf{k}_2 + \overline{\mathbf{w}}_2 = 0 \tag{6.26}$$

k_1 can be obtained by solving Equation 6.25 and k_2 from Equation 6.26. Then $r_1 = W_1^T \cdot k_1$ and $r_2 = \bar{W}_2^T \cdot k_2$. Correction numbers of $W \cdot r + w = 0$ are $r = r_1 + r_2$.

6.1.4 Successive Independent Conditional Adjustment

If conditional equations consist of two sets,

$$\begin{pmatrix} W_1 \\ W_2 \end{pmatrix} \cdot r + \begin{pmatrix} w_1 \\ w_2 \end{pmatrix} = 0 \tag{6.27}$$

and solving them as a whole, then the correction numbers can be expressed as

$$r = P^{-1} \begin{pmatrix} W_1^T & W_2^T \end{pmatrix} \begin{pmatrix} k_1 \\ k_2 \end{pmatrix} \tag{6.28}$$

where k_1 and k_2 are two vectors of connection numbers of the two conditional equation sets (Equation 6.27).

For independent conditional adjustment, P is a symmetric and positive definite matrix and the normal equations can be formulated as

$$\begin{pmatrix} W_1 \\ W_2 \end{pmatrix} P^{-1} \begin{pmatrix} W_1^T & W_2^T \end{pmatrix} \begin{pmatrix} k_1 \\ k_2 \end{pmatrix} + \begin{pmatrix} w_1 \\ w_2 \end{pmatrix} = 0 \tag{6.29}$$

or

$$\begin{pmatrix} W_1 \cdot P^{-1} \cdot W_1^T & W_1 \cdot P^{-1} \cdot W_2^T \\ W_2 \cdot P^{-1} \cdot W_1^T & W_2 \cdot P^{-1} \cdot W_2^T \end{pmatrix} \begin{pmatrix} k_1 \\ k_2 \end{pmatrix} + \begin{pmatrix} w_1 \\ w_2 \end{pmatrix} = 0 \tag{6.30}$$

Equations 6.30 can be rewritten as

$$S_{11} \cdot k_1 + S_{12} \cdot k_2 + w_1 = 0 \tag{6.31}$$

$$S_{21} \cdot k_1 + S_{22} \cdot k_2 + w_2 = 0 \tag{6.32}$$

where $S_{11} = W_1 \cdot P^{-1} \cdot W_1^T$; $S_{12} = W_1 \cdot P^{-1} \cdot W_2^T$; $S_{21} = W_2 \cdot P^{-1} \cdot W_1^T$; and $S_{22} = W_2 \cdot P^{-1} \cdot W_2^T$.

According to Equation 6.31,

$$k_1 = -S_{11}^{-1} (S_{12} \cdot k_2 + w_1) \tag{6.33}$$

Substituting Equation 6.33 into Equation 6.32 yields

$$-\mathbf{S}_{21} \cdot \mathbf{S}_{11}^{-1}\left(\mathbf{S}_{12} \cdot \mathbf{k}_2 + \mathbf{w}_1\right) + \mathbf{S}_{22} \cdot \mathbf{k}_2 + \mathbf{w}_2 = 0 \tag{6.34}$$

then

$$\mathbf{k}_2 = \left(\mathbf{S}_{22} - \mathbf{S}_{21} \cdot \mathbf{S}_{11}^{-1} \cdot \mathbf{S}_{12}\right)^{-1} + \left(\mathbf{w}_2 - \mathbf{S}_{21} \cdot \mathbf{S}_{11}^{-1} \cdot \mathbf{w}_1\right) \tag{6.35}$$

If the equation set (Equation 6.27) is successively solved, the correction numbers of the first equation set can be expressed as

$$\mathbf{r}_1' = \mathbf{Q}_{LL} \cdot \mathbf{W}_1^{T} \cdot \mathbf{k}_1' \tag{6.36}$$

where $\mathbf{Q}_{LL} = \mathbf{P}^{-1}$ is the inverse matrix of weight matrix $\mathbf{P}$.
The correction numbers, $\mathbf{k}_1'$, are the solutions of the following normal equations:

$$\mathbf{S}_{11} \cdot \mathbf{k}_1' + \mathbf{w}_1 = 0 \tag{6.37}$$

Then,

$$\mathbf{k}_1' = -\mathbf{S}_{11}^{-1} \cdot \mathbf{w}_1 = -\mathbf{S}_{11}^{-1}\left(\mathbf{W}_1 \cdot \mathbf{v} + \mathbf{w}_1^{0}\right) \tag{6.38}$$

The first correction numbers $\mathbf{r}_1'$ are added to the observed values $\mathbf{v}$ and we can obtain the calculation results of the first equation set:

$$\mathbf{v}' = \mathbf{v} + \mathbf{r}_1' = \mathbf{v} + \mathbf{Q}_{LL} \cdot \mathbf{W}_1^{T} \cdot \mathbf{k}_1' = \mathbf{v} - \mathbf{Q}_{LL} \cdot \mathbf{W}_1^{T} \cdot \mathbf{S}_{11}^{-1}(\mathbf{W}_1 \cdot \mathbf{v} + \mathbf{W}_1^{0})$$
$$= \left(\mathbf{I} - \mathbf{Q}_{LL} \cdot \mathbf{W}_1^{T} \cdot \mathbf{S}_{11}^{-1} \cdot \mathbf{W}_1\right) \cdot \mathbf{v} - \mathbf{Q}_{LL} \cdot \mathbf{W}_1^{T} \cdot \mathbf{S}_{11}^{-1} \cdot \mathbf{W}_0 \tag{6.39}$$

In terms of co-factor propagation, the inverse weight-matrix can be formulated as

$$\mathbf{Q}_{L'L'} = \left(\mathbf{I} - \mathbf{Q}_{LL} \cdot \mathbf{W}_1^{T} \cdot \mathbf{S}_{11}^{-1} \cdot \mathbf{W}_1\right)\mathbf{Q}_{LL}\left(\mathbf{I} - \mathbf{Q}_{LL} \cdot \mathbf{W}_1^{T} \cdot \mathbf{S}_{11}^{-1} \cdot \mathbf{W}_1\right)^{T}$$
$$= \mathbf{Q}_{LL} - \mathbf{Q}_{LL} \cdot \mathbf{W}_1^{T} \cdot \mathbf{S}_{11}^{-1} \mathbf{W}_{1} \mathbf{Q}_{LL} \tag{6.40}$$

According to Equation 6.34,

$$\left(\mathbf{S}_{22} - \mathbf{S}_{21} \cdot \mathbf{S}_{11}^{-1} \cdot \mathbf{S}_{12}\right)\mathbf{k}_2 + \left(\mathbf{w}_2 - \mathbf{S}_{21} \cdot \mathbf{S}_{11}^{-1} \cdot \mathbf{w}_1\right) = 0 \tag{6.41}$$

or

$$W_2 \cdot \left(P^{-1} - P^{-1} \cdot W_1^T \cdot S_{11}^{-1} \cdot W_1 \cdot P^{-1}\right) \cdot W_2^T \cdot k_2$$

$$+ \left(\left(W_2 \cdot v + w_2^0\right) - W_2 \cdot P^{-1} \cdot W_1^T \cdot S_{11}^{-1}\left(W_1 \cdot v + w_1^0\right)\right) = 0 \qquad (6.42)$$

Combining Equations 6.40, 6.39, and 6.42, leads to

$$W_2 \cdot Q_{L'L'} \cdot W_2^T \cdot k_2 + \left(W_2 \cdot v' + w_2^0\right) = 0 \qquad (6.43)$$

The following expression can be obtained by combining Equations 6.28, 6.33, 6.36, and 6.40:

$$r = Q_{LL}(W_1^T \cdot k_1 + W_2^T \cdot k_2)$$

$$= Q_{LL}\left(W_1^T\left(-S_{11}^{-1}\left(S_{12} \cdot k_2 + w_1\right)\right)\right) + Q_{LL} \cdot W_2^T \cdot k_2$$

$$= -Q_{LL} \cdot W_1^T \cdot S_{11}^{-1} \cdot w_1 - Q_{LL} \cdot W_1^T \cdot S_{11}^{-1} \cdot S_{12} \cdot k_2 + Q_{LL} \cdot W_2^T \cdot k_2 \qquad (6.44)$$

$$= Q_{LL} \cdot W_1^T \cdot k_1' - Q_{LL} \cdot W_1^T \cdot S_{11}^{-1} \cdot W_1 \cdot Q_{LL} \cdot W_2^T \cdot k_2 + Q_{LL} \cdot W_2^T \cdot k_2$$

$$= r_1' + \left(Q_{LL} - Q_{LL} \cdot W_1^T \cdot S_{11}^{-1} \cdot W_1 \cdot Q_{LL}\right)W_2^T \cdot k_2$$

$$= r_1' + Q_{L'L'} \cdot W_2^T \cdot k_2 = r_1' + r_2''$$

$$r_2'' = Q_{L'L'} \cdot W_2^T \cdot k_2 \qquad (6.45)$$

In brief, certain steps must be followed in order to achieve successive independent conditional adjustment: (a) divide the conditional equations into two sets and calculate r', k_1', v', and $Q_{L'L'}$ in terms of Equations 6.36, 6.38, 6.39, and 6.40; (b) set-up the second equation set (Equation 6.43) on the basis of the first adjusted values v_1'; (c) solve connection numbers k_2 according to the second conditional equation set and calculate the second adjustment numbers r_2'' according to Equations 6.45; (d) calculate adjustment values $\hat{v} = v' + r_2'' = v + r_1' + r_2''$.

6.2 Adjustment Computation of HASM

It is usually convenient to consider the sampled values as arising from the underlying function $f(x, y)$, which is unnecessarily known (Franke, 1982).

The sampled values are not assumed to satisfy any particular condition as to spacing or density. Suppose the coordinates of grid cells are (l, k) where the sampled points are located, the corresponding values are $\bar{f}_{l,k}$, and the grid-cell size is h; then the finite-difference equation set of HASM could be formulated as (Yue et al., 2007, 2010)

$$
\left\{
\begin{aligned}
\frac{f_{i+1,j} - 2f_{i,j} + f_{i-1,j}}{h^2} &= (\Gamma^1_{11})_{i,j}\, \frac{f_{i+1,j} - f_{i-1,j}}{2h} + (\Gamma^2_{11})_{i,j}\, \frac{f_{i,j+1} - f_{i,j-1}}{2h} \\
&\quad + \frac{L_{i,j}}{\sqrt{E_{i,j} + G_{i,j} - 1}} \\
\frac{f_{i,j+1} - 2f_{i,j} + f_{i,j-1}}{h^2} &= (\Gamma^1_{22})_{i,j}\, \frac{f_{i+1,j} - f_{i-1,j}}{2h} + (\Gamma^2_{22})_{i,j}\, \frac{f_{i,j+1} - f_{i,j-1}}{2h} \\
&\quad + \frac{N_{i,j}}{\sqrt{E_{i,j} + G_{i,j} - 1}} \\
\frac{f_{i+1,j+1} - f_{i-1,j+1} + f_{i-1,j-1} - f_{i+1,j+1}}{4h^2} &= (\Gamma^1_{12})_{i,j}\, \frac{f_{i+1,j} - f_{i-1,j}}{2h} + (\Gamma^2_{12})_{i,j} \\
&\quad \times \frac{f_{i,j+1} - f_{i,j-1}}{2h} + \frac{M_{i,j}}{\sqrt{E_{i,j} + G_{i,j} - 1}} \\
f_{l,k} &= \bar{f}_{l,k}
\end{aligned}
\right.
\tag{6.46}
$$

where $(\Gamma^1_{11})_{i,j}$, $(\Gamma^1_{12})_{i,j}$, $(\Gamma^1_{22})_{i,j}$, $(\Gamma^2_{11})_{i,j}$, $(\Gamma^2_{12})_{i,j}$, and $(\Gamma^2_{22})_{i,j}$ are the definite difference of the Christoffel symbols of the second kind; $E_{i,j}$, $F_{i,j}$, and $G_{i,j}$ are the first fundamental coefficients; and $L_{i,j}$, $M_{i,j}$, and $N_{i,j}$ are the second fundamental coefficients. They can be formulated as follows:

$$
E_{i,j} = 1 + \left(\frac{f_{i+1,j} - f_{i-1,j}}{2h} \right)^2
$$

$$
G_{i,j} = 1 + \left(\frac{f_{i,j+1} - f_{i,j-1}}{2h} \right)^2
$$

$$
F_{i,j} = \left(\frac{f_{i+1,j} - f_{i-1,j}}{2h} \right)\left(\frac{f_{i,j+1} - f_{i,j-1}}{2h} \right)
$$

$$L_{i,j} = \frac{\dfrac{f_{i+1,j} - 2f_{i,j} + f_{i-1,j}}{h^2}}{\sqrt{1 + \left(\dfrac{f_{i+1,j} - f_{i-1,j}}{2h}\right)^2 + \left(\dfrac{f_{i,j+1} - f_{i,j-1}}{2h}\right)^2}}$$

$$N_{i,j} = \frac{\dfrac{f_{i,j+1} - 2f_{i,j} + f_{i,j-1}}{h^2}}{\sqrt{1 + \left(\dfrac{f_{i+1,j} - f_{i-1,j}}{2h}\right)^2 + \left(\dfrac{f_{i,j+1} - f_{i,j-1}}{2h}\right)^2}}$$

$$M_{i,j} = \frac{\left(\dfrac{f_{i+1,j+1} - f_{i+1,j-1}}{4h^2}\right) - \left(\dfrac{f_{i-1,j+1} - f_{i-1,j-1}}{4h^2}\right)}{\sqrt{1 + \left(\dfrac{f_{i+1,j} - f_{i-1,j}}{2h}\right)^2 + \left(\dfrac{f_{i,j+1} - f_{i,j-1}}{2h}\right)^2}}$$

$$(\Gamma_{11}^1)_{i,j} = \frac{G_{i,j} \cdot (E_{i+1,j} - E_{i-1,j}) - 2F_{i,j} \cdot (F_{i+1,j} - F_{i-1,j}) + F_{i,j} \cdot (E_{i,j+1} - E_{i,j-1})}{4(E_{i,j} \cdot G_{i,j} - (F_{i,j})^2)h}$$

$$(\Gamma_{11}^2)_{i,j} = \frac{2E_{i,j} \cdot (F_{i+1,j} - F_{i-1,j}) - E_{i,j} \cdot (E_{i,j+1} - E_{i,j-1}) - F_{i,j} \cdot (E_{i+1,j} - E_{i-1,j})}{4(E_{i,j} \cdot G_{i,j} - (F_{i,j})^2)h}$$

$$(\Gamma_{22}^1)_{i,j} = \frac{2G_{i,j} \cdot (F_{i,j+1} - F_{i,j-1}) - G_{i,j} \cdot (G_{i+1,j} - G_{i-1,j}) - F_{i,j} \cdot (G_{i,j+1} - G_{i,j-1})}{4(E_{i,j} \cdot G_{i,j} - (F_{i,j})^2)h}$$

$$(\Gamma_{22}^2)_{i,j} = \frac{E_{i,j} \cdot (G_{i,j+1} - G_{i,j-1}) - 2F_{i,j} \cdot (F_{i,j+1} - F_{i,j-1}) + F_{i,j}(G_{i+1,j} - G_{i-1,j})}{4(E_{i,j} \cdot G_{i,j} - (F_{i,j})^2)h}$$

$$(\Gamma_{12}^1)_{i,j} = \frac{G_{i,j} \cdot (E_{i,j+1} - E_{i,j-1}) - F_{i,j} \cdot (G_{i+1,j} - G_{i-1,j})}{4(E_{i,j} \cdot G_{i,j} - (F_{i,j})^2)h}$$

$$(\Gamma_{12}^2) = \frac{E_{i,j} \cdot (G_{i+1,j} - G_{i-1,j}) - F_{i,j} \cdot (F_{i,j+1} - F_{i,j-1})}{4(E_{i,j} \cdot G_{i,j} - (F_{i,j})^2)h}$$

The finite-difference equation set means that 5×5 grid cells are necessary for a solution and there is a maximum of $3 + m$ constraint conditions. Here, m represents the number of sampling points in the 5×5 grid cells and

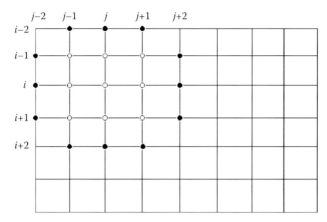

FIGURE 6.1
The 9 inner grid cells (circles) and 12 boundary grid cells (black dots).

$0 \leq m \leq 9$. The nine inner grid cells among the 5×5 grid cells are used for the solving process and the 12 grid cells on the boundary are used only to calculate the first and second fundamental coefficients (Figure 6.1).

The correction vector, $\underset{n \times 1}{\mathbf{r}}$, can be expressed as

$$\underset{9 \times 1}{\mathbf{r}} = \begin{bmatrix} r_{i-1,j-1} \\ r_{i-1,j} \\ r_{i-1,j+1} \\ r_{i,j-1} \\ r_{i,j} \\ r_{i,j+1} \\ r_{i+1,j-1} \\ r_{i+1,j} \\ r_{i+1,j+1} \end{bmatrix} \rightarrow \begin{bmatrix} r_0 \\ r_1 \\ r_2 \\ r_3 \\ r_4 \\ r_5 \\ r_6 \\ r_7 \\ r_8 \end{bmatrix} \tag{6.47}$$

The coefficient matrix can be rewritten as

$$\underset{(3+m) \times 9}{\mathbf{W}} = \begin{bmatrix} \underset{3 \times 9}{\mathbf{V}} \\ \underset{m \times 9}{\mathbf{S}} \end{bmatrix} \tag{6.48}$$

where, $\underset{3\times9}{\mathbf{V}}$ is determined by the Gauss equation set and $\underset{m\times9}{\mathbf{S}}$ is determined by sampling points.

The equation set (Equation 16.1) can be reformulated as

$$
\begin{cases}
\dfrac{(f_{i+1,j} + r_7) - 2(f_{i,j} + r_4) + (f_{i-1,j} + r_1)}{h^2} = (\Gamma^1_{11})_{i,j}\, \dfrac{(f_{i+1,j} + r_7) - (f_{i-1,j} + r_1)}{2h} \\[6pt]
\quad + \left(\Gamma^2_{11}\right)_{i,j} \dfrac{(f_{i,j+1} + r_5) - (f_{i,j-1} + r_3)}{2h} + \dfrac{L_{i,j}}{\sqrt{E_{i,j} + G_{i,j} - 1}} \\[10pt]
\dfrac{(f_{i,j+1} + r_5) - 2(f_{i,j} + r_4) + (f_{i,j-1} + r_3)}{h^2} = (\Gamma^1_{22})_{i,j}\, \dfrac{(f_{i+1,j} + r_7) - (f_{i-1,j} + r_1)}{2h} \\[6pt]
\quad + \left(\Gamma^2_{22}\right)_{i,j} \dfrac{(f_{i,j+1} + r_5) - (f_{i,j-1} + r_3)}{2h} + \dfrac{N_{i,j}}{\sqrt{E_{i,j} + G_{i,j} - 1}} \\[10pt]
\dfrac{(f_{i+1,j+1} + r_8) - (f_{i-1,j+1} + r_2) + (f_{i-1,j-1} + r_0) - (f_{i+1,j=1} + r_6)}{4h^2} \\[6pt]
= (\Gamma^1_{12})_{i,j}\, \dfrac{(f_{i+1,j} + r_7) - (f_{i=1,j} + r_1)}{2h} + \left(\Gamma^2_{12}\right)_{i,j} \dfrac{(f_{i,j+1} + r_5) - (f_{i,j=1} + r_3)}{2h} \\[6pt]
\quad + \dfrac{M_{i,j}}{\sqrt{E_{i,j} + G_{i,j} - 1}} \\[10pt]
\bar{f}_{m,n} + r_q = \bar{f}_{m,n}
\end{cases}
\tag{6.49}
$$

Then, we have

$$
\underset{3\times9}{\mathbf{V}} =
\begin{bmatrix}
0 & \dfrac{(2 + h(\Gamma^1_{11})_{i,j})}{2h^2} & 0 & \dfrac{(\Gamma^2_{11})_{i,j}}{2h} \\[10pt]
0 & \dfrac{(\Gamma^1_{22})_{i,j}}{2h} & 0 & \dfrac{(2 + h(\Gamma^2_{22})_{i,j})}{2h^2} \\[10pt]
\dfrac{1}{4h^2} & \dfrac{(\Gamma^1_{12})_{i,j}}{2h} & -\dfrac{1}{4h^2} & \dfrac{(\Gamma^2_{12})_{i,j}}{2h}
\end{bmatrix}
$$

$$
\begin{bmatrix}
-\dfrac{2}{h^2} & -\dfrac{(\Gamma^2_{11})_{i,j}}{2h} & 0 & \dfrac{(2 - h(\Gamma^1_{11})_{i,j})}{h^2} & 0 \\[10pt]
-\dfrac{2}{h^2} & \dfrac{(2 - h(\Gamma^2_{22})_{i,j})}{2h^2} & 0 & -\dfrac{(\Gamma^1_{22})_{i,j}}{2h} & 0 \\[10pt]
-\dfrac{(\Gamma^2_{12})_{i,j}}{2h} & -\dfrac{1}{4h^2} & \dfrac{(\Gamma^1_{12})_{i,j}}{2h} & \dfrac{1}{4h^2}
\end{bmatrix}
\tag{6.50}
$$

$$
\mathbf{w}_{(3+u)\times1} =
\begin{bmatrix}
(\Gamma_{11}^1)_{i,j}\dfrac{f_{i+1,j}-f_{i-1,j}}{2h} + (\Gamma_{11}^2)\dfrac{f_{i,j+1}-f_{i,j-1}}{2h} + \dfrac{L_{i,j}}{\sqrt{E_{i,j}+G_{i,j}-1}} \\[2ex]
-\dfrac{f_{i+1,j}-2f_{i,j}+f_{i-1,j}}{h^2} \\[2ex]
(\Gamma_{22}^1)_{i,j}\dfrac{f_{i+1,j}-f_{i-1,j}}{2h} + (\Gamma_{22}^2)_{i,j}\dfrac{f_{i,j+1}-f_{i,j-1}}{2h} + \dfrac{N_{i,j}}{\sqrt{E_{i,j}+G_{i,j}-1}} \\[2ex]
-\dfrac{f_{i,j+1}-2f_{i,j}+f_{i,j-1}}{h^2} \\[2ex]
(\Gamma_{12}^1)_{i,j}\dfrac{f_{i+1,j}-f_{i-1,j}}{2h} + (\Gamma_{12}^2)_{i,j}\dfrac{f_{i,j+1}-f_{i,j-1}}{2h} + \dfrac{M_{i,j}}{\sqrt{E_{i,j}+G_{i,j}-1}} \\[2ex]
-\dfrac{f_{i+1,j+1}-f_{i-1,j+1}+f_{i-1,j-1}-f_{i+1,j-1}}{4h^2} \\[2ex]
0 \\
\vdots \\
0
\end{bmatrix}
\tag{6.51}
$$

The number of rows of $\underset{m\times9}{\mathbf{S}}$ is determined by the number of sampling points of m ($m \le 9$). The element of every row vector of $\underset{m\times9}{\mathbf{S}}$ takes its value of 1 when it corresponds to an element of $\underset{9\times1}{\mathbf{r}^{\mathrm{T}}} = [r_0, r_1, \ldots, r_8]$, or else it takes 0.

When the solution procedure is completed, the spatial distribution of simulation error Δ can be expressed as

$$
\underset{9\times9}{\Delta^2} = \sigma_0^2 \cdot \underset{9\times9}{\mathbf{Q}}
\tag{6.52}
$$

where

$$
\sigma_0^2 = \frac{\underset{9\times1}{\mathbf{r}^{\mathrm{T}}} \cdot \underset{9\times9}{\mathbf{P}} \cdot \underset{9\times1}{\mathbf{r}}}{3+m} = -\frac{\underset{(3+u)\times1}{\mathbf{w}^{\mathrm{T}}} \cdot \underset{(3+u)\times1}{\mathbf{k}}}{3+m};
$$

$$
\underset{9\times9}{\mathbf{Q}} = \underset{9\times9}{\mathbf{P}^{-1}} - \underset{9\times9}{\mathbf{P}^{-1}} \cdot \underset{9\times(3+m)}{\mathbf{W}^{\mathrm{T}}} \cdot \underset{(3+m)\times(3+m)}{\mathbf{S}^{-1}} \cdot \underset{(3+m)\times9}{\mathbf{W}} \cdot \underset{9\times9}{\mathbf{P}^{-1}};
$$

$\underset{9\times1}{\mathbf{r}}$ are correction values; $\underset{9\times9}{\mathbf{P}}$ is weight matrix; $\underset{(3+m)\times1}{\mathbf{k}}$ are connection numbers; $\underset{(3+m)\times1}{\mathbf{w}} = - \underset{(3+m)\times9}{\mathbf{W}} \cdot \underset{9\times1}{\mathbf{v}}$, $\underset{(3+m)\times9}{\mathbf{W}}$ is the coefficient matrix of conditional equations; and $\underset{(3+m)\times(3+m)}{\mathbf{S}} = \underset{(3+m)\times9}{\mathbf{W}} \cdot \underset{9\times9}{\mathbf{P}^{-1}} \cdot \underset{9\times(3+m)}{\mathbf{W}^{\mathrm{T}}}$.

If I and J are respectively the number of computational units consisting of 5×5 grid cells in latitudinal direction and longitudinal direction, the number of computational units in a computational domain would be $15(I-1) \cdot (J-1) + I \cdot J$. It should be noted that the computational unit will simply be termed "unit" from this point onward. The number of equations to be solved is $15(I-1) \cdot (J-1) + I \cdot J + m$, in which m is the number of sampling points. If the computational domain becomes too big, this may mean that computation can no longer be conducted. To solve this problem, huge equation sets can be grouped into a number of small equation sets and every small equation set is successively solved by means of the method for successive independent conditional adjustments. The solution vector of each small equation set, which is their initial values plus their correction values, can be taken as virtually observed values for solving the next small equation set. This procedure is continued until the final solution is obtained. The solution of holistic computation equals that of grouped successive computation.

The solution vector in unit Ψ is independent of the one in its neighbor unit Θ of which the distance from the center of unit Ψ is 4 grid cells. Thus, all these independent units can form one group. One computational domain can be classified into a maximum of 16 such groups (Figure 6.2). In each group, the distance between the center of the neighbor units is equal to 4 grid cells, and so the units in the group can be dealt with according to the location of the starting unit center and its shift.

In every group, the normal equation set can be expressed as

$$\underset{q \times q}{\mathbf{S}} \cdot \underset{q \times 1}{\mathbf{k}} + \underset{q \times 1}{\mathbf{w}} = 0 \qquad (6.53)$$

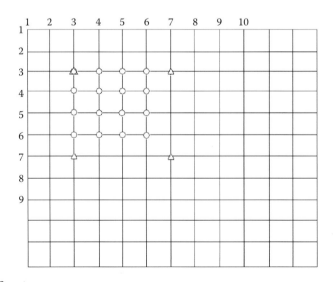

FIGURE 6.2

The starting unit centers of all the 16 groups are marked by ○, and the centers of the neighbor units in one given group are marked by △.

where

$$S_{q \times q} = \begin{bmatrix} Q_1 \cdots \cdots \cdots \\ \cdots Q_2 \cdots \cdots \\ \cdots \cdots Q_3 \cdots \cdots \\ \cdots \cdots \cdots \cdots \\ \cdots \cdots \cdots Q_k \cdots \end{bmatrix}$$

and Q_1, Q_2, Q_3, ..., Q_k are the matrices corresponding to the independent computational units.

According to the principle of matrix partitioning,

$$S_{q \times q}^{-1} = \begin{bmatrix} Q_1^{-1} \cdots \cdots \cdots \\ \cdots Q_2^{-1} \cdots \cdots \\ \cdots \cdots Q_3^{-1} \cdots \cdots \\ \cdots \cdots \cdots \cdots \\ \cdots \cdots \cdots Q_k^{-1} \end{bmatrix}.$$

Thus, the whole solving process is transformed into successively solving the 16 groups and the solving process of each group is transformed into solving the equation set in the independent units successively. According to this principle, the whole solving process is a repeat of the solving process for the unit of 5×5 grid cells and the maximum number of equations to be solved is 12. Thus, computation becomes very small, and the small storage space is utilized. The computational time of HASM-AC can be calculated by $t = \sum_{i=1}^{q} t_i$ where $q = 15(I - 1) \cdot (J - 1) + I \cdot J$ and t_i is the computation time of the ith unit.

6.3 Tests of Accuracy and Computing Speed

The accuracy and speed of IDW, kriging, spline, and HASM-AC, respectively, are comparatively calculated, whereby a higher accuracy means lower error. The classical methods are performed by the module of 3D analyst in ArcGIS 9.2. The default parameters are employed by the software. For IDW, the power is 2, the search radius is variable, and the maximum number of searched points is 12. For kriging, the ordinary method is selected, the model of semivariogram is spherical, the search radius is variable, and the maximum number of search points is 12. For spline, the regularized option is used, the weight is 0.1, and the number of searched points is 12.

HASM-AC should be run using six steps: (1) computing the initial values of all grid cells in the computational domain in terms of the sampled data, (2) flagging the sampled points as control points to transform the control points into grid coordinates, (3) extending the control point that might not locate at a grid-cell center to its closest lattice center by means of the Taylor expansion method, (4) setting a weight threshold for the control points; (5) creating a weight matrix in terms of the weight threshold, (6) starting the simulation on the basis of the initial values and the weight matrix; (7) outputting the simulated results, including simulated surface, the histogram of absolute error, error surface, *RMSE*, and computational time.

6.3.1 Errors

An error is defined as the difference between the sampled value and the simulated value. In the test, absolute error (AE) and *RMSE* are calculated. They are respectively formulated as

$$AE = f_{i,j} - Sf_{i,j} \tag{6.54}$$

and

$$RMSE = \left(\frac{1}{I \times J} \sum_{i=1}^{I} \sum_{j=1}^{J} \left(f_{i,j} - Sf_{i,j} \right)^2 \right)^{-1/2} \tag{6.55}$$

where $Sf_{i,j}$ is the simulated value of $f(x, y)$ at lattice (i, j); $f_{i,j}$ is the true value or the sampled value of $f(x, y)$ at the lattice (i, j), $i = 1, 2, ..., I, j = 1, 2, ..., J$.

6.3.1.1 Numerical Test

A mathematical surface (Figure 2.1) is selected to test the accuracies of HASM-AC and the classical methods so that the true value can be predetermined to avoid uncertainty caused by uncontrollable data errors. The mathematical surface has the following formulation:

$$z(x, y) = 3 + 2\sin(2\pi x) \sin(2\pi y) + \exp(-15(x - 1)^2 - 15(y - 1)^2)$$

$$+ \exp(-10x^2 - 15(y - 1)^2) \tag{6.56}$$

The computational domain of the test surface is normalized as $[0,1] \times [0,1]$ and is orthogonally divided into 100×100 grid cells, that is, the grid spacing is 0.01. The sampling interval is defined as 0.02, which means that the sampling proportion is 25%.

The error histograms indicate the general patterns of absolute errors from different methods (Figure 6.3). The vertical coordinate represents frequency, that is, the number of grids at which there is an equal absolute error. The abscissa axis represents absolute error. The absolute error ranges of IDW, kriging, and spline are [−0.330   0.332], [−0.168   0.168], and [−0.065   0.065], respectively, while HASM-AC is [−0.008   0.008]. For IDW, kriging, and spline the number of grids where absolute error is bigger than 0.008, are 4591, 4599, and 4602, respectively; those where the absolute error is less than 0.008, are 4597, 4596, and 4606. In other words, only 8.21%, 8.05%, and 7.92% of grid cells

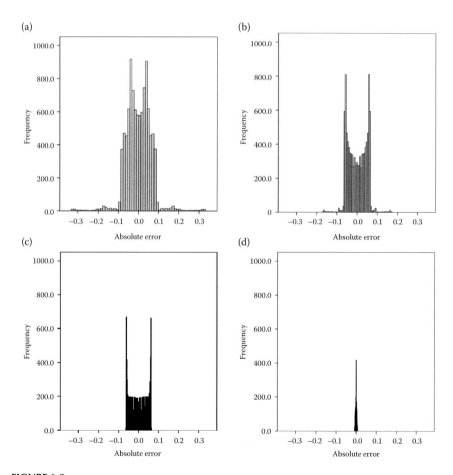

FIGURE 6.3
Histograms of absolute errors when the sampling proportion is 25%: (a) from IDW, (b) from spline, (c) from kriging, and (d) from HASM-AC.

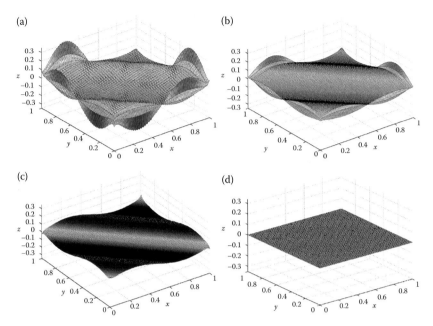

FIGURE 6.4
Error surfaces when the sampling proportion is 25%: (a) from IDW, (b) from kriging, (c) from spline, and (d) from HASM-AC.

simulated by IDW, kriging, and spline, respectively, are as accurate as those simulated by HASM-AC.

RMSEs of IDW, kriging, and spline are 0.062, 0.047, and 0.044, respectively, while RMSE of HASM-AC are 0.003. In other words, the accuracy of HASM-AC is 20.67, 15.67, and 14.67 times higher than IDW, kriging, and spline, respectively.

The error surfaces indicate that IDW, kriging, and spline have similar error magnitudes and distribution patterns (Figure 6.4a, b, and c). The biggest absolute error appears in the peak and hollow areas. These areas evidently have peak-cutting and hollow-filling problems. Systematic errors are created in the simulation processes of IDW, kriging, and spline, but HASM-AC's error (Figure 6.4d) are more evenly distributed and have smaller magnitudes than those of IDW, spline, or kriging.

The sampling proportion is increased to 34% by adding sample points at the grid cells where both their row number and column number can be exactly divided by 3. Then, the numerical test indicates that the absolute error ranges of IDW, kriging, spline, and HASM-AC become [−0.078 0.079], [−0.011 0.011], [−0.00015 0.00014], and [−0.00011 0.00009], respectively. RMSEs decrease to 0.012, 0.002, 0.0002, and 0.0001. In other words, an increase of sample points makes all these methods more accurate (Figure 6.5).

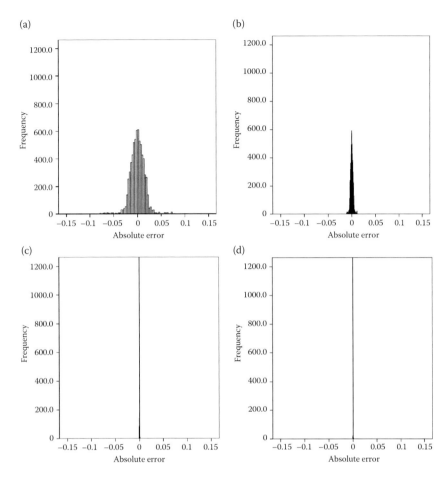

FIGURE 6.5
Histograms of absolute errors when the sampling proportion is 34%: (a) from IDW, (b) from kriging, (c) from spline, and (d) from HASM-AC.

6.3.1.2 Real-World Test of Accuracy

A rectangular area in the Tongzhou district of Beijing is selected as the real-world test area (Figure 6.6). Its length is 0.5 km east to west and its width is 0.4 km north to south. The North Canal of Beijing flows through the rectangular area from north to south. This test area is relatively flat—an effective way to reveal accuracy differences among surface-modeling methods. Its elevation is greater than 14 m and less than 22 m. Total station instruments are used to randomly sample height points in the test area. About 395 height points are sampled, that is, roughly one sample for each 600 m² or approximately one sampling interval of 22 m on an average. Around 347 sampled points are used to simulate the elevation surface and the remaining 48 height

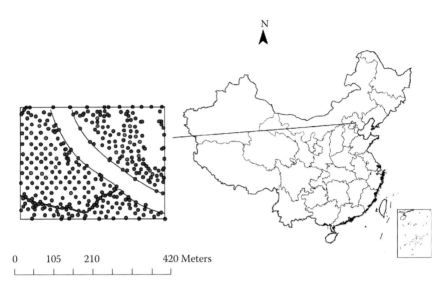

FIGURE 6.6
Tongzhou test area for accuracy, in which black dots represent the sampled points.

points are employed to cross-validate the simulated surface. The cross-validation comprises three steps: (1) 48 height points are removed from the 395 sampled points randomly; (2) the elevation surface is simulated using the remaining data, and (3) the removed data points are returned. This process continues until all the data points have been simulated in this manner and the simulation errors have been analyzed.

We first test the effect of different spatial resolutions, 45×45 m, 35×35 m, 25×25 m, 15×15 m, and 5×5 m, on the accuracy of the simulated surface by every method of surface modeling. When spatial resolution is changed from 45×45 m to 5×5 m, RMSE of HASM-AC is monotonously decreased from 0.40 m to 0.11 m and reaches its minimum AE amplitude in a spatial resolution of 5×5 m (Table 6.1). However, spline, kringing, and IDW have their minimum RMSE in a spatial resolution of 15×15 m. Spline presents its minimum AE amplitude in a spatial resolution of 5×5 m, kriging in a spatial resolution of 35×35 m, and IDW in 25×25 m. HASM-AC shows the highest accuracy in almost all spatial resolutions compared to the classic methods, except for that of IDW on spatial resolutions of 25×25 m and 35×35 m. The accuracy of HASM-AC has stable change trends, but those of the classical methods are unstable.

We then test the effect of sampling intervals on the simulation accuracies of all methods in a spatial resolution of 15×15 m. The test results indicate that HASM-AC accuracy becomes lower monotonously when sampling points become sparser, while accuracies of the classic methods see an oscillating change when sampling points become sparser (Table 6.2). All the

TABLE 6.1

Effect of Grid Spacing on Errors

Methods	Spatial Resolution (m)	Minimum AE (m)	Maximum AE (m)	AE Amplitude (m)	RMSE (m)
Spline	45 × 45	−2.47	1.64	4.11	0.79
	35 × 35	−2.35	1.32	3.67	0.61
	25 × 25	−1.65	1.70	3.35	0.51
	15 × 15	−1.56	1.02	2.58	0.34
	5 × 5	−0.50	1.38	1.88	0.35
Kriging	45 × 45	−1.35	1.05	2.4	0.42
	35 × 35	−0.89	0.91	1.8	0.33
	25 × 25	−0.96	1.34	2.3	0.36
	15 × 15	−1.09	0.77	1.86	0.28
	5 × 5	−0.74	0.75	1.49	0.36
IDW	45 × 45	−1.78	1.25	3.03	0.49
	35 × 35	−0.80	0.68	1.48	0.28
	25 × 25	−0.44	0.84	1.28	0.25
	15 × 15	−0.39	0.99	1.38	0.21
	5 × 5	−1.01	1.16	2.17	0.24
HASM-AC	45 × 45	−0.76	1.66	2.42	0.40
	35 × 35	−1.02	0.56	1.58	0.30
	25 × 25	−0.77	0.95	1.72	0.25
	15 × 15	−0.66	1.06	1.72	0.20
	5 × 5	−0.29	0.38	0.67	0.11

classic methods are unstable, while HASM-AC is very stable. When the sampling intervals are on an average shorter than 26 m, HASM-AC has the highest accuracy but when the sampling intervals are longer than 26 m, IDW has a higher accuracy than HASM-AC. However, the accuracy of HASM-AC is always higher than that of spline and kriging for all sampling intervals. When the sampling points are reduced from 347 to 299 in the simulation process, AE amplitudes of all methods reach their smallest value, which means that around 48 sampling points among the sampled data have a significantly higher impact on the whole data structure. As a result, AE amplitudes of HASM-AC and kriging become monotonously larger when sampling intervals increase, while those of spline oscillate. In other words, HASM-AC is the most accurate and stable method, while spline is the least accurate and the most unstable one.

The simulated surfaces (Figure 6.7) indicate that the HASM-AC result fits the real terrain in terms of our on-the-spot investigation. Spline has a great oscillatory surge at the right corner, which makes its error much higher than those of IDW and kriging. The North Canal is not clearly reflected by

TABLE 6.2

Effect of Sampling Intervals on Errors

Methods	Number of Sampled Height Points	Sampling Interval on an Average (m)	Minimum AE (m)	Maximum AE (m)	AE Amplitude (m)	RMSE (m)
Spline	347	24	−1.56	1.02	2.58	0.34
	299	26	−1.04	0.75	1.79	0.40
	251	28	−2.12	1.60	3.72	0.52
	203	32	−2.15	1.03	3.18	0.45
	155	36	−2.03	3.01	5.04	0.64
Kriging	347	24	−1.09	0.77	1.86	0.28
	299	26	−0.94	0.55	1.49	0.32
	251	28	−1.12	0.72	1.84	0.35
	203	32	−1.26	0.60	1.86	0.39
	155	36	−1.88	0.34	2.22	0.48
IDW	347	24	−0.39	0.99	1.38	0.21
	299	26	−0.64	0.61	1.25	0.21
	251	28	−1.16	0.17	1.33	0.23
	203	32	−1.21	0.24	1.45	0.22
	155	36	−1.67	0.32	1.99	0.36
HASM-AC	347	24	−0.66	1.06	1.72	0.20
	299	26	−0.66	0.46	1.12	0.21
	251	28	−0.67	0.54	1.21	0.24
	203	32	−1.15	0.73	1.88	0.29
	155	36	−1.61	1.44	3.05	0.43

IDW or kriging simulations. This is demonstrated in the result simulated by kriging in which a branch of the left side of the North Canal does not appear at all.

6.3.2 Real-World Test of Computing Speed

An area with 6000×6000 grid cells is selected to test the computational speeds of the various methods, because the data set sizes of the mathematical surface and the rectangular area of the Tongzhou district in Beijing are not big enough to discriminate the differences in computational speeds. The area for testing computational speed is located in 100°E–105°E and 30°N–35°N, where Qinghai province, Gansu province, and Sichuan province meet. The dataset has a spatial resolution of 3′ × 3′. This area has an elevation range between 333 and 5739 m, which includes various landforms such as plateaus, ridgelines, hills, and plains. To the west of this area, is the Tibet plateau. The middle part of the area has an undulating topography including crisscross

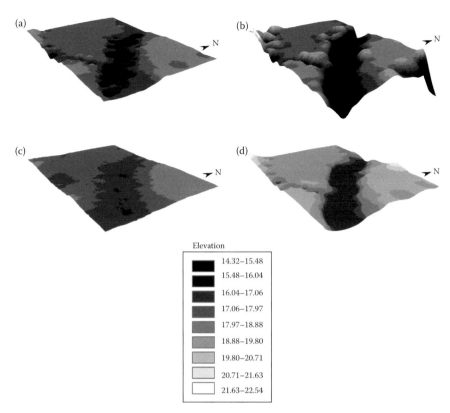

FIGURE 6.7
Simulated surfaces: (a) from IDW, (b) from spline, (c) from kriging, and (d) from HASM-AC.

valleys and widely distributed basins. The southeast is a transition area from Hengduan Mountains to the plain of western Sichuan province. The northeast is a transition area from the plateau of southern Gansu province to Loess Plateau (Figure 6.8).

HASM is run by means of a successive independent conditional adjustment. The holistic computational process is transformed and so it is able to compute the independent computational units in 16 different groups. HASM-AC dynamically reading and storing data, reduces the occupancy of computer resources and shortens the computational time. The tests of different methods are conducted using the same computer with Intel Core Duo, 2.67 GHz of CPU, DDR2 800/2G of memory, P35 chipsets, a SATA2/320G hard disk, a Windows XP, and NET Framework 2.0 operating system. The test results demonstrate that it takes HASM-AC 1.6 h to simulate the test area, while it takes spline, IDW, and kriging 12.84, 17.06, and 906.27 h, respectively. This means that the computing speed of HASM-AC is 8, 11, and 563 times higher than spline, IDW, and kriging, respectively.

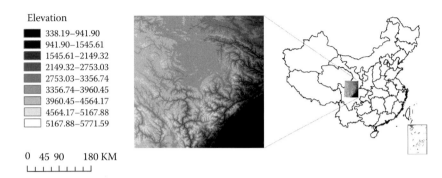

Elevation
- 338.19–941.90
- 941.90–1545.61
- 1545.61–2149.32
- 2149.32–2753.03
- 2753.03–3356.74
- 3356.74–3960.45
- 3960.45–4564.17
- 4564.17–5167.88
- 5167.88–5771.59

0   45 90   180 KM

FIGURE 6.8
An area for testing computational speed.

6.4 Conclusions and Discussion

6.4.1 Conclusions

The advantages of HASM-AC can be summarized as follows (Yue and Wang, 2010): (1) the finite difference of HASM can be used to establish condition equations and conduct constraint evaluations in a computational domain under the principle of linear unbiased estimates with minimum error variance; (2) the principle of linear unbiased estimates with minimum error variance ensures that the simulation approaches the most probable surface; (3) the successive least-squares method based on grouping computational units reduces the occupancy of computer resources considerably, which greatly shortens the computing time and makes HASM-AC able to process voluminous data; (4) data error in a computational unit not only impacts the simulation in its own computational unit, but also propagates among all computational units; HASM-AC ensures very high accuracy while distributing its error impact evenly in spacial terms.

The numerical test indicates that HASM-AC has the highest accuracy. Spline accuracy is always higher than IDW and kriging, and improves much faster than both when the density of sample points is increased. In the real-world test of the Tongzhou district of Beijing, HASM-AC is the most accurate and most stable method. Spline is the least accurate and the most unstable method because of its great oscillatory surge at the right corner. The real-world test of computing speed in the area, where Qinghai, Gansu, and Sichuan meet, shows that HASM-AC has the highest computing speed. The computational time of spline is shorter than IDW. Kriging computation takes longer time than all other methods. In brief, HASM-AC performs best in both the numerical and the real-world tests. Spline, IDW, and kriging have their own advantages and drawbacks.

6.4.2 Discussion

HASM-AC is theoretically based on the fundamental theorem of surfaces. In terms of this theorem, a surface is uniquely defined by the first fundamental coefficients and the second fundamental coefficients (Henderson, 1998). The first fundamental coefficients of a surface yield information about some geometric properties of the surface, by which we can calculate the length of curves, the angle of tangent vectors, the area of regions, and the geodesics on the surface. These geometric properties and objects that can be determined only in terms of the first fundamental coefficients of a surface are called the intrinsic geometric properties. These properties and objects form the subject of the intrinsic geometry of a surface, which reveals those of its properties that do not depend on the shape of the surface, but instead depend only on measurements that we carry out while on a surface itself (Toponogov, 2006). The second fundamental coefficients reflect local warping of the surface—its deviation from the tangent plane at the point under consideration (Liseikin, 2004), which can be observed from outside Earth. HASM-AC is theoretically perfect, which ensures that it performs best.

The classical methods such as IDW, kriging, and spline have various theoretical shortcomings. IDW uses an inverse distance weighting function of sampling values to determine the interpolation value for any given point within the calculated area, but fails to incorporate the spatial structure and ignores information beyond the neighborhood of the sampling values (Magnussen et al., 2007). Kriging tries to have the mean error equal to zero and aims at minimizing the variance of errors, but its goals are practically unattainable since the mean error and the variance of errors are always unknown (Isaaks and Srivastava, 1989). Spline uses univariate cubic basis-splines to approximately simulate surfaces, but a few types of surfaces fit the formulation of univariate cubic basis-splines (Watt, 2000). Therefore, errors of all these classical methods are difficult to control in the process of surface modeling, because they are not based on the intrinsic property of surfaces.

HASM-AC conducts its computation in 16 groups, and data fusion among the groups must be implemented in the simulation process, which may take some time. Therefore, HASM-AC needs to be further improved. Theoretically, a complete simulation of a surface needs to conduct the 16 grouping computations one by one. The accuracy improvement decreases as the computation process continues. However, it may not be necessary to conduct all the grouping computations once the accuracy requirement has been satisfied. In further research work, configuration of the groups will be studied to find out the dominant connection groups and rationally plan for each of them, by which HASM-AC could become more applicable and efficient.

The next logical step would be to extend HASM-AC to a parallel algorithm. Most of the parallel systems currently used fall into a single data stream (SIMD) or multiple instruction/multiple data streams (MIMD) (Kontoghiorghes, 2006). In SIMD, all processors are given the same instruction; each processor operates on different data, and processors may sit out a sequence of instructions. In MIMD, each processor runs its own instruction sequence, works on a different part of the problem, communicates data to other parts, and may have to wait for other processors for access to data. The parallel algorithm of HASM-AC would be based on SIMD.

References

Chilani, C.D. and Wolf, P.R. 2006. *Adjustment Computations*. NJ: John Wiley & Sons.

Franke, R. 1982. Scattered data interpolation: Tests of some methods. *Mathematics of Computation* 38: 181–200.

Henderson, D.W. 1998. *Differential Geometry*. London: Prentice-Hall, Inc.

Isaaks, E.H. and Srivastava, R.M. 1989. *Applied Geostatistics*. NY: Oxford University Press.

Kolman, B. and Trench, W.F. 1971. *Elementary Multivariable Calculus*. New York: Academic Press.

Kontoghiorghes, E.J. 2006. *Handbook of Parallel Computing and Statistics*. Boca Raton, FL: Chapman & Hall/CRC.

Liseikin, V.D. 2004. *A Computational Differential Geometry Approach to Grid Generation*. Berlin: Springer.

Magnussen, S., Næsset, E., and Wulder, M.A. 2007. Efficient multiresolution spatial predictions for large data arrays. *Remote Sensing of Environment* 109(4): 451–463.

Toponogov, V.A. 2006. *Differential Geometry of Curves and Surfaces*. NY: Birkhaeuser Boston.

Watt, A. 2000. *3D Computer Graphics*. NY: Addison-Wesley.

Yue, T.X. and Wang, S.H. 2010. Adjustment computation of HASM: A high accuracy and speed method. *International Journal of Geographical Information Sciences* (in press).

Yue, T.X., Du, Z.P., Song, D.J., and Gong, Y. 2007. A new method of surface modeling and its application to DEM construction. *Geomorphology* 91(1–2): 161–172.

Yue, T.X., Song, D.J., Du, Z.P., and Wang, W. 2010. High accuracy surface modeling and its application to DEM generation. *International Journal of Remote Sensing* 31(8): 2205–2226.

7

Optimal Control Method of HASM*

7.1 Introduction

Digital elevation models (DEMs) are representations of terrain elevation as a function of geographical location (Raaflaub and Collins, 2006), and are typically represented in one of the two formats: contour maps where the surface is represented by lines of constant elevation at even intervals or point heights where the surface elevation is sampled on either regular or irregular grids. In this chapter, the method of HASM (Yue et al., 2007) is combined with optimal control theory to make the accuracy of DEM as high as possible.

The optimal control theory determines the control signals that cause a process to satisfy the physical constraints, simultaneously minimizing the performance criterion (Kirk, 1970). It is a common objective in many research fields to develop control functions for commanding dynamic systems to get a desired output or for augmenting the system's stability (Stengel, 1994). If the control objective can be formulated as a quantitative criterion, then optimization of this criterion establishes a feasible design structure for the control function. Optimal control, concerns the properties of control functions that give solutions which minimize the measure of performance. The differential equation describes the dynamic response of the mechanism to be controlled, which depends on the control function (Vinter, 2000).

There have been a number of crucial theoretical advances in optimal control theory since the late 1930s (Tan and Bennett, 1984). For example, it was discovered that the mixture of message and noise could be filtered by using a linear operator to minimize the noise (Wiener, 1950)—a development which solved many control problems. Dynamic programming and maximum principle provided the basis for the development of modern control theory. Dynamic programming was designed to treat multistage processes possessing certain invariant aspects; it provided a solution to the multistage control problem from the initial to the final stage; a single N-dimensional problem was reduced to a sequence of N one-dimensional problems (Bellman, 1954, 1957). The maximum principle could be deduced using the principle of dynamic programming (Pontraygin et al., 1962). The maximum principle

---

* Dr. Dun-Jiang Song and Dr. Zheng-Ping Du are major collaborators of this chapter.

was a set of necessary conditions for a control function to be optimal, which was appropriate for handling the deterministic control problem. For the certainty equivalence principle, it was concluded that the optimal control setting of linear models with additive random terms could be found in the same way as for deterministic models (Simon, 1956; Theil, 1957). This was a significant development where it recognized the control problem as the inverse of the estimation problem (Stolz, 1960). The development of nonsmooth analysis and the viscosity method in the 1970s was also an important breakthrough. It demonstrated that much of the optimization and analysis which had evolved under traditional smoothness assumptions could be developed in a general nonsmooth setting (Clarke, 1983).

The optimal control theory has since been widely applied in many fields with the development of computer science and technology. For instance, the optimal control theory was used to calculate minimum-fuel rocket trajectories to successfully navigate the first moon landing (Breakwell and Dixon, 1975). Thin plate spline (TPS) for spatial interpolation was then developed by minimizing bending energy (Duchon, 1977). An optimal control model was used in greenhouse climate management during the cultivation of a lettuce crop (van Henten, 2003). Optimal control theory was also employed to develop integrating simulation models for the operation and control of urban wastewater systems (Butler and Schutze, 2005). An algorithm for optimum control of groundwater conditions was derived for the purpose of uniformly distributing the groundwater level (Bobarykin and Latyshev, 2007). An optimal control function was derived to asymptotically stabilize chaotic rotations and minimize the required like-energy cost (El-Gohary, 2009). An optimal control model was also used to determine the optimized energy losses in multiboiler steam systems (Bujak, 2009). The existence and uniqueness of the optimal controller was established by using Ekeland's principle to deal with an optimal control problem for an age-dependent biological population system (Chen and He, 2009). Optimal control was applied to optimal harvesting of renewable resources with alternative use (Piazza and Rapaport, 2009). An optimal control model was designed for the load-shifting problem in energy management to improve energy efficiency through the control of conveyor belts (Middelberg et al., 2009). The optimal purification of a polluted section of the river was formulated as a hyperbolic optimal control problem with control constraints, where the state system was given by shallow-water equations coupled with the pollutant concentration equation; the control was the flux of injected water, and the objective function was related to the total quantity of injected water and the pollution thresholds (Alvarez-Vazquez et al., 2009). However, measuring the accuracy of the relevant variables, as well as various computational difficulties associated with determining an optimal control often impeded its economical implementation (Athans and Falb, 2007).

In this chapter, taking the DEM as an example, an optimal control method of HASM (HASM-OC) is developed and validated to meet the specific requirements of extremely high accuracy.

7.2 HASM-OC

7.2.1 Equality Constraint of HASM

Suppose, $\{(x_i, y_j) \mid x_i = i \cdot h, y_j = j \cdot h, 0 \le i \le I + 1, 0 \le j \le J + 1\}$ is an orthogonal division of computational domain Ω and h is the grid cell size of the division. Then, the iterative formulation of HASM can be formulated as (Yue et al., 2007, 2010),

$$\begin{cases} \mathbf{A} \cdot \mathbf{z}^{(n+1)} = \mathbf{d}^{(n)} \\ \mathbf{B} \cdot \mathbf{z}^{(n+1)} = \mathbf{q}^{(n)} \end{cases} \tag{7.1}$$

where $\mathbf{z}^{(n+1)} = (f_{1,1}^{(n+1)}, \ldots, f_{1,J}^{(n+1)}, \ldots, f_{I,1}^{(n+1)}, \ldots, f_{I,J}^{(n+1)})^T = (z_1^{(n+1)}, \ldots, z_J^{(n+1)}, \ldots, z_{(I-1) \cdot J+1}^{(n+1)}$ $, \ldots, z_{I \cdot J}^{(n+1)})^T$, $z_{(i-1) \cdot J+j}^{(n+1)} = f_{i,j}^{(n+1)}$; $\mathbf{A}$ and $\mathbf{B}$ respectively, represent the coefficient matrixes of the first equation and the second equation in equation set (Equation 7.1); $\mathbf{d}^{(n)}$ and $\mathbf{q}^{(n)}$ are respectively the right-hand vectors of equation set (Equation 7.1).

If the kth sampling point is located at the lattice (x_i, y_j) in the computational domain, the simulation value should be equal or approximate to the sampling value at this lattice so that a constraint equation set is added to the equation set (Equation 7.1). HASM could be formulated as

$$\begin{cases} \min \left\| \begin{bmatrix} \mathbf{A} \\ \mathbf{B} \end{bmatrix} \cdot \mathbf{z}^{(n+1)} - \begin{bmatrix} \mathbf{d}^{(n)} \\ \mathbf{q}^{(n)} \end{bmatrix} \right\| \\ s.t. \quad \mathbf{S} \cdot \mathbf{z}^{(n+1)} = \mathbf{k} \end{cases} \tag{7.2}$$

where $\mathbf{S} \in \mathbb{R}^{S_p \times (I \times J)}$ represents the sampling matrix, S_p is the total number of sampling points; the nonzero element of the sample matrix can be expressed as $S_{p,(i-1) \times I + j} = 1$; $\mathbf{k} \in \mathbb{R}^{S_p \times 1}$ represents the sampling vector and the nonzero element of the sample vector $k_p = \bar{f}_{i,j}$.

This method based on the equality constraint makes the simulated surface have its minimum error under the condition that the simulated value equals the sampled value at a sampled point. In the iterative simulation process, iterant of $f(x,y)$ always equals its initial value on the boundary of the computational domain, that is, $f_{i,j}^n = f_{i,j}^0$ for $i = 0$ or $j = 0$ where, $n > 0$.

7.2.2 Inequality Constraint of HASM

Contour lines with a shared boundary divide a computational domain into different subdomains. Contour lines enclosing the subdomain determine the

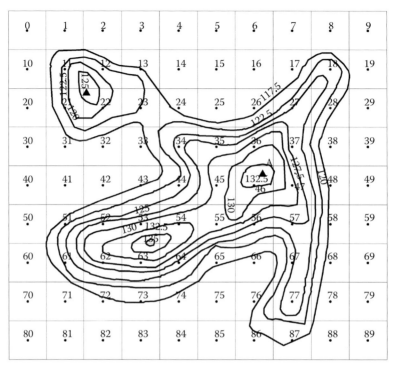

FIGURE 7.1
A schematic diagram for establishing the inequality constraint condition.

range of value of $f(x, y)$ at every lattice within it. For example, the numbered dots represent central points of lattices in Figure 7.1. The contour interval is 2.5 m. The elevation of a contour line is marked by a number along the contour line. The central point numbered 46 is surrounded by a mountain-summit contour line of 132.5 m. Meanwhile, a sampled point with an elevation of 134 m, A, is located within the mountain-summit contour line. This sampled point is approximately regarded as the maximum value of the mountain summit. Then, the elevation of the lattice numbered 46 must satisfy the inequality, $lb(46) < f(46) < ub(46)$ where, $lb(46) = 132.5$ and $ub(46) = 134$. If the sampled point did not exist, $ub(46)$ could approximately take its value of 135 m, which is the elevation of the mountain-summit contour line plus one contour interval. The lattice numbered 51 is located between the contour lines with an elevation of 117.5 m and 120 m, respectively. Obviously, $lb(51) < f(51) < ub(51)$ where, $lb(51) = 117.5$ and $ub(51) = 120$. This kind of inequality can be established for every lattice in the computational domain in the iterative process, that is,

$$\mathbf{l}_b < \mathbf{z}^{(n+1)} < \mathbf{u}_b \tag{7.3}$$

$$\mathbf{l}_b = (lb_1, \ldots, lb_J, \ldots, lb_{(I-1) \cdot J+1}, \ldots, lb_{I \cdot J})^{\mathrm{T}} \tag{7.4}$$

$$\mathbf{u}_b = (ub_1, \ldots, ub_j, \ldots, ub_{(I-1) \cdot J+1}, \ldots, ub_{I \cdot J})^{\mathrm{T}} \tag{7.5}$$

where $lb_{(i-1) \cdot J + j}$ and $ub_{(i-1) \cdot J + j}$ are respectively the lower and upper bounds of $f(x, y)$ at lattice (x_i, y_j), in which $1 \leq i \leq I$ and $1 \leq j \leq J$.

It should be noted that when a contour line passes through the central point of a lattice, then this central point would be considered a sampled point. When a contour line is close to the central point of a lattice and the distance between the contour line and the central point is smaller than the threshold value, σ, the central point could approximately be regarded as a sampled point. In Figure 7.1, these kinds of points include those, numbered 21, 22, 27, 43, 56, 57, and 62. When contour lines are dense and the simulated DEM has a coarse spatial resolution, there might not be any central point of a lattice between two neighboring contour lines. Then, the number of the retrieved contour lines from the then simulated DEM would be less than the originals.

HASM-OC can be formulated as

$$\begin{cases} \min \left\| \begin{bmatrix} \mathbf{A} \\ \mathbf{B} \end{bmatrix} \cdot \mathbf{z}^{(n+1)} - \begin{bmatrix} \mathbf{d}^{(n)} \\ \mathbf{q}^{(n)} \end{bmatrix} \right\| \\ s.t. \\ \mathbf{S} \cdot \mathbf{z}^{(n+1)} = \mathbf{k} \\ \mathbf{l}_b < \mathbf{z}^{(n+1)} < \mathbf{u}_b \end{cases} \tag{7.6}$$

The values of $f(x, y)$ at the lattices on the boundary of the computational domain do not change in the iterative process for solving the differential equation of HASM-OC, that is, $f_{i,j}^{(n)} = f_{i,j}^{(0)}$ where $n > 0$, $i = 0$ or $j = 0$. The optimal control faces a considerable challenge, because an error on the boundary of the computational domain might violate the inequality constraint condition. However, the computational domain can be made bigger so that the lattices on the boundary become inner lattices of the extended computational domain. Then, the violation of the inequality constraints can be avoided.

If the computational domain Ω is extended M lattices outwards, the new lattices in the extended computational domain can be described as

$$\{(\bar{x}_i, \bar{y}_j) \mid 0 \leq i \leq I + 2M, 0 \leq j \leq J + 2M\} \tag{7.7}$$

Then correspondences between the initial and the new lattices can be expressed as

$$(x_i, y_j) = (\bar{x}_{i+M}, \bar{y}_{j+M}), \quad (0 \leq i \leq I, 0 \leq j \leq J) \tag{7.8}$$

Suppose, $\overline{f}_{i,j}^{(n)}$ is the iterant of $f(x, y)$ at the nth iteration step in the extended computational domain, then, for $n \geq 0$

$$
\begin{aligned}
\mathbf{z}^{n+1} &= \left(\overline{f}_{1,1}^{(n+1)}, \ldots, \overline{f}_{1,J+2M}^{(n+1)}, \ldots, \overline{f}_{I+2M,1}^{(n+1)}, \ldots, \overline{f}_{I+2M,J+2M}^{(n+1)} \right)^{\mathrm{T}} \\
&= \left(\overline{z}_{1}^{(n+1)}, \ldots, \overline{z}_{J+2M}^{(n+1)}, \ldots, \overline{z}_{(I+2M-1)\cdot J+1}^{(n+1)}, \ldots, \overline{z}_{(I+2M)\cdot(J+2M)}^{(n+1)} \right)^{\mathrm{T}}
\end{aligned}
$$
(7.9)

where $\overline{z}_{(i-1)\cdot(J+2M)+j}^{(n+1)} = \overline{f}_{i,j}^{(n+1)}$.

In the extended computational domain, the optimal control formulation can be expressed as

$$
\begin{cases}
\min \left\| \begin{bmatrix} \overline{\mathbf{A}} \\ \overline{\mathbf{B}} \end{bmatrix} \cdot \overline{\mathbf{z}}^{(n+1)} - \begin{bmatrix} \overline{\mathbf{d}}^{(n)} \\ \overline{\mathbf{q}}^{(n)} \end{bmatrix} \right\| \\[2ex]
s.t. \\[1ex]
\overline{\mathbf{S}} \cdot \overline{\mathbf{z}}^{(n+1)} = \overline{\mathbf{k}} \\[1ex]
\overline{\mathbf{l}}_{b} < \overline{\mathbf{z}}^{(n+1)} < \overline{\mathbf{u}}_{b}
\end{cases}
$$
(7.10)

In formulation (Equation 7.10), if a lattice (x_i, y_j) is located outside the initial computational domain, the lower and upper bounds of $f(x, y)$ respectively, take negative and positive infinity when there is no information about elevation. The accuracy of elevation at the lattices outside the initial computational domain is very low, so the simulation value at the lattices, inside the initial computational domain cannot be further improved when iteration reaches the mth step. Thus, the simulation process returns to the initial computational domain from the extended computational domain and continues the iterative process under the optimized boundary conditions, in which the optimized values of $f(x,y)$ at the lattices, on the boundary of the initial computational domain are used, that is, $f_{i,j}^{(0)} = \overline{f}_{i+M,j+M}^{(m)}$ where $0 \leq i \leq I, 0 \leq j \leq J$.

The iteration in the extended computational domain is called preiteration. The iteration in the initial computational domain after its boundary condition is optimized, is termed as postiteration. The preiteration makes the values of $f(x, y)$ at the lattices on the boundary, $f_{i,j}^{(n)}$ where $n \geq 0$ and $i = 0, i = I$, $j = 0$ or $j = J$, satisfy the constraint condition so that the accuracy of postiteration is improved considerably.

The optimal control method of HASM can be as achieved using the following steps: (1) determine the lower and upper bounds of $f(x,y)$ at every lattice in terms of contour lines on the basis of an orthogonal division of the computational domain Ω, (2) set up the sampling matrix and sampling vector, (3) extend the computational domain outward, (4) establish inequality constraints, a sampling matrix and sampling vector on the extended computational domain, (5) optimize the boundary condition, using the preiteration process, (6) return to the initial computational domain by cutting out the extended part of the simulated surface, (7) conduct the postiteration in which $f_{i,j}^{(0)} = \overline{f}_{i+M,j+M}^{(m)}$ where $0 \leq i \leq I$, $0 \leq j \leq J$, and (8) stop the iteration and output the final simulated surface.

7.2.3 Determination of the Range of Elevation at the Central Point of Every Lattice

A contour tree is an abstraction of a scalar field that encodes the nesting relationships of isosurfaces (Carr et al., 2010). The contour tree (Chen et al., 2004) is used to determine the range of elevation at the central point of every lattice, which involves three steps: (1) sample the scattered points and contour lines passing through or almost passing through the central points of the lattices to establish the equality constraints of elevations; (2) retrieve the lower and upper bounds (contour elevation) of the unsampled lattices on the basis of the contour tree; (3) transform the mixed constraints into boundary constraints to simplify the simulation by slightly perturbing the sampled elevations. In other words, HASM-OC includes equality constraints and boundary constraints; step (3) allows HASM-OC to deal with boundary constraints alone.

Step 1: Sample the scattered points and contour lines passing through or almost passing through the central points of lattices to establish the equality constraints of elevations; Grid(i, j) represents a DEM lattice; it has elevation, lower, and upper bounds; C(n) stores n contour lines; CInterval is interval of contour lines; SamData stores elevations of the sampled data of scattered points and contour lines.

```
Delta = 1e-3; //threshold
For each Grid(i,j)
  If there exists any scattered point, scatteredPoint(m), located in Grid(i,j)
       Record the Grid(i,j) and the height of scatteredPoint(m)to SamData;
     Else if there is a contour C(k) passing through Grid(i,j)
      If the distance of Grid(i,j) from the contour C(k)
      is less than the threshold (Delta *cellsize)
          Record the Grid(i,j) and the height of contour C(k) to SamData;
      End If
  End If
End For
```

Step 2: To get the lower and upper bounds (contour elevations) of unsampled lattices on the basis of contour trees

```
For each Grid(i,j)
  While (1)
    If Grid(i,j) is not a sampled lattice
    Find any contour C(k) which contains Grid(i,j) in contour lines data set C(n);
     Let Hk be the elevation of contour C(k);
      If contour C(k) is a leaf node
      Grid(i,j).Lower = Hk - CInterval;
      Grid(i,j).Upper = Hk + CInterval;
      Break;
      Else
   Search among C(k)'s children for the child contour C(w) which contains Grid(i,j);
```

```
    If contour C(w) is empty
        Grid(i,j).Lower = Hk - CInterval;
        Grid(i,j).Upper = Hk + CInterval;
         Break;
    Else
      Let the child contour C(w) be the father contour C(k);
    End If
    End If
   End If
  End While
End For
```

Step 3: To transform mixed constraints into boundary constraints

```
For each Grid(i,j)
  If Grid(i,j) is a sampled lattice
      //calculate the lower and upper bounds of this sampled lattice
    Find the corresponding sampling data SamData(k);
      Grid(i,j).Upper = SamData(k) + Delta * CInterval;
      Grid(i,j).Lower =SamData(k) - Delta * CInterval;
   Else
    Grid(i,j).Upper = Grid(i,j).Upper - Delta * CInterval;
    Grid(i,j).Lower = Grid(i).Lower + Delta * CInterval;
   End If
End For
```

7.3 Validation of HASM-OC

It is believed that TPS provides accurate, operationally straightforward and computationally efficient solutions to the problem of spatial interpolation (Hutchinson, 1995). TPS can be used to calculate regular grid DEMs from arbitrarily large topographic data sets. It automatically removes spurious sinks or pits by imposing a drainage enforcement condition on the fitted grid values and eliminates one of the main weaknesses of elevation grids produced by interpolation techniques for general purposes. This greatly improves the utility of DEMs for hydrological applications (Hutchinson, 1989).

The basic function of thin TPS can be formulated as (Bookstein, 1989),

$$\varphi\,(r) = r^2 \cdot \log r \tag{7.11}$$

where r is the Euclidean distance between two points.

Given a set of control points $\{(x_i, y_i), i = 1, 2, \ldots, n\}$, a weighted combination of TPS centered around each data point gives the interpolation function

$$f(x,y) = \sum_{i=1}^{n} c_i \cdot \varphi(r) \tag{7.12}$$

where $r = \sqrt{(x - x_i)^2 - (y - y_i)^2}$.

The interpolation function passes through the points exactly, whilst minimizing the bending energy that is defined as

$$I\left[f(x,y)\right] = \iint_{R^2} \left(f_{xx}^2 + 2f_{xy}^2 + f_{yy}^2\right)dx\,dy \qquad (7.13)$$

In other words, TPS is a kind of optimal control method; a two-dimensional analog of cubic spline in one dimension. TPS fits a spline function to the observations, for example, the fitted function agrees with the observation values at the observation points.

The numerical tests and real-world studies (Yue et al., 2007, 2010; Yue and Song, 2008; Yue and Wang, 2010) demonstrate that the accuracy of HASM is much higher than the classical methods such as inverse distance weight (IDW), triangulated irregular network (TIN), OK, and cubic spline. In this chapter, HASM-OC is validated by comparing it with TPS.

7.3.1 Numerical Test

A Gaussian synthetic surface is selected as a numerical test surface. Its computational domain is $[-3, 3] \times [-3, 3]$. The test surface has the following formulation:

$$f(x,y) = 3(1-x)^2 e^{-x^2-(y+1)^2} - 10(x/5 - x^3 - y^5)e^{-x^2-y^2} - e^{-(x+1)^2-y^2}/3 \qquad (7.14)$$

where $-6.5510 < f(x,y) < 8.1062$.

The numerical test process can be achieved in five steps: (1) create a DEM with 2001×2001 grid cells by using Equation 7.14 of the Gaussian synthetic surface, (2) convert the DEM into contour lines (Figure 7.2), regarded as the original contour lines, (3) simulate the DEM using the data set of the original contour lines by means of both HASM-OC and TPS, (4) convert the simulated DEM into contour lines that are termed retrieved contour lines, and (5) compare the errors created by HASM-OC and TPS, in which an error is defined as the difference between the original contour lines at step (2) and the retrieved contour lines at step (4).

At step (2), different contour intervals of 2, 1, 0.5, 0.2, and 0.1 m are selected to analyze the effect of the densities of contour lines on simulation errors. At step (3), various spatial resolutions of 0.12×0.12, 0.06×0.06, 0.03×0.03, 0.015×0.015, and 0.0075 m $\times 0.0075$ m are chosen to simulate the impact of different spatial resolutions on accuracy. In total, 25 numerical tests are conducted. The input parameters of TPS are a default value of ArcGIS 9.0. Three error indexes are involved at step (5). They are the contour-line-number (CLN) difference between the original and the retrieved contour lines, the ratio of intersection area

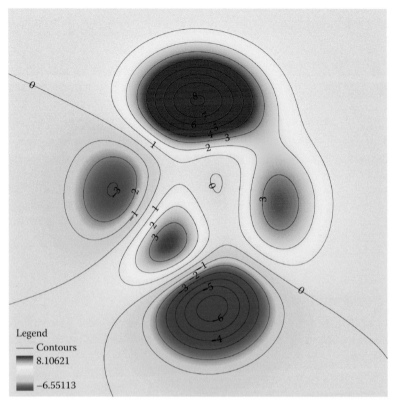

FIGURE 7.2
Contour lines of the Gaussian synthetic surface, of which the contour interval is 1 m.

between the original and the retrieved contour lines to the total length of the original contour lines (RIAL), and *RMSE*.

The contour interval, 1 m, is selected to display the CLN difference between the original and the retrieved contour lines (Figures 7.3 and 7.4), because it is difficult to compare the two when the contour lines are very dense. Figure 7.3 indicates that it is almost impossible to identify the difference on the finer spatial resolutions of 0.03, 0.015, and 0.0075 m, except to say that there is little difference between the original contour lines and those retrieved by HASM-OC on spatial resolutions of 0.12 and 0.06 m. When the contour interval is 0.1 m, two of the original contour lines, cannot be retrieved on spatial resolution of 0.12 m by HASM-OC. This is because the spatial resolutions are too coarse, so the contour lines do not enclose any central point of the lattice. When the spatial resolution becomes fine enough, this problem can be solved (Table 7.1).

However, Figure 7.4 and Table 7.2 demonstrate that the contour lines retrieved by TPS oscillationally vary with the spatial resolution becoming

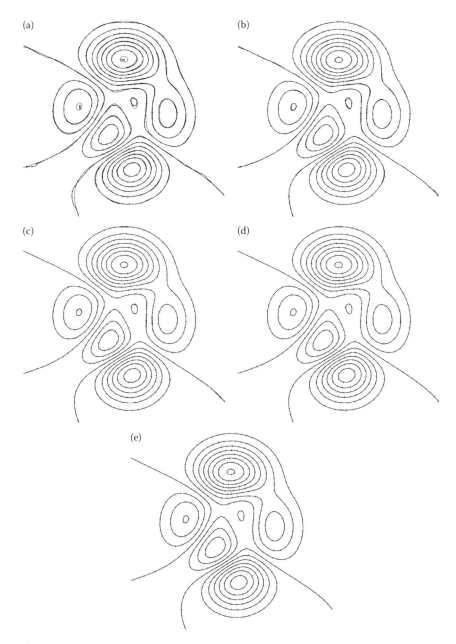

FIGURE 7.3
Comparison between the original and the retrieved contour lines by HASM-OC on different spatial resolutions, in which real line represents the original contour lines and the line with cross bar the retrieved contour lines: (a) on a spatial resolution of 0.12 m, (b) 0.06 m, (c) 0.03 m, (d) 0.15 m, and (e) 0.0075 m.

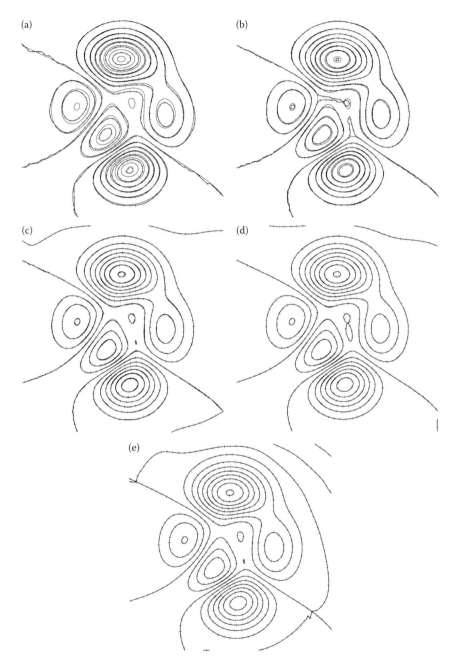

FIGURE 7.4
Comparison between the original and the contour lines from DEM simulated by TPS on different spatial resolutions, in which real line represents the original contour lines and the line with cross bar the contour lines from DEM simulated by TPS: (a) on a spatial resolution of 0.12 m, (b) 0.06 m, (c) 0.03 m, (d) 0.015 m, and (e) 0.0075 m.

TABLE 7.1

CLN Comparison between Original and Retrieved Contour Lines by HASM-OC

Spatial Resolution/ Contour Interval	2	1	0.5	0.2	0.1
0.12	12	24	47	113	224
0.06	12	24	47	113	225
0.03	12	24	47	113	226
0.015	12	24	47	113	226
0.0075	12	24	47	113	226
The number of the original contour lines	12	24	47	113	226

finer and the simulated CLN almost always different from the original. The fact that the retrieved CLN equals the original, when the spatial resolution is 0.06 m and the contour intervals are 2 and 0.5 m, is the only exception; perhaps a mere coincidence, because TPS failed to retrieve one of the original contour lines and instead added a false contour line. Thus, the simulation results of TPS are very unstable.

We find that by comparing Table 7.3 with Table 7.4, (1) RIAL of HASM-OC is always much smaller than that of TPS, (2) *RMSE* and RIAL of HASM-OC monotonically decrease with the spatial resolution becoming finer for fixed contour intervals, while the *RMSE* and RIAL of TPS show a great deal of oscillatory change, (3) there are three exceptions in which the *RMSE* of HASM-OC is bigger than that of TPS; an unusual phenomenon that might be caused by the great deal of oscillation shown by *RMSE* of TPS, (4) the ratio of HASM-OC RIAL to TPS RIAL is greater than 3 and has a maximum value of 593.5, (5) the ratio of HASM-OC RIAL to spatial resolution is 6.42% on an average, which means that the average distance between the retrieved contour lines and the original ones is less than 6.5% of lattice width. The results show that DEM with a fixed spatial resolution can be simulated by

TABLE 7.2

CLN Comparison between Original and Retrieved Contour Lines by TPS

Spatial Resolution/ Contour Interval	2	1	0.5	0.2	0.1
0.12	11	21	42	107	215
0.06	12	23	47	119	239
0.03	15	27	52	129	249
0.015	14	27	56	131	261
0.0075	15	29	60	140	257
The number of the original contour lines	12	24	47	113	226

operating HASM-OC on data with different contour-line densities. The DEM accuracy of HASM-OC increases considerably with the contour interval decrease, while the accuracy of TPS is very unstable.

7.3.2 SRTM Test

Seventeen areas (Figure 7.5) scattered across China are selected on the basis of Shuttle Radar Topography Mission (SRTM) surfaces (van Zyl, 2001), in which various landform types such as plains, plateaus, hills, basins, and karst are included. In terms of the location of every selected area, the SRTM surface is transformed into the Beijing geodetic coordinate system 54. Taking the central point of every selected region as the center, a square with sides of

TABLE 7.3

Errors of Retrieved Contour Lines by HASM-OC

ID	Contour Interval (m)	Spatial Resolution (m)	Total Length of the True Contour Lines (m)	*RMSE* (m)	RIAL (m)
1	2	0.12	47	0.145	1.26×10^{-2}
2	1		97	0.056	9.19×10^{-3}
3	0.5		197	0.037	7.45×10^{-3}
4	0.2		494	0.028	5.68×10^{-3}
5	0.1		991	0.015	4.37×10^{-3}
6	2	0.06	47	0.294	4.74×10^{-3}
7	1		97	0.124	3.68×10^{-3}
8	0.5		197	0.063	3.62×10^{-3}
9	0.2		494	0.033	3.14×10^{-3}
10	0.1		991	0.02	2.50×10^{-3}
11	2	0.03	47	0.46	2.26×10^{-3}
12	1		97	0.217	1.91×10^{-3}
13	0.5		197	0.105	1.82×10^{-3}
14	0.2		494	0.039	1.84×10^{-3}
15	0.1		991	0.019	1.51×10^{-3}
16	2	0.015	47	0.536	1.15×10^{-3}
17	1		97	0.245	1.11×10^{-3}
18	0.5		197	0.113	1.06×10^{-3}
19	0.2		494	0.04	9.49×10^{-4}
20	0.1		991	0.018	8.76×10^{-4}
21	2	0.0075	47	0.563	5.53×10^{-4}
22	1		97	0.257	5.15×10^{-4}
23	0.5		197	0.117	4.92×10^{-4}
24	0.2		494	0.041	4.72×10^{-4}
25	0.1		991	0.018	4.25×10^{-4}

TABLE 7.4

Errors of Retrieved Contour Lines by TPS

ID	Contour Interval (m)	Spatial Resolution (m)	Total Length of the True Contour Lines (m)	RMSE (m)	RIAL (m)
1	2	0.12	47	0.56	5.23×10^{-2}
2	1		97	0.19	2.83×10^{-2}
3	0.5		197	0.161	1.47×10^{-1}
4	0.2		494	0.188	1.06×10^{-1}
5	0.1		991	0.182	7.61×10^{-2}
6	2	0.06	47	0.569	1.70×10^{-1}
7	1		97	0.158	1.33×10^{-2}
8	0.5		197	0.084	8.86×10^{-2}
9	0.2		494	0.13	7.07×10^{-2}
10	0.1		991	0.126	4.60×10^{-2}
11	2	0.03	47	0.541	1.18×10^{-1}
12	1		97	0.149	1.12×10^{-1}
13	0.5		197	0.15	2.05×10^{-1}
14	0.2		494	0.178	1.15×10^{-1}
15	0.1		991	0.134	5.47×10^{-2}
16	2	0.015	47	0.611	5.30×10^{-2}
17	1		97	0.173	4.02×10^{-2}
18	0.5		197	0.243	2.39×10^{-1}
19	0.2		494	0.217	1.42×10^{-1}
20	0.1		991	0.18	1.06×10^{-1}
21	2	0.0075	47	0.552	5.56×10^{-2}
22	1		97	0.309	1.67×10^{-1}
23	0.5		197	0.393	2.92×10^{-1}
24	0.2		494	0.274	1.46×10^{-1}
25	0.1		991	0.213	6.08×10^{-2}

90 km is extracted by cutting into the transformed SRTM surface. Data sets of contour lines with contour intervals of 200, 150, 100, and 50 m are generated from the 17 extracted squares that are termed as test regions. A primary analysis of the extracted data sets indicates that there is a lower density of contour lines in plains and a higher density in karst. Mountainous areas have a bigger standard error (Table 7.5).

Error 010277, for example, too many points in the contour polyline, is reported in the TPS simulation processes of 8 test regions. These are regions 3, 4, 7, 8, 9, 10, 13, and 16. But when HASM-OC is operated on the extracted data sets of the 17 test regions, all the simulation processes are completed successfully. This phenomenon demonstrates that HASM-OC is more robust than TPS.

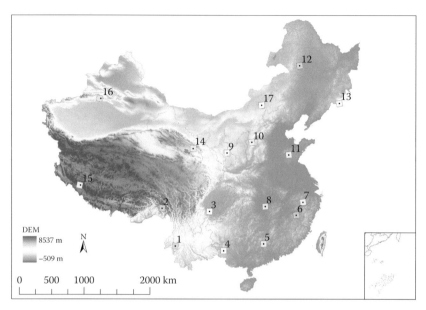

FIGURE 7.5
Locations of the 17 test regions.

Test regions 1, 2, 5, 6, 11, 12, 14, 15, and 17 (Figure 7.6) are selected to compare the difference between the CLNs of the original DEMs and the simulated DEMs. The comparison (Table 7.6) indicates that 88% of the original contour lines were successfully retrieved by HASM-OC, while only 54% were captured by TPS on an average. We take region 1 as an example to visualize overlaying the original contour lines with the ones, respectively simulated by HASM-OC and TPS. The impression drawings show that the contour lines simulated by HASM-OC almost match the original contour lines (Figure 7.7), but many spilths appear in the results simulated by TPS (Figure 7.8).

7.3.3 Real-World Test of Qian-Yan-Zhou in Jiangxi Province of China

Qian-Yan-Zhou is an ecological experiment station run by the Chinese Academy of Sciences situated in the red earth hilly area. It is 2.04 km² in area, and lies in central Jiangxi Province. It is located at 115°04′E and 26°44′N. Its elevation ranges between 60 and 150 m. It consists of 80 small hills and 9 gullies. The annual mean temperature is 17.8°C. The annual mean precipitation is 1360 mm. The original vegetation used to be evergreen broad-leaved forest, but due to degradation this was replaced by planted forest and grassland, shrub, and other secondary vegetation. The planted pine forest, citrus garden, farmland, and fish pond are main land-use types. The red soil, paddy soil, and meadow soil are main types of soil.

TABLE 7.5

Primary Analysis of the 17 Extracted Squares, in which Bold Number Means Error 010277 is Reported in the Simulation Processes of TPS

ID of Test Region	Latitude	Longitude	Spatial Resolution (m)	Area (m²) (×10⁹)	Minimum Elevation (m)	Maximum Elevation (m)	Standard Error (m)	Total Number of Contour Lines with Different DNCL			
								50 m	100 m	150 m	200 m
1	24°20′24″N	99°11′10″E	91.84	8.10	544	3326	441.2	7078	3557	2442	1850
2	29°12′32″N	96°45′59″E	88.69	8.10	1911	6548	728.5	7052	3509	2399	1746
3	29°4′54″N	104°12′32″E	88.45	8.11	225	1994	253.3	11178	3677	2064	2567
4	23°51′30″N	106°21′49″E	89.99	8.10	94	1601	266.9	**16173**	**8113**	**5301**	**4193**
5	23°54′10″N	111°34′44″E	90.21	8.11	25	1629	283.4	11723	5879	3857	2983
6	25°49′59″N	115°46′31″E	88.45	8.11	27	2155	370.8	7993	3922	2684	1908
7	30°57′1″N	118°13′21″E	88.45	8.11	0	1494	226.2	10955	5994	3228	2222
8	29°27′58″N	113°4′39″E	88.69	8.10	−98	893	58.4	2760	1130	507	**337**
9	33°38′51″N	110°50′19″E	84.21	8.10	1169	2032	138.2	**8447**	**4216**	**2837**	**2022**
10	38°2′18″N	111°3′52″E	84.47	8.09	974	2799	254.8	**6518**	**3376**	**2167**	1709
11	35°39′17″N	116°2′21″E	84.21	8.10	95	1125	125.8	3972	1929	1498	1024
12	47°18′34″N	122°10′36″E	78.33	8.10	193	1237	172.9	3389	1691	1108	821
13	42°16′36″N	124°10′12″E	81.74	8.10	507	2478	295.5	**3264**	**1669**	**1048**	**825**
14	37°18′59″N	101°16′40″E	84.21	8.10	2420	5210	441.1	4126	2113	1359	1020
15	30°34′14″N	83°15′12″E	86.78	8.10	4578	6453	402.1	4133	2364	1677	1000
16	42°17′32″N	83°29′25″E	81.74	8.10	1071	4509	795.8	**6263**	3199	2121	1618
17	42°48′55″N	113°53′44″E	81.74	8.10	973	1449	64.1	3104	1402	1590	350

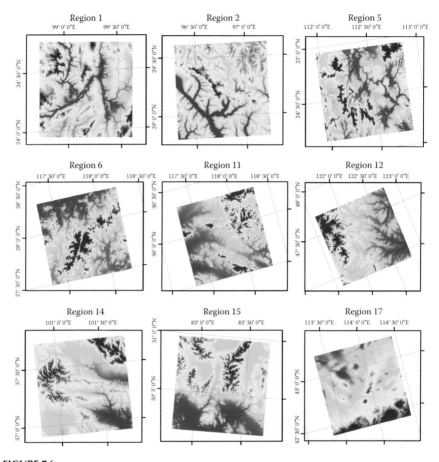

FIGURE 7.6
Data matrixes of the nine test regions, in which the longitudinal axis represents rows and the abscissa axis is the column of data matrixes (red means higher elevation and blue means lower elevation).

The topographical map of Beijing geodetic coordinate system 54 is digitalized; its contour interval is 2.5 m (Figure 7.9). The data set from digitizing the map includes control points of elevation and contour lines. DEMs on spatial resolutions of 2, 3, 4, 5, 6, 8, 10, 15, and 20 m are simulated in order to analyze the effects of various spatial resolutions on the simulation results. The results demonstrate that the contour lines retrieved by HASM-OC almost match the originals when the spatial resolution varies from 2 to 6 m (Table 7.7). When the spatial resolution is coarser than 6 m, some spilths appear in the results simulated by HASM-OC. The ratio of spilths created by HASM-OC to original contour lines reaches its maximum value, 3%, when

TABLE 7.6

Comparison between Retrieved CLNs and Original CLN in the Nine Test Regions

ID of Test Region	Contour Interval (m)	CLN Retrieved by HASM-OC	CLN Retrieved by TPS	Original CLN
1	200	1503	851	1850
2		1442	1002	1746
5		2643	1237	2983
6		1615	1045	1908
11		899	855	1024
12		774	722	821
14		887	628	1020
15		866	561	1000
17		194	125	350
1	150	1998	1139	2422
2		2018	1370	2399
5		3446	1613	3857
6		2321	1332	2685
11		1379	894	1499
12		1101	1110	1108
14		1167	822	1359
15		1494	775	1677
17		1294	932	1590
1	100	3011	1688	3557
2		3046	2128	3510
5		5349	2411	5879
6		3462	1859	3923
11		1793	1134	1929
12		1565	1160	1691
14		1936	1215	2113
15		2129	1192	2365
17		1147	335	1403
1	50	5979	3448	7079
2		6362	4008	7053
5		10,883	4918	11,714
6		7263	3852	7994
11		3797	2071	3973
12		3207	1918	3390
14		3922	2499	4127
15		3864	2353	4134
17		2781	805	3105

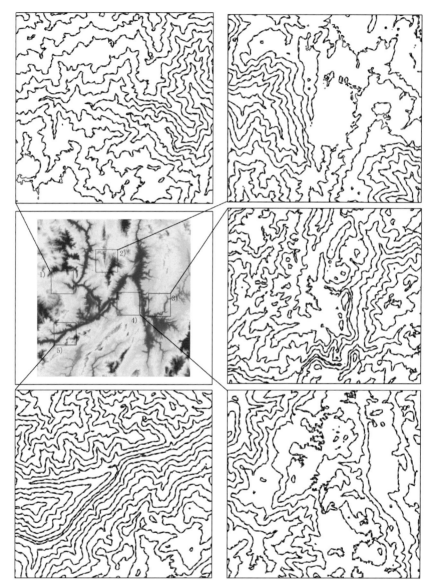

FIGURE 7.7
Impression drawing of the overlaying original contour lines with the simulated contour lines, in which the blue contour lines represent original, broken lines retrieved by HASM-OC.

the spatial resolution is 20 m. By contrast, TPS generates seven spilths when the spatial resolution is 2 m. Thus, the number of contour lines simulated by TPS, drops sharply when the spatial resolution becomes coarser. When the spatial resolution is 20 m, the contour lines produced by TPS, account for only 60% of the original contour lines.

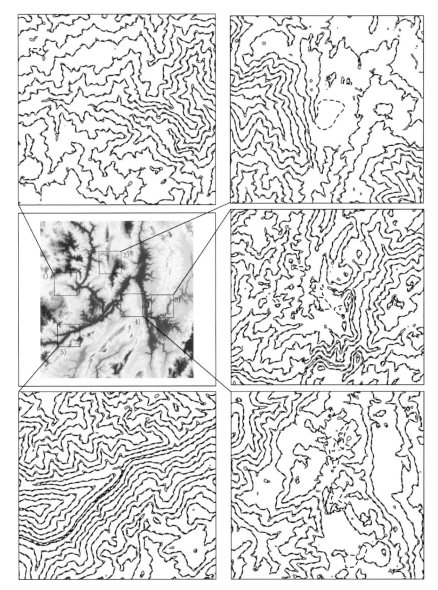

FIGURE 7.8
Impression drawing of the overlaying original contour lines with the simulated contour lines, in which the blue contour lines represent original, broken lines retrieved by TPS.

7.4 Conclusions and Discussion

A new method for the optimal control of HASM has now been developed to further improve the accuracy of HASM. We conducted a numerical test on a

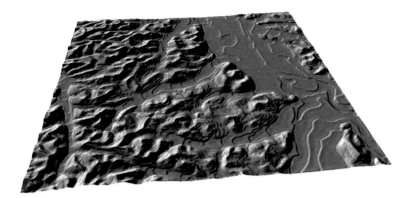

FIGURE 7.9
Shaded relief of DEM on spatial resolution of $2\,m \times 2\,m$ simulated by HASM-OC (its altitude angle is 45 and its azimuth angle is 315).

synthetic Gaussian surface, a test of an SRTM surface on national level, and a real-world test in Qian-Yan-Zhou of Jiangxi province in China to assess the accuracy of HASM-OC by comparing its simulation results with those produced by TPS, another kind of optimal control method. The tests indicate that the contour lines retrieved by HASM-OC match their corresponding original contour lines very well on appropriate spatial resolutions. HASM-OC gives much better simulation results than TPS in almost all cases. Even when TPS reported error in its simulation processes in several test regions, which meant the simulation was unable to continue, HASM-OC successfully completed its simulation in these test regions. The lower and upper bounds for the central point of every lattice were always fixed on different resolutions, which prove that the simulation accuracy of HASM-OC is stable on various resolutions.

TABLE 7.7

Comparison of the Original CLN with the Ones Retrieved Respectively by HASM-OC and TPS

Spatial Resolution (m)	CLN Retrieved by TPS	CLN Retrieved by HASM-OC	The Original CLN
2	671	664	664
3	658	664	664
4	658	665	664
5	639	665	664
6	626	664	664
8	590	676	664
10	559	676	664
15	464	681	664
20	401	685	664

The contour lines retrieved by HASM-OC were much closer to the original contour lines compared with those produced by TPS. The deviation of the contour lines retrieved by HASM-OC was less than 6.5% of the lattice width on average in the numerical tests. A DEM on a specific spatial resolution could be simulated by operating HASM-OC on data sets with various densities of contour lines. The accuracy of the DEM simulated by HASM-OC increased considerably when the contour lines became denser. But the DEM simulated by TPS had a very unstable accuracy.

HASM-OC in this chapter is in a very early stage of development. Its constraints are very simple. HASM-OC will be further improved by adding sufficient constraints, such as terrain characteristic points, lines, and surfaces. The improved HASM-OC would have the following formulation:

$$
\begin{cases}
\min \left\| \begin{bmatrix} \mathbf{A} \\ \mathbf{B} \end{bmatrix} \cdot \mathbf{z}^{(n+1)} - \begin{bmatrix} \mathbf{d}^{(n)} \\ \mathbf{q}^{(n)} \end{bmatrix} \right\| \\
s.t. \\
\mathbf{S} \cdot \mathbf{z}^{(n+1)} = \mathbf{k} \\
\mathbf{l}_b < \mathbf{z}^{(n+1)} < \mathbf{u}_b \\
\mathbf{C}_l \cdot \mathbf{z}^{(n+1)} \le \mathbf{b}_l
\end{cases}
\tag{7.15}
$$

The computational speed will be a major challenge faced by HASM-OC because of its complex formulation. We may be able to find a solution to this challenge by employing the multigrid method, adaptive algorithm, and parallel computing technology.

References

Alvarez-Vazquez, L.J., Martinez, A., Vazquez-Mendez, M.E., and Vilar, M.A. 2009. An application of optimal control theory to river pollution remediation. *Applied Numerical Mathematics* 59 (5): 845–858.

Athans, M. and Falb, P.L. 2007. *Optimal Control: An Introduction to the Theory and Its Applications*. New York, NY: Dover Publications, Inc.

Bellman, R.E. 1954. The Theory of Dynamic Programming. *Bulletin of the American Mathematical Society* 60: 503–515.

Bellman, R.E. 1957. *Dynamic Programming*. Princeton, NJ: Princeton University Press.

Bobarykin, N.D. and Latyshev, K.S. 2007. Optimum control of the ground-water level with account for a rain or snow fall-out. *Journal of Engineering Physics and Thermophysics* 80(2): 370–373.

Bookstein, F.L. 1989. Principal warps: Thin plate splines and the decomposition of deformations. *IEEE Transactions on Pattern Analysis and Machine Intelligence* 11: 567–585.

Breakwell, J.V. and Dixon, J.F . 1975. *Journal of Optimization Theory and Applications* 17: 465–975.

Bujak, J. 2009. Optimal control of energy losses in multi-boiler steam systems. *Energy* 34(9): 1260–1270.

Butler, D. and Schutze, M. 2005. Integrating simulation models with a view to optimal control of urban wastewater systems. *Environmental Modelling and Software* 20(4): 415–426.

Carr, H., Snoeyink, J. and van de Panne, M. 2010. Flexible isosurfaces: Simplifying and displaying scalar topology using the contour tree. *Computational Geometry* 43: 42–58.

Chen, J.J. and He, Z.R. 2009. Optimal control for a class of nonlinear age-distributed population systems. *Applied Mathematics and Computation* 214(2): 574–580.

Chen, J., Qiao, C.F. and Zhao, R.L. 2004. A Voronoi interior adjacency-based approach for generating a contour tree. *Computers and Geosciences* 30: 355–367.

Clarke, F.H. 1983. *Optimization and Nonsmooth Analysis.* New York, NY: Wiley-Interscience.

Duchon, J. 1977. Splines minimizing rotation invariant seminorms in sobolev spaces. *Lecture Notes in Mathematics* 571: 85–100.

El-Gohary, A. 2009. Chaos and optimal control of steady-state rotation of a satellite-gyrostat on a circular orbit. *Chaos, Solitons and Fractals* 42(5): 2842–2851.

Hutchinson, M.F. 1989. A new method for gridding elevation and streamline data with automatic removal of pits. *Journal of Hydrology* 106: 211–232.

Hutchinson, M.F. 1995. Interpolating mean rainfall using thin plate smoothing splines. *International Journal of Geographical Information Science* 9: 385–403.

Kirk, D.E. 1970. *Optimal Control Theory: An Introduction.* Englewood Cliffs, NJ: Prentice-Hall, Inc.

Middelberg, A., Zhang, J.F., and Xia, X.H. 2009. An optimal control model for load shifting-with application in the energy management of a colliery. *Applied Energy* 86(7–8): 1266–1273.

Piazza, A. and Rapaport, A. 2009. Optimal control of renewable resources with alternative use. *Mathematical and Computer Modelling* 50(1): 260–272.

Pontraygin, L.S., Boltyanski, V.G., Gamkrelidge, R.V., and Mishchenko, E.F. 1962. *The Mathematic Theory of Optimal Processes.* New York: Wiley-Interscience.

Raaflaub, L.D. and Collins, M.J. 2006. The effect of error in gridded digital elevation models on the estimation of topographic parameters. *Environmental Modelling and Software* 21(5): 710–732.

Simon, H.A. 1956. Dynamic programming under uncertainty with a quadratic criterion function. *Econometrica* 24: 74–81.

Stengel, R.F. 1994. *Optimal Control and Estimation.* New York: Dover Publications, INC.

Stolz Jr., G. 1960. Numerical solutions to an inverse problem of heat conduction for simple shapes. *Journal of Heat Transfer* 82: 20–26.

Tan, K.C. and Bennett, R.J. 1984. Optimal Control of Spatial Systems. London: George Allen & Unwin.

Theil, H. 1957. A note on certainty equivalence in dynamic planning. *Econometrica* 25: 346–349.

van Henten, E.J. 2003. Sensitivity analysis of an optimal control problem in green-house climate management. *Biosystems Engineering* 85(3): 355–364.

van Zyl, J.J. 2001. The Shuttle Radar Topography Mission (SRTM): A breakthrough in remote sensing of topography. *Acta Astronautica* 48(5–12): 559–565.

Vinter, R. 2000. *Optimal Control*. Boston: Birkhaeuser.

Wiener, N. 1950. *Extrapolation, Interpolation, and Smoothing of Stationary Time Series*. New York, NY: Technology Press of M.I.T. and Wiley.

Yue, T.X., Du, Z.P., Song, D.J., and Gong, Y. 2007. A new method of surface modeling and its application to DEM construction. *Geomorphology* 91(1–2): 161–172.

Yue, T.X. and Song, Y.J. 2008. The YUE-HASM method. In *Accuracy in Geomatics, Proceedings of the 8th International Symposium on Spatial Accuracy Assessment in Natural Resources and Environmental Sciences*, Shanghai, 2008, eds D. Li, Y. Ge, and G.M. Foody, 148–153. Liverpool: World Academic Union Ltd.

Yue, T.X., Song, D.J., Du, Z.P., and Wang, W. 2010. High accuracy surface modeling and its application to DEM generation. *International Journal of Remote Sensing* 31(8): 2205–2226.

Yue, T.X. and Wang, S.H. 2010. Adjustment computation of HASM: A high accuracy and speed method. *International Journal of Geographical Information Sciences* (in press).

8

Basic Approach of HASM to Dynamic Simulation*

8.1 Introduction

Dynamic simulation is defined as creating time-dependent models for physical systems (Fishwick, 2007). Various approaches to the dynamic simulation of earth surface systems have been developed and most of these are based on geographical information systems (GIS). For instance, in the 1970s, spatial interaction modeling was proposed to test ideas concerning the development and evolution of urban spatial structures (Harris and Wilson, 1978). A cellular model for residential site selection behavior was designed as an experiment into dynamic modeling using GIS (Gimblett, 1989). A spatial modeling workstation was developed to apply parallel processors to spatial ecosystem modeling by linking GIS with a general dynamic simulation system (Costanza and Maxwell, 1991). A spatial dynamic simulation and assessment system was described for the evaluation and simulation of environmental and ecological processes (Bali and Gimblett, 1992). A spatio-temporal interpolation method for GIS temporal modeling for urban expansion processes was presented to model missing information on changes that take place between consecutive snapshots (Dragicevic and Marceau, 2000). A dynamic information architecture system was used to build a suite of models for the purpose of assessing the ecological impacts of military land use and land management practices (Sydelko et al., 2001). The current status of real-time hydrological models was assessed for flood nowcasting and hazard mitigation (Al-Sabhan et al., 2003). A Web-based geographic information system that could delineate watersheds in real time was developed to support hydrologic model operation on the internet (Choi and Engel, 2003). Cellular automata were applied to representations of the future evolution of cities (Barredo et al., 2003). A real-time GIS-driven surveillance pilot system was studied to enhance West Nile virus dead bird surveillance in Canada (Shuai et al., 2006). A generic dynamic crop model was developed to meet the requirements of a variety of agricultural stakeholders on estimates of yield, pest-related losses, soil carbon, nitrogen dynamics, and emissions of various greenhouse

---

* Dr. Zheng-Ping Du is a major collaborator of this chapter.

gases (Aggarwal et al., 2006). Dynamic simulation was conducted by linking an erosion model with a GIS and then developing the resulting spatial information into visualizations of the evolving coastal environment (Brown et al., 2006). A promising first step in the effort to develop tangible geospatial modeling environments allowed users to interact with 3D landscape data by coupling a tangible physical model with GIS (Mitasova et al., 2006). A dynamic modeling approach using cellular automata was also explored to assess the regional distribution patterns of rock-glaciers (Frauenfelder et al., 2008). A real-time automatic interpolation system was presented to better understand natural variability and, improve our ability to detect radiological accidents (Hiemstra et al., 2009). An integrated software system was designed for the dynamic simulation of fires following an earthquake (Zhao, 2010). As demonstrated above, it is likely that GIS use will extend beyond mapping, toward a richer use of its dynamic simulation capabilities.

However, most simulations integrating GIS with modeling do not have a dynamic capability and require events to be premodeled. The utility of GIS for real-time decision making is questionable owing to a number of practical and implementation impediments (Zerger and Smith, 2003). GIS has great difficulty in handling change over time, largely due to the map sheet approach to topology building. It is not easy to add or remove one feature at a time during the updating process. Time is usually treated as a series of "snapshots" perhaps at 10-year intervals, while what is needed for updating is an incremental approach. Thus, spatial operations that fall outside traditional GIS need to be considered; in particular, simulation of one form or another where the objects move in space and the topology needs to be dynamically updated (Gold and Mostafavi, 2000).

Recently, as mobile devices have rapidly developed and wireless internet is increasingly used, interest in mobile GIS has gradually grown (Yun et al., 2006). The location data that are used by mobile GIS are usually dynamic data that are frequently updated, not static data that are updated less frequently. Therefore, it is quite inefficient to use the existing GIS and spatial indexing to manage the location data of moving objects that should be processed by mobile GIS. To offer mobile GIS such as location-based services efficiently, there must be a real-time GIS that can deal with the dynamic status of moving objects and a location index that can deal with the characteristics of location data. However, current GIS technology usually deals with static data, hence it is not suitable for real-time GIS.

Till date, most surface-modeling efforts have been focused on static snapshots that reduce the profoundly dynamic nature of the real world to simpler and abstracted perspectives that are fixed or stationary in some way (Yuan and Hornsby, 2008). A new generation of surface models is required to move beyond the snapshot view to a more realistic and dynamic view. The University Consortium for Geographic Information Science (UCGIS) has identified the visualization of geographic dynamics as one of its eight research challenges. In this chapter, HASM (Yue et al., 2007, 2010; Yue and Song, 2008; Yue and Wang, 2010) is employed to find a solution for the challenges faced by current GIS.

8.2 Basic Formulation

8.2.1 Solving Procedure of HASM

There are two modes of simulation: static and dynamic. The former is largely concerned with the pattern and structure while the latter deals with processes and subsequent change of patterns (Clarke et al., 1998). The dynamic simulation can be mathematically described as a simulation process when new information is added or removed from the grid cells of the computational domain.

The matrix formulation of HASM can be expressed as

$$\begin{cases} \mathbf{A} \cdot \mathbf{z}^{(n+1)} = \mathbf{d}^{(n)} \\ \mathbf{B} \cdot \mathbf{z}^{(n+1)} = \mathbf{q}^{(n)} \end{cases} \quad (8.1)$$

Let $\mathbf{C} = \begin{bmatrix} \mathbf{A} \\ \mathbf{B} \end{bmatrix}$, $\mathbf{u}^{(n)} = \begin{bmatrix} \mathbf{d}^{(n)} \\ \mathbf{q}^{(n)} \end{bmatrix}$, $\mathbf{S}$ is location matrix of the sampled points and $\mathbf{k}$ represents the vector of sampled values. Then we can get

$$\begin{cases} \min \left\| \mathbf{C} \cdot \mathbf{z}^{(n+1)} - \mathbf{u}^{(n)} \right\|_2 \\ \text{s.t. } \mathbf{S} \cdot \mathbf{z}^{(n+1)} = \mathbf{k} \end{cases} \quad (8.2)$$

where $\mathbf{C} \in \mathbb{R}^{(2I \cdot J) \times (I \cdot J)}$, $\mathbf{u}^{(n)} \in \mathbb{R}^{(2I \cdot J) \times 1}$, $\mathbf{S} \in \mathbb{R}^{S_p \times (2I \cdot J)}$, $\mathbf{k} \in \mathbb{R}^{S_p \times 1}$, and $I \cdot J$ is the total number of inner grid cells in the computational domain.

A necessary and sufficient condition for the solution of the least-squares problem under equality constraint (Equation 8.2) is $rank(\mathbf{S}) = S_p$ and $rank([\mathbf{C}^T \quad \mathbf{S}^T]^T) = I \cdot J$.

Owing to $\min \left\| \mathbf{C} \cdot \mathbf{z}^{(n+1)} - \mathbf{u}^{(n)} \right\|_2 = \min \left((\mathbf{C} \cdot \mathbf{z}^{(n+1)} - \mathbf{u}^{(n)})^T (\mathbf{C} \cdot \mathbf{z}^{(n+1)} - \mathbf{u}^{(n)}) \right)$ (Equation 8.2) is equivalent to

$$\begin{cases} \min \left((\mathbf{z}^{(n+1)})^T \cdot \mathbf{C}^T \cdot \mathbf{C} \cdot \mathbf{z}^{(n+1)} - 2(\mathbf{u}^{(n)})^T \cdot \mathbf{C} \cdot \mathbf{z}^{(n+1)} + (\mathbf{u}^{(n)})^T \cdot \mathbf{u}^{(n)} \right) \\ \text{s.t. } \quad \mathbf{S} \cdot \mathbf{z}^{(n+1)} = \mathbf{k} \end{cases} \quad (8.3)$$

In terms of the Kuhn–Tucker condition $\mathbf{z}^{(n+1)} \in \mathbb{R}^{(I \cdot J) \times 1}$, $\boldsymbol{\lambda} \in \mathbb{R}^{S_p \times 1}$ must satisfy the following equation set:

$$\begin{cases} \mathbf{C}^T \cdot \mathbf{C} \cdot \mathbf{z}^{(n+1)} - \mathbf{C}^T \cdot \mathbf{u}^{(n)} = \mathbf{S}^T \cdot \boldsymbol{\lambda} \\ \mathbf{S} \cdot \mathbf{z}^{(n+1)} = \mathbf{k} \end{cases} \tag{8.4}$$

or

$$\begin{bmatrix} \mathbf{C}^T \cdot \mathbf{C} & -\mathbf{S}^T \\ \mathbf{S} & \mathbf{0} \end{bmatrix} \begin{bmatrix} \mathbf{z}^{(n+1)} \\ \boldsymbol{\lambda} \end{bmatrix} = \begin{bmatrix} \mathbf{C}^T \cdot \mathbf{u}^{(n)} \\ \mathbf{k} \end{bmatrix} \tag{8.5}$$

If $rank(\mathbf{S}) = K$ and $rank([\mathbf{C}^T \quad \mathbf{S}^T]^T) = I \cdot J$, the coefficient matrix of Equation 8.5 can be factorized as

$$\begin{bmatrix} \mathbf{C}^T \cdot \mathbf{C} & -\mathbf{S}^T \\ \mathbf{S} & \mathbf{0} \end{bmatrix} = \begin{bmatrix} \mathbf{S}_w & \mathbf{0} \\ \mathbf{P} & \mathbf{S}_c \end{bmatrix} \begin{bmatrix} \mathbf{S}_w^T & -\mathbf{P}^T \\ \mathbf{0} & \mathbf{S}_c^T \end{bmatrix} \tag{8.6}$$

and Equation 8.5 can be formulated as

$$\begin{bmatrix} \mathbf{S}_w & \mathbf{0} \\ \mathbf{P} & \mathbf{S}_c \end{bmatrix} \begin{bmatrix} \mathbf{S}_w^T & -\mathbf{P}^T \\ \mathbf{0} & \mathbf{S}_c^T \end{bmatrix} \begin{bmatrix} \mathbf{z}^{(n+1)} \\ \boldsymbol{\lambda} \end{bmatrix} = \begin{bmatrix} \mathbf{C}^T \cdot \mathbf{u}^{(n)} \\ \mathbf{k} \end{bmatrix} \tag{8.7}$$

where $\mathbf{S}_w \in \mathbb{R}^{(I \cdot J) \times (I \cdot J)}$ and $\mathbf{S}_c \in \mathbb{R}^{S_p \times S_p}$ are the lower triangular matrix; $\mathbf{P} \in \mathbb{R}^{S_p \times (I \cdot J)}$.

Thus, we can get $\mathbf{z}^{(n+1)}$ by solving the following lower triangular equation set and upper triangular equation set successively, that is,

$$\begin{bmatrix} \mathbf{S}_w & \mathbf{0} \\ \mathbf{P} & \mathbf{S}_c \end{bmatrix} \begin{bmatrix} \hat{\mathbf{z}}_1 \\ \hat{\mathbf{z}}_2 \end{bmatrix} = \begin{bmatrix} \mathbf{C}^T \cdot \mathbf{u}^{(n)} \\ \mathbf{k} \end{bmatrix} \tag{8.8}$$

$$\begin{bmatrix} \mathbf{S}_w^T & -\mathbf{P}^T \\ \mathbf{0} & \mathbf{S}_c^T \end{bmatrix} \begin{bmatrix} \mathbf{z}^{(n+1)} \\ \boldsymbol{\lambda} \end{bmatrix} = \begin{bmatrix} \hat{\mathbf{z}}_1 \\ \hat{\mathbf{z}}_2 \end{bmatrix} \tag{8.9}$$

The solving procedure can be summarized into four steps: (1) $\mathbf{S}_w \cdot \hat{\mathbf{z}}_1 = \mathbf{C}^T \cdot \mathbf{u}^{(n)}$; (2) $\mathbf{S}_c \cdot \hat{\mathbf{z}}_2 = \mathbf{k} - \mathbf{P} \cdot \hat{\mathbf{z}}_1$; (3) $\mathbf{S}_c^T \cdot \boldsymbol{\lambda} = \hat{\mathbf{z}}_2$; (4) $\mathbf{S}_w^T \cdot \mathbf{z}^{(n+1)} = \hat{\mathbf{z}}_1 + \mathbf{P}^T \cdot \boldsymbol{\lambda}$.

8.2.2 Factorization of the HASM Coefficient Matrix

In terms of equation set (Equation 8.6), we get

$$
\begin{bmatrix} \mathbf{C}^T \cdot \mathbf{C} & -\mathbf{S}^T \\ \mathbf{S} & 0 \end{bmatrix} = \begin{bmatrix} \mathbf{S}_w & 0 \\ \mathbf{P} & \mathbf{S}_c \end{bmatrix} \begin{bmatrix} \mathbf{S}_w^T & -\mathbf{P}^T \\ 0 & \mathbf{S}_c^T \end{bmatrix} = \begin{bmatrix} \mathbf{S}_w \cdot \mathbf{S}_w^T & -\mathbf{S}_w \cdot \mathbf{P}^T \\ \mathbf{P} \cdot \mathbf{S}_w^T & \mathbf{S}_c \cdot \mathbf{S}_c^T - \mathbf{P} \cdot \mathbf{P}^T \end{bmatrix} \quad (8.10)
$$

so that $\mathbf{C}^T \cdot \mathbf{C} = \mathbf{S}_w \cdot \mathbf{S}_w^T$, $\mathbf{S} = \mathbf{P} \cdot \mathbf{S}_w^T$, $\mathbf{S}_c \cdot \mathbf{S}_c^T = \mathbf{P} \cdot \mathbf{P}^T$ Thus, $\mathbf{S}_w$ is the Cholesky factor of the symmetric and positive-definite matrix $\mathbf{C}^T \cdot \mathbf{C}$. $\mathbf{P} = \mathbf{S} \cdot (\mathbf{S}_w^T)^{-1}$ is a full row rank matrix and $\mathbf{P} \cdot \mathbf{P}^T$ is a symmetric and positive definite matrix. There exists a lower triangular matrix $\mathbf{S}_c \in \mathbb{R}^{S_p \times S_p}$ which assures that $\mathbf{S}_c \cdot \mathbf{S}_c^T = \mathbf{P} \cdot \mathbf{P}^T$.

In practice, we first conduct QR factorization of $\mathbf{C}$ so that $\mathbf{C} = \mathbf{Q} \cdot \begin{bmatrix} \mathbf{R} \\ 0 \end{bmatrix}$ in which $\mathbf{Q} \in \mathbb{R}^{2(I \cdot J) \times 2(I \cdot J)}$ is an orthogonal matrix and $\mathbf{R} \in \mathbb{R}^{(I \cdot J) \times (I \cdot J)}$ is an upper triangular matrix with principal diagonal elements bigger than zero. Let $\mathbf{S}_w = \mathbf{R}^T$; then,

$$
\mathbf{S}_w \cdot \mathbf{S}_w^T = \mathbf{R}^T \cdot \mathbf{R} = [\mathbf{R}^T \ 0]\begin{bmatrix} \mathbf{R} \\ 0 \end{bmatrix} = [\mathbf{R}^T \ 0](\mathbf{Q}_{2(I \cdot J)}^T \cdot \mathbf{Q}_{2(I \cdot J)})\begin{bmatrix} \mathbf{R} \\ 0 \end{bmatrix} = \mathbf{C}^T \cdot \mathbf{C} \quad (8.11)
$$

The solving procedure of Equations 8.2, the least-squares problem under equality constraint, can be updated as

1. To make QR factorization of $\mathbf{C}$ to obtain $\mathbf{R}$
2. Let $\mathbf{S}_w = \mathbf{R}^T$; then $\mathbf{P} = \mathbf{S} \cdot (\mathbf{S}_w^T)^{-1}$
3. To find $\mathbf{S}_c$ so that $\mathbf{S}_c \cdot \mathbf{S}_c^T = \mathbf{P} \cdot \mathbf{P}^T$
4. To successively solve equation sets $\begin{bmatrix} \mathbf{S}_w & 0 \\ \mathbf{P} & \mathbf{S}_c \end{bmatrix}\begin{bmatrix} \hat{\mathbf{z}}_1 \\ \hat{\mathbf{z}}_2 \end{bmatrix} = \begin{bmatrix} \mathbf{C}^T \cdot \mathbf{u}^{(n)} \\ \mathbf{k} \end{bmatrix}$ and

$\begin{bmatrix} \mathbf{S}_w^T & -\mathbf{P}^T \\ 0 & \mathbf{S}_c^T \end{bmatrix}\begin{bmatrix} \mathbf{z}^{(n+1)} \\ \lambda \end{bmatrix} = \begin{bmatrix} \hat{\mathbf{z}}_1 \\ \hat{\mathbf{z}}_2 \end{bmatrix}$, for the solution of $\mathbf{z}^{(n+1)}$.

We refer to this procedure as HASM-F. Suppose, $\mathbf{P} = (p_{i,j})$, $\mathbf{S} = (s_{i,j})$, $\mathbf{S}_w = ((s_w)_{i,j})$, and $\mathbf{S}_c = ((s_c)_{i,j})$; then,

$$
p_{i,j} = \left(s_{i,j} - \sum_{k=1}^{j-1} p_{i,k} \cdot (s_w)_{j,k} \right) \bigg/ (s_w)_{j,j}, \quad i = 1, 2, \ldots, S_p \quad \text{and} \quad j = 1, 2, \ldots, I \cdot J - 1, I \cdot J
$$

(8.12)

$$
(s_c)_{i,j} = \left(\sum_{k=1}^{I \cdot J} p_{i,k} \cdot p_{j,k} - \sum_{l=1}^{j-1} (s_c)_{i,l} \cdot (s_c)_{j,l} \right) \bigg/ (s_c)_{j,j}, \quad i = 1, 2, \ldots, S_p \quad \text{and} \quad j = 1, 2, \ldots, i
$$

(8.13)

8.2.3 Formulation of HASM-F by Dynamically Adding Information

If $\hat{S}_p$ new sampling points are added and the new sampling location matrix

is expressed as $\tilde{\mathbf{S}} = \begin{bmatrix} \mathbf{S} \\ \hat{\mathbf{S}} \end{bmatrix}$, in which $\hat{\mathbf{S}} \in \mathbb{R}^{\hat{S}_p \times (I \cdot J)}$, then

$$\tilde{\mathbf{P}} = \tilde{\mathbf{S}} \cdot (\mathbf{S}_w^T)^{-1} = \begin{bmatrix} \mathbf{S} \cdot (\mathbf{S}_w^T)^{-1} \\ \hat{\mathbf{S}} \cdot (\mathbf{S}_w^T)^{-1} \end{bmatrix} \tag{8.14}$$

Let $\hat{\mathbf{P}} = \hat{\mathbf{S}} \cdot (\mathbf{S}_w^T)^{-1}$; then,

$$\tilde{\mathbf{P}} = \begin{bmatrix} \mathbf{S} \cdot (\mathbf{S}_w^T)^{-1} \\ \hat{\mathbf{S}} \cdot (\mathbf{S}_w^T)^{-1} \end{bmatrix} = \begin{bmatrix} \mathbf{P} \\ \hat{\mathbf{P}} \end{bmatrix} \tag{8.15}$$

$$\tilde{\mathbf{P}} \cdot \tilde{\mathbf{P}}^T = \begin{bmatrix} \mathbf{P} \\ \hat{\mathbf{P}} \end{bmatrix} \begin{bmatrix} \mathbf{P}^T & \hat{\mathbf{P}}^T \end{bmatrix} = \begin{bmatrix} \mathbf{P} \cdot \mathbf{P}^T & \mathbf{P} \cdot \hat{\mathbf{P}}^T \\ \hat{\mathbf{P}} \cdot \mathbf{P}^T & \hat{\mathbf{P}} \cdot \hat{\mathbf{P}}^T \end{bmatrix} \tag{8.16}$$

Owing to $\mathbf{S}_c \cdot \mathbf{S}_c^T = \mathbf{P} \cdot \mathbf{P}^T$, $\tilde{\mathbf{P}} \cdot \tilde{\mathbf{P}}^T$ can be factorized as

$$\tilde{\mathbf{P}} \cdot \tilde{\mathbf{P}}^T = \begin{bmatrix} \mathbf{S}_c & 0 \\ \mathbf{V} & \mathbf{S}_a \end{bmatrix} \begin{bmatrix} \mathbf{S}_c^T & \mathbf{V}^T \\ 0 & \mathbf{S}_a^T \end{bmatrix} \tag{8.17}$$

then,

$$\begin{bmatrix} \mathbf{P} \cdot \mathbf{P}^T & \mathbf{P} \cdot \hat{\mathbf{P}}^T \\ \hat{\mathbf{P}} \cdot \mathbf{P}^T & \hat{\mathbf{P}} \cdot \hat{\mathbf{P}}^T \end{bmatrix} = \begin{bmatrix} \mathbf{S}_c & 0 \\ \mathbf{V} & \mathbf{S}_a \end{bmatrix} \begin{bmatrix} \mathbf{S}_c^T & \mathbf{V}^T \\ 0 & \mathbf{S}_a^T \end{bmatrix} = \begin{bmatrix} \mathbf{S}_c \cdot \mathbf{S}_c^T & \mathbf{S}_c \cdot \mathbf{V}^T \\ \mathbf{V} \cdot \mathbf{S}_c^T & \mathbf{V} \cdot \mathbf{V}^T + \mathbf{S}_a \cdot \mathbf{S}_a^T \end{bmatrix} \tag{8.18}$$

Thus, $\mathbf{V} \cdot \mathbf{S}_c^T = \hat{\mathbf{P}} \cdot \mathbf{P}^T$ and $\mathbf{V} = \hat{\mathbf{P}} \cdot \mathbf{P}^T (\mathbf{S}_c^T)^{-1}$ can be calculated by

$$\mathbf{S}_a \cdot \mathbf{S}_a^T = \hat{\mathbf{P}} \cdot \hat{\mathbf{P}}^T - \mathbf{V} \cdot \mathbf{V}^T \tag{8.19}$$

so that we can achieve a new lower triangular matrix

$$\tilde{\mathbf{S}}_c = \begin{bmatrix} \mathbf{S}_c & 0 \\ \mathbf{V} & \mathbf{S}_a \end{bmatrix} \tag{8.20}$$

The factorization of the updated coefficient matrix of HASM, when new sampling points are added, can be formulated as

$$
\begin{bmatrix} \mathbf{C}^T \cdot \mathbf{C} & -\tilde{\mathbf{S}}^T \\ \tilde{\mathbf{S}} & 0 \end{bmatrix} = \begin{bmatrix} \mathbf{S}_w & 0 \\ \tilde{\mathbf{P}} & \tilde{\mathbf{S}}_c \end{bmatrix} \begin{bmatrix} \mathbf{S}_w^T & -\tilde{\mathbf{P}}^T \\ 0 & \tilde{\mathbf{S}}_c^T \end{bmatrix} = \begin{bmatrix} \mathbf{S}_w & 0 & 0 \\ \mathbf{P} & \mathbf{S}_c & 0 \\ \hat{\mathbf{P}} & \mathbf{V} & \mathbf{S}_a \end{bmatrix} \begin{bmatrix} \mathbf{S}_w^T & -\mathbf{P}^T & -\hat{\mathbf{P}}^T \\ 0 & \mathbf{S}_c^T & \mathbf{V}^T \\ 0 & 0 & \mathbf{S}_a^T \end{bmatrix}
$$

$$(8.21)$$

Equation 8.5 can be updated as

$$
\begin{bmatrix} \mathbf{C}^T \cdot \mathbf{C} & -\tilde{\mathbf{S}}^T \\ \tilde{\mathbf{S}} & 0 \end{bmatrix} \begin{bmatrix} \mathbf{z}^{(n+1)} \\ \tilde{\lambda} \end{bmatrix} = \begin{bmatrix} \mathbf{S}_w & 0 & 0 \\ \mathbf{P} & \mathbf{S}_c & 0 \\ \hat{\mathbf{P}} & \mathbf{V} & \mathbf{S}_a \end{bmatrix} \begin{bmatrix} \mathbf{S}_w^T & -\mathbf{P}^T & -\hat{\mathbf{P}}^T \\ 0 & \mathbf{S}_c^T & \mathbf{V}^T \\ 0 & 0 & \mathbf{S}_a^T \end{bmatrix} \begin{bmatrix} \mathbf{z}^{(n+1)} \\ \lambda \\ \hat{\lambda} \end{bmatrix} = \begin{bmatrix} \mathbf{C}^T \cdot \mathbf{u}^{(n)} \\ \mathbf{k} \\ \hat{\mathbf{k}} \end{bmatrix}
$$

$$(8.22)$$

where $\hat{\lambda} \in \mathbb{R}^{\hat{S}_p \times 1}$; $\hat{\mathbf{k}} \in \mathbb{R}^{\hat{S}_p \times 1}$ is the vector of sampled values of the newly added sampling points.

Comparing Equation 8.22 with Equation 8.5, the only difference between them is that $\hat{\mathbf{P}}$, $\mathbf{V}$, and $\mathbf{S}_a$ need to be calculated additionally in equation set (Equation 8.22) when new sampling points are added.

8.2.4 Formulation of HASM-F by Dynamically Removing Information

Suppose, $\check{S}_p$ information points are removed and the sampling location matrix is updated as

$$
\mathbf{S} = \begin{bmatrix} \hat{\mathbf{S}} \\ \check{\mathbf{S}} \end{bmatrix}
$$

$$(8.23)$$

where $\hat{\mathbf{S}} \in \mathbb{R}^{(S_p - \check{S}_p) \times (I \cdot J)}$ and $\check{\mathbf{S}} \in \mathbb{R}^{\check{S}_p \times (I \cdot J)}$.

Let $\check{\mathbf{P}} = \check{\mathbf{S}} \cdot (\mathbf{S}_w^T)^{-1}$; then,

$$
\mathbf{P} = \mathbf{S} \cdot (\mathbf{S}_w^T)^{-1} = \begin{bmatrix} \hat{\mathbf{S}}(\mathbf{S}_w^T)^{-1} \\ \check{\mathbf{S}}(\mathbf{S}_w^T)^{-1} \end{bmatrix} = \begin{bmatrix} \hat{\mathbf{P}} \\ \check{\mathbf{P}} \end{bmatrix}
$$

$$(8.24)$$

$$
\mathbf{P} \cdot \mathbf{P}^T = \begin{bmatrix} \hat{\mathbf{P}} \\ \check{\mathbf{P}} \end{bmatrix} \begin{bmatrix} \hat{\mathbf{P}}^T & \check{\mathbf{P}}^T \end{bmatrix} = \begin{bmatrix} \hat{\mathbf{P}} \cdot \hat{\mathbf{P}}^T & \hat{\mathbf{P}} \cdot \check{\mathbf{P}}^T \\ \check{\mathbf{P}} \cdot \hat{\mathbf{P}}^T & \check{\mathbf{P}} \cdot \check{\mathbf{P}}^T \end{bmatrix} = \mathbf{S}_c \cdot \mathbf{S}_c^T
$$

$$(8.25)$$

Let $\mathbf{S}_c = \begin{bmatrix} \hat{\mathbf{S}}_c & 0 \\ \breve{\mathbf{V}} & \breve{\mathbf{S}}_c \end{bmatrix}$; then,

$$\mathbf{P} \cdot \mathbf{P}^T = \mathbf{S}_c \cdot \mathbf{S}_c^T = \begin{bmatrix} \hat{\mathbf{S}}_c & 0 \\ \breve{\mathbf{V}} & \breve{\mathbf{S}}_c \end{bmatrix} \begin{bmatrix} \hat{\mathbf{S}}_c^T & \breve{\mathbf{V}}^T \\ 0 & \breve{\mathbf{S}}_c^T \end{bmatrix} = \begin{bmatrix} \hat{\mathbf{S}}_c \cdot \hat{\mathbf{S}}_c^T & \hat{\mathbf{S}}_c \cdot \breve{\mathbf{V}}^T \\ \breve{\mathbf{V}} \cdot \hat{\mathbf{S}}_c^T & \breve{\mathbf{V}} \cdot \breve{\mathbf{V}}^T + \breve{\mathbf{S}}_c \cdot \breve{\mathbf{S}}_c^T \end{bmatrix} \qquad (8.26)$$

Thus,

$$\hat{\mathbf{P}} \cdot \hat{\mathbf{P}}^T = \hat{\mathbf{S}}_c \cdot \hat{\mathbf{S}}_c^T \qquad (8.27)$$

Therefore, $\hat{\mathbf{P}}$ consists of elements in the first $(S_p - \breve{S}_p)$ rows and $\hat{\mathbf{S}}_c$ consists of $\left(S_p - \breve{S}_p\right) \times \left(S_p - \breve{S}_p\right)$ elements in the upper corner on the left of $\mathbf{S}_c$ when $\breve{S}_p$ information points are removed.

The procedure for dynamic simulation with removed information points can be achieved using the following four steps:

1. Construct $\hat{\mathbf{P}}$ by taking the first $(S_p - \breve{S}_p)$ rows of $\mathbf{P}$.
2. Construct $\hat{\mathbf{S}}_c$ by taking the elements in the first $(S_p - \breve{S}_p)$ rows and the first $(S_p - \breve{S}_p)$ columns of $\mathbf{S}_c$.
3. Select the surface, before removing information points, as $\mathbf{z}^{(0)}$ for calculating $\mathbf{u}^{(0)}$.
4. Solve the equation sets $\begin{bmatrix} \mathbf{S}_w & 0 \\ \hat{\mathbf{P}} & \hat{\mathbf{S}}_c \end{bmatrix} \begin{bmatrix} \hat{\mathbf{z}}_1 \\ \hat{\mathbf{z}}_2 \end{bmatrix} = \begin{bmatrix} \mathbf{C}^T \cdot \mathbf{u}^{(n)} \\ \mathbf{k} \end{bmatrix}$ and $\begin{bmatrix} \mathbf{S}_w^T & -\hat{\mathbf{P}}^T \\ 0 & \hat{\mathbf{S}}_c^T \end{bmatrix} \begin{bmatrix} \mathbf{z}^{(n+1)} \\ \lambda \end{bmatrix} = \begin{bmatrix} \hat{\mathbf{z}}_1 \\ \hat{\mathbf{z}}_2 \end{bmatrix}$ successively to simulate the surface after removing the information points where $n \geq 0$.

It is noted that, for dynamical simulation with added information points, one iteration step can be conducted if there is no special requirement for simulation accuracy; if the accuracy needs to be improved, the iteration step can be continuously repeated until the requirement for accuracy is satisfied. However, it may be sufficient for the dynamical simulation with removed information points to conduct only one iteration step, because its accuracy may be reduced if additional iteration steps are conducted.

8.3 Numerical Test

$f(x, y) = (2x + y) \sin(2\pi x) \sin(2\pi y) + 1$ is selected as the numerical test surface (Figure 8.1). Its computational domain is $[0\ 1] \times [0\ 1]$. A uniform grid cell size

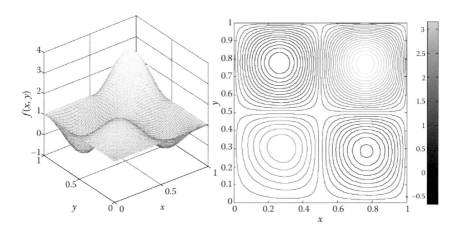

FIGURE 8.1
Test surface and its contour map.

is designed for the dynamic simulation with 20 iteration steps. New sampling points are randomly added to the dynamical simulation with added information points, of which there are initially 40, with 20 more sampling points added each time (Figure 8.2). The dynamic simulation with removed information points requires the opposite procedure.

We first compare the simulation errors of HASM-F and HASM-B, respectively. The mean absolute error (MAE) and mean relative error (MRE) are formulated as

$$MAE = \frac{1}{(I+2)\cdot(J+2)}\sum_i\sum_j \left|f_{i,j} - Sf_{i,j}\right| \tag{8.28}$$

$$MRE = \frac{1}{(I+2)\cdot(J+2)}\sum_i\sum_j \left|\frac{f_{i,j} - Sf_{i,j}}{f_{i,j}}\right| \times 100\% \tag{8.29}$$

where $f_{i,j}$ is the true value of $f(x, y)$ and $Sf_{i,j}$ is the simulated value of $f(x, y)$ at (x_i, y_j); I and J are respectively the number of grid cells in directions x and y.

The comparison shows that there is almost no difference in errors (Table 8.1), but HASM-F requires a longer computational time because its simulation process must calculate the inverse matrixes of the triangular matrixes many times, which makes HASM-F much more complex than HASM-B.

Here we introduce two terms: direct simulation and dynamic simulation. Direct simulation means that Equation 8.7 is conducted on all sampled points. Dynamic simulation includes dynamic simulation by adding information and dynamic simulation by removing information. Dynamic simulation by

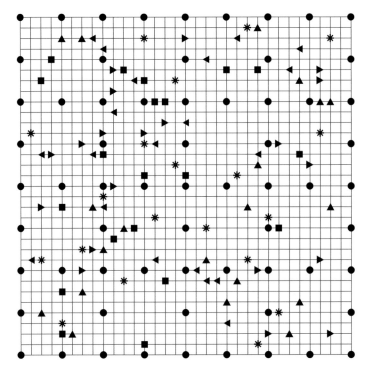

FIGURE 8.2
Dynamically adding information points: ● represents initial sampling points: ■, ▲, ✱, ▶, and ◀ respectively represent additional sampling points.

adding information is the procedure described in Section 8.2.3. Dynamic simulation for removing information is shown in Section 8.2.4.

Dynamic simulation by adding information starts with 40 sampling points. To begin with, the direct simulation and the dynamic simulation have the same procedure and accuracy (Table 8.2). The dynamic simulation conducts 20 iteration steps for every new data set of sampling points that is added. The dynamic simulation also becomes more accurate every time a data set is added, and thus achieves higher accuracy overall than direct simulation.

Dynamic simulation by removing information starts with 140 sampling points, and then removes 20 sampling points each time until there are only 20 sampling points left. To begin with, dynamic simulation by removing sampling points, dynamic simulation by adding sampling points, and direct simulation have the same accuracy. However, direct simulation's accuracy decreases much faster than dynamic simulation when sampling points are removed in the systematic way described above (Table 8.3).

Figure 8.3 indicates the change processes of simulation errors in a numerical test when sampling points are first added and then removed successively. In this numerical test, the total number of grid cells selected is 1089.

TABLE 8.1

Comparison between HASM-F Errors and HASM-B Errors

Total Number of Grid Cells	Number of Sampling Points	Mean Absolute Error		Mean Relative Error	
		HASM-B	HASM-F	HASM-B	HASM-F
289	33	1.029×10^{-2}	1.028×10^{-2}	1.256×10^{-2}	1.256×10^{-2}
	63	5.012×10^{-3}	5.010×10^{-3}	1.569×10^{-3}	1.569×10^{-3}
625	85	1.944×10^{-3}	1.944×10^{-3}	2.346×10^{-3}	2.345×10^{-3}
	135	1.312×10^{-3}	1.312×10^{-3}	1.616×10^{-3}	1.610×10^{-3}
1089	33	1.005×10^{-2}	1.005×10^{-2}	2.304×10^{-3}	2.304×10^{-3}
	133	1.964×10^{-3}	1.964×10^{-3}	4.596×10^{-3}	4.595×10^{-3}
1681	56	3.478×10^{-3}	3.478×10^{-3}	6.258×10^{-3}	6.258×10^{-3}
	206	8.549×10^{-4}	8.548×10^{-4}	1.195×10^{-3}	1.195×10^{-3}
2401	85	1.840×10^{-3}	1.840×10^{-3}	2.069×10^{-3}	2.068×10^{-3}
	285	5.155×10^{-4}	5.154×10^{-4}	5.823×10^{-4}	5.821×10^{-4}

TABLE 8.2

Comparison between Errors of Dynamic Simulation by Adding Information and Direct Simulation

Number of Sampling Points	Mean Absolute Error		Mean Relative Error	
	Direct Simulation	Dynamic Simulation	Direct Simulation	Dynamic Simulation
40	6.014×10^{-2}	6.014×10^{-2}	6.147×10^{-2}	6.147×10^{-2}
60	1.582×10^{-2}	1.363×10^{-2}	2.170×10^{-2}	1.769×10^{-2}
80	8.922×10^{-3}	6.971×10^{-3}	1.411×10^{-2}	1.009×10^{-2}
100	5.679×10^{-3}	4.897×10^{-3}	1.269×10^{-2}	7.991×10^{-3}
120	4.591×10^{-3}	3.710×10^{-3}	9.990×10^{-3}	6.669×10^{-3}
140	3.465×10^{-3}	2.646×10^{-3}	5.939×10^{-3}	3.817×10^{-3}

TABLE 8.3

Comparison between Errors of Dynamic Simulation by Removing Information and Direct Simulation

Number of Sampling Points	Mean Absolute Error		Mean Relative Error	
	Direct Simulation	Dynamic Simulation	Direct Simulation	Dynamic Simulation
140	3.335×10^{-3}	3.335×10^{-3}	5.023×10^{-3}	5.023×10^{-3}
120	3.967×10^{-3}	3.358×10^{-3}	5.680×10^{-3}	5.090×10^{-3}
100	5.082×10^{-3}	3.410×10^{-3}	6.536×10^{-3}	5.165×10^{-3}
80	7.778×10^{-3}	3.495×10^{-3}	7.483×10^{-3}	5.119×10^{-3}
60	1.412×10^{-2}	3.554×10^{-3}	1.619×10^{-2}	5.315×10^{-3}
40	6.014×10^{-2}	4.014×10^{-3}	6.147×10^{-2}	5.464×10^{-3}

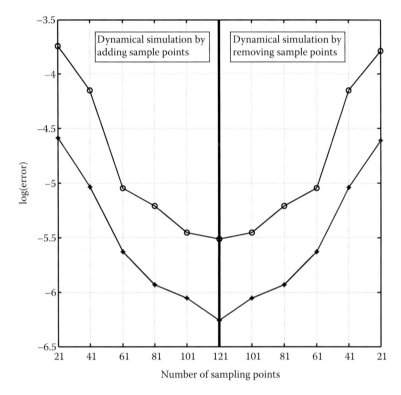

FIGURE 8.3
Error change of the dynamic simulation by adding and removing sampling points: ✶ represents relative error and ○ absolute error.

The dynamic simulation starts with 21 sampling points and 20 extra sampling points are added each time until there are 121 sampling points in all. The dynamic simulation then starts from 121 sampling points and 20 sampling points are removed each time until there are only 21 sampling points left. Once again, $f(x, y) = (2x + y) \sin(2\pi x) \sin(2\pi y) + 1$ is selected as the test surface. In Figure 8.3, the vertical axis is the natural logarithm of the mean absolute error and mean relative error; abscissa indicates the changing number of sampling points. Figure 8.3 shows that the error curve is symmetric when sample points increase to 121 and then decrease to 21 successively, which means that the dynamically simulated surface is unique.

8.4 Conclusions and Discussion

The HASM-F-based dynamic simulation method (HASM-FDS) is developed on the basis of matrix factorization. The dynamic simulation by adding sample points, which tries to describe how a system changes in time, only needs to conduct factorization of the submatrix created by the added sample points. The dynamic simulation by removing sample points, which is designed to find the error sources and delete them, only needs to calculate the submatrix of the remaining sample points. There is no significant difference in the computational complexity or simulation accuracy of either dynamic or direct simulation.

It is important that we develop fast and accurate simulation methods that are dynamic in time; for example, results will vary according to when the simulation is conducted, in order to effectively assimilate continuous streams of data into running simulations, and track and steer remotely distributed simulations to interact with computations as a means of dealing with noisy data (Douglas and Efendiev, 2006). HASM-FDS is an alternative way to use dynamic data to drive a real-time simulation. It can be concluded that HASM-FDS is applicable to the dynamic simulation of real-time issues if its computational speed can be improved enough.

References

Aggarwal, P.K., Kalra, N., Chander, S., and Pathak, H. 2006. InfoCrop: A dynamic simulation model for the assessment of crop yields, losses due to pests, and environmental impact of agro-ecosystems in tropical environments. *Agricultural Systems* 89: 1–25.

Al-Sabhan, W., Mulligan, M., and Blackburn, G.A. 2003. A real-time hydrological model for flood prediction using GIS and the WWW. *Computers, Environment and Urban Systems* 27: 9–32.

Bali, G.L. and Gimblett, R. 1992. Spatial dynamic emergent hierarchies simulation and assessment system. *Ecological Modelling* 62: 107–121.

Barredo, J.I., Kasanko, M., McCormick, N., and Lavalle, C. 2003. Modelling dynamic spatial processes: Simulation of urban future scenarios through cellular automata. *Landscape and Urban Planning* 64: 145–160.

Brown, I., Jude, S., Koukoulas, S., Nicholls, R., Dickson, M., and Walkden, M. 2006. Dynamic simulation and visualization of coastal erosion. *Computers, Environment and Urban Systems* 30: 840–860.

Choi, J.Y. and Engel, B.A. 2003. Real-time watershed delineation system using web-GIS. *Journal of Computing in Civil Engineering* 17(3): 189–196.

Clarke, G., Langley R., and Cardwell, W. 1998. Empirical applications of dynamic spatial interaction models. *Computers, Environment and Urban Systems* 22(2): 157–184.

Costanza, R. and Maxwell, T. 1991. Spatial ecosystem modelling using parallel processors. *Ecological Modelling* 58: 159–183.

Douglas, C.C. and Efendiev, Y. 2006. A dynamic data-driven application simulation framework for contaminant transport problems. *Computers and Mathematics with Applications* 51: 1633–1646.

Dragicevic, S. and Marceau, D.J. 2000. An application of fuzzy logic reasoning for GIS temporal modeling of dynamic processes. *Fuzzy Sets and Systems* 113: 69–80.

Fishwick, P.A. 2007. *Handbook of Dynamic System Modeling*. Boca Raton, FL: Chapman & Hall/CRC Taylor & Francis Group.

Frauenfelder, R., Schneider, B., and Kääb, A. 2008. Using dynamic modelling to simulate the distribution of rock-glaciers. *Geomorphology* 93: 130–143.

Gimblett, R. 1989. Linking perception research, visual simulations and dynamic modeling within a GIS framework: The ball state experience. *Computers, Environment and Urban Systems* 13: 109–123.

Gold, C. and Mostafavi, M.A. 2000. Towards the global GIS. *ISPRS Journal of Photogrammetry and Remote Sensing* 55: 150–163.

Harris, B. and Wilson, A.G. 1978. Equilibrium values and dynamics of attractiveness terms in production-constrained spatial interaction models. *Environment and Planning A* 10: 371–388.

Hiemstra, P.H., Pebesma, E.J., Twenho, C.J.W., and Heuvelink, G.B.M. 2009. Real-time automatic interpolation of ambient gamma dose rates from the Dutch radioactivity monitoring network. *Computers and Geosciences* 35: 1711–1721.

Mitasova, H., Mitas, L., Ratti, C., Ishii, H., Alonso, J., and Harmon, R.S. 2006. Real-time landscape model interaction using a tangible geospatial modeling environment. *IEEE Computer Graphics and Applications* 26(4): 55–63.

Shuai, J.P., Buck, P., Sockett, P., Aramini J., and Pollari, F. 2006. A GIS-driven integrated real-time surveillance pilot system for national West Nile virus dead bird surveillance in Canada. *International Journal of Health Geographics* 5: 17, doi:10.1186/1476–072X-5-17.

Sydelko, P.J., Hlohowskyj, I., Majerus, K., Christiansen, J., and Dolph, J. 2001. An object-oriented framework for dynamic ecosystem modeling: Application for integrated risk assessment. *The Science of the Total Environment* 274: 271–281.

Yuan, M. and Hornsby, K.S. 2008. *Computation and Visualization for Understanding Dynamics in Geographic Domain*. NY: CRC Press.

Yue, T.X. and Song, Y.J. 2008. The YUE-HASM Method. In *Accuracy in Geomatics: Proceedings of the 8th International Symposium on Spatial Accuracy Assessment in Natural Resources and Environmental Sciences*, Shanghai, June 25–27, 2008, pp. 148–153.

Yue, T.X. and Wang, S.H., 2010. Adjustment computation of HASM: A high accuracy and speed method. *International Journal of Geographical Information Sciences* (in press).

Yue, T.X., Du, Z.P., Song, D.J., and Gong, Y. 2007. A new method of surface modeling and its application to DEM construction. *Geomorphology* 91(1–2): 161–172.

Yue, T.X., Song, D.J., Du, Z.P., and Wang, W. 2010. High accuracy surface modeling and its application to DEM generation. *International Journal of Remote Sensing* 31(8): 2205–2226.

Yun, J.K., Kim, D.O., Hong, D.S., Kim, M.H., and Han, K.J. 2006. A real-time mobile GIS based on the HBR-tree for location based services. *Computers and Industrial Engineering* 51: 58–71.

Zerger, A. and Smith, D.I. 2003. Impediments to using GIS for real-time disaster decision support. *Computers, Environment and Urban Systems* 27: 123–141.

Zhao, S.J. 2010. GisFFE—an integrated software system for the dynamic simulation of fires following an earthquake based on GIS. *Fire Safety Journal* 45(2): 83–97.

Section II

Surface Modeling of Environment Components of Earth Surface

9

Surface Modeling of DEM*

9.1 Introduction

A digital elevation model (DEM) is a representation of terrain elevation as a function of geographic location. DEMs are typically represented in two formats: contour maps, where the surface is represented by lines of constant elevation at even intervals; or point heights, where the elevation surface is sampled on either a regular or irregular basis (Raaflaub and Collins, 2006). Elevation is a distinctive variable in GIS, because it is not only a simple, inexpensive, and intensively sampled variable, but also the most crucial and probably the most fundamental of all the variables (Atkinson, 2002). DEMs are playing an increasingly important role in the area of remote sensing image-based data capture, especially with the development of high spatial resolution satellite images (Zhu et al., 2005). DEM has become an indispensable quantitative environmental variable in most of the research topics of remote sensing (San and Suezen, 2005). DEM is now commonly used for rectifying satellite images, and the accuracy of these images is highly dependent on the accuracy of the DEM.

In the past, DEMs were generated using stereo pair data sets from the same sensors. Due to the high spatial resolution of recent satellite sensors in the visible spectrum, many researchers around the world have since investigated the extraction of elevation, mostly by means of clinometry or stereoscopy (Wolf, 1974; Buchroithner, 1989; Petrie et al., 1997). Recently, elevation data have been derived from sets of synthetic aperture radar (SAR) images (Li et al., 2003) and nighttime thermal infrared (NTI) images (Saraf et al., 2005). DEMs are considered to be the most permanent and reusable georelated data sets over time and the most important data structures for geospatial analysis. Unfortunately, DEMs's ability to provide usable information is still not available for the whole earth, and where they are available, the information they provide is not always sufficiently accurate (Toutin, 2001).

Various techniques of surface modeling have been used for compiling DEMs (Etzelmuller, 2000; Schneider, 2005). DEMs have been used for identifying and classifying landforms (Hayakawa and Oguchi, 2006; Prima et al., 2006), analyzing terrain characteristics (Gao, 1997; Ganas et al., 2005; Glenn et al., 2006;

---

* Dr. Chuan-Fa Chen and Dr. Dun-Jiang Song are major collaborators of this chapter.

Lin and Oguchi, 2006), and studying geomorphic processes (Bolongaro-Crevenna et al., 2005; Mitasova et al., 2005). Any DEM, even in a very high quality, is an approximation of a real-world continuous surface (Carter, 1988). The sources of DEM errors include the quality of source data, the capability of data-capturing equipment, the transformation of control points, the mathematical model for constructing the surface, and grid resolution and orientation (Zhou and Liu, 2002; Zhu et al., 2005). Such DEM errors can be propagated through various simulation processes to affect the quality of the final product (Huang and Lees, 2005; Oksanen and Sarjakoski, 2005).

A large number of studies have been focused on finding solutions for DEM errors. For instance, Goodchild (1982) introduced the fractal Brownian process as a terrain simulation model for improving DEM accuracy. Walsh et al. (1987) found that total errors could be minimized by recognizing inherent errors within input products and operational errors created by combinations of input data. Hutchinson and Dowling (1991) introduced a drainage enforcement algorithm that tried to remove the spurious pits in order to yield a DEM that would reflect the natural drainage structure. Unwin (1995) suggested that it would be useful if general tools for examining how errors propagate through GIS operations could be made available. Wise (2000) argued that a clear distinction needed to be made between the lattice and pixel models when current GIS is used; the value stored in a pixel relates to the whole extent of the grid, while the value stored in a lattice relates solely to the central point of the grid. The United States Geological Survey (1997) adopted statistical accuracy tests, data editing, the verification of logical and physical formats of data files, and visual inspection as the principle items of its DEM quality control system (Lo and Yeung 2002). Florinsky (2002) suggested four alternative methods to reduce DEM errors caused by the Gibbs phenomenon. Shi et al. (2005) investigated high-order interpolation algorithms to reduce DEM errors. Podobnikar (2005) proposed to produce high-quality DEMs by using all available data sources, even lower quality data sets and data sets without a height attribute. Chaplot et al. (2006) evaluated the performance of various classic techniques for DEM generation. However, it is clear that more studies still need to be undertaken to attack the error problem at its root.

In this chapter, HASM is introduced to develop high-accuracy DEMs. HASM-PCG is then employed to fill the voids in the SRTM DEM. HASM-B is applied to developing DEMs by respectively using a topological map and height-sampling data.

9.2 SRTM Void Filling

Data sets of the SRTM, which was derived from the Space Shuttle Endeavour in February 2000, have become a useful source of elevation data and thus critical to modern imagery analysis and geospatial intelligence requirements (van Zyl, 2001). The highest spatial resolution of the DEM data sets, which is publicly available, is 1 arc second, approximately 30 m. The SRTM DEM has an absolute vertical accuracy of ±16 m, a relative vertical accuracy of ±10 m for 90% of the data, and an absolute horizontal accuracy of ±20 m (Walker et al., 2007). All accuracies were quoted at the 90% confidence level, consistent with National Mapping Accuracy Standards (Rabus et al., 2003).

SRTM's accuracy has been assessed since it was released (Helm et al., 2002; Smith and Sandwell, 2003; Falorni et al., 2005). Various validations have demonstrated the applicability of SRTM data. For instance, the surface height measured independently by SRTM and SLA-02 was cross-validated (Sun et al., 2003), and it was inferred from the standard deviation of the height differences that the absolute accuracy of SRTM height at a low vegetation area was in fact better than the SRTM mission specifications (16 m). A validation work in Southeast Michigan indicated that SRTM data exceeded the requirement for height accuracy (Brown et al., 2005). Rodriguez et al. (2006) presented an assessment of the SRTM topographic data using a variety of globally distributed data sets; their results demonstrated that SRTM data accuracy exceeded its mission goal of 16 m. A validation of the SRTM DEM data set, using a unique database of completely independent height measurements derived from satellite altimeter echoes, showed that SRTM data are generally very good indeed (Berry et al., 2007).

However, SRTM data has various-sized voids, resulting in incomplete data sets (Grohman et al., 2006). These voids account for 0.3% of the total data set analyzed in a study by the United States, but they can amount to up to 30% of rugged terrain. Most voids are less than five pixels in size. Voids may occur for different reasons: water surfaces produce radar signal scattering, which makes it impossible for the interferometer to detect meaningful reflections; for surface inclinations above 20°, the frequency of data voids increases because of radar shadowing; while in desert areas, a void is more likely to occur due to a complex dielectric constant (Luedeling et al., 2007; Reuter et al., 2007). Small voids can be successfully filled by means of the interpolation of values around the edges. Large voids however, cannot be simply interpolated from the edges and need to be filled with topographic information from other sources.

Any areas of missing data that exist in the SRTM data are defined as voids. Void-filling is the process of all no-data areas being filled using the best-performing interpolation algorithm available (Reuter et al., 2007). Many approaches have been developed to fill voids in SRTM DEM such as Fill and Feather (FF), IDW-based Delta Surface Fill (DSF), TIN-based delta surface (TDS), kriging, spline, and the advanced spline method (ANUDEM). The FF method replaces a void with the most accurate digital elevation source available and then smoothes the transition to mitigate any abrupt change, in which the void-specific bias of the alternative surface is removed by adding a constant, and then feathered at the edges to provide a seamless transition. FF corrupts the presumably correct SRTM surface at the void edges and cannot account for varying vertical biases within the void (Luedeling et al., 2007). The DSF process replaces the void with fill source posts that are adjusted to the SRTM values found at the void interface, in which inverse distance weighted interpolation is used (Grohman et al., 2006). TDS is a method similar to DSF, in which voids are filled by TINs that connect the points bordering the voids (Luedeling et al., 2007).

All interpolation methods for void-filling use the elevation data surrounding the void if auxiliary data are unavailable. If auxiliary sources of elevation are available, then some of these algorithms can incorporate this information to improve the accuracy of the interpolation (Reuter et al., 2007). The quality of different methods has been extensively evaluated. The void-filling errors of spline, kriging, and linear estimation were comparatively analyzed in hilly, mountainous, and planar areas respectively; the results indicated that there was no significant difference in the results obtained using any of the three methods mentioned above (Katzil and Doytsher, 2000). Fisher and Tate (2006) concluded that no single interpolation method is more accurate than any other method for the interpolation of terrain data after they reviewed the source and nature of errors in digital models of elevation, and in the derivatives of these models.

But many other studies have demonstrated the large differences in results that can appear when different methods are used, and that potential improvement that can be achieved with good quality auxiliary information (Grohman et al., 2006). The tests of IDW, kriging, multiquadratic radial basis function (MRBF), and spline in the mountainous region of northern Laos and in western France showed that a few differences exist between the interpolation methods at a higher sampling density, while the accuracy was more dependent on the choice of interpolation techniques at lower sampling density; kriging yielded the better estimations if the spatial structure of altitude was strong; IDW and spline performed better when the spatial structure of height was weak, and MRBF performed well in the mountainous areas (Chaplot et al., 2006). Wilson and Gallant (2000) indicated that geostatistical methods such as kriging performed better than other methods among the many existing interpolation techniques. However, Aguilar et al. (2005) claimed that neighborhood approaches such as IDW were as accurate as kriging or even better. Jarvis et al. (2008) concluded that kriging or IDW might be better employed to fill small- or medium-sized voids in relatively flat low-lying areas, spline for small- or medium-sized voids in high altitude or dissected terrain, TIN or IDW for large voids in very flat areas, and ANUDEM for large voids in other terrains.

In this section, HASM is applied to void-filling. Its performance is compared with those of kriging, IDW, spline, and ANUDEM, taking void-filling in different terrain types of China as examples.

9.2.1 HASM

HASM as seen in Chapter 4, can be expressed as (Yue et al., 2007, 2010)

$$\mathbf{W}_h \cdot \mathbf{z}^{(n+1)} = \mathbf{v}_h^{(n)} \tag{9.1}$$

where $\mathbf{W}_h = \begin{bmatrix} \mathbf{A}_h^T & \mathbf{B}_h^T & \lambda \cdot \mathbf{S}^T \end{bmatrix} \begin{bmatrix} \mathbf{A}_h \\ \mathbf{B}_h \\ \lambda \cdot \mathbf{S} \end{bmatrix}$, $\mathbf{v}_h^{(n)} = \begin{bmatrix} \mathbf{A}_h^T & \mathbf{B}_h^T & \lambda \cdot \mathbf{S}^T \end{bmatrix} \begin{bmatrix} \mathbf{d}_h^{(n)} \\ \mathbf{q}_h^{(n)} \\ \lambda \cdot \mathbf{k} \end{bmatrix}$

and h is the grid cell size.

In previous HASM applications (Yue et al., 2007, 2010; Shi et al., 2009), λ was a real-number parameter. It is here extended to a vector, $\lambda = \bar{\lambda}(\lambda_1, \lambda_2, \ldots, \lambda_k, \ldots, \lambda_{K-1}, \lambda_K)^T$, in the process of filling SRTM voids for weighting the different sampled points that have various accuracy and confidence degrees, in which $\bar{\lambda}$ is a real-number coefficient and λ_k is the weight of the kth sampled point. If a sampled point had higher accuracy or higher degree of confidence, it will have a bigger weight value. If all the sampled points had almost the same accuracy, the weight value of a sampled point would be determined by the variance of elevation data surrounding it. The neighborhood of every sampled point can be determined by searching adjacent points.

A Delaunay triangulation is regarded as usable for the purpose of interpolation if its triangles are nearly equiangular (Sibon, 1978). But in practice with arbitrarily placed points, a close approach to equiangularity is seldom possible. Thus, resultant triangles are made as nearly equiangular as possible. From an interpolation point of view, Delaunay triangulation is optimal (Correc and Chapuis, 1987). The points are usually connected according to Delaunay triangulation, a procedure that joins the centers of neighboring Thiessen polygons. A Thiessen is the dual of a Delaunay. Thiessen polygons are generated from a set of points. The boundaries of the Thiessen polygons define the area that is closest to each point relative to all other points (http://en.mimi.hu/gis/thiessen_polygon.htm).

For the computational domain Ω, a set of sampled points is used to construct a Delaunay triangulation by the procedure of Thiessen polygons $\{TRK_k \mid [(x_{k_1}, y_{k_1})(x_{k_2}, y_{k_2})(x_{k_3}, y_{k_3})], (x_{k_i}, y_{k_i}) \in \Phi, i = 1, 2, 3\}$. For every sampled point, all other sampled points that have an edge connecting with it are defined as its adjoining points. Suppose the number of adjoining points is k_w, obviously $k_w \geq 2$. V_k is the elevation variance of the sampled points with its adjoining points. For all sampled points, we can achieve a sequence of variances, $\{V_k\}$, in which $V_{max} = \max \{V_k\}$ and $V_{min} = \min \{V_k\}$. If λ_{max} and λ_{min} are respectively the maximum and minimum values of elements of $(\lambda_1, \lambda_2, \ldots, \lambda_k, \ldots, \lambda_{K-1}, \lambda_K)^T$, the weight of a sampled point can be formulated as

$$\lambda_k = -\frac{\lambda_{max} - \lambda_{min}}{V_{max} - V_{min}} \cdot V_k + \frac{\lambda_{max} \cdot V_{max} - \lambda_{min} \cdot V_{min}}{V_{max} - V_{min}} \tag{9.2}$$

In other words, the sampled point has more weight if the elevation around it has less variety. If the elevation change is slighter, the sampled point has a greater weight. The maximum variance corresponds to the minimum weight and the minimum variance corresponds to the maximum weight.

9.2.2 Materials and Verification Procedures

China is a mountainous country and has five major mountainous ranges (Zhao, 1986). The first is the east–west mountain range; from north to south, there are three major subranges. These are: the Tianshan-Yinshan-Yanshan mountain range; the Kunlun-Qinling-Dabie mountain range; and the Nanling mountain range. The second is the north–southward mountain range, which includes Helan mountain, Liupan mountain, and the Hengduan mountain subrange. The third is the north–eastward mountain range, which includes the Dahinggan mountains, the Taihang mountains and the Wushan mountains. The fourth is the northwestward mountain range, which includes the Altay Mountains and the Qilian mountains. The fifth is the arc mountain range, which includes the Himalaya Mountains and the Taiwan mountains. China's land can be approximately divided into three areas: the first is from Qinghai-Xizang plateau with its mean elevation of over 4000 m; the second is from the eastern margin of Qinghai-Xizang plateau extending eastward to the Da Hinggan-Taihang-Wushan line with elevations from 2000 to 1000 m; the third is from the Da Hinggan-Taihang-Wushan line eastward toward the coast.

Voids account for 0.15% of the total data set of China. Many areas have voids more than 30%, such as southeast of Qinghai-Xizang plateau, areas around Sichuan basin and areas around Tianshan mountain (Figure 9.1). It is necessary to conduct void-filling if SRTM data are used.

Nine verification regions without voids are selected in three different geomorphologic types; hilly, plateau, and mountainous areas. In every geomorphologic type, three verification regions are determined according to void-area size (Figure 9.2). The terrain parameters are described in Table 9.1.

The DEM of China is calculated in terms of elevation above Yellow Sea level, while SRTM data is based on Earth Gravitational Model 1996 (EGM96), an improved spherical harmonic model of the Earth's gravitational potential to 360°. The Global Positioning System (GPS) uses WGS 84 as its reference coordinate system, dating from 1984 and last revised in 2004. It comprises a standard coordinate frame for the Earth, a standard spheroid reference surface (the datum or reference ellipsoid) for raw altitude data, and a gravitational equipotent surface (geoids) that defines the nominal sea level. However, we cannot realize the transformation between the coordinate systems of China DEM and EGM96; data sampled using GPS is also inapplicable in this case, because it does not provide the potential coefficients of the gravitational potential model for China. Therefore, we design a verification procedure as follows: (1) SRTM3 data of the nine verification regions are downloaded from

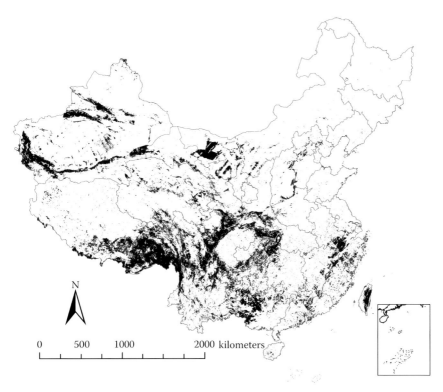

FIGURE 9.1
Spatial distribution of voids in China.

the Web site, http://srtm.csi.cgiar.org/, in which there are no voids; (2) voids are artificially made by cleaning out the SRTM3 data in the verification regions; the cleaned-out data are then selected as the "true" values for verification; (3) 100 pixels are extended from the borders of the artificially made voids and used to simulate values in the voids respectively by means of HASM, TIN, IDW, spline, kriging, and ANUDEM; (4) SRTM30 data (http://www.dgadv.com/srtm30/) are selected as the auxiliary data to fill the voids by means of HASM, TIN, IDW, spline, kriging, and ANUDEM; (5) the void-filling results by HASM, TIN, IDW, spline, kriging, and ANUDEM, with auxiliary data and without auxiliary data, are respectively compared with the "true" values; and (6) errors of void-filling results are comparatively analyzed in terms of *RMSE* and *MAE*. The lower errors mean higher accuracy.

MAE and *RMSE* are respectively formulated as

$$MAE = \frac{1}{I \times J} \sum_{i=1}^{I} \sum_{j=1}^{J} (f_{i,j} - Sf_{i,j}) \qquad (9.3)$$

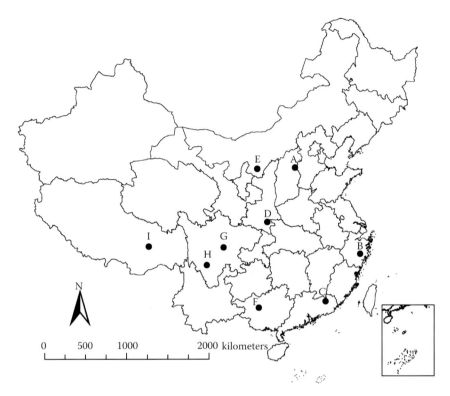

FIGURE 9.2
The selected regions for verification.

TABLE 9.1

Terrain Parameters in the Research Regions

Geomorphologic Type	Verification Region	Minimum Elevation (m)	Maximum Elevation (m)	Elevation on an Average (m)	Standard Deviation (m)	Void Count
Hilly area	A	469	1816	1023	263	520
	B	71	1060	343	199	177
	C	113	673	193	61	442
Plateau	D	276	2324	1187	455	712
	E	1310	1510	1385	36	3230
	F	−25	1143	466	214	84,891
Mountainous area	G	567	5922	2161	405	2109
	H	1225	5125	3041	742	104,514
	I	2519	6841	4620	670	197,708

and

$$RMSE = \left(\frac{1}{I \times J} \sum_{i=1}^{I} \sum_{j=1}^{J} (f_{i,j} - Sf_{i,j})^2 \right)^{-1/2} \tag{9.4}$$

where $Sf_{i,j}$ is the interpolated value of $f(x, y)$ at the lattice (i, j); $f_{i,j}$ is the true value of $f(x, y)$ at the lattice (i, j), $i = 1, 2, …, I, j = 1, 2, …, J$.

RMSE is equal to the standard deviation of the error if the mean error is (or is assumed to be) zero. *RMSE* is a widely used measure of conformity between a set of estimates and actual values, and has become a standard measure of map accuracy (Fisher and Tate, 2006).

For the classical methods, TIN, IDW, spline, kriging, and ANUDEM, default values of their input parameters are used under the assumption that the default values would provide reasonable results under most conditions because they are all well-known and long-standing implementations (Reuter et al., 2007).

9.2.3 Comparative Analyses of Void-Filling Results

If we use the standard deviation as a measure of landform complexity, region A, has the biggest landform complexity and region C, has the smallest complexity among the three regions of the hilly area (Table 9.1). In region A, located in 38°29'10.67"N–38°30'38.75"N and 113°22'55.48"E–113°25'08.72"E, accuracies of the simulated results without auxiliary data by all methods are much higher than the ones when the auxiliary data are added; MAEs of TIN, IDW, spline, kriging, ANUDEM, and HASM are increased by 38.7%, 14.9%, 40.9%, 32.6%, 26.6%, and 19.2%, respectively; RMSEs are increased by 54%, 27.6%, 56.2%, 46%, 35.1%, and 20.2% (Table 9.2). Spline and TIN produce the most inaccurate results while HASM produces the most accurate ones.

In region B, located in 28°41'12.90"N–28°42'12.32"N and 120°35'40.58"E–120°36'51.42"E, both MAEs and RMSEs of TIN, ANUDEM, and HASM are increased when the auxiliary data are introduced; MAEs are respectively increased by 8.1%, 0.5%, and 3.7%, RMSEs by 8.2%, 0.6%, and 2.6%. MAE and RMSE of Kriging are respectively decreased by 1.6% and 0.7% because of the introduction of the auxiliary data. MAEs of IDW and spline are respectively increased by 0.7% and 2.7%, while their RMSEs are decreased by 0.2% and 1.1%. TIN performs worst and HASM best.

In region C, located in 24°15'11.32"N–24°16'3.77"N and 115°41'27.09"E–115°43'10.44"E, the introduction of the auxiliary data lessens the simulation errors of all methods. MAEs of TIN, IDW, spline, kriging, ANUDEM, and HASM are decreased respectively by 17%, 4.7%, 34.5%, 4.5%, 24.1%, and 4.8%; their RMSEs are reduced by 18.3%, 3.6%, 38.4%, 5.5%, 22.7%, and 3.8%. Spline produces the highest error and HASM the lowest.

TABLE 9.2

Accuracies of Different Methods in Hilly Regions

Region	Method	MAE (m) Without Auxiliary Data	MAE (m) With Auxiliary Data	RMSE (m) Without Auxiliary Data	RMSE (m) With Auxiliary Data
A	TIN	57.1	79.2	75.7	116.6
	IDW	47.7	54.8	64.1	81.8
	SPLINE	57.4	80.9	78.3	122.3
	KRIGING	44.2	58.6	60.4	88.2
	ANUDEM	49.7	62.9	68.6	92.7
	HASM	38.5	45.9	51.6	62.0
B	TIN	45.4	49.1	57.1	61.8
	IDW	41.2	41.5	52.4	52.3
	SPLINE	25.4	26	36.6	36.2
	KRIGING	44.8	44.1	55.3	54.9
	ANUDEM	40.7	40.9	53.5	53.8
	HASM	24.6	25.5	34.0	34.9
C	TIN	4.7	3.9	6.0	4.9
	IDW	4.3	4.1	5.5	5.3
	SPLINE	11.3	7.4	16.4	10.1
	KRIGING	4.4	4.2	5.5	5.2
	ANUDEM	5.4	4.1	6.6	5.1
	HASM	4.2	4.0	5.3	5.1

In hilly areas, it may be concluded that (1) the simulation accuracy of every method is closely related to landform complexity, (2) spline has the most sensitive response to adding the auxiliary data and this response is negative in rugged areas and positive in relatively flat areas, and (3) HASM has the highest accuracy compared to the classic methods.

In the plateau, region D has the highest landform complexity and region E the lowest, while region F has the largest void area and region D the smallest (Table 9.1). In region D, located in 33°00′19.14″N–33°03′19.05″N and 109°12′04.35″E–109°14′11.09″E, the introduction of SRTM30 data makes MEAs of TIN, spline, and ANUDEM increase respectively by 41.9%, 36%, and 16.3%, and RMSEs by 58.7%, 41.1%, and 22.1%; but SRTM30 reduces MEAs and RMSEs of IDW and HASM; IDW has the lowest accuracy and HASM has the highest accuracy. In region E, located in 38°36′15.96″N–38°40′48.81″N and 108°05′59.05″E–108°12′42.31″E and region F, in 23°46′53.75″N–24°14′03.07″N and 107°40′10.28″E–108°02′10.36″E, SRTM30 causes a great decrease of MEAs and RMSEs. MEA and RMSE of spline respectively decrease by 51.9% and 55.8% in region E, by 88.6%, and 88.7% in region F. HASM is the most accurate method and spline the least accurate.

In general, there are far fewer MEAs and RMSE in region E than in regions D and F (Table 9.3); the biggest MEAs and RMSEs appear in region F. This means that simulation accuracy is not only related to landform complexity, but also to the size of the void area. The accuracy of HASM simulation is always higher than the classic methods. The method employed plays an important role in using SRTM30 to improve void-filling accuracy.

In the mountainous area, the highest landform complexity appears in region H (27°58′47.40″N–28°43′06.28″N, 101°32′01.24″E–101°51′27.33″E) and the lowest in region G (30°23′52.87″N–30°27′19.94″N, 103°30′41.67″E–103°33′42.65″E). The biggest void area is in region I (30°03′53.96″N–30°41′14.57″N, 93°54′06.49″E–94°32′12.40″E) and the smallest is in region G. The introduction of SRTM30 in region G increases *RMSE* of ANUDEM by 55.6%, HASM by 11.5%, TIN by 5.9%, and IDW by 1.4, while it reduces RMSEs of spline, and kriging by 2.7% and 2.8%, respectively. HASM produces the most accurate results (Table 9.4). In regions H and I, the auxiliary data greatly improve the simulation results of methods and all RMSEs are decreased by more than 50%. For all methods, the biggest void-filling errors are in region I, whether the auxiliary data are used or not. HASM is always more accurate than all other methods.

TABLE 9.3

Accuracies of Different Methods in Plateau

		MAE (m)		RMSE (m)	
Plateau	Method	Without Auxiliary Data	With Auxiliary Data	Without Auxiliary Data	With Auxiliary Data
D	TIN	30.3	43.0	39.7	63.0
	IDW	87.1	83.0	113.1	108.5
	SPLINE	36.9	50.2	65.7	92.7
	KRIGING	53.8	53.8	77.9	79.0
	ANUDEM	30.1	35.0	39.0	47.6
	HASM	22.3	21.3	30.2	29.2
E	TIN	5.3	3.3	7.5	4.5
	IDW	7.3	5.4	9.8	7.1
	SPLINE	10.6	5.1	16.5	7.3
	KRIGING	6.3	3.6	8.8	4.9
	ANUDEM	4.9	3.0	7.4	4.0
	HASM	4.9	3.0	7.2	4.0
F	TIN	54.4	41.5	68.0	50.2
	IDW	70.0	37.4	93.5	46.3
	SPLINE	454.0	51.7	759.7	86.2
	KRIGING	69.6	36.9	92.5	45.6
	ANUDEM	48.7	40.1	63.3	48.8
	HASM	48.6	36.3	63.1	44.9

TABLE 9.4

Accuracies of Different Methods in Mountainous Areas

Mountainous Area	Method	MAE (m) Without Auxiliary Data	MAE (m) With Auxiliary Data	RMSE (m) Without Auxiliary Data	RMSE (m) With Auxiliary Data
G	TIN	98.7	108.8	138.7	146.9
	IDW	109.2	110.1	148.4	150.5
	SPLINE	114.8	120.9	168.0	163.5
	KRIGING	113.1	111.1	151.2	146.9
	ANUDEM	68.1	103.3	90.3	140.5
	HASM	65.8	73.1	86.0	95.9
H	TIN	196.9	88.2	292.9	115.9
	IDW	312.3	109.3	448.9	140.6
	SPLINE	435.6	104.3	758.4	145.6
	KRIGING	312.1	140.4	447.3	181.7
	ANUDEM	239.5	78.4	359.5	104.0
	HASM	187.4	77.7	277.7	103.4
I	TIN	364.9	168.4	503.5	221.1
	IDW	487.9	202.1	666.6	258.6
	SPLINE	1175.6	172.7	2242.4	247.5
	KRIGING	492.1	251.8	668.3	318.9
	ANUDEM	408.5	160.4	574.6	208.3
	HASM	331.6	136.1	481.4	182.2

9.3 Surface Modeling of DEM Based on Height-Sampling Points

Five sets of DEMs for the Da-Fo-Si coal mine in Shaan-Xi Province, China, were constructed using HASM, kriging, TIN, spline, and IDW, respectively, and their errors were comparatively analyzed. The coal mine (Figure 9.3) is located at N35°05′ and E108°00′ in northwestern Xian-Yang City, and its area is 1.4 km². The great topographical variety of the area makes it very suitable for the study of DEM accuracy.

Twelve GPS control points were allocated within the coal mine. The Leica TC403 total station was then used to randomly collect 3856 elevation points, of which 771 points (Figure 9.4a) were randomly taken to construct DEMs, while the other 3085 points (Figure 9.4b) were used to inspect the accuracy of the constructed DEMs.

Coal mine DEMs with a spatial resolution of 1×1 m were constructed by means of HASM, kriging, TIN, spline, and IDW. Shaded relief maps of the DEMs were used to represent the terrain relief of the coal mine, which was

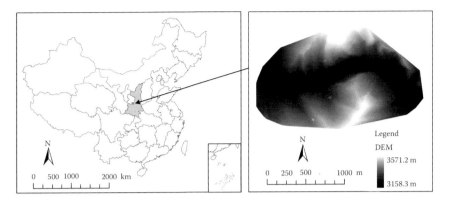

FIGURE 9.3
Location and topography of the Da-Fo-Si coal mine.

lit up from a single direction. The Azimuth angle of the light source is 315°. The azimuth is expressed in positive degrees from 0° to 360° and measured clockwise from the north. The altitude angle of the light source above the horizon is 45°. The slope is expressed in positive degrees, with 0° at the horizon and 90° directly overhead. All the shaded relief maps are made in the same azimuth and altitude angle (Figure 9.5).

Figure 9.5 represents the fact that the kriging simulation makes the DEM surface fragmentary (Figure 9.5b). The surface of the TIN simulation within each triangle is a plane passing through three vertices. Each facet has one value for gradient and one for aspect, while the curvature is zero everywhere (Hugentobler et al., 2005). These properties of TIN cause the so-called peak-truncation and pit-fill problem (Figure 9.5c). The spline simulation has a higher oscillation on the boundary of the DEM (Figure 9.5d). The IDW simulation produces many "bull's eyes" on the DEM surface (Figure 9.5e).

9.4 Surface Modeling of DEM Based on Topological Maps

The Qian-Yan-Zhou Experimental Station for Red Soil and Hilly Land of the Chinese Academy of Sciences is selected as the test area. The Qian-Yan-Zhou Experimental Station is located in the middle of the Jiang-Xi province of China. The geographic coordinates of its central point is 26°44′44″N and 115°03′44″E (Figure 9.6). Its area is 204.16 ha, including about 80 hillocks and 9 gullies. Its elevation varies from 60 to 150 m. It includes four geomorphologic types: hillocks, terraces, higher flood-plains, and lower flood-plains. The hillocks have round summit surfaces; their relative heights vary from 20 to 50 m; and their slopes vary from 10° to 30°. The terraces have even surfaces; their relative heights are about 2 m; and their slopes vary from 1° to 2°.

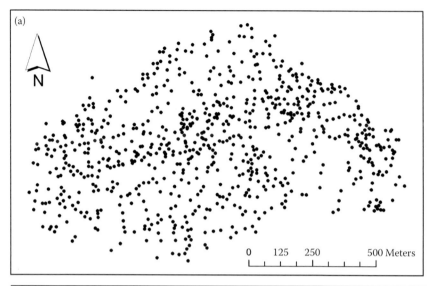

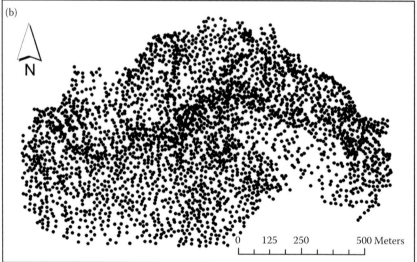

FIGURE 9.4

Location of height-sampling points. (a) Points for DEM construction and (b) points for inspecting DEM accuracy.

The surface slopes of the higher flood-Plains vary from 1° to 3° and their relative heights from 0.6 to 1 m. The lower flood-plains are slightly undulant and their relative heights vary from 0.5 to 0.6 m. The Qian-Yan-Zhou has great topographical variety and so is highly suitable for testing the efficiency of all methods of surface modeling.

The DEM of the Qian-Yan-Zhou Experimental Station is simulated by means of HASM, spline, TIN, kriging, and IDW, respectively, for which a

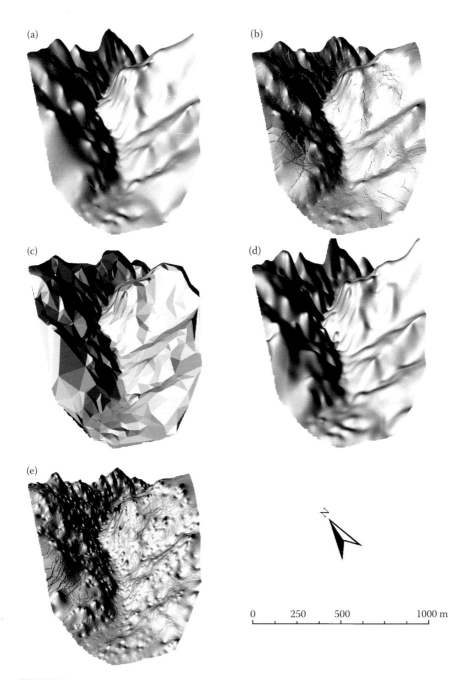

FIGURE 9.5
Shaded relief maps of constructed DEMs by (a) HASM, (b) kriging, (c) TIN, (d) spline, and (e) IDW.

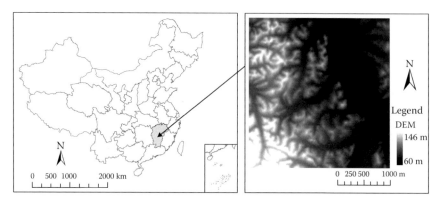

FIGURE 9.6
Location of Qian-Yan-Zhou Experimental Station for Red Soil and Hilly Land.

scanned topographical map on a spatial scale of 1:5000 is vectorized. Shaded relief maps of the simulated DEMs are developed to represent the terrain relief of the Qian-Yan-Zhou. All the shaded relief maps are made in the same azimuth and altitude angle (Figure 9.7). The Azimuth angle of the light source is 180°, which is from south to north. The azimuth is expressed in positive degrees from 0° to 360° and measured clockwise from the north. The altitude angle of the light source above the horizon is 45°. The slope is expressed in positive degrees, with 0° at the horizon and 90° directly overhead.

The 3D shaded relief maps show that HASM has a better simulation result (Figure 9.7a) than those of the classical methods. Spline has a serious oscillation problem (Figure 9.7b), especially in the right bottom area (Figure 9.7b). TIN has an obvious peak truncation and pit-fill problem (Figure 9.7c). IDW and kriging make the surface of the Qian-Yan-Zhou very coarse and the hillocks like sand dunes (Figure 9.7d and e).

9.5 Discussion and Conclusions

9.5.1 SRTM Void Filling

The comparative analyses indicates that the introduction of SRTM30 data as the auxiliary data might create great errors when void areas have a high landform complexity or sizes of void areas are insufficiently big. HASM results always have the highest accuracy compared with all the classic methods, whether SRTM30 data are added or not, landform complexity is higher or lower, or void area is larger or smaller in all the nine regions of the three topographic types. However, HASM accuracy is also impacted by both landform complexity and the size of the void area (Table 9.5).

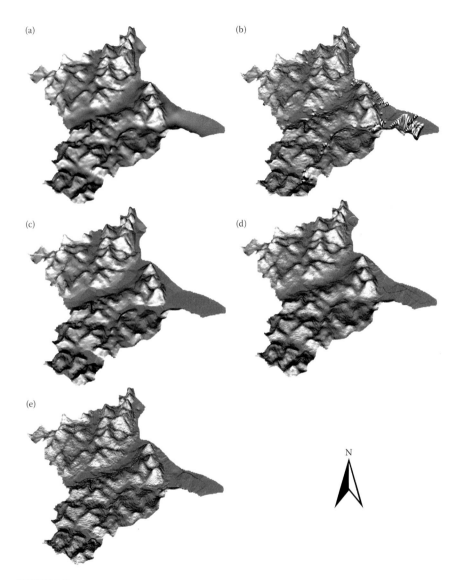

FIGURE 9.7
A comparison of DEMs developed by different methods: (a) by HASM, (b) by spline, (c) by TIN, (d) by kriging, and (e) by IDW.

We design a simple impact index of landform complexity and size of void area

$$SI_i = w \cdot \frac{SD_i}{\sum\limits_{i=1}^{9} SD_i} + (1-w) \cdot \frac{SV_i}{\sum\limits_{i=1}^{9} SV_i} \qquad (9.5)$$

TABLE 9.5

Accuracy Analyses of HASM Void-Filling

Geomorphologic Type	Verification Region	Region Number	Standard Deviation (m)	Void Count	RMSE (m) Without SRTM30	RMSE (m) With SRTM30
Hilly area	A	1	263	520	51.6	62.0
	B	2	199	177	34.0	34.9
	C	3	61	442	5.3	5.1
Plateau	D	4	455	712	30.2	29.2
	E	5	36	3230	7.2	4.0
	F	6	214	84,891	63.1	44.9
Mountainous area	G	7	405	2109	86.0	95.9
	H	8	742	104,514	277.7	103.4
	I	9	670	197,708	481.4	182.2

where $i = 1, 2, \ldots, 9$ are the numerals corresponding to regions A, B, ..., I; SD_i and SV_i are respectively the standard deviation and the size of void area in the ith region; w is a weight and $w \leq 1$.

Our calculation (Table 9.5) shows that for HASM, results without SRTM30 when $w = 0.5$, SI_i and $RMSE_i$ has the highest correlation coefficient, $R_i^2 = 0.93$. They have the following regression relation:

$$RMSE_i = 1272.1 SI_i - 26.2 \tag{9.6}$$

For HASM results with SRTM30, SI_i and $RMSE_i$ reach the highest correlation coefficient, $R_i^2 = 0.80$, when $w = (2/3)$. The linear regression equation can be formulated as

$$RMSE_i = 486.36 SI_i + 8.36 \tag{9.7}$$

This means that landform complexity and the size of the void area have equal impact on the accuracy of HASM void-filling when the data surrounding the void are only used to fill the void. When the auxiliary data are added, landform complexity has twice the impact on the void-filling error than the size of the void area.

In addition to landform complexity and the size of the void area, the slope aspect has a major influence on the elevation error (Bourgine and Baghdadi, 2005). The errors have a strong correlation with slope and certain aspect values, taking into account the fact that slope and aspect considerably improved the accuracy of the SRTM3 data for terrain with slope values greater than 10° (Gorokhovich and Voustianiouk, 2006). SRTM elevation data exhibited a significant sensitivity to the vertical structure of vegetation and SRTM held considerable potential for developing statistically significant inversion models for deriving estimates of vegetation canopy height (Kellndorfer et al., 2004). A significant positive bias in the data was observed in areas where extensive tree and/or shrub cover was present (Walker et al., 2007). In vegetated terrain, the DEMs represented neither the "bald earth" surface nor the canopy top surface, but some elevation between the two (Hofton et al., 2006).

The combined impact of landform complexity, size of void area, slope and aspect, as well as vegetation will be studied after we have obtained the potential coefficients of the gravitational potential model in China.

9.5.2 DEMs Based on Height-Sampling Points and on Topological Maps

Although the accuracy of HASM was higher than that of TIN, spline, and IDW, the difference was less distinct than that of the numerical tests. In other words, HASM lost accuracy in the practical applications, which may have been caused by location differences between the sampling points and the corresponding central points of the lattices of the simulated surfaces. In the real world example, most of the sampling points were not located at the centers of their corresponding lattices, and the nearest neighbor method was used to neglect the location differences. This accuracy loss problem could be resolved in two alternative ways: (1) sampling heights at the centers of the lattices through a careful design or (2) establishing a continuous Taylor expansion of DEM, $z = f(x, y)$, on the basis of topographical characteristics.

In short, the highly accurate results obtained through the successful operation of HASM on SRTM elevation data, height-sampling points, and the topological map demonstrate that HASM can provide an alternative solution to the DEM error problem.

References

Aguilar, F.J., Agüera, F., Aguilar, M.A., and Carvajal, F. 2005. Effects of terrain morphology, sampling density, and interpolation methods on grid DEM accuracy. *Photogrammetric Engineering and Remote Sensing* 71: 805–816.

Atkinson, P.M. 2002. Surface modelling: What's the point? *Transactions in GIS* 6(1): 1–4.

Berry, P.A.M., Garlick, J.D., and Smith, R.G. 2007. Near-global validation of the SRTM DEM using satellite radar altimetry. *Remote Sensing of Environment* 106: 17–27.

Bolongaro-Crevenna, A., Torres-Rodríguez, V., Sorani, V., Frame, D., and Ortiz, M.A. 2005. Geomorphometric analysis for characterizing landforms in Morelos State, Mexico. *Geomorphology* 67: 407–422.

Bourgine, B., and Baghdadi, N. 2005. Assessment of C-band SRTM DEM in a dense equatorial forest zone. *Comptes Rendus Geoscience* 337: 1225–1234.

Brown, C.G., Sarabandi, K., and Pierce, L.E. 2005. Validation of the shuttle radar topography mission height data. *IEEE Transactions on Geoscience and Remote Sensing* 43 (8): 1707–1715.

Buchroithner, M. 1989. Stereo-viewing from space. *Advances in Space Research* 19: 29–40.

Carter, J.R. 1988. Digital representations of topographic surfaces. *Photogrammetric Engineering and Remote Sensing* 54: 1577–1580.

Chaplot, V., Darboux, F., Bourennane, H., Leguedois, S., Silvera, N., and Phachomphon, K. 2006. Accuracy of interpolation techniques for the derivation of digital elevation models in relation to landform types and data density. *Geomorphology* 77: 126–141.

Correc, Y. and Chapuis, E. 1987. Fast computation of Delaunay triangulations. *Advanced Engineering Software* 9(2): 77–83.

Etzelmuller, B. 2000. On the quantification of surface changes using grid-based digital elevation models (DEMs). *Transitions in GIS* 4: 129–143.

Falorni, G., Teles, V., Vivoni, E. R., Bras, R. L., and Amaratunga, K. 2005. Analysis and characterization of the vertical accuracy of digital elevation models from the shuttle radar topography mission. *Journal of Geophysical Research* 110: F02005.

Fisher, P.F. and Tate, N.J. 2006. Causes and consequences of error in digital elevation models. *Progress in Physical Geography* 30(4): 467–489.

Florinsky, I.V. 2002. Errors of signal processing in digital terrain modeling. *International Journal of Geographical Information Science* 16: 475–501.

Ganas, A., Pavlides, S., and Karastathis, V. 2005. DEM-based morphometry of range-front escarpments in Attica, central Greece, and its relation to fault slip rates. *Geomorphology* 65: 301–319.

Gao, J. 1997. Resolution and accuracy of terrain representation by grid DEMs at a micro-scale. *International Journal of Geographical Information Science* 11: 199–212.

Glenn, N.F., Streutker, D.R., Chadwick, D.J., Thackray, G.D., and Dorsch, S.J. 2006. Analysis of LiDAR-derived topographic information for characterizing and differentiating landslide morphology and activity. *Geomorphology* 73: 131–148.

Goodchild, M.F. 1982. The fractal Brownian process as a terrain simulation model. *Modelling and Simulation* 13: 1133–1137.

Gorokhovich, Y. and Voustianiouk, A. 2006. Accuracy assessment of the processed SRTM-based elevation data by CGIAR using field data from USA and Thailand and its relation to the terrain characteristics. *Remote Sensing of Environment* 104: 409–415.

Grohman, G., Kroenung, G., and Strebeck, J. 2006. Filling SRTM voids: The delta surface fill method. *Photogrammetric Engineering and Remote Sensing* 72(3): 213–216.

Hayakawa, Y.S. and Oguchi, T. 2006. DEM-based identification of fluvial knickzones and its application to Japanese Mountain Rivers. *Geomorphology* 78: 90–106.

Helm, A., Braun, A., Eickschen, S., and Schune, T. 2002. Calibration of the shuttle radar topography mission X-SAR instrument using a synthetic altimetry data model. *Canadian Journal of Remote Sensing* 28(4): 573–580.

Hofton, M., Dubayah, R., Blair, J. B., and Rabine, D. 2006. Validation of SRTM elevations over vegetated and non-vegetated terrain using medium footprint Lidar. *Photogrammetric Engineering and Remote Sensing* 72(3): 279–285.

Hugentobler, M., Purves, R.S., and Schneider, B. 2005. Evaluating methods for interpolating continuous surfaces from irregular data: A case study. In *Developments in Spatial Data Handling*, ed. P.F. Fisher, 109–124. Berlin: Springer.

Huang, Z. and Lees, B. 2005. Representing and reducing error in natural-resource classification using model combination. *International Journal of Geographical Information Science* 19(5): 603–621.

Hutchinson, M.F. and Dowling, T.I. 1991. A continental hydrological assessment of a new grid-based digital elevation model of Australia. *Hydrological Processes* 5: 45–58.

Jarvis, A., Reuter, H.I., Nelson, A., and Guevara, E. 2008. *Hole-filled SRTM for the globe Version 4*. Available from the CGIAR-CSI SRTM 90 m Database: http://srtm.csi.cgiar.org.

Katzil, Y. and Doytsher, Y. 2000. Height estimation methods for filling gaps in gridded DTM. *Journal of Surveying Engineering* 126(4): 145–162.

Kellndorfer, J., Walker, W., Pierce, L., Dobson, C., Fites, J.A., Hunsaker, C., Vona, J., and Clutter, M. 2004. Vegetation height estimation from Shuttle Radar Topography Mission and National Elevation Data sets. *Remote Sensing of Environment* 93: 339–358.

Li, X.W., Guo, H.D., Wang, C.L., Li, Z., and Liao, J.J. 2003. DEM generation in the densely vegetated area of Hotan, north-west China using SIR-C repeat pass polarimetric SAR interferometry. *International Journal of Remote Sensing* 24(14): 2997–3003.

Lin, Z. and Oguchi, T. 2006. DEM analysis on longitudinal and transverse profiles of steep mountainous watersheds. *Geomorphology* 78: 77–89.

Lo, C.P. and Yeung, A.K.W. 2002. *Concepts and Techniques of Geographic Information Systems*. Upper Saddle River, New Jersey: Prentice-Hall.

Luedeling, E., Siebert, S., and Buerkert, A. 2007. Filling the voids in the SRTM elevation model—a TIN-based delta surface approach. *ISPRS Journal of Photogrammetry and Remote Sensing* 62(4): 283–294.

Mitasova, H., Overton, M., and Harmon, R.S. 2005. Geospatial analysis of a coastal sand dune field evolution: Jockey's Ridge, North Carolina. *Geomorphology* 72: 204–221.

Oksanen, J. and Sarjakoski, T. 2005. Error propagation of DEM-based surface derivatives. *Computers and Geosciences* 31: 1015–1027.

Petrie, G., Al-Roussan, N., El-Niweiri, A. H. A., Li, Z., and Valadan Zoej, M. J. 1997. Topographic mapping of arid and semi-arid areas in the Red Sea region from stereo space imagery. *Advances in Remote Sensing* 5: 11–29.

Prima, O.D.A., Echigo, A., Yokoyama, R., and Yoshida, T. 2006. Supervised landform classification of Northeast Honshu from DEM-derived thematic maps. *Geomorphology* 78: 373–386.

Podobnikar, T. 2005. Production of integrated digital terrain model from multiple data sets of different quality. *International Journal of Geographical Information Science* 19: 69–89.

Raaflaub, L.D. and Collins, M.J. 2006. The effect of error in gridded digital elevation models on the estimation of topographic parameters. *Environmental Modelling and Software* 21(5): 710–732.

Rabus, B., Eineder, M., Roth, A., and Bamler, R. 2003. The Shuttle Radar Topography Mission—a new class of digital elevation models acquired by space-borne radar. *ISPRS Journal of Photogrammetry and Remote Sensing* 57: 241–262.

Reuter, H.I., Nelson, A., and Jarvis, A. 2007. An evaluation of void-filling interpolation methods for SRTM data. *International Journal of Geographical Information Science* 21(9): 983–1008.

Rodriguez, E., Morris, C.S., and Belz, J.E. 2006. A global assessment of the SRTM performance. *Photogrammetric Engineering and Remote Sensing* 72(3): 249–260.

San, B.T. and Suezen, M.L. 2005. Digital elevation model (DEM) generation and accuracy assessment from ASTER stereo data. *International Journal of Remote Sensing* 26(22): 5013–5027.

Saraf, A.K., Mishra, B.P., Choudhury, S., and Ghosh, P. 2005. Digital elevation model (DEM) generation from NOAA-AVHRR night-time data and its comparison with USGS-DEM. *International Journal of Remote Sensing* 26(18): 3879–3887.

Schneider, B. 2005. Extraction of hierarchical surface networks from bilinear surface patches. *Geographical Analysis* 37: 244–263.

Shi, W.J., Liu, J.Y., Song, Y.J., Du, Z.P., Chen, C.F., and Yue, T.X. 2009. Surface modeling of soil pH. *Geoderma* 150(1–2): 113–119.

Shi, W.Z., Li, Q.Q., and Zhu, C.Q. 2005. Estimating the propagation error of DEM from higher-order interpolation algorithms. *International Journal of Remote Sensing* 26: 3069–3084.

Sibon, R. 1978. Locally equiangular triangulations. *The Computer Journal* 21(3): 243–245.

Smith, B. and Sandwell, D. 2003. Accuracy and resolution of shuttle radar topography mission data. *Geophysical Research Letters* 30(9): 1467.

Sun, G., Ranson, K.J., Khairuk, V.I., and Kovacs, K. 2003. Validation of surface height from shuttle radar topography mission using shuttle laser altimeter. *Remote Sensing of Environment* 88(4): 401–411.

Toutin, Th. 2001. Elevation modeling from satellite visible and infrared (VIR) data. *International Journal of Remote Sensing* 22(6): 1097–1125.

United States Geological Survey. 1997. *Standards for Digital Terrain Models.* Reston, Virginia: United States Geological Survey.

Unwin, D.J. 1995. Geographical information systems and the problems of 'error and uncertainty'. *Progress in Human Geography* 19: 549–558.

van Zyl, J.J. 2001. The Shuttle Radar Topography Mission (SRTM): A breakthrough in remote sensing of topography. *Acta Astronautica* 48(5–12): 559–565.

Walker, W.S., Kellndorfer, J.M., and Pierce, L.E. 2007. Quality assessment of SRTM C- and X-band interferometric data: Implications for the retrieval of vegetation canopy height. *Remote Sensing of Environment* 106: 428–448.

Walsh, S.J., Lightfoot, D.R., and Butler, D.R. 1987. Recognition and assessment of error in geographical information systems. *Photogrammetric Engineering and Remote Sensing* 53: 1423–1430.

Wilson, J.P. and Gallant, J.C. 2000. *Terrain Analysis: Principles and Applications.* New York: Wiley.

Wise, S. 2000. GIS data modeling—lessons from the analysis of DTMs. *International Journal of Geographical Information Science* 14: 313–318.

Wolf, P.R. 1974. *Elements of Photogrammetry.* New York: McGraw-Hill.

Yue, T.X., Du, Z.P., Song, D.J., and Gong, Y. 2007. A new method of surface modeling and its application to DEM construction. *Geomorphology* 91(1–2): 161–172.

Yue, T.X., Song, D.J., Du, Z.P., and Wang, W. 2010. High accuracy surface modeling and its application to DEM generation. *International Journal of Remote Sensing* 31(8): 2205–2226.

Zhao, S.Q. 1986. *Physical Geography of China.* New York: John Wiley & Sons.

Zhou, Q.M. and Liu, X.J. 2002. Error assessment of grid-based flow routing algorithms used in hydrological models. *International Journal of Geographical Information Science* 16(8): 819–842.

Zhu, C.Q., Shi, W.Z., Li, Q.Q., Wang, G.X., Cheung, T.C.K., Dai, E.F., and Shea, G.Y.K. 2005. Estimation of average DEM accuracy under linear interpolation considering random error at the nodes of TIN model. *International Journal of Remote Sensing* 26(24): 5509–5523.

10

Surface Modeling of Population Distribution*

10.1 Introduction

Reliable information on population distribution is essential for a wide range of applications in both the science and policy domains. Without an adequate knowledge of where people live and spend their time, it is all but impossible to model human activities, to plan service provision, to estimate pressures on the environment, or to assess human exposures and risks to human health (Briggs et al., 2007).

The population distribution analysis method represents a particularly influential line of enquiry (Woods and Rees, 1986). It traces back the statistical revolution that began in 1750 in Europe and 1790 in America (Jefferson, 1909), which was the result of combining the regular population censuses with vital registration data organized and manipulated within a framework of administrative units (Cullen, 1975; Cassedy, 1984). Approaches to estimating the population distribution can be divided into four categories: (1) even-distribution, (2) areal interpolation, (3) kernel models, and (4) surface modeling.

Most publicly available demographic data sets such as those generated by the population statistics and census are aggregated to administrative areal units (Mennis, 2003). The even-distribution method transformed population data from census to grid under the assumption that the population was evenly distributed within each census district (Tobler et al., 1997). Even distribution of population using administrative areal units gives the impression that the population is distributed homogeneously throughout each areal unit, even when portions of the region are actually uninhabited. It cannot sufficiently represent the underlying geographical distribution of the population, because it is reported through aggregating individual population counts in irregular areal units, which means the results in some cases are geographically meaningless (Wu and Murray, 2005). Another problem with even-distribution population data is related to incompatible spatial information layers (Goodchild et al., 1993). Different departments and agencies collect and distribute data in varying zonal arrangements. As a consequence, issues arise with regional analysis and modeling, in which multiple data

---

* Dr. Ying-An Wang is a major collaborator of this chapter.

sources must be integrated before analysis can be implemented. The modifiable areal unit problem may exist when utilizing such data in geographical applications, while the relationship between variables may only be valid for one particular zonal arrangement and scale.

Areal interpolation between one partitioning of geographical space and another remains an important issue, particularly in terms of population counts and related statistics, which are often required in order to compute incidence ratios (Langford, 2006, 2007). Areal interpolation is the transformation of data between different sets of areal units (Goodchild and Lam, 1980). The set of zones for which data are available is termed source zones. The second set of zones, for which estimates need to be derived, is termed target zones (Goodchild et al., 1993; Moxey and Allanson, 1994). However, these methods fail to recognize that population patterns are mostly disjunct with settlement centers separated by widely dispersed vegetation areas.

Areal interpolation is referred to as dasymetric mapping when ancillary information is used (Wright, 1936). Dasymetric mapping is well placed to overcome the inertia associated with the even-distribution method. In its most basic form, remotely sensed imagery is used to distinguish populated areas from unpopulated areas. Census population counts are then redistributed internally within each source zone so that they lie only within populated subzones, thereby addressing the improbable uniform distribution assumption (Langford, 2007). Dasymetric mapping, for example, redistributing census with the help of satellite imagery, is a significant improvement on simple areal interpolation methods. A regression model was developed by examining the correlation between population counts from census and landcover types (Yuan et al., 1997; Lo, 2008). Impervious surface, for example, any material prohibiting the infiltration of water into soil, is a major component of urban infrastructure. Impervious surface fraction, calculated as the proportion of impervious surface to investigation region over a small area, has been found to reveal more information about built-up areas than land-use and land-cover classification (Ji and Jensen, 1999). The relationship between built-up surface and population density were quantified by integrating remote sensing and GIS technologies, which showed that the proportion of built-up surface was closely related to population density (Mesev, 1998; Yin et al., 2005). Urban population distribution was estimated by means of the co-kriging method combined with residential impervious surface fraction, a method derived from remotely sensed imagery (Wu and Murray, 2005). Area-to-point residual kriging was proposed to improve interpolation accuracy by disaggregating population from census units to the land-use zones within them (Liu et al. 2008). The dasymetric-based approach integrates multispectral satellite imagery with a recently built mailing information database with which residential areas can be identified to simulate urban population distribution (Langford et al., 2008).

Kernel models such as monocentric models (Clark, 1951; Newling, 1969; Parr, 1985; Bracken and Martin, 1989; Wang, 1998; Baumont et al., 2004; Harris and Chen, 2005) and polycentric models (Gordon et al., 1986; Small and Song, 1994; McMillen and McDonald, 1998; Wu, 1998; Loibl and Toetzer, 2003) involve the redistribution of census data, which is governed by a distance decay function within a locally adaptive kernel and transforms vector point data to a near-continuous raster surface of cells (Cressman, 1959; Martin, 1996). The monocentric urban form, agglomerating industry, and employment in a single center and packing the population around the center and along radiating transport network, was popular prior to the 1950s, but has since been eroded by suburbanization and decentralization (Berry and Kim, 1993). The development of urban regions is no longer based on a single development pole such as the city centre. Recent developments are most obviously characterized by the emergence of new centers within the suburban areas themselves (Mueller and Rohr-Zaenker, 1995). The method of turning monocentric urban regions into polycentric urban structures that contain at least two kernels has become much more universal since 1960 (Hall, 1999). Across the developed world, metropolitan regions have become increasingly polycentric (Bontje, 2005). A number of cubic-degree polynomials, termed cubic-spline density functions, were used to describe these polycentric urban structures (Anderson, 1982, 1985; Zheng, 1991). A spatial regression approach was employed to simulate population density in monocentric and polycentric cities using a Minkowskian distance metric (Griffith and Wong, 2007).

Surface modeling provides a grid data structure for the explicitly spatial analysis of population distribution (Martin, 1998). Surface modeling of population distribution (SMPD) is aimed at formulating population in a regular grid system, in which each grid cell contains an estimate of the total population representative of that particular location (Tobler, 1979; Goodchild et al., 1993; Langford and Unwin, 1994; Yue et al., 2003, 2005a,b, 2010). Population distribution is expressed as a continuously varying surface by surface modeling and can be measured at any given location. Representing data in grid form has at least four advantages: (1) surface modeling may offer a more accurate cartographic representation of population distribution than the conventional choropleth maps (Langford and Unwin, 1994); (2) regular grids can be easily re-aggregated to any areal arrangement required; (3) producing population data in grid form is one way of ensuring compatibility between heterogeneous data sets; and (4) converting data into grid form avoids some of the problems imposed by artificial political boundaries (Martin and Bracken, 1991; Deichmann, 1996). Surface modeling consists of three basic steps: (1) a surface of weighting factors is created in a regular grid system for the study areas; (2) the basic weights derived in the first step are adjusted by using ancillary data sources, and (3) the total population in the study areas is distributed to the corresponding grids in proportion to the weights constructed in the previous steps.

In recent years, Grid Population of the World (GPW), LandScan 2000 Global Population Database and the co-Kriging method were developed to reference population in a uniform coordinate system. GPW simply allocates the total population on an administrative unit over grid cells proportionally under the assumption that the population is distributed evenly over the administrative unit (Tobler et al., 1997). LandScan 2000 distributes best available census counts to grid cells in terms of probability coefficients based on publicly available databases offering worldwide coverage of roads, slopes, land cover, and nighttime lights (Dobson et al., 2000; Sutton et al., 2003). The Co-kriging method was developed to interpolate population density by modeling the spatial correlation and cross-correlation of population on impervious surface fraction (Wu and Murray, 2005). However, GPW's weakness is the assumption that the uniformly distributed population is within the administrative unit; this smoothes local variability and renders the results geographically meaningless. LandScan 2000 usually loses detailed biophysical information in producing land-cover data from remotely sensed images, and the limited land-cover types are too coarse to estimate detailed population density. The Co-kriging method is based on quantifying impervious surface from remotely sensed data, which may include a great deal of nonresidential information such as roads and public squares.

In this chapter, we introduce a newly developed method for SMPD. It is applied to simulating change trends and developing scenarios of population distribution on national and local levels.

10.2 SMPD Method

Sir Isaac Newton propounded his Law of Universal Gravitation in 1687, that is, any two bodies attract each other in proportion to the product of their masses and in inverse proportion to the square of their distance. In analogy to the physical gravity model, the concept of potential population distribution was developed, which is a measure of average accessibility of a given location with respect to the size and location of other features (Plane and Rogerson, 1994; Deichmann, 1996). The influence of a city upon a grid cell is assumed to be proportional to the city's size, weighted inversely by the distance of separation between the grid cell and the city. Within a given threshold distance, the potential population distribution is formulated as

$$p_{i,j} = \sum_{k=1}^{M} \frac{S_k}{(d_{i,j,k})^a} \qquad (10.1)$$

where $p_{i,j}$ is the population in grid cell (i, j); S_k is the size of city k; $d_{i,j,k}$ is the distance between grid cell (i, j) and city k; M is the total number of cities within the given threshold distance; and a is the exponent to be simulated.

In fact, the spatial distribution of population is greatly influenced by natural factors. For instance, population distribution in China shows three obvious characteristics: (1) that most people inhabit the area along the coast, (2) that most people inhabit areas with an elevation lower than 500 m, and (3) that most people inhabit areas with a warm-wet climate (Zhang, 1997). The influence of each of these factors is not straightforward. It depends on other factors such as the spatial distribution of transport infrastructures and where the cities are located. For instance, when a railway traverses the Qinghai-Xizang plateau of China, or when a new city appears, the distribution of population is immediately modified. The growth of various means of transport is fundamental to the distribution of population. In China, there is an evident contrast between regions already reached by modern transportation and those less affected by it. The former develops rapidly, in which great cities expand at an accelerating pace. Transport clearly favors the growth of cities. Cities are one of the major elements in the distribution of population. In general, it is possible to distinguish between a central and more densely peopled nucleus and a peripheral residential area as these are extended star-like along the lines of transportation. The influence of cities on the distribution of population is exercised directly through the concentration of people in cities and also indirectly through the swarms of suburbanites scattered over their peripheral countryside in dormitory suburbs or satellite towns (Deichmann, 1996).

Therefore, in addition to the size of city k and the distance between grid cell (i, j) and city k embodied in the formulation (10.1), land cover, net primary productivity (NPP), transport infrastructures, and elevation are involved in SMPD. SMPD is formulated as a transformation between computational domain (i, j) and physical domain $(i, j, MSPD_{ij}(t))$ (Yue et al., 2003, 2005a,b),

$$SMPD_{i,j}(t) = G(n,t) \cdot W_{i,j}(t) \cdot f_1(Tran_{i,j}(t)) \cdot f_2(NPP_{i,j}(t))$$

$$\times f_3(DEM_{i,j}(t)) \cdot f_4(Ct_{i,j}(t)) \tag{10.2}$$

where t is the time variable; $G(n, t)$ is a parameter determined by total population in administrative division n where grid cell (i, j) is located; $W_{i,j}(t)$ is an indicative factor of water area; $f_1(Tran_{i,j}(t))$ is a function determined by the condition of transport infrastructures of grid cell (i, j); $f_2(NPP_{i,j}(t))$ is a function determined by the condition of the NPP of grid cell (i, j); $f_3(DEM_{i,j}(t))$ is a function determined by the elevation of grid cell (i, j); $f_4(Ct_{i,j}(t))$ is a function determined by the contribution of cities to population density in grid cell (i, j); $Ct_{i,j}(t) = \sum_{k=1}^{M(t)}((S_k(t))^{a_1}/d_{i,j,k}(t))$; $S_k(t)$ is size of the kth city; a_1 is an exponent to be simulated; $M(t)$ is the total number of cities in a defined search radius; $d_{i,j,k}(t)$ is the distance from grid cell (i, j) to the core grid cell that has the highest population density in the kth city; $SMPD_{i,j}(t)$ is a residence-based measure, which represents the population density that should be there in terms of the natural, economic, and social environments where the grid cell (i, j) is located.

10.3 SMPD-Based Population Simulation on a National Level of China

10.3.1 Background of Major Data Layers

The major data layers that are matched with their corresponding SMPD variables include NPP, elevation, water systems, city distribution and urbanization, transport infrastructure condition and planning, and total population in every province, and population projection in China. The elevation is a natural factor and has a slow change with time so that it is a spatial variable and could be regarded as a temporal constant within the last 100 years. Although NPP is based on climate and soil, it could be modified by human activities so that it becomes a spatial and temporal variable. Urbanization and transport infrastructure distribution are spatial and temporal variables, which are determined by both natural factors and human activities, and have changed rapidly within China in the last 100 years.

10.3.1.1 Elevation

Elevation is an important variable of human population distribution because most human settlements occur on lower elevation in China. For instance, the area of plains and hills with an elevation lower than 500 m accounts for about 28% of the total land area of China, and yet 74% of the total population inhabit these areas (Zhang, 1997). The terrestrial parts of China are broadly divided into three areas from the Qinghai-Xizang Plateau eastwards (Zhao, 1986). The lofty and extensive Qinghai-Xizang Plateau is the first great topographic area. Its eastern and northern borders roughly coincide with the 3000 m contour line. Its elevation is between 4000 and 5000 m, hence it's titled, the roof of the world.

From the eastern margin of Qinghai-Xizang Plateau eastward to the Da Hinggan-Taihang-Wushan mountains lies the second great topographic area. It mainly comprises of plateaus and basins with elevations between 1000 and 2000 m, such as the Nei Mongol, Ordos, Loess, and Yunnan-Guizhou plateaus and the Tarim, Junggar, and Sichuan basins.

From the eastern margin of the second area eastward to the coast is the third great topographic area. The largest plains of China, the North China Plain, the Northeast China Plain, and the middle and lower Changjiang Plain are located in this area. All three generally lie at an elevation of below 200 m (Figure 10.1).

10.3.1.2 Water Systems

Water systems in China that include rivers and lakes (Figure 10.2) have a considerable effect on human population distribution. Rivers can be divided

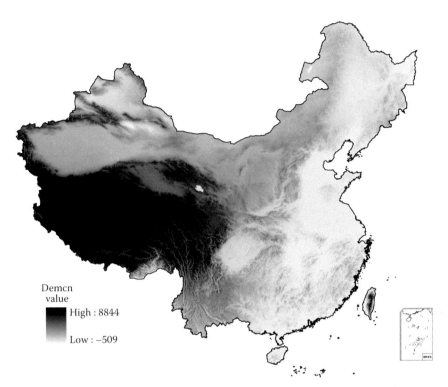

Demcn
value

High : 8844

Low : −509

FIGURE 10.1
Digital elevation model of China.

into those that discharge into oceans and inland rivers that begin in mountainous areas and disappear in conoplain or flow into inland lakes. The oceanic system can be subdivided into Pacific, Indian, and Arctic drainage basins, accounting for 64% of the total land area in China. The inland system includes a few perennial rivers and has large tracts with no runoff, accounting for 36% of the total land area in China (Zhao, 1986).

The Pacific oceanic subsystem accounts for 88.9% of the total oceanic system; rivers flowing into the Pacific ocean include the Heilongjiang river, the Liaohe river, the Luanhe river, the Haihe river, the Yellow river, the Yangtze river, the Qiantangjiang river, the Zhujiang river, the Yuanjiang river, and the Lancangjiang river. The Indian oceanic subsystem accounts for 10.3% of the total oceanic system; rivers flowing into the Indian Ocean have only their upper reaches located in China, including the Nujiang river, the Longchuan river, the Yarlu Zangbujiang river, and the Shiquan river. The Arctic oceanic subsystem has only one large river, the Ertix river. The Ertrix river is a tributary of the Ob river, and accounts for 0.8% of the total oceanic system.

The inland river system is mostly located in arid and semiarid Northwest China, the northwestern Qinghai-Xizang Plateau, and the central Ordos plateau. The great divide between inland and oceanic river systems runs

FIGURE 10.2
Water systems of China.

essentially northeast to southwest, starting from the southern end of the Da Hinggan mountains, passing through Yanshan, Helanshan, the eastern Qilian, Bayan Har, Nyainqentanglha, and Gangdise mountains, and up to the southwestern margin of the Qinghai-Xizang Plateau.

Five major lake regions can be identified: the northeast lake region; the northwest lake region; the Qinghai-Xizang lake region; the eastern lake region; and the southwest lake region. They include approximately 2800 natural lakes, each with an area greater than 1 sq. km, and many reservoirs that serve as artificial lakes. In the northeast lake region, there are large tracts of marsh and numerous small lakes with a total lake area of 3722 sq. km. The northwest lake region is located in extensive arid and semiarid northwest China; there are numerous inland lakes with a total area of 22,500 sq. km. In the Qinghai-Xizang lake region, the lakes are fed by melting snow and the lake basins are mainly controlled by tectonic structure, with a total lake area of 30,974 sq. km. The eastern lake region is mostly located in the middle and lower reaches of Changjiang, Huaihe, and Zhujiang rivers, with a total lake area of 22,161 sq. km. The southwest lake region is located in the Yunnan-Guizhou plateau, mostly in karsts area, with a total lake area of 1188 sq. km.

10.3.1.3 Land Use and Net Primary Productivity

Land use is a good indicator of spatial human population distribution (Tian et al., 2005). In most regions, population ranges from extremely low density in desert, water, wetlands, ice, or tundra land, to high density in developed or urban land, between which cultivated lands would range (Dobson et al., 2000). The land-use database of China developed during the 1980s and 1990s (Figure 10.3), is derived from Landsat Thematic Mapper (TM) imagery at 30-m resolution (Liu et al., 2003).

NPP is another good indicator of spatial population distribution. NPP is the difference between accumulative photosynthesis and accumulative auto-trophic respiration by green plants per unit time and space (Lieth and Whittaker, 1975). Terrestrial ecosystem modeling (Liu, 2003) is employed for analyzing the spatial distribution of NPP in China. It integrates different types of data that include land-use change data, daily climatic data and soil data. The analysis results show that the mean annual NPP of terrestrial eco-systems in China was 3.588×10^{15} g C/year in the 1990s, far greater than 0.094×10^{15} g C/year, the mean annual NPP of terrestrial ecosystems in the 1980s. In other words, the mean annual NPP increased by 0.49 g C m^{-2}/year during the 20 years.

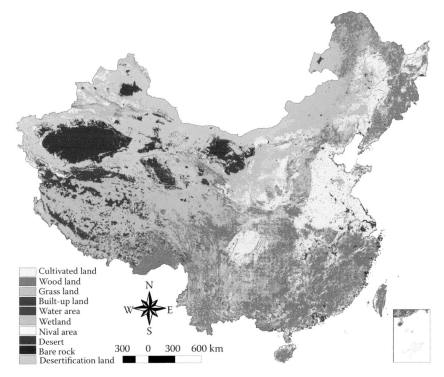

FIGURE 10.3
Land use in 2000.

For China, the general trend is that from southeast to northwest the NPP becomes smaller and smaller gradually (Figure 10.4). Most of the NPP is distributed in the East of the rainfall line where the annual precipitation is 410 mm, with the exception of the southern slopes of the Tianshan Mountains and the Altai Mountains in Xinjiang, where the NPP is higher. The maximum NPP appears in Xiaoxinganling mountain and Changbai mountain in the northeast of China, Yunnan-Guizhou plateau, Guangxi, Hainan, Chongqin, and the provinces along the middle and lower reaches of the Yangtze river.

In terms of land-use types, on average, the NPP of shrub and open forest is 1071 g C m^{-2} year^{-1}, the NPP of evergreen broad-leaved forests is 975 g C m^{-2} year^{-1}, the NPP of deciduous broad-leaved forests is 928 g C m^{-2} year^{-1}, the NPP of coniferous and broad-leaved mixed forests is 870 g C m^{-2} year^{-1}, the NPP of farmland systems is 752 g C m^{-2} year^{-1}, the NPP of evergreen coniferous forests is 587 g C m^{-2} year^{-1}, the NPP of deciduous coniferous forests is 585 g C m^{-2} year^{-1}, and the NPP of grassland is 271 g C m^{-2} year^{-1} (Figure 10.4) (Liu, 2003).

10.3.1.4 Spatial Distribution of Cities and Urbanization

Urbanization describes the process of the population of cities becoming concentrated. Spatial distribution of cities and proximity to cities are essential

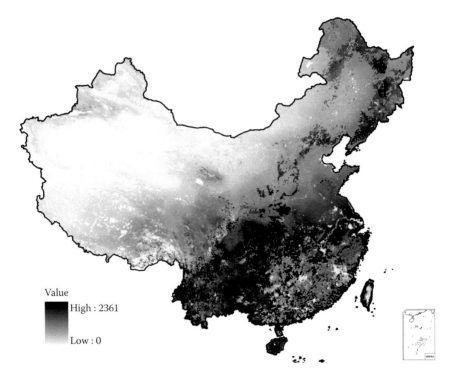

Value

High : 2361

Low : 0

FIGURE 10.4
Spatial distribution of the mean NPP in the 1990s (unit: g C m^{-2}/year).

factors for human population distribution in China. In recent times, the density of population in cities in eastern China has become much higher than that of cities in western China. Between 1843 and 1893, the proportion of urban population in China grew relatively slowly, gradually increasing from 5.1% to 6.0% on an average; in the lower reaches of the Yangtze river the proportion of urban population increased from 7.4% to 10.6%; in the coastal area of South China the proportion increased from 7.0% to 8.7%; while in the inland area the proportion of urban population oscillated between 4.0% and 5.0%. From 1895 to 1931, in areas along the coast and Yangtze River, North and Northeast China cities developed rapidly, while inland area cities developed much slowly, sometimes even coming to a standstill. In the early 1930s, the proportion of urban population was about 9.2% in China. From 1931 to 1949, the turbulent and unstable political situation led to slow population growth in China and the urban population proportion only increased to 10.6% (Zhang, 1997). Cities in China spatially concentrate in coastal areas, especially the Yangtze River Delta, the Pearl River Delta, and the Beijing-Tianjin-Tangshan area. In 2000, 42.1% of the 667 major cities of China were distributed in eastern China where area accounts for 9.5% of the whole area of China; 34% were distributed in middle China where the area accounts for 17.4%; and 23.8% were distributed in western China where the area accounts for 70.4% (National Bureau of Statistics of the People's Republic of China, 2001). The distribution densities of cities in eastern and middle China were respectively 13.1 times and 5.8 times that of western China (Figure 10.5). The urban population proportion was 36.22% in 2000.

According to the National Report of China Urban Development (2001–2002) (Editorial Committee of National Report of China Urban Development of China Mayor Association, 2003), the proportion of urban population in China should be 46.22% and 56.22%, respectively, in the years 2010 and 2020. In terms of statistics, the annual increase rate of the proportion of urban population has been 1.44% in the last five years (National Bureau of Statistics of the People's Republic of China, 2003), judging by which the urban population proportion should respectively be 50.62% and 65.02% in the years 2010 and 2020. According to a report on a Senior Forum organized by the Ministry of Construction of the People's Republic of China in October 2004, the annual increase rate of the proportion of urban population has increased by 1.88% on average in recent years (http://www.sdinfo.net.cn/daily/2004/10/20041018135647.htm), judging by which the urban population proportion should respectively be 55.02% and 73.82% in the years 2010 and 2020.

10.3.1.5 Spatial Distribution of Transport Infrastructure in China

Transport infrastructure is a primary indicator of human population distribution (Dobson et al., 2000). Roads and railways are especially indicative because of their vital role in human activities. The construction of the railroad from Tang-Shan city to Feng-Nan County in 1881, which was about

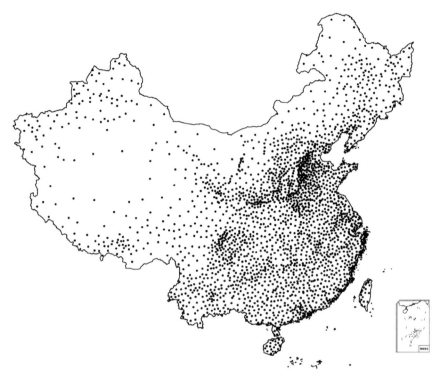

FIGURE 10.5
Spatial distribution of cities in China in the year 2000 (unit: thousand persons grouped by nonagricultural population in urban district).

9.7 km in length, signaled the beginning of railroad development in China. There was 14,411 km of railroad in China in 1930 (Figure 10.6) and 21,800 km in 1949 (Figure 10.7). The total length of railways in operation was 68,700 km in 2000 (Figure 10.8) (Year Book House of China Transportation and Communication, 2001). According to medium- and long-term planning of rail network construction in China, the total length of railroad across the country should reach 85,000 and 100,000 km in the years 2010 and 2020, respectively (Figure 10.9).

In 1902, the first automobile was imported into China, and in 1906 the first piece of road was constructed. In 1949, the total length of highways usable to automobiles was 80,700 km (Figure 10.10). After 50 years of construction, the total length of highways in China (Figure 10.11) came to about 1.4 million km in 2000 (Year Book House of China Transportation and Communication, 2001). According to China's development plan for the national trunk-road network, the trunk line system of national highways, seven east–west main trunk roads and five south–north main trunk roads would be completed around 2010, of which the total length would be about 35,000 km. A major framework of trunk roads would be completed around 2020. The major

FIGURE 10.6
Spatial distribution of railways in the 1930 in China.

framework includes Beijing radiation highways, south–north trunk roads, and east–west trunk roads. The major framework also includes the five vertical trunk roads and seven horizontal trunk roads that link the municipalities directly under central government with the provincial capitals of China, covering all metropolises with a population of more than 1 million and 93% of big cities with population of more than 0.5 million people (Figure 10.12).

10.3.2 Population Growth

10.3.2.1 Change Trend of Population

The population in China has increased by about 2.76 times (Table 10.1) from 1930 to 2000. Hu's result (1983) shows, that the total population of China was 452.8 million in 1930, and 541.67 million in 1949. Between 1930 and 1949, both the fertility rate and death rate were higher so the natural rate of population growth was, therefore, lower. Between 1950 and 2000, the total population increased by 725.74 million, which means the annual mean growth rate was 2.7%. Population growth underwent a rapid increase from 1950 to 1973. After the birth control policy of only one child per couple was introduced in China in 1973, both the fertility rate and death rate declined. Although the birth

FIGURE 10.7
Spatial distribution of railways in 1949 in China.

control policy has restrained rapid population growth, according to recent figures there are still 9.5 million babies born in China every year, because of the huge base number.

10.3.2.2 Population Projection

Population projection can be defined as the numerical outcome of a set of assumptions made about future trends (Smith, 1984). Forecasts can be reviewed as specific types of projections, which are likely to provide accurate predictions about the future population. The most commonly used projection methods can be classified as mathematic extrapolation (ME), Ratio (RA), Cohort Component (CC), and Economic-Demographic (ED). The ME method extrapolates collected population data from a particular base period into the future by using mathematical models such as linear regression, geometric growth rates, and logistical curves, which assume that past trends fitted to a particular mathematic curve will continue into the future (Pritchett, 1891; Pearl and Reed, 1920). The RA method expresses population in smaller areas as ratios of a population in larger areas where a projection already exists that keeps the whole equal to the sum of its parts. The CC method (Whelpton, 1928) projects separate births, deaths, and migrations for each

FIGURE 10.8
Spatial distribution of railways in the year 2000 in China.

age group of the population, which provides projections of age, sex, racial composition, as well as total size. The ED method focuses on the interplay between economic and demographic variables and the effects of economic factors on births, deaths, and immigration (Lee and Hong, 1974).

The CC method is used to make population projections of China in this chapter (Feeney and Yu, 1987; Ma, 1989). Parity progression fertility is employed to project the number of births, which includes the following four steps.

Step 1: To calculate the total fertility rate

$$f(t) = \sum_i f_i(t) \tag{10.3}$$

where $f_i(t) = \sum_{a=15}^{49} f_{a,i}(t)$ represents the proportion of women who have an ith birth during time period t, $f_{a,i}(t) = (b_{a,i}(t)/w_a(t))$ is the age-order-specific birth rate calculated in terms of census, $w_a(t)$ denotes the number of women aged a in completed years at the beginning of the year t, and $b_{a,i}(t)$ is the number of ith births these women have had, aged a;

FIGURE 10.9
Spatial distribution of railways in the year 2020 in China.

Step 2: To project the age-order-specific birth rate during time period $t + n$

$$pf_{a,i}(t + n) = g_{a,i}(t) \times f_i(t + n) \tag{10.4}$$

where $g_{a,i}(t) = (f_{a,i}(t)/f_i(t))$

Step 3: To calculate the age-order-specific birth probability $p_{a,i}(t + 1)$ in terms of the age-order-specific birth rate $f_{a,i}(t + 1)$ as follows:

$$f_{15,1}(t + 1) = p_{15,1}(t + 1) \tag{10.5}$$

for $a \geq 15$

$$f_{a+1,1}(t + 1) = \frac{b_{a+1,1}(t + 1)}{w_{a+1}(t + 1)} = \frac{b_{a+1,1}(t + 1)}{w_{a+1,0}(t + 1)} \cdot \frac{w_{a+1,0}(t + 1)}{w_{a+1}(t + 1)} \tag{10.6}$$

$$= p_{a+1,0}(t + 1) \prod_{k=15}^{a} (1 - p_{k,1}(t + 1))$$

$$f_{15,2}(t + 1) = p_{15,2}(t + 1) = 0 \tag{10.7}$$

FIGURE 10.10
Spatial distribution of roads in 1949 in China.

for $a \geq 16$

$$f_{a+1,2}(t+1) = \frac{b_{a+1,2}(t+1)}{w_{a+1}(t+1)} = \frac{b_{a+1,2}(t+1)}{w_{a+1,1}(t+1)} \cdot \frac{w_{a+1,1}(t+1)}{w_{a+1}(t+1)}$$

$$= p_{a+1,1}(t+1) \cdot \left(\frac{w_{a,1}(t+1)}{w_{a+1}(t+1)} - f_{a,2}(t+1) + f_{a,1}(t+1) \right) \quad (10.8)$$

$$= p_{a+1,1}(t+1) \cdot \left(-\sum_{k=16}^{a} f_{k,2}(t+1) + \sum_{k=15}^{a} f_{k,1}(t+1) \right)$$

We can find the other probability using a similar method: $p_{a,i}(t+n)$.

Step 4: To project births

$$B(t+1) = \sum_{i=1}^{5} \sum_{a=15}^{49} w_a(t+1) \cdot p_{a,i}(t+1) \quad (10.9)$$

FIGURE 10.11
Spatial distribution of roads in the year 2000 in China.

The number of deaths in the year t is calculated in terms of life tables by means of the following formulation:

$$D(t) = \sum_{a=0}^{115} D(a,t) \tag{10.10}$$

$$D(a,t) = P(a,t) \cdot d(a,t) \tag{10.11}$$

where $D(a, t)$ is the number of age-specific deaths; $P(a, t)$ is the mid-year number of people at age a in the year t; and $d(a, t)$ is the mortality of people at age a in the year t.

The population balance equation is formulated as

$$P(t + 1) = P(t) + B(t + 1) - D(t + 1) + I(t + 1) \tag{10.12}$$

where $P(t)$ is the population in the year t; $B(t + 1)$, $D(t + 1)$, and $I(t + 1)$ are, respectively, the numbers of births, deaths, and immigrants in the year $t + 1$.

The projection results (Table 10.2 through 10.4) show that the total population of China should be respectively 1420.4, 1432.3, 1489.8 million in 2020 under low, medium, and high projections (Research Group of National Population Strategy for Development of China, 2007).

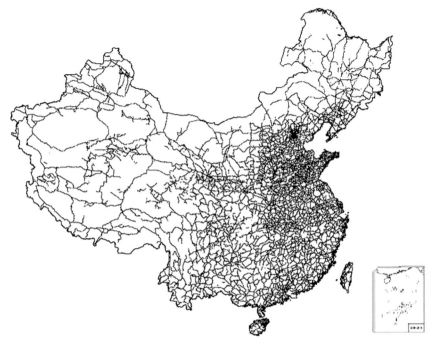

FIGURE 10.12
Spatial distribution of national highways in the year 2020 in China.

10.3.3 SMPD Operation

SMPD in China is formulated as (Yue et al., 2003, 2005a,b)

$$SMPD_{ij} = G(n) \cdot \frac{p_{ij}}{\sum p_{ij}} \tag{10.13}$$

$$p_{ij} = W_{ij} \cdot (NPP_{ij})^{a_5} \cdot (DEM_{ij})^{a_4} (Tran_{ij})^{a_3} \cdot \left(\sum_{k=1}^{M} \frac{(S_k)^{a_1}}{d_{ijk}} \right)^{a_2} \tag{10.14}$$

where $G(n)$ is the total population in province n, in which grid cell (i, j) is located; W_{ij} is the indicative factor of water area, when grid cell (i, j) is located in water area $W_{ij} = 0$, or else $W_{ij} = 1$; $Tran_{ij}$ is the transport infrastructure factor of grid cell (i, j); NPP_{ij} is the factor of NPP of grid cell (i, j); DEM_{ij} is the elevation factor of grid cell (i, j); S_k is the size of the kth city; M is the total number of cities in the defined search radius; d_{ijk} is the distance from grid cell (i, j) to the core grid cell that has the highest population density in the kth city; and a_k $(k = 1, 2, 3, 4, 5)$ are exponents to be simulated.

The transport infrastructure factor, the factor of NPP, and the elevation factor, by comparing the effects of different functional forms of each

TABLE 10.1

Change Trend of the Provincial Population in China Excluding Taiwan, Hong Kong, and Macao Temporarily

Region	Area (km²)	Population Size (Million Persons)			Population Density (Person/km²)		
		1930	1949	2000	1930	1949	2000
Western China	**6,725,746**	**110.63**	**174.57**	**354.60**	**16**	**26**	**53**
Inner Mongolia	1,143,327	4.40	37.88	23.01	4	33	20
Guangxi	236,544	11.50	18.42	47.24	49	78	200
Chongqing	82,390	Belong to Sichuan	Belong to Sichuan	30.91	Belong to Sichuan	Belong to Sichuan	375
Sichuan	483,759	51.34	57.30	84.07	106	118	174
Guizhou	176,109	11.03	14.16	36.77	63	80	209
Yunnan	383,101	11.52	15.95	40.77	30	42	106
Tibet	1,201,653	0.78	1.00	2.51	1	1	2
Shaanxi	205,732	10.39	13.17	35.72	50	64	174
Gansu	404,622	5.49	9.68	25.34	14	24	63
Qinghai	716,677	1.28	1.48	4.80	2	2	7
Ningxia	51,785	0.39	1.20	5.54	8	23	107
Xinjiang	1,640,111	2.51	4.33	17.92	2	3	11
Middle China	**1,670,726**	**150.15**	**161.41**	**419.41**	**90**	**97**	**251**

Shanxi	156,563	11.30	12.81	31.96	72	82	204
Anhui	140,165	21.92	27.86	62.78	156	199	448
Jiangxi	166,960	17.16	12.68	41.64	103	76	249
Henan	165,619	31.92	41.74	95.27	193	252	575
Hubei	185,950	25.94	25.36	59.36	140	136	319
Hunan	211,815	29.54	29.87	65.15	139	141	308
Jilin	191,093	7.82	10.09	26.27	41	53	137
Helongjiang	452,561	4.55	1.01	36.98	10	2	82
Eastern China	**1,203,528**	**187.23**	**205.68**	**462.71**	**156**	**171**	**384**
Beijing	16,386	1.52	4.14	11.14	93	253	680
Tianjin	11,620	1.47	3.99	9.19	126	344	791
Hebei	188,111	30.29	30.86	66.71	161	164	355
Liaoning	146,316	16.08	18.31	41.35	110	125	283
Shanghai	8013	3.91	5.06	13.22	488	632	1650
Jiangsu	103,405	30.29	35.12	70.69	293	340	684
Zhejiang	103,196	20.07	20.83	45.01	194	202	436
Fujian	122,468	13.99	11.88	33.05	114	97	270
Shandong	157,119	36.67	45.49	89.75	233	290	571
Guangdong	179,776	32.93	30.00	74.99	183	167	417
Hainan	40,070			7.61			190
	Belong to Guangdong	Belong to Guangdong	Belong to Guangdong	Belong to Guangdong	Belong to Guangdong	Belong to Guangdong	

TABLE 10.2

Total Population under Low Projection

Year	Total Population (Million)	Male (Million)	Female (Million)	Births (Million)	Deaths (Million)	Fertility (‰)	Mortality (‰)	Natural Increase Rate (‰)	Total Fertility Rate	Child Ratio (%)	Labor Ratio (%)	Elder Ratio (%)	Labor Number (Million)
2010	1347.9	693.7	654.1	18	9.96	13.4	7.41	5.99	1.73	19.02	73.01	7.97	984
2011	1356.1	697.8	658.4	18.36	10.08	13.58	7.46	6.12	1.72	18.78	73.06	8.16	991
2012	1364.5	701.9	662.7	18.61	10.21	13.68	7.5	6.18	1.72	18.61	73.01	8.38	996
2013	1372.9	706.	667	18.71	10.34	13.67	7.55	6.12	1.71	18.49	72.9	8.61	1001
2014	1381.1	709.9	671.2	18.65	10.46	13.54	7.6	5.95	1.71	18.4	72.67	8.93	1004
2015	1389	713.8	675.2	18.44	10.59	13.32	7.65	5.67	1.7	18.34	72.39	9.28	1005
2016	1396.4	717.3	679	18.12	10.71	13.01	7.69	5.32	1.7	18.29	72.05	9.66	1006
2017	1403.3	720.6	682.6	17.72	10.84	12.66	7.74	4.92	1.69	18.27	71.62	10.11	1005
2018	1409.6	723.6	686	17.29	10.98	12.3	7.81	4.49	1.69	18.26	71.15	10.59	1003
2019	1415.3	726.3	689	16.85	11.13	11.93	7.88	4.05	1.69	18.25	70.63	11.13	1000
2020	1420.4	728.6	691.8	16.4	11.28	11.57	7.96	3.61	1.69	18.13	70.24	11.63	998
2021	1424.9	730.7	694.3	15.96	11.45	11.22	8.05	3.17	1.68	18.01	69.91	12.08	996
2022	1428.8	732.3	696.5	15.52	11.63	10.88	8.15	2.73	1.68	17.86	69.57	12.57	994
2023	1432.1	733.7	698.4	15.1	11.82	10.56	8.27	2.29	1.68	17.68	69.39	12.93	994
2024	1434.8	734.7	700	14.69	12.02	10.25	8.39	1.86	1.68	17.46	69.4	13.14	996
2025	1436.8	735.4	701.4	14.3	12.24	9.96	8.53	1.43	1.67	17.19	69.54	13.27	999
2026	1438.3	735.8	702.5	13.94	12.49	9.7	8.69	1.01	1.67	16.89	69.68	13.43	1002
2027	1439.2	735.8	703.3	13.64	12.75	9.48	8.86	0.62	1.66	16.56	69.41	14.03	999

2028	1439.5	735.6	704.	13.4	13.01	9.31	9.04	0.27	1.66	16.21	68.89	14.91	992
2029	1439.5	735.1	704.4	13.24	13.29	9.2	9.24	-0.04	1.65	15.85	68.46	15.69	985
2030	1439.1	734.4	704.6	13.17	13.6	9.15	9.45	-0.3	1.65	15.51	68.05	16.44	979
2031	1438.3	733.6	704.7	13.15	13.92	9.14	9.68	-0.54	1.65	15.19	67.68	17.13	973
2032	1437.2	732.6	704.6	13.18	14.27	9.17	9.93	-0.76	1.65	14.9	67.31	17.79	967
2033	1435.8	731.4	704.4	13.24	14.63	9.22	10.18	-0.97	1.65	14.65	66.79	18.56	959
2034	1434.1	730.1	704.	13.33	14.98	9.29	10.44	-1.16	1.64	14.44	66.21	19.35	95
2035	1432.2	728.7	703.5	13.42	15.36	9.36	10.72	-1.36	1.64	14.26	65.65	20.08	94
2036	1430.	727.2	702.8	13.5	15.73	9.44	10.99	-1.55	1.64	14.13	65.13	20.74	931
2037	1427.5	725.5	702.	13.57	16.07	9.5	11.25	-1.75	1.64	14.02	64.65	21.32	923
2038	1424.7	723.7	701.	13.59	16.41	9.53	11.5	-1.97	1.63	13.96	64.2	21.84	915
2039	1421.5	721.8	699.8	13.58	16.71	9.54	11.75	-2.21	1.63	13.92	63.83	22.26	907
2040	1418.	719.6	698.4	13.51	17	9.51	11.97	-2.46	1.62	13.9	63.53	22.56	901
2041	1414.2	717.4	696.8	13.39	17.23	9.46	12.17	-2.71	1.62	13.91	63.3	22.79	895
2042	1410.	714.9	695.1	13.23	17.42	9.37	12.34	-2.96	1.62	13.93	63.11	22.96	89
2043	1405.4	712.2	693.2	13.04	17.63	9.26	12.53	-3.27	1.61	13.96	62.92	23.12	884
2044	1400.4	709.3	691.1	12.81	17.77	9.13	12.66	-3.54	1.61	13.98	62.73	23.29	878
2045	1395.1	706.2	688.9	12.56	17.87	8.98	12.78	-3.8	1.61	14	62.58	23.42	873
2046	1389.4	702.9	686.5	12.29	17.98	8.83	12.92	-4.09	1.61	14	62.36	23.64	866
2047	1383.2	699.3	683.9	12.01	18.21	8.66	13.14	-4.47	1.6	13.98	62.09	23.92	859
2048	1376.2	695.4	680.8	11.73	18.8	8.5	13.62	-5.12	1.6	13.95	61.96	24.08	853
2049	1368.2	691.	677.1	11.45	19.48	8.35	14.2	-5.85	1.6	13.91	61.87	24.22	846
2050	1359.4	686.3	673.	11.19	19.98	8.2	14.65	-6.45	1.59	13.84	61.67	24.49	838

Source: Adapted from Research Group of National Population Strategy for Development of China. 2007. *National Population Strategy for Development of China.* Beijing: China Population Press (in Chinese).

TABLE 10.3

Total Population under Medium Projection

Year	Total Population (Million)	Male (Million)	Female (Million)	Births (Million)	Deaths (Million)	Fertility (‰)	Mortality (‰)	Natural Increase Rate (‰)	Total Fertility Rate	Child Ratio (%)	Labor Ratio (%)	Edler Ratio (%)	Labor Number (Million)
2010	1354.5	697.2	657.3	18.55	9.98	13.74	7.39	6.35	1.78	19.41	72.65	7.93	984
2011	1363.3	701.6	661.8	18.9	10.1	13.91	7.43	6.47	1.78	19.2	72.68	8.12	991
2012	1372.3	705.9	666.3	19.16	10.23	14.01	7.48	6.53	1.77	19.05	72.63	8.33	997
2013	1381.2	710.3	670.9	19.26	10.36	13.99	7.52	6.47	1.76	18.92	72.51	8.56	1001
2014	1389.9	714.5	675.3	19.2	10.48	13.86	7.56	6.3	1.76	18.84	72.29	8.87	1005
2015	1398.3	718.6	679.6	18.99	10.61	13.62	7.61	6.01	1.75	18.77	72.02	9.21	1007
2016	1406.2	722.5	683.7	18.66	10.73	13.31	7.65	5.66	1.75	18.73	71.67	9.59	1008
2017	1413.6	726	687.6	18.26	10.86	12.95	7.7	5.25	1.74	18.72	71.25	10.04	1007
2018	1420.4	729.3	691.1	17.83	11	12.58	7.76	4.82	1.74	18.7	70.79	10.51	1006
2019	1426.7	732.2	694.4	17.39	11.14	12.21	7.83	4.39	1.74	18.66	70.3	11.04	1003
2020	1432.3	734.9	697.5	16.95	11.3	11.86	7.9	3.95	1.73	18.54	69.93	11.53	1002
2021	1437.4	737.2	700.2	16.53	11.47	11.52	7.99	3.53	1.73	18.4	69.62	11.98	1001
2022	1441.9	739.2	702.7	16.11	11.64	11.19	8.09	3.11	1.73	18.25	69.29	12.46	999
2023	1445.7	740.8	704.9	15.71	11.84	10.88	8.2	2.68	1.73	18.07	69.12	12.81	999
2024	1449.	742.2	706.8	15.32	12.04	10.59	8.32	2.27	1.72	17.85	69.14	13.02	1002
2025	1451.7	743.2	708.5	14.95	12.26	10.31	8.45	1.85	1.72	17.59	69.28	13.13	1006
2026	1453.8	743.9	709.9	14.62	12.51	10.06	8.61	1.45	1.71	17.29	69.43	13.29	1009
2027	1455.4	744.3	711.1	14.34	12.77	9.86	8.78	1.08	1.71	16.96	69.16	13.88	1007

2028	1456.5	744.4	712.1	14.12	13.03	9.7	8.95	0.75	1.7	16.62	68.65	14.74	1000
2029	1457.1	744.3	712.8	13.98	13.31	9.6	9.14	0.46	1.7	16.27	68.23	15.5	994
2030	1457.4	744.	713.4	13.92	13.62	9.55	9.35	0.21	1.7	15.94	67.82	16.24	988
2031	1457.4	743.6	713.9	13.91	13.94	9.55	9.57	-0.02	1.69	15.63	67.46	16.91	983
2032	1457.1	742.9	714.1	13.94	14.29	9.57	9.81	-0.24	1.69	15.36	67.09	17.55	978
2033	1456.4	742.2	714.3	14.01	14.65	9.62	10.05	-0.44	1.69	15.12	66.59	18.3	970
2034	1455.5	741.3	714.3	14.1	15	9.68	10.3	-0.62	1.69	14.92	66.01	19.07	961
2035	1454.3	740.2	714.1	14.19	15.38	9.75	10.57	-0.82	1.69	14.75	65.47	19.78	952
2036	1452.9	739.1	713.8	14.28	15.75	9.82	10.83	-1.01	1.68	14.62	64.96	20.42	944
2037	1451.1	737.8	713.3	14.34	16.09	9.88	11.08	-1.2	1.68	14.53	64.49	20.98	936
2038	1449.1	736.4	712.6	14.37	16.43	9.91	11.33	-1.42	1.68	14.47	64.05	21.48	928
2039	1446.7	734.9	711.8	14.35	16.73	9.91	11.56	-1.65	1.67	14.43	63.69	21.88	921
2040	1443.9	733.1	710.8	14.28	17.02	9.88	11.77	-1.89	1.67	14.42	63.41	22.17	916
2041	1440.9	731.2	709.6	14.16	17.25	9.82	11.96	-2.14	1.66	14.43	63.2	22.38	911
2042	1437.4	729.2	708.3	14	17.44	9.73	12.12	-2.39	1.66	14.44	63.03	22.53	906
2043	1433.6	726.9	706.7	13.8	17.65	9.61	12.29	-2.68	1.66	14.47	62.86	22.67	901
2044	1429.3	724.3	705	13.57	17.78	9.48	12.42	-2.95	1.65	14.49	62.68	22.83	896
2045	1424.8	721.6	703.1	13.31	17.88	9.33	12.53	-3.2	1.65	14.5	62.56	22.94	891
2046	1419.8	718.7	701.1	13.05	18	9.17	12.66	-3.48	1.65	14.5	62.36	23.14	885
2047	1414.4	715.5	698.9	12.77	18.23	9.01	12.86	-3.85	1.64	14.48	62.12	23.41	879
2048	1408.1	711.9	696.1	12.49	18.82	8.85	13.33	-4.48	1.64	14.44	62.01	23.55	873
2049	1400.8	708	692.8	12.22	19.5	8.7	13.89	-5.18	1.64	14.39	61.94	23.67	868
2050	1392.7	703.7	689.1	11.96	20	8.56	14.32	-5.76	1.63	14.32	61.77	23.91	860

Source: Adapted from Research Group of National Population Strategy for Development of National Population Strategy for Development of China. 2007. *National Population Strategy for Development of China.* Beijing: China Population Press (in Chinese).

TABLE 10.4

Total Population under High Projection

Year	Total Population (Million)	Male (Million)	Female (Million)	Births (Million)	Deaths (Million)	Fertility (‰)	Mortality (‰)	Natural Increase Rate (‰)	Total Fertility Rate	Child Ratio (%)	Labor Ratio (%)	Edler Ratio (%)	Labor Number (Million)
2010	1382. 6	711.9	670.7	22.09	10.11	16.05	7.34	8.71	2.13	21.05	71.18	7.77	984
2011	1394.6	717.9	676.8	22.24	10.23	16.01	7.37	8.65	2.1	21.02	71.05	7.93	991
2012	1406.6	723.8	682.8	22.35	10.35	15.96	7.39	8.57	2.08	21.03	70.85	8.12	997
2013	1418.5	729.7	688.8	22.37	10.47	15.84	7.41	8.42	2.06	21.07	70.6	8.33	1001
2014	1430.2	735.5	694.6	22.26	10.59	15.63	7.44	8.19	2.04	21.13	70.25	8.62	1005
2015	1441.5	741.1	700.4	22.02	10.72	15.34	7.47	7.87	2.03	21.21	69.85	8.93	1007
2016	1452.3	746.5	705.8	21.67	10.84	14.98	7.49	7.48	2.03	21.33	69.39	9.28	1008
2017	1462.6	751.5	711.1	21.25	10.97	14.58	7.53	7.06	2.02	21.45	68.85	9.69	1007
2018	1472.3	756.3	716	20.8	11.11	14.17	7.57	6.6	2.02	21.57	68.29	10.13	1005
2019	1481.3	760.7	720.7	20.33	11.25	13.77	7.62	6.15	2.02	21.68	67.7	10.62	1003
2020	1489.8	764.8	725	19.87	11.41	13.37	7.68	5.69	2.02	21.27	67.66	11.07	1008
2021	1497.6	768.5	729.1	19.41	11.58	13	7.75	5.25	2.02	20.91	67.6	11.49	1012
2022	1504.9	771.9	732.9	18.97	11.75	12.63	7.83	4.81	2.02	20.59	67.49	11.92	1016
2023	1511.4	775	736.4	18.53	11.95	12.29	7.93	4.36	2.01	20.28	67.48	12.24	1020
2024	1517.4	777.7	739.7	18.12	12.15	11.96	8.03	3.94	2.01	19.96	67.62	12.41	1026
2025	1522.8	780.1	742.6	17.75	12.38	11.68	8.14	3.53	2.01	19.63	67.87	12.5	1034
2026	1527.6	782.2	745.4	17.47	12.62	11.45	8.28	3.18	2	19.28	68.1	12.62	1040
2027	1532.1	784.1	747.9	17.33	12.89	11.33	8.42	2.9	1.98	18.92	67.92	13.16	1041

2028	1536.3	785.9	750.4	17.37	13.16	11.32	8.58	2.75	1.97	18.56	67.49	13.95	1037
2029	1540.5	787.6	752.9	17.65	13.45	11.48	8.74	2.73	1.97	18.23	67.13	14.63	1034
2030	1544.8	789.4	755.4	18.09	13.76	11.73	8.92	2.81	1.97	17.95	66.76	15.29	1031
2031	1549.3	791.3	758	18.57	14.09	12	9.11	2.9	1.98	17.71	66.41	15.88	1029
2032	1553.9	793.2	760.6	19.01	14.45	12.25	9.31	2.94	1.99	17.53	66.05	16.42	1026
2033	1558.4	795.1	763.3	19.37	14.81	12.45	9.52	2.93	2	17.4	65.53	17.06	1021
2034	1562.9	797	765.9	19.65	15.17	12.59	9.72	2.87	2	17.32	64.96	17.72	1015
2035	1567.2	798.9	768.3	19.85	15.56	12.68	9.94	2.74	2	17.28	64.4	18.31	1009
2036	1571.3	800.6	770.7	20.01	15.92	12.75	10.15	2.6	2.01	17.29	63.88	18.83	1004
2037	1575.1	802.2	772.9	20.12	16.27	12.79	10.34	2.44	2.01	17.32	63.4	19.28	999
2038	1578.7	803.7	775	20.17	16.61	12.79	10.53	2.26	2.01	17.4	62.94	19.66	994
2039	1581.9	805.1	776.8	20.16	16.92	12.76	10.71	2.05	2.01	17.49	62.56	19.95	990
2040	1584.8	806.3	778.5	20.1	17.2	12.69	10.87	1.83	2.01	17.62	62.25	20.13	987
2041	1587.3	807.3	780.1	19.98	17.44	12.6	10.99	1.6	2.01	17.75	62	20.25	984
2042	1589.5	808.1	781.4	19.81	17.63	12.47	11.1	1.37	2	17.89	61.81	20.3	982
2043	1591.3	808.7	782.6	19.6	17.84	12.32	11.22	1.1	2	18.01	61.63	20.35	981
2044	1592.7	809.1	783.5	19.35	17.98	12.15	11.29	0.86	2	18.11	61.48	20.41	979
2045	1593.6	809.3	784.4	19.07	18.08	11.97	11.35	0.62	2	18.17	61.4	20.43	978
2046	1594.2	809.2	785	18.77	18.2	11.78	11.42	0.36	2	18.18	61.3	20.53	977
2047	1594.2	808.9	785.4	18.46	18.43	11.58	11.56	0.02	2	18.15	61.18	20.67	975
2048	1593.4	808.1	785.3	18.15	19.02	11.39	11.93	-0.54	2	18.09	61.19	20.71	975
2049	1591.5	807	784.6	17.86	19.7	11.21	12.37	-1.16	2	18.01	61.26	20.74	975
2050	1588.9	805.5	783.4	17.58	20.2	11.05	12.7	-1.65	2	17.9	61.24	20.86	973

Source: Adapted from Research Group of National Population Strategy for Development of China. 2007. *National Population Strategy for Development of China.* Beijing: China Population Press (in Chinese).

influencing factor on the corresponding simulation results, can be respectively formulated as

$$Tran_{ij} = \frac{ra_{ij} + ro_{ij}}{\underset{i,j}{Max}\{ra_{ij} + ro_{ij}\}} \tag{10.15}$$

$$NPP_{ij} = \exp\left\{-\frac{(MNPP_{ij} - 752)^2}{10^6}\right\} \tag{10.16}$$

$$DEM_{ij}(t) = \begin{cases} \dfrac{500}{\left(dem_{ij}(t)\right)^2} & dem_{ij}(t) \geq 3700\,\text{m} \\[3mm] \dfrac{500}{dem_{ij}(t)} & 500\,\text{m} < dem_{ij}(t) < 3700\,\text{m} \\[3mm] 1 & dem_{ij}(t) \leq 500\,\text{m} \end{cases} \tag{10.17}$$

where ra_{ij} and ro_{ij} respectively represent rail density and road density at grid cell (i, j); $MNPP_{ij}$ is the mean annual NPP in the 1990s at grid cell (i, j); dem_{ij} is the elevation at grid cell (i, j).

In the SMPD process, province-level census is used to simulate spatial population distribution and the county-level census is used to determine the exponents. First, let $a_2 = a_3 = a_4 = a_5 = 1$ and change the value of a_1. It is found that when $a_1 = 1$, the correlation coefficient between the county-level census and the mean simulated population on the county-level reaches the maximum value. Second, let $a_3 = a_4 = a_5 = 1$ and change the value of a_2, it is found that when $a_2 = 1.2$, the correlation coefficient reaches the maximum value. Similarly, $a_3 = 1.3$, $a_4 = 0.7$, and $a_5 = 0.0001$ are respectively determined. Because the simulation priority of a_k $(k = 1,2,3,4,5)$ may have an effect on the simulation results and $\{a_1, a_2, a_3, a_4, a_5\}$ has 120 possible permutations, another 119 similar simulation processes have to be conducted by considering the simulation priority of each a_k $(k = 1,2,3,4,5)$ in different permutations of $\{a_1, a_2, a_3, a_4, a_5\}$. The results show that the correlation coefficient between the county-level census and the mean simulated population on county-level has the biggest value if the order of the simulation priority of the exponents $\{a_1, a_2, a_3, a_4, a_5\}$ is taken up in terms of the subscripted label order from 1 to 5. Therefore, p_{ij} is finally formulated as

$$p_{ij} = W_{ij} \cdot \left(NPP_{ij}\right)^{0.0001} \cdot \left(DEM_{ij}\right)^{0.7} \cdot \left(Tran_{ij}\right)^{1.3} \cdot \left(\sum_{k=1}^{M} \frac{S_k}{d_{ijk}}\right)^{1.2} \tag{10.18}$$

The major auxiliary tools include ArcInfo GIS and Delphi computer language. Eight data layers are involved, which are NPP, DEM, WA (water area), Grid Rail (railway network), Grid Road (the road network), Chbnd (the administrative boundary), Chzh (the urban area) and Cityshp (the geographical coordinates of the city). The data are first preprocessed as follows: (1) by converting NPP into vector data, (2) by overlaying Chbnd with GridRoad and GridRail using intersect and creating a data layer, ChBndNew, (3) by adding fields; CityFlag for urban code and rural code and CityArea for urban districts, in Chzh, (4) by overlaying Chzh with ChBndNew by using intersect and creating a data layer, ChCity, (5) by overlaying NPP with ChCity by using intersect and creating a data layer, NppNew, (6) by overlaying DEM with NppNew by using intersect and creating a data layer, DNpp, (7) by overlaying WA with DNpp by using intersect and creating a data layer, WDNpp.

Every grid cell in 1 × 1 km resolution is generated on the basis of WDNpp and CityShp, which includes six steps: (1) to read the attribute values of natural and socioeconomic indicators at every SMPD grid cell, (2) to simulate the contribution of NPP and elevation to SMPD, (3) to define the search radius of the SMPD grid cell and search cities and transport infrastructures that have considerable effects on the SMPD grid cell, (4) to simulate the contribution of the searched cities and transport infrastructures to the SMPD grid cell, (5) to operate the SMPD, and (6) to convert the text file of the calculated result into point vector data and create the grid data from the point vector data.

10.3.4 Change Trends in Population Distribution

According to the current ecological and economical situation, China can be geographically analyzed in three regions: western, middle, and eastern China. The western region of China consists of five provinces in southwest China, five provinces in northwest China, the Inner Mongolia Autonomous region, and the Guangxi Zhuang Autonomous region. The five provinces in southwest China are Sichuan province, Chongqing city, Yunnan province, Guizhou province, and the Tibet Autonomous region. The five provinces in northwest China are Shaanxi province, Gansu province, the Ningxia Hui Autonomous region, the Xinjiang Uygur Autonomous region, and Qinghai province. The area of the western region of China is about 6.7546 million square kilometers, accounting for 70% of the whole of China. The middle region of China consists of eight provinces; Shanxi, Anhui, Jiangxi, Henan, Hubei, Hunan, Jilin, and Helongjiang, of which the total area is 1.67 million sq. km, accounting for 17.4% of the whole of China. The eastern region of China consists of 11 provinces; Beijing, Tianjin, Hebei, Liaoning, Shanghai, Jiangsu, Zhejiang, Fujian, Shandong, Guangdong, and Hainan, of which the total area is 1.2 million sq. km, accounting for 12.5% of the whole of China.

A comparison of the simulation results shows that the ratio of population in the western region of China to that of the total population of China was 24% in 1930 (Figure 10.13), 32% in 1949 (Figure 10.14), and 29% in 2000 (Figure 10.15); whereas the ratio of population in the middle region of China was 33% in 1930, 30% in 1949, and 34% in 2000; and the ratio of population in the eastern region of China was 41% in 1930, 38% in 1949, and 37% in 2000. Human population had a slanting trend from the eastern region to the western and middle regions of China during the period from 1930 to 2000. From 1930 to 1949, on an average, the annual growth rate of the population was 3% in the western region, 0.4% in the middle region and 0.5% in the eastern region; from 1949 to 2000, the annual growth rates of the population were, respectively, 2%, 3.1%, and 2.5% in the western, middle, and eastern regions (Table 10.1).

The change of spatial population distribution trends in China from the 1930s to the 1990s is likely to have been caused by the following five factors: (1) in the 1930s and 1940s, the China–Japan war and China civil war led to a large number of deaths and the migration of scores of inhabitants of the eastern region of China; (2) in the 1950s, the newly established government of China organized a series of massive immigrations to bring unused land under cultivation in the western region of China. Examples include the production and construction corps in the Xinjiang Uygur autonomous region, and the immigration and reclamation bureau in Qinghai province; (3) in the 1960s, a large number of factories, scientific research institutions, colleges, and universities were moved from the east coast to the western region, meaning

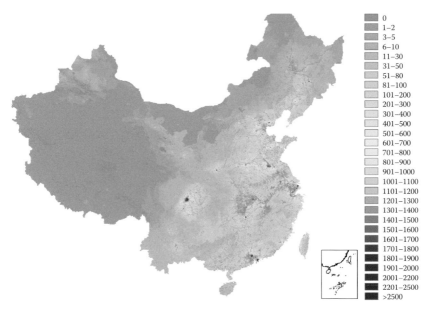

FIGURE 10.13
Human population distribution of China in 1930 (unit: persons/sq. km).

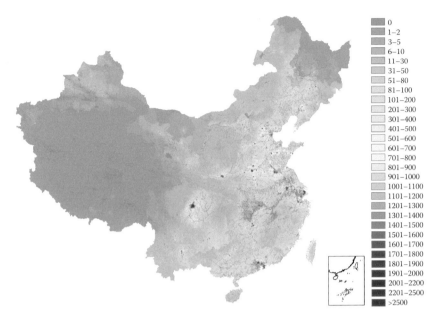

FIGURE 10.14
Human population distribution of China in 1949 (unit: persons/sq. km).

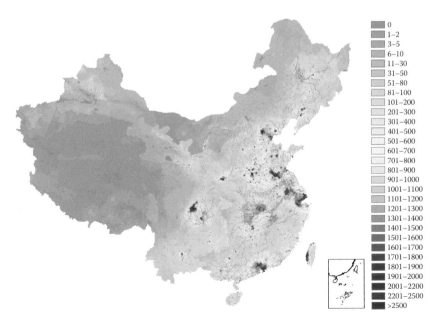

FIGURE 10.15
Human population distribution of China in 2000 (unit: persons/sq. km).

that about 17 million middle-school students in cities were sent to rural areas in order to gain a peasant education; (4) since the 1970s, the birth control policy which allows only one child per couple has been rigorously enforced in the eastern region of China, and loosely in the western region; (5) in recent years, implementation of the western development strategy and massive construction of infrastructures in the western region of China have contributed to an increase in the ratio of population in the western region to that of the total population of China.

10.3.5 Scenarios of Population Distribution

The creation of scenarios as instruments for problem identification and decision making dates back to the 1940s, when nuclear physicists used computer simulations to tackle probabilities and uncertainties, allowing military strategists to help military organizations to plan operations based on the results (Yue et al., 1999). In the late 1960s, scenarios began to be used outside of a military context for the first time (Kahn and Wiener, 1967), and by the early 1970s, many companies and governmental organizations started using scenarios for their strategic decision-making and planning (Jefferson, 1983). Since the late 1970s, scenarios have been extensively used in the context of sustainable development (Haefele et al., 1981; World Energy Council, 1993; Schoute et al., 1995; Gallopin and Rijsberman, 1999; Nakicenovic et al., 2000; Raskin, 2000; Alcamo and Ribeiro, 2001; Organization of Economic Cooperation and Development, 2001; Putting and Bakkes, 2003). In the 1980s and 1990s, physical planning was brought into the public sphere in order to promote public discussion, prompting the creation of more integrated scenarios with various alternatives (Wack, 1985; van de Klundert, 1995; World Business Council on Sustainable Development, 1997).

Scenarios can be defined as plausible alternative futures, each being an example of what might happen as the result of particular assumptions (Millennium Ecosystem Assessment, 2003). However, there are many other definitions of scenarios. For instance, the Intergovernmental Panel on Climate Change described scenarios as alternative futures that are neither predictions nor forecasts but an alternative image of how the future might unfold (Nakicenovic et al., 2000). Alcamo et al. (1995) defined scenarios as projections of the future state of society and the environment based on specific assumptions on key determinants such as population, economic growth, technological change, or environmental policies. Veeneklaas and van den Berg (1995) defined scenarios as a description of a possible or desirable future state as well as a series of events that could lead from the current state of affairs to this future state. Godet (1987) defined scenarios as a description of a future situation together with the progression of events leading from the current situation to the future situation.

Each of the SMPD scenarios is a plausible alternative future under the particular assumptions of elevation, water systems, NPP, urbanization, transport

TABLE 10.5

Assumptions of the Three Scenarios

Scenarios	Elevation	Water Systems	NPP Increase Rate (g C·m⁻²/yr)	Urbanization (%)	Rail (km)	Road (Million km)	Population (Million)
I	2020	No change	No change	0.49	73.82	114030	1.540	1597.68
II	2020	No change	No change	0.49	65.02	100000	1.470	1508.84
III	2020	No change	No change	0.49	56.22	85980	1.435	1456.39

infrastructure development, and population growth. The three scenarios, which are broken down into I, II, and III, are developed under the general assumption that railway construction planning has been successfully carried out. They show that the increase rate of NPP would be 0.49 g C m⁻²/year, and that water systems and elevation would change very little on the national level (Table 10.5).

In scenario I, it is supposed that the proportion of urban population, the total length of highway, the total length of railway, and the total population would respectively be 73.82%, 1.54 million km, 114,030 km, and 1489.8 million in the year 2020.

In scenario II, the proportion of urban population would be 65.02%, the total length of highway would be 1.47 million km, the total length of railway would be 100,000 km, and total population would be 1432.3 million in the year 2020.

In scenario III, the proportion of urban population would be 56.22%, the total length of highway would be 1.47 million km, the total length of railway would be 85,980 km and the total population would be 1420.4 million in the year 2020.

All scenarios show that, if there were no limits on population migration in China, large swaths of the population would likely migrate from the western and middle regions of China to the eastern region. In fact, the simulation of population trends between the years 1930 and 2000 implies that population migration was in fact limited by provincial administrations

TABLE 10.6

SMPD Scenarios in Different Regions of China

Years		The Western Region Population	Ratio (%)	The Middle Region Population	Ratio (%)	The Eastern Region Population	Ratio (%)
2000		354.6	29	419.41	34	462.71	37
2020	Scenario I	230.47	15.47	487.02	32.69	772.31	51.84
	Scenario II	223.30	15.59	468.36	32.7	740.64	51.71
	Scenario III	224.00	15.77	464.47	32.7	731.79	51.52

(Yue et al., 2003). If the population could freely migrate within the whole of China, the balanced ratios of population in the western region, the middle region, and the eastern region to that of the total population in the whole China would be about 16%, 33%, and 52%, respectively, in 2020 (Table 10.6). As the three scenarios of population distribution in the year 2020 have a very similar spatial pattern, we will only present scenario II in 2020 as an example in this chapter (Figure 10.16).

10.3.6 Brief Summary

In terms of the SMPD scenarios, 16% of the population would live in the western region, 33% in the middle region and 52% in the eastern region, if the population could freely migrate within the whole of China. This spatial pattern of population distribution in China under current economic and infrastructure conditions would hasten the development of the Encircling-Bohai-Sea, the Yangtze-River-Delta, and the Pear-River-Delta urban agglomerations in the eastern region of China (Wang, 2003), and make the rapid development of these urban agglomerations an inexorable trend in the near future. The Encircling-Bohai-Sea urban agglomeration, taking Beijing, Tianjin, and Dalian as its cores, embodies five prefectural cities in Hebei province which are Tangshan, Baoding, Langfang, Qinhuangdao, and Cangzhou, eight prefectural cities in Shandong province which include

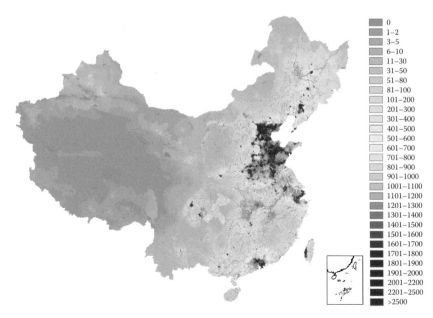

FIGURE 10.16
Scenario II of human population distribution in the year 2020 (unit: persons/sq. km).

Jinan, Dongying, Yantai, Weihai, Qingdao, Weifang, Zibo, and Binzhou, and nine prefectural cities in Liaoning province which consist of Shenyang, Dalian, Benxi, Anshan, Yingkou, Fushun, Panjin, Jinzhou, and Huludao. The Yangtze-River-Delta urban agglomeration, taking Shanghai, Suzhou, Wuxi, Nanjing, Hangzhou, and Ningbo as its cores, embodies eight prefectural cities in Jiangsu province including Nanjing, Suzhou, Wuxi, Changzhou, Yangzhou, Zhenjiang, Nantong, and Taizhou, and six prefectural cities in Zhejiang province, Hangzhou, Ningbo, Huzhou, Jiaxing, Shaoxing, and Zhousha. The Pear-River-Delta urban agglomeration taking Guangzhou, Shenzhen, Zhuhai, Dongguan, and Zhongshan as its cores, embodies nine prefectural cities in Guandong province, that is, Guangzhou, Dongguan, Shenzhen, Zhongshan, Zhuhai, Foshan, Jiangmen, Zhaoqing, and Huizho (Editorial Committee of National Report of China Urban Development of China Mayor Association, 2004) (Figure 10.17).

A great immigration of population from the western and middle regions to the eastern region of China and the development of three urban agglomerations in the eastern region would lead to many serious environmental problems, such as a shortage of natural resources and energy resources, and the exacerbation of environmental pollution and land degradation. These environmental problems pose as a significant challenge to the inhabitants of these areas.

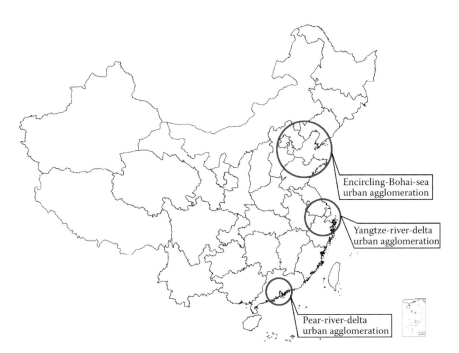

FIGURE 10.17
Location of the three urban agglomerations in the eastern region of China.

10.4 SMPD in Beijing

The simulation of urban population density began with the pioneering work of Clark (1951). He believed that the falling off of population density followed a simple mathematical equation of exponential decline. Since then, a range of monocentric models and polycentric models have been proposed. However, many urban spatial patterns, such as the Beijing master plan from 2004 to 2020, challenge these polycentric models, in which population is distributed across widely dispersed kernels and corridors, and in myriad clusters throughout the urban landscape. Monocentric models have also been criticized; it has been argued that they need to be revamped beyond polycentricity concepts into an even more general spatial framework (Waddell and Shukla, 1993). In other words, both monocentric models and polycentric models are unlikely to be able to comprehend the more-than-polycentric nature of urban spatial structures, or capture the dynamics of rapid restructuring in Beijing in the near future.

More reliable models of population distribution require more sophisticated techniques (Briggs et al., 2007). In addition to the classical methods, SMPD was also implemented by integrating census, land use, settlements, road networks, and slopes to simulate urban dynamics (Schneiderbauer and Ehrlich, 2005). For instance, a model for urban dynamics was developed to spatially analyze the evolution of a population living in a fixed region with fixed employment (Ghordaf et al., 2009); and a dynamic activity-based model was developed to improve travel behavior analysis and predict transport modes (Beckx et al., 2009). In this chapter, SMPD is specified to simulate changes in trends and to develop scenarios of population distribution in Beijing by integrating data sets of census, digital elevation models, transportation systems, and residential distribution.

10.4.1 Specification of SMPD

SMPD integrates these related factors together using a grid generation method, generally formulated as Equations 10.13 and 10.14 on the national level of China (Yue et al., 2005a,b, 2009), in which a factor is a function of a variable while a parameter is a constant to be estimated.

For an urban area, $NPP_{i,j}(t)$ is replaced by a greenery-area indicator, $GR_{i,j}(t)$. When grid cell (i, j) is located in the greenery area, $GR_{i,j}(t) = 0$ or else $GR_{i,j}(t) = 1$, because the greenery patches are guaranteed by the regulation based on the land-use plan.

The formulation (Equation 10.14) is adjusted as

$$p_{i,j}(t) = W_{i,j}(t) \cdot GR_{i,j}(t) \cdot (DEM_{i,j}(t))^{a_3} \cdot (Tran_{i,j}(t))^{a_2} \cdot (Ct_{i,j}(t))^{a_1} \quad (10.19)$$

First, a search radius of 1 km is defined to formulate the road contribution to the transportation factor. Taking grid cell (i, j) as the search center, of which population density is being calculated, all roads are searched in a range of 1 km away from the search center. If no road is found in the search range, the search is repeated by increasing the search radius by 1 km until a road is found. All railways are searched in a range of 10 km away from the search center. The simulation results demonstrate that the transportation factor, $Tran_{i,j}(t)$, can be formulated as

$$Tran_{i,j}(t) = \left(\frac{Ra_{i,j}(t)}{\underset{i,j}{Max}\{Ra_{i,j}(t)\}} + \frac{Ro_{i,j}(t)}{\underset{i,j}{Max}\{Ro_{i,j}(t)\}} \right)$$

$$\times \left(\underset{i,j}{Max}\left\{ \frac{Ra_{i,j}(t)}{\underset{i,j}{Max}\{Ra_{i,j}(t)\}} + \frac{Ro_{i,j}(t)}{\underset{i,j}{Max}\{Ro_{i,j}(t)\}} \right\} \right)^{-1} \tag{10.20}$$

where $Ro_{i,j}(t) = \sum_{k=1}^{Nro_{i,j}(t)}(1/dro_{i,j,k}(t))$, $Nro_{i,j}(t)$ is the total number of roads found in the searching process, $dro_{i,j,k}(t)$ is the distance (km) from grid cell (i, j) to the kth road; $Ra_{i,j}(t) = \sum_{k=1}^{Nra_{i,j}(t)}(1/dra_{i,j,k}(t))$, $Nra_{i,j}(t)$ is the total number of railways found in the searching process, $dro_{i,j,k}(t)$ is the distance (km) from grid cell (i, j) to the kth railway and t is the time variable.

Taking grid cell (i, j) as the search center, all residential areas are searched in a range of 20 km away from the search center. If $M_{i,j}(t)$ is the total number of residential areas found in this search, our analyses shows that the residential factor and census have the biggest correlation coefficient when the residential factor at grid cell (i, j), $Ct_{i,j}(t)$, is formulated as

$$Ct_{i,j}(t) = \frac{RE_{i,j}(t)}{\underset{i,j}{Max}\{RE_{i,j}(t)\}} \tag{10.21}$$

where $RE_{i,j}(t) = \sum_{k=1}^{M_{i,j}(t)}(1/d_{i,j,k}(t))$, $d_{i,j,k}(t)$ is the distance (km) from grid cell (i, j) to the core grid cell that has the highest population density in the kth residential area.

Statistical analysis demonstrates that more than 80% of the population of Beijing inhabits an area where the elevation is lower than 80 m. The elevation factor and the census have the biggest correlation coefficient when the elevation factor is expressed as

$$DEM_{i,j} = dem_{i,j} \cdot \left(\underset{i,j}{Max}\{dem_{i,j}\} \right)^{-1} \tag{10.22}$$

where $dem_{i,j} = 160/(ele_{i,j} + 80)$, $ele_{i,j}$ is elevation (m) at grid cell (i, j).

The input data sets for running the specified SMPD include railway and road systems, residential-area distribution, DEM, township boundaries, county boundaries, and land use supplying data on water areas and greenery areas (Figure 10.18). These data sets are used to estimate the parameters, a_k ($k = 1,2,3$), by means of an exhaustion method (Layton, 1951; Kouremenos, 1997). First, let $a_3 = a_2 = 1$ and change the value of a_1, so the value starts at 0 and the increase step is 0.1. It is found that when $a_1 = 1.8$, the correlation coefficient between the census and the simulated population in the 18 administrative regions on a county-level reaches the maximum value. If we let $a_3 = 1$, and change the value of a_2, it is found that when $a_2 = 1.5$, the correlation coefficient reaches the maximum value. Using a similar method, we determine $a_3 = 0.8$. Because the simulation priority of a_k ($k = 1, 2, 3$) may have an effect on the simulation results, and { a_1, a_2, a_3} has six possible permutations, the other five similar simulation processes are also conducted by considering the simulation priority of each a_k ($k = 1, 2, 3$) in different permutations. The results show that the correlation coefficient between the county-level census and the mean simulated population on county-level has the biggest value if the order of the simulation priority of the exponents ($k = 1, 2, 3$) is taken up in terms of the subscripted label order from 1 to 3. Finally, the Beijing population distribution model is specified as

$$p_{i,j}(t) = W_{i,j}(t) \cdot GR_{i,j}(t) \cdot \left(DEM_{i,j}\right)^{0.8} \cdot \left(Tran_{i,j}(t)\right)^{1.5} \cdot \left(Ct_{i,j}(t)\right)^{1.8} \quad (10.23)$$

$$SMPD_{i,j}(t) = G_k(t) \cdot \frac{p_{i,j}(t)}{\sum p_{i,j}(t)} \quad (10.24)$$

where $G_k(t)$ is total the population of the census in the kth county of Beijing at time t; $p_{i,j}(t)$ is the weight of population at grid cell (i, j).

10.4.2 Descriptions Related to Data Sets

10.4.2.1 Baseline Situation of Beijing in 2000

Beijing is the capital city of China. In the year 2000, its total population was 13.57 million (Office for the Fifth Census of Beijing, 2002). It consisted of 18 counties, including 287 townships, and 4521 residential areas. Its road network comprised of 12 Beijing sections of national roads, 64 municipal roads, 238 county roads, and lots of township roads. The length of the 12 radial national roads starting from Beijing was 859 km in total, not including overlap sections. The 64 municipal roads, 238 county roads, and the township roads were, respectively, 1424 km, 2242 km, and 4879 km in 2000 (Editorial Board of Beijing Annals, 2000). The total road length in Beijing was 13,597 km at the end of 2000, including highways of 267 km, class-1 roads of 298 km, and class-2 roads of 1444 km (Qu, 2002). Class-1 roads are designed to have a

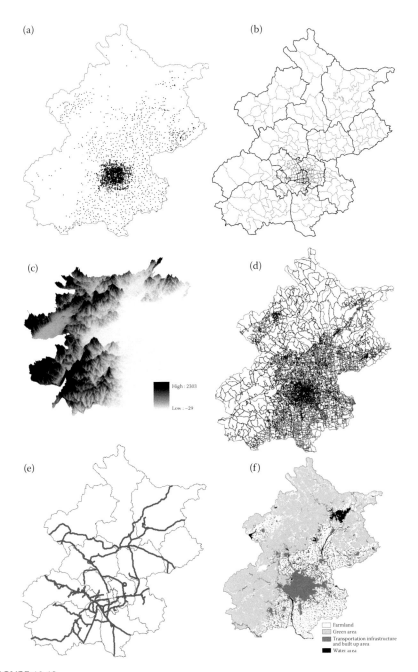

FIGURE 10.18
Baseline situation data sets of Beijing in 2000: (a) spatial distribution of residential areas, (b) boundaries of 18 counties and 287 townships, (c) digital elevation model (its azimuth angle of the light source is 315° and its altitude angle of the light source above the horizon is 30°), (d) road network, (e) rail network, and (f) land use.

traffic flow of between 15,000–30,000 cars per day and night. Class-2 roads are designed to have a traffic flow of between 3000 and 7500 cars per day and night. In 2000, there were 10 trunk railways radiating from Beijing to the northeast, the northwest, the south, the southeast, and the southwest of China (Editorial Board of Beijing Annals, 2004). In addition, Beijing had 11 branch railways and 32 linking railways. The Beijing sections of the trunk railways, the branch railways and the linking railways form the Beijing hub, one of the largest railway hubs in China.

The five major rivers that flow through Beijing territory include the Chao-Bai river, the Bei-Yun river, the Ji-Yun river, the Yong-Ding river, and the Ju-Ma river. The area of Beijing as a whole is 16,410 km^2 (Beijing Municipal Bureau of Statistics, 2007), in which plain accounts for 38.6%, and mountainous areas account for 61.4%. The west, north, and northeast of Beijing is surrounded by mountains. The southeast is plain, which slants down toward the Bohai sea. The elevation of the plain area is about 40 m on average. The elevation in the mountainous area is about 1250 m on average. The highest mountain peak, Dong-Ling mountain, where Beijing and the Hebei province meet, has an elevation of 2303 m (Figure 10.18).

10.4.2.2 Beijing Master Plan

The Beijing master plan from 2004 to 2020 has made a huge difference to the original urban spatial pattern of isotropic development around a single center (Figure 10.19b). The newly designed urban spatial pattern consists of two axes and two poly centers (Beijing municipality, 2005). The two axes refer to the east–west axis along Chang-An street, while the traditional south–north axis passes through the Olympic Center in northern Beijing and Nan-Yuan in southern Beijing.

The poly-centers refer to several functional centers to be constructed in Beijing that will serve the whole of China and be linked to the rest of the world. They include the Zhong-Guan-Cun Core Area of Sciences and Technologies, the Olympic Center, the Central Business District, the Scientific and Technological Innovation Center in the Hai-Dian Mountainous Area, the Shun-Yi Base for Modern Manufacturing Industry, the Tong-Zhou Comprehensive Service Center, the Yi-Zhuang Development Center for Sophisticated Technology Industry, and the Shi-Jing-Shan Comprehensive Service Center.

In terms of the Beijing master plan, Beijing can be divided into three districts: the old city, the central city, and the suburban region. The old city consists of Dong-Cheng, Xi-Cheng, Chong-Wen, and Xuan-Wu (Figure 10.20); the central city consists of Shi-Jing-Shan, Hai-Dian, Chao-Yang, and Feng-Tai as well as Hui-Long-Guan and the northern area of Bei-Yuan; while the suburban region consists of the planned new cities, Tong-Zhou, Shun-Yi, Yi-Zhuang, Da-Xing, Fang-Shan, Chang-Ping, Huai-Rou, Mi-Yun, Ping-Gu, Yan-Qing, and Men-Tou-Gou. Under the Beijing master plan, the population

(a)

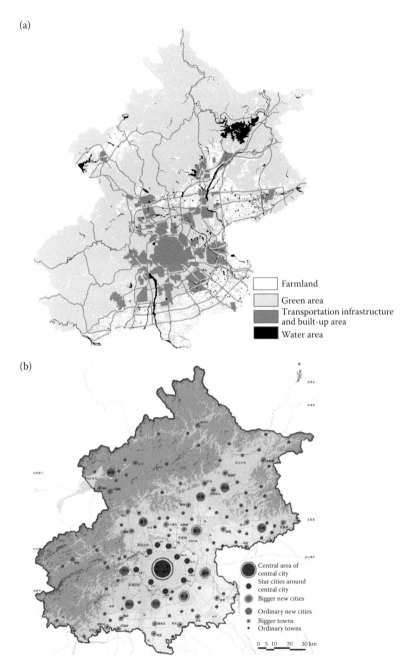

Farmland

Green area

Transportation infrastructure
and built-up area

Water area

(b)

Central area of
central city

Star cities around
central city

Bigger new cities

Ordinary new cities

Bigger towns

Ordinary towns

0 5 10   20   30 km

FIGURE 10.19
Relative plans: (a) the land-use plan, (b) the master plan, (c) the road plan, and (d) the railway
plan.

(c)

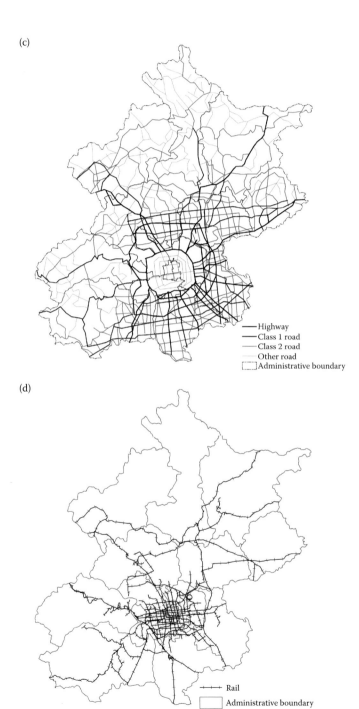

(d)

FIGURE 10.19
Continued.

FIGURE 10.20
Administrative units of Beijing on a county level.

of the old city would be limited to 1.1 million, the population of the central city would be limited to 7.4 million, and the total population of Beijing would be limited to 18 million in 2020. The annual population growth rate would therefore be 1.4% on an average. The population of the star cities around the central city would be 2.7 million, in the new cities it would be 5.7 million, and in the towns around the new cities it would be, 1.8 million. The proportion of urban population would be about 90%. The Master Plan will pilot the redistribution of population in Beijing, with population and industries likely to be decentralized from the old city and central city to the new cities and towns.

10.4.2.3 Land-Use Plan

In terms of the land-use plan of Beijing (Ouyang et al., 2005), land-use types include farmland, green areas, water areas, built-up areas, and transportation infrastructure. The spatial pattern of Beijing's greenbelt landscape that could be described as forest includes its greenbelt matrix, its rivers, and transport routes, and the built-up areas inlaid of the matrix (Figure 10.19a). Under Beijing's land-use plan, the ratio of greenery patches, including forest, grassland, water areas, wetland, farmland, and orchards, would be bigger than 60%, 70% of which would be made up of forest. The ratio of built-up areas, such as transportation routes, industrial areas, residential areas, and airports, would be limited to 40%. An appropriate quantity of residential quarters would be distributed in the greenbelt, but the total population in every residential quarter would not be allowed to exceed 0.1 million, the residential area would be limited to 3 km², and the distance between residential quarters would be required to be longer than 2 km. Greenery patches would account for 40% more than previously allowed in each residential quarter.

10.4.2.4 Beijing Transportation Infrastructure Plan

In 2020, the total planned length of road networks in Beijing is estimated to reach 22,000 km, of which the total length of trunk road networks, consisting of Beijing's national and municipal roads, would reach 3000 km. Road density would be 1.34 km/km². The national road system would include three trunk national roads and eight national roads (Figure 10.19c).

There are two systems of urban railways currently in operation in Beijing (Figure 10.19d); the first is the downtown railway transport system (DRTS), the second is the suburban railway transport system (SRTS). The DRTS would serve the central city and new cities, and include 16 underground railways and six subways. Its total length would be 693 km. The SRTS would serve the new cities and the areas along the railways between the new cities and the central city. The SRTS as planned would include six trunk railways, and its total length would be 429 km.

10.4.3 Simulation of Change Trends and Scenarios

10.4.3.1 Validation of the Specified SMPD

The censuses of Beijing in 2000 are classified into a training data set and a testing data set. The training data set is the census on a county level. The testing data set is the census on a township level. The training data set is used to simulate the spatial distribution of population on a spatial resolution of 100 × 100 m in Beijing (Figure 10.21) by operating the specified SMPD on data sets in 2000 (Figure 10.18). The testing data set is used to validate the model by calculating the difference between the census of every township and the sum of population on every grid cell on which central point is located in the administrative boundary of the corresponding township. If the census on the township level is regarded as the true value, the correlation coefficient between the census of every township and the sum of population on every grid cell of the township is $R^2 = 0.72$, which indicates that the simulated

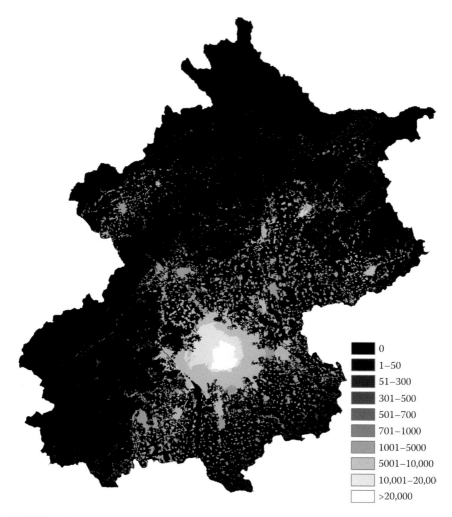

■	0
■	1–50
▦	51–300
▨	301–500
▨	501–700
▨	701–1000
▨	1001–5000
▨	5001–10,000
▨	10,001–20,00
□	>20,000

FIGURE 10.21
Simulated population distribution in the year 2000.

population basically matches the fifth census on the township level (Figure 10.22), that is, the specified SMPD can be applied to simulating population distribution in Beijing.

10.4.3.2 Changing Trends

Data sets of land-use (supplying information on water areas, greenery land, and transport systems) (Figure 10.23) and residential-area distribution (Figure 10.24) in 1985, 1995, and 2005 are used as inputs to simulate a change of population distribution trends under the assumption that DEM saw almost no change by running the specified SMPD.

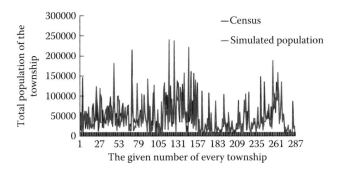

FIGURE 10.22
Comparison of census and simulated population on a township level.

The simulation results (Table 10.7 and Figure 10.25) indicate that the population in the old city had a decreasing trend. The old city had a population of 2.502 million in 1985, which reduced to 2.052 million in 2005. Its population decreased by 6600 people annually during the period from 1985 to 1995 and by 38,400 people annually from 1995 to 2005 on an average. The proportion

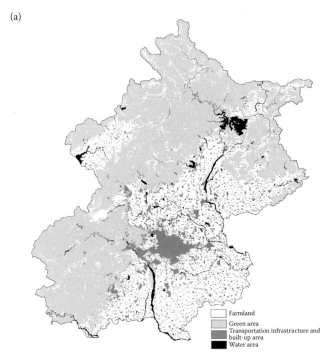

FIGURE 10.23
Land-use changes of Beijing in (a) 1985, (b) 1995, and (c) 2005. (Adapted from Data Center for Resources and Environmental Sciences, Chinese Academy of Sciences.)

(b)

(c)

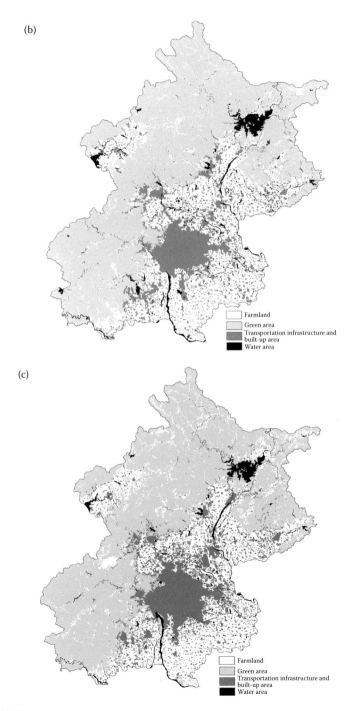

FIGURE 10.23
Continued.

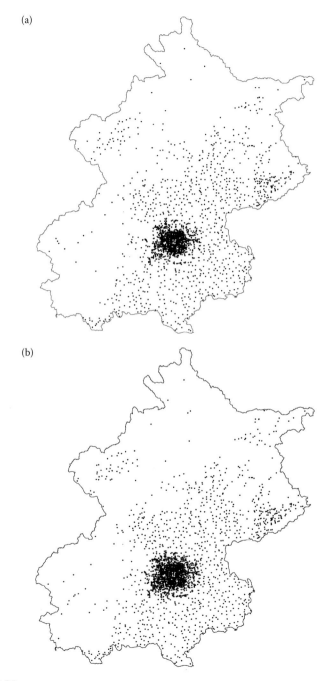

FIGURE 10.24
Residential-area distribution in (a) 1985, (b) 1995, and (c) 2005. (Adapted from Data Center for Resources and Environmental Sciences, Chinese Academy of Sciences.)

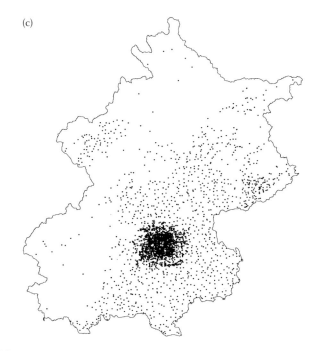

(c)

FIGURE 10.24
Continued.

of population in the old city to that of the whole of Beijing was reduced from 25.3% in 1985 to 13.34% in 2005.

The population in the central city increased considerably, from 3.306 million in 1985 to 7.48 million in 2005. Its increase rate was 17.42% during the period from 1985 to 1995 and 92.68% from 1995 to 2005. In the second decade,

TABLE 10.7

Changing Trend of Population Distribution in Beijing

Year	Region	Population	
		Million Persons	**Proportion (%)**
1985	Old city	2.502	25.30
	Central city	3.306	33.42
	Suburban region	4.083	41.28
1995	Old city	2.436	22.76
	Central city	3.882	36.27
	Suburban region	4.385	40.97
2005	Old city	2.052	13.34
	Central city	7.48	48.63
	Suburban region	5.848	38.03

(a)

(b)

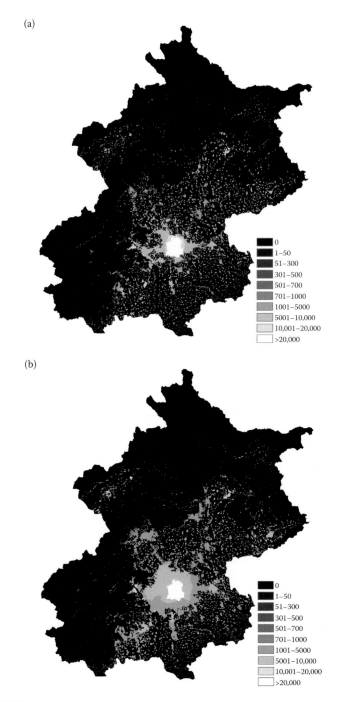

FIGURE 10.25
Change of population distribution trends in Beijing in (a) 1985, (b) 1995, and (c) 2005.

(c)

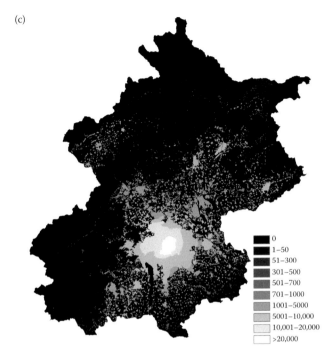

■	0
■	1–50
■	51–300
■	301–500
■	501–700
▦	701–1000
▦	1001–5000
▦	5001–10,000
▢	10,001–20,000
▢	>20,000

FIGURE 10.25
Continued.

the increase rate was much faster than that of the first decade. The proportion of population in the central city to that of the entire population of Beijing increased from 33.42% in 1985 to 48.63% in 2005.

The population of the suburban region increased from 4.083 million in 1985 to 5.848 million in 2005. There was an annual growth rate of 30,200 during the period from 1985 to 1995 (the first decade) and 146,300 from 1995 to 2005 (the second decade) on average. But the proportion of population in the suburban region to that of the entire population of Beijing had a decreasing trend. The decrease rate was 0.03% in the first decade and 0.3% in the second decade.

In short, much of the population of Beijing has been evacuated from the old city to its peripheral area since 1985. Although the population in both the central city and in the suburban region has seen an increase, the population in the central city has grown much faster than that of the suburban region. This means that the population has been congregating to central city at an increasing rate.

10.4.3.3 Scenarios

Three scenarios of Beijing population distribution (Figure 10.26 and Table 10.8) are developed by running the specified SMPD on the data of

(a)

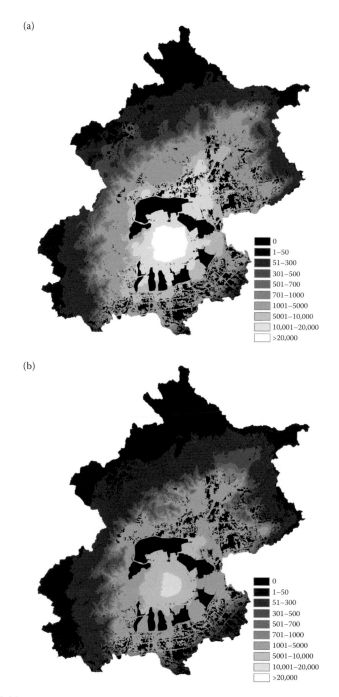

(b)

FIGURE 10.26
Scenarios of Beijing population distribution in the year 2020: (a) scenario I, (b) scenario II, and (c) scenario III.

(c)

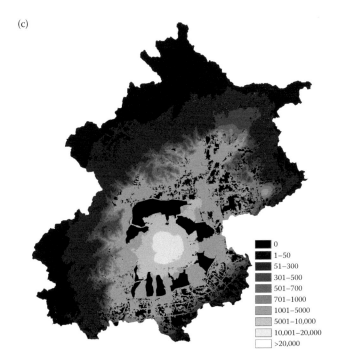

■	0
■	1–50
■	51–300
■	301–500
■	501–700
■	701–1000
■	1001–5000
■	5001–10,000
□	10,001–20,000
□	>20,000

FIGURE 10.26
Continued.

the land-use plan, the master plan, and the transportation plan (Figure 10.19) as well as related assumptions. Scenario I assumes that the total population of Beijing would be 46.6 million in 2020 if people were allowed to migrate freely in China (Yue et al., 2005b). Scenario II assumes that the total population of Beijing would be limited to 18 million in 2020 as specified by the Beijing master plan if inhabitants could migrate freely within Beijing. Scenario III is developed under the assumption that the Beijing master plan and greenbelt planning have been successfully carried out; and concludes that the total population of Beijing would be 18 million, 7.4 million of whom would be distributed in the central city and 1.1 million in the old city.

In scenario I, the population in all districts and counties of Beijing in 2020 would see considerable growth compared to that of the year 2000. The biggest annual growth rates would appear in Shun-Yi, Huai-Ru, and Chang-Ping; these growth rates would be 46.05%, 42.28%, and 30.2%, respectively. The old city would have 2.827 million inhabitants. The central city would have 16.574 million inhabitants. 27.199 million inhabitants would live in the suburban region.

In scenario II, the population of the old city would be 1.09 million and the population of the central city would be 6.4 million in 2020 (Table 10.9). The

TABLE 10.8

Population in 2000 and Population Scenarios in 2020 in Administrative Regions on the County Level

Administrative Regions on the County Level	Census in 2000 (Million Persons)	Scenarios I in 2020 (Million Persons)	Scenarios II in 2020 (Million Persons)	Scenarios III in 2020 (Million Persons)
Xuan-Wu	0.526	0.573	0.221	0.223
Xi-Cheng	0.707	1.008	0.389	0.392
Chong-Wen	0.346	0.455	0.176	0.177
Dong-Cheng	0.536	0.791	0.305	0.308
Shi-Jing-Shan	0.489	1.077	0.416	0.481
Chao-Yang	2.290	5.609	2.167	2.504
Hai-Dian	2.240	5.348	2.066	2.388
Feng-Tai	1.369	4.541	1.754	2.027
Tong-Zhou	0.674	2.367	0.914	0.827
Da-Xing	0.672	2.848	1.100	0.995
Shun-Yi	0.637	6.504	2.512	2.272
Chang-Ping	0.615	4.330	1.673	1.512
Ping-Gu	0.397	0.755	0.291	0.264
Fang-Shan	0.814	3.337	1.289	1.165
Mi-Yun	0.420	2.255	0.871	0.788
Men-Tou-Gou	0.267	1.646	0.636	0.575
Huai-Rou	0.296	2.799	1.081	0.978
Yan-Qing	0.275	0.359	0.139	0.125
The whole of Beijing	13.57	46.6	18	18

TABLE 10.9

Comparisons of SMPD with GPW and LandScan

Method	SMPD	LandScan	GPW
Spatial Resolution	1 km	1 km	5 km
Input Variables	• Census • Land cover • Transport infrastructures • DEM • Spatial distribution of cities • NPP • Meteorological data • Soil data • Satellite image	• Census • Land cover roads • DEM • Night-time lights	• Census

population in all districts of the old city would see annual decrease rates of 2.44% on an average. In the central city, the population of Shi-Jing-Shan, Hai-Dian, and Chao-Yang would see an annual decrease rate of 0.75%, 0.39%, and 0.27%, respectively. In the suburban region, the population of Yan-Qing and Ping-Gu would considerably decrease at an annual rate of 2.9% and 1.33%, respectively. In other districts and counties, the population would increase.

In scenario III, 1.015 million of the population of the old city would be decentralized between the years 2000 and 2020. The districts and counties, for which the population would increase, include Hai-Dian, Chao-Yang, Tong-Zhou, Fang-Shan, Feng-Tai, Da-Xing, Mi-Yun, Men-Tou-Gou, Chang-Ping, Huai-Rou, and Shun-Yi. Shun-Yi and Huai-Rou would see the greatest growth, at an annual rate of 12.83% and 11.51%, respectively, because Shun-Yi has been developing a base for Modern Manufacturing Industry and the most important highways and railways from Beijing to northeast China passing through Huai-Rou. Feng-Tai is a crucial part of the Beijing transportation hub, while in Chao-Yang both an Olympic Center and a Central Business District are being developed. The Yi-Zhuang Development Center for Sophisticated Technology Industry is currently under construction in Da-Xing. In the future, Beijing will have more of an urban structure than a polycentric one.

10.4.4 Brief Summary

During the period from 1985 to 2005, the population congregated into the central city from the old city and the suburban region. Clearly, population congregation was speeding up. The proportion of the population in the old city, the central city, and the suburban region to that of the whole of Beijing was respectively 13.34%, 48.63%, and 38.03%. The master plan of Beijing has changed this congregation trend. Three scenarios of Beijing population distribution in 2020 show that the old city and the central city would respectively have 1.09 million and 6.4 million inhabitants in 2020 if the total population of Beijing was limited to 18 million in 2020, and if its inhabitants could migrate freely within Beijing. The proportions of the population in the old city, the central city, and the suburban region to that of the whole of Beijing was respectively 6.07%, 35.57%, and 58.36%. If the Beijing master plan was successfully carried out, the total population in Beijing would be 18 million, 7.4 million of whom would be distributed in the central city and 1.1 million of which would be distributed in the old city. It is noted that the total population of Beijing would be 46.6 million in 2020 if people were allowed to migrate freely in China in terms of our previous result (Yue et al., 2005b).

This result proves that SMPD is not only able to simulate both mono-centric and polycentric urban population densities, but can also express the more-than-polycentric urban structure and capture the dynamics of rapid restructuring as designed in the Beijing master plan. In addition, SMPD not only pays attention to the situation of the related elements at the site of the generated grid cell itself, but also calculates the contributions of other grid cells by searching the surrounding environment of the generated grid cell. The search function of SMPD allows micro-scale simulation results to be integrated into large-scale patterns of population distribution. SMPD provides a solution for across-scale problems in simulating urban population distribution.

However, it is a hard work to formulate the factors and to estimate the parameters of SMPD. The processes of specifying SMPD would have to be considerably improved for different regions on different spatial levels for future results to be easily obtainable.

10.5 Conclusions and Discussion

When comparing SMPD with Landscan and GPW (as seen in Table 10.9), it is clear that SMPD is based on much more essential data such as land use, climate, soil, elevation, water systems, urbanization, transport infrastructure, and the census. When planning of land-cover, transport infrastructures, and cities are available for the SMPD, the scenarios of spatial population distribution can be developed on the basis of total population projections.

Human population distribution is mainly determined by geographical location factors such as distance from cities, transport infrastructures, and environmental conditions, for example, NPP and elevation. Research results of wildlife population distributions (Ji and Jeske, 2000; Krivan, 2003; Westerberg and Wennergren, 2003) show that distributions of wildlife population are characterized by geographical location and seasonal variations of movement patterns. The movement behavior of wildlife populations is impacted by environmental conditions and land uses. In addition to predator–prey dynamics and interspecific competition, physical texture, food resources, weather conditions, and landscape structure have a direct effect on wildlife population distributions. Differences in the magnitude of these factors also play a part. Therefore, wildlife population distribution is generally determined by geographical locations and environmental conditions. But some specific factors such as urbanization and transportation infrastructures, which might negatively impact some wildlife population distributions, have a positive effect on human population distribution; while other factors such as NPP have a similar effect on both human population and wildlife population. From this it seems that the grid generation method employed in SMPD could also be used to simulate wildlife population distribution.

References

Alcamo, J., Bouwman, A., Edmonds, J., Gruebler, A., Monita, T., and Sugandhy, A. 1995. An evaluation of the IPCC IS92 emission scenarios. In *Climate Change 1994*, ed., IPCC, 251–304. Cambridge: Cambridge University Press.

Alcamo, J. and Ribeiro, T. 2001. *Scenarios as Tools for International Environmental Assessments.* Luxembourg: Office for Official Publications of the European Communities.

Anderson, J.E. 1982. Cubic-spline urban-density functions. *Journal of Urban Economics* 12: 155–167.

Anderson, J.E. 1985. The changing structure of a city: Temporal changes in cubic-spline urban density patterns. *Journal of Regional Science* 25: 413–425.

Baumont, C., Ertur, C., and Gallo, J.L. 2004. Spatial analysis of employment and population density: The case of the agglomeration of Dijon 1999. *Geographical Analysis* 36(2): 146–176.

Beckx, C., Panis, L.I., Arentze, T., Janssens, D., Torfs, R., Broekx, S., and Wets, G. 2009. A dynamic activity-based population modelling approach to evaluate exposure to air pollution: Methods and application to a Dutch urban area. *Environmental Impact Assessment Review* 29: 179–185.

Beijing Municipality. 2005. Beijing Master Plan (from 2004 to 2020). *Beijing City Planning and Construction Review* 101 (2): 5–51 (in Chinese).

Beijing Municipal Bureau of Statistics. 2007. *Beijing Statistical Yearbook.* Beijing: China Statistics Press.

Berry, B.L.L. and Kim, H.M. 1993. Challenges to the Monocentric Model. *Geographical Analysis* 25 (1): 1–4.

Bontje, M. 2005. Edge Cities, European-style: Examples from Paris and the Randstad. *Cities* 22 (4): 317–330.

Bracken, I. and Martin, D. 1989. The generation of spatial population distribution from census centroid data. *Environment and Planning A* 21: 537–543.

Briggs, D.J., Gulliver, J., Fecht, D., and Vienneau, D.M. 2007. Dasymetric mapping of small-area population distribution using land-cover and light emissions data. *Remote Sensing of Environment* 108: 451–466.

Cassedy, J.H. 1984. *American Medicine and Statistical Thinking, 1800–1860.* Cambridge: Harvard University Press.

Clark, C. 1951. Urban population densities. *Journal of Royal Statistical Society* 114: 490–496.

Cressman, G.P. 1959. An operational objective analysis system. *Monthly Weather Review* 87: 367–374.

Cullen, J. 1975. *The Statistical Movement in Early Victorian Britain: The Foundation of Empirical Social Science.* Brighton: Hassocks.

Deichmann, U. 1996. *A Review of Spatial Population Database Design and Modeling.* Technical Report 96–3, National Center for Geographic Information and Analysis, USA.

Dobson, J.E., Bright, E.A., Coleman, P.R., Durfee, R.C., and Worley, B.A. 2000. LandScan: A global population database for estimating populations at risk. *Photogrammetric Engineering and Remote Sensing* 66(7): 849–857.

Editorial Board of Beijing Annals. 2000. *Beijing Annals of Road Transportation.* Beijing: Beijing Press (in Chinese).

Editorial Board of Beijing Annals. 2004. *Beijing Annals of Railway Transportation*. Beijing: Beijing Press (in Chinese).

Editorial Committee of National Report of China Urban Development of China Mayor Association. 2003. *National Report of China Urban Development (2001–2002)*. Beijing: Xiyuan Press (in Chinese).

Editorial Committee of National Report of China Urban Development of China Mayor Association. 2004. *National Report of China Urban Development (2002–2003)*. Beijing: Xiyuan Press (in Chinese).

Feeney, G. and Yu, J.Y. 1987. Period parity progression measures of fertility in China. *Population Studies* 41(1): 77–102.

Gallopin, W. and Rijsberman, F. 1999. *Three Global Scenarios*. Paris: World Water Council.

Ghordaf, J.E., Hbid, M.L., Sánchez, E., and Langlais, M. 2009. On the evolution of spatially distributed urban populations: Modelling and mathematical analysis. *Nonlinear Analysis: Real World Applications* 10(5): 2945–2960.

Godet, M. 1987. *Scenarios and Strategic Management*. London: Butterworth Scientific Ltd.

Goodchild, M.F., Anselin, L., and Deichmann, U. 1993. A framework for the areal interpolation of socioeconomic data. *Environment and Planning A* 25: 383–397.

Goodchild, M.F. and Lam, N.S.N. 1980. Areal interpolation: A variant of the traditional spatial problem. *Geo-Processing* 1: 297–312.

Gordon, P., Richardson, H.W., and Wong, H.L. 1986. The distribution of population and employment in a polycentric: The case of Los Angeles. *Environment and Planning A* 18: 161–173.

Griffith, D.A. and Wong, D.W. 2007. Modeling population density across major US cities: A polycentric spatial regression approach. *Journal of Geographical Systems* 9: 53–75.

Haefele, W., Anderer, J., Mcdonald, A., and Nakicenovic, N. 1981. *Energy in a Finite World: Paths to a Sustainable Future*. Cambridge: Ballinger.

Hall, P. 1999. The future of cities. *Computers, Environment and Urban Systems* 23: 173–185.

Harris, R. and Chen, Z.Q. 2005. Giving dimension to point locations: Urban density profiling using population surface models. *Computers, Environment and Urban Systems* 29: 115–132.

Hu, H.Y. 1983. *Discussion on Distribution of Population in China*. Shanghai: Eastern China Normal University Press (in Chinese).

Jefferson, M. 1909. The anthropography of some great cities: A study in distribution of population. *Bulletin of American Geographical Society* 41(4): 537–566.

Jefferson, M., 1983. Economic uncertainty and business decision-making. In *Beyond Positive Economics?* Ed., J. Wiseman, 122–159. London: MacMillan Press.

Ji, M. and Jensen, J.R. 1999. Effectiveness of subpixel analysis in detecting and quantifying urban imperviousness from Landsat Thematic Mapper imagery. *Geocarto International* 14(4): 31–39.

Ji, W. and Jeske, C. 2000. Spatial modeling of the geographical distribution of wildlife populations: A case study in the lower Mississippi River region. *Ecological Modelling* 132: 95–104.

Kahn, H. and Wiener, A.J. 1967. *The Year 2000*. New York: MacMillan.

Kouremenos, T. 1997. Mathematical rigor and the origin of the exhaustion method. *Centaurus* 39: 230–252.

Krivan, V. 2003. Idea free distributions when resources undergo population dynamics. *Theoretical Population Biology* 64: 25–38.

Langford, M. and Unwin, D.J. 1994. Generating and mapping population density surfaces within a geographical information system. *The Cartographic Journal* 31: 21–26.

Langford, M. 2006. Obtaining population estimates in noncensus reporting zones: An evaluation of the 3-class dasymetric method. *Computers, Environment and Urban Systems* 30: 161–180.

Langford, M. 2007. Rapid facilitation of dasymetric-based population interpolation by means of raster pixel maps. *Computers, Environment and Urban Systems* 31: 19–32.

Langford, M., Higgs, G., Radcliffe, J., and White, S. 2008. Urban population distribution models and service accessibility estimation. *Computers, Environment and Urban Systems* 32: 66–80.

Layton, W. 1951. The relationship between the method of successive residuals and the method of exhaustion. *Psychometrika* 16(1): 51–56.

Lee, J. and Hong, W. 1974. *Metropolitan Demographic Projections: 1960–1985*. Washington, DC: National Planning Association.

Lieth, H. and Whittaker, R.H. 1975. *Primary Productivity of the Biosphere*. New York: Springer-Verlag.

Liu, M.L. 2003. *Dynamic Responses of Terrestrial Ecosystems to Carbon Cycle under Impact of Climate Change and Land-use Change in China*. Post-Doctoral Thesis of Institute of Geographical Sciences and Natural Resources Research of CAS (in Chinese).

Liu, J.Y., Liu, M.L., Zhuang, D.F., Zhang, Z.X., and Deng, X.Z. 2003. Study on spatial pattern of land-use change in China during 1995–2000. *Science in China (Series D)* 46: 373–384.

Liu, X.H., Kyriakidis, P.C., and Goodchild, M.F. 2008. Population-density estimation using regression and area-to-point residual kriging. *International Journal of Geographical Information Science* 22(4–5): 431–447.

Lo, C.P. 2008. Population estimation using geographically weighted regression. *GIS and Remote Sensing* 45(2): 131–148.

Loibl, W. and Toetzer, T. 2003. Modeling growth and densification processes in suburban regions—Simulation of landscape transition with spatial agents. *Environmental Modelling and Software* 18: 553–563.

Ma, Y.T. 1989. *Population Statistics*. Beijing: Hongqi Press (in Chinese).

Martin, D. 1996. An assessment of surface and zonal models of population. *International Journal of Geographical Information Systems* 10(8): 973–989.

Martin, D. 1998. Automatic neighbourhood identification from population surfaces. *Computers, Environment and Urban Systems* 22(2): 107–120.

Martin, D. and Bracken, I. 1991. Techniques for modeling population-related raster databases. *Environment and Planning A* 23: 1069–1075.

Mcmillen, D.P. and Mcdonald, J.F. 1998. Population density in Chicago: A bid rent approach. *Urban Studies* 7: 1119–1130.

Mennis, J. 2003. Generating surface models of population using dasymetric mapping. *The Professional Geographer* 55(1): 31–42.

Mesev, V. 1998. The use of census data in urban image classification. *Photogrammetric Engineering and Remote Sensing* 64(5): 431–438.

Millennium Ecosystem Assessment. 2003. *Ecosystems and Human Well-being*. Washington: Island Press.

Moxey, A. and Allanson, P. 1994. Areal interpolation of spatially extensive variables: A comparison of alternative techniques. *International Journal of Geographical Information Systems* 8(5): 479–487.

Mueller, W. and Rohr-Zaenker, R. 1995. Neue Zentren in den Verdichtungsraeumen der USA. *Raumforschung und Raumordnung* 53(6): 436–443.

Nakicenovic, N., Alcamo, J., Davis, G. et al., 2000. *Special Report on Emissions Scenarios*. Cambridge: Cambridge University Press.

National Bureau of Statistics of the People's Republic of China. 2001. *China Statistical Yearbook*. Beijing: China Statistics Press.

National Bureau of Statistics of the People's Republic of China. 2003. *China Statistical Yearbook*. Beijing: China Statistics Press.

Newling, B.E. 1969. The spatial variation of urban population densities. *Geographical Review* 59: 242–252.

Office for the Fifth Census of Beijing. 2002. *Census of Beijing in 2000*. Beijing: China Statistics Press (in Chinese).

Organization of Economic Cooperation and Development. 2001. *OECD Environmental Outlook*. Paris: OCED.

Ouyang, Z.Y., Wang, R.S., Li, W.F., Paulussen, J., Li, D.H., Xiao, Y., and Wang, X.K. 2005, Ecological planning on greenbelt surrounding mega city, Beijing. *Acta Ecologia Sinica* 25(5): 965–971 (in Chinese).

Parr, J.B. 1985. The form of the regional density function. *Regional Studies* 19: 535–546.

Pearl, R. and Reed, L.J. 1920. On the rate of growth of the population of the United States since 1790 and its mathematical representation. *Proceedings of the National Academy of Sciences* 6(6): 275–288.

Plane, D.A. and Rogerson, P.A. 1994. *The Geographical Analysis of Population: with Applications to Planning and Business*. New York: John Wiley & Sons.

Pritchett, H.S. 1891. A formula for predicting the population of the United States. *Quarterly Publication of the American Statistical Association* 2(14): 278–286.

Putting, J. and Bakkes, J. 2003. *The GEO-3 Scenarios 2002–2032*. Nairobi: UNEP.

Qu, F. 2002. Road planning and construction of Beijing. *Beijing City Planning and Construction Review* 82(1): 32–34 (in Chinese).

Raskin, P. 2000. *Regional Scenarios for Environmental Sustainability: A Review of the Literature*. Nairobi: UNEP.

Research Group of National Population Strategy for Development of China. 2007. *National Population Strategy for Development of China*. Beijing: China Population Press (in Chinese).

Schneiderbauer, S. and Ehrlich, D. 2005. Population density estimations for disaster management: Case study rural Zimbabwe. In *Geo-information for Disaster Management*, eds., P. van Oosterom, S. Zlatanova, and E.M. Fendel, 901–921. Berlin: Springer.

Schoute, J.F.T., Finke, P.A., Veeneklaas, F.R., and Wolfert, H.P. 1995. *Scenario Studies for the Rural Environment*. London: Kluwer Academic Publishers.

Small, K. and Song, S. 1994. Population and employment densities: Structure and change. *Journal of Urban Economics* 36: 292–313.

Smith, S.K. 1984. *Population Projections: What Do We Really Know?* Gainesville: Board of Regents of the State of Florida.

Sutton, P., Elvidge, C., and Obremski, T. 2003. Building and evaluating models to estimate ambient population density. *Photogrammetric Engineering & Remote Sensing* 69(5): 545–553.

Tian, Y.Z., Yue, T.X., Zhu, L.F., and Clinton, N. 2005. Modeling population density using land-cover data. *Ecological Modelling* 189: 72–88.

Tobler, W.R. 1979. Smooth pycnophylactic interpolation for geographical regions. *Journal of the American Statistical Association* 74: 519–530.

Tobler, W., Deichmann, U., Gottsegen, J., and Maloy, K. 1997. World population in a grid of spherical quadrilaterals. *International Journal of Population Geography* 3: 203–225.

van de Klundert, A.F. 1995. The future's future: Inherent tensions between research, policy and the citizen in the use of future oriented studies. In *Scenario Studies for the Rural Environment*, eds., J.F.T. Schoute, P.A. Finke, F.R. Veeneklaas, and H.P. Wolfert, 25–32. London: Kluwer Academic Publishers.

Veeneklaas, F.R. and Van Den Berg, L.M. 1995. Optimization of regional water management through scenarios analysis. In *Scenario Studies for the Rural Environment*, eds., J.F.T. Schoute, P.A. Finke, F.R. Veeneklaas, and H.P. Wolfert, 11–13. London: Kluwer Academic Publishers.

Wack, P. 1985. Scenarios: Shooting the rapids. *Harvard Business Review* 63: 135–150.

Waddell, P. and Shukla, V. 1993. Employment dynamics, spatial restructuring, and the business cycle. *Geographical Analysis* 25(1): 35–52.

Wang, F. 1998. Urban population distribution with various road networks: A simulation approach. *Environment Planning B* 25: 265–278.

Wang, G.T. 2003. Issues on urbanization of China. *City Planning Review* 27(4): 11–14 (in Chinese).

Westerberg, L. and Wennergren, U. 2003. Predicting the spatial distribution of a population in a heterogeneous landscape. *Ecological Modelling* 166: 53–65.

Whelpton, P.K. 1928. Population of the United States, 1925 to 1975. *American Journal of Sociology* 34(2): 253–270.

Woods, R. and Rees, P. 1986. Spatial demography: Themes, issues and progress. In *Population Structures and Models: Developments in Spatial Demography*, eds. R. Woods and P. Rees, 1–3. London: Allen & Unwin.

World Business Council on Sustainable Development. 1997. *Exploring Sustainable Development*. Geneva: WBCSD.

World Energy Council. 1993. *Energy for Tomorrow's World*. London: Kogan Page Ltd.

Wright, J. 1936. A method of mapping densities of population with Cape Cod as an example. *Geographical Review* 26: 519–536.

Wu, F. 1998. Polycentric urban development and land-use change in a transitional economy: The case of Guangzhou. *Environment and Planning A* 30: 1077–1100.

Wu, C.S. and Murray, A.T. 2005. A cokriging method for estimating population density in urban areas. *Computers, Environment and Urban Systems* 29: 558–579.

Year Book House of China Transportation and Communications. 2001. *Year Book of China Transportation and Communications*. Beijing: Year Book House of China Transportation and Communications (in Chinese).

Yin, Z.Y., Stewart, D.J., Bullard, S., and Maclachlan, J.T. 2005. Changes in urban built-up surface and population distribution patterns during 1986–1999: A case study of Cairo, Egypt. *Computers, Environment and Urban Systems* 29: 595–616.

Yuan, Y., Smith, R.M., and Limp, W.F. 1997. Remodelling census population with spatial information from Landsat TM imagery. *Computers Environment and Urban Systems* 21: 245–258.

Yue, T.X., Haber, W., Grossmann, W.D., and Kasperidus, H.D. 1999. A method for strategic management of land. In *Environmental Indices—Systems Analysis*

Approach, eds., Y.A. Pykh, D.E. Hytt, and R.J.M. Lenz, 181–201. Oxford: EOLSS Publishers Co Ltd.

Yue, T.X., Wang, Y.A., and Fan, Z.M. 2010. Surface modelling of population distribution. In *Handbook of Ecological Modelling and Informatics*, Sven Erik Jørgensen, ed., Southampton: WIT Press.

Yue, T.X., Wang, Y.A., Chen, S.P., Liu, J.Y., Qiu, D.S., Deng, X.Z., Liu, M.L., and Tian, Y.Z. 2003. Numerical simulation of population distribution in China. *Population and Environment* 25(2): 141–163.

Yue, T.X., Wang, Y.A., Liu, J.Y., Chen, S.P., Qiu, D.S., Deng, X.Z., Liu, M.L., Tian, Y.Z., and Su, B.P. 2005a. Surface modelling of human population distribution in China. *Ecological Modelling* 181(4): 461–478.

Yue, T.X., Wang, Y.A., Liu, J.Y., Chen, S.P., Tian, Y.Z., and Su, B.P. 2005b. SMPD scenarios of spatial distribution of human population in China. *Population and Environment* 26(3): 207–228.

Zhang, S.Y. 1997. *Population Geography of China*. Beijing: Business Press House (in Chinese).

Zhao, S.Q. 1986. *Physical Geography of China*. New York, NY: John Wiley & Sons.

Zheng, X.P. 1991. Metropolitan spatial structure and its determinants: A case-study of Tokyo. *Urban Studies* 28(1): 87–104.

11

*Surface Modeling of Human Carrying Capacity**

11.1 Introduction

11.1.1 Definitions of Carrying Capacity

Discussions about carrying capacity can be traced back to Malthus' treatise on human population growth which is based on the assumption that human population increases exponentially and that food is the sole limiting factor on its growth (Malthus, 1789). Although Multhus' theory has been criticized, his treatise, due to its great influence on Darwin's concept of natural selection and its part in developing the incipient science of human demography, provides a solid basis for the concept of carrying capacity (Seidl and Tisdell, 1999; Price, 1999).

Many definitions of carrying capacity, based on different aims and various conditions, have been proposed. In 1922, based on observing the effects of the introduction of populations of reindeer in Alaska, carrying capacity was defined as the number of stock that a range can support, without injuring that range (Hawden and Palmer, 1922). As a planning tool, carrying capacity is defined as the ability of a natural or artificial system to support the demands of various uses. It subsequently refers to inherent limits in the systems beyond which instability, degradation, or irreversible damage occurs (Godschalk and Park, 1975). In tourism, carrying capacity is defined as the maximum number of visitors that can be tolerated without unacceptable deterioration of the physical environment and without considerably diminishing user satisfaction (Mathieson and Wall, 1982). Human carrying capacity (HCC) is defined as the maximum level of exploitation of a renewable resource, imposing limits on a specific type of land-use, which can be sustained without causing irreversible land degradation within a given area (Kessler, 1994). Human carrying capacities are contingent on technology, preferences, and structures of production and consumption (Arrow et al., 1995). As a basis of sustainable development, carrying capacity is defined as the ability to produce desired outputs from a constrained resource base to achieve a higher and more equitable quality of life, while maintaining the desired environmental

---

* Dr. Yong-Zhong Tian is a major collaborator of this chapter.

quality and ecological health (Khanna et al., 1999). The definition of carrying capacity for protected areas focuses on the acceptability of natural resources and the human impact of visitation, and considers biophysical characteristics of a protected area, social factors, and management policies to be more important determinants of carrying capacity than the number of visitors (Prato, 2001). In planning and managing urban development, carrying capacity is defined as the level of human activity, population growth, land-use, and physical development that can be sustained by the urban environment without causing serious degradation or irreversible damage (Oh et al., 2005).

In general, carrying capacity can be defined as the maximum human, livestock, or wildlife population size that a given habitat can support without being permanently damaged (Odum, 1989; Rees, 1992; Kessler, 1994; Hui, 2006; Haraldsson and Ólafsdóttir, 2006). The HCC of terrestrial ecosystems in China is calculated by surface modeling under this general definition.

11.1.2 Estimates of Human Carrying Capacity on a Global Level

At least 42 estimates of HCC on a global level can be found in scientific literature (Cohen, 1995). Estimated HCCs vary from 2.026 billion to 1022 billion people (Figure 11.1), while the projected range of world population is from 7.8 billion to 12.5 billion in 2050 and from 6 billion to 20 billion at the end of the twenty-first century (Conway, 1997). Overall, 57.14% of estimates range between 7 billion and 30 billion people. The methods include categorical assertion, generalization from observed population sizes, and estimation in terms of limiting factors. For instance, according to Ravenstein (1891), the Earth could support 5.994 billion people. Penck (1925) described his calculation as the number of people on the earth equals the productive area times its average production per unit of area, divided by the average nutritional requirement per individual. His calculation showed that the highest conceivable number of inhabitants would be 15.9 billion. De Wit (1967) concluded that 1022 billion people could live on the Earth if photosynthesis is the limiting factor; if people wanted a million kilocalories a year, plus 750 square meters per person for

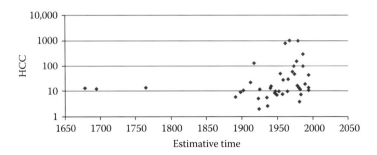

FIGURE 11.1
Various estimates of HCC (billion persons) in chronological order.

living and recreation, then 146 billion could be supported on total land of 13.1 billion hectares; if 200 g of meat was added to the daily diet, HCC would be reduced from 146 billion to 126 billion. Hulett (1970) estimated that the optimum population of the world must be less than 1 billion people when world population approached 3.7 billion people in 1970. According to Revelle's estimates (1976), the total potential gross cropped area is 4.23 billion of hectares, on which 40 billion could be supported in pest losses and nonfood uses could be kept to 10% of the harvest. On the world's total land area of 12.3 billion hectares, excluding ice and tundra, the world's potential agricultural and forest land could supply the needs of 157 billion people living at Japanese standards of food consumption and Asian standards of timber requirements (Clark, 1977). On the basis of the overlay of FAO (Food and Agriculture Organization of the United Nations) maps of climate on FAO maps of soil at the 1:5,000,000 scale, the potential productivity of each crop was estimated on each agro-ecological cell of land by combining multiple factors, in which the amount of each crop could be reduced to calories. The results indicated that on a supporting capacity with high inputs, the Earth could potentially provide for 32.8 billion people on a subsistence diet similar to that of the developing world, on 6.495 billion hectares of land area excluding the areas mapped as water bodies (Higgins et al., 1983; United Nations Fund for Population Activities, 1993). Millman et al. (1991) estimated that the primary food supply could potentially feed 5.9 billion people on a basic diet, 3.9 billion people on an improved diet, and 2.9 billion people on a full-but-healthy diet.

According to Cohen's comments (Cohen, 1995), Penck's procedure for estimating maximum population density represented an advancement in theory, but not in method, over Ravenstein's. The major assumptions underlying Penck's equations are that human numbers are limited by food alone and individual food requirements are constant in every place and time. De Wit's procedure omitted the physical problem of supplying soils everywhere with water and minerals, the biological problem of preventing all forms of damage and spoilage before final consumption and the economic problem of providing infrastructure and preparing soils for maximal productivity. Hulett's 15.7 ratio between the caloric value of animal feed and the caloric value of consumed animal products is much higher than the 4.5 used by Revelle and the 10 used by De Wit. Revelle assumed that the amount of cultivated land could be increased more than twice, the present impact of agriculture on environmental systems could be greatly increased without penalty and food would be perfectly equitably distributed among people without grain being diverted to livestock. Clark paid little attention to the possible loss of food between its growth in the field and its final consumption and his estimates are sensitive to the details of his assumptions. Although Higgins et al. recognized that soil, water, and climate can limit photosynthesis before sunlight does, their calculation excluded East Asia and the developed world. The Hunger Reports assumed that if food were uniformly distributed to all people and the number of projected producers differed from the actual number of producers, then the total food supply would remain equal to the actual food supply, which seems unlikely. All these estimates inadequately picture human–planetary interactions that are intrinsically dynamic because knowledge of the past, present, and future is highly imperfect.

11.2 Method

11.2.1 Ecological Thresholds

In the physical sciences, a threshold is described as a point or zone where there is a dramatic change in the state matter or system (Luck, 2005). In the literature of ecology, many definitions of ecological thresholds have been put forward (Huggett, 2005). For instance, Friedel (1991) considered thresholds as boundaries in space and time between two different states. Muradian (2001) regarded thresholds as the critical values of an independent variable, around which a change from one stable state to another occurs. Meyers and Walker (2003) defined an ecological threshold as a bifurcation point between alternate states that, when passed, causes a system to "flip" to a different state.

Ecological thresholds can be classified into extinction thresholds and carrying capacity thresholds. In a conservation biology context, extinct thresholds of native vegetation cover are defined as abrupt and nonlinear changes that occur in some measure, such as the rate of loss of species across a small amount of habitat loss (With and King, 1999; Lindenmayer et al., 2005). In population ecology, an extinction threshold is a population size below which extinction risk becomes so high that it is considered unacceptable (Shaffer, 1981; Hildenbrandt et al., 2006).

If the population density in an ecosystem exceeds its carrying capacity, the ecosystem can be damaged due to overload, and the overused ecosystem may change to a different ecosystem, which is understood as ecological discontinuity. Ecological discontinuity is generally defined as a sudden change in any property of an ecological system as a consequence of smooth and continuous change in an independent variable (Muradian, 2001). Ecological discontinuities imply critical values of the independent variable, around which the system mutates from one stable state to another. The critical values are carrying-capacity thresholds. Carrying-capacity thresholds for natural conservation areas have been proposed to ensure the sound development of natural areas; these capacities range from 10% to 30% of the agricultural area (Haber, 1972; Andrén, 1994; Fahrig, 1998; Luck, 2005). These carrying-capacity thresholds have been embraced by political parties as a means of protecting the environment, although the thresholds are still a matter of extended scientific discussion (Herrmann et al., 2003).

After summarizing the related research achievements (Beverton and Holt, 1957; Ricker, 1958; Slobodkin, 1962; Silliman, 1969; Odum, 1971), Odum (1983) concluded that: (1) the optimum density of population sustainable over a long period of time in the face of environmental uncertainties is perhaps 50% lower than the theoretical maximum carrying capacity K; (2) from a sustainable development standpoint, the range between maximum carrying capacity and optimum density seems to represent the sustainable growth range of population; (3) the sustainable growth range of population is often skewed to the left (taking the $n = 0.5\ K$ as its central axis).

Here, we answer the following three questions using the theory of dissipative structures: (1) under what condition the sustainable growth range of population $[0.5K, K]$ is valid (2) under what condition the sustainable growth range is skewed to the left (3) if the sustainable growth range is skewed to the left, how far it will be skewed and what will be its upper and lower limits.

11.2.1.1 Entropy, Entropy Production, and Excess Entropy Production

The total entropy change in a system can be formulated as

$$dS = d_eS + d_iS, \tag{11.1}$$

where d_eS is the transfer of entropy across the boundaries of the system; d_iS the entropy produced within the system.

According to the second law of thermodynamics, the entropy production within the system is positive (Prigogine, 1968, 1980). The total entropy change with time in a system, can be expressed as (Glausdorff and Prigogine, 1971)

$$\frac{dS}{dt} = \int_V \frac{\partial S}{\partial t} \cdot dV = -\int_\Sigma d\Sigma n \cdot J_s + \int_V \sigma \cdot dV \tag{11.2}$$

where J_s is the exchange rate of entropy through a unit area, that is, entropy flow; σ is the rate of entropy production in a unit volume, that is, entropy production.

Therefore,

$$\frac{d_iS}{dt} = \int_V \sigma \cdot dV = P \tag{11.3}$$

$$\frac{d_eS}{dt} = \int_\Sigma d\Sigma n \cdot J_s \tag{11.4}$$

where V is volume; Σ is the surface area encircling the volume.

Then, entropy production can be formulated as (Onsager, 1931a,b)

$$\sigma = \sum_k J_k \cdot X_k \tag{11.5}$$

where J_k represents the generalized flow of the kth irreversible process, which relates to speed; X_k represents the generalized forces of the kth irreversible process, which relates to motive force.

For an open system, when the critical conditions compel the system to leave the equilibrium state, a macroirreversible process begins. Because the generalized force is the cause of the generalized flow, the generalized flow can be regarded as a specific function of the generalized force. If this function exists and can be expanded on the basis of the equilibrium state, in which the generalized force and the generalized flow are equal to zero in terms of the Taylor series, then

$$J(X) = J(X_0) + \left(\frac{\partial J}{\partial X}\right)_0 \cdot (X - X_0) + \frac{1}{2} \cdot \left(\frac{\partial^2 J}{\partial X^2}\right) \cdot (X - X_0)^2 + \cdots$$

$$= \left(\frac{\partial J}{\partial X}\right)_0 \cdot X + \frac{1}{2} \cdot \left(\frac{\partial^2 J}{\partial X^2}\right) \cdot X^2 + \cdots \tag{11.6}$$

When the generalized force is very weak and the system has a small deviation from the equilibrium state, the higher powers of X can be eliminated and then

$$J(X) = \left(\frac{\partial J}{\partial X}\right)_0 \cdot X \tag{11.7}$$

The nonequilibrium state with relation to Equation 11.7 is called the linear region of it.

When the generalized force is stronger and the system is far away from the equilibrium state, the expanding formula of $J(X)$ includes the higher powers of X so that the generalized flow is a nonlinear function of the generalized force, which is called the nonlinear region of the nonequilibrium state. Under this condition, the entropy S and entropy production P can be expanded near the equilibrium state and the following formula can be given through strictly deductive reasoning:

$$\frac{d\left(\frac{1}{2}(\delta^2 S)\right)}{dt} = \int \sum_k \delta J_k \cdot \delta X_k \cdot dV = \delta_X P \tag{11.8}$$

In other words, the derivative of $(1/2)(\delta^2 S)$ with respect to time is the excess entropy production $\delta_X P$, in which suffixal variable X is the generalized force and δ is a variational symbol.

In short, in the linear region of nonequilibrium, the stability of the system can be judged by the entropy production $P \geq 0$ and $(dP/dt) \leq 0$; in the nonlinear region of nonequilibrium, the excess entropy production $\delta_X P$, greater than, smaller than, and equal to zero, respectively, correspond to stable, unstable, and critical states.

11.2.1.2 Logistic Equation of Population Growth

All populations have the characteristics of growth. These characteristics can be classified into two types, the *r*-type and the *k*-type growth (McCullough, 1979). The individuals of the *r*-type population are smaller in size; they mature earlier, have a larger reproductive allocation, and tend to have more offsprings (Begon et al., 1990). The *r*-type growth can be formulated as an exponential equation of

$$\frac{dn(t)}{dt} = r \cdot n(t) \tag{11.9}$$

where *r* is the intrinsic growth rate of the population and $n(t)$ is the number of individuals in the population of the unit area.

The individuals of the *k*-type population are larger in size; they tend to delay reproduction, have a lower reproductive allocation, and produce fewer offsprings with more parental care. The *k*-type growth can be formulated as a logistic equation

$$\frac{dn(t)}{dt} = r \cdot n(t) \cdot \left(1 - \frac{n(t)}{K}\right) \tag{11.10}$$

where *K* is the maximum carrying capacity in the unit area.

11.2.1.3 Theoretically Reasoning the Range of Sustainable Growth Population

Haber classified ecosystems into natural, near-natural, seminatural, anthropogenic (biotic), and techno-ecosystems in terms of increasing artificiality (Haber, 1990). If the natural, near-natural, and seminatural ecosystems are regarded as the linear region of the nonequilibrium state, and the anthropogenic biotic and anthropogenic technical ecosystems are considered as the nonlinear region of the nonequilibrium state, then for the linear region of the nonequilibrium state, the generalized flow $J(t)$ and the generalized force $X(t)$ can be respectively written as

$$J(t) = r \cdot n(t) \tag{11.11}$$

and

$$X(t) = R \cdot \left(1 - \frac{n(t)}{K}\right) \tag{11.12}$$

where R is a dimensional constant greater than zero and t is the time variable.

Then, the entropy production can be respectively expressed as

$$P(t) = R \cdot r \cdot n(t) \cdot \left(1 - \frac{n(t)}{K}\right) \tag{11.13}$$

and

$$\frac{dP(t)}{dt} = R \cdot r \cdot \left(1 - \frac{2n(t)}{K}\right) \cdot \frac{dn(t)}{dt} \tag{11.14}$$

Combining Equations 11.10 and 11.14, we find that

$$\frac{dP(t)}{dt} = R \cdot r^2 \cdot n(t) \cdot \left(1 - \frac{n(t)}{K}\right) \cdot \left(1 - \frac{2n(t)}{K}\right) \tag{11.15}$$

When $(K/2) \le n(t) \le K$, we have $P(t) \ge 0$ and $(dP(t)/dt) \le 0$. Therefore, when $(K/2) \le n(t) \le K$, the ecosystems are locally stable.

For the nonlinear region of the nonequilibrium state, the generalized flow and generalized force can be respectively written as

$$J(t) = r \cdot n(t) \tag{11.16}$$

and

$$X(t) = R \cdot r^2 \cdot n(t) \cdot \left(1 - \frac{n(t)}{K}\right) \cdot \left(\frac{2 \cdot n(t)}{K} - 1\right) \tag{11.17}$$

In fact, in the nonlinear region of the nonequilibrium state, any appearance of a new order can be considered as the reference state of a certain disorder losing its stability. Therefore, the generalized force in the nonlinear region of the nonequilibrium state can be formulated as $X(t) = -(dP(t)/dt)$; then,

$$\delta_X P(t) = -\frac{6 \cdot r^3 \cdot R}{K^2} \cdot \left(n(t) - (1 + \sqrt{3}) \cdot \frac{K}{2}\right) \cdot \left(n(t) - (1 - \sqrt{3}) \cdot \frac{K}{2}\right) \cdot (\delta_X n(t))^2 \tag{11.18}$$

According to Equation 11.18, when $\left(1 - \sqrt{3}\right) \cdot (K / 2) < n(t) < \left(1 + \sqrt{3}\right) \cdot (K / 2)$, $\delta_X P(t) > 0$, the ecosystems are globally stable; when $n(t) = \left(1 - \sqrt{3}\right) \cdot (K / 2)$ or $n(t) = \left(1 + \sqrt{3}\right) \cdot (K / 2)$, $\delta_X P(t) = 0$, the ecosystems are critical; and when $n(t) < \left(1 - \sqrt{3}\right) \cdot (K / 2)$ or $n(t) > \left(1 + \sqrt{3}\right) \cdot (K / 2)$, $\delta_X P(t) > 0$, the ecosystems are globally unstable.

11.2.1.4 Brief Summary

According to this model of ecological thresholds, the sustainable growth range of k-type growth population is $[0.5K, K]$ for less disturbed ecosystems such as natural ecosystems, near-natural ecosystems, and seminatural ecosystems, and approximately $[0.211K, 0.789K]$ for more disturbed ecosystems such as anthropogenic biotic ecosystems and anthropogentic technical systems, where K is the maximum carrying capacity of the ecosystems. In other words, for weakly disturbed ecosystems, the extinction threshold is $0.5K$ and the carrying-capacity threshold is K; for strongly disturbed ecosystems, the extinction threshold is $0.211K$ and the carrying-capacity threshold is $0.789K$ (Yue, 2000). In this chapter, 78.9% is adopted as the critical value in calculating the HCC of terrestrial ecosystems in China.

11.2.2 Models for Food Provision Capacity of Terrestrial Ecosystems

Here, we focus on four types of terrestrial ecosystems: cropland, grassland, woodland, and aquatic. The food provision capacity of cropland ecosystems is formulated as (Tian, 2004; Dang et al., 1999)

$$Y_{\text{grain}} = Y(Q) \cdot f(T) \cdot f(W) \cdot f(S) \cdot f(M) \cdot CA \cdot HI \cdot A_{\text{food}} \qquad (11.19)$$

where Y_{grain} is the grain yield per unit area (kg/hm^2); CA is the impact coefficient (%) of natural disasters; HI is the harvest index (%); $A_{\text{food}} = A_{\text{gross}} \cdot C_{\text{net}} \cdot C_{\text{food}}$ is the effective area (hm^2) of cultivated land; A_{gross} is the gross cultivated area (hm^2); C_{net} is the net cultivation coefficient (%); C_{food} is the effective cultivation coefficient (%); $Y(Q) = 0.219 \times Q$ is photosynthetic potential (kg/hm^2); Q is solar radiation (j/cm^2); $f(T)$, $f(W)$, $f(S)$, and $f(M)$ are effective coefficients (%) of temperature, soil moisture content, soil fertility, and agricultural production input, respectively; $f(S) = C_{\text{SFI}} \cdot C_{\text{erode}}$, C_{SFI} is the modification coefficient (%) of soil fertility, C_{erode} is the soil erosion coefficient (%); $f(M) = 0.6$; $f(T) = (1/1 + e^{2.052 - 0.161\,T})$, when $T < -10$, $f(T) = 0$; $f(W) = K(0.7C_{\text{slope}}C_{\text{DEM}} + 0.3C_{\text{slopewater}})$; K is the humidity coefficient (%); C_{slope} is the moisture modification coefficient (%) of surface slope; C_{DEM} is the moisture modification coefficient (%) of elevation; and $C_{\text{slopewater}}$ is the hydrological modification coefficient (%).

In addition to grain, crop straw could be converted into food, which is formulated as

$$Y_{\text{mutton}} = Y_{\text{straw}} \cdot C_{\text{fodder}} \cdot C_{\text{mutton}} \qquad (11.20)$$

where Y_{mutton} represents food from the converted crop straw (mutton unit); Y_{straw} represents the yield (kg) of the crop straw; C_{fodder} represents the coefficient (%) of the crop straw converted into fodder; and C_{mutton} represents the conversion coefficient (%) between mutton and fodder.

The food provision capacity of grassland ecosystems is formulated as follows (Chen, 2001):

$$Y_{grassland} = Y(Q) \cdot f(T) \cdot f(W) \cdot f(S) \cdot f(M) \cdot A_{grass} \cdot C_{use} \cdot C_{grassmutton} \qquad (11.21)$$

$$Y(Q) = PAR \cdot AB \cdot CL \cdot G \cdot CH \cdot E \cdot (1 - B) \cdot (1 - C)/F \qquad (11.22)$$

where $Y_{grassland}$ is the food provision capacity of grassland ecosystems (mutton unit); C_{use} is the utilization ratio (%) of forage grass; $C_{grassmutton}$ is the conversion coefficient (%) between forage grass and mutton; A_{grass} is the grassland area (hm²); PAR is effective photosynthetic radiation (j/cm²); AB is the maximum absorption ratio (%) of PAR; CL is the modification coefficient (%) of vegetation coverage; G is the modification coefficient (%) of growth rate; CH is the harvest coefficient (%) of forage grass; E is the quantum conversion ratio (%); B is the modification coefficient (%) of vegetation absorption; F is the heat quantity of dry vegetation matter (J/g); C is the ash ratio of forage grass (%); the calculation of $f(T)$ and $f(W)$ is the same as the calculation for the food provision capacity of cropland ecosystems; $f(W) = K = R/E_0 = (R/0.0018(25 + T)^2(100 - F))$, K is the monthly humidity index (%), where R is the monthly precipitation (mm), E_0 is the monthly evaporation (mm), T is the monthly mean temperature (°C), and F is the monthly mean relative humidity on average (%); $f(M) = C_{Pop} \cdot C_{Scale}$, C_{Pop} is the modification coefficient (%) of population density, and C_{Scale} is the modification coefficient (%) of grassland size.

The food provision capacity of woodland ecosystems is formulated as (Tian, 2004; Liu et al., 2005a,b)

$$P_{ij} = A_i \cdot N_{ij} \cdot CT_{ij} \cdot CP_{ij} \cdot CS_{ij} \cdot CR_{ij} \cdot CV_{ij} \cdot CO_{ij}O \qquad (11.23)$$

where P_{ij} is the potential of the jth type of food provisioned by the ith forest type (kg/hm²); A_i is the area of the ith forest type (hm²); N_{ij} is the baseline yield of the jth type of food of the ith forest type (kg/hm²); and CT_{ij}, CP_{ij}, CS_{ij}, CR_{ij}, CV_{ij}, and CO_{ij} are the modification coefficients (%) of temperature, precipitation, soil, solar radiation, vegetation, and others, respectively.

The aquatic ecosystem model for food provision capacity is formulated as

$$P_i = \sum_{j=i}^{4} PA_{ij} \cdot A_{ij} \cdot C_{ij} \qquad (11.24)$$

where P_i is the artificial fish potential (kg/hm²) of the ith pixel; PA_{ij} is the natural fish productivity (kg/hm²) of the jth type of water body within the ith pixel; A_{ij} is the area (hm²) of the jth type of water body within the ith pixel;

C_{ij} is the coefficient (%) of the increasing fish production from artificial input; $j = 1, 2, 3$, and 4 represents the fishery paddy field, reservoir, lake, and river.

In terms of the simulation results, natural fish productivity could be formulated as (Tian, 2004; Liu et al., 2005a,b)

$$PA_{i1} = 0.042 \cdot (\ln SFI_i)^{2.4} \cdot (\ln P_{mi})^{1.6} \cdot (\ln PD_i)^{0.72} \cdot ((1 + t_i)/100)^{3.4} \qquad (11.25)$$

$$PA_{i2} = 0.334 \cdot (\ln SFI_i)^{0.72} \cdot (\ln P_{mi})^{0.83} \cdot (\ln PD_i)^{0.45} \cdot (\ln SR_i)^{1.31}$$
$$\times ((1 + t_i)/100)^{2.42} \qquad (11.26)$$

$$PA_{i3} = 3.94 \cdot (\ln SFI_i)^{0.55} \cdot (\ln P_{mi})^{1.21} \cdot (\ln PD_i)^{0.47} \cdot (\ln NDVI_i)^{0.18}$$
$$\times (\ln S_i)^{-0.25} \cdot ((1 + t_i)/100)^{3.4} \qquad (11.27)$$

$$PA_{i4} = 0.179 \cdot (\ln SFI_i)^{0.81} \cdot (\ln P_{mi})^{1.87} \cdot (\ln PD_i)^{0.56} \cdot (\ln NDVI_i)^{0.21}$$
$$\times ((1 + t_i)/100)^{1.78} \qquad (11.28)$$

where *SFI* is the soil fertility index (%); P_m is the precipitation coefficient (mm); t is the temperature (°C); *PD* is the population density (individuals/km²); *S* is the slope gradient; and normalized difference vegetation index (*NDVI*) is the vegetation index.

The total food provision capacity of all types of ecosystems is calculated by

$$TN_i = \sum_{j=1}^{m} P_j \cdot N_{ij} \qquad (11.29)$$

where TN_i is the potential of the *i*th kind of nutrients ($i = 1, 2,$ and 3 represents calories, protein, and fat, respectively); P_j is the potential of the *j*th type of food; *m* is the number of food types; and N_{ij} is the content of the *i*th kind of nutrients from the *j*th type of food.

11. 3 Data Sets

11.3.1 Data Sets for Ecosystem Classification of Terrestrial Ecosystems

Ecosystem classification is essential for spatially analyzing the food provision capacities of terrestrial ecosystems. The data set was developed from

Landsat ETM scenes with a spatial resolution of 30×30 m. In the late 1990s, 512 ETM scenes were used for ecosystem classification. These Landsat ETM images were geo-referenced and orthorectified by using field-collected ground control points and high-resolution digital elevation models. The interpretation of ETM images and land-cover classifications were validated against extensive field surveys. During 1999 and 2000, ground truth checking was conducted for more than 75,000 km of transects across China, and more than 8000 photographs were taken by cameras equipped with a global positioning system. A hierarchical classification system of 25 land-cover classes was applied to the data. The 25 classes of land-cover were grouped further into six smaller aggregated classes: croplands, woodlands, grasslands, water bodies, unused land, and built-up areas. This database, for which the spatial resolution is scaled up to 1×1 km, is used to calculate the HCC of terrestrial ecosystems in China.

11.3.2 Data Sets for the Calculation of Food Provision Capacity

The major variables of the models for food provision capacity include solar radiation, temperature, precipitation, soil fertility, terrain, population density, and NDVI. NDVI is formulated by the spectral reflectance of the visible red and the near-infrared bands and has been widely used in describing the relationships between vegetation characteristics (Yue et al., 2002). Data sets corresponding to the variables include a 1:1000000 vegetation cover (Editorial Board of Vegetation Map of China, 2001), climatic data on a resolution of 1×1 km (Yue et al., 2005a), a DEM based on a resolution of 1×1 km (Yue et al., 2005b), NDVI data on a resolution of 1×1 km (http://www.resdc.cn), soil data (http://www.resdc.cn), land use and land cover (Liu et al., 2005a,b), human population distribution (Yue et al. 2003, 2005c), a data set on natural conservation areas, and statistical data collected from counties, provinces, the State Forestry Administration (2005), the National Bureau of Statistics of the People's Republic of China (2005), and the Ministry of Agriculture of the People's Republic of China (2005).

11.4 Food Provision Capacity of Terrestrial Ecosystems

11.4.1 Cropland

Although cropland accounts for only 14.9% of the total land area of China, 86% of food consumed is grown in cropland ecosystems. Cropland ecosystems are of great importance to food provision. In the recent years, the major grain crops in China have included wheat, rice, maize, soybean, yam, oats, rye, barley, millet, and Chinese sorghum, among which the production of wheat, rice, and maize accounts for 85% of total grain production. The sown area of these three crops accounts for 93% of the total cropland area. Rice is mainly distributed in southern China, of which the mean yield is 6200 kg/hm² and the biggest yield reaches 18,000 kg/hm². Wheat and maize are mainly distributed in northern China. The mean yield of wheat is 3700 kg/hm² and reaches 5000 kg/hm² in Tibet. The mean yield of maize is 4600 kg/hm² and reaches 7000 kg/hm² in Xinjiang and Shanghai.

According to the model of the food provision capacity of cropland ecosystems, grain yield is mainly determined by solar radiation, temperature, soil moisture content, soil fertility, agricultural production input, the harvest index, and the impact coefficient of natural disasters. The data for solar radiation, temperature, and soil moisture content were selected from the 735 weather observation stations scattered across China, which were interpolated on 1 × 1 km surfaces, as input data of the model, by combining high-accuracy surface modeling (Yue et al., 2010) with a zonal variation pattern in the climate (Yue et al., 2006). The effective coefficient of soil fertility is calculated in terms of soil texture, organic matter, total nitrogen, total phosphorus, readily available phosphorus, readily available potassium, and pH value, which are modified by elevation, topographical factors, and soil erosion intensity. Agricultural production input is calculated in terms of fertilizer, irrigation, technical progress, and management.

In terms of the FAO's project (FAO, 1995), food production has been increasing in the various soil types, due to the improvement of ecosystem management and technical progress, which have their roots in the beginning of the agricultural revolution 10,000 years ago (Higgins et al., 1983). According to the research results of Wang and Zhao (2004), the harvest index has increased from about 0.2 in the 1950s to 0.6 in recent years. However, natural disasters still have a considerable impact on food production and have caused a yearly drop of 11% in grain production on an average since 1950.

In the recent years, 25% of crop straw has been used as fodder (Sun et al., 2003). According to a food research group concerned with middle and long-term development, the utilization ratio of crop straw could potentially be as high as 50% (Lu and Liu, 1993). Crop straw, however, is an important source of organic matter. Using too much crop straw as fodder causes soil degradation, because the organic matter in the soil cannot be efficiently supplemented (Wu, 2003). The maximum proportion of crop straw utilization as fodder should be 40%.

The result of the operating model, taking into consideration of the information presented above, show that cropland ecosystems could produce 118,000 million kg of grain (Figure 11.2), which amounts to 411,910,000 million kilocalories, 114,839 million kg of protein, and 33,047 million kg of fat. In addition, cropland ecosystems could produce 812,000 million kg of crop straw, of which a 40% utilization of fodder amounts to 15,157 million kg of mutton. The total amount of calories, protein, and fat that could be provisioned by cropland ecosystems is 4,155,620,000 million kilocalories, 116,643 million kg, and 36,246 million kg, respectively.

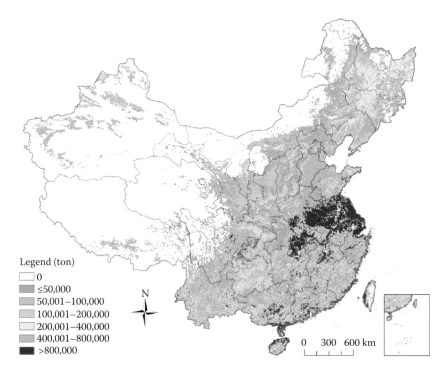

Legend (ton)

0

≤50,000

50,001–100,000

100,001–200,000

200,001–400,000

400,001–800,000

>800,000

0   300   600 km

FIGURE 11.2
Grain potential of cropland ecosystems in China.

11.4.2 Grassland

According to remotely sensed data, there are about 303 million hectares of grassland in China, which accounts for 31.9% of its total land area (Liu et al., 2005a,b). Of this grassland, 92.7% is distributed in western China, 4.4% in middle China, and 2.9% in eastern China. The grassland can be classified into meadow steppe, pasture, desertification grassland, and herbosa steppe. Meadow steppe is mainly found in northeastern China, east Inner Mongolia, and on the Qinhai-Xizang Plateau. Pasture is mainly found on the Inner Mongolia plateau, the Loess Plateau, Qinhai-Xizang Plateau, and in the mountainous areas of northern China. Grass coverage, yield, and quality of pasture are all higher than those of desertification grassland, but inferior to those of meadow steppe. Desertification grassland is mainly found in north-western China, where dominant vegetation consists of sparse xerophytic shrubs that have poor productivity. Herbosa steppe is scattered through-out agricultural areas in southeastern China. Due to the warm and moist climate of that area, grass coverage, yield, and the quality of pasture are relatively high.

In terms of the research results (Xu, 1992; Chen, 2001; Liu et al., 2002; Yang, 2005), the maximum absorption ratio of PAR, $AB = 0.8$; the modification coefficient of growth rate, $G = 0.5$; the harvest coefficient of forage grass, $CH = 0.4$; the quantum conversion ratio $E = 0.2625$; the modification coefficient of vegetation absorption, $B = 0.33$; the heat quantity of dry vegetation matter, $F = 17.8$ kj/g; the ash ratio of forage grass, $C = (12.622 - 0.2382T)\%$; and the 300 kg of dry forage grass amounts to one sheep unit, about 14 kg of mutton. A 100 g of mutton amounts to 241 calories, 11.9 g of protein and 21.1 g of fat (FAO, 1953).

Based on the above, the results of the model of the food provision capacity of grassland ecosystems indicate that grassland ecosystems in China could produce 17,970 million kg of mutton (Figure 11.3), which amounts to 43,320,000 million kilocalories, 2139 million kg of protein and 3793 million kg of fat. The grassland ecosystems, which have a higher production potential, are found in eastern Inner Mongolia, southern Tibet, eastern and western Yunan, southern Gansu, eastern Qinghai, western Sichuan, and western Xiinjiang. The six provinces that have the highest grassland production are Inner Mongolia (3680 million kg of mutton), Tibet (3060 million kg of mutton), Yunnan (1980 million kg), Sichuan (1760 million kg), Qinghai (110 million kg), and Xinjiang (930 million kg). The production of these six provinces accounts for 69.73% of the total production in China.

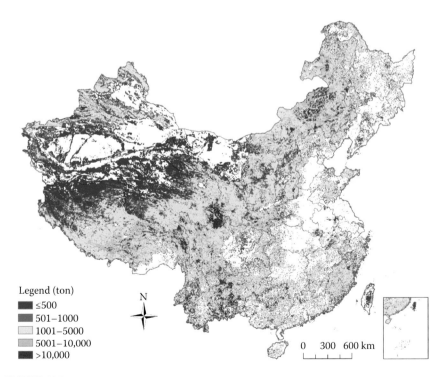

Legend (ton)
■ ≤500
■ 501–1000
□ 1001–5000
▨ 5001–10,000
■ >10,000

N

0   300   600 km

FIGURE 11.3
Food potential of grassland ecosystems in China.

11.4.3 Woodland

The woodland area of China, including artificial economic forests, comprises of approximately 226.74 million hectares (Liu et al., 2005a,b). The calculation of the food provision capacity of woodland takes into account meat, forage, oil-bearing seed, fruits, and vegetables from nonconservation areas such as economic forests.

Woodland is the habitat of many wild animals and, with the exception of species protected by wildlife conservation law, these animals can be considered a source of food. Leaves and branches of trees and grasses in woodland can be used as forage for domestic animals. Tea-oil seed and walnuts are the main edible oils supplied by the woodlands of China, and they account for 99% of total edible oil from woodland. Fruits mainly include apples, oranges, pears, jujubes, persimmons, bananas, and pineapples. Vegetables refer to edible wild vegetation that could be classified into six types: root vegetables, stem vegetables, leaf vegetables, flower vegetables, fruit vegetables, and mushrooms, among which bamboo shoots—a kind of stem vegetable, has the highest yield.

Based on the above, the results of the model of the food provision capacity show that woodland ecosystems could provide 1850 million kg of meat

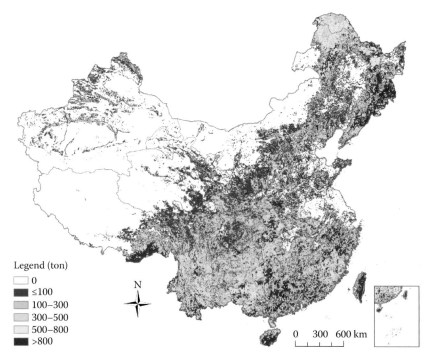

Legend (ton)
0
≤100
100–300
300–500
500–800
>800

N

0   300   600 km

FIGURE 11.4
Meat potential of woodland ecosystems.

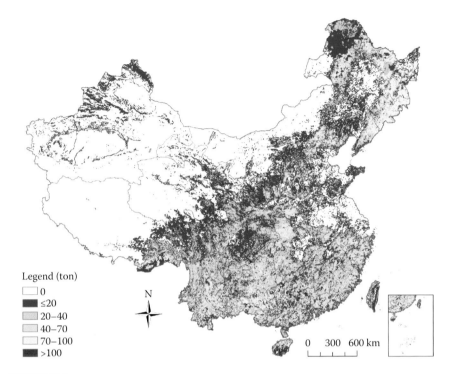

FIGURE 11.5
Forage potential of woodland ecosystems.

(Figure 11.4), 171,080 million kg of forage (Figure 11.5), 25,050 million kg of vegetables (Figure 11.6), 20,380 million kg of oil-bearing seed (Figure 11.7), and 475,440 million kg of fruits (Figure 11.8). This amounts to 229,780,000 million kilocalories, 4299 million kg of protein and 8200 million kg of fat in total.

11.4.4 Aquatic Ecosystems

Terrestrial aquatic ecosystems include rivers, lakes, reservoirs, and paddy fields. The aquatic ecosystems provide various kinds of aquatic products such as fish, shrimp, and crab. In this chapter, all aquatic products are converted into fish to easily calculate the food provision capacity of aquatic ecosystems.

11.4.4.1 Paddy Fields

Rivers and lakes have been described in detail in Chapter 10. We give here a brief introduction to paddy fields as an aquatic ecosystem type. Rice-field fish cultivation, which combines organic rice production and aquaculture, is a major source of aquatic products for the local inhabitants of many regions

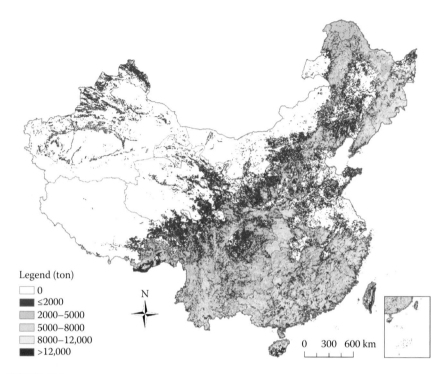

Legend (ton)
- ☐ 0
- ■ ≤2000
- ☐ 2000–5000
- ☐ 5000–8000
- ☐ 8000–12,000
- ■ >12,000

N

0   300   600 km

FIGURE 11.6
Vegetable potential of woodland ecosystems.

of China. For instance, in the Jiangsu province, rice-field fish cultivation accounts for 90% of the land area; in the Guizhou province, rice-field fish cultivation covers 123,132 ha and its production accounts for 40% of all aquatic products. However, rice fields that are used for fish cultivation must meet the essential requirements of rich water resources, good water quality, fertile soil, and sufficient sunshine. Rice fields in Northwest China and the Qinghai-Xizang Plateau cannot be used for fish cultivation. Rice fields, suitable for fish cultivation are mainly found in Sichuan, Chongqing, Jiangsu, Anhui, Hunan, Jiangxi, and Hubei. There were 1.53 million hectares of rice-field fish cultivated in the year 2000 in China, which yielded 748.5 million kg of fish (Ministry of Agriculture of the People's Republic of China, 2001).

11.4.4.2 Fish Productivity

Fish productivity can be classified into natural fish productivity and artificial potential. Natural fish productivity occurs under natural conditions, without investment. Artificial potential is maximum fish productivity based on investment. In terms of the simulation results of the aquatic ecosystem models for natural fish productivity, the total natural fish production in China could reach 3665.1 million kg, of which paddy field, river, lake, and

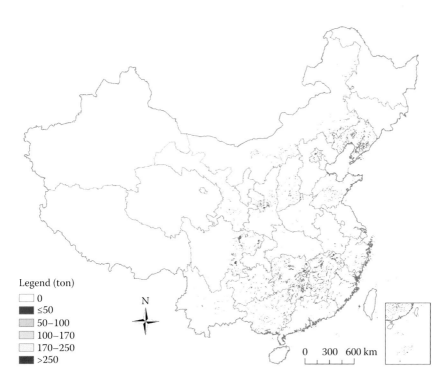

FIGURE 11.7
Oil-bearing seed potential of woodland ecosystems.

reservoir production would account for 38.34%, 7.08%, 25%, and 29.58%, respectively. The natural fish productivities of paddy fields, rivers, lakes, and reservoirs are 123, 75, 142, and 263 kg/hm², respectively.

However, according to statistics (Ministry of Agriculture of the People's Republic of China, 2002), the real fish productivities of paddy fields, rivers, lakes, and reservoirs in 2001 were 487, 1756, 1043, and 2910 kg/hm², respectively. Based on experimental studies, fish productivity could be increased to 3700 kg/hm² in paddy fields (Lin, 1996), 14,530 kg/hm² in reservoirs (Yao, 1992), 5426 kg/hm² in lakes, and 4405 kg/hm² in rivers through fertilization, fishing bait input, and increased oxygen.

It should be noted that only a minority of waters are suitable for artificial fish-farming, as the slope must be smaller than 1.5° and the annual mean temperature must be greater than 0°C. The results of these calculations show that artificial fish production of aquatic ecosystems could reach 38,586.3 million kg (Figure 11.9), of which 16,274.1, 2093.9, 4971.2, and 15,247 million kg could be supplied by paddy-fields, rivers, lakes, and reservoirs, respectively. According to FAO standards, 100 g of fish provides 132 kcal, 18.8 g of protein, and 5.7 g of fat. Aquatic ecosystems in China could supply 5.093×10^7 million kcal, 7255 million kg of protein, and 2200 million kg of fat in total.

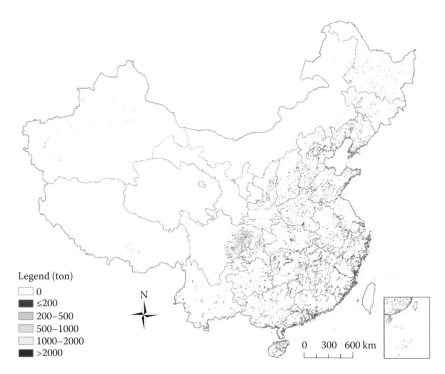

FIGURE 11.8
Fruit potential of woodland ecosystems.

11.4.5 Human Carrying Capacity

In general, terrestrial ecosystems could potentially supply 4,479,650,000 million kilocalories, 130,336 million kg of protein, and 50,439 million kg of fat (Tables 11.1 and 11.2). In view of the threshold of HCC (Yue, 2000) and the 11% production drop caused by natural disasters (Yue et al., 2008), 3,101,800,000 million kcal, 90,247 million kg of protein, and 34,925 million kg of fat could be made available for human consumption.

In terms of the food development strategy and food security goals of China (Lu and Liu, 1993; Lu, 2003), the living standards during different development stages of China can be classified as primary well-to-do life, full well-to-do life, and well-off life. Under the living standards of the primary well-to-do life, one person would be expected to consume 2289 kilocalories, 77 g of protein and 67 g of fat daily. Under living standards of the full well-to-do life, one person would be expected to consume 2295 kcal, 81 g of protein and 67.5 g of fat daily. Under living conditions of the well-off life, one person would be expected to consume 2347 kcal, 86 g of protein and 72 g of fat daily. Under the primary well-to-do life, full well-to-do life, and well-off life, the HCC of calories would be 3765, 3755, and 3672 million individuals, protein would be 3265, 3096, and 2916 million individuals, and fat

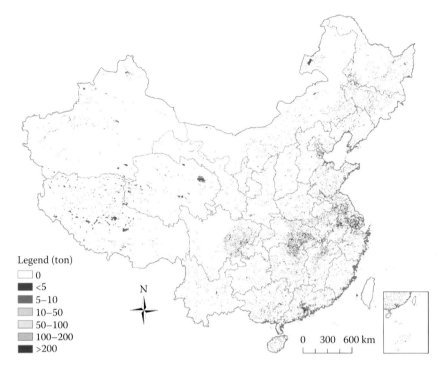

FIGURE 11.9
Artificial fish potential of aquatic ecosystems.

TABLE 11.1

Food Provision Capacity of Terrestrial Ecosystems in China

Nutrient Type	Cropland	Woodland	Grassland	Aquatic Ecosystems	Totals
Calorie (million kilocalories)	4,155,620,000	229,780,000	43,320,000	50,930,000	4,479,650,000
Protein (million kg)	116,643	4299	2139	7255	130,336
Fat (million kg)	36,246	8200	3793	2200	50439

TABLE 11.2

Proportion of Nutrients Provisioned by Cropland, Grassland, Woodland, and Aquatic Ecosystems

Nutrient Type	Cropland (%)	Woodland (%)	Grassland (%)	Aquatic Ecosystems (%)
Calorie (million kilocalories)	92.77	0.97	5.13	1.14
Protein (million kg)	89.49	1.64	3.30	5.57
Fat (million kg)	71.86	7.52	16.26	4.36

would be 1448, 1438, and 1348 million individuals, respectively (Table 11.3 and Figure 11.9).

The results of the calculations show that nutrients provided by terrestrial ecosystems would be imbalanced under the current agricultural production structure. If the agricultural production structure was improved, and the nutrients were balanced, the HCC under each living standard could be formulated as follows

$$\begin{cases} NPF_1 = PF_1(1 - x) \\ NPF_2 = PF_2 + PF_1 \times x \times y \\ P_{cal} = (NPF_1 \times U_{cal1} + NPF_2 \times U_{cal2})/(UD_{cal} \times 365) \\ P_{pro} = (NPF_1 \times U_{pro1} + NPF_2 \times U_{pro2})/(UD_{pro} \times 365) \\ P_{fat} = (NPF_1 \times U_{fat1} + NPF_2 \times U_{fat2})/(UD_{fat} \times 365) \\ P_{cal} = P_{pro} = P_{fat} \end{cases} \qquad (11.30)$$

where x is the proportion of food on the first trophic level for supporting the production of food on the second trophic level to the total food on the first trophic level, $x \in (0,1)$; $y = 1/3.6$ is the conversion rate from the food on the first trophic level to the food on the second trophic level (Lu and Liu, 1993); PF_1 and PF_2 are respectively foods on the first and second level before the nutrients are balanced; NPF_1 and NPF_2 are respectively foods on the first and second trophic levels after the nutrients have been balanced; P_{cal}, P_{pro}, and P_{fat} are respectively human carrying capacities in terms of calories, protein, and fat; U represents the nutritive contents of one kilogram of one kind of food; UD represents one kind of nutrient needed by one person on a daily basis.

TABLE 11.3

Human Carrying Capacity of Terrestrial Ecosystems in China

Nutrient Type	Primary Well-to-do Life	Full Well-to-do Life	Well-off Life
Imbalanced Nutrients			
Calorie	3765	3755	3672
Protein	3256	3096	2916
Fat	1448	1438	1348
Balanced Nutrients			
All nutrients	2029	1914	1794

Note: Unit: 1 million individuals.

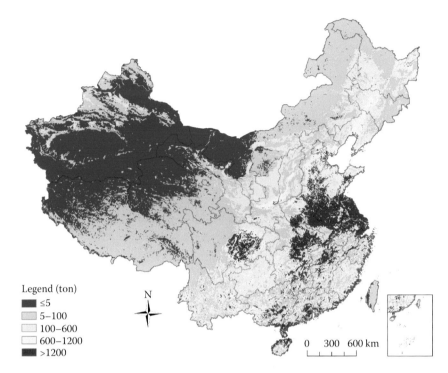

FIGURE 11.10
Human carrying capacity of terrestrial ecosystems under the living standards of the well-off life in China.

After the nutrients were balanced by improving the agricultural production structure, the HCC would be 2029, 1914, and 1794 million individuals under the living standards of primary well-to-do life, full well-to-do life, and well-off life, respectively, taking into account the threshold of HCC and the 11% production drop caused by natural disasters (Figure 11.10).

11.5 Discussion and Conclusions

In terms of food availability, the HCC of terrestrial ecosystems in China has been hotly debated since Brown (1994, 1995) argued that large Chinese grain imports could drive world food prices to dangerously high levels by the year 2030. Brown asserted that terrestrial ecosystems in China would bear an overload of between 450 million and 802 million individuals by the year 2030, under the assumption of a grain consumption of 460 kg per capita annually. However, Harris (1996) argued that China's yield growth of 1.5–2.0% per annum enables China to remain almost self-sufficient in grain production. According to an ecosystem approach based on the adaptation of crops to the physical environment, and an energy/mass balance between the physical environment, primary producer (crops), and secondary producer (livestock), the HCC of China's agro-ecosystems would be 1720 million individuals (Cao et al., 1995).

According to research results produced by the National Situation Analysis Group of the Chinese Academy of Sciences (NSAG, 1997), the grain production of China in the year of 2030 could be 700,000 million kg under a high grain production scenario, 660,000 million kg under a medium scenario, and 630,000 million kg under a low scenario. If the annual grain consumption were 460 kg per capita, the HCC of grain would be 1522, 1435, and 1370 million individuals under the high, medium, and low grain production scenarios, respectively. Our results show that, under the three living standards of primary well-to-do life, full well-to-do life, and well-off life, the HCC of calories would be 3765, 3755, and 3672 million individuals, protein would be 3265, 3096, and 2916 million individuals, and fat would be 1448, 1438, and 1348 million individuals, respectively. Overall, there is structural shortage of food as per the current agricultural production structure in China.

The water diversion project will have a huge effect on HCC in China, as the spatial distribution of water resources is currently considerably uneven. Precipitation shows a tendency to decrease from the southeastern coastal area to the northwestern inland area. Water resources in northern China have a volume of 521.3 billion m^3 (Table 11.4), accounting for only 18.6% of the total water resources of China (Qian and Zhang, 2001), but cultivated land in northern China accounts for 64% of China's total cultivated land (Zhang et al., 1999). Northern China can be divided into six regions: the Yellow River, the Huaihe River, the Haihe River, the Helongjiang River, and the Liaohe River, as well as the inland water systems in northwestern China. Southern China can be divided into four regions: the Yangtze River, the Zhujiang River, the rivers in the southeastern coastal area, and the rivers in southwestern China. Overall, northern China has a serious water deficit, while southern China has sufficient surplus water resources that could be transferred to alleviate the deficit in northern China.

The south-to-north water diversion project (SNWDP) has been under consideration by institutes such as the Ministry of Water Resources, the Ministry of Science and Technology, and the National Development and Reform Commission of the People's Republic of China since the 1950s. The project consists of three water diversion-routes, namely the East Route, transferring water from the lower reaches of the Yangtze River to northern China, the Middle Route, transferring water from the middle reaches of the Yangtze River to northern China, and the West Route, transferring water from the upper reaches of the Yangtze River to the upper reaches of the Yellow River (Figure 11.11). The East and Middle Routes were started in 2003 and will be completed by 2020. A feasible work plan for the West Route has been

TABLE 11.4

Spatial Distribution of Annual Water Volume, on an Average, in China

Region	Area (km²)	Water Resources (Billion m³)	Water Resources per Unit (m³/hm²)
Northern China			
Rivers in Northeast China	1,248,445	178.3	1428
Haihe river	318,161	42.1	1323
Yellow river	794,712	74.4	936
Huaihe river	329,211	96.1	2919
Inland rivers	3,374,443	130.4	386
Total in northern China	6,064,972	521.3	860
Southern China			
Yangtze river	1,808,500	961.3	5315
Rivers in southeastern coastal area	239,803	259.2	10809
Zhujiang river	580,641	470.8	8108
Rivers in southwestern China	851,406	585.3	6875
Total in southern China	3,480,350	2276.6	6541

proposed. It will be launched in the 2020s and will be completed before 2050. The East Route will supply water volume of 18 billion m³ to northern China, 15.7 billion m³ of which will be used for the urbanization and industrialization of Tianjin and Shandong. The remaining 2.3 billion m³ will be used for the agriculture and shipping industry along the trunk canal. The Middle Route will supply a water volume of 22 billion m³, 10 billion m³ of which will be used to meet the urban and industrial demands of Beijing, Tianjin, Hebei, Henan, and Hubei. The remaining 12 billion m³ will be used for agriculture and ecosystems along the trunk canal (Zhang et al., 1999; Qian and Zhang, 2001). The West Route will supply a water volume of 17 billion m³, 5.3 billion m³ of which will be used for ecological purposes, 3.5 billion m³ for cities, 5.2 billion m³ for industries, and 3 billion m³ for agriculture, forestry, and livestock farming in the Qingai, Gansu, Ningxia, Inner Mongolia, Shaanxi, and Shanxi provinces (Tan et al., 2004).

Thus, 17.3 billion m³ of water could be replenished into agriculture in northern China. Almost all agricultural land needs to be irrigated in the arid and semiarid areas of northern China, because precipitation is unable to meet the agricultural demands for water. The yield of the nonirrigated land is 2100 kg/hm², on average. The rice yield is 7500 kg/hm² and other irrigated

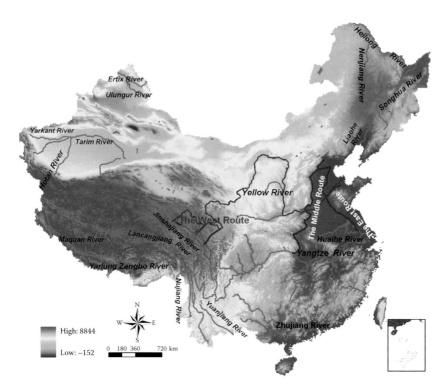

FIGURE 11.11
The south-to-north water diversion project.

land yields are 4500 kg/hm². Additionally, 1.23 m³ of water is needed for an increase of 1 kg of grain (Shen and Su, 1998), enabling SNWDP to achieve an increase of 14 billion kg of grain that, in northern China, could feed 29, 26, and 23 million individuals under the primary well-to-do life, full well-to-do life, and well-off life, respectively, in terms of national food and essential nutrients.

In short, the HCC would be 2029, 1914, and 1794 million individuals under the living standards of primary well-to-do life, full well-to-do life, and well-off life, respectively, if the agricultural production structure were improved so as to result in balanced nutritional value. If the contribution of SNWDP is considered, the HCC could be 2058, 1940, and 1817 million individuals under the three living standards, respectively.

The model in this chapter provides an upper limit of population that could be supported by the food provision services of terrestrial ecosystems in China, which are independent of economic and social factors. Further work on scenarios of HCC in China will need focus on climate change, interactions between the industrial and the agricultural sector, industrialization, and the import of food.

References

Andrén, H. 1994. Effects of habitat fragmentation on birds and mammals in landscapes with different proportions of suitable habitat: A review. *Oikos* 71: 355–366.

Arrow, K., Bolin, B., Constanza, R., Dasgupta, P., Folke, C., Holling, C.S., Jansson, B.O., Levin, S., Maeler, K.G., Perrings, C., and Pimental, D. 1995. Economic growth, carrying capacity, and the environment. *Science* 268: 520–521.

Begon, M., Harper, J.L., and Townsend, C.R. 1990. *Ecology: Individuals, Populations and Communities*. Cambridge, USA: Blackwell Scientific Publications.

Beverton, R.J.H. and Holt, S.J. 1957. *On the Dynamics of Exploited Fish Populations*. London: H.M. Stationery Off.

Brown, L.R. 1994. Who will feed China? *Worldwatch* 7: 10–19.

Brown, L.R. 1995. *Who Will Feed China? Wake-up Call for a Small Planet*. New York, NY: W.W. Norton and Co.

Cao, M.K., Ma, S.J., and Han, C.R. 1995. Potential productivity and human carrying capacity of an agro-ecosystem: An analysis of food production potential of China. *Agricultural Systems* 47: 387–414.

Chen, B.M. 2001. *Productivity and Human Carrying Capacity of Agricultural Resources in China*. China, Beijing: Meteorological Press (in Chinese).

Clark, C. 1977. *Population Growth and Land Use*, 2nd edn. London: Macmillan.

Conway, G. 1997. *The Doubly Green Revolution: Food for All the Twenty-First Century*. New York, NY: Cornell University Press.

Cohen, J.E. 1995. *How Many People Can the Earth Support?* New York, NY: W.W. Norton & Company.

Dang, A.R., Yan, S.J., and Zhou, Y. 1999. A GIS-based study on grain potential in China. *Journal of Remote Sensing* 3(3): 225–229 (in Chinese).

De Wit, C.T. 1967. Photosynthesis: Its relationship to overpopulation. In *Harvesting the Sun: Photosynthesis in Plant Life*, eds., A.S. Pietro, F.A. Greer, and T.J. Army, 315–320. New York, NY: Academic Press.

Editorial Board of Vegetation Map of China. 2001. *Vegetation Atlas of China 1:1000000*. Beijing: Science Press (in Chinese).

Fahrig, L. 1998. When does fragmentation of breeding habitat affect population survival? *Ecological Modelling* 105: 273–292.

Food and Agriculture Organization of the United Nations (FAO). 1953. *Food Composition Tables for International Use*, 2nd edn. Rome: FAO.

Food and Agriculture Organization of the United Nations (FAO). 1995. *FAO Agrostat-PC, Computer Disk*. Rome: FAO.

Friedel, M.H. 1991. Range condition assessment and the concept of thresholds: A view point. *Journal of Range Management* 44: 422–426.

Glausdorff, P. and Prigogine, I. 1971. *Thermodynamic Theory of Structure, Stability and Fluctuations*. New York, NY: Wiley-Interscience.

Godschalk, D.R. and Park, F.H. 1975. Carrying capacity: A key to environmental planning. *Journal of Soil Water Conservation* 30: 160–165.

Haber, W. 1972. Grundzuege einer oekologischen Theorie der Landnutzung. *Innere Kolonisation* 21: 294–298.

Haber, W. 1990. Using landscape ecology in planning and management. In *Changing Landscapes: An Ecological Perspective*, eds., I.S. Zonneveld and R.T.T. Forman, 217–231, New York, NY: Springer-Verlag.

Haraldsson, H.V. and Ólafsdóttir, R. 2006. A novel modelling approach for evaluating the preindustrial natural carrying capacity of human population in Iceland. *Science of the Total Environment* 372: 109–119.

Harris, J.M. 1996. World agricultural futures: Regional sustainability and ecological limits. *Ecological Economics* 17: 95–115.

Hawden, S. and Palmer, L.J. 1922. Reindeer in Alaska. *Bulletin of the U.S. Department of Agriculture* 1089: 1–70.

Herrmann, S., Dabbert, S., and Raumer, H.G.S. 2003. Threshold values for nature protection areas as indicators for bio-diversity: A regional evaluation of economic and ecological consequences. *Agriculture, Ecosystems and Environment* 98: 493–506.

Higgins, G.M., Kassam, A.H. Naiken, L. Fischer, G., and Shah, M.M. 1983. Potential population supporting capacities of lands in the developing world. *Technical Report of Project INT/75/P13, Land Resources for Populations of the Future, FPA/INT/513*. Rome: Food and Agricultural Organization of the United Nations.

Hildenbrandt, H., Mueller, M.S., and Grimm, V. 2006. How to detect and visualize extinction thresholds for structured PVA models. *Ecological Modelling* 191(3–4): 545–550.

Huggett, A.J. 2005. The concept and utility of "ecological threshold" in biodiversity conservation. *Biological Conservation* 124: 301–310.

Hui, C. 2006. Carrying capacity, population equilibrium, and environment's maximal load. *Ecological Modelling* 192: 317–320.

Hulett, H.R. 1970. Optimum world population. *BioScience* 20: 160–161.

Kessler, J.J. 1994. Usefulness of the human carrying capacity concept in assessing ecological sustainability of land use in semi-arid regions. *Agriculture, Ecosystems and Environment* 48(3): 273–284.

Khanna, P., Babu, P.M., and George, M.S. 1999. Carrying-capacity as a basis for sustainable development: A case study of National Capital Region in India. *Progress in Planning* 52: 101–166.

Lin, Z.H. 1996. Environmental effects of rice-field fish cultivation. *Agriculture and Environmental Protection* 15(4): 177–181 (in Chinese).

Lindenmayer, D.B., Fischer, J., and Cunningham, R.B. 2005. Native vegetation cover thresholds associated with species responses. *Biological Conservation* 124: 311–316.

Liu, J.Y., Tian, H.Q., Liu, M.L., Zhuang, D.F., and Melillo, J.M. 2005a. China's changing landscape during the 1990s: Large-scale land transformations estimated with satellite data. *Geophysical Research Letters* 32: 1–5.

Liu, J.Y., Yue, T.X., Ju, H.B., Wang, Q., and Li, X.B. 2005b. *Integrated Ecosystem Assessment of Western China*. Beijing: China Meteorological Press.

Liu, L.M., Zhang, F.R., and Zhao, Y.W. 2002. An analysis of production potential of grassland in China and its utilization counter measures. *China Population, Resources and Environment* 12(4): 100–105 (in Chinese).

Lu, L.S. 2003. *Food and Nutrient for Development in China*. Beijing: China Agriculture Press (in Chinese).

Lu, L.S. and Liu, Z.C. 1993. *The Middle- and Long-Term Development Strategy of Food in China*. Beijing: China Agiriculture Press (in Chinese).

Luck, G.W. 2005. An introduction for thresholds. *Biological Conservation* 124: 299–300.

Malthus, T.R. 1789. *An Essay on the Principle of Population*. London: Pickering.

Mathieson, A. and Wall, G. 1982. *Tourism: Economic, Physical, and Social Impacts*. Harlow, UK: Longman.

McCullough, D.R. 1979. *The George Reserve Deer Herd: Population Ecology of a k-Selected Species*. Ann Arbor: The University of Michigan Press.

Meyers, J. and Walker, B.H. 2003. Thresholds and alternate states in ecological and social-ecological systems: Thresholds database. *Resilience Alliance*, http://www.resalliance.org.au.

Millman, S.R., Chen, R.S., Emlen, J., Haarmann, V., Kasperson, J.X., and Messer, E. 1991. *The Hunger Report: Update 1991*. Alan Shawn Feinstein World Hunger Program, Brown University.

Ministry of Agriculture of the People's Republic of China. 2001. *Statistics of China Agriculture 2000*. Beijing: China Agriculture Press (in Chinese).

Ministry of Agriculture of the People's Republic of China, 2002. *Statistics of China Fishery 2001*. Beijing: China Agriculture Press (in Chinese).

Ministry of Agriculture of the People's Republic of China. 2005. *Statistics of China Agriculture 2004*. Beijing: China Agriculture Press (in Chinese).

Muradian, R. 2001. Ecological thresholds: A survey. *Ecological Economics* 38: 7–24.

National Bureau of Statistics of the People's Republic of China. 2005. *China Statistical Yearbook*. Beijing: China Statistics Press.

National Situation Analysis Group of Chinese Academy of Sciences. 1997. *The National Situation Report No.5: Agriculture and Development*. Shenyang: Liaoning People's Press (in Chinese).

Odum, E. P. 1971. *Fundamentals of Ecology*, Philadelphia: W. B. Saunders Company.

Odum, E. P. 1983. *Basic Ecology*. Philadelphia: Saunders College Publishing.

Odum, E. 1989. *Ecology and Our Endangered Life Support Systems*. Sunderland: Sinauer Associates.

Oh, K., Jeong, Y., Lee, D., Lee, W., and Choi, J. 2005. Determining development density using the Urban Carrying Capacity Assessment System. *Landscape and Urban Planning* 73(1): 1–15.

Onsager, L. 1931a. Reciprocal relation in irreversible processes I. *Physical Review* 37: 405–426.

Onsager, L. 1931b. Reciprocal relation in irreversible processes II. *Physical Review* 38: 2265–2279.

Penck, A. 1925. Das Hauptproblem der physischen Anthropogeographie. *Zeitschrift fuer Geopolitik* 2: 330–348.

Prato, T. 2001. Modeling carrying capacity for national parks. *Ecological Economics* 39: 321–331.

Price, D. 1999. Carrying capacity reconsidered. *Population and Environment* 21(1): 5–26.

Prigogine, I. 1968. *Introduction to Thermodynamics of Irreversible Processes*, New York, NY: Interscience Pub.

Prigogine, I. 1980. *From Being to Becoming*, San Francisco: W. H. Freeman and Comp.

Qian, Z.Y. and Zhang, G.D. 2001. *A Summing-up Research Report on Water Resources Strategy for Sustainable Development of China*. Beijing: China Water Power Press (in Chinese).

Ravenstein, E.G. 1891. Lands of the global still available for European settlement. *Proceedings of the Royal Geographical Society* 13: 27–35.

Rees, W.E. 1992. Ecological footprints appropriated carrying capacity: What urban economics leaves out. *Environment and Urbanization* 4: 121–130.

Revelle, R. 1976. The resources available for agriculture. *Scientific American* 235(3): 165–178.

Ricker, W.E. 1958. Handbook of computation for biological statistics of fish popula-
    tions. *Bulletin of the Fisheries Research Board of Canada* 119: 1–300.
Seidl, I. and Tisdell, C.A. 1999. Carrying capacity reconsidered: From Malthus'
    population theory to cultural carrying capacity. *Ecological Economics* 31:
    395–408.
Shaffer, M.L. 1981. Minimum population sizes for species conservation. *BioScience* 31:
    131–134.
Shen, Z.R. and Su, R.Q. 1998. *Research on Countermeasures to Water Shortage for
    Agriculture in China.* Beijing: China Agricultural Science and Technology Press
    (in Chinese).
Silliman, R.P. 1969. Population models and test populations as research tools.
    Bio-Science 19: 524–529.
Slobodkin, L.B. 1962. *Growth and Regulation of Animal Populations.* New York: Rinehart
    and Winston.
State Forestry Administration. 2005. Statistics of Forest Resources of China. Beijing:
    China Forestry Press (in Chinese).
Sun, M.H., Shan, H.G., and Nan, F. 2003. Comprehensive utilization of crop straw.
    Solar Energy 24: 8–11 (in Chinese).
Tan, Y.W., Liu, X., and Cui, Q. 2004. *West-Line Project Drawing Water from Southern
    China to Northern China.* Zhengzhou: Yellow River Water Conservancy Press.
Tian, Y.Z. 2004. *An Assessment of Food Provisioning Services of Terrestrial Ecosystems in
    China.* Doctoral Thesis of Institute of Geographical Sciences and Natural
    Resources Research of CAS, Beijing (in Chinese).
United Nations Fund for Population Activities. 1993. *Population Issues: Briefing kit
    1993.* New York, NY: United Nations Fund for Population Activities.
Wang, S.A. and Zhao. M. 2004. The way of taping crop production potential. [Online]
    2004–10–4, available at: http://www.bjsp.org.cn/kjlt/xny/wj.htm.
With, K.A. and King, A.W. 1999. Extinction thresholds for species in fractal land-
    scapes. *Conservation Biology* 13: 314–326.
Wu, R.G. 2003. An analysis of effects of crop straw utilization on environment. *Yunnan
    Environmental Science* 22: 86–87 (in Chinese).
Xu, Z.R. 1992. Climatic production potential of grassland in Sichuan province. *Rural
    Eco-Environment* 8(1): 19–23 (in Chinese).
Yang, Y.X. 2005. *China Food Composition 2004.* Beijing: Peking University Medical Press
    (in Chinese).
Yao, H.L. 1992. Specific principles and techniques of comprehensive fish engineering
    in China. *Aquaculture* 2: 24–29 (in Chinese).
Yue, T.X. 2000. Stability analysis on the sustainable growth range of population.
    Progress in Natural Science 10(8): 631–636.
Yue, T.X., Chen, S.P., Xu, B., Liu, Q.S., Li, H.G., Liu, G.H., and Ye, Q.H. 2002. A curve-
    theorem based approach for change detection and its application to Yellow
    River delta. *International Journal of Remote Sensing* 23(11): 2283–2292.
Yue, T.X., Fan, Z.M., and Liu, J.Y. 2005a. Changes of major terrestrial ecosystems in
    China since 1960. *Global and Planetary Change* 48: 287–302.
Yue, T.X., Wang, Y.A., Liu, J.Y., Chen, S.P., Qiu, D.S., Deng, X.Z., Liu, M.L., Tian,Y.Z.
    and Su, B. 2005b. Surface modelling of human population distribution in China.
    Ecological Modelling 181(4): 461–478.
Yue, T.X., Fan, Z.M., Liu, J.Y., and Wei, B.X. 2006. Scenarios of major terrestrial eco-
    systems in China. *Ecological Modelling* 199: 363–376.

Yue, T.X., Song, D.J., Du, Z.P., and Wang, W. 2010. High accuracy surface modeling and its application to DEM generation. *International Journal of Remote Sensing* 31(8): 2205–2226.

Yue, T.X., Tian, Y.Z., Liu, J.Y., and Fan, Z.M. 2008. Surface modeling of human carrying capacity of terrestrial ecosystems in China. *Ecological Modelling* 214: 168–180.

Yue, T.X., Wang, Y.A., Chen, S.P., Liu, J.Y., Qiu, D.S., Deng, X.Z., Liu, M.L., and Tian, Y.Z. 2003. Numerical simulation of population distribution in China. *Population and Environment* 25(2): 141–163.

Yue, T.X., Wang, Y.A., Liu, J.Y., Chen, S.P., Tian,Y.Z., and Su, B.P. 2005c. MSPD scenarios of spatial distribution of human population in China. *Population and Environment* 26(3): 207–228.

Zhang, D.Z., Luo, X.L., and Yu, C.S. 1999. *Drawing Water from Southern China to Northern China*. Beijing: China Water Power Press (in Chinese).

12

Changing Trends of Food Provision in China*

12.1 Introduction

According to surface modeling results of the carrying capacity of terrestrial ecosystems in China (Yue et al., 2008), the maximum amounts of calories, protein, and fat which could be produced by farmland ecosystems are 4,155,620,000 million kcal, 116,643 million kg, and 36,246 million kg, respectively. Grassland ecosystems in China could produce 17,970 million kg of mutton, which amounts to 43,320,000 million kcal of calories, 2139 million kg of protein and 3793 million kg of fat. Woodland ecosystems could provide 229,780,000 million kcal of calories, 4299 million kg of protein, and 8200 million kg of fat. Aquatic ecosystems in China could supply 50,930,000 million kcal of calories, 7255 million kg of protein, and 2200 million kg of fat. In view of the threshold of carrying capacity (Yue, 2000), 78.9% of the food produced by ecosystems could be used for human demand. A 11% production drop has been caused by natural disasters according to statistical analysis. Thus, 3,101,800,000 million kcal of calories, 90,247 million kg of protein, and 34,925 million kg of fat would be available for human consumption.

The surface modeling of human carrying capacity provides an upper limit of population that could be supported by the food provision services of terrestrial ecosystems in China. These are independent of economic and social factors. Changes in food production trends in China deal with the effect of climate change, land-use change, Grain for Green programs, fisheries along the Chinese coast, and food imports.

12.2 Farmland Change

Statistics published by the State Statistical Bureau have underreported the total area of farmland in China. The State Statistical Bureau reported 96.846 million hectares in 1985, 94.974 million hectares in 1995 and 130.039 million hectares in 2005. However, farmland estimates derived from remote sensing of satellites were respectively 183.676, 181.301, and 179.688 million hectares in 1985, 1995, and 2005. Annual surveys produced by the Ministry of Land and

---

* Dr. Qing Wang and Dr. Yi-Min Lu are major collaborators of this chapter.

TABLE 12.1

Farmland Area from Different Sources (in Million Hectares)

Year	Interpretation from Satellite Image	Investigation Data of Agricultural Ministry	Statistics
1985	183.676	154.651	96.846
1995	181.301	143.196	94.974
2005	179.688	156.988	130.039

Resources since 1985 also exceed the State Statistical Bureau's statistics by 34% on an average (Table 12.1). The comprehensive land survey conducted in 1996 by the Ministry of Land and Resources reported 130.039 million hectares of farmland and 10.0 million hectares of horticultural land (Lichtenberga and Ding, 2008). However, the total farmland area of China as reported by the Ministry of Agriculture was 149.265 in 1996, 9.226 million hectares more than the figures stated in the Ministry of Land and Resources' survey for the same year, with an overall difference of 6.2%. The interpretation of China's total farmland from remote sensing of satellites differed from the Ministry of Agriculture's survey by 21.9% on an average. From this, we can conclude that the remotely sensed data of farmland had an error of 21.9%, if we take the Ministry of Agriculture's annual survey as the "true" data; for example, there were 154.651 million hectares of farmland in 1985, 143.196 million hectares in 1995, and 156.988 million hectares in 2005.

12.2.1 Grain for Green Program

The Grain for Green program is one of the world's largest environmental set-aside programs, of which the main objective is to restore the country's forests and grasslands to prevent soil erosion. The program designers have made the steepness of the slope one of the main criteria on which plots are selected for inclusion in the Grain for Green program (Xu et al., 2006). The steepness criterion means that the program targets land with a slope of more than 25°.

The Grain for Green program was divided into three phases. The experimental phase was from 1999 to 2001; the construction phase was from 2002 to 2010 and the consolidation phase will commence in 2011 and conclude in 2020. By the end of 2001, 1.206 million hectares of farmland were converted into forestland or grassland and 1.097 million hectares of barren land had been afforested. Before 2011, 14.667 million hectares of farmland, including almost all farmland with a slope of more than 25°, will be converted to forestland or grassland; 2.667 million hectares of cultivated desertification land will be converted to grassland; and 17.333 million hectares of barren land will be afforested. During the consolidation phase, converted land and afforested land will have to be carefully managed in order to meet Grain for Green's target.

TABLE 12.2

Converted Farmland and Afforested Barren Land in the Grain for Green Program in China (in Million Hectares)

Year	Converted Farmland	Afforestation on Bare Land	Total
1999	0.381	0.066	0.448
2000	0.405	0.468	0.872
2001	0.420	0.563	0.983
2002	2.647	3.082	5.729
2003	3.367	3.767	7.133
2004	0.667	3.333	4.000
2005	1.114	1.321	2.435
1999–2010	14.667	17.333	32

In 1999, Grain for Green's pilot program was first experimented in three provinces, Gansu, Shaanxi, and Sichuan; 0.381 million hectares of farmland was converted into forestland and 0.066 million hectares of barren land was afforested. In 2000, the experimentation area was expanded to 17 provinces; the converted farmland and the afforested barren land were respectively 0.41 million hectares and 0.449 million hectares. In 2001, 20 provinces were involved in the experiment; 0.42 million hectares of farmland was converted into forestland and 0.563 million barren land was afforested (Table 12.2). In 2002, the Grain for Green program was launched across China. By 2005, 9.001 million hectares had been withdrawn from farmland and planted with trees or converted to permanent grassland.

Farmers receive compensation of RMB1575 per hectare in South China and RMB1050 per hectare in North China for the land they dedicate to soil conservation measures. Participating households enjoy a faster increase in assets such as livestock. In some local areas, agricultural income increases due to more intensive agricultural production on nonprogram plots, because better seed stock can be used, multicropping can replace single cropping and more livestock can be produced.

12.2.2 Urbanization

The statistics from the Ministry of Construction show that urbanization levels in China increased from 17.9% in 1978 to 40.5% in 2003, which demonstrates a growth twice as fast as the world average in the same period (Chen, 2007). Recognizing urbanization's central role in further economic growth and social development, China will continue to give high priority to urbanization in the coming decades. According to the National Report of China Urban Development (2001–2002) (Editorial Committee of National Report of China Urban Development of China Mayor Association, 2003), the urban population proportion will be 56.22% in 2020. Urbanization is characterized by an extensive establishment of development zones in and near cities, and a drastic expansion of rural construction driven by the development of township–village enterprises.

Satellite images are much more accurate at interpreting built-up areas than they are at interpreting other land-use types such as farmland, grassland, and forestland. The interpretation of remotely sensed data from Advanced Very High Resolution Radiometer (AVHRR) indicates that there were 5.291 million hectares of built-up area in China in 1985, 6.579 million hectares in 1995 and 19.136 million hectares in 2005. The built-up area increased by 13.845 million hectares during the period 1985 to 2005. The losses of cultivated area to urban uses have been concentrated in the coastal and central provinces. Net losses of farmland in these provinces appear to be in the region of 2 million hectares between 1985 and 1995. That rate of loss accelerated to 3.5 million hectares between 1996 and 2003. Urbanization, industrialization, infrastructure, and other nonagricultural developments appear to be the main causes of farmland loss in these rapidly industrializing coastal provinces (Lichtenberga and Ding, 2008).

However, much farmland has been newly reclaimed in North China. For instance, during the period 2000 to 2005, 0.79 million hectares of farmland was newly reclaimed in the Xinjiang Uygur Autonomous region. Its annual increasing area is 0.132 million hectares on an average, in which saline and alkaline land, desert, and bare land, respectively, account for 10.47%, 8.95%, and 3.44% according to our investigation.

12.3 Yield Change and Food Production in Farmland

12.3.1 Yield

Over the last 30 years, grain yield has gone up from 2528 kg/hm^2 in 1978 to 4748 kg/hm^2 in 2007 (Table 12.3). The increasing trend can be formulated into a linear regression equation with the correlation coefficient of $R^2 = 0.92$:

$$Y_g = 71.44t - 138476 \tag{12.1}$$

where Y_g is the grain yield and t is the corresponding year.

Rice is the most important crop in China, being the staple food of more than 65% of Chinese people. During the last three decades, rice farming accounted for about 27–29% of the total area sown with grain, and rice production represented 41–45% of total grain production in the country. Since 1970, China has had nearly one-fourth of the total rice-sown area and more than one-third of the total rice production in the world (Tao et al., 2008). With the improvement of paddy field management and the introduction of newly

TABLE 12.3

Yields of Major Crops in China in the Recent 30 Years (kg/hm^2)

Year	Grain	Rice	Wheat	Maize	Bean	Oil Crop
1978	2528	3975	1845	2805	1058	855
1979	2783	4245	2138	2985	1028	908
1980	2738	4133	1890	3075	1095	903
1981	2828	4320	2108	3045	1163	1080
1982	3124	4886	2449	3266	1073	1116
1983	3306	5096	2802	3623	1290	1135
1985	3483	5097	2937	3607	1394	1304
1989	3632	5509	3043	3878	1269	1125
1990	3933	5726	3194	4524	1403	1386
1991	3876	5640	3100	4578	1379	1347
1992	4004	5803	3331	4533	1426	1324
1993	4131	5848	3519	4963	1156	1516
1994	4063	5831	3426	4693	1160	1551
1995	4240	6025	3541	4917	1540	1670
1996	4483	6212	3734	5203	1770	1713
1997	4377	6319	4102	4387	1765	1663
1998	4502	6366	3685	5268	1782	1752
1999	4493	6345	3947	4945	1841	1832
2000	4261	6272	3738	4598	1577	1842
2001	4267	6163	3806	4698	1582	1849
2002	4399	6189	3777	4924	1893	1889
2003	4333	6061	3932	4813	1772	1700
2004	4620	6311	4252	5120	1815	1988
2005	4642	6260	4275	5287	1705	1974
2006	4716	6232	4550	5394	1721	2072
2007	4748	6433	4608	5167	1454	2114

bred seed, rice has a mean yield of 5665 kg/hm^2 and has seen an annual increase rate of 2.1% over the last 30 years. Its changing trend with the correlation coefficient $R^2 = 0.86$ can be regressively expressed as

$$Y_{RI}(t) = 78.91t - 151640 \tag{12.2}$$

where Y_{RI} is the rice yield and t is the corresponding year.

Wheat is the second most important grain in China. It had a mean yield of 3374 kg/hm^2 and has seen an annual increase rate of 5% over the last 30 years. The change trend can be formulated into the following regression equation with a correlation coefficient of $R^2 = 0.94$

$$Y_w(t) = 84.62t - 165305 \tag{12.3}$$

Maize is the third most important grain after wheat and rice in China. An increasing proportion of maize is being used as feed grain as a result of rising demand for meat, population growth, and rapid economic development. The mean yield of maize was 4396 kg/hm² during the period 1978 to 2007 and the annual increase rate was 2.8%. The increasing trend can be formulated as follows, of which correlation coefficient is $R^2 = 0.85$:

$$Y_M(t) = 83.18t - 161413 \tag{12.4}$$

Other dominant crops include bean and oil crop. They have mean yields of 1466 and 1523 kg/hm², respectively (Table 12.3). Their increasing trends with correlation coefficients of $R^2 = 0.68$ and $R^2 = 0.95$ can be respectively expressed as

$$Y_{bean}(t) = 25.65t - 49672 \tag{12.5}$$

$$Y_{oil} = 41.27t - 80752 \tag{12.6}$$

12.3.2 Production

The production of farmland is calculated in terms of calories, protein, and fat using the following formulation:

$$CN_i(t) = \sum_j P_j(t) \cdot C_{i,j} \tag{12.7}$$

where $CN_i(t)$ is the total nutrients from farmland ; $i = 1$, 2, and 3, respectively represent calories (million kcal), protein (million kg), and fat (million kg); $P_j(t)$ is the total production (million kg) of crop j; $C_{i,j}$ is the conversion factor from crop j to nutrient i.

According to data produced by the Chinese Academy of Agricultural Sciences (Table 12.4), since 1980, agricultural production has generally increased, with a certain degree of oscillation. The production of calories, protein, and fat (Table 12.5) can be formulated as the following regression equations with correlation coefficients of $R^2 = 0.69$, $R^2 = 0.76$, and $R^2 = 0.91$, respectively:

$$CN_1(t) = 21673485t - 41969847292 \tag{12.8}$$

$$CN_2(t) = 730t - 1418099 \tag{12.9}$$

$$CN_3(t) = 441t - 864723 \tag{12.10}$$

TABLE 12.4

Nutrient Contents in 100 Grams of Food

Crops	Calorie (kcal)	Protein (g)	Fat (g)
Rice	346	7.4	0.8
Wheat	317	11.9	1.3
Maize	335	8.7	3.8
Bean	359	35.0	16.0
Oil crop	298	12.0	25.4

Source: Adapted from Yang, Y.X. 2005. *China Food Composition 2004.* Beijing: Peking University Medical Press (in Chinese).

Owing to the 11-year cyclicity of climate change in terms of droughts and floods (Yue et al., 2005), it is necessary to analyze the change of farmland provisions first in the period between 1985 and 1995 and second in the period between 1995 and 2005, dependent on the availability of data. The annual mean production of farmland in the first period was 1,179,629,173 million kcal

TABLE 12.5

Total Production of Farmland in Terms of Major Nutrients in China

Year	Calorie (Million kcal)	Protein (Million kg)	Fat (Million kg)
1980	826,200,957	24,264.6	7372.2
1981	872,359,099	25,740.6	8194.6
1983	968,246,183	28,983.9	8907.0
1985	1,054,288,320	32,032.3	10,706.3
1986	1,039,763,722	31,998.9	10,546.5
1987	1,088,636,935	33,318.1	11,438.4
1988	1,107,327,802	33,873.4	10,917.7
1989	1,169,404,140	34,978.9	8123.7
1990	1,323,728,223	39,006.5	12,550.5
1991	1,187,764,119	35,562.9	12,365.2
1992	1,216,146,710	36,613.0	11,990.4
1993	1,216,187,430	37,200.4	12,591.1
1994	1,207,047,938	36,762.9	12,848.2
1995	1,365,625,568	41,576.2	15,191.3
1996	1,501,655,704	45,426.4	16,177.3
1997	1,538,585,251	47,168.7	16,400.0
1998	1,485,573,785	45,410.0	16,632.8
1999	1,370,741,536	41,922.2	15,947.4
2000	1,314,481,390	40,851.0	16,547.6
2001	1,331,843,249	41,249.2	16,835.2
2002	1,273,466,903	39,993.8	16,763.6
2003	1,221,664,463	38,794.8	16,665.9
2004	1,401,775,069	43,549.8	17,919.5
2005	1,478,976,303	45,741.7	18,557.1

TABLE 12.6

Change of Farmland Provision

Nutrient	Calorie	Protein	Fat
Annual mean production from 1985 to 1995 (million kcal)	1,179,629,173	35,720	11,752
Annual mean production from 1995 to 2005 (million kcal)	1,389,489,929	42,880	16,694
Increased production (million kcal)	209,860,756	7160	4943
Increased rate (%)	18	20	42

of calories, feeding 1408 million individuals, 35,720 million kg of protein, feeding 1208 million individuals and 11,752 million kg of fat, feeding 477 million individuals under full well-to-do life on an average. The annual mean production in the second period was 1,389,489,929 million kcal of calories, 42,880 million kg of protein, and 16,694 million kg of fat, which could meet the demands of 1659 million individuals, 1450 million individuals, and 678 million individuals, respectively, under a full well-to-do life. The production of calories increased by 209,860,756 million kcal, protein by 7160 million kg and fat by 4943 million kg annually in the second period compared to those produced in the first period on an average (Table 12.6).

During both the first and the second periods, the highest yields of calories and protein appeared in the North China Plain, the middle and lower reaches of the Yangtze river, the Sichuan Basin, and the Northeast China Plain (Figure 12.1a, b, d, and e). The North China Plain and Northeast China Plain had the highest yield of fat (Figure 12.1c and f).

During the second period, a drop in production took place in the Eastern Coastal Area, and middle and lower reaches of the Yangtze River where there had been rapid urbanization over the past 20 years, and in the Sichuan Basin where many farmers went to cities for nonagricultural work, which meant farmland was cropped less efficiently (Figure 12.2). Production increase appeared in Northwest China and Northeast China where low-yield farmland was converted into grassland and forest for the Grain for Green program, and the western North China Plain which benefited from being on the Middle Route of the South-to-North Water Diversion Project, which transfers water from the middle reaches of the Yangtze River to northern China (Yue et al., 2008).

12.4 Grassland Productivity

A data set of bimonthly normalized difference vegetation index (NDVI) in a spatial resolution of 8 km × 8 km is downloaded from the Global Inventory

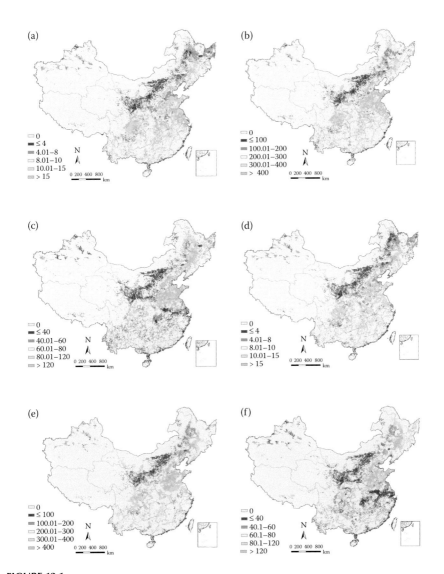

FIGURE 12.1

Annual mean yield of farmland: (a) calories (million kcal/hm²) during the first period (1985–1995), (b) protein (kg/hm²) during the first period, (c) fat (kg/hm²) during the first period, (d) calories (million kcal/hm²) during the second period (1995–2005), (e) protein (kg/hm²) during the second period, and (f) fat (kg/hm²) during the second period.

Modeling and Mapping Studies, which spans a 25-year period 1981 to 2006 (http://glcf.umiacs.umd.edu/data/gimms). The NDVI data set in a spatial resolution of 1 km × 1 km is available from the period 1992 to 1996. The data set is derived from imagery obtained using the AVHRR and has been corrected for calibration, view geometry, and volcanic aerosols.

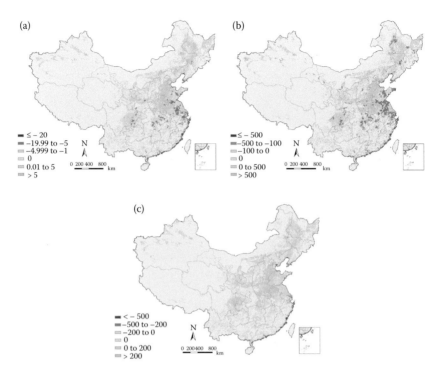

FIGURE 12.2
Difference of annual mean farmland production between the two periods: (a) calories (million kcal/hm²), (b) protein (kg/hm²), and (c) fat (kg/hm²).

The peak value of above ground biomass of grass communities in north China usually appears at the end of August. 268 samples were taken at the end of August between 1992 and 1994, and 851 samples during the same period between 2002 and 2004 (Figure 12.3). The sample plots sized 1 km × 1 km consist of 10 smaller typical quadrats. The size of the typical quadrat is 1m × 1m for herbaceous plants and 2 m × 2 m for shrubs. Biomass in a sample plot is calculated by weighting the biomass in the 10 typical quadrats in terms of the proportion of quadrat types in the sample plot. The location of the sample plot center is determined by means of a global positioning system (GPS).

The first step of biomass calculation is fusing remotely sensed data with sampled data. The NDVI data set in spatial resolution 8 km × 8 km has a long-time series, but the data set in spatial resolution 1 km × 1 km has a much shorter time series although it temporarily matches the sampled data during the period 1992 to 1994.

The relationship between values at pixels of the 1 km × 1 km NDVI, corresponding to the 268 samples, and the mean values on 1 km² of the

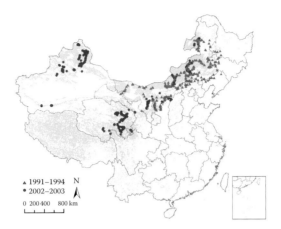

FIGURE 12.3
Location of the sample plots in grassland.

8 km × 8 km pixels that take the 1 km × 1 km pixels as their centers can be simulated as the following regressive equation (Xin et al., 2009):

$$NDVI_{8\times8} = 1.01NDVI_{1\times1} \qquad (12.11)$$

where $NDVI_{1\times1}$ is the value at the 1 km × 1 km pixel corresponding to one of the 268 plots during the period 1992 to 1994; $NDVI_{8\times8}$ is the mean value on 1 km² of a 8 km × 8 km pixel taking the 1 km × 1 km pixel as its center; the correlation coefficient is $R^2 = 0.9751$.

The regressive Equation 12.11 means that the sampled data can be used to simulate the relationship between biomass and NDVI in a spatial resolution of 8 km × 8 km. 787 plots are randomly selected from the 1119 plots in a size of 1 km × 1 km during the periods from 1992 to 1994 and from 2002 to 2004 to simulate the relationship between biomass and NDVI. The remaining 332 plots are used to validate the simulated results.

The simulation results demonstrate that in temperate steppe the relationship between biomass and NDVI can be formulated as

$$Biomass = 37.25 \cdot \exp\{1.73NDVI\} \qquad (12.12)$$

in temperate meadow,

$$Biomass = 169.38 \cdot \exp\{1.06NDVI\} \qquad (12.13)$$

in temperate desert grassland,

$$Biomass = 35.85 \cdot \exp\{1.48NDVI\} \qquad (12.14)$$

in frigid grassland,

$$\text{Biomass} = 29.44 \cdot \exp\{1.49NDVI\} \tag{12.15}$$

in temperate desert,

$$\text{Biomass} = 126.57NDVI + 34.97 \tag{12.16}$$

in lowland meadow,

$$\text{Biomass} = 55.9\exp\{1.79NDVI\} \tag{12.17}$$

in temperate montane meadow,

$$\text{Biomass} = 41.06\exp\{2.77NDVI\} \tag{12.18}$$

in frigid meadow,

$$\text{Biomass} = 39.37\exp\{1.5NDVI\} \tag{12.19}$$

and in sandy grassland

$$\text{Biomass} = 70.34.37e \quad x\{1\ p9NDVI\} \tag{12.20}$$

The validation indicates that the simulated biomass and the sampled biomass have a correlation coefficient of 0.75. In terms of the formulations between biomass and NDVI, the annual grassland provision was 318,964.1 million kg of dry biomass on an average during the period 1982 to 1992, which can be converted into 14,885 million kg of mutton and amounts to 30,216,535 million kcal of calories, 2828.15 million kg of protein and 2098.78 million kg of fat. The food provision of grassland was 29,917,240 million kcal of calories, 2800.14 million kg of protein, and 2078 million kg of fat on an average during the period 1992 to 2002 (Table 12.7).

It is found that climate change in terms of droughts and floods has a small cyclicity of 11 years (Yue et al., 2005). Thus, we comparatively analyze the provision change between the first period 1982 to 1992 (Figure 12.4a) and the second period 1992 to 2002 (Figure 12.4b). There was a drop of 316,000 billion kg in biomass production in the second period compared to the first period. The biggest drop, over 900 million kg, happened in the Yunnan province and Sichuan province. Drop of over 200 million kg appeared in the Guangxi Zhuang autonomous region, the Hubei, the Hunan, the Guizhou, the Gansu, and the Guandong provinces. However, there was considerable

TABLE 12.7

Food Provision of Grassland on an Average in China

Production	During the Period 1982 to 1992 on an Average	During the Period 1992 to 2002 on an Average	Increase
Biomass (million kg)	318,964.10	315,804.80	−3,159.3
Mutton (million kg)	14,885.00	14,737.60	−147.4
Calorie (million kcal)	30,216,535.00	29,917,240.00	−299,295.2
Protein (million kg)	2828.15	2800.14	−28.0
Fat (million kg)	2098.78	2078.00	−20.8

increase in the Xinjiang Uygur, the Tibet, and the Inner Mongolian autonomous regions, and the Hebei province, of 750, 551, 367, and 120 million kg, respectively (Table 12.8 and Figure 12.5).

12.5 Aquatic Products

Aquatic products include fish, shellfish, seashells, algae, and cephalopods from inland aquiculture, inland fish, marine cultures, and marine fish. The total production of aquatic food can be calculated by the following formulation:

$$AN_i(t) = \sum_j P_j(t) \cdot C_{i,j} \tag{12.21}$$

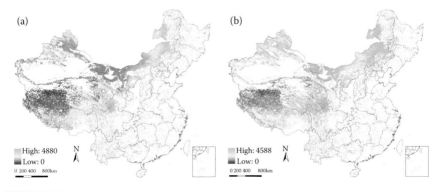

(a)

(b)

High: 4880
Low: 0
0 200 400 800km

High: 4588
Low: 0
0 200 400 800km

FIGURE 12.4

Biomass productivity (kg/hm²) of grassland on an average: (a) from 1982 to 1992 and (b) from 1992 to 2002.

TABLE 12.8

Annual Mean Biomass Change of Grassland Comparing those
in the First Period with the Second Period

Production Drop (Million kg)		Production Increase (Million kg)	
Province	Production Change	Province	Production Change
Yunnan	−993.90	Xinjiang	750.10
Sichuan	−923.57	Tibet	551.47
Guangxi	−523.20	Inner Mongolia	366.73
Hubei	−436.66	Hebei	119.73
Hunan	−375.61	Jilin	54.78
Guizhou	−279.22	Helongjiang	51.89
Gansu	−240.81	Ningxia	24.90
Guangdong	−226.70	Qinghai	7.74
Henan	−193.55	Tianjin	4.84
Jiangxi	−183.54	Jiangsu	3.83
Shanxi	−160.62		
Shaanxi	−111.72		
Shandong	−108.10		
Fujian	−86.75		
Liaoning	−77.95		
Zhejiang	−67.74		
Anhui	−61.18		
Hainan	−18.16		
Taiwan	−12.60		
Beijing	−2.51		

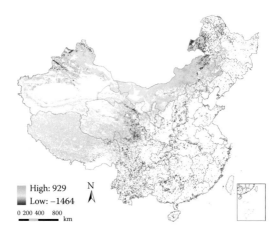

High: 929
Low: −1464

0 200 400   800
km

N

FIGURE 12.5

Biomass productivity change (kg/hm²) of grassland on an average between the two periods.

TABLE 12.9

Contents of Calorie, Protein, and Fat per 100 g of Aquatic Products

Aquatic Foods	Calorie (kcal)	Protein (g)	Fat (g)
Marine fishes	112.71	18.93	3.51
Marine shellfishes	105.00	19.98	1.48
Marine seashells	66.50	9.63	1.10
Marine algae	134.33	17.83	0.97
Marine cephalopods	85.80	15.82	0.82
Other marine products	61.67	8.73	0.27
Inland fishes	103.00	16.60	3.30
Inland shellfishes	95.00	16.95	2.50
Inland seashells	53.67	9.63	0.80
Inland algae	272.50	42.45	1.80
Other inland products	105.50	19.15	2.75

Source: Adapted from Yang, Y.X. 2005. *China Food Composition 2004*. Beijing: Peking University Medical Press (in Chinese).

where $AN_i(t)$ is the total aquatic production in terms of major nutrients; $i = 1$, 2, and 3 represent calories (million kcal), protein (million kg) and fat (million kg), respectively; $P_j(t)$ is total production (million kg) of the aquatic product j; $C_{i,j}$ is the conversion factor from aquatic product j to nutrient i.

The total aquatic production has seen a considerable increase since 1998 (Fishery Administration of Ministry of Agriculture of the People's Republic of China, 1998–2008) and follows a linear regression equation with $R^2 = 0.9952$:

$$AP(t) = 1674t + 34741 \tag{12.22}$$

where $AP(t)$ represents the total aquatic production in the year t.

In 2006, the total aquatic production reached 52,904 million kg, which could be converted into 50,851.5 billion kcal of calories, 8245.4 million kg of protein, and 1333.1 million kg of fat in terms of China Food Composition (Tables 12.9 and 12.10).

12.6 Import and Export of Food

In terms of available data during the period 1982 to 2006 (Compiling Committee for Almanac of China's Foreign Economic Relations and Trade, 1984–2007), imports and exports of food and animal feed included grain, bean, oil crop, meat, aquatic products, animal feed, vegetables, and fruits. During this time, the import and export of grain fluctuated considerably.

TABLE 12.10

Total Aquatic Production

Year	Total Aquatic Production (Million kg)	Calorie (Billion kcal)	Protein (Million kg)	Fat (Million kg)
1998	39,066.5	37,914.7	6146.0	1010.8
1999	41,224.3	39,910.6	6460.9	1058.1
2000	42,790.0	41,229.9	6673.9	1087.8
2001	43,821.0	42,073.2	6806.6	1107.2
2002	45,651.8	45,222.4	7346.4	1164.5
2003	47,061.1	45,298.2	7345.3	1184.3
2004	49,017.7	47,129.5	7643.3	1227.7
2005	51,016.5	49,072.6	9955.5	1283.2
2006	52,904.0	50,851.5	8245.4	1333.1

Exports reached their lowest point when imports reached their highest, and vice versa (Figure 12.6a). For instance, the value of grain imports peaked in 1982, from 1987 to 1990, 1995 and 2004, while exports steadily declined; however, when imports were at their lowest in 1985, 1986, 1993, and 2003, exports were at peak value.

Imports of oil crop and bean have seen the most rapid increase since 1994. Oil crop imports were over 4000 million kg in 1994 and reached 37,400 million kg in 2006 (Figure 12.6b). Bean imports were at more than 10,000 million kg in 2000 and reached 28,638 million kg in 2006 (Figure 12.6c). But since 1982, far less bean has been exported. Before 1995, China exported much more animal feed than it imported, but since 1995, imports have risen and exports have shrunk. In 1997, the value of imports of animal feed peaked and then started to decrease. Since 2003, imports have increased once again and in 2006 it has reached 6712 million kg (Figure 12.6d).

Data of the imports and exports of meat and aquatic products were unavailable before 1995. Since 1995, meat exports have risen while imports have declined. Exports showed a valley in 1999 and 2000 and then started increasing rapidly. Exports in 2006 reached 5680 million kg (Figure 12.6e). Imports and exports of aquatic products have also in turn increased. China now imports more than it exports. The import and export of aquatic products in 2006 were 1429 and 1811 million kg, respectively (Figure 12.6f).

If net imports, which are defined as imports subtracting from exports, are converted into calories, protein, and fat, we find that net imports were bigger than zero during the period 1995 to 2006 (Table 12.11). In 2004, the net import of calories, protein and fat were 121460766 million kcal, 7546 million kg, and 5880 million kg, respectively. Under a full well-to-do life, the net import of calories could meet the demands of 145 million people; protein could meet the demands of 255 million people; and fat could meet the demands of 239 million people.

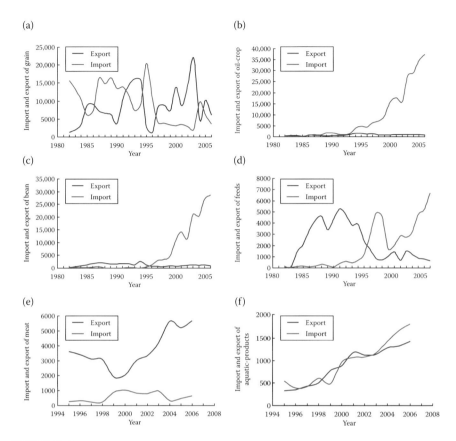

FIGURE 12.6
Imports and exports (million kg): (a) grain, (b) oil crop, (c) bean, (d) feed, (e) meat, and (f) aquatic products.

The net import of calories since 1982 can be formulated as, of which $R^2 = 0.58$,

$$IN_1(t) = 40720t^3 - 243277520t^2 + 484482798086t - 321612522017247 \quad (12.23)$$

The net import of protein since 1982 with correlation coefficient $R^2 = 0.92$:

$$IN_2(t) = 36t^2 - 144287t + 143451639 \quad (12.24)$$

The net import of fat since 1982 with correlation coefficient $R^2 = 0.88$:

$$IN_3(t) = 16t^2 - 62613t + 62181184 \quad (12.25)$$

where $t = 1982, 1983, \ldots, 2005$ or 2006 represents time.

TABLE 12.11

Net Imports in Terms of Calorie, Protein, and Fat

Year	Calorie (Million kcal)	Protein (Million kg)	Fat (Million kg)
1982	−8,125,740	−314.20	−396.04
1983	−11,357,966	−460.81	−580.55
1984	−21,079,688	−1118.81	−689.49
1985	−42,783,797	−1833.47	−982.20
1986	−16,330,068	−1611.92	−564.88
1987	17,187,953	−870.98	−226.59
1988	21,940,269	−939.81	−258.42
1989	34,897,629	−416.65	1179.51
1990	19,166,035	−1187.49	752.78
1991	−10,116,500	−1783.03	−278.98
1992	−34,086,199	−2019.03	−485.78
1993	−44,564,599	−1895.10	−1009.60
1994	−29,187,908	−2579.01	−163.49
1995	77,965,417	1512.48	1827.53
1996	47,388,461	2046.37	1386.85
1997	13,186,590	2313.22	1111.84
1998	12,699,335	2367.40	1502.41
1999	8,895,110	1280.83	1872.38
2000	5,599,330	2984.32	2278.13
2001	35,433,303	4439.88	2154.48
2002	4,527,890	2600.14	2012.48
2003	24,428,852	5177.95	4301.61
2004	121,460,766	7546.39	5880.29
2005	100,173,412	9287.76	5352.83
2006	118,455,946	10,161.36	5506.84

12.7 Population That Food Could Support

Total food provisions during the period 1998 to 2004 have been calculated, because the availability of data is limited, that is, the production data of farmland are only available for the period 1980 to 2005, of grassland from 1992 to 2004, of aquatic products from 1998 to 2006, and of imports and exports from 1982 to 2006. The biggest provision appeared in 2004, of which calories was 1,601,217,474 million kcal, protein was 61,627 million kg, and fat was 27,170 million kg during the period 1998 to 2004 (Table 12.12).

In terms of demand for calories, protein and fat under different living standards (Table 12.13), China supplied calories for 1917 million people under primary well-to-do life, 1912 million people under a full well-to-do life and

TABLE 12.12

Total Food Provisions

Year	Calorie (Million kcal)	Protein (Million kg)	Fat (Million kg)
1998	1,564,538,245	56,577	21,115
1999	1,449,633,505	52,480	20,968
2000	1,390,347,195	53,227	21,930
2001	1,439,907,348	55,356	22,219
2002	1,353,649,048	52,789	22,054
2003	1,322,033,475	54,186	24,280
2004	1,601,217,474	61,627	27,170

1869 million people under a well-off life. It supplied protein for 2193, 2084, and 1963 million people, and fat for 1111, 1103, and 1034 million people, respectively, under the three different living standards in 2004 (Table 12.14).

In 2004, China's total population was 1300 million. In terms of calories and protein, there was sufficient food for a full well-to-do life and well-off life in China, while there was a great shortage in terms of fat. The fat shortage was caused by imbalanced nutrient provisions. Thus, a model for balancing nutrients (MBN) has been developed, in which foods are classified into grain, bean, oil crop, meat, and aquatic products. MBN can be formulated as follows:

$$
\begin{cases}
PA_{11} = PB_{11} \cdot (1 - x_1) \\
PA_{12} = PB_{12} \cdot (1 - x_2) \\
PA_{13} = PB_{13} \cdot (1 - x_3) \\
PA_{21} = PB_{21} + (PB_{11} \cdot x_1 + PB_{12} \cdot x_2 + PB_{13} \cdot x_3) \cdot y \\
PA_{22} = PB_{22} \\
P_{cal} = (PA_{11} \cdot CAL_{11} + PA_{12} \cdot CAL_{12} + PA_{13} \cdot CAL_{13} + PA_{21} \cdot CAL_{21} + PA_{22} \cdot CAL_{22}) \\
\qquad \div (365 \cdot DD_{cal}) \\
P_{pro} = (PA_{11} \cdot PRO_{11} + PA_{12} \cdot PRO_{12} + PA_{13} \cdot PRO_{13} + PA_{21} \cdot PRO_{21} + PA_{22} \cdot PRO_{22}) \\
\qquad \div (365 \cdot DD_{pro}) \\
P_{fat} = (PA_{11} \cdot FAT_{11} + PA_{12} \cdot FAT_{12} + PA_{13} \cdot FAT_{13} + PA_{21} \cdot FAT_{21} + PA_{22} \cdot FAT_{22}) \\
\qquad \div (365 \cdot DD_{fat}) \\
P_{cal} = P_{pro} = P_{fat}
\end{cases}
$$

$$(12.26)$$

TABLE 12.13

Nutrients Consumed Daily by an Individual under Different Living Standards

Living Standard	Calorie (kcal)	Protein (g)	Fat (g)
Primary well-to-do	2289	77	67
Full well-to-do	2295	81	67.5
Well-off	2347	86	72

where x_1, x_2, and x_3, respectively, represent the proportion of grain, bean, and oil crop converted into food on the second trophic level such as meat, obviously $x_i \in (0,1)$, $i = 1, 2, 3$; $y = 1/3.6$ is the conversion rate from food on the first trophic level, such as grain, bean and oil crops, to food on the second trophic level (Yue et al., 2008); PB_{11}, PB_{12}, and PB_{13} are productions of grain, bean, and oil crops before they are balanced and PA_{11}, PA_{12}, and PA_{13} are productions of grain, bean, and oil crop after they are balanced; PA_{21} and PA_{22} are respectively the production of meat and aquatic products after the balance; PB_{21} and PB_{22} are productions of meat and aquatic products before the balance; P_{cal}, P_{pro}, and P_{fat}, respectively, represent the population supported by calories, protein, and fat; CAL, PRO, and FAT, respectively, represent calories, protein, and fat content per kg of a kind of food; DD is one kind of nutrient needed daily by one individual.

The MBN result (Table 12.15) shows that China could supply food for 1534 million people, 1500 million people, and 1411 million people under primary well-to-do life, full well-to-do life, and well-off life, respectively, in 2004. In addition to meeting the demands of China's 1300 million population in 2004, China had food for 234 million people, 172 million people, and 111 million people under the three different living standards.

TABLE 12.14

Population (Million People) That Food Could Support before the Balance of Nutrients

	Primary Well-to-Do Life			Well-to-Do Life			Well-Off Life		
Year	Calorie	Protein	Fat	Calorie	Protein	Fat	Calorie	Protein	Fat
1998	1873	2013	863	1868	1914	857	1826	1802	803
1999	1735	1867	857	1731	1775	851	1692	1672	798
2000	1664	1894	897	1660	1800	890	1623	1696	834
2001	1723	1970	909	1719	1872	902	1681	1763	845
2002	1620	1878	902	1616	1786	895	1580	1682	839
2003	1582	1928	993	1578	1833	985	1543	1726	924
2004	1917	2193	1111	1912	2084	1103	1869	1963	1034
On an average	1731	1963	933	1726	1866	926	1688	1758	868

TABLE 12.15

Population That Food Could Support after the Balance of Nutrients

Year	Primary Well-to-Do Life (Million People)	Well-to-Do Life (Million People)	Well-Off Life (Million People)
1998	1290.1	1262.0	1187.6
1999	1225.2	1198.5	1128.1
2000	1228.1	1201.3	1130.5
2001	1371.6	1341.8	1262.1
2002	1312.0	1283.5	1207.3
2003	1385.3	1355.2	1274.3
2004	1533.5	1500.2	1410.8
On an average	1335.1	1306.1	1228.7

12.8 Conclusions and Discussion

Terrestrial ecosystems including farmland, grassland, and inland aquatic ecosystems provisioned 1,454,237,000 million kcal of calories, 49,958 million kg of protein, and 20,739 million kg of fat in 2004 (Table 12.16). The maximum food provisioning capacity of terrestrial ecosystems would be 4,479,650,000 million kcal of calories, 130,336 million kg of protein, and 50,439 million kg of fat (Yue et al., 2008). Food provision of terrestrial ecosystems in 2004 reached 32%, 38%, and 41% of the maximum provisioning capacities of calories, protein, and fat, respectively.

Food from farmland accounts for a much greater proportion of overall food than grassland, aquatic ecosystems, or net imports. The proportion of food supplied by farmland shows a decreasing trend, from 84.66% in 1998 to 74.72% in 2004. Net imports have generally increased, from 4.04% in 1998 to 13.82% in 2004, and the increase is speeding up (Table 12.17).

In terms of calories, 92.94% was supplied by farmland, 2.08% by grassland, 2.97% by aquatic products, and 2.01% by net imports on an average during

TABLE 12.16

Food Provision of Terrestrial Ecosystems

Year	Calorie (Million kcal)	Protein (Million kg)	Fat (Million kg)
1998	1,529,617,890	50,611	19,096
1999	1,417,561,778	47,455	18,564
2000	1,361,143,845	46,434	19,118
2001	1,380,708,520	47,087	19,532
2002	1,323,300,606	46,001	19,485
2003	1,272,759,723	44,994	19,435
2004	1,454,237,000	49,958	20,739

TABLE 12.17

Proportions (%) of Foods from Farmland, Grassland, Aquatic Ecosystems, and Net Imports

Year	Farmland	Grassland	Aquatic Ecosystems	Net Import
1998	84.66	5.28	6.02	4.04
1999	83.50	5.80	6.70	3.99
2000	82.25	5.46	6.82	5.47
2001	80.93	5.61	6.73	6.73
2002	81.95	5.74	7.51	4.79
2003	77.55	5.46	7.29	9.71
2004	74.72	4.83	6.62	13.82

the period 1998 to 2004 (Table 12.18). 75.63% of protein was provisioned by farmland, 5.1% by grassland, 12.56% by aquatic products, and 6.71% by net imports. 73.81% of fat was supplied by farmland, 9.18% by grassland, 4.92% by aquatic products, and 12.09% by net imports on an average during the period 1998 to 2004. Food, especially fat, is becoming increasingly dependent on imports in China.

On an average, during the period 1998 to 2004, China was annually able to provide calories for 1731 million people, protein for 1963 million people, and fat for 933 million people under primary well-to-do-life; calories for 1726 million people, protein for 1866 million people and, fat for 926 million people under a full well-to-do life; and calories for 1688 million people, protein for

TABLE 12.18

Proportion Changes of Calorie, Protein, and Fat from Farmland, Grassland, Aquatic Products, and Net Imports

Year	Food Sources	1998	1999	2000	2001	2002	2003	2004	On an Average
Calorie (%)	Farmland	94.96	94.56	94.55	92.50	94.08	92.41	87.55	92.94
	Grassland	1.81	2.07	2.09	2.12	2.25	2.32	1.93	2.08
	Aquatic products	2.42	2.75	2.97	2.92	3.34	3.43	2.94	2.97
	Net imports	0.81	0.61	0.40	2.46	0.33	1.85	7.58	2.01
Protein (%)	Farmland	80.26	79.88	76.75	74.52	75.76	71.60	70.67	75.63
	Grassland	4.69	5.37	5.11	5.17	5.40	5.29	4.69	5.10
	Aquatic products	10.86	12.31	12.54	12.30	13.92	13.56	12.40	12.56
	Net imports	4.18	2.44	5.61	8.02	4.93	9.56	12.25	6.71
Fat (%)	Farmland	78.77	76.06	75.46	75.77	76.01	68.64	65.95	73.81
	Grassland	9.33	9.97	9.20	9.55	9.58	8.77	7.89	9.18
	Aquatic products	4.79	5.05	4.96	4.98	5.28	4.88	4.52	4.92
	Net imports	7.12	8.93	10.39	9.70	9.13	17.72	21.64	12.09

1758 million people, and fat for 868 million people under well-off-life. There was a considerable shortage of fat, that is, the nutrient structure was quite imbalanced in China. The balanced results indicate that food produced by China could support 1335.1, 1306.1, and 1228.7 million people under primary well-to-do life, well-to-do life, and well-off life, respectively, on an average (Yue et al., 2010).

Since 1978, China has introduced the "household responsibility system" and had started to dismantle the people's commune system. When introduced, the "household responsibility system" linked remuneration to output and mobilized peasants' enthusiasm for production. The mean yield of grain was 3943 kg/hm² during the 30 year period 1978 to 2007, while its yield was 1547 kg/hm² during the 30 year period 1949 to 1978. The improvements to farmland management and the introduction of new seed-breeding technologies have made it possible to greatly increase the yields of crops. For instance, in terms of our investigation, the maximum yield of rice reached 18,000 kg/hm² in the Yunan province, but the mean yield was 5665 km/hm² over the last 30 years; grain yield in the North China Plain was 20,100 kg/hm², as reported in the Scientific and Technological Project for High Yield of Grain by the Ministry of Science and Technology of the People's Republic of China in 2008, while in 2007 the yield was 4748 kg/hm² on an average. In other words, there is a great deal of potential for raising food provisions in China.

References

Chen, J. 2007. Rapid urbanization in China: A real challenge to soil protection and food security. *Catena* 69: 1–15.

Compiling Committee for Almanac of China's Foreign Economic Relations and Trade. 1984. *Almanac of China's Foreign Economic Relations and Trade*. Beijing: Publishing Company of China's Foreign Economic Relations and Trade (in Chinese).

Compiling Committee for Almanac of China's Foreign Economic Relations and Trade. 1985. *Almanac of China's Foreign Economic Relations and Trade*. Beijing: Publishing Company of Water Conservancy and Electric Power (in Chinese).

Compiling Committee for Almanac of China's Foreign Economic Relations and Trade. 1986. *Almanac of China's Foreign Economic Relations and Trade*. Beijing: China Expectation Press (in Chinese).

Compiling Committee for Almanac of China's Foreign Economic Relations and Trade. 1987. *Almanac of China's Foreign Economic Relations and Trade*. Beijing: Publishing Company of China's Foreign Economic Relations and Trade (in Chinese).

Compiling Committee for Almanac of China's Foreign Economic Relations and Trade. 1988. *Almanac of China's Foreign Economic Relations and Trade*. Beijing: China Expectation Press (in Chinese).

Compiling Committee for Almanac of China's Foreign Economic Relations and Trade. 1989. *Almanac of China's Foreign Economic Relations and Trade*. Beijing: China Expectation Press (in Chinese).

Compiling Committee for Almanac of China's Foreign Economic Relations and Trade. 1990–1995. *Almanac of China's Foreign Economic Relations and Trade*. Beijing: China Social Press (in Chinese).

Compiling Committee for Almanac of China's Foreign Economic Relations and Trade. 1996–1998. *Almanac of China's Foreign Economic Relations and Trade*. Beijing: China Economic Press (in Chinese).

Compiling Committee for Almanac of China's Foreign Economic Relations and Trade. 1999–2007. *Almanac of China's Foreign Economic Relations and Trade*. Beijing: Publishing Company of China's Foreign Economic Relations and Trade (in Chinese).

Editorial Committee of National Report of China Urban Development of China Mayor Association. 2003. *National Report of China Urban Development (2001–2002)*. Beijing: Xiyuan Press (in Chinese).

Fishery Administration of Ministry of Agriculture of the People's Republic of China. 1998–2008. *China Fishery Yearbook*. Beijing: China Agriculture Press (in Chinese).

Lichtenberga, E. and Ding, C.R. 2008. Assessing farmland protection policy in China. *Land Use Policy* 25: 59–68.

Tao, F.L., Hayashi, Y., Zhang, Z., Sakamoto, T., and Yokozawa, M. 2008. Global warming, rice production, and water use in China: Developing a probabilistic assessment. *Agricultural and Forest Meteorology* 148: 94–110.

Xin, X.P., Zhang, H.B., Li, G., Chen, B.R., and Yang, G.X. 2009. Variation in spatial pattern of grassland biomass in China from 1982 to 2003. *Journal of Natural Resources* 24(9): 1582–1592 (in Chinese).

Xu, Z.G., Xu, J.T., Deng, X.Z., Huang, J.K., Uchida, E., and Rozelle, S. 2006. Grain for Green versus Grain: Conflict between food security and conservation set-aside in China. *World Development* 34(1): 130–148.

Yang, Y.X. 2005. *China Food Composition 2004*. Beijing: Peking University Medical Press (in Chinese).

Yue, T.X. 2000. Stability analysis on the sustainable growth range of population. *Progress in Natural Science* 10(8): 631–636.

Yue, T.X., Fan, Z.M. and Liu, J.Y. 2005. Changes of major terrestrial ecosystems in China Since 1960. *Global and Planetary Change* 48: 287–302.

Yue, T.X., Tian, Y.Z., Liu, J.Y., and Fan, Z.M. 2008. Surface modeling of human carrying capacity of terrestrial ecosystems in China. *Ecological Modelling* 214: 168–180.

Yue, T.X., Wang, Q., Lu, Y.M., Xin, X.P., Zhang, H.B., and Wu, S.X. 2010. Change trends of food provisions in China. *Global and Planetary Change* 72: 118–130.

13

Surface Modeling of Ecological Diversity*

13.1 Introduction

Species diversity has two components: richness, also called species density, based on the total number of species present, and evenness, based on the relative abundance of species and the degree of its dominance thereof (Odum, 1983; Hamilton, 2005). This concept has been used to formulate ecotope diversity (Yue et al., 2001, 2003, 2005, 2007). An ecotope is the smallest holistic land unit, characterized by homogeneity of at least one land attribute of the geosphere—namely, atmosphere, vegetation, soil, rock, water, and so on, with nonexcessive variations in other attributes (Naveh and Lieberman, 1994). An ecotope commonly includes three characteristics: (1) The smallest homogenous mapable unit of land, (2) homogenous in general substrate condition, potential natural vegetation, and potential ecosystem functioning, and (3) the composition of patches in different successional stages of land-use (Forman, 1995). Species diversity can be distinguished into individual-counting diversity and biomass-based diversity. Ecotope diversity can be classified into individual ecotope-counting diversity and ecotope-area-based diversity.

13.1.1 Debates on the Relationship between Ecological Diversity and Ecosystem Functions

Prior to the 1970s, scientists attempted to develop a general theory linking stability and diversity to species observation (Odum, 1953; MacArthur, 1955; Elton, 1958; Hutchinson, 1959). Since Gardner and Ashby (1970) and May (1972) challenged the conventional wisdom that stability increases with species diversity, the thinking of some scientists gradually started to change. Two schools of thought developed on the diversity–stability hypothesis. Some scientists argued that their research results did not support the conventional wisdom of natural historians: diversity begets stability (Gilpin, 1975; Woodward, 1994; Beeby and Brennan, 1997; Naeem, 2002; Lhomme and Winkel, 2002; Pfisterer and Schmid, 2002). However, many scientists still believed that diversity begets stability (Odum, 1971; Watt, 1973; McNaughton,

---

* Dr. Qiquan Li and Dr. Shengnan Ma are major collaborators of this chapter.

1978; Glowka et al., 1994; Pennist, 1994; McGrady-Steed et al., 1997; Naeem and Li, 1997; Tilman et al., 1997). Several studies have indicated that diversity can be expected, on an average to give rise to the ecosystem stability (Bengtsson et al., 2000; Chapin III et al., 2000; McCann, 2000; Tilman, 2000; Wolfe, 2000). The long-term stability of an ecosystem service depends on biodiversity. Biodiversity could therefore, be an important element for the reliable and sustainable provisioning of ecosystem services (Tilman et al., 2006). The results of a decade-long experiment indicated that biological diversity makes ecosystems both stable and productive (Bakalar, 2007). Species richness enhances the stability of algal community biomass across a range of environmental settings in marine macroalgal communities (Boyer et al., 2009).

The relationship between diversity and productivity has been a central, but contentious issue within ecology (Schmid, 2002). In terms of Darwin's result, the ecological diversity of communities is due to niche diversification of the co-occurring species, and such diversification would lead to greater community productivity due to more effective resource exploitation (Darwin, 1872). In 1968, evidence from California grasslands showed that net productivity was inversely related to species diversity (McNaughton, 1978), which challenged Darwin's result. Since the 1990s, there have been ardent debates on the diversity–productivity relationship. New York successional analyses suggested that the average net productivity is negatively related to species diversity (McNaughton, 1994). Johnson et al. (1996) argued that attempts to unveil the relationship between ecological diversity and ecosystem productivity continue to generate contradictory conclusions. Rusch and Oesterheld (1997) claimed that diversity has a negative effect on productivity. Tilman et al. (1997) concluded that species composition, rather than species diversity, is the main determinant of ecosystem productivity. Hooper and Vitousek (1997), in relation to their experiment in grassland of California, argued that primary productivity does not correlate with increasing functional group richness, but the composition explains much more variance than richness. Grime (1997) stated that dominant plants rather than ecological diversity controls ecosystem productivity. Huston et al. (2000) concluded that species richness per se has no significant statistical or biological effect on plant productivity. Benedetti-Cecchi and Gurevitch (2005) reported that an increase in spatial variability of biodiversity may cause dramatic decreases in the mean productivity of the system. However, many ecologists still believe Darwin's result. The experimental results at eight European field sites show that each halving of the number of plant species reduced productivity by approximately 80 g/m^2 on an average (Hector et al., 1999). Tilman (2000) reviewed recent experimental, theoretical and observational studies and stated that on an average greater diversity leads to greater productivity in plant communities, greater nutrient retention in ecosystems, and greater ecosystem stability. Purvis and Hector (2000) summarized that 95% of the experimental studies support a positive relationship between diversity and ecosystem functioning. Many ecologists (Loreau, 2000; Loreau and Hector, 2001; Tilman et al., 2001) think that productivity may be greater at higher diversity because of niche complementarity among particular combinations of species and the greater chance of occurrence of such combinations at higher diversity. Pfisterer and Schmid (2002), in their combinatorial ecological diversity experiments, found that higher diversity tends to lead to higher productivity. Striebel et al. (2009) observed a positive relationship between diversity and productivity in their investigations of terrestrial and aquatic plant communities. Marquard et al. (2009) found positive effects of species richness on above-ground productivity, due to increased plant density.

On the scale of ecotope observation (Naveh and Lieberman, 1994; Forman, 1995), Odum (1969) proposed the strategy of ecosystem development and stated that the most pleasant and certainly the safest landscape to live in is one containing a variety of crops, forests, lakes, streams, roadsides, marshes, seashores, and "waste places." Haber (1971) applied this strategy in land utilization systems and proposed the concepts of differentiated land use (Haber, 1979) and differentiated land-use strategy (Haber, 1990). Numerous authors have stressed the favorable effect of diversity on agroecosystem functions such as ecosystem stability and productivity (Altieri, 1991; Prinsley, 1992; Ripl, 1995; Burel, 1996; Lenz and Haber, 1996; OECD, 1997; Palma et al., 2007).

13.1.2 Relationships between Species Diversity and Ecotope Diversity

An understanding of the relationship between species diversity and ecotope diversity, including their response to past, current, and future human activities, is essential to developing models for sustainable land management. Therefore, since the early 1960s, many scientists have paid particular attention to this relationship, and two camps have appeared on the relationship hypothesis. Some scientists found that an increase of ecotope diversity leads to an increase in species diversity, but others' findings did not support this hypothesis.

13.1.2.1 Hypothesis That Ecotope Diversity Positively Relates to Species Diversity

In America in 1961, MacArthur and MacArthur (1961) statistically analyzed the relationship between species diversity and landscape diversity by means of Shannon's index. The results showed that both bird species diversity and plant species diversity were closely related to landscape diversity and their correlation coefficients were 0.887 and 0.817, respectively. Tilman (1982) indicated that increased species diversity is often associated with greater environmental heterogeneity. Huston (1994) suggested that the spatial heterogeneity of an area is strongly correlated with the number of species that are found there. The decreased diversity of habitats associated with traditional management has resulted in a decline of plant species richness in European landscapes (Kienast, 1993; Ihse, 1995; Poschlod and Bonn, 1998). Greenberg et al.

(1995) demonstrated that bird species richness and diversity were positively correlated with increasing vertical and horizontal habitat structures. According to heterogeneous and homogenous treatments in experiments by Vivian-Smith (1997), heterogeneity significantly influenced temporal changes in species richness and more rare species were present in heterogeneous treatments; components of diversity, richness, and evenness were greater in the environment with greater heterogeneity. Heterogeneity promotes diversity at different scales and rare species appear to benefit more than common species, in which habitat heterogeneity is quantified as the variability in the principal component analysis. An evaluation of biodiversity at two spatial scales indicated that species diversity in an area increased with habitat variability and with habitat heterogeneity, in which habitat variability was defined as the number of biotope types per unit area and habitat heterogeneity was defined as the number of patches and the length of ecotones per unit area (Duelli, 1997). Species diversity should be higher in more diverse surroundings, because more species from ecologically different neighboring patches immigrate than in more uniform areas. Species diversity increases with the number of the mosaic patches. These matrix effects have been demonstrated in many case studies (Jonsen and Fahrig, 1997; Miller et al., 1997; Burel et al., 1998; Thies and Tscharntke, 1999; Weibull et al., 2000). Therriault and Kolasa (2000) agreed that both spatial and temporal habitat heterogeneity affects the structure and dynamics of ecological communities by increasing species diversity in terrestrial and aquatic systems. Jongman and Bunce (2000) found that landscape diversity was a major cause for species diversity within European regions. Therriault and Kolasa (2000) concluded that species diversity has a significant influence on the level of habitat heterogeneity. Species extinctions might additionally reduce heterogeneity and encourage further extinction.

Holt et al. (2002) found that structurally complex landscapes could enhance local species richness possibly due to a higher species pool in the complex landscapes. Maintaining habitat heterogeneity was proposed as a means of conserving species richness in habitats threatened by human activities. A rich diversity of habitats that were home to large numbers of plant diversity was developed as the result of varying natural conditions and different land use. Dauber et al. (2003) concluded that landscape diversity might service as a useful indicator for species richness at the landscape scale in terms of their study carried out at 20 managed grassland sites in Germany. An increase in the number of habitats led to an increase in species diversity in a landscape because of an expansion in the number of partitionable niche dimensions. Kostylev et al. (2005) concluded that complex habitats allowed more species to co-exist in a given area after they assessed the effects of surface complexity and area on species richness and abundance. Ma (2008) analyzed the landscape composition around each sampling site on the spatial scale of 4 ha by using ArcGIS Desktop 9.0, the result showed that the diversity of land-cover types was positively correlated with species richness.

13.1.2.2 Evidence against the Hypothesis That Ecotope Diversity Positively Relates to Species Diversity

Hoover and Parker (1991) studied six investigation areas across Georgia, two each from the Appalachian Highlands, Piedmont, and the Coastal Plain. They calculated species diversity using the Shannon–Weiner index and community diversity using the modified Simpson index in every investigation area. The calculation results showed that species diversity decreased from the Appalachian Highlands to the Coastal Plain, while community diversity increased from the Appalachian Highlands to the Coastal Plain. In other words, species diversity and landscape diversity had opposing trends in Georgia. Several studies also report the negative effects of intensive land use on species richness (Luoto, 2000; Zechmeister and Moser, 2001). Overgaard et al. (2003) stated that anopheline species diversity was negatively related to landscape diversity. Forest fragmentation resulting from human economic activities often increased ecotope diversity, which could result in a reduction in anopheline species diversity. Biodiversity is dependent on the natural richness of different parts of the continent, and also depends on the impact of humans and the way they change natural landscapes into cultural landscapes. Costamagma et al. (2004) found that landscape complexity had no positive effect on species richness. Cramer and Willig (2005) concluded that there was no evidence to support the hypothesis that habitat heterogeneity would enhance the diversity of a landscape by increasing the number of species in an area. According to a study in an area of 4 km^2 near the village of Friedeberg (10°34′E, 45°12′N) of Central Germany (Baessler and Klotz, 2006), the ecotope diversity of land use in 1957, 1979, and 2000 was respectively 1.02, 1.05, and 1.04, while the weed species diversity was 0.47, 0.37, and 0.38 (Tables 13.1 and 13.2). The correlation coefficient was −0.92. The less intense agricultural land use during the period 1945 to 1952 promoted a species rich weed flora. There was still high species diversity during the period 1952 to 1968. The enlargement of arable fields and the increase in agricultural

TABLE 13.1

Ecotope Diversity of Land Use in the Dry Region of Central Germany in Terms of the Scaling Diversity Index

Proportion of Land-Use Types (%)	1957	1979	2000
Woodland	17.05	13.97	12.03
Meadow	3.72	6.04	5.92
Dry grassland	0.85	4.64	5.38
Arable field	63.44	58.31	61.42
Built-up area	8.21	8.57	10.83
River Saale	4.03	5.48	2.61
Others	2.71	2.98	1.81
Ecotope diversity	**1.02**	**1.05**	**1.04**

TABLE 13.2

Weed Species Diversity in the Dry Region of Central Germany
in Terms of the Scaling Diversity Index

Weed Species	1957	1979	2000
Amaranthus retrofiexus	0.01	0.87	1.90
Arenaria serpyllifolia	0.49	0	0
Capsella bursa-pastoris	0.42	0.08	0.05
Consolida regalis	1.09	0.03	0.04
Descurainia sophia	0.39	0.03	0.16
Euphorbia esula	0.02	0	0
Euphorbia exigua	0.47	0.03	0.19
Euphorbia peplus	0.19	0.07	0.02
Galium aparine	0.21	4.34	0.39
Lamium amplexicaule	0.35	0.13	0.02
Lithospermum arvense	0.21	0.01	0.01
Mentha arvensis	1.17	0.05	0.08
Papaver rhoeas	0.60	0.03	0.34
Plantago intermedia	0	0.05	0
Senecio vulgaris	0.15	0.01	0.02
Silene noctiflora	0.33	0.21	0.02
Sinapis arvensis	0.63	0.17	0
Sonchus oleraceus	0.39	0.22	0.02
Stellaria media	3.43	1.76	0.55
Thlaspi arvense	0.25	0.02	0.01
Veronica agrestis	0.40	0	0
Veronica hederifolia	0.99	0.18	0
Veronica polita	0	0.18	0.35
Species diversity	**0.47**	**0.37**	**0.38**

intensification during the period 1968 to 1989 led to a decrease in weed species diversity. In 2000, ecotope diversity was lower and weed species diversity was higher when compared with 1979. Vollhardt et al. (2008) studied the
landscape complexity and species diversity of cereal aphid parasitoids; their
results did not support the hypothesis that increasing landscape diversity
could enhance local species diversity in farmland.

13.2 Ecological Diversity Index

Measuring ecological diversity has become a growth industry because of the
great significance of diversity to studies relating to ecosystems, ecosystem services and their changes (Williams and Humphries, 1996; Ibanez et al., 2005).
Numerous indices of ecological diversity have been proposed. Unsatisfying

TABLE 13.3

Unsatisfying Diversity Indices

Ordinal Number	Formula	Explanation of Parameters and Variables	Reference
1	$$H = -\sum_{i=1}^{m} p_i \ln p_i$$	p_i is the proportion of individuals found in the species i or the proportion of ecotope number in type i; and m is the total number of species or ecotope types.	Odum, 1969
2	$$HB = \frac{\ln N! - \sum \ln n_i!}{N}$$	n_i is the number of individuals in the species i; N is the total number of individuals	Pielou, 1966
3	$$d = \left(\sum_{i=1}^{m} p_i^2\right)^{-1}$$	p_i is the proportion of individuals or biomass that contributes to the total in the sample; m is the total number of species in the community; d is Simpson's index	Harper and Hawksworth, 1996
4	$$HS = 1 - \sum p_i^2$$	p_i is the proportion of individuals found in the species i	Simpson, 1949
5	$$N_a = \left(\sum_{i=1}^{m} (p_i)^a\right)^{1/(1-a)}$$	N_a is the ath "order" of diversity; p_i is the proportional abundance of the species i	Hill, 1973
6	$$D = \frac{\sum_{i=1}^{m} p_i \ln p_i}{\ln m}$$	D is the measure of ecotope diversity; p_i is the proportion of the landscape in type i; and m is the total number of ecotope types	Mladenoff et al., 1997
7	$$d_{Mg} = \frac{m-1}{\ln N}$$	m is the number of species; N is the total number of individuals summed over all m species	Margalef, 1957
8	$$d_{Mn} = \frac{m}{N^{1/2}}$$	m is the number of species; N is the total number of individuals summed over all m species	Whittaker, 1977
9	$$d = \frac{m}{\log N}$$	m is the number of species and N is the number of individuals	McNaughton, 1994
10	$$d = \frac{N - \left(\sum n_i^2\right)^{1/2}}{N - N^{1/2}}$$	n_i is the number of individuals in the species i; N is the total number of individuals	McIntosh, 1967
11	$$d = \frac{N_{max}}{N}$$	N_{max} is the number of individuals in the most abundant species; N is the total number of individuals	Berger and Parker, 1970

diversity indices (Table 13.3) have been criticized by many ecologists (Barrett, 1968; Odum, 1969; Pimm, 1994; Harper and Hawksworth, 1996; Beeby and Brennan, 1997; Mladenoff et al., 1997; Yue et al., 1998; Ricotta, 2002; Hoffmann and Greef, 2003; Ricotta et al., 2004; Roy et al., 2004; Scholes and Biggs, 2005). Odum (1969) stated that the Shannon–Weiner index (Formula 1 in Table 13.3) may obscure the behavior of the different aspects of diversity, richness, and evenness. For example, in field experiments, an acute stress from insecticide reduces the number of species of insects relative to the number of individuals but increases the evenness in the relative abundance of the surviving species (Barrett, 1968). Thus, in this case, the "richness" and "evenness" components would tend to cancel each other. Harper and Hawksworth (1996) pointed out that the Shannon–Weiner index and Simpson's index (Formula 3 in Table 13.3) are inadequate for some purposes because it is possible for high richness but low evenness to have a lower index overall than in cases where the opposite occurs. Beeby and Brennan (1997) assert that while various indices attempt to measure diversity, no single measurement of diversity has yet been adopted as being the most effective under all circumstances.

Pimm's research (1994) reviewed the history of the relationship between diversity and stability and concluded that many diversity indices ignore the evenness of species and looks only at the species list. Many measures focus on the richness aspect of diversity. For instance, most policy-makers are used to seeing species diversity simply as the changing number of species on a species list (Mace, 2005). Many studies (Yoshida, 2003; Haberl et al., 2004; Uys et al., 2004; Hanski, 2005; Hodgson et al., 2005; Ibanez et al., 2005) equate richness to diversity. Richness, though necessary, is not sufficient to support the components of ecological diversity that underlie the key functions and benefits of an ecosystem (Mace, 2005).

Ecological diversity is the result of ecological processes acting at various spatial and temporal scales (Alados et al., 2004). Studies on scaling issues are burgeoning because of the increasing need for ecological modeling and simulation. They are driven by progress in remote sensing technologies to obtain data on various resolutions and by the integration of geo-referenced data collected at various scales (Martin et al., 2005). However, all diversity indices in Table 13.3 ignore the important parameter of scale.

A useful measure of ecological diversity should be theoretically sound and sensitive to changes at policy-relevant spatial scales; it should allow comparison with a baseline situation and policy target, be usable in the simulation of scenarios, and be amenable to aggregation and disaggregation on local, national, regional, and international levels (Scholes and Biggs, 2005).

13.2.1 Theoretical Analysis of Diversity Indices

13.2.1.1 Drawbacks of the Diversity Indices

None of the diversity indices in Table 13.3 bear any relation to the investigation area, and so are unable to formulate the richness aspect. Because the

Shannon–Weiner index has been most widely used to formulate diversity in recent years (Alados et al., 2004; Mueller, 2004; Roy et al., 2005; Sandstroem et al., 2006), we will take it as a detailed example for analyzing the drawbacks of the unsatisfying models.

If the Shannon–Weiner index was used, individuals of every species or every ecotope type would be greater than 100. Suppose, n_i is the individual number of species i or ecotope type i, m the total species number or total ecotope types, and N the total individual number of all species or total number of all ecotopes, then,

$$N = \sum_{i=1}^{m} n_i \tag{13.1}$$

and

$$R = \frac{N!}{\prod_{i=1}^{m} n_i!} \tag{13.2}$$

then

$$H = \frac{\ln R}{N} = \frac{1}{N}\left(\ln N! - \sum_{i=1}^{m} \ln n_i!\right) \tag{13.3}$$

According to Stirling's Formula

$$n! = \left(\frac{n}{e}\right)^n (2\pi n)^{\frac{1}{2}} e^{w(n)} \tag{13.4}$$

$$H = \frac{1}{N}\left(\ln\left(\left(\frac{N}{e}\right)^N (2\pi n)^{\frac{1}{2}} e^{w(N)}\right) - \sum_{i=1}^{m} \ln\left(\left(\frac{n_i}{e}\right)^{n_i} (2\pi n_i)^{\frac{1}{2}} e^{w(n_i)}\right)\right)$$

$$\tag{13.5}$$

$$= -\sum_{i=1}^{m} p_i \ln p_i + \varepsilon(n_1, n_2, \ldots, n_m)$$

where,

$$\varepsilon(n_1, n_2, \ldots, n_m) = \frac{1}{2N}\left(\ln(2\pi N) - \sum_{i=1}^{m} \ln(2\pi n_i)\right) + \frac{w(N) - \sum_{i=1}^{m} w(n_i)}{N};$$

$e = 2.7183$; $p_i = (n_i/N)$; $\pi = 3.1415$; and

$$\frac{1}{12(n + 0.5)} < w(n) < \frac{1}{12n}.$$

When $n_i \geq 100$, we can get an approximate formulation (Haken, 1983), that is,

$$H \approx -\sum_{i=1}^{m} p_i \ln p_i \tag{13.6}$$

Shannon and Weaver (1962) gave this formulation the name entropy in relation to the mathematical form, which expresses uncertainty (Peters, 1975). This Shannon–Weiner index has been widely used to formulate biodiversity by ecologists.

The Shannon–Weiner index does not include any information about the area under investigation. However, the species number is closely associated with the area or spatial scale (MacArthur and Wilson, 1967; Williamson, 1981); "You will find more species if you sample a larger area" (Rosenzweig, 1995). The Global Biodiversity Assessment (Bisby, 1995) states that "the central single measure of ecological diversity is species richness" and "species richness is related to area in a complicated way, so we must exercise caution in comparing the diversities of areas that differ greatly in size." The relationship between species number and area has been formulated as (Arrhennius, 1921; Preston, 1960, 1962; Gorman, 1979; Browne, 1981)

$$\frac{m}{C} = \left(\frac{1}{A}\right)^{-D_0} \tag{15.7}$$

where m is the number of species; A the area; C the constant; and D_0 the Hausdorff dimension.

A research report compiled by the World Conservation Monitoring Center in collaboration with the Natural History Museum, the World Conservation Union, the United Nations Environment Program, the World Wide Fund for Nature, and the World Resources Institute, indicates that "a ten-fold decrease in area leads to a loss of half the species present" (Groombridge, 1992). In this case, it is easy to get a calculation result, $D_0 = 0.301$.

Formulation (Equation 15.7) shows that species number, taking $(1/A)$ as the scaling factor, has statistical self-similarity. The self-similarity is the property that at every scale of observation new details are revealed, yet these details are reminiscent of details elsewhere in the structure of the object or in the same part of the object but at a different scale (Iannaccone and Khokha, 1996). The mathematician, Felix Hausdorff (1919), statistically defined such self-similarity as $D_0 = -\lim_{\varepsilon \to 0}(\ln N / \ln \varepsilon)$. When $(\ln N/\ln \varepsilon)$ is a constant, $N = (\varepsilon)^{-D_0}$ where ε is the scaling factor, D_0 the Hausdorff dimension, and N the fractal object. Obviously, area has an essential effect on the number of species, but is not included in the Shannon–Weiner index and other unsatisfying diversity indices (Table 13.3).

The Shannon–Weiner index cannot express the "richness" component of diversity. For the Shannon–Weiner index, the essential function $f(x) = -x \ln x$ does not strictly increase because $df(x)/dx = -(\ln x + 1)$. Thus, $df(x)/dx$ is respectively greater than zero, smaller than zero and equals zero when x is smaller than 0.3679, greater than 0.3679, and equal to 0.3679. In other words, $f(x) = -x \ln x$ increases when x is smaller than 0.3679; it decreases when x is greater than 0.3679; and it reaches the maximum value when x equals 0.3679 (Table 13.4). Therefore, if the individual proportion of species i to total individuals of all species in the investigation region, p_i, is smaller than 0.3679, the contribution of p_i to the Shannon–Weiner index would increase with an increase of p_i. When p_i is greater than 0.3679, contribution of p_i to Shannon–Weiner index would decrease with an increase of p_i. The contribution of p_i to Shannon–Weiner index reaches maximum when p_i equals 0.3679. The Shannon–Weiner index has its maximum value, $\ln m$, when $p_i = (1/m)$ (Alados et al., 2004).

TABLE 13.4

Relation between the Function $f(x) = - x \ln x$ and Its Independent Variable x

x	$f(x) = -x \ln x$
0	0
0.00001	0.0001
0.0001	0.0009
0.001	0.0069
0.01	0.0461
0.1	0.2303
0.2	0.3219
0.3679	0.3679
0.4	0.3665
0.5	0.3466
0.6	0.3065
0.7	0.2497
0.8	0.1785
0.9	0.0948
1	0

13.2.1.2 Scaling Diversity Index

The scaling diversity index is expressed as (Yue et al., 2001, 2003, 2004)

$$D(\varepsilon,r,t) = -\frac{\ln\left(\sum_{i=1}^{m(\varepsilon,r,t)} \left(p_i(\varepsilon,r,t)\right)^{1/2}\right)^2}{\ln \varepsilon} \tag{13.8}$$

where $p_i(\varepsilon, r, t)$ is the probability of the ith ecotope or of the ith species; $m(\varepsilon, r, t)$ is total number of species or ecotopes under investigation; t represents time; $\varepsilon = (e + A)^{-1}$, A is the area of the investigation region measured by hectare; r is the spatial resolution of the data set used; and e equals 2.71828.

In terms of the Method of Lagrange Multipliers (Kolman and Trench, 1971), the necessary condition under which $D(\varepsilon, t)$ reaches the maximum value is

$$\frac{\partial D(\varepsilon,r.t)}{\partial p_j(\varepsilon,r,t)} + \lambda \cdot \frac{\partial k\left(p_1(\varepsilon,r,t),\, \ldots,\, p_m(\varepsilon,r,t)\right)}{\partial p_j(\varepsilon,r,t)} = 0 \tag{13.9}$$

where $j = 1, 2, \ldots, m$;

$$k(p_1(\varepsilon,r,t),\, \ldots,\, p_m(\varepsilon,r,t)) = 1 - \sum_{i=1}^{m(\varepsilon,r,t)} p_i(\varepsilon,r,t)$$

and λ is an arbitrary constant.

The solution of the differential equation 13.9 is $p_j(\varepsilon,r,t) = (1/m(\varepsilon,r,t))$. In other words, when every investigation species or ecotope has an equal proportion, $D(\varepsilon, r, t)$ reaches its maximum value, $-(\ln m(\varepsilon,r,t)/\ln \varepsilon)$, under constraint condition, $\sum_{i=1}^{m(\varepsilon,r,t)} p_i(\varepsilon,r,t) = 1$. Thus, the scaling diversity index can express the "evenness" aspect of diversity.

$D(\varepsilon, r, t)$ is a strictly increasing function of $p_j(\varepsilon, r, t)$, because

$$\frac{\partial D(\varepsilon,r,t)}{\partial p_j(\varepsilon,r,t)} = \frac{1}{\left(p_j(\varepsilon,r,t)\right)^{1/2} \cdot \sum_{i=1}^{m} \left(p_i(\varepsilon,r,t)\right)^{1/2}} > 0$$

for $p_j(\varepsilon,t) > 0 (j = 1, 2, \ldots, m)$. Therefore, the scaling diversity can express the "richness" aspect of diversity.

13.3 Effects of Different Spatial Scales on the Relationship between Species Diversity and Ecosystem Diversity

The effects of different spatial scales on ecological diversity and on relationships between ecological diversity and ecosystem functions have been discussed by many scientists. For instance, Noordwijk's results (2002) showed that intensification of the crop-fallow system is likely to decrease the average species diversity, but ecotope diversity may initially increase; while further intensification is likely to reduce all aspects of ecological diversity. Enquist and Niklas (2001) found that organizing principles are needed to link ecological diversity across spatial and temporal scales. Crawley and Harral (2001) demonstrated that different processes might determine plant diversity at different spatial scales. Ritchie and Olff (1999) proposed that the spatial scaling of resources use by species of different sizes may explain many species' diversity patterns across a range of spatial scales, which provides a basis for the development of ecological theories that are trans-scalar in geographical space (Whittaker, 1999).

Many scientists note that relationships between ecological diversity and ecosystem functions require particular attention to be paid to various scales such as local, landscape, and regional scales. For instance, Chase and Leibold (2002) concluded that the shape of the productivity–diversity relationship depends on spatial scale; when data were viewed among ponds, the relationship between species diversity and productivity was hump-shaped, whereas when the same data were viewed among watersheds, the relationship was positively linear. Noordwijk (2002) stated that trade-offs between productivity and ecological diversity depend on the scale of model application. Loreau, et al. (2001) stated that generally the relative effects of individual species and species richness may be expected to be the greatest at small-to-intermediate spatial scales, but these biological factors should be less important as predictors of ecosystem processes at regional scales, where environmental heterogeneity is greater. Purvis and Hector (2000) found that the relationship between plant diversity and productivity changes with spatial scales. Gaston (2000) stated that observed patterns may vary with spatial scales while processes at regional scales influence patterns observed at local ones. The results of plant diversity and productivity experiments in European grasslands highlight the importance of considering scale when studying relationship between diversity and productivity (Hector et al., 1999).

13.4 Case Studies

13.4.1 Case Study in Fukang of the Xinjiang Uygur Autonomous Region

Fukang of Xinjiang Uygur Autonomous Region is located in the northern foot of the eastern part of the Tianshan Mountains, with an area of 8767 km². The ranges of geographical coordinates of Fukang are 43°45′–45°30′N and 87°465′–88°44′E (Figure 13.1). Fukang is 140 km long from south to north and

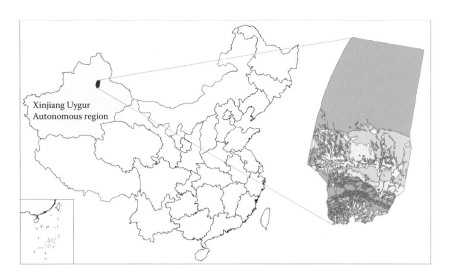

FIGURE 13.1
Location of Fukang.

75 km wide from west to east. The terrain of Fukang, which slants from southeast to northwest, can be divided into three geomorphologic units: mountainous area, plain, and desert. Bogda peak in the mountainous area has an elevation of 5445 m, which is the highest peak in the eastern part of the Tianshan Mountains; the south edge of the desert has an elevation lower than 460 m. The elevation difference is about 5000 m. The area proportions of mountainous area, plain, and desert to the total area of Fukang are, respectively 50.74%, 29.00%, and 20.26%. The Fukang region from the Bogda peak to the north edge of the desert, about 80 km, has a clear vertical spectrum of natural zones: the alpine nival zone, the alpine and sub-alpine meadow zone, the forest zone, the hilly grassland zone, the piedmont semidesertification zone, the desertification zone in alluvial and diluvial plain, and the fixed and semifixed sand dunes in the desert zone.

13.4.1.1 Data Acquisition

The Landsat TM image of Fukang was taken in the autumn of 2002 on a spatial resolution of 30×30 m. We applied unsupervised classification and generated clusters using ISODATA clustering analysis. Investigation data were used to identify training classes and to label and merge clusters for identification. Signatures of identified clusters were subsequently used in a maximum likelihood classification to generate a land-cover map. The land-cover types include woodland, sparse woodland, shrub land, other woodland, dense grassland, moderately dense grassland, sparse grassland, paddy field, dry farmland, lake, reservoir and water pond, urban area,

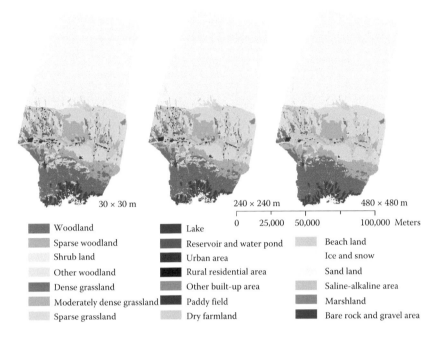

FIGURE 13.2
Land-cover maps of Fukang on different resolutions: left map on resolution 30 × 30 m, middle map on resolution 240 × 240 m, and right map on resolution 480 × 480 m.

rural residential area, other built-up area, nival area, beach land, sand land, saline-alkaline area, marshland, and bare rock and gravel area (Figure 13.2). The land-cover data set on a spatial resolution of 30 × 30 m are transformed into a series of land-cover data sets on spatial resolutions of $(k \times 30) \times (k \times 30)$m ($k = 1,2,3,...,16$) by an upscaling process. The land-cover type in each pixel of the new land-cover data set is derived from the dominant land-cover type of the transformed pixels. When a new data set is created, the boundary of every newly created ecotope, topological relationship, and attribute of the new data are rectified. Finally, the newly created data set is exported to a vector polygon file of Coverage of Arc/Info.

Classification accuracy of the land-cover was assessed by means of PCI Geomatica Focus. 396 samples were stratified and randomly selected from the land-cover maps. Each sampled land-cover type of the classified image was identified on the basis of a topographical map on 1:50,000 and land-cover map on 1:100,000 of Fukang. The classification accuracies for final products were evaluated in error matrices and summarized in Table 13.5. The overall accuracy achieved was 95.75%. The producer and user accuracies and the conditional kappa coefficient for each individual land-cover type were also reported.

TABLE 13.5

Classification Accuracy and Conditional Kappa Statistics for Each Land-Cover Type

Land-Cover Type	Reference Totals	Classified Totals	Number Correct	Producer's Accuracy (%)	User's Accuracy (%)	Kappa Statistics
Woodland	15	14	13	86.67	92.86	0.93
Shrub land	9	8	8	88.89	88.89	0.89
Sparse woodland	8	6	6	75.00	100.00	1.00
Other woodland	1	1	1	100.00	100.00	1.00
Dense grassland	34	37	32	94.12	86.49	0.85
Moderately dense grassland	37	39	37	100.00	94.87	0.94
Sparse grassland	52	50	50	96.15	100.00	1.00
Lake	6	6	6	100.00	100.00	1.00
Reservoir and water pond	9	9	9	100.00	100.00	1.00
Nival land	7	7	7	100.00	100.00	1.00
Beach land	7	5	5	71.43	100.00	1.00
Urban area	6	6	6	100.00	100.00	1.00
Rural residential area	7	6	6	85.71	100.00	1.00
Other built-up area	5	6	5	100.00	83.33	0.83
Sand land	145	144	144	99.31	100.00	1.00
Saline–alkaline area	4	4	4	100.00	66.67	0.66
Marshland	2	2	2	100.00	100.00	1.00
Bare rock and gravel area	14	15	14	93.33	93.33	0.93
Paddy field	3	3	3	75.00	100.00	1.00
Dry farmland	25	28	25	92.59	89.29	0.89
Totals	396	396	383			
Overall accuracy (%)						95.75
Overall kappa statistics (%)						0.95

13.4.1.2 Results

The statistic analysis of ecotope diversity on different spatial resolutions shows that the ecotope diversity of Fukang landscape, which is calculated by the scaling diversity index, increases on an average and converges to 0.210 with the spatial resolution becoming finer (Table 13.6). Ecotope diversity change (Figure 13.3) can be formulated as the following regression equation:

$$D = 0.21 + 0.04r - 3.73r^2 + 37.09r^3 - 162.47r^4 + 260.07r^5 \qquad (13.10)$$

TABLE 13.6

Changes of Relative Indices with Spatial Resolution Becoming Finer

Spatial Resolution (km²)	Ecotope Number	Ecotope Density (Ecotope Number/km²)	Ecotope Type	Ecotope Diversity
0.2304	547	0.12673	18	0.186
0.2025	583	0.12581	18	0.189
0.1764	622	0.12239	17	0.192
0.1521	654	0.12228	19	0.193
0.1296	737	0.12205	19	0.199
0.1089	749	0.11931	19	0.199
0.0900	829	0.11098	20	0.201
0.0729	856	0.10494	18	0.203
0.0576	920	0.09764	18	0.206
0.0441	973	0.09456	19	0.207
0.0324	1046	0.08543	20	0.208
0.0225	1070	0.08407	20	0.210
0.0144	1072	0.07460	20	0.210
0.0081	1111	0.07095	20	0.210
0.0036	1103	0.06650	20	0.210
0.0009	1073	0.06239	20	0.210

where D is ecotope diversity and r is the spatial resolution of the data set. The correlation coefficient of the regression equation is $R^2 = 0.9922$.

The change of ecotope density with spatial resolution (Figure 13.4) can be formulated as

$$Ri = 0.1273e^{-3.2645r} \qquad (13.11)$$

where Ri is the ecotope density and r is the spatial resolution of the data set. The correlation coefficient of the regression equation is $R^2 = 0.9885$.

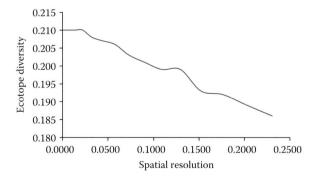

FIGURE 13.3
Ecotope diversity change with spatial resolution becoming finer.

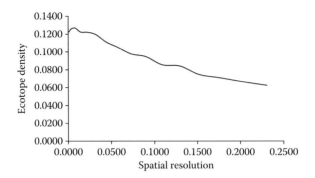

FIGURE 13.4
Relationship between ecotope density and spatial resolution.

The relationship between ecotope diversity and ecotope density (Figure 13.5) can be statistically expressed as

$$D = 0.2984Ri^{0.1673} \tag{13.12}$$

where D is ecotope diversity and Ri is ecotope density. The correlation coefficient of the regression equation is $R^2 = 0.9861$.

The results show that more ecotope types and ecotopes can be found in a given investigation region with spatial resolution becoming finer. Ecotope diversity has a closely positive relationship with ecotope density. Regression Equation 13.12 can be reformulated as $D/0.2984 = ((1/Ri))^{-0.1673}$. Ecotope diversity, taking $(1/Ri)$ as the scaling factor, has statistical self-similarity and the Hausdorff dimension is 0.1673. In other words, the scaling diversity index can express the richness aspect of diversity.

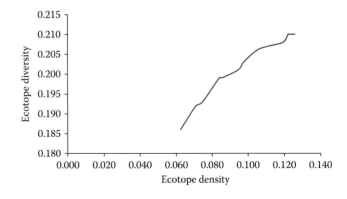

FIGURE 13.5
Relationship between ecotope diversity and ecotope density with spatial resolution becoming finer.

13.4.2 Case Study of Bayin Xile Grassland in the Inner Mongolia Autonomous Region

The investigated region with a total area of 97,200 ha is located in 43°29′44″ – 43°46′19″N and 116°25′3″ – 116°49′33″E, within the Bayin Xile Grassland of the Xilin Gol League of Inner Mongolia Autonomous Region. It has a semiarid and continental grassland climate in the temperate zone. The annual average temperature is 0.2°C, with a highest temperature of 38.5°C in summer and a lowest temperature of –42.8°C in winter. The annual average precipitation is 350 mm. The most common species are *Aneurolepidium chinensis* (Trin.) Kitag. and *Stipa grandis* P. Smirn., of which biomass accounts for 60.4% of the total amount. In precise terms, *Aneurolepidium chinensis* (Trin.) Kitag. accounts for 48.8% and *Stipa grandis* P. Smirn. 11.6%. Grass plants return green at the end of April and senescence in early October. The plant-growing period lasts about 150 days.

13.4.2.1 Data Acquisition of Land Cover

Landsat TM/ETM images, taken on July 31, 1987, August 11, 1991, September 27, 1997, and May 23, 2000, were analyzed by applying digital image processing techniques to the six visible/near-infrared bands (bands 1, 2, 3, 4, 5, and 7). Ancillary data included a vegetation map of Bayin Xile, a soil map of the Xilin River Basin, a topographical map, and biomass data sampled in the field. Using the six atmospherically corrected bands as input, 48 spectral classes were generated by unsupervised classification procedure of ISODATA (ERDAS/Imagine 8.4 package). Cropland and wetland could not be separated as unique spectral classes, and so supervised classification was used to define training samples of cropland and wetland. The supervised training samples and the 48 unsupervised spectral classes were combined and the whole image was classified again using the maximum likelihood classification procedure. Using the ancillary data as a reference, the spatial relationships between spectral classes and land-cover types were established. The final land-cover classification map had 14 identical classes: *Filifolium sibiricum* steppe, *Stipa baicalensis* steppe, *Aneurolepidium chinensis* + forbs steppe, *Aneurolepidium chinensis* + bunchgrasss steppe, *Aneurolepidium chinensis* + *Artemisia* frigidas steppe, *Stipa grandis* + *Aneurolepidium chinensis* steppe, *Stipa grandis* + bunchgrass steppe, *Stipa krylovii* steppe, *Artemisia* frigida steppe, cropland, wetland, desertification land, saline–alkaline land, and water area (Figures 13.6 through 13.9).

13.4.2.2 Data Acquisition of Maximum Aboveground Biomass in Terms of Species

The peak value of aboveground biomass usually appears at the end of August, which is considered as the grassland productivity. In order to analyze the relationship between species biomass diversity and the grassland productivity, 57 sampling quadrates 1 m × 1 m were randomly selected.

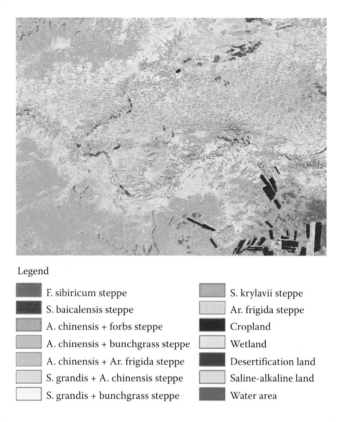

Legend

■	F. sibiricum steppe	■	S. krylavii steppe
■	S. baicalensis steppe	□	Ar. frigida steppe
■	A. chinensis + forbs steppe	■	Cropland
■	A. chinensis + bunchgrass steppe	□	Wetland
■	A. chinensis + Ar. frigida steppe	■	Desertification land
□	S. grandis + A. chinensis steppe	□	Saline-alkaline land
□	S. grandis + bunchgrass steppe	■	Water area

FIGURE 13.6
Land cover in 1987 (30 × 30 m).

The sampling process included five steps: (1) cutting grass to the roots at the end of August, (2) classifying the cut grass in terms of species, (3) drying the cut grass in terms of species at 60°C, and (4) weighting the dried grass in terms of species. The Inner Mongolia Grassland Ecosystem Research Station, which was founded in 1979 and listed as a key project demonstration station by UNESCO's MAB program in 1988, has been repeating the sampling process in the study region at the end of August every year since 1980. We will only pay attention to the data sampled in the periods 1986 to 1988, 1990 to 1992, 1996 to 1998, and 1999 to 2001 and to the averages of biomass of each species in the four periods (Table 13.7) in order to make calculation results; species diversity have an identical time phase with that of ecotope diversity based on remotely sensed data.

13.4.2.3 Results

The land-cover data set on a 30 × 30 m spatial resolution is transformed into 20 more data sets using an upscaling process (Tables 13.8 through 13.10).

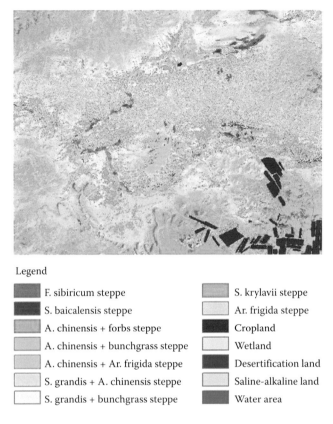

Legend

▨	F. sibiricum steppe	▨	S. krylavii steppe
▨	S. baicalensis steppe	▨	Ar. frigida steppe
▨	A. chinensis + forbs steppe	▨	Cropland
▨	A. chinensis + bunchgrass steppe	▨	Wetland
▨	A. chinensis + Ar. frigida steppe	▨	Desertification land
▨	S. grandis + A. chinensis steppe	▨	Saline-alkaline land
▨	S. grandis + bunchgrass steppe	▨	Water area

FIGURE 13.7
Land cover in 1991 (30 × 30 m).

The pixel side of every new land-cover data set is 30 m bigger than the transformed data set in both width and height. The land-cover type in each pixel of the new land-cover data is derived from the dominant land-cover type of the transformed pixels. When every new data set has been created, it can be exported to a vector polygon file such as Coverage of Arc/Info.

The analysis results show that resolution change causes a nonlinear change of the correlation coefficients of ecotope diversity with grassland productivity and desertification area. When the resolution is 210 × 210 m, the correlation coefficients between ecotope diversity and grassland productivity reach maximum values for the scaling index, Shannon's index, and Simpson's index, which are 0.52, 0.75, and 0.89, respectively. But the correlation coefficients between ecotope diversity and desertification area reach maximum values for both the scaling index and Shannon's index when the pixel size is 30 × 30 m, and for Simpson's index when the pixel size is 60 × 60 m. Therefore, on an average for the three diversity indexes, ecotope diversity seems to have a close relationship with productivity at a spatial resolution of 210 × 210 m;

Legend

■ F. sibiricum steppe		▨ S. krylavii steppe	
■ S. baicalensis steppe		▫ Ar. frigida steppe	
▨ A. chinensis + forbs steppe		■ Cropland	
▨ A. chinensis + bunchgrass steppe		▫ Wetland	
▨ A. chinensis + Ar. frigida steppe		■ Desertification land	
▫ S. grandis + A. chinensis steppe		▫ Saline-alkaline land	
▫ S. grandis + bunchgrass steppe		■ Water area	

FIGURE 13.8
Land cover in 1997 (30×30 m).

and there is a relationship between ecotope diversity and desertification area on a resolution of 30×30 m. So we assert that if the correlation coefficient is less than 0.5 they have no relationship; if the correlation coefficient is located in the range between 0.5 and 0.6 they have little relationship; if the correlation coefficient is located in the range between 0.6 and 0.8 they have some relationship; and if the correlation coefficient is located in the range between 0.8 and 1.0 they have a close relationship.

In terms of scaling index, with the exception of ecotope diversity bearing little relation to productivity on a special resolution of 210×210 m, ecotope diversity has no relationship with productivity on all other resolutions (Table 13.8). In terms of Simpson's index, ecotope diversity has no relationship with desertification area on any resolutions other than 30×30 m, 60×60 m, and 90×90 m (Table 13.9). In terms of Shannon's index, ecotope diversity has a relationship with desertification area on a resolution of 480×480 m, while it has no relationship with desertification area on resolutions of 600×600 m, 570×570 m, 540×540 m, 510×510 m, 450×450 m, 420×420 m, 390×390 m, and 360×360 m (Table 13.10).

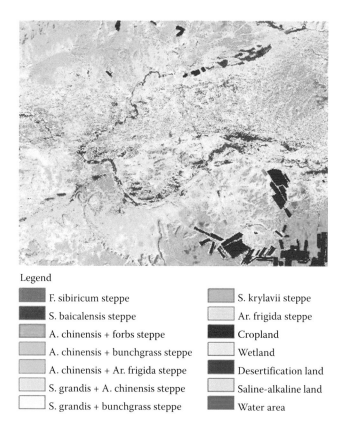

Legend

■	F. sibiricum steppe	■	S. krylavii steppe
■	S. baicalensis steppe	☐	Ar. frigida steppe
☐	A. chinensis + forbs steppe	■	Cropland
☐	A. chinensis + bunchgrass steppe	☐	Wetland
☐	A. chinensis + Ar. frigida steppe	■	Desertification land
☐	S. grandis + A. chinensis steppe	☐	Saline-alkaline land
☐	S. grandis + bunchgrass steppe	■	Water area

FIGURE 13.9
Land cover in 2000 (30 × 30 m).

On the scale of ecotope observation, calculation results of the three diversity indices on the data at a 30 × 30 m resolution (Figures 13.6 through 13.9) show that scaling diversity and Shannon's diversity have no relationship with grassland productivity, but the correlation coefficient between Simpson's diversity and grassland productivity is 0.73 (Table 13.11). Statistically, the calculation results for the three diversity indices have different trends from 1987 to 2000. Shannon's diversity and scaling diversity show an increasing trend from 1991 to 1997, but Simpson's diversity demonstrates a decreasing trend during this period. From 1997 to 2000, Shannon's diversity increases very little, Simpson's diversity increases a great deal while scaling diversity does not change (Table 13.12).

On the given resolution 210 × 210 m, the correlation coefficient between ecotope diversity and desertification area is 0.73 for the scaling index, 0.51 for Shannon's index, and 0.18 for Simpson's index; the correlation coefficient between ecotope diversity and grassland productivity is 0.52 for the scaling index, 0.75 for Shannon's index, and 0.89 for Simpson's index

TABLE 13.7

Averages of Sampling Data of Maximum Aboveground Biomass during the Four
Periods (g/m²)

Species	Average of Biomass from 1986 to 1988	Average of Biomass from 1990 to 1992	Average of Biomass from 1996 to 1998	Average of Biomass from 1999 to 2000
Aneurolepidium chinensis (Trin.) Kitag.	53.96	87.78	64.15	66.37
Stipa grandis P. Smirn.	13.86	26.88	6.58	20.70
Achnathrum sibiricum (L.) Keng	3.70	7.85	6.47	11.41
Caragana microphylla Lam.	11.72	3.21	0.83	2.48
Agropyron michnoi Roshev	2.95	24.19	3.01	1.91
Artemisia commutata Bess.	13.46	5.29	0.91	0.90
Carex korshinshyi Kom.	3.78	1.88	4.92	23.64
Artemisia scoparia Wald. et Kit.	2.38	1.70	0.17	0.29
Salsola collina Pall.	1.63	1.70	0.04	0.33
Kochia prostrata (L.) Schrad.	1.81	1.60	0.82	4.22
Serratula centeuroides L.	1.94	1.06	1.25	3.18
Artemisia frigida Willd.	0.80	1.58	1.34	2.46
Cleistogenes squarrosa (Trin.) Keng	0.59	2.38	2.92	2.09
Koeleria cristata (L.)Pers.	0.61	2.36	1.78	7.82
Heteropappus altaicus (Willd.) Novopokr.	1.69	0.72	0.88	0.19
Poa palustris L.	0.17	3.30	0.44	0.68
Allium ramosum L.	0.01	0.10	2.32	0.15
Allium senescens L.	2.19	3.82	0.00	3.75
Potentilla tanacetifolia Willd. Ex Schlecht.	0.49	1.36	0.61	2.96
Melissitus ruthenica (L.) Peschkova	0.25	0.09	1.82	2.35
Orostachys fimbriatus (Turcz.) Berger	0.58	2.10	0.62	0.00
Allium tenuissimum L.	0.38	1.78	0.26	2.68
Potentilla acaulis L.	0.10	0.41	0.68	2.86
Allium bidentatum Fisch. ex Prokh.	0.08	0.09	0.00	3.91
Dontostemon micranthus C. A. Mey	0.81	0.52	0.00	0.00
Allium condensatum Turcz.	0.01	0.43	0.22	0.36
Saposhnikovia divaricala (Turca.) Schischk.	0.19	0.31	0.25	1.14
Artemisia sieversiana Willd.	2.47	0.00	0.00	0.00

TABLE 13.7   (continued)

Averages of Sampling Data of Maximum Aboveground Biomass during the Four Periods (g/m²)

Species	Average of Biomass from 1986 to 1988	Average of Biomass from 1990 to 1992	Average of Biomass from 1996 to 1998	Average of Biomass from 1999 to 2000
Potentilla bifurica L.	0.55	0.01	0.26	0.76
Allium anisopodium Ldb.	0.06	0.75	0.00	1.98
Oxytropis myriophylla (Pall.) DC.	0.29	0.86	0.08	0.00
Elymus dahuricus var. *tangutorum* Roshev.	0.00	0.55	0.11	1.23
Astragalus adsurgens Pall.	0.00	0.02	0.06	0.01
Thalictrum petaloideum var.	0.31	0.52	0.03	0.05
Chenopodium glaucum L.	0.11	0.25	0.18	0.02
Pulsatilla tenuiloba (Turcz.) Tuz.	0.08	0.06	0.00	2.02
Thermopsis lanceolata R. Br.	0.00	0.02	0.00	0.08
Adenophora stenanthina (Ldb.) Kitag.	0.09	0.02	0.00	1.04
Haplophyllum dauricum Juss.	0.02	0.07	0.00	0.01
Iris tenuifolia Pall.	0.00	0.15	0.00	0.10
Melandrium brachypetalium (Horn) Fenzl.	0.07	0.00	0.00	0.00
Potentilla verticillaris Steph. ex Willd.	0.01	0.17	0.00	0.40
Cymbaria dahurica L.	0.05	0.00	0.00	1.24
Adenophora crispata (Korsh.) Kitag.	0.00	0.00	0.20	0.46
Gueldenstaedtia verna (Georgi) A. Bor.	0.00	0.01	0.02	0.00
Astragalus galactites Pall.	0.05	0.38	0.00	0.42
Chenopodium aristatum L.	0.01	0.04	0.00	0.00
Gentiana squarrosa Ldb.	0.00	0.03	0.00	0.00
Others	0.33	2.11	1.34	2.42

(Tables 13.8, 13.9, and 13.10). In terms of the scaling index, we can conclude that ecotope diversity has a strong relationship with desertification area but only a small connection to productivity. In terms of Simpson's index, ecotope diversity bears no relation to desertification area, but is closely related to productivity. In terms of Shannon's index, ecotope diversity bears little relation to desertification area, but has a strong relationship with ecosystem productivity.

TABLE 13.8

Effect of Different Spatial Scales on Correlation Coefficient of Scaling Diversity with Grassland Productivity and Desertification Area

Ordinal Number	Pixel Size	Year				Correlation Coefficient	
		1987	1991	1997	2000	With Productivity	With Desertification Area
1	30 × 30 m	0.53	0.54	0.55	0.55	−0.08	0.96
2	60 × 60 m	0.49	0.49	0.50	0.50	0.08	0.90
3	90 × 90 m	0.44	0.46	0.46	0.46	0.32	0.82
5	120 × 120 m	0.41	0.43	0.43	0.43	0.39	0.79
6	150 × 150 m	0.39	0.41	0.40	0.41	0.46	0.76
7	180 × 180 m	0.37	0.39	0.38	0.39	0.43	0.75
8	210 × 210 m	0.35	0.37	0.37	0.38	0.52	0.73
9	240 × 240 m	0.34	0.36	0.35	0.36	0.42	0.72
10	270 × 270 m	0.32	0.35	0.34	0.35	0.43	0.69
11	300 × 300 m	0.31	0.34	0.33	0.34	0.41	0.70
12	330 × 330 m	0.30	0.33	0.32	0.33	0.35	0.69
13	360 × 360 m	0.29	0.32	0.31	0.32	0.38	0.75
14	390 × 390 m	0.28	0.31	0.30	0.31	0.36	0.66
15	420 × 420 m	0.27	0.30	0.29	0.30	0.46	0.64
16	450 × 450 m	0.27	0.29	0.29	0.30	0.37	0.69
17	480 × 480 m	0.26	0.28	0.29	0.29	0.25	0.79
18	510 × 510 m	0.25	0.29	0.28	0.28	0.38	0.62
19	540 × 540 m	0.25	0.28	0.27	0.28	0.48	0.62
20	570 × 570 m	0.24	0.27	0.26	0.27	0.40	0.64
21	600 × 600 m	0.24	0.27	0.26	0.27	0.38	0.48

It is one of the major aims of this chapter to explore self-similarity among ecotope diversity on different resolutions and species diversity. In order not to ignore the effect of species diversity in 1986 on ecotope diversity in 1987 and the effect of ecotope diversity in 1987 on species diversity in 1988 as well as the similar interactions between species diversity and ecotope diversity during other three periods from 1990 to 1992, 1996 to 1998, and 1999 to 2001, the biomass averages of each species during the four periods are calculated respectively (Table 13.7). The three diversity indices are then operated on the averages, in which the area of the sampling quadrate is 1×1 m and the smallest crown diameter of the sampled individuals is 0.001×0.001 m. The calculation results (Table 13.13) show that the correlation coefficients of scaling diversity, Shannon's diversity, and Simpson's diversity with grassland productivity are respectively 0.77, 0.67, and 0.63. In other words, species biomass diversity calculated using the scaling index shows a closer relationship with grassland productivity than either Shannon's or Simpson's diversity.

TABLE 13.9

Effect of Different Spatial Scales on Correlation Coefficient between Simpson's Diversity and Grassland Productivity

		Year				Correlation Coefficient	
Ordinal Number	Pixel Size	1987	1991	1997	2000	With Productivity	With Desertification Area
1	30 × 30 m	86.35	422.20	298.18	585.14	0.68	0.71
2	60 × 60 m	85.32	284.89	312.16	444.36	0.48	0.89
3	90 × 90 m	69.25	227.20	124.91	322.94	0.82	0.60
5	120 × 120 m	60.78	182.26	72.16	248.79	0.89	0.47
6	150 × 150 m	53.57	162.16	51.76	167.87	0.85	0.26
7	180 × 180 m	46.16	147.56	43.75	125.68	0.76	0.12
8	210 × 210 m	43.18	126.99	24.72	134.59	0.89	0.18
9	240 × 240 m	40.81	108.86	37.66	126.70	0.89	0.33
10	270 × 270 m	38.01	102.35	20.30	101.85	0.88	0.12
11	300 × 300 m	33.57	93.39	19.37	93.57	0.88	0.14
12	330 × 330 m	29.89	84.60	19.74	80.04	0.85	0.12
13	360 × 360 m	31.13	77.65	16.52	60.21	0.76	−0.09
14	390 × 390 m	24.55	82.62	15.86	47.67	0.57	−0.19
15	420 × 420 m	27.31	70.79	16.47	55.59	0.76	−0.06
16	450 × 450 m	21.48	67.28	13.75	45.40	0.66	−0.11
17	480 × 480 m	19.51	58.13	17.52	49.07	0.76	0.09
18	510 × 510 m	18.93	67.99	13.65	47.78	0.68	−0.05
19	540 × 540 m	19.43	63.17	12.71	38.19	0.60	−0.17
20	570 × 570 m	18.20	55.95	9.88	35.45	0.64	−0.17
21	600 × 600 m	17.20	59.43	10.60	32.23	0.54	−0.22

13.4.3 Case Study of Dongzhi Tableland in the Loess Plateau

Dongzhi Tableland is geographically north of the Jin-He river, between the Ma-Lian-He river, and the Pu-He river. Dongzhi tableland is located in 35°15′28″ ~ 36°3′55″N and 107°27′40″ ~ 107°57′45″E, in the hinterland of the Loess Plateau (Figure 13.10). Its total area is 2778 km², of which 960 km² is flat land. Serious soil erosion is Dongzhi tableland's major ecological problem, and is the main restriction of local economic development. The density of gullies is 2.17 km/sq. km on an average. The soil erosion area is 2724 sq. km, accounting for 98.1% of the total area of Dongzhi tableland. Water erosion and gravity erosion are the major soil erosion types. The elevation of Dongzhi tableland is 1350 m on an average. Its elevation is lower in the south at about 882 m, and higher in the north at about 1576 m. Dongzhi tableland includes four counties, Xifeng, Qingcheng, Ninxian, and Heshui, and 301 villages. There were 0.636 million people in Dongzhi tableland in 2007.

The area featured, has an inland monsoon climate with a short, cold winter season and a rainy autumn season. Its mean annual precipitation is 548 mm.

TABLE 13.10

Effect of Different Spatial Scales on Correlation Coefficient of Shannon's Diversity with Grassland Productivity and Desertification Area

Ordinal Number	Pixel Size	Year				Correlation Coefficient	
		1987	1991	1997	2000	With Productivity	With Desertification Area
1	30 × 30 m	8.76	9.37	9.71	9.77	0.17	0.93
2	60 × 60 m	8.13	8.78	8.88	9.02	0.28	0.85
3	90 × 90 m	7.30	8.09	7.91	8.31	0.52	0.73
5	120 × 120 m	6.73	7.57	7.30	7.78	0.58	0.69
6	150 × 150 m	6.32	7.17	6.75	7.33	0.67	0.59
7	180 × 180 m	5.91	6.85	6.39	6.97	0.64	0.57
8	210 × 210 m	5.64	6.55	5.96	6.73	0.75	0.51
9	240 × 240 m	5.42	6.29	5.85	6.44	0.67	0.58
10	270 × 270m	5.21	6.12	5.50	6.22	0.74	0.46
11	300 × 300 m	4.96	5.94	5.35	6.05	0.70	0.50
12	330 × 330 m	4.77	5.78	5.24	5.82	0.63	0.52
13	360 × 360 m	4.67	5.57	5.02	5.63	0.69	0.48
14	390 × 390 m	4.50	5.52	4.87	5.37	0.60	0.37
15	420 × 420 m	4.44	5.38	4.73	5.37	0.70	0.41
16	450 × 450 m	4.23	5.26	4.64	5.23	0.65	0.44
17	480 × 480 m	4.07	5.07	4.69	5.15	0.55	0.61
18	510 × 510 m	4.08	5.16	4.47	5.05	0.63	0.39
19	540 × 540 m	4.08	5.04	4.32	4.91	0.68	0.31
20	570 × 570 m	3.89	4.89	4.11	4.79	0.71	0.31
21	600 × 600 m	3.86	4.95	4.10	4.67	0.63	0.22

Its mean annual temperature is 11°C. It can be climatically divided into two areas; temperate grassland and temperate deciduous broadleaf forest. The temperate grassland area is in the north and the temperate deciduous broadleaf forest is in the south. Dongzhi tableland is covered by black humus soil and loess soil. The current land-cover types include forest land, grassland,

TABLE 13.11

Correlation Coefficient of Productivity with Species Biomass Diversity and Ecotope Diversity

Diversity Index Name	Species Biomass Diversity	Ecotope Diversity (30 × 30 m Resolution)
Scaling index	0.77	0.10
Shannon's index	0.67	0.22
Simpson's index	0.63	0.73

TABLE 13.12

Desertification Area and Ecotope Diversity

Year	Desertification Area (hectare)	Ecotope Diversity (30 × 30 m Resolution)		
		Scaling Index	Shannon's Index	Simpson's Index
1987	357.12	0.53	8.76	86.35
1991	930.78	0.54	9.37	422.20
1997	2503.98	0.55	9.71	298.18
2000	2986.56	0.55	9.77	585.14

crop land, water area, built-up area, and bare land. The forest land is dominated by locusts, pines, and poplar. The economic trees mainly include apple and almond. The grassland is dominated by white-sheep grass, Altai puppy-flowers, Xing'an bush-clover, holy sagebrush, scabrous hide-seed-grass, discolor cinquefoil, thin-leaf milkwort, virgate sagebrush, white-leaf sage-brush, China ixeris, narrow-leaf rice-bag, girald sagebrush, two-color crazyweed, hard bluegrass, and crested wheat-grass. The bush is dominated by sallow thorn. Dongzhi tableland has 115,120 hm² of cultivated land, 38,568 hm² of forest land, and 85,600 hm² of grassland (Figure 13.11).

13.4.3.1 Species Data Acquisition and Species Diversity

The quadrate for trees was 10 × 10 m in size and for grasses 1 × 1 m. The records of every quadrate included longitude, latitude, slope, aspect, elevation, and ecological circumstance. The records of trees detailed species name and individual number of each species, as well as crown cover, height, and pectoral diameter of every species on an average. The records of grass detailed species name, individual number of each species, fresh weight, height, and coverage

TABLE 13.13

Grassland Productivity and Species Biomass-Based Diversity

Period	Biomass (g/m²)	Species Biomass-Based Diversity		
		Scaling Index	Shannon's Index	Simpson's Index
Average of biomass from 1986 to 1988	124.64	0.20	2.16	4.45
Average of biomass from 1990 to 1992	190.50	0.21	2.08	3.95
Average of biomass from 1996 to 1998	105.57	0.19	1.76	2.61
Average of biomass from 1999 to 2001	181.10	0.22	2.44	5.74

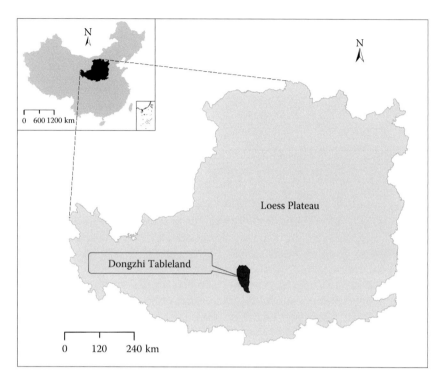

FIGURE 13.10
Location of Dongzhi tableland.

of every species. The aboveground parts of grass were taken to laboratory to calculate biomass, while 54 standard trees were cut down for analyzing biomass. The statistically regressive average biomass, height, and pectoral diameter of all tree species were calculated on the basis of analyzing the biomass of the trunks, branches, and leaves of every standard tree. Tree discs, tree branches, tree leaves, bush branches, bush leaves and grasses were dried at 80°C for one hour using a thermostatic container; the temperature was then regulated to 105 °C until a constant weight was reached.

For calculating the relationship between individual-counting diversity and biomass-based diversity, we selected 98 sample quadrates in Dongzhi tableland, including 19 sample quadrates of woodland and 79 grassland quadrates (Figure 13.12). In all, 117 species can be found in the woodland quadrates (Supplementary Tables 13.14 and 13.15) and 104 species in the grassland quadrates (Supplementary Tables 13.16 and 13.17).

The calculation results of sampling data on woodland indicate that the correlation coefficient between individual-counting diversity and biomass-based diversity is 0.286 and the significance level is 0.235. In other words, individual-counting and biomass-based diversities have no significant dependency.

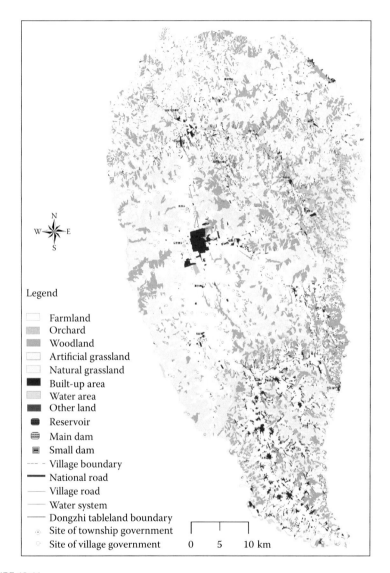

FIGURE 13.11
Land cover of Dongzhi tableland in 2007.

However, the results in grassland show that individual-counting and biomass-based diversities are significantly correlated and the regressive equation can be formulated as

$$y = 0.854x + 0.399 \tag{13.13}$$

where y represents biomass-based diversity, x is individual-counting diversity, the correlation coefficient is 0.934, and the significance level is less than 0.0001.

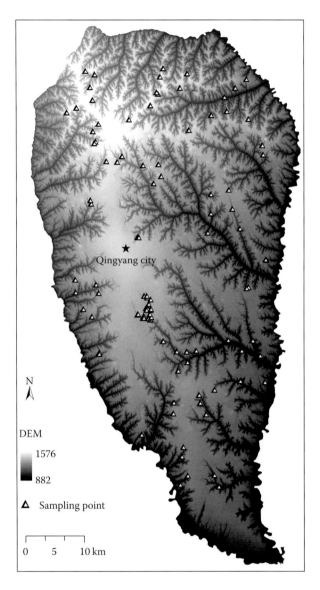

FIGURE 13.12
Spatial distributions of the sample quadrats.

13.4.3.2 Ecotope Data Acquisition and Ecotope Diversity

The 197 topographical maps on a spatial scale of 1:10,000 were used to investigate land cover (Figure 13.13). The boundary of every ecotope was drawn
on the corresponding topographical map in terms of land-cover types and
landform, in which the village, the land-cover types belonged to were
marked. Every ecotope was numbered. After the fieldwork was completed,

TABLE 13.14

Individual Number of Species in the Quadrates of Woodland

Quandrant Code	S33	S35	S41	S42	S43	S44	S47	S48	S51
Stipa bungeana	114	809	487		1082	230	155	579	197
Bothriochloa ischaemum	418	13	404	103	22	17	322	164	59
Heteropappus altaicus	6	3	3	1	6	4	113		
Lespedeza davurica	9		108	12	89	113	79	18	
Artemisia leucophylla		85	22	13	16	48	3	1	
Cleistogenes squarrosa	36		9	7	9	32	6		617
Poa sphondylodes		3		978	17	116	30		3
Robinia pseudoacacia		8	14					61	
Artemisia sacrorum	4		57	51	29	22	42	25	8
Artemisia scoparia	3			2	6	15	15	1	
Ixeris chinensis	4	1	9		2	35	8		
Gueldenstaedtia stenophylla		1			6	13	11	1	9
Agropyron cristatum		34				4	3		1
Ixeris sonchifolia	7	4	1	6	5	5	2		
Potentilla discolor	56		37	5			14	12	14
Oxytropis bicolor	6		2		13		22	1	2
Galium verum	29		28	11	47				2
Carum buriaticum		2	15	10			2		3
Chrysanthemum indicum		55		11	8	2			
Vicia amoena		1		9		1	12		
Medicago falcata				5			4		5
Potentilla reptans var. *sericophylla*	5		26						
Leontopodium leontopodioides		43	40	28	2			38	
Setaria viridis								1	
Lespedeza juncea	7								12
Carex duriuscula				13		23			
Artemisia giraldii	58							52	49
Prunus americana	5			6	15		2		
Pinus tabulaeformis				14		14	6		48
Thymus mongolicus	26		17		106				
Lespedeza potaninii									105
Leymus secalinus									
Hippophae rhamnoides				3					
Pyrus betulaefolia								8	
Populus simonii									
Lycium barbarum		4						4	
Viola philippica				2		7	1		
Potentilla multifida		1				5			4
Others	8	21	68	9	105	15	13	46	21
Total species	**19**	**20**	**30**	**26**	**29**	**21**	**25**	**16**	**27**
Total individual number	**796**	**1050**	**1326**	**1335**	**1623**	**712**	**872**	**966**	**1197**

continued

TABLE 13.14   (continued)

Individual Number of Species in the Quadrates of Woodland

Quadrant Code	S57	S60	S79	S83	S87	S89	S90	S97	S99	S100
Stipa bungeana	169	16	33	133	91	64	228	378	798	273
Bothriochloa ischaemum	62				7	931	156	553	274	23
Heteropappus altaicus	3		89		23	35	15	69	6	3
Lespedeza davurica			48	5	29	11	19	52	34	44
Artemisia leucophylla	19		252	39	37	3	24	21		86
Cleistogenes squarrosa	25		106	392	24		6	240	27	
Poa sphondylodes	4	3	139		213	92	1			66
Robinia pseudoacacia	30	16	8	12	3	4	6	3	7	3
Artemisia sacrorum	5			11					76	91
Artemisia scoparia	13		1		10			9	42	1
Ixeris chinensis	2				16	4	10	2	6	
Gueldenstaedtia stenophylla	2		20		1			3	38	7
Agropyron cristatum	90	18	15		62	4	8			155
Ixeris sonchifolia	2	3			2					2
Potentilla discolor	177							13	172	1
Oxytropis bicolor						27	3	4	6	
Galium verum					8	5	88			14
Carum buriaticum		9	4		18	1				
Chrysanthemum indicum				23	88	5				98
Vicia amoena		1	2	2	19					
Medicago falcata		5	8	29		5				
Potentilla reptans var. *sericophylla*		3	85	4	328					121
Leontopodium leontopodioides						12				3
Setaria viridis		1			93			5	14	3
Lespedeza juncea			1			9	2		36	
Carex duriuscula		190	380		10					
Artemisia giraldii								8		
Prunus armeniaca										
Pinus tabulaeformis										
Thymus mongolicus										
Lespedeza potaninii	30									
Leymus secalinus							147			46
Hippophae rhamnoides										
Pyrus betulaefolia										
Populus simonii		5								
Lycium barbarum	1									
Viola philippica	1	11			22	6		1		
Potentilla multifida					23	12	5			2
Others	16	42	9	12	53	73	19	21	2	55
Total species	**23**	**23**	**19**	**15**	**28**	**26**	**19**	**19**	**15**	**29**
Total individual number	**651**	**323**	**1200**	**662**	**1180**	**1303**	**737**	**1382**	**1538**	**1097**

TABLE 13.15

Species Biomass in the Quadrates of Woodland

Quadrant Code	S33	S35	S41	S42	S43	S44	S47	S48	S51
Stipa bungeana	4.84	32.56	21.63		23.25	6.67	5.79	47.15	9.02
Bothriochloa ischaemum	82.25	0.35	52.68	9.39	2.33	0.88	32.84	21.53	4.96
Heteropappus altaicus	1.75	2.45	1.58	0.29	1.17	0.58	27.24		
Lespedeza davurica	3.43		12.66	3.51	18.05	25.73	19.52	3.51	
Artemisia leucophylla		36.00	11.00	6.47	3.00	14.20	1.60	0.80	
Cleistogenes squarrosa	0.40		0.83	0.53	0.33	1.03	0.30		8.63
Poa sphondylodes		0.18		40.88	1.73	4.73	1.13		0.30
Robinia pseudoacacia		160928	59920					122000	
Artemisia sacrorum	1.17		22.75	19.05	5.92	9.82	8.78	6.57	5.27
Artemisia scoparia	0.60			0.47	2.07	3.33	2.00	0.87	
Ixeris chinensis	0.05	0.02	0.18		0.05	0.33	0.13		
Gueldenstaedtia stenophylla		0.07			0.28	1.08	0.68	0.24	0.32
Agropyron cristatum		6.40				1.00	0.87		0.33
Ixeris sonchifolia	0.92	0.12	0.19	0.31	0.46	0.73	0.19		
Potentilla discolor	29.40		7.07	1.47			9.17	7.91	4.62
Oxytropis bicolor	0.45		0.55		1.00		1.65	0.25	0.30
Galium verum	3.77		2.27	1.31	1.11				0.34
Carum buriaticum		0.19	0.98	0.41			0.38		0.25
Chrysanthemum indicum		15.12		1.24	1.08	0.36			
Vicia amoena		0.22		0.99		0.18	1.28		
Medicago falcata				0.85			0.85		0.65
Potentilla reptans var. *sericophylla*		0.70		1.20					
Leontopodium leontopodioides			2.82	1.99	1.65	0.26			2.43
Setaria viridis								0.34	
Lespedeza juncea	0.90								2.50
Carex duriuscula				0.82		0.77			
Artemisia giraldii	30.78							33.31	14.00
Prunus armeniaca	55492				466450	160753.2	122660		
Pinus tabulaeformis				42000		49000	18000		360000
Thymus mongolicus	0.68		0.77		1.79				
Lespedeza potaninii									9.10
Leymus secalinus									
Hippophae rhamnoides				7500					
Pyrus betulaefolia							472.00		
Populus simonii									
Lycium barbarum		3.72						9.40	
Viola philippica			0.31		0.50	0.19			
Potentilla multifida		0.09			0.68				0.68
Others	0.78	2.12	12.34	1.99	6.48	3.75	2.28	6.75	13.20
Total species	**19**	**20**	**30**	**26**	**29**	**21**	**25**	**16**	**27**
Total biomass (g)	**55654**	**161028**	**60071**	**516043**	**160826**	**49076**	**141249**	**122139**	**360077**

continued

TABLE 13.15　(continued)

Species Biomass in the Quadrates of Woodland

Quadrant Code	S57	S60	S79	S83	S87	S89	S90	S97	S99	S100
Stipa bungeana	8.73	1.10	1.50	18.00	35.00	5.00	11.50	17.00	47.00	10.00
Bothriochloa ischaemum	6.59				0.70	123.55	28.00	86.10	68.60	3.85
Heteropappus altaicus	1.93		12.24		11.56	7.14	3.40	11.56	2.72	1.36
Lespedeza davurica			4.66	1.23	4.90	8.33	37.24	10.29	8.82	7.11
Artemisia leucophylla	16.47		104.80	31.20	20.00	0.40	6.80	13.00		29.20
Cleistogenes squarrosa	0.50		6.20	31.00	0.80		0.20	2.60	0.60	
Poa sphondylodes	0.38	0.41	7.20		10.80	9.45	0.36			2.48
Robinia pseudoacacia	168000	585600	155200	232800	6000	8000	12000	6000	15000	61200
Artemisia sacrorum	7.35			3.51					50.31	22.62
Artemisia scoparia	12.07		0.62		7.44			5.43	32.55	0.78
Ixeris chinensis	0.05				0.21	0.04	0.26	0.04	0.06	
Gueldenstaedtia stenophylla	0.32		0.84		0.12			0.24	0.84	0.48
Agropyron cristatum	26.60	2.73	2.80		5.60	0.40	3.60			15.20
Ixeris sonchifolia	0.31	0.37			0.23					0.35
Potentilla discolor	8.75							2.31	31.92	0.42
Oxytropis bicolor						3.72	0.31	0.31	0.47	
Galium verum					1.74	0.29	11.89			1.02
Carum buriaticum		1.74	0.19		0.48	0.02				
Chrysanthemum indicum				12.90	32.40	2.70				30.60
Vicia amoena		0.18	0.22	0.18	0.66					
Medicago falcata		0.85	0.45	4.20		1.20				
Potentilla reptans var. sericophylla		0.28	5.40	0.18	16.92					5.40
Leontopodium leontopodioides						1.30				0.26
Setaria viridis		0.27			1.80			0.10	1.00	0.20
Lespedeza juncea			0.45			2.25	11.70		9.45	
Carex duriuscula		6.57	6.38		0.29					
Artemisia giraldii								12.68		
Prunus armeniaca										
Pinus tabulaeformis										

TABLE 13.15 (continued)

Species Biomass in the Quadrates of Woodland

Quadrant Code	S57	S60	S79	S83	S87	S89	S90	S97	S99	S100
Thymus mongolicus										
Lespedeza potaninii	7.40									
Leymus secalinus							35.64			4.95
Hippophae rhamnoides										
Pyrus betulaefolia										
Populus simonii		134550								
Lycium barbarum	1.14									
Viola philippica	0.19	0.50			1.84	0.58		0.02		
Potentilla multifida					3.77	6.96	11.60			0.58
Others	2.98	21.52	19.83	4.45	8.71	31.91	12.71	17.13	0.08	16.24
Total species	**23**	**23**	**19**	**15**	**28**	**26**	**19**	**19**	**15**	**29**
Total biomass (g)	**168,102**	**720,187**	**155,374**	**232,907**	**6166**	**8205**	**12,175**	**6179**	**15,254**	**61,353**

the investigated data were checked and corrected and then digitized by means of ArcGIS. 8 land-cover types and 35 ecotope types were identified and stored in vector format. This work started in August 2007 and finished in July 2008, during which the authors of this chapter, worked with 119 local technicians. In all, 6 months were spent on fieldwork, 2 months on checking and correcting the investigated data, and 3 months on digitizing the maps.

In order to analyze the relationship between ecotope diversity and species diversity as well as individual ecotope-counting diversity and ecotope-area-based diversity, 10 square regions were selected randomly in Dongzhi table-land. Every region is 8×8 km in size. In each square region, 6 quadrates were designed in different slope aspects and landform types (Figure 13.14).

Overall, 32 ecotope types were identified in the 10 square regions, including slope farmland, terrace farmland, dam farmland, apple, orchard on slope, pear, China pine, acacia, poplar, osier, sallow thorn, Chinese date, haw, walnut, almond, Chinese prickly ash, a nursery of young plants, natural grassland, clover blossom, river, reservoir, pond, low land, reed land, urban area, rural residential area, industrial and mining area, road, water conservancy facilities, and other built-up areas (Tables 13.18 and 13.19). The calculation result shows that individual ecotope-counting diversity and ecotope-area-based diversity have a significant correlation, which can be formulated as

$$y = 1.205x + 0.046 \tag{13.15}$$

where y represents ecotope-area-based diversity, x is individual ecotope-counting diversity, the correlation coefficient is 0.802 and the significance level is 0.00036.

TABLE 13.16

Individual Number of Species in the Quadrates of Grassland

Quadrant Code	S1	S2	S3	S4	S5	S6	S7	S8	S9	S10
Stipa bungeana	40	168	480	340	474	165	592	361	290	182
Bothriochloa ischaemum	3800	86	510	20	258	227	380	311	1706	
Artemisia sacrorum		154		18	45	6		206	26	75
Heteropappus altaicus			18	4	40	3	6	22		
Lespedeza davurica	18	93	64	7	110	16	68	5	6	6
Cleistogenes squarrosa		22			36	109	35	6	8	592
Potentilla discolor			23		37	120	31	6	88	
Artemisia giraldii	30	126	10			200	69	162	3	44
Polygala tenuifolia		7	5	6				10	1	
Artemisia scoparia	7	2	35	15	11	38	28	1	4	
Ixeris chinensis		8	4	12	13	4	3	3		11
Gueldenstaedtia stenophylla	37	5	4	1	1	2	4			
Artemisia leucophylla				33						28
Oxytropis bicolor	6		21	1		4	17			
Agropyron cristatum		5	61	183						6
Poa sphondylodes		148		14	149			13		
Allium ramosum			20	11		38		9	15	
Adenophora polyantha				1						
Ixeris sonchifolia			2	2	19					5
Galium verum				26				3		4
Viola philippica		2	1	3	5		2			7
Bupleurum angustissimum										
Medicago falcata		2		18	4					
Setaria viridis		25								
Potentilla reptans var. sericophylla		470		1	95					100
Carum buriaticum		17	31	4	26					
Potentilla multifida				7						1
Viola dissecta		18								14
Lespedeza juncea										
Leontopodium leontopodioides										54
Carex duriuscula		5			4					60
Artemisia japonica										53
Chrysanthemum indicum										
Lespedeza potaninii										
Carduus nutans										2
Plantago asiatica		16		41	2					1
Thymus mongolicus										
Leymus secalinus										
Others	5	57	43	126	3	6	3	4	0	12
Total species	**8**	**24**	**18**	**37**	**19**	**16**	**13**	**17**	**10**	**22**
Total individual number	**3943**	**1436**	**1332**	**894**	**1332**	**938**	**1238**	**1122**	**2147**	**1257**

TABLE 13.16 (continued)

Individual Number of Species in the Quadrates of Grassland

Quadrant Code	S11	S12	S13	S14	S15	S16	S17	S18	S19	S20
Stipa bungeana	61	575	1033	254	233	110	358	8	510	170
Bothriochloa ischaemum	736	543	221		166	1318	378	1716		24
Artemisia sacrorum	3	4	3	261	64	73		53	5	27
Heteropappus altaicus	21		4	1	11		3		71	9
Lespedeza davurica	162	89					88	15	43	25
Cleistogenes squarrosa	3		35		153					4
Potentilla discolor	55	87	10	20	4	3	33		2	66
Artemisia giraldii		1	95	33	335	4		48		71
Polygala tenuifolia	12	82	5	12				11	2	19
Artemisia scoparia					1	2	1		7	8
Ixeris chinensis	1	2		13			5	4	2	14
Gueldenstaedtia stenophylla		38	1	1	3		3		1	61
Artemisia leucophylla	70	73	5	10			19	27	127	
Oxytropis bicolor	9	1		9			55		16	49
Agropyron cristatum	3		6					2	81	
Poa sphondylodes				13					10	
Allium ramosum			7			7			1	
Adenophora polyantha		15	5	7			2		15	1
Ixeris sonchifolia	27			7					5	
Galium verum		1		42			10		12	
Viola philippica	9			13			8			6
Bupleurum angustissimum	14	2		1			3		1	
Medicago falcata			11	76					61	
Setaria viridis				1					19	
Potentilla reptans var. *sericophylla*				42					78	
Carum buriaticum									22	
Potentilla multifida		8					26		34	
Viola dissecta				14				1		
Lespedeza juncea			36		1	11				
Leontopodium leontopodioides				111			243		1	
Carex duriuscula				116						
Artemisia japonica				339			12			
Chrysanthemum indicum	30								273	
Lespedeza potaninii			75			14				
Carduus nutans	5									
Plantago asiatica				5					3	
Thymus mongolicus							113			418
Leymus secalinus	3	2								
Others	90	49	10	37		64	25	2	59	127
Total species	**21**	**21**	**18**	**28**	**10**	**11**	**23**	**11**	**31**	**19**
Total individual number	**1314**	**1572**	**1562**	**1438**	**971**	**1606**	**1385**	**1887**	**1461**	**1099**

continued

TABLE 13.16 (continued)

Individual Number of Species in the Quadrates of Grassland

Quadrant Code	S21	S22	S23	S24	S25	S26	S27	S28	S29	S30
Stipa bungeana	73	428	63	393	37	240	535	170	374	340
Bothriochloa ischaemum	492	770	466	370	56	371	240	3774		562
Artemisia sacrorum	25	7	10	37			8		21	
Heteropappus altaicus	7		25	5	34	7	8			
Lespedeza davurica	58	11					13	114	56	60
Cleistogenes squarrosa		7	16		70	171	21	57		14
Potentilla discolor		87	80		27	17	4	8		1
Artemisia giraldii		58	21	79	74	117	2	12	21	6
Polygala tenuifolia	1		5		4		6	25	24	
Artemisia scoparia		29	11			6	7	8		7
Ixeris chinensis			9				4	5		1
Gueldenstaedtia stenophylla	2	6	1	35						1
Artemisia leucophylla	63			153			44		76	109
Oxytropis bicolor			181		8	3	27	3		
Agropyron cristatum	1			9			35		340	
Poa sphondylodes	316						7		16	
Allium ramosum					2	1	12	7	2	
Adenophora polyantha	3								10	
Ixeris sonchifolia						1	9		4	
Galium verum							42		45	
Viola philippica							4			
Bupleurum angustissimum	2						13		9	
Medicago falcata	2		17				10			
Setaria viridis				14						1
Potentilla reptans var. *sericophylla*	35									
Carum buriaticum	20								22	
Potentilla multifida						2	10			
Viola dissecta						3	13		6	
Lespedeza juncea			26		22	43	16			
Leontopodium leontopodioides										
Carex duriuscula	24									
Artemisia japonica										
Chrysanthemum indicum	81								6	
Lespedeza potaninii			58	11	59	28				
Carduus nutans								2	1	
Plantago asiatica									2	
Thymus mongolicus										
Leymus secalinus										
Others	2	1	8	2	7	2	26	8	73	8
Total species	**18**	**10**	**17**	**12**	**12**	**15**	**30**	**12**	**24**	**14**
Total individual number	**1207**	**1404**	**997**	**1108**	**400**	**1012**	**1118**	**4191**	**1108**	**1110**

TABLE 13.16   (continued)

Individual Number of Species in the Quadrates of Grassland

Quadrant Code	S31	S32	S34	S38	S39	S40	S45	S46	S49	S50
Stipa bungeana	186	115	43	315	304	192	98	590	1027	550
Bothriochloa ischaemum	167	258	345	543	717	352	54	61	195	308
Artemisia sacrorum		67		17		91	47	57	12	55
Heteropappus altaicus	9	30	5	12			4	14	137	11
Lespedeza davurica	3			60	74		124	172	20	
Cleistogenes squarrosa	13	167	193	30	22	12	11	99	544	165
Potentilla discolor	5			3	29	1	7	21	79	27
Artemisia giraldii	162	4		105		70	168		46	
Polygala tenuifolia	8	2	1	3	1		5	16	5	2
Artemisia scoparia	18	20			99		2	21	13	3
Ixeris chinensis	3	4	4	26	5		1	3	9	32
Gueldenstaedtia stenophylla	1		1	7	90		3	1	9	4
Artemisia leucophylla		2	18	26	36	10	150	78	8	
Oxytropis bicolor		32		6	79			31		1
Agropyron cristatum			7				10			36
Poa sphondylodes			2			90	196	105		16
Allium ramosum	1			31			1	1	2	
Adenophora polyantha		2	19	4		8				10
Ixeris sonchifolia		3	15	3		1	1	1		20
Galium verum			29	44		1		40		82
Viola philippica			22	5			1	2		
Bupleurum angustissimum		6	6	13				4		1
Medicago falcata				4						15
Setaria viridis	2			1	7			2	2	
Potentilla reptans var. sericophylla								18		
Carum buriaticum			7			5	1	2		29
Potentilla multifida		1	20	55		1				4
Viola dissecta		2	4							
Lespedeza juncea	1		74							
Leontopodium leontopodioides		5	61	33		6		39		31
Carex duriuscula						9				106
Artemisia japonica			68			12				
Chrysanthemum indicum			21			18				
Lespedeza potaninii		58	65			88				157
Carduus nutans				3				5		2
Plantago asiatica										
Thymus mongolicus			17							
Leymus secalinus										
Others	0	46	140	19	0	82	6	9	22	21
Total species	**14**	**22**	**33**	**28**	**12**	**22**	**22**	**29**	**19**	**30**
Total individual number	**579**	**824**	**1187**	**1368**	**1463**	**1049**	**890**	**1392**	**2130**	**1688**

continued

TABLE 13.16 (continued)

Individual Number of Species in the Quadrates of Grassland

Quadrant Code	S52	S53	S54	S55	S56	S58	S59	S61	S62	S63
Stipa bungeana	326	142	263	916	50	164	621	289	739	81
Bothriochloa ischaemum	254	468		72	446	86	3	50	2012	695
Artemisia sacrorum	34	5	19	29	47	139	230	66	3	
Heteropappus altaicus	1	53	28	70	30	92	5	3	3	
Lespedeza davurica						238		7	29	275
Cleistogenes squarrosa	155	13	170	20	12	2		4		
Potentilla discolor	26	106	17	16	9	20		5	18	3
Artemisia giraldii		79	12			21		223	74	
Polygala tenuifolia	4	14	1	10	8	8		20	8	2
Artemisia scoparia		14	15	6	7	13		8	108	6
Ixeris chinensis	1	4	1	8	4	73	4			7
Gueldenstaedtia stenophylla	1	3	3		2	5		1	1	
Artemisia leucophylla	118	30	53	117	51	18				
Oxytropis bicolor	4	23	6	47	8	14		1	38	16
Agropyron cristatum		14	36	16	58	2	31	158		
Poa sphondylodes	6		45	10	136	41	7	25		
Allium ramosum	1	7		2			1			
Adenophora polyantha					3	2		6		
Ixeris sonchifolia		6	1	10	5	18	21			
Galium verum				76	54	22	4	140		
Viola philippica		1		1	4	2	6			3
Bupleurum angustissimum			1	5	2	2	1			
Medicago falcata		2	1	9	28	2	38			
Setaria viridis			1	4		4	4			
Potentilla reptans var. sericophylla			72	14	205		169			
Carum buriaticum	4			15	3	1	6			
Potentilla multifida				11	1	11				
Viola dissecta						1				
Lespedeza juncea	19	4			27					
Leontopodium leontopodioides				231	37	83				
Carex duriuscula			165		9					
Artemisia japonica					7	22	69			
Chrysanthemum indicum										
Lespedeza potaninii	73	15	27	107	30		1			
Carduus nutans	1						1			
Plantago asiatica				8	3		34			
Thymus mongolicus										
Leymus secalinus										
Others	9	12	23	55	20	18	80	29	0	3
Total species	**22**	**23**	**26**	**38**	**34**	**34**	**28**	**19**	**11**	**10**
Total individual number	**1037**	**1015**	**960**	**1885**	**1306**	**1124**	**1336**	**1035**	**3033**	**1091**

TABLE 13.16 (continued)

Individual Number of Species in the Quadrates of Grassland

Quadrant Code	S64	S65	S66	S67	S68	S69	S70	S71	S73	S74
Stipa bungeana	242	808	167	280	856	141	89	24	263	277
Bothriochloa ischaemum		19		767		819		38		148
Artemisia sacrorum	89		86		10	51	56	55	165	115
Heteropappus altaicus	4		3	10	7	1			2	1
Lespedeza davurica	42	55	27	37	7	6		109	2	10
Cleistogenes squarrosa	46	245	13	102	48	67	17			91
Potentilla discolor			1	67		4			14	56
Artemisia giraldii		62				116	13		33	169
Polygala tenuifolia			5	17				4	4	4
Artemisia scoparia		2	4	1			1		3	
Ixeris chinensis			2				5			
Gueldenstaedtia stenophylla		14	2	1						2
Artemisia leucophylla	29		357	36			18		22	
Oxytropis bicolor			18	21		1				1
Agropyron cristatum	4	1			7		50			
Poa sphondylodes			55		3	58			87	
Allium ramosum									1	2
Adenophora polyantha	10		1		2	10		5	5	
Ixeris sonchifolia							2			
Galium verum	40		169					53		
Viola philippica			2							
Bupleurum angustissimum	22					1		5	2	
Medicago falcata	4			27						
Setaria viridis		16	1				2			
Potentilla reptans var. sericophylla	93		332		62		2	31		
Carum buriaticum	11		5		50					
Potentilla multifida	1		1			2				
Viola dissecta							8	15	19	
Lespedeza juncea			6			93				
Leontopodium leontopodioides	6							153		
Carex duriuscula	448					96		1080	198	
Artemisia japonica	98						18	166		
Chrysanthemum indicum	42		26				52		10	
Lespedeza potaninii										
Carduus nutans	1						1			2
Plantago asiatica			2							
Thymus mongolicus										
Leymus secalinus			1		225		184	101		
Others	29	34	7	0	15	75	61	73	32	11
Total species	**30**	**13**	**29**	**11**	**15**	**21**	**25**	**25**	**24**	**15**
Total individual number	**1261**	**1256**	**1293**	**1339**	**1319**	**1541**	**579**	**1912**	**862**	**889**

continued

TABLE 13.16 (continued)

Individual Number of Species in the Quadrates of Grassland

Quadrant Code	S75	S76	S7	S8	S80	S81	S82	S84	S85	S86
Stipa bungeana	330	223	197	163	291	202	376	927	83	493
Bothriochloa ischaemum		32	202	1240		983	193	373		815
Artemisia sacrorum	2	55	103		38		17	17	3	
Heteropappus altaicus		13	2	26	1	22	1	3	27	9
Lespedeza davurica	14	17	7	54	54	135	18	187	32	45
Cleistogenes squarrosa	9			56	6	116	23	85		63
Potentilla discolor		12	18	44				26	3	130
Artemisia giraldii		173	50	91	9	8			20	4
Polygala tenuifolia	3	6	5	1	3	32		17	4	
Artemisia scoparia								18	3	7
Ixeris chinensis	1			2	1	4		2	1	7
Gueldenstaedtia stenophylla			2		2			8	6	2
Artemisia leucophylla	25		34	13	16	12	2	22	94	
Oxytropis bicolor				24		2	2	3		15
Agropyron cristatum	2	130	5			26	71	3	30	
Poa sphondylodes			64		101	27	37	10	114	
Allium ramosum				2	5	3	1	15	2	
Adenophora polyantha	34				5	1	2			
Ixeris sonchifolia					2		2		18	
Galium verum	103		35		41	70			114	
Viola philippica	27				1				4	
Bupleurum angustissimum	29		4		12		8			
Mcdicago falcata	77		121			18	21		5	
Setaria viridis	1	1				2		5	25	1
Potentilla reptans var. sericophylla	11				8		27		124	
Carum buriaticum	2		4			3				
Potentilla multifida					6		2			
Viola dissecta	12				17		7			
Lespedeza juncea			3	5		11				
Leontopodium leontopodioides						9	142			
Carex duriuscula	74				178		18		3	
Artemisia japonica	106				63	51	132			
Chrysanthemum indicum					28				91	
Lespedeza potaninii							54			
Carduus nutans										
Plantago asiatica	4						3			
Thymus mongolicus										
Leymus secalinus					17	114				
Others	167	2	35	1	28	26	13	2	5	14
Total species	**33**	**12**	**19**	**14**	**30**	**24**	**28**	**18**	**24**	**15**
Total individual number	**1033**	**664**	**891**	**1722**	**933**	**1877**	**1172**	**1723**	**811**	**1605**

TABLE 13.16 (continued)

Individual Number of Species in the Quadrates of Grassland

Quadrant Code	S88	S91	S93	S94	S95	S96	S98	S101	S102
Stipa bungeana	165	432	275	372	176	117	420	161	776
Bothriochloa ischaemum	573		16	787	8	1240	48	2	115
Artemisia sacrorum		36	67		64	1	4	70	1
Heteropappus altaicus	4	7	1	7		16			37
Lespedeza davurica	44	74	114	6	30	10	4	34	34
Cleistogenes squarrosa	1		48	234	7		30	95	
Potentilla discolor	29			38		15			131
Artemisia giraldii		1		3		4	3	43	54
Polygala tenuifolia	5	10		1	3				26
Artemisia scoparia	1			6		33			73
Ixeris chinensis		20				9	1		7
Gueldenstaedtia stenophylla				1	2	30			34
Artemisia leucophylla	18		19	19	15		111	97	34
Oxytropis bicolor	15			7	2	8			75
Agropyron cristatum		2		42	39		71	66	2
Poa sphondylodes		188	4		71		17	25	
Allium ramosum			15	28		6	1	4	
Adenophora polyantha	4	3			4		1	6	
Ixeris sonchifolia	3	6	3		13		11	2	
Galium verum	76		11		17		146	19	
Viola philippica	4	1	2		11		6	1	
Bupleurum angustissimum	1	2			26		1		
Medicago falcata	17	4							
Setaria viridis							10		2
Potentilla reptans var. *sericophylla*			125		112		392	265	
Carum buriaticum								7	
Potentilla multifida	9				1				2
Viola dissecta			10		11	2	4	14	2
Lespedeza juncea	9			1		1			
Leontopodium leontopodioides	15	21	141		152				
Carex duriuscula			199		448		1	6	
Artemisia japonica		25	29		20		33	30	
Chrysanthemum indicum		7	42	41	91		48	16	40
Lespedeza potaninii									
Carduus nutans	1			10	5				
Plantago asiatica			21						
Thymus mongolicus	275								
Leymus secalinus	182		46					120	
Others	0	64	72	57	75	0	106	46	26
Total species	**22**	**21**	**25**	**22**	**31**	**14**	**29**	**29**	**24**
Total individual number	**1451**	**903**	**1260**	**1660**	**1403**	**1492**	**1469**	**1129**	**1471**

TABLE 13.17

Species Biomass in the Quadrates of Grassland

Quadrant Code	S1	S2	S3	S4	S5	S6	S7	S8	S9	S10
Stipa bungeana	0.44	2.64	28.60	15.84	20.68	5.72	16.72	20.68	9.68	8.80
Bothriochloa ischaemum	154.00	3.50	49.70	0.70	34.83	60.20	21.00	20.65	121.80	
Artemisia sacrorum		46.41		8.58	7.41	1.17		29.64	35.10	50.70
Heteropappus altaicus			12.25	2.80	2.98	1.05	1.23	4.90		
Lespedeza davurica	0.98	11.27	14.70	1.47	9.31	2.45	5.88	0.49	1.47	1.47
Cleistogenes squarrosa		0.50			0.40	2.20	0.40	0.06	0.10	13.00
Potentilla discolor			11.76		8.40	21.00	4.62	1.26	14.49	
Artemisia giraldii	16.72	22.80	3.04			38.00	16.53	47.12	0.57	13.49
Polygala tenuifolia		0.72	0.18	0.36				0.72	0.18	
Artemisia scoparia	0.82	0.62	6.80	5.80		6.40	5.20	0.40	0.80	
Ixeris chinensis		0.23	0.08	0.15	0.10	0.05	0.04	0.03		0.05
Gueldenstaedtia stenophylla	0.84	0.36	0.12	0.12	0.12	0.19	0.12			
Artemisia leucophylla				8.80						14.40
Oxytropis bicolor	0.60		1.80	0.15		0.15	0.90			
Agropyron cristatum		0.32	19.60	42.80						0.60
Poa sphondylodes		3.60		0.90	13.05			0.23		
Allium ramosum			1.58	0.75		0.75		0.15	0.45	
Adenophora polyantha				0.27						
Ixeris sonchifolia			0.69	0.35	0.81					0.35
Galium verum				5.80				0.09		0.15
Viola philippica		0.23	0.35	0.23	0.46		0.23			0.35
Bupleurum angustissimum										
Medicago falcata		0.60		17.40	0.90					
Setaria viridis		0.50								
Potentilla reptans var. sericophylla		37.44		0.18	2.52					5.40
Carum buriaticum		0.76	0.29	0.10	0.38					
Potentilla multifida				4.06						0.29
Viola dissecta		0.52								0.30
Lespedeza juncea										
Leontopodium leontopodioides										2.86
Carex duriuscula		0.15			0.15					1.45
Artemisia japonica										8.23
Chrysanthemum indicum										
Lespedeza potaninii										
Carduus nutans										2.85
Plantago asiatica		1.10		1.10	0.18					0.11
Thymus mongolicus										
Leymus secalinus										
Others	0.25	1.68	6.56	43.49	1.00	2.57	0.27	1.24	0.00	3.86
Total species	**8**	**24**	**18**	**37**	**19**	**16**	**13**	**17**	**10**	**22**
Total biomass	**175**	**136**	**158**	**162**	**104**	**142**	**73**	**128**	**185**	**129**

TABLE 13.17   (continued)

Species Biomass in the Quadrates of Grassland

Quadrant Code	S11	S12	S13	S14	S15	S16	S17	S18	S19	S20
Stipa bungeana	1.50	10.12	43.56	13.20	7.92	4.84	13.20	0.22	14.08	2.20
Bothriochloa ischaemum	26.60	29.75	21.00		7.00	78.75	40.25	81.20		0.53
Artemisia sacrorum	30.42	14.43	0.70	58.50	19.89	18.33		25.74	2.34	5.07
Heteropappus altaicus	0.18	0.35	0.70	0.18	1.75		0.53		11.20	1.05
Lespedeza davurica	1.05						14.70	3.68	6.86	2.45
Cleistogenes squarrosa	0.10		0.60		1.70					0.10
Potentilla discolor	9.24	9.24	3.78	3.36	0.84	0.84	3.36		0.21	5.67
Artemisia giraldii	14.44	14.82	27.36	14.82	65.36	0.38		12.54		21.28
Polygala tenuifolia		0.14	0.90	0.72				0.72	0.18	0.90
Artemisia scoparia	1.00	11.40			0.60	0.80	0.20		4.80	1.00
Ixeris chinensis	0.03	0.01		0.08			0.03	0.03	0.01	0.05
Gueldenstaedtia stenophylla		0.48	0.05	0.12	0.12		0.10		0.02	0.72
Artemisia leucophylla			2.80	1.60			1.60	4.00	23.20	
Oxytropis bicolor	0.30	0.06		0.90			2.85		0.60	1.20
Agropyron cristatum			1.80						0.40	4.80
Poa sphondylodes	0.18			0.45						0.23
Allium ramosum	0.60		0.75			0.53				0.02
Adenophora polyantha	5.94	0.27	2.16	0.81			1.35		3.51	0.27
Ixeris sonchifolia				0.23					0.23	
Galium verum		1.31		2.90			0.44		0.87	
Viola philippica		0.12		0.69			0.23			0.23
Bupleurum angustissimum	1.50			0.53			0.35		0.25	
Medicago falcata			3.60	6.90					9.90	
Setaria viridis				0.04					0.40	
Potentilla reptans var. sericophylla				2.52					5.04	
Carum buriaticum									0.86	
Potentilla multifida							5.80		8.12	
Viola dissecta				0.38				0.08		
Lespedeza juncea		1.75	9.90		0.23	1.35				
Leontopodium leontopodioides				3.38			9.10		0.05	
Carex duriuscula				2.90						
Artemisia japonica	10.50			30.80			2.80			
Chrysanthemum indicum									33.12	
Lespedeza potaninii			5.61	18.87			2.04			
Carduus nutans										
Plantago asiatica				0.22					0.11	
Thymus mongolicus							6.63			10.88
Leymus secalinus										
Others	11.91	1.66	3.27	3.53		9.41	2.22	0.15	3.65	8.07
Total species	**21**	**21**	**18**	**28**	**10**	**11**	**23**	**11**	**31**	**19**
Total biomass	**116**	**102**	**142**	**150**	**105**	**117**	**106**	**129**	**135**	**62**

continued

TABLE 13.17   (continued)

Species Biomass in the Quadrates of Grassland

Quadrant Code	S21	S22	S23	S24	S25	S26	S27	S28	S29	S30
Stipa bungeana	2.20	8.80	1.10	13.20	2.42	10.12	19.80	5.28	23.32	25.52
Bothriochloa ischaemum	49.70	28.70	42.00	34.30	8.40	32.90	44.45	82.60		61.25
Artemisia sacrorum	9.95	1.56	1.56	17.16			14.82		7.61	
Heteropappus altaicus	2.80		3.15	1.93	6.30	0.70	1.75			
Lespedeza davurica	18.13	0.98					4.66	28.42	11.27	24.01
Cleistogenes squarrosa		0.10	0.10		2.00	5.80	0.20	1.00		0.40
Potentilla discolor		8.82	18.90		23.52	4.62	2.52	5.88		0.63
Artemisia giraldii		38.00	9.12	26.22	63.84	58.33	0.57	3.61	8.74	2.28
Polygala tenuifolia	0.18		0.72		0.90		0.36	1.08	1.44	
Artemisia scoparia		3.20	2.40	0.20		4.40	0.60	1.80		6.40
Ixeris chinensis			0.05				0.05	0.02		0.03
Gueldenstaedtia stenophylla	0.12	0.12	0.12	0.72						0.12
Artemisia leucophylla	23.60			74.00			15.00		14.80	32.00
Oxytropis bicolor		9.90			0.60	0.30	1.20	0.60		
Agropyron cristatum	0.20			1.80			4.80		31.20	
Poa sphondylodes	15.53						0.68		1.35	
Allium ramosum					0.23	0.08	0.60	0.30	0.15	
Adenophora polyantha	0.81								4.05	
Ixeris sonchifolia						0.09	0.46		0.35	
Galium verum							2.32		4.06	
Viola philippica							0.35			
Bupleurum angustissimum	0.35						4.55		3.33	
Mcdicago falcata	0.30		9.00				4.05			
Setaria viridis				0.18						0.20
Potentilla reptans var. sericophylla	4.32									
Carum buriaticum	0.29								0.76	
Potentilla multifida						1.16	12.09			
Viola dissecta						0.23	0.15		0.12	
Lespedeza juncea			6.30		6.53	1.80	4.50			
Leontopodium leontopodioides										
Carex duriuscula	0.58									
Artemisia japonica										
Chrysanthemum indicum	15.36								2.04	
Lespedeza potaninii			19.13	2.30	57.12	12.75				
Carduus nutans							0.67		0.19	
Plantago asiatica									0.11	
Thymus mongolicus										
Leymus secalinus										
Others	0.11	0.08	5.39	2.54	5.55	0.30	10.92	1.64	14.24	9.32
Total species	**18**	**10**	**17**	**12**	**12**	**15**	**30**	**12**	**24**	**14**
Total biomass	**144.5**	**90.4**	**128.9**	**174.5**	**177.4**	**133.6**	**152.1**	**132.2**	**129.1**	**162.2**

TABLE 13.17 (continued)

Species Biomass in the Quadrates of Grassland

Quadrant Code	S31	S32	S34	S38	S39	S40	S45	S46	S49	S50
Stipa bungeana	18.04	6.16	2.86	10.12	7.92	10.56	4.47	20.90	22.15	18.70
Bothriochloa ischaemum	35.35	52.50	47.95	42.53	60.20	36.40	3.56	8.58	12.89	11.49
Artemisia sacrorum		23.79		2.73		31.59	11.57	9.82	3.06	17.03
Heteropappus altaicus	8.23	4.38	0.53	1.23			1.63	2.98	10.56	1.93
Lespedeza davurica	1.96			16.91	17.15		24.58	49.25	3.68	
Cleistogenes squarrosa	0.70	2.30	8.20	0.20	0.30	0.40	0.30	1.77	3.73	1.47
Potentilla discolor	5.88			1.05	3.78	0.71	1.75	5.11	11.97	3.08
Artemisia giraldii	70.68	4.56		25.46		40.28	27.93		10.58	
Polygala tenuifolia	0.90	0.18	0.11	0.22	0.11		0.84	1.44	0.54	0.30
Artemisia scoparia	7.60	4.80			26.20		0.37	2.00	2.13	0.47
Ixeris chinensis	0.03	0.05	0.03	0.13	0.05		0.03	0.06	0.10	0.82
Gueldenstaedtia stenophylla	0.05		0.07	0.18	1.20		0.28	0.24	0.28	0.44
Artemisia leucophylla		1.60	11.60	5.80	17.00	5.20	56.60	21.40	2.33	
Oxytropis bicolor		2.10		0.15	2.25			1.65		0.25
Agropyron cristatum			1.40				1.27			4.60
Poa sphondylodes			0.23			6.30	17.33	6.60		1.13
Allium ramosum	0.05			0.38			0.13	0.18	0.20	
Adenophora polyantha		1.89	2.97	1.89		2.16				2.12
Ixeris sonchifolia		0.35	0.46	0.07		0.02	0.23	0.23		5.10
Galium verum			3.48	2.61		0.06		2.95		3.72
Viola philippica			0.92	0.23			0.19	0.19		
Bupleurum angustissimum		0.70	1.23	2.63				0.76		0.23
Medicago falcata				0.60						2.25
Setaria viridis	0.21			0.12	0.41			0.34	0.34	
Potentilla reptans var. sericophylla								1.62		
Carum buriaticum			0.38			0.38	0.16	0.16		0.41
Potentilla multifida		3.48	4.64	11.02		0.29				0.92
Viola dissecta		0.11	0.08							
Lespedeza juncea	0.23		16.65							
Leontopodium leontopodioides		0.91	2.08	1.17		0.39		3.34		0.95
Carex duriuscula						0.29				3.05
Artemisia japonica			10.15			3.85				
Chrysanthemum indicum			2.16			2.16				
Lespedeza potaninii		46.92	14.79			41.82				18.02
Carduus nutans				0.38				0.76		0.41
Plantago asiatica										
Thymus mongolicus			0.51							
Leymus secalinus										
Others		15.68	10.34	2.65	0.00	57.74	1.28	6.46	3.87	1.73
Total species	**14**	**22**	**33**	**28**	**12**	**22**	**22**	**29**	**19**	**30**
Total biomass	**149.9**	**172.4**	**143.8**	**130.4**	**136.6**	**240.6**	**154.5**	**148.8**	**88.4**	**100.6**

continued

TABLE 13.17   (continued)

Species Biomass in the Quadrates of Grassland

Quadrant Code	S52	S53	S54	S55	S56	S58	S59	S61	S62	S63
Stipa bungeana	20.31	4.18	12.69	32.12	1.98	3.74	21.49	13.00	24.50	1.50
Bothriochloa ischaemum	30.04	38.68		3.56	48.48	5.66	0.53	5.25	64.40	119.4
Artemisia sacrorum	19.70	3.71	6.70	6.05	21.52	35.58	45.44	24.18	0.39	
Heteropappus altaicus	0.41	8.63	2.74	10.91	4.03	12.89	1.81	1.02	0.34	
Lespedeza davurica						20.99		1.96	1.72	28.91
Cleistogenes squarrosa	3.83	0.23	2.63	0.43	0.17	0.17		0.20		
Potentilla discolor	7.21	15.33	3.50	3.22	1.58	1.75		1.05	1.26	1.89
Artemisia giraldii		29.45	8.93			6.14		64.22	37.83	
Polygala tenuifolia	0.48	1.44	0.42	0.90	1.08	0.78		4.92	0.72	0.29
Artemisia scoparia		4.34	1.60	2.60	2.73	1.53		2.17	12.40	1.55
Ixeris chinensis	0.04	0.08	0.05	0.08	0.04	0.21	0.05			0.05
Gueldenstaedtia stenophylla	0.20	0.30	0.24		0.24	0.32		0.07	0.07	
Artemisia leucophylla	48.60	8.60	25.53	35.80	19.40	7.53				
Oxytropis bicolor	0.45	1.50	0.45	3.00	0.65	0.60		0.16	1.24	0.93
Agropyron cristatum		1.40	13.87	1.33	9.00	0.47	3.13	23.80		
Poa sphondylodes	0.68		2.70	0.60	8.03	1.95	0.45	2.48		
Allium ramosum	0.18	0.73		0.20			0.13			
Adenophora polyantha					0.59	0.50		0.81		
Ixeris sonchifolia		0.42	0.15	0.77	0.38	0.38	1.15			
Galium verum				4.16	5.61	1.35	0.34	6.40		
Viola philippica		0.15		0.15	0.31	0.19	0.38			0.23
Bupleurum angustissimum			0.48	1.58	0.76	0.52	0.51			
Medicago falcata			0.45	0.35	1.75	7.10	0.45	3.00		
Setaria viridis			0.34	0.41		0.41	0.34			
Potentilla reptans var. sericophylla			3.06	1.88	11.58		7.62			
Carum buriaticum	0.13			1.43	0.25	0.13	0.35			
Potentilla multifida				1.84	0.24	1.60				
Viola dissecta						0.10				
Lespedeza juncea	11.63	8.90			21.25					
Leontopodium leontopodioides				9.36	0.07	2.50				
Carex duriuscula			3.96		0.24					
Artemisia japonica					1.69	3.21	9.98			
Chrysanthemum indicum										
Lespedeza potaninii	32.73	3.06	3.92	16.07	9.78		0.34			
Carduus nutans	0.54						1.81			
Plantago asiatica				0.59	0.34		1.87			
Thymus mongolicus										
Leymus secalinus										
Others	2.05	2.32	13.31	6.85	2.74	2.26	7.40	1.55	0.00	1.53
Total species	**22**	**23**	**26**	**38**	**34**	**34**	**28**	**19**	**11**	**10**
Total biomass	**179.2**	**133.9**	**107.6**	**147.6**	**181.9**	**113.9**	**108.1**	**153.2**	**144.9**	**156.2**

TABLE 13.17 (continued)

Species Biomass in the Quadrates of Grassland

Quadrant Code	S64	S65	S66	S67	S68	S69	S70	S71	S73	S74
Stipa bungeana	22.00	55.50	11.00	11.00	71.75	8.00	5.50	2.50	15.50	14.0
Bothriochloa ischaemum		3.15		63.70		39.20		8.40		4.90
Artemisia sacrorum	32.37		38.22		38.42	17.55	88.14	22.82	107.3	52.65
Heteropappus altaicus	0.68		0.51	3.40	1.36	0.68			0.51	0.68
Lespedeza davurica	18.62	22.05	22.05	5.39	3.43	0.98		25.48	0.25	0.98
Cleistogenes squarrosa	1.00	12.40	0.60	1.80	3.60	0.50	0.80			1.10
Potentilla discolor			0.21	19.32		0.63			5.88	7.98
Artemisia giraldii		54.60				28.08	3.90		24.96	50.70
Polygala tenuifolia			0.54	4.14				0.36	0.36	0.36
Artemisia scoparia		1.24	0.62	0.31			0.31		0.93	
Ixeris chinensis			0.03				0.12			
Gueldenstaedtia stenophylla		2.16	0.48	0.12						0.12
Artemisia leucophylla	13.60		120.0	10.00			3.20		10.00	
Oxytropis bicolor			1.55	1.24		0.31				0.31
Agropyron cristatum	1.80	0.60			3.20		2.40			
Poa sphondylodes			2.70		0.23	2.70			7.20	
Allium ramosum									0.08	0.05
Adenophora polyantha	4.86		1.22		1.35	3.78		0.81	0.95	
Ixeris sonchifolia							0.12			
Galium verum	4.79		14.00					3.50		
Viola philippica			0.23							
Bupleurum angustissimum	1.12					0.23		2.76	5.98	
Medicago falcata	5.40				5.70					
Setaria viridis		1.00	0.06				0.10			
Potentilla reptans var. sericophylla	2.16		17.64		4.68		0.18	0.90		
Carum buriaticum	0.19		0.19		1.33					
Potentilla multifida	0.87		0.29			0.44				
Viola dissecta							1.20	0.51	1.19	
Lespedeza juncea			0.90			22.50				
Leontopodium leontopodioides	0.52							7.28		
Carex duriuscula	28.13					6.67		52.20	14.21	
Artemisia japonica	19.95						4.73	27.30		
Chrysanthemum indicum	6.36		3.60				21.15		2.70	
Lespedeza potaninii										
Carduus nutans	0.19						7.22			1.14
Plantago asiatica			0.11							
Thymus mongolicus										
Leymus secalinus			0.83		19.14		16.17	15.84		
Others	22.80	27.94	4.27	0.00	3.28	18.47	45.05	20.75	38.54	12.00
Total species	**30**	**13**	**29**	**11**	**15**	**21**	**25**	**25**	**24**	**15**
Total biomass	**187.4**	**180.6**	**241.8**	**120.4**	**157.5**	**150.7**	**200.3**	**191.4**	**236.5**	**147.0**

continued

TABLE 13.17 (continued)

Species Biomass in the Quadrates of Grassland

Quadrant Code	S75	S76	S77	S78	S80	S81	S82	S84	S85	S86
Stipa bungeana	16.50	8.50	10.00	5.00	23.50	13.00	46.00	47.00	20.00	16.50
Bothriochloa ischaemum		1.70	14.00	34.30		123.2	32.20	41.30		61.95
Artemisia sacrorum	0.59	13.26	61.62		16.97		9.95	8.58	1.17	
Heteropappus altaicus		1.36	1.02	2.04	0.34	2.04	0.34	0.68	2.55	1.36
Lespedeza davurica	1.47	1.47	1.96	9.80	19.60	22.54	1.96	22.54	4.41	1.96
Cleistogenes squarrosa	0.20			0.40	0.16	2.20	0.80	2.20		0.80
Potentilla discolor		2.10	5.04	8.40				5.46	0.42	25.20
Artemisia giraldii		70.20	21.84	22.23	9.17	1.17			19.11	4.29
Polygala tenuifolia	0.04	0.36	0.36	0.18	0.36	2.88		0.90	0.29	
Artemisia scoparia								9.83	0.47	3.88
Ixeris chinensis	0.02			0.06	0.02	0.06		0.06	0.01	0.06
Gueldenstaedtia stenophylla			0.24		0.12			0.36	0.60	0.12
Artemisia leucophylla	2.60		24.00	2.80	4.80	1.60	0.40	17.60	30.00	
Oxytropis bicolor					1.09		2.17	0.25	0.31	0.78
Agropyron cristatum	0.40	20.00	1.60			3.60	9.60	1.20	5.40	
Poa sphondylodes			4.50		5.18	1.35	2.70	0.68	4.95	
Allium ramosum				0.15	0.45	0.15	0.15	1.65	0.30	
Adenophora polyantha	5.13				0.81	0.27	0.41			
Ixeris sonchifolia					0.23		0.12		2.30	
Galium verum	5.25		3.23		3.57	4.93			11.31	
Viola philippica	0.58				0.12				0.35	
Bupleurum angustissimum	4.14		1.84		4.37		2.53			
Medicago falcata	5.10		18.90			4.65	1.95		0.60	
Setaria viridis	0.04	0.04				0.10		0.20	0.20	0.04
Potentilla reptans var. sericophylla	0.36				0.58		1.98		12.96	
Carum buriaticum	0.10		0.29			0.10				
Potentilla multifida					2.03		1.74			
Viola dissecta	0.68				0.51		0.26			
Lespedeza juncea			2.48	0.45		1.35				
Leontopodium leontopodioides						0.52	7.54			
Carex duriuscula	1.16				9.28		0.29		0.15	
Artemisia japonica	10.85				8.05	14.00	14.35			
Chrysanthemum indicum					4.65				15.00	
Lespedeza potaninii							49.47			
Carduus nutans										
Plantago asiatica	0.33						0.22			
Thymus mongolicus										
Leymus secalinus					0.99	14.19				
Others	29.32	1.46	2.54	0.24	3.08	2.08	2.22	0.05	0.35	2.54
Total species	**33**	**12**	**19**	**14**	**30**	**24**	**28**	**18**	**24**	**15**
Total biomass	**84.8**	**120.5**	**175.5**	**87.1**	**118.9**	**218.1**	**187.4**	**160.6**	**132.9**	**119.5**

TABLE 13.17 (continued)

Species Biomass in the Quadrates of Grassland

Quadrant Code	S88	S91	S93	S94	S95	S96	S98	S101	S102
Stipa bungeana	8.00	40.00	20.50	23.00	13.00	3.50	30.00	12.00	22.50
Bothriochloa ischaemum	70.35		2.10	63.70	0.35	121.1	10.15	0.35	9.10
Artemisia sacrorum		20.67	34.13		17.94	0.78	0.78	8.19	0.39
Heteropappus altaicus	0.34	2.38	0.27	1.36		1.36			8.50
Lespedeza davurica	10.29	33.32	27.93	1.47	6.86	1.23	2.45	22.54	8.33
Cleistogenes squarrosa	0.06		1.60	2.60	0.20		1.00	3.00	
Potentilla discolor	4.62			6.30		2.52			16.38
Artemisia giraldii		1.56		0.98		1.76	1.37	24.96	42.12
Polygala tenuifolia	0.72	1.26		0.14	0.36				1.26
Artemisia scoparia	0.09			2.48		4.65			9.61
Ixeris chinensis		0.77				0.09	0.03		0.08
Gueldenstaedtia stenophylla				0.12	0.24	0.48			0.24
Artemisia leucophylla	7.20		13.20	16.00	4.80		55.20	41.60	13.20
Oxytropis bicolor	2.17			3.10	0.31	1.55			6.82
Agropyron cristatum		0.60		6.40	4.00		9.20	6.80	0.40
Poa sphondylodes		9.00	0.45		2.25		0.45	1.71	
Allium ramosum			0.45	1.28		0.23	0.15	0.15	
Adenophora polyantha	3.24	0.81			1.35		0.54	1.89	
Ixeris sonchifolia	0.35	0.35	0.23		0.92		0.58	0.18	
Galium verum	6.67		1.02		1.16		15.95	3.19	
Viola philippica	0.46	0.23	0.12		0.58		0.69	0.05	
Bupleurum angustissimum	0.46	0.69			4.60		0.69		
Medicago falcata	6.30	1.50							
Setaria viridis							0.20		0.04
Potentilla reptans var. sericophylla			8.28		5.40		26.64	10.62	
Carum buriaticum								0.19	
Potentilla multifida	5.22				0.87				1.16
Viola dissecta			0.34		0.26	0.17	0.60	0.60	0.17
Lespedeza juncea	29.70			0.45		0.90			
Leontopodium leontopodioides	1.69	1.04	7.02		11.78				
Carex duriuscula			9.28		24.65		0.09	0.29	
Artemisia japonica		9.45	4.90		3.15		11.20	8.40	
Chrysanthemum indicum		4.20	18.30	12.30	25.80		33.00	5.40	3.00
Lespedeza potaninii									
Carduus nutans	2.66			2.28	1.33				
Plantago asiatica			2.64						
Thymus mongolicus	4.93								
Leymus secalinus	20.79		5.61					16.50	
Others		16.34	5.74	30.76	7.27	0.00	73.47	9.21	3.37
Total species	**22**	**21**	**25**	**22**	**31**	**14**	**29**	**29**	**24**
Total biomass	**186.3**	**144.2**	**164.1**	**174.7**	**139.4**	**140.3**	**274.4**	**177.8**	**146.7**

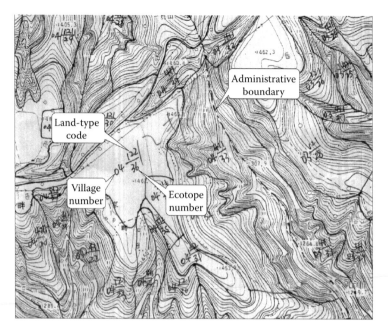

FIGURE 13.13
Data acquisition of ecotopes.

In the six quadrats of square region 1 (Table 13.20), 55 species were found, in which the dominant species included *Stipa bungeana*, *Bothriochloa ischaemum*, and *Artemisia sacrorum*. In square region 2 (Table 13.21), 61 species were sampled, in which the dominant species included *Artemisia sacrorum*, *Bothriochloa ischaemum*, *Carex duriuscula*, *Stipa bungeana*, *Cleistogenes squarrosa*, and *Hippophae rhamnoides*. In square region 3 (Table 13.22), 46 species were found in the sampled quadrats; the species with the largest individual number and biggest biomass were *Bothriochloa ischaemum*, *Artemisia giraldii*, *Stipa bungeana*, and *Agropyron cristatum*. In square region 4 (Table 13.23), 71 species, dominated by *Stipa bungeana*, *Artemisia sacrorum*, *Bothriochloa ischaemum*, *Artemisia japonica*, *Carex duriuscula*, and *Leontopodium leontopodioides*, were found. In square region 5 (Table 13.24), 45 species with dominants of *Robinia pseudoacacia*, *Bothriochloa ischaemum*, *Stipa bungeana*, and *Cleistogenes squarrosa* were identified. In square region 6 (Table 13.25), 50 species with dominants of *Stipa bungeana*, *Bothriochloa ischaemum*, and *Artemisia leucophylla* appeared. In square region 7 (Table 13.26), 52 species with dominants of *Robinia pseudoacacia*, *Bothriochloa ischaemum*, and *Stipa bungeana* were found. In square region 8 (Tables 13.27), 53 species with dominants of *Bothriochloa ischaemum*, *Stipa bungeana*, *Carex duriuscula*, and *Cleistogenes squarrosa* were identified. In square region 9 (Table 13.28), 53 species dominated by *Robinia pseudoacacia*, *Bothriochloa ischaemum*, *Stipa bungeana*, *Potentilla reptans* var. *sericophylla*, and *Cleistogenes*

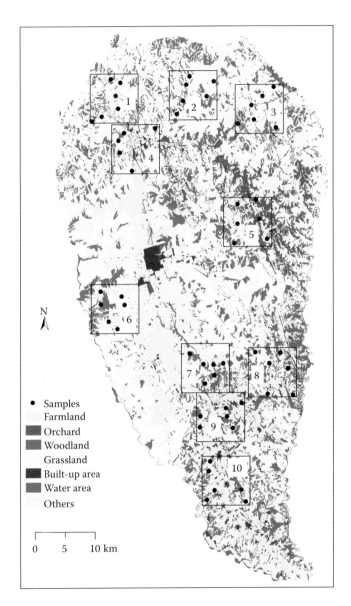

FIGURE 13.14
Location of the 10 square regions and corresponding quadrates.

squarrosa were found. Finally, in square region 10 (Table 13.29), 56 species, dominated by *Prunus armeniaca, Robinia pseudoacacia, Bothriochloa ischaemum, Stipa bungeana,* and *Potentilla reptans* var. *sericophylla* were identified.

The analyses in terms of Supplementary Tables 13.17 through 13.29, demonstrates that the correlation coefficient between individual-counting

TABLE 13.18

Ecotope Number of Every Ecotope Type in 2007

Region Code	1	2	3	4	5	6	7	8	9	10
Slope farmland	130	223	207	88	208	75	56	35	10	18
Terrace farmland	138	82	124	96	107	102	74	187	224	216
Dam farmland			2							
Apple	53	7	19	80	134	31	11	23	14	
Pear	1		1		1					
China pine			23			20	27	6		
Locust	86	49	63	47	82	49	70	150	85	84
Poplar		1		1	1		7	5	9	2
Osier										6
Sallow thorn	1	32	15	5	2			19	1	
Chinese date								1		
Haw		1							1	
Walnut					1					
Almond	36	17	12	21	6		11	9	9	
Chinese prickly ash									2	
Nursery of young plants						1				
Natural grassland	354	325	287	302	246	356	261	209	232	119
Artificial grassland	26	41	9	8	2	4	22	9	11	4
River			4		1	1	3	2		1
Reservoir	6		18			4			5	25
Pond							2			
Low land			4			2	1	2		1
Reed land							5			
Urban area	1	5	5	15	8	3				2
Rural residential area	111	43	55	171	168	52	56	145	154	141
Industrial and mining area				7						2
Road	1	1	1	1	1	2	3	3	1	1
Water conservancy facilities					2					
Total number of ecotopes	**944**	**827**	**849**	**842**	**970**	**702**	**628**	**787**	**757**	**622**
Total ecotope types	**13**	**13**	**17**	**13**	**16**	**14**	**16**	**15**	**13**	**14**

diversity and individual ecotope-counting diversity was −0.381 and that the significance level was 0.178. Individual-counting diversity and ecotope-area-based diversity had a correlation coefficient of −0.253 and a significance level of 0.481. Biomass-based diversity and individual ecotope-counting diversity had a correlation coefficient of 0.19 and a significance of 0.599. The correlation coefficient between biomass-based diversity and ecotope-area-based diversity was 0.188 and its significance level was 0.604. In short, species diversity and ecotope diversity have no significant correlation in Dongzhi tableland.

TABLE 13.19

Area of Every Ecotope Type in 2007 (Unit: km^2)

Region Code	1	2	3	4	5	6	7	8	9	10
Slope farmland	8.980	18.123	15.521	6.804	14.489	10.485	9.673	2.042	0.222	1.273
Terrace farmland	16.057	7.906	12.333	21.728	8.613	14.020	15.389	21.833	31.998	31.880
Dam farmland			0.045							
Apple	0.787	0.141	0.425	1.417	11.009	0.218	0.204	0.311	0.553	
Pear	0.036		0.002		0.030					
China pine			1.609			2.544	2.719	0.439		
Locust	6.805	7.654	6.354	4.132	9.755	8.429	7.479	13.734	5.732	8.148
Poplar		0.028		0.038	0.027		0.350	0.399	0.337	0.145
Osier										0.406
Sallow thorn	0.047	2.848	1.640	0.567	0.120		2.018	0.012		
Chinese date								0.244		
Haw		0.040							0.043	
Walnut					0.019					
Almond	1.690	1.328	0.701	1.787	0.226		1.189	0.739	0.438	
Chinese prickly ash									0.089	
Nursery of young plants						0.029				
Natural grassland	26.976	22.794	22.324	21.683	16.497	25.558	22.379	19.254	16.543	14.437
Artificial grassland	0.657	1.710	0.632	0.258	0.061	0.205	0.957	0.347	0.336	0.472
River			0.121		0.008	0.002	0.017	0.077		0.002
Reservoir	0.068		0.100			0.030			0.045	0.201
Pond							0.001			
Low land			0.200			0.017	0.016	0.075		0.003
Reed land							0.202			
Urban area	0.003	0.053	0.029	0.749	0.196	0.051				0.373
Rural residential area	0.888	0.530	0.944	3.856	2.377	1.916	0.962	3.530	6.429	4.763
Industrial and mining area				0.103						0.289
Road	1.006	0.844	1.021	0.876	0.554	0.496	0.445	0.967	1.233	1.608
Water conservancy facilities					0.020					
Total area	64.0	64.0	64.0	64.0	64.0	64.0	64.0	64.0	64.0	64.0
Total ecotope types	13	13	17	13	16	14	16	15	13	14

13.5 Discussion and Conclusions

13.5.1 Definitions

Biodiversity definitions vary within the discipline (Holt, 2006). Diversity has been used by landscape architects, ecologists, policy analysts, politicians,

TABLE 13.20

Individual Number and Biomass of Every Species in Square Region 1

Quadrant Code	Individual Number						Biomass (g)					
	SQ15	SQ16	SQ73	SQ74	SQ75	SQ76	SQ15	SQ16	SQ73	SQ74	SQ75	SQ76
Bothriochloa ischaemum	166	1318		148		32	7.00	78.75		4.90		1.70
Stipa bungeana	233	110	263	277	330	223	7.92	4.84	15.50	14.00	16.50	8.50
Artemisia giraldii	335	4	33	169		173	65.36	0.38	24.96	50.70		70.20
Artemisia sacrorum	64	73	165	115	2	55	19.89	18.33	107.25	52.65	0.59	13.26
Carex duriuscula			198		74				14.21		1.16	
Cleistogenes squarrosa	153			91	9		1.70			1.10	0.20	
Agropyron cristatum					2	130					0.40	20.00
Artemisia japonica					106						10.85	
Galium verum					103						5.25	
Potentilla discolor	4	3	14	56		12	0.84	0.84	5.88	7.98		2.10
Poa sphondylodes			87						7.20			
Medicago falcata					77						5.10	
Artemisia frigida		63						9.00				
Viola pseudo bambusetorum			1		59				0.05		1.12	
Artemisia leucophylla			22		25				10.00		2.60	
Lespedeza davurica			2	10	14	17			0.25	0.98	1.47	1.47
Others	16	35	77	23	232	22	2.695	5.125	51.18	14.655	39.599	3.22
Total species	**10**	**11**	**24**	**15**	**33**	**12**	**10**	**11**	**24**	**15**	**33**	**12**
Total	971	1606	862	889	1033	664	105.41	117.27	236.47	146.97	84.834	120.45

TABLE 13. 21

Individual Number and Biomass of Every Species in Square Region 2

Quadrant Code	Individual Number						Biomass (g)					
	SQ10	SQ11	SQ69	SQ70	SQ71	SQ72	SQ10	SQ11	SQ69	SQ70	SQ71	SQ72
Hippophae rhamnoides						13						8125
Artemisia sacrorum	75	162	51	56	55	14	50.7	30.42	17.55	88.14	22.82	6.24
Bothriochloa ischaemum		736	819		38	876		26.6	39.2		8.4	98
Carex duriuscula	60		96		1080		1.45		6.67		52.2	
Artemisia giraldii	44	70	116	13	166		13.49	14.44	28.08	3.9		
Artemisia japonica	53	30		18	109	60	8.225	10.5		4.725	27.3	
Lespedeza davurica	6	21	6		24	271	1.47	1.045	0.98		25.48	12.25
Stipa bungeana	182	61	141	89	101		8.8	1.5	8	5.5	2.5	11
Leymus secalinus				184	153					16.17	15.84	
Leontopodium leontopodioides	54						2.86				7.28	
Lespedeza juncea			93			26			22.5			5.85
Potentilla discolor		55	4			7		9.24	0.63			1.89
Artemisia leucophylla	28			18		17	14.4			3.2		3.6
Chrysanthemum indicum				52						21.15		
Glycyrrhiza uralensis				2		3				6.67		22.62
Cleistogenes squarrosa	592	3	67	17			13	0.1	0.5	0.8		
Others	163	176	148	130	186	37	14.29	21.63	26.6	50.03	29.59	3.85
Total species	**22**	**21**	**21**	**25**	**25**	**13**	**22**	**21**	**21**	**25**	**25**	**13**
Total	**1257**	**1314**	**1541**	**579**	**1912**	**1324**	**128.7**	**115.5**	**150.7**	**200.3**	**191.4**	**8290**

TABLE 13.22

Individual Number and Biomass of Every Species in Square Region 3

Quadrant Code	Individual Number						Biomass (g)					
	SQ3	SQ4	SQ8	SQ61	SQ62	SQ63	SQ3	SQ4	SQ8	SQ61	SQ62	SQ63
Bothriochloa ischaemum	510	20	311	50	2012	695	49.70	0.70	20.65	5.25	64.40	119.35
Artemisia giraldii	10	340	162	223	74		3.04		47.12	64.22	37.83	
Stipa bungeana	480		361	289	739	81	28.60	15.84	20.68	13.00	24.50	1.50
Agropyron cristatum	61	183		158			19.60	42.80		23.80		
Artemisia sacrorum		18	206	66	3			8.58	29.64	24.18	0.39	
Lespedeza davurica	64	7	5	7	29	275	14.70	1.47	0.49	1.96	1.72	28.91
Artemisia scoparia	35	15	1	8	108	6	6.80	5.80	0.40	2.17	12.40	1.55
Heteropappus altaicus	18	4	22	3	3		12.25	2.80	4.90	1.02	0.34	
Astragalus melilotoides Pall.		10				3		15.99				1.53
Medicago falcata		18						17.40				
Potentilla discolor	23		6	5	18	3	11.76		1.26	1.05	1.26	1.89
Galium verum		26	3	140				5.80	0.09	6.40		
Phragmites australis		7						11.44				
Artemisia leucophylla		33						8.80				
Polygala tenuifolia	5	6	10	20	8	2	0.18	0.36	0.72	4.92	0.72	0.29
Potentilla bifurca	39						5.61					
Others	87	207	35	66	39	26	5.835	24.41	1.704	5.262	1.312	1.21
Total species	**18**	**37**	**17**	**19**	**11**	**10**	**18**	**37**	**17**	**19**	**11**	**10**
Total	**1332**	**894**	**1122**	**1035**	**3033**	**1091**	**158.08**	**162.19**	**127.65**	**153.23**	**144.87**	**156.23**

TABLE 13.23

Individual Number and Biomass of Every Species in Square Region 4

Quadrant Code	Individual Number						Biomass (g)					
	SQ12	SQ13	SQ14	SQ17	SQ19	SQ64	SQ12	SQ13	SQ14	SQ17	SQ19	SQ64
Stipa bungeana	575	1033	254	358	510	242	10.12	43.56	13.20	13.20	14.08	22.00
Artemisia sacrorum	89	3	261		5	89	14.43	0.70	58.50		2.34	32.37
Bothriochloa ischaemum	543	221		378			29.75	21.00		40.25		
Artemisia giraldii	73	95	33				14.82	27.36	14.82			
Artemisia japonica			339	12		98			30.80	2.80		19.95
Artemisia leucophylla		5	10	19	127	29		2.80	1.60	1.60	23.20	13.60
Lespedeza davurica				88	43	42				14.73	6.86	18.62
Chrysanthemum indicum					273	42					33.12	6.36
Carex duriuscula			116			448			2.90			28.13
Medicago falcata	36	11	76		61	4		3.60	6.90		9.90	5.40
Lespedeza potaninii	87	75					5.61	18.87				
Potentilla discolor	82	10	20	33	2		9.24	3.78	3.36	3.36	0.21	
Artemisia scoparia	82			1	7		11.40			0.20	4.80	
Galium verum	15		42	10	12	40	1.31		2.90	0.44	0.87	4.79
Potentilla reptans var. *sericophylla*			42		78	93			2.52		5.04	2.16
Leontopodium leontopodioides			111	243	1	6			3.38	9.10	0.05	0.52
Others	72	109	134	243	342	128	4.834	20.128	8.855	20.072	34.17	33.512
Total species	21	18	28	23	31	30	21	18	28	23	31	30
Total	1572	1562	1438	1385	1461	1261	101.51	141.8	149.74	105.72	134.64	187.41

TABLE 13.24

Individual Number and Biomass of Every Species in Square Region 5

Quadrant Code	Individual Number						Biomass (g)					
	SQ5	SQ6	SQ23	SQ24	SQ83	SQ84	SQ5	SQ6	SQ23	SQ24	SQ83	SQ84
Robinia pseudoacacia					12						232800	
Bothriochloa ischaemum	258	227	466	370		373	34.83	60.2	42	34.3		41.3
Artemisia leucophylla				153	39	22				74	31.2	17.6
Stipa bungeana	474	165	63	393	133	927	20.68	5.72	1.1	13.2	18	47
Artemisia giraldii		200	21	79				38	9.12	26.22		
Potentilla discolor	37	120	80			26	8.4	21	18.9			5.46
Artemisia sacrorum	45	6	10	37	11	17	7.41	1.17	1.56	17.16	3.51	8.58
Cleistogenes squarrosa	36	109	16		392	85	0.4	2.2	0.1		31	2.2
Lespedeza davurica	110	16			5	187	9.31	2.45			1.225	22.54
Lespedeza potaninii	11		58	11					19.13	2.295		
Artemisia scoparia		38	11			18		6.4	2.4	0.2		9.83
Medicago falcata	4		17		29		0.9		9		4.2	
Poa sphondylodes	149					10	13.05					0.675
Chrysanthemum indicum					23						12.9	
Oxytropis bicolor		4	181			3		0.15	9.9			0.31
Heteropappus altaicus	40	3	25	5		3	2.975	1.05	3.15	1.925		0.68
Others	168	50	49	60	18	52	5.706	3.56	12.58	5.235	4.806	4.418
Total species	**19**	**16**	**17**	**12**	**15**	**18**	**19**	**16**	**17**	**12**	**15**	**18**
Total	**1332**	**938**	**997**	**1108**	**662**	**1723**	**103.66**	**141.90**	**128.94**	**174.54**	**232906.84**	**160.59**

TABLE 13.25

Individual Number and Biomass of Every Species in Square Region 6

Quadrant Code	Individual Number						Biomass (g)					
	SQ29	SQ30	SQ65	SQ66	SQ67	SQ68	SQ29	SQ30	SQ65	SQ66	SQ67	SQ68
Stipa bungeana	374	340	808	167	280	856	23.32	25.52	55.50	11.00	11.00	71.75
Artemisia leucophylla	76	109		357	36		14.80	32.00		120.00	10.00	
Bothriochloa ischaemum		562	19		767			61.25	3.15		63.70	
Lespedeza davurica	56	60	55	27	37	7	11.27	24.01	22.05	22.05	5.39	3.43
Artemisia sacrorum	21			86		10	7.61			38.22		38.42
Artemisia giraldii	21	6	62				8.74	2.28	54.60			
Agropyron cristatum	340		1			7	31.20		0.60			3.20
Limonium bicolor			9			7			22.62			1.95
Potentilla reptans var. *sericophylla*				332		62				17.64		4.68
Potentilla discolor		1		1	67			0.63		0.21	19.32	
Leymus secalinus				1		225				0.83		19.14
Cleistogenes squarrosa		14	245	13	102	48		0.40	12.40	0.60	1.80	3.60
Galium verum	45			169			4.06			14.30		
Artemisia scoparia		7	2	4	1			6.40	1.24	0.62	0.31	
Polygala tenuifolia	24			5	17		1.44			0.54	4.14	
Chrysanthemum indicum	6			26			2.04			3.60		
Others	145	11	55	105	32	97	24.64	9.667	8.484	12.525	4.76	11.3
Total species	**24**	**14**	**13**	**29**	**11**	**15**	**24**	**14**	**13**	**29**	**11**	**15**
Total	**1108**	**1110**	**1256**	**1293**	**1339**	**1319**	**129.1**	**162.2**	**180.6**	**241.83**	**120.4**	**157.5**

TABLE 13.26

Individual Number and Biomass of Every Species in Square Region 7

Quadrant Code	Individual Number						Biomass (g)					
	SQ39	SQ40	SQ89	SQ90	SQ91	SQ92	SQ39	SQ40	SQ89	SQ90	SQ91	SQ92
Robinia pseudoacacia			4	6					8000	12000		
Bothriochloa ischaemum	717	352	931	156		573	60.2	36.4	123.6	28		47.6
Lespedeza davurica	74		11	19	74	22	17.15		8.33	37.24	33.32	6.37
Stipa bungeana	304	192	64	228	432	297	7.92	10.56	5	11.5	40	7
Caragana microphylla		76						56.84				
Artemisia sacrorum		91			36	7		31.59			20.67	2.73
Artemisia giraldii		70			1			40.28			1.56	
Lespedeza potaninii		88						41.82				
Leymus secalinus				147						35.64		
Artemisia leucophylla	36	10	3	24		9	17	5.2	0.4	6.8		2
Heteropappus altaicus			35	15	7	130			7.14	3.4	2.38	17
Triarrhena			23	5					19.08	10.6		
Artemisia scoparia	99					31	26.2					2.11
Poa sphondylodes		90	92	1	188	127		6.3	9.45	0.36	9	18.27
Potentilla discolor	29	1	12				3.78	0.714	6.96			
Potentilla multifida		1		5				0.29		11.6		
Others	204	78	128	131	165	388	4.318	10.61	25.32	30.07	37.23	12.27
Total species	**12**	**22**	**26**	**19**	**21**	**21**	**12**	**22**	**26**	**19**	**21**	**21**
Total	**1463**	**1049**	**1303**	**737**	**903**	**1584**	**136.6**	**240.6**	**8205**	**12175**	**144.2**	**115.3**

TABLE 13.27

Individual Number and Biomass of Every Species in Square Region 8

Quadrant Code	Individual Number						Biomass (g)					
	SQ31	SQ32	SQ93	SQ94	SQ95	SQ96	SQ31	SQ32	SQ93	SQ94	SQ95	SQ96
Bothriochloa ischaemum	167	258	16	787	8	1240	35.35	52.5	2.1	63.7	0.35	121.1
Stipa bungeana	186	115	275	372	176	117	18.04	6.16	20.5	23	13	3.5
Artemisia giraldii	162	4		3		4	70.68	4.56		0.975		1.755
Artemisia sacrorum		67	67	41	64	1		23.79	34.13		17.94	0.78
Chrysanthemum indicum			42		91				18.3	12.3	25.8	
Lespedeza potaninii		58						46.92				
Lespedeza davurica	3		114	6	30	10	1.96		27.93	1.47	6.86	1.225
Artemisia leucophylla		2	19	19	15			1.6	13.2	16	4.8	
Carex duriuscula			199		448				9.28		24.65	
Leontopodium leontopodioides		5	141		152			0.91	7.02		11.78	
Artemisia scoparia	18	20		6		33	7.6	4.8		2.48		4.65
Thalictrum petaloideum				16						18.08		
Heteropappus altaicus	9	30	1	7		16	8.225	4.375	0.272	1.36		1.36
Potentilla discolor	5			38		15	5.88			6.3		2.52
Potentilla reptans var. sericophylla			125		112				8.28		5.4	
Agropyron cristatum				42	39					6.4	4	
Others	29	265	261	323	268	56	2.148	26.83	23.09	22.65	24.84	3.415
Total species	**14**	**22**	**25**	**22**	**31**	**14**	**14**	**22**	**25**	**22**	**31**	**14**
Total	**579**	**824**	**1260**	**1660**	**1403**	**1492**	**149.9**	**172.4**	**164.1**	**174.7**	**139.4**	**140.3**

TABLE 13.28

Individual Number and Biomass of Every Species in Square Region 9

Quadrant Code	Individual Number						Biomass (g)					
	SQ35	SQ36	SQ97	SQ98	SQ99	SQ100	SQ35	SQ36	SQ97	SQ98	SQ99	SQ100
Robinia pseudoacacia	8		3		7	3	160928		6000		15000	61200
Ziziphus jujuba var. *spinosa*		25						1320				
Bothriochloa ischaemum	13	108	553	48	274	23	0.35	10.5	86.1	10.15	68.6	3.85
Stipa bungeana	809	535	378	420	798	273	32.56	22.88	17	30	47	10
Artemisia leucophylla	85	12	21	111		86	36	7.2	13	55.2		29.2
Lespedeza potaninii		104						126				
Artemisia sacrorum		15		4	76	91		13.26		0.78	50.31	22.62
Chrysanthemum indicum	55			48		98	15.12			33		30.6
Artemisia argy				84						65.1		
Artemisia scoparia		50	9		42	1		7.2	5.425		32.55	0.775
Potentilla discolor			13		172	1			2.31		31.92	0.42
Potentilla reptans var. *sericophylla*	5			392		121	0.7			26.64		5.4
Agropyron cristatum	34			71		155	6.4			9.2		15.2
Heteropappus altaicus	3	37	69		6	3	2.45	10.85	11.56		2.72	1.36
Lespedeza davurica			52	4	34	44			10.29	2.45	8.82	7.105
Artemisia japonica				33		13				11.2		8.4
Others	38	193	284	254	129	185	6.714	9.322	33.12	30.69	12.495	18.142
Total species	**20**	**13**	**19**	**29**	**15**	**29**	**20**	**13**	**19**	**29**	**15**	**29**
Total	**1050**	**1079**	**1382**	**1469**	**1538**	**1097**	**161028**	**1527**	**6179**	**274.4**	**15254**	**61353**

TABLE 13.29

Individual Number and Biomass of Every Species in Square Region 10

Quadrant Code	Individual Number						Biomass (g)					
	SQ33	SQ34	SQ85	SQ86	SQ87	SQ88	SQ33	SQ34	SQ85	SQ86	SQ87	SQ88
Prunus armeniaca	5						55491.5					
Robinia pseudoacacia					3						6000	
Bothriochloa ischaemum	418	345		815	7	573	82.25	47.95		61.95	0.7	70.35
Stipa bungeana	114	43	83	493	91	165	4.84	2.86	20	16.5	35	8
Artemisia leucophylla		18	94		37	18		11.6	30		20	7.2
Potentilla discolor	56		3	130			29.4		0.42	25.2		
Artemisia giraldii	58		20	4		29	30.78		19.11	4.29		4.62
Chrysanthemum indicum		21	91		88			2.16	15		32.4	
Lespedeza juncea	7	74				9	0.9	16.65				29.7
Potentilla reptans var. *sericophylla*			124		328				12.96		16.92	
Galium verum	29	29	114		8	76	3.77	3.48	11.31		1.74	6.67
Lespedeza davurica	9		32	45	29	44	3.43		4.41	1.96	4.9	10.29
Leymus secalinus						182						20.79
Heteropappus altaicus	6	5	27	9	23	4	1.75	0.525	2.55	1.36	11.56	0.34
Poa sphondylodes		2	114		213			0.225	4.95		10.8	
Lespedeza potaninii		65						14.79				
Others	94	585	109	109	353	351	5.05	43.555	12.177	8.213	31.945	28.348
Total species	**19**	**33**	**24**	**15**	**28**	**22**	**19**	**33**	**24**	**15**	**28**	**22**
Total	**796**	**1187**	**811**	**1605**	**1180**	**1451**	**55653.7**	**143.8**	**132.887**	**119.47**	**6165.97**	**186.31**

artists, and so on. Each discipline has provided its own understanding and definition. For instance, the multiplicative concept expressed gamma diversity as the product of the mean alpha diversity and beta diversity (Whittaker, 1972), in which diversity was equated to species richness (Wagner et al., 2000). The additive partitioning of species diversity proposed that the diversity between two samples was equal to the combined diversity of the two samples minus the average within-sample diversity (MacArthur et al., 1966). The additive version treated alpha diversity as the average within-sample diversity, beta diversity as the average amount of diversity not found in a single, randomly-chosen sample, and gamma diversity as alpha diversity plus beta diversity (Veech et al., 2002; Chen et al., 2008). Zamora et al. (2007) defined the alpha diversity as local species richness, beta diversity as species turnover and gamma diversity as landscape richness.

The relationships between diversity and ecosystem stability, between diversity and productivity, and between species diversity and ecotope diversity were long debated using differing languages of diversity (Roper-Lindsay et al., 2003; Ewers and Rodrigues, 2006), which made these debates meaningless.

Here, we adopt the definition of individual-counting diversity that combines richness and evenness. Richness is based on the total number of species present and evenness is based on the relative abundance of the species and the degree of its dominance thereof. This concept is then generalized to formulate biomass-based diversity, individual ecotope-counting diversity, and ecotope-area-based diversity.

13.5.2 Diversity Indexes

Different diversity indexes have different calculation results. In addition to the scaling diversity index that has recently been developed, 11 traditional diversity indices can be found in scientific literature. Our study demonstrates that some of these diversity indices are unable to formulate the richness aspect, while others are unable to express the evenness aspect. By contrast, the scaling diversity index can formulate both the evenness and richness aspects of diversity. It is able to combine factors of spatial resolution, spatial scale, and temporal scale with both the richness and evenness components. It is scientifically sound and can be operated at affordable cost.

13.5.3 Effects of Spatial Resolutions on the Relationship between Diversity and Ecosystem Functions

The case study in Fukang of the Xinjiang Uygur Autonomous Region shows that the finer the spatial resolution, the more ecotopes are found in the given region and the higher the ecotope diversity is on an average. The case study in Bayin Xile Grassland shows that ecotope diversity calculated by scaling the diversity index strictly increases as the resolution increases.

According to the design of the scaling diversity index and fractal theory (Falconer, 2003),

$$\lim_{\varepsilon \to 0} d(\varepsilon, t) = -\lim_{\varepsilon \to 0} \frac{\ln\left(\sum_{i=1}^{m(\varepsilon,t)} \left(p_i(\varepsilon,t)\right)^{1/2}\right)^2}{\ln(\varepsilon)}$$

must have its limit, $D_{1/2}(t)$, when the resolution approaches zero.

Spatial resolution of data has a nonlinear effect on the conclusions of the relationships between ecological diversity and ecosystem functions such as productivity and stability. For a given diversity index, we may get different conclusions on different resolutions of data. On a given resolution or scale of given observation, different diversity indexes might give different conclusions. In other words, debates on relationships between ecological diversity and ecosystem functions are meaningless if they do not make clear what kind of diversity index is used, how large the spatial scale or how high spatial resolution is in terms of the data their calculations are based on.

13.5.4 Species Diversity and Ecotope Diversity

First, we try to answer the question of whether it is necessary to distinguish species diversity into individual-counting diversity and biomass-based diversity, and ecotope diversity into individual ecotope-counting diversity and ecotope-area-based diversity. The analysis results show that individual-counting diversity and biomass-based diversity have no significant correlation in woodland ecosystems. In grassland ecosystems, individual-counting diversity is significantly correlated with biomass-based diversity. Individual ecotope-counting diversity and ecotope-area-based diversity also have a significant correlation. Therefore, it is unnecessary to distinguish ecotope diversity into individual ecotope-counting diversity and ecotope-area-based diversity and to discriminate between individual-counting diversity and biomass-based diversity for grassland ecosystems for studies that have no special requirement for accuracy. But it is necessary in the case of woodland ecosystems to divide species diversity into individual-counting diversity and biomass-based diversity. In conclusion, we find that individual-counting diversity and species-biomass have no significant correlation with ecotope diversity, according to our analysis of the results produced by data sampled in Dongzhi tableland.

References

Alados, C.L., Elaich, A., Papanastasis, V.P., Ozbek, H., Navarro, T., Freitas, H., Vrahnakis, M., Larrosi, D., and Cabezudo, B. 2004. Change in plant spatial patterns and diversity along the successional gradient of Mediterranean grazing ecosystems. *Ecological Modelling* 180: 523–535.

Altieri, M.A. 1991. How best can we use biodiversity in agro-ecosystems? *Outlook on Agriculture* 20: 15–23.

Arrhennius, O. 1921. Species and area. *Journal of Ecology* 9: 95–99.

Baessler, C. and Klotz, S. 2006. Effects of changes in agricultural land-use on landscape structure and arable weed vegetation over the last 50 years. *Agriculture, Ecosystems and Environment* 115: 43–50.

Bakalar, N. 2007. Biodiversity aids productivity. *Discover* 28(1): 53.

Barrett, G.W. 1968. The effects of an acute insecticide stress on a semi-enclosed grassland ecosystem. *Ecology* 49: 1019–1035.

Beeby, A. and Brennan, A.M. 1997. *First Ecology*. London: Chapman & Hall.

Benedetti-Cecchi, L., and Gurevitch, J. 2005. Unanticipated impacts of spatial variance of biodiversity on plant productivity. *Ecology Letters* 8(8): 791–799.

Bengtsson, J., Nilsson, S.G., France, A., and Menozzi, P. 2000. Biodiversity, disturbances, ecosystem function and management of European forests. *Forest Ecology and Management* 132: 39–50.

Berger, W.H. and Parker, F.L. 1970. Diversity of planktonic foraminifera in deep sea sediments. *Science* 168: 1345–1347.

Bisby, F.A. 1995. Characterization of biodiversity. In *Global Biodiversity Assessment*, eds., V.H. Heywood and R.T. Watson, 21–106. Great Britain: Cambridge University Press.

Boyer, K.E., Kertesz, J.S., and Bruno, J.F. 2009. Biodiversity effects on productivity and stability o f marine macroalgal communities: The role of environmental context. *Oikos* 118(7): 1062–1072.

Browne, R.A. 1981. Lakes as islands: Biogeographic distribution, turnover rates, and species composition in the lakes of central New York. *Journal of Biogeography* 8: 75–83.

Burel, F. 1996. Hedgerows and their role in agricultural landscapes. *Critical Reviews in Plant Sciences* 15: 169–190.

Burel, F., Baudry, J., Butet, A., Clergeau, P., Delettre, Y., LeCoeur, D., Dubs, F. et al., 1998. Comparative biodiversity along a gradient of agricultural landscapes. *Acta Oecologica* 19: 47–60.

Chapin III, F.S., Zavaleta, E.S., Eviner, V.T., Naylor, R.L., Vitousek, P.M., Reynolds, H.L., Hooper, D.U. et al., 2000. Consequences of changing biodiversity. *Nature* 405: 234–242.

Chase, J.M. and Leibold, M.A. 2002. Spatial scale dictates the productivity–biodiversity relationship. *Nature* 416: 427–430.

Chen, X., Li., B.L., and Zhang, X. 2008. Using spatial analysis to monitor tree diversity at a large scale: A case study in Northeast China Transect. *Journal of Plant Ecology* 1(2): 137–141.

Costamagma, A.C., Menalled, F.D., and Landis, D.A. 2004. Host density influences parasitism of the armyworm *Pseudaletia unipuncta* in agricultural landscapes. *Basic and Applied Ecology* 4: 347–355.

Cramer, M.J. and Willig, M.R. 2005. Habitat heterogeneity, species diversity and null models. *Oikos* 108: 209–218.

Crawley, M.J. and Harral, J.E. 2001. Scale dependence in plant biodiversity. *Science* 291: 864–868.

Darwin, C. 1872. *The Origin of Species*. Chicago: Thompson and Thomas.

Dauber, J., Hirsch, M., Simmering, D., Waldhardt, R., Otte, A., and Wolters, V. 2003. Landscape structure as an indicator of biodiversity: Matrix effects on species richness. *Agriculture, Ecosystems and Environment* 98: 321–329.

Duelli, P. 1997. Biodiversity evaluation in agricultural landscapes, an approach at two different scales. *Agriculture, Ecosystems and Environment* 62: 81–91.

Elton, C.S. 1958. *The Ecology of Invasions by Animals and Plants*. London: Methuen and Co Ltd.

Enquist, B.J. and Niklas, K.J. 2001. Invariant scaling relations across tree-dominated communities. *Nature* 410: 655–660.

Ewers, R.M. and Rodrigues, A.S.L. 2006. Speaking different languages on biodiversity. *Nature* 443: 506.

Falconer, K. 2003. *Fractal Geometry*. Chichester: Join & Sons Ltd.

Forman, R.T.T. 1995. *Land Mosaics: The Ecology of Landscapes and Regions*. New York, NY: Cambridge University Press.

Gardner, M.R. and Ashby, W.R. 1970. Connections of large dynamic systems: Critical values for stability. *Nature* 228: 784.

Gaston, K.J. 2000. Global patterns in biodiversity. *Nature* 405: 220–227.

Gilpin, M.E. 1975. Stability of feasible predator–prey systems. *Nature* 254: 137–139.

Glowka, L., Burhenne-Guilmin, F., and Synge, H. 1994. *A Guide to the Convention on Biological Diversity*. IUCN-The World Conservation Union. Cambridge, UK: The Burlington Press.

Gorman, M.L. 1979. *Island Ecology*. London: Chapman & Hall.

Greenberg, C.H., Harris, L.D., and Neary, D.G. 1995. A comparison of bird communities in burned and salvage-logged, clearcut, and forested Florida sand pine scrub. *The Wilson Bulletin* 107: 40–54.

Grime, J.P. 1997. Biodiversity and ecosystem function: The debate deepens. *Nature* 277: 1260–1263.

Groombridge, B. 1992. *Global Biodiversity: Status of the Earth's Living Resources*. London: Chapman & Hall.

Haber, W. 1971. Landscaftspflege durch differenzierte Bodennutzung. *Bayerisches Landwirtschaftliches Jahrbuch* 48, 19–35.

Haber, W. 1979. Raumordnungs-Konzepte aus der Sicht der Ökosystemforschung. *Forschungs-und Sitzungsberichte Akademie f. Raumforschung und Landesplanung* 131: 12–24.

Haber, W. 1990. Using landscape ecology in planning and management. In *Changing Landscapes: An Ecological Perspective*, eds., I.S. Zonneveld and R.T.T. Forman, 217–231. New York, NY: Springer-Verlag.

Haberl, H., Schulz, N.B., Plutzar, C., Erb, K.H., Krausmann, F., Loibl W., Moser, D. et al., 2004. Human appropriation of net primary production and species diversity in agricultural landscapes. *Agriculture, Ecosystems and Environment* 102: 213–218.

Haken, H. 1983. *Synergetics*. Berlin: Springer-Verlag.

Hamilton, A.J. 2005. Species diversity or biodiversity? *Journal of Environmental Management* 75: 89–92.

Hanski, I. 2005. Landscape fragmentation, biodiversity loss and the societal response. *EMBO* 6(5): 388–392.

Harper, J.L. and Hawksworth, D.L. 1996. Preface. In *Biodiversity: Measurement and Estimation*, ed., D.L. Hawksworth, 5–12. London: Chapman & Hall.

Hausdorff, F. 1919. Dimension und aeusseres Mass. Mathematische Annalen 19: 157–159.

Hector, A., Schmid, B., Beierkuhnlein, C. et al., 1999. Plant diversity and productivity experiments in European grasslands. *Science* 286: 1123–1127.

Hill, M.O. 1973. Diversity and evenness: A unifying notation and its consequences. *Ecology* 54: 427–431.

Hodgson, J.G. and other 30 authors. 2005. How much will it cost to save grassland diversity? *Biological Conservation* 122: 263–273.

Hoffmann, J. and Greef, J.M. 2003. Mosaic indicators—theoretical approach for the development of indicators for species diversity in agricultural landscapes. *Agriculture, Ecosystems and Environment* 98: 387–394.

Hooper, D.U. and Vitousek, P.M. 1997. The effects of plant composition and diversity on ecosystem processes. *Science* 277: 1302–1305.

Holt, A. 2006. Biodiversity definitions vary within the discipline. *Nature* 444: 146.

Holt, A.R., Gaston, K.J., and He, F. 2002. Occupy–abundance relationships and spatial distribution: A review. *Basic and Applied Ecology* 3: 1–13.

Hoover, S.R. and Parker, A.J. 1991. Spatial components of biotic diversity in landscapes of Georgia, USA. *Landscape Ecology* 5(3): 125–136.

Huston, M. 1994. *Biological Diversity: The Coexistence of Species on Changing Landscapes*. Cambridge, UK: Cambridge University Press.

Huston, M.A., Aarssen, L.W., Austin M.P., and Cade, B.S. 2000. No consistent effect of plant diversity on productivity. *Science* 289: 1255–1257.

Hutchinson, G.E. 1959. Homage to Santa Rosalia or why are there so many kinds of animals? *The American Naturalist* 93: 145–159.

Iannaccone, P.M. and Khokha, M.K. 1996. Fractal geometry. In *Fractal Geometry in Biological Systems*, eds., P.M. Iannaccone and M.K. Khokha, 3–11. New York, NY: CRC Press.

Ibanez, J.J., Caniego, J., San Jose, F., and Carrera, C. 2005, Pedodiversity—area relationships for islands. *Ecological Modelling* 182: 257–269.

Ihse, M. 1995. Swedish agricultural landscapes: Patterns and changes during the last 50 years, studied by aerial photos. *Landscape Urban Plan* 31: 21–37.

Johnson, K.H., Vogt, K.A., Clark, H.J., Schmitz, O.J., and Vogt, D.J. 1996. *Trends in Ecology and Evolution* 11(9): 372–377.

Jongman, R.H.G. and Bunce, R.G.H. 2000. Landscape classification, scales and biodiversity in Europe. In *Consequences of Land Use Changes, Advances in Ecological Sciences 5*, eds., R.H.G. Jongman and Ü. Mander, 11–38. Southampton, Boston: Computational Mechanics Publications.

Jonsen, I.D. and Fahrig, L. 1997. Response of generalist and specialist insect herbivores to landscape spatial structure. *Landscape Ecology* 12: 185–197.

Kienast, F. 1993. Analysis of historic landscape patterns with Geographical Information System: A methodological outline. *Landscape Ecology* 8: 103–118.

Kolman, B. and Trench, W.F. 1971. *Elementary Multivariable Calculus*. New York, NY: Academic Press.

Kostylev, V.E., Erlandsson, J., Ming, M.Y., and Williams, G.A. 2005. The relative importance of habitat complexity and surface area in assessing biodiversity: Fractal application on rocky shores. *Ecological Complexity* 2: 272–286.

Lenz, R.J.M. and Haber, W. 1996. Classification theory of ecological systems: Are the generalizations only the exceptions? *Bulletin of the Ecological Society of America* 77: 62–64.

Lhomme, J.P. and Winkel, T. 2002. Diversity–stability relationships in community ecology: Re-examination of the portfolio effect. *Theoretical Population Biology* 62: 271–279.

Loreau, M. 2000. Biodiversity and ecosystem functioning: Recent theoretical advances. *Oikos* 91: 3–17.

Loreau, M. and Hector, A. 2001. Partitioning selection and complementarity in biodiversity experiments. *Nature* 412: 72–76.

Loreau, M., Naeem, S., Inchausiti, P., Bengtsson, J., Grime, J.P., Hector, A., Hooper, D.U. et al., 2001. Biodiversity and ecosystem functioning: Current knowledge and future challenges. *Science* 294: 804–808.

Luoto, M. 2000. Modelling of rare plant species richness by landscape variables in an agriculture area in Finland. *Plant Ecology* 149: 157–168.

Ma, M.H. 2008. Multi-scale responses of plant species diversity in semi-natural buffer strips to agricultural landscapes. *Applied Vegetation Science* 11(2): 269–278.

MacArthur, R. 1955. Fluctuations of animal populations, and a measure of community stability. *Ecology* 36(3): 533–536.

MacArthur, R.H. and MacArthur, J.W. 1961. On bird species diversity. *Ecology* 58: 594–598.

MacArthur, R., Recher, H., and Cody, M. 1966. On the relation between habitat selection and species diversity. *The American Naturalist* 100: 319–332.

Macarthur, R.H. and Wilson, E.O. 1967. *The Theory of Island Biogeography*. Princeton, NJ: Princeton University Press.

Mace, G.M. 2005. An index of intactness. *Nature* 434: 32–33.

Margalef, R. 1957. La teoria de la informacion en ecologia. *Mem. Real Acad. Ciencias by Artes de Barcelona* 32: 373–449.

Marquard, E., Weigelt, A., Roscher, C., Gubsch, M., Lipowsky, A., and Schmid, B. 2009. Positive biodiversity–productivity relationship due to increased plant density. *Journal of Ecology* 97: 696–704.

Martin, M.A., Pachepsky, Y.A., and Perfect, E. 2005. Scaling, fractals and diversity in soils and ecohydrology. *Ecological Modelling* 182: 217–220.

May, R.M. 1972. Will a large complex system be stable? *Nature* 238, 413–414.

McCann, K.S. 2000. The diversity–stability debate. *Nature* 405: 228–233.

McGrady-Steed, J., Harries, P., and Morin, P.J. 1997. Biodiversity regulates ecosystem predictability. *Nature* 390: 162–165.

Mcintosh, R.P. 1967. An index of diversity and the relation of certain concepts to diversity. *Ecology* 48: 392–404.

McNaughton, S.J. 1978. Stability and diversity of ecological communities. *Nature* 274: 251–253.

McNaughton, S.J. 1994. Biodiversity and function of grazing ecosystems. In *Biodiversity and Ecosystem Function*, eds., E.D. Schulze and H.A. Mooney, 361–383. Berlin: Springer-Verlag.

Miller, J.N., Brooks, R.P., and Croonquist, M.J. 1997. Effects of landscape patterns on biotic communities. *Landscape Ecology* 12: 137–153.

Mladenoff, D.J., Niemi, G.J., and White, M.A. 1997. Effects of changing landscape pattern and USGS land-cover data variability on eco region discrimination across a forest-agriculture. *Landscape Ecology* 12: 379–396.

Mueller, C., Berger, G., and Glemnitz, M. 2004. Quantifying geomorphological heterogeneity to assess species diversity of set-aside arable land. *Agriculture, Ecosystems and Environment* 104: 587–594.

Naeem, S. 2002. Biodiversity equals instability? *Nature* 416: 23–24.

Naeem, S. and Li, S. 1997. Biodiversity enhances ecosystem reliability. *Nature* 390: 507–509.

Naveh, Z. and Lieberman, A. S. 1994. *Landscape Ecology: Theory and Application*. New York, NY: Springer-Verlag.

Noordwijk, M.V. 2002. Scaling trade-offs between crop productivity, carbon stocks and biodiversity in shifting cultivation landscape mosaics: The FALLOW model. *Ecological Modelling* 149: 113–126.

Odum, E.P. 1953. *Fundamentals of Ecology*. Philadelphia: Saunders College Publishing.

Odum, E.P. 1969. The strategy of ecosystem development. *Science* 164: 262–270.

Odum, E.P. 1971. *Fundamentals of Ecology*. Philadelphia: W. B. Saunders Company.

Odum, E. P. 1983. *Basic Ecology*. Philadelphia: Saunders College Publishing.

Organization for Economic Co-operation and Development (OECD). 1997. *Environmental Indicators for Agriculture*. Paris: OECD Publications.

Overgaard, H.J., Ekbom, B., Suwonkerd, W., and Takagi, M. 2003. Effect of landscape structure on anopheline mosquito density and diversity in northern Thailand: Implications for malaria transmission and control. *Landscape Ecology* 18: 605–619.

Palma, J.H.N., Graves, A.R., Burgess, P.J., Keesman, K.J., van Keulen, H., Mayus, M., Reisner, Y., and Herzog, F. 2007. Methodological approach for the assessment of environmental effects of agroforestry at the landscape scale. *Ecological Engineering* 29: 450–462.

Pennist, E. 1994. Biodiversity helps keep ecosystems healthy. *Science News* 145: 84.

Peters, J. 1975. Entropy and information: Conformities and controversies. In *Entropy and Information in Science and Philosophy*, eds., L. Kubat and J. Zeman, 61–81. New York, NY: Elsevier Scienctific Publishing Company.

Pfisterer, A.B. and Schmid, B. 2002. Diversity-dependent production can decrease the stability of ecosystem functioning. *Nature* 416: 84–86.

Pielou, E.C. 1966. Species diversity and pattern diversity in the study of ecological succession. *Journal of Theoretical Biology* 10: 370–383.

Pimm, S.L. 1994. Biodiversity and the balance of nature. In *Biodiversity and Ecosystem Function*, eds., E.D. Schulze and H.A. Mooney, 347–359. Berlin: Springer-Verlag.

Poschlod, P. and Bonn, S. 1998. Changing dispersal processes in the central European landscape since the last ice age: An explanation for the actual decrease of plant species richness in different habitats? *Acta Botanica Neerlandica* 47(1): 27–44.

Preston, F.W. 1960. Time and space and the variation of species. *Ecology* 41: 785–790.

Preston, F.W. 1962. The canonical distribution of commonness and rarity. *Ecology* 43: 185–215 and 410–432.

Prinsley, R.T. 1992. The role of trees in sustainable agriculture—an overview. *Agroforestry Systems* 20: 87–115.

Purvis, A. and Hector, A. 2000. Getting the measure of biodiversity. *Nature* 405: 212–219.

Ricotta, C. 2002. Bridging the gap between ecological diversity indices and measures of biodiversity with Shannon's entropy: Comment to Izsák and Papp. *Ecological Modelling* 152: 1–3.

Ricotta, C., Chiarucci, A., and Avena, G. 2004, Quantifying the effects of nutrient addition on community diversity of serpentine vegetation using parametric entropy of type α. *Acta Oecologica* 25: 61–65.

Ripl, W. 1995. Management of water cycle and energy flow for ecosystem control: The energy-transport-reaction (ETR) model. *Ecological Modelling* 78: 61–76.

Ritchie, M.E. and Olff, H. 1999. Spatial scaling laws yield a synthetic theory of biodiversity. *Nature* 400: 557–560.

Roper-Lindsay, J., Simmons, E., Solon, J., Jongman, R., Degorski, M., and Miller, C. 2003. Biodiversity and landscape diversity. In *Multifunctional Landscapes, Vol. II, Monitoring, Diversity and Management*, eds. J. Brandt and H. Vejre, 155–159. Southampton, UK: WIT Press.

Rosenzweig, M.L. 1995. *Species Diversity in Space and Time*. Great Britain: Cambridge University Press.

Roy, P.S., Padalia, H., Chauhan, N., Porwal, M.C., Gupta, S., Biswas, S., and Jagdale, R. 2005. Validation of geospatial model for biodiversity characterization at landscape level: a study in Andaman & Nicobar Islands, India. *Ecological Modelling* 185, 349–369.

Roy, A., Tripathi, S.K., and Basu, S.K. 2004. Formulating diversity vector for ecosystem comparison. *Ecological Modelling* 179: 499–513.

Rusch, G.M. and Oesterheld, M. 1997. Relationship between productivity, species and functional group diversity in grazed and nongrazed Pampas grassland. *Oikos* 76: 519–526.

Sandstroem, U.G., Angelstam, P., and Mikusinski, G. 2006. Ecological diversity of birds in relation to the structure of urban green space. *Landscape and Urban Planning* 77(1–2): 39–53.

Schmid, B. 2002. The species richness–productivity controversy. *TRENDS in Ecology and Evolution* 17(3): 113–114.

Scholes, R.J. and Biggs, R. 2005. A biodiversity intactness index. *Nature* 434: 45–49.

Shannon, C.E. and Weaver, W. 1962. *The Mathematical Theory of Communication*. Urbana, IL: The University of Illinois Press.

Simpson, E. H. 1949. Measurement of diversity. *Nature* 163: 688.

Striebel, M., Behl, S., and Stibor, H. 2009. The coupling of biodiversity and productivity in phytoplankton communities: Consequences for biomass stoichiometry. *Ecology* 90(8): 2025–2031.

Therriault, T.W. and Kolasa, J. 2000. Explicit links among physical stress, habitat heterogeneity and biodiversity. *Oikos* 89: 387–391.

Thies, C. and Tscharntke, T. 1999. Landscape structure and biological control in agroecosystems. *Science* 285: 893–895.

Tilman, D. 1982. *Resource Competition and Community Structure*. Princeton, NJ: Princeton University Press.

Tilman, D. 2000. Causes, consequences and ethics of biodiversity. *Nature* 405: 208–211.

Tilman, D., Knops, J., Wedin, D., Reich P., Ritchie, M., and Siemann, E. 1997. The influence of functional diversity and composition on ecosystem processes. *Science* 277: 1300–1302.

Tilman, D., Naeem, S., Knops, J., Reich, P., Siemann, E., Wedin, D., Ritchie, M., et al. 1997. Biodiversity and ecosystem properties. *Science* 278: 1865–1869.

Tilman, D., Reich, P.B., and Knops, J.M.H. 2006. Biodiversity and ecosystem stability in a decade long grassland experiment. *Nature* 441: 629–632.

Tilman, D., Reich, P.B., Knops, J., Wedin, D., Mielke, T., and Lehman, C. 2001. Diversity and productivity in long-term grassland experiment. *Science* 294: 843–845.

Uys, R.G., Bond, W.J., and Everson, T.M. 2004. The effect of different fire regimes on plant diversity in southern African grasslands. *Biological Conservation* 118: 489–499.

Veech, J.A., Summerville, K.S., Crist, T.O., and Gering, J.C. 2002. The additive partitioning of species diversity: Recent revival of an old idea. *Oikos* 99: 3–9.

Vivian-Smith, G. 1997. Microtopographic heterogeneity and floristic diversity in experimental wetland communities. *Journal of Ecology* 85: 71–82.

Vollhardt, I.M.G., Tscharntke, T., Waeckers, F.L., Bianchi, F.J.J.A., and Thies, C. 2008. Diversity of cereal aphid parasitoids in simple and complex landscapes. *Agriculture, Ecosystems and Environment* 126: 289–292.

Wagner, H.H., Wildi, O., and Ewald, K.C. 2000. Additive partitioning of plant species diversity in an agricultural mosaic landscape. *Landscape Ecology* 15: 219–227.

Watt, K.E.F. 1973. *Principles of Environmental Science*. New York, NY: McGraw-Hill.

Weibull, A.C., Bengtsson, J., and Nohlgren, E. 2000. Diversity of butterflies in the agricultural landscape: The role of farming system and landscape heterogeneity. *Ecography* 23: 743–750.

Whittaker, R.H. 1972. Evolution and measurement of species diversity. *Taxon* 21: 213–251.

Whittaker, R.H. 1977. Evolution of species diversity in land communities. In *Evolutionary Biology* 10, eds., M.K. Hecht, W.C. Steere and B. Wallace, 1–67. New York, NY: Plenum.

Whittaker, R.J. 1999. Scaling, energetics and diversity. *Nature* 401: 865–866.

Williamson, M. 1981. *Island Population*. Oxford: Oxford University Press.

Williams, P.H. and Humphries, C.J. 1996. Comparing character diversity among biota. In *Biodiversity: A Biology of Numbers and Difference*, ed., K.J. Gaston, 54–76. Cambridge: Blackwell Science.

Wolfe, M.S. 2000. Crop strength through diversity. *Nature* 406: 681–682.

Woodward, F.I. 1994. How many species are required for a functional ecosystem? In *Biodiversity and Ecosystem Function*, eds., E.D. Schulze and H.A. Mooney, 271–291. Berlin: Springer-Verlag.

Yoshida, K. 2003. Evolutionary dynamics of species diversity in an interaction web system. *Ecological Modelling* 163: 131–143.

Yue, T.X., Haber, W., Grossmann, W.D., and Kasperidus, H.D. 1998, Towards the satisfying models for biological diversity. *Ekologia (Bratislava)* 17: 129–141.

Yue, T.X., Liu, J.Y., Chen, S.Q., Li, Z.Q., Ma, S.N., Tian, Y.Z., and Ge, F. 2005. Considerable effects of diversity indices and spatial scales on conclusions relating to ecological diversity. *Ecological Modelling* 188: 418–431.

Yue, T.X., Liu, J.Y., Jørgensen, S.E., and Ye, Q.H. 2003. Landscape change detection of the newly created wetland in Yellow River Delta. *Ecological Modelling* 164: 21–31.

Yue, T.X., Liu, J.Y., Jørgensen, S.E., Gao, Z.Q., Zhang, S.H., and Deng, X.Z. 2001. Changes of Holdridge life zone diversity in all of China over a half century. *Ecological Modelling* 144: 153–162.

Yue, T.X., Ma, S.N., Wu, S.X., and Zhan, J.Y. 2007. Comparative analyses of the scaling diversity index and its applicability. *International Journal Remote Sensing* 28(7): 1611–1623.

Yue, T.X., Xu, B., and Liu, J.Y. 2004. A patch connectivity index and its change on a newly born wetland at the Yellow River Delta. *International Journal of Remote Sensing* 25(21): 4617–4628.

Zamora, J., Verdu, J.R., and Galante, E. 2007. Species richness in Mediterranean agroecosystems: Spatial and temporal analysis for biodiversity conservation. *Biodiversity Conservation* 134(1): 113–121.

Zechmeister, H.G. and Moser, D. 2001. The influence of agricultural land-use intensity on bryophyte species richness. *Biodiversity and Conservation* 10: 1609–1625.

14

Change Detection

14.1 Introduction

One of the major applications of remotely sensed data obtained from Earth-orbiting satellites is change detection (Anderson, 1977; Nelson, 1983). Change detection, defined as the process of identifying differences in the state of an object or phenomenon by observing it at different times (Singh, 1989), is useful in extracting environmental changes.

Change detection was first used to detect changes in position by applying, the "blink" principle (Menzel, 1970). This approach was applied to detect differences in multitemporal Landsat MSS images (Masry et al., 1975). Since then, many approaches to change detection have been developed. For instance, postclassification methods (Weismiller et al., 1977) were successfully applied to regional deforestation studies (Skole and Tucker, 1993). Composite analysis was the earliest semiautomated computer-assisted approach used to generate land-cover and land-use change maps from satellite data (Weismiller et al., 1977). Image differencing was based on image subtraction and threshold (Nelson, 1983). Principle components analysis was a powerful data transformation technique for information extraction in multidimensional remote sensing data (Lillesand and Kiefer, 1979). Change vector analysis (Engvall et al., 1977) was developed for application to land-cover changes by using MODIS (Lunetta, 1998). Fuzzy set operation combined change information from different image channels into a single-image channel (Gong, 1993). Spectral mixture analysis was introduced for high spectral resolution images (Adams et al., 1995). A computational model motivated by human cognitive processing and selective attention was proposed for detecting critical changes in environments (Fang et al., 2003). It was found that the temporally invariant cluster was a simple, effective and repeatable method to create radiometrically comparable data sets for the remote detection of landscape change (Chen et al., 2005). A digital technique for change detection, named radiometric rotation controlled by no-change axis (RCNA), was proposed for allowing the use of satellite images without complex atmospheric corrections of similar spectral bands by combining information from field surveys (Maldonado et al., 2007). Landsat TM and SRTM-DEM derived variables were integrated with decision tree classifiers to detect biodiversity and ecosystem services in complex wet tropical environments (Sesnie et al., 2008). A rank-order change detection approach was developed for applications involving multitemporal data sets where problems may exist due to image normalization, cross-sensor radiometric calibration, or unavailability of a desired sensor type (Wulder et al., 2008). A method for the timely extraction of reliable land-cover change information was developed on the basis of cross-correlogram spectral matching (CCSM) with the aim of identifying interannual land-cover changes from time series NDVI data (Wang et al., 2009). The change detection analysis in natural resource remote sensing projects can be summarized by four general steps: image and reference data acquisition, preprocessing, analysis, and evaluation (Kennedy et al., 2009). A new method was proposed for the change detection of buildings in urban environments from very high spatial resolution images (VHSR) using existing digital cartographic data (Bouziani et al., 2010).

In this chapter, we detect landscape change by combining the scaling diversity index, the patch connectivity index, the model for mean center of land-cover, and the model for human impact intensity. A curve-theorem-based method is developed to detect land-cover change in the Yellow River Delta.

14.2 Landscape Change Detection

Landscape changes can be distinguished into conversions from one land-cover type into another and transformations within a given land-cover type. Land-cover mapping represents a major step forward in the application of remote sensing data to landscape change detection (Oetter et al., 2001). The first coarse-scale land-cover map was created in 1991, and developed by composite data sets of AVHRR polar orbiting satellites and NDVI composites of the EROS Data Center (Loveland et al., 1991). In order to generate products of landscape change detection, the North American Landscape Characterization program began to compile MSS data sets for the 1970s, 1980s, and 1990s in 1992 (Lunetta et al., 1993). In 2000, a land-cover and land-use database was compiled by the Multiresolution Land Characteristics Program.

We will detect the landscape change of the newly created wetland of the Yellow River Delta by generating a land-cover database.

14.2.1 Newly Created Wetlands of the Yellow River Delta

The newly created wetland of the Yellow River Delta consists of the mouth of the Yellow River, currently on course to the Bohai sea, and the area close to the mouth (Figure 14.1). It is located in 37°35–37°54′N and 118°43–119°20′E. The current course of the Yellow River was formed artificially by changing its previous course from the Diaokou river to the Qingshui gully in 1976. Due to the deposition of a large amount of sand and mud transported by the

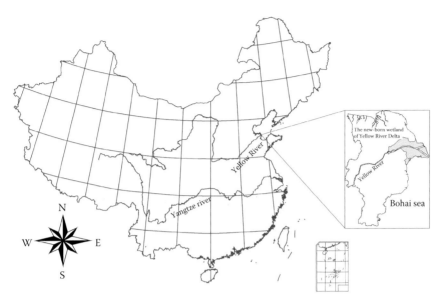

FIGURE 14.1
Location of the newly created wetland of the Yellow River Delta.

Yellow River, the area of newly created wetland enlarges by 32.4 sq. km each year (Zhao, 1997). The salt content in the soil of the newly deposited land is more than 3%, on which *Suaeda salsa* Pall. is partially distributed. These increase organic matter in the soil, which transforms the area into *Tamarix chinensis* Lour or *Aeluropus liitorallis* (Gouan) Parl. var. *sinensis* Debeanx Land. With *Tamarix chinensis* Lour and *Aeluropus liitorallis* (Gouan) Parl. var. *sinensis* Debeanx secreting salt, and their dead branches and leaves accumulating in the soil, the salt content of the soil is reduced and its fertility is increased, which turns the area into rank grassland. The low-lying land evolves into *Phragmitas communis* Trin. Land. With the gradual raise of terrain and drench of rainfall, the land is desalted and evolves into land suitable for forestry and agriculture (Tian et al., 1999). In other words, the low-lying land along the Yellow River and its gullies has full moisture content and damp soil, in which *Phragmitas communis* Trin. is dominant. The beach near the sea has a low and flat terrain subject to sea tide erosion, in which the salt content of soil is higher and *Suaeda salsa* Pall. and *Tamarix chinensis* Lour are dominant. The higher land has lower salt content, in which *Aeluropus liitorallis* (Gouan) Parl. var. *sinensis* Debeanx, *imperata cylindrica* (L.) Beauv. var. *major* (Nees) C. B. Hubb. and *Setaria viridis* (L.) Beauv. are dominant. The area where the salt content of soil is lower than 0.3% has now been partially cultivated into artificial forestland, wheat land, and soybean land (Figure 14.2).

The newly created wetland of the Yellow River Delta is an ecosystem typical of littoral wetland in estuaries, containing rich wetland vegetation, and hydrobios. It is an important transfer station, wintering habitat, and breeding

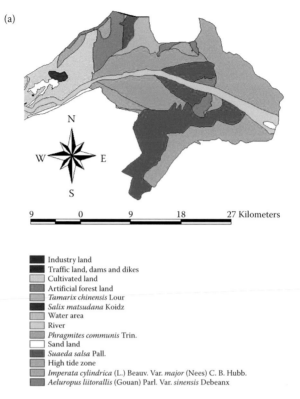

(a)

Industry land
Traffic land, dams and dikes
Cultivated land
Artificial forest land
Tamarix chinensis Lour
Salix matsudana Koidz
Water area
River
Phragmites communis Trin.
Sand land
Suaeda salsa Pall.
High tide zone
Imperata cylindrica (L.) Beauv. Var. *major* (Nees) C. B. Hubb.
Aeluropus liitorallis (Gouan) Parl. Var. *sinensis* Debeanx

FIGURE 14.2
Landscape of the newly created wetland of the Yellow River Delta in: (a) 1984.

farm for birds in the northeast Asian inland and in the Pacific area. The newly created wetland of the Yellow River Delta has three obvious characteristics: (1) it is one of the fastest-growing wetland ecosystems in estuary in the world; (2) its vegetation is in the early stages of development; and (3) with land extending into the Yellow Sea and vegetation developing toward the seashore, there is frequent succession among various vegetation communities. Because of these characteristics, the newly created wetland is very fragile.

14.2.2 Methods

14.2.2.1 Patch Connectivity Index

Since Menger's Theorem came into being in 1927, connectivity has become one of the most important aspects of graph theory. Since the early 1960s, connectivity has evolved into many research fields as a mathematical tool, and has solved a wide variety of problems. In the early 1980s, the term connectivity was first applied to studies of landscape ecology (Merriam, 1984; Risser et al., 1984).

(b)

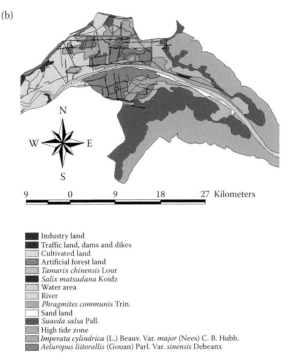

9   0   9   18   27 Kilometers

- Industry land
- Traffic land, dams and dikes
- Cultivated land
- Artificial forest land
- *Tamarix chinensis* Lour
- *Salix matsudana* Koidz
- Water area
- River
- *Phragmites communis* Trin.
- Sand land
- *Suaeda salsa* Pall.
- High tide zone
- *Imperata cylindrica* (L.) Beauv. Var. *major* (Nees) C. B. Hubb.
- *Aeluropus liitorallis* (Gouan) Parl. Var. *sinensis* Debeanx

FIGURE 14.2
(*Continued*) (b) 1991.

Haber and his institute adopted connectivity as an important research step in their holistic approach to land planning (Haber, 1984, 1986, 1987, 1988, 1990; Haber and Burhardt, 1986, 1988). Connectivity was regarded as a parameter of landscape function that measures the process by which sub-populations of a landscape are interconnected into a demographic function unit (Baudry and Merriam, 1988; Mcdonnell and Pickett, 1988). Risser et al. (1984) stated that the movement of migrants or propagules among elements of the land mosaic could be expressed in graph theoretic measures of connectivity. Forman and Godron (1986) defined it as a measure of how connected or spatially continuous a corridor is in terms of the mathematical concept of connectivity in topology, known as the concept of network connectivity. Schreiber (1988) stated that connectivity in landscape ecology includes the entire complex of relationships in and between ecological systems, that is, not only the interrelationships in communities and between organisms, but also the network of interactions and flows between the biotic and abiotic compartments of an ecosystem. Janssens and Gulinck (1988) believed that connectivity in landscape is a combination of contiguity and proximity. Haber (1990) stated that connectivity is an assessment of spatial interrelations among all ecotope types or ecotope assemblages of a regional natural unit, with special emphasis on connectedness and mutual dependence.

(c)

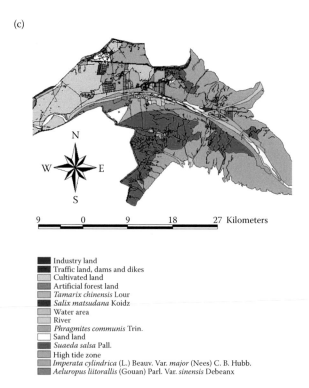

Industry land
Traffic land, dams and dikes
Cultivated land
Artificial forest land
Tamarix chinensis Lour
Salix matsudana Koidz
Water area
River
Phragmites communis Trin.
Sand land
Suaeda salsa Pall.
High tide zone
Imperata cylindrica (L.) Beauv. Var. *major* (Nees) C. B. Hubb.
Aeluropus liitorallis (Gouan) Parl. Var. *sinensis* Debeanx

FIGURE 14.2
(*Continued*) (c) 1996.

Taylor et al. (1993) defined landscape connectivity as the degree to which the landscape facilitates or impedes movement among resource patches. Forman (1995) defined connectivity as a measure of how connected or spatially continuous a corridor, network, or matrix is. With et al. (1997) described landscape connectivity as the functional relationship among habitat patches, owing to the spatial contagion of habitat and the movement responses of organisms to the landscape structure.

The models for landscape connectivity can be correspondingly classified into those used for line connectivity (Menger, 1927; Swart and Lawes, 1996), those used for vertex connectivity (Garrison, 1960; Taaffe and Gauthier, Jr. 1973; Lowe and Moryadas, 1975), those used for network connectivity (Shimbel, 1953; Shimbel and Katz, 1953; Haggett and Chorley, 1969) and those used for patch connectivity. The first three have long been studied. Mladenoff et al. (1997) applied the gravity model to an index for the connectivity of forest patches. In our case, the edge-to-edge distance of patches is either equal to zero or too small to be put on the denominator of the formula. Thus, we introduce another index of patch connectivity.

Some animals such as small herbivores primarily move in habitat interiors. A variety of carnivores and mammalian large herbivores prefer to move along forest–grassland edges. Winds may lift seeds and deposit them along these edges. According to Forman (1995), a compact patch should contain higher species richness than an elongated patch with its fewer interior species, if the substrate is homogeneous. It is widely considered to be true that an elongated patch is less effective in conserving internal resources and species than a round patch. In other words, the perimeter and the maximum radial distance of a polygon are important parameters of landscape.

Patch connectivity can be defined as the average movement efficiency of migrants or propagules in the patches of the region being considered. The efficiency of movement starting from a competing center can be assessed by measuring the distance from the center to the outlying parts within the territory (Haggett and Chorley, 1969). Efficiency of boundaries can be measured by the length of the territory's perimeter. For these minimum energy (or shortest distance) criteria, Coxeter (1961) proposed three geometrical principles: (1) regular polygons are more economical shapes than irregular polygons; (2) circles are the most economical of the regular polygons; and (3) hexagons are the regular polygons that allow the greatest amount of packing into an area, consistent with minimizing movement and boundary costs (minimizing distance). For instance, if an individual moves from the center to the furthest point, within a regular square, with an area of 1 sq. km, the distance is 0.7071 km; if an individual moves from the center to the furthest point, within a rectangle, with an area of 1 sq. km, of which two sides are three times as long as the others, the distance is 1.291 km.

For a regular k-gon such as a triangle (3-gon), square (4-gon), pentagon (5-gon) or hexagon (6-gon), if r_k is the radius of the k-gon, the maximum radial distance, the area A_k and the perimeter Pr_k can be respectively formulated as (Yue et al., 2003, 2004)

$$
\begin{aligned}
A_k &= k \cdot \left(r_k \cdot \sin\left(\frac{360}{2k} \cdot \frac{\pi}{180}\right) \cdot r_k \cdot \mathrm{con}\left(\frac{360}{2k} \cdot \frac{\pi}{180}\right)\right) \\
&= \frac{k \cdot r_k^2}{2} \cdot \sin\left(\frac{2\pi}{k}\right)
\end{aligned}
\tag{14.1}
$$

$$
\begin{aligned}
\mathrm{Pr}_k &= 2 \cdot k \cdot r_k \cdot \sin\left(\frac{360}{2k} \cdot \frac{\pi}{180}\right) \\
&= 2 \cdot k \cdot r_k \cdot \sin\left(\frac{\pi}{k}\right)
\end{aligned}
\tag{14.2}
$$

$$
A_\infty = \lim_{k \to \infty} A_k = \pi \cdot r_\infty^2
\tag{14.3}
$$

$$
\mathrm{Pr}_\infty = \lim_{k \to \infty} \mathrm{Pr}_k = 2 \cdot \pi \cdot r_\infty
\tag{14.4}
$$

When $A_k = 1$,

$$r_k = \left(\frac{2}{k}\right)^{1/2} \cdot \left(\sin\left(\frac{2\pi}{k}\right)\right)^{-1/2} \tag{14.5}$$

$$\text{Pr}_k = 2 \cdot \sin\left(\frac{\pi}{k}\right) \cdot (2k)^{1/2} \cdot \left(\sin\left(\frac{2\pi}{k}\right)\right)^{-1/2} \tag{14.6}$$

Obviously, the perimeter is directly proportional to the maximum radial distance; and the perimeter and maximum radial distance are inversely proportional to the number of the sides.

Combining the information above, we take hexagon (6-gon) as a comparative standard and introduce an index of movement efficiency of migrants or propagules in the patch (i, j),

$$S_{ij}(t) = \frac{8\sqrt{3} \cdot A_{ij}(t)}{(\text{Pr}_{ij}(t))^2} \tag{14.7}$$

where A_{ij} and Pr_{ij} are the area and the perimeter of the jth patch in the ith land-cover type, respectively; t is the variable of time. The coefficient $8\sqrt{3}$ is the ratio of the square of perimeter to the area of a hexagon. When it is a circle, S_{ij} has a maximum value of $(2\sqrt{3}/\pi)$, approximately 1.1.

The model for patch connectivity can be formulated as

$$CO(t) = \sum_{i=1}^{m(t)} \sum_{j=1}^{n_i(t)} p_{ij}(t) \cdot S_{ij}(t) \tag{14.8}$$

where p_{ij} is the proportion of the area of the jth patch in the ith type to the total area under investigation; t is the variable of time. When all patches have the shape of a hexagon (6-gon), $CO(t) = 1.0$.

If it is necessary to consider the degree of difficulty or the ease of movement in the patches, we only need introduce a corresponding parameter. In other words, we can formulate the model as follows:

$$CO(t) = \sum_{i=1}^{m(t)} \sum_{j=1}^{n_i(t)} df_{ij}(t) \cdot p_{ij}(t) \cdot S_{ij}(t) \tag{14.9}$$

where $df_{ij}(t)$ is the degree of difficulty or ease of movement for animal migrants or plant propagules in the jth patch of the ith type.

14.2.2.2 Diversity Index

The diversity index has been is expressed in Chapter 13 as

$$d(t) = -\frac{\ln\left(\sum_{i=1}^{m(\varepsilon)} (p_i(t))^{\frac{1}{2}}\right)^2}{\ln(\varepsilon)} \tag{14.10}$$

where $p_i(t)$ is the probability of the ith ecotope; $m(\varepsilon)$ is the total number of the investigated ecotopes; t represents the variable of time; $\varepsilon = 1/(e + A)$; A is the area of the studied region measured by hectare; and e equals 2.71828.

This index is applicable to various spatial scales such as species, ecotopes and Holdridge life zones (HLZ). In this section, the model is used on the ecotope observation scale. The term ecotope is used instead of the term landscape, so that the term is comparable with the terms of species and HLZ.

14.2.2.3 Model for Human Impact Intensity

The model for human impact intensity is formulated as (Yue et al., 1998)

$$HU(t) = \sum_{i=1}^{m(t)} p_i(t) \cdot h_i(t) \tag{14.11}$$

where $p_i(t)$ is the proportion of the area of land-cover type i; $h_i(t)$ is the impact intensity of human activities on the land-cover type i; t is the variable of time; $m(t)$ is the total number of land-cover types.

14.2.2.4 Model for the Mean Center of Land Cover

The model of the mean center was introduced in the US census of population in 1870 (Shaw and Wheeler, 1985). Such spatial statistics have been used in geography since the 1950s (Hart, 1954; Warntz and Neft, 1960; Ebdon, 1978; Shaw and Wheeler, 1985; Gao et al., 1998). The major applications were those concerned with central tendency, especially the weight mean center of spatial distribution. The model is formulated as

$$x_j(t) = \sum_{i=1}^{I_j} \frac{s_{ij}(t) \cdot X_{ij}(t)}{S_j(t)} \tag{14.12}$$

$$y_j(t) = \sum_{i=1}^{I_j} \frac{s_{ij}(t) \cdot Y_{ij}(t)}{S_j(t)} \tag{14.13}$$

where t is the variable of time; $I_j(t)$ is the patch number of land-cover type j; $s_{ij}(t)$ is area of ith patch of land-cover type j; $S_j(t)$ is the total area of land-cover type j; $(X_{ij}(t), Y_{ij}(t))$ is the longitude and latitude coordinate of the geometric center of the ith patch of land-cover type j; $(x_j(t), y_j(t))$ is the mean center of the land-cover type j.

14.2.3 Materials and Results

14.2.3.1 Data Acquisition

In general, the data from 1984, 1991, and 1996 were produced using processing Landsat TM images of the newly created wetland in all four seasons of the year. First, the synthetic images were made into templates by unsupervised classification. Then, the training samples selected in the newly created wetland were added to the templates, and the templates were edited and evaluated until their recognition was satisfactory. Finally, the image was zoned and classified by means of maximum likelihood classification. In other words, the unsupervised and the supervised classifications were jointly used in order to improve classification accuracy.

The different growing seasons of various plants in the newly created wetland led to a seasonal change in land cover. Different plants growing in a single land unit led to mixed pixels. As a result, the spectral characteristics of every land unit are different, and varying landscape types cannot easily be identified by means of spectral characteristics alone. However, it is very easy to differentiate artificial cultivation traces, and farmland division is distinct. Farmland is first divided by visual interpretation. Agricultural land-use types are interpreted according to the spectral characteristics, the way in which a crop is growing, and the growing season. The farmland feature is extracted and made to mask. The farmland feature is then taken from an image from the period when its major natural vegetation was growing well. Combining with investigation on the spot, major quadrate selection and template edition, images from various periods are classified by ISODATA cluster analysis. The results interpreted primarily by computer are finally improved by a second investigation, using information on the texture of the images, the interrelation of surface features, geographical law, and so on.

14.2.3.2 Results

By operating the models corresponding to formulation (Equations 14.9 through 14.11) on the data from Figure 14.2, the following results can be reached (Table 14.1):

$$CO(t) = -1.3049 \ d(t) + 0.6322 \qquad (14.14)$$

$$CO(t) = -1.5134 \ HU(t) + 0.5522 \qquad (14.15)$$

TABLE 14.1

Change of Patch Connectivity and Ecotope Diversity of the Newly Created
Wetland of Yellow River Delta

Period	1984	1991	1996
Patch connectivity $CO(t)$	0.3258	0.1959	0.0113
Ecotope diversity d(t)	0.2791	0.2880	0.4779
Human impact intensity $HU(t)$	0.1384	0.2750	0.3291

where t is the variable of time; $CO(t)$ is the index of patch connectivity; d(t) is the index of ecotope diversity; and $HU(t)$ is the index of human impact intensity. The correlation coefficient between $CO(t)$ and d(t) is −0.9272 and the one between $CO(t)$ and $HU(t)$ is −0.9411.

The research results show that human impact intensity and ecotope diversity had an increasing trend in the newly created wetland of the Yellow River Delta during the period from 1984 to 1996, while patch connectivity had a decreasing trend. The main driving forces of patch connectivity change were rapid population growth, especially nonpeasant population growth in the Yellow River Delta, and the environmental evolution of the newly created wetland.

In 1984, the total population of the Yellow River Delta was 1.4275 million and the ratio of nonpeasant population to total population was 14.76%, of which the annual growth rate was 4.12% on an average from 1978 to 1984. In 1991, the total population was 1.5852 million and the ratio of nonpeasant population was 23.06%, of which the annual growth rate was 10.5% on an average from 1984 to 1991. In 1996, the total population reached 1.6577 million and the ratio of nonpeasant population was 39.83%, of which the annual growth rate of nonpeasant population was 16.12% from 1991 to 1996 on an average. The total staff and workers of the Shenli oil fields accounted for 72.7% of the nonpeasant population in the Yellow River Delta. Development of the Shenli oil fields greatly promoted economic growth, but disturbed the landscape of the newly created wetland. Since the development of the Shenli oil fields about 30 years, oil and gas wells can be found scattered all over the Yellow River Delta.

Owing to the effects of both human activity and natural environmental evolution, the weighted mean centers of all land-cover types saw a great change. From 1984 to 1991, the weighted mean center of spatial distribution in the high-tide zone moved 8015 m toward the southeast; the zone mainly consisting of *Suaeda salsa* Pall., and *Tamarix chinensis* Lour moved 2442 m toward the southeast, with land extending into the Yellow Sea; the zone mainly consisting of *Aeluropus liitorallis* (Gouan) Parl. var. *sinensis* Debeanx, and *imperata cylindrica* (L.) Beauv. var. *major* (Nees) C. B. Hubb. moved 3854 m toward the northwest; the zone mainly consisting of *Phragmitas communis* Trin. moved 5076 m toward the southeast, distributing across low-lying land along the course of the Yellow River; the forestland zone moved 1711 m toward the southeast; the river, and water zone moved 2442 m toward the southeast; the farmland zone moved 3854 m toward the northeast; and the traffic, industrial and residential zones moved 9794 m toward the northeast. From 1991 to 1996, the weighted mean center of spatial distribution in the high-tide zone moved 5134 m toward the southeast; the zone mainly consisting of *Suaeda salsa* Pall.–*Tamarix chinensis* Lour community moved 1742 m toward the southwest; the zone mainly consisting of *Aeluropus liitorallis* (Gouan) Parl. var. *sinensis* Debeanx–*imperata cylindrica* (L.) Beauv. var. *major* (Nees) C. B. Hubb. moved 779 m toward the northwest; the zone mainly consisting of *Phragmitas communis* Trin. moved 2053 m toward the southeast; the forestland zone moved 4986 m toward the northeast; the river and water zone moved 2372 m toward the southeast; the farmland zone moved 446 m toward the northwest; and the traffic, industrial, and residential zones moved 2663 m toward the northeast (Table 14.2).

TABLE 14.2

Central Tendency of Spatial Distribution of Each Land-Cover Type in the Newly Created Wetland of Yellow River Delta

Land-Cover Type	From 1984 to 1991		From 1991 to 1996	
	Movement Distance of the Mean Center (m)	Movement Direction	Movement Distance of the Mean Center (m)	Movement Direction
High-tide zone	8015	Toward southeast	51,342	Toward southeast
Suaeda salsa Pall.–*Tamarix chinensis* Lour community	2442	Toward southeast	1742	Toward southwest
Aeluropus liitorallis (Gouan) Parl. var. *sinensis* Debeanx–*imperata cylindrica* (L.) Beauv. var. *major* (Nees) C. B. Hubb. community	3854	Toward northwest	779	Toward northwest
Phragmitas communis Trin. community	5076	Toward southeast	2053	Toward southeast
Forestland	1711	Toward southeast	4986	Toward northeast
River and water area	2442	Toward southeast	2372	Toward southeast
Farmland	3854	Toward northeast	446	Toward northwest
Industrial and traffic land and residential area	9794	Toward northeast	2663	Toward northeast

14.3 Change Detection of Land Cover in the Yellow River Delta

Remote sensing has been used to estimate the extent of specific land-cover types and to detect changes in land cover that have occurred in the past. For instance, Hall et al. (1991) detected large-scale patterns of forest succession. Bauer et al. (1994) detected forest cover in Minnesota. Stone et al. (1994) mapped the vegetation of South America. Fuller et al. (1994) detected changes in land cover in Great Britain. Jensen et al. (1995) detected inland wetland changes in the Everglades Water Conservation in Florida. Landsat data were used to estimate vegetation types and wildlife habitats and to detect changes in land cover in the Northern Forest region of Vermont, New Hampshire, and western Maine (Miller et al., 1998; Bryant et al., 1993). Sobrino and Raissouni (2000) monitored changes in land cover in Moroco. Fung and Siu (2000) detected changes in land cover in Hong Kong. An automated change detection of land-cover was explored by using 250 m multitemporal MODIS NDVI 16-day composite data (Lunetta et al., 2006). A noncontextual model, a contextual model based on spatial smoothing, and a contextual model based on Markov random fields were used to analyze error propagation in land-cover change detection (Liu and Chun, 2009). Land-cover and land-use changes in the Mediterranean region of Turkey were analyzed over approximately 30 years using Landsat Multispectral Scanner (MSS) data (1975), and Landsat Thematic Mapper (TM) data (1987, 1995, and 2003) by image classification techniques (Onur et al., 2009). The National Land-Cover Database (NLCD) 2001, released by the US Geological Survey, was updated to a nominal date of 2006 by using both Landsat imagery and data from NLCD 2001 as the baseline in which change areas had been identified (Xian et al., 2009).

In this section, a curve-theorem based method is used to analyze land-cover change in the Yellow River Delta.

14.3.1 Methods

14.3.1.1 Curve Theorem in the Plane and the Approach for Change Detection

The curve theorem: Let $k:\{S_0, S\} \to \Re$ be continuous. Then there is a curve, $c:\{S_0, S\} \to \Re$, parameterized by arc-length, whose curvature at s is $k(s)$ for all $s \in \{S_0, S\}$. Moreover, if c_1 and c_2 are two such curves, then $c_1 = \alpha \cdot c_2$ where α is a proper Euclidean motion; that is, a translation followed by a rotation (Spivak, 1979).

According to the curve theorem, the overall difference between the two plane curves can be simulated as the following (Yue and Ai, 1990; Yue, 1994):

$$CD = \frac{1}{S - S_0} \int_{S_0}^{S} \left((c_1(S_0) - c_2(S_0))^2 + (\alpha_1(s) - \alpha_2(s))^2 + (k_1(s) - k_2(s))^2 \right) ds \quad (14.16)$$

where $k_i(s)$ and $\alpha_i(s)$ are the curvature and slope of the plane curve L_i, respectively; and $c_i(S_0)$ is the initial value ($i = 1, 2$).

It can be proven (Yue et al., 1999) that $CD(c_1, c_2)$ has the following three properties: (a) $CD(c_1, c_2) \geq 0$; $CD(c_1, c_2) = 0$ if and only if $c_1 = c_2$; (b) $CD(c_1, c_2) = CD(c_2, c_1)$; (c) $CD(c_1, c_3) \leq CD(c_1, c_2) + CD(c_2, c_3)$. In functional analysis, $CD(c_1, c_2)$ is a kind of distance in metric space of curves (Taylor, 1958).

If curves c_i could be simulated as

$$y = f_i(x) \tag{14.17}$$

then, α_i and k_i can be respectively formulated as

$$\alpha_i(x) = \frac{df_i(x)}{dx} \tag{14.18}$$

$$k_i(x) = \frac{d\alpha_i(x)}{dx} \cdot (1 + \alpha_i^2(x))^{-3/2} \tag{14.19}$$

$$ds = (1 + \alpha^2(x))^{1/2} dx \tag{14.20}$$

where x is abscissa and s is arc length.

Suppose that curve $f_2(x) = 0$ is considered as an intended-goal-function and $f_1(x)$ is an arbitrary function. The general index can be formulated as

$$CD = \frac{1}{X - X_0} \int_{X_0}^{X} (\alpha_1^2(x) + \kappa_1^2(x) + f_1^2(X_0))(1 + \alpha_1^2(x))^{\frac{1}{2}} dx \tag{14.21}$$

where $CD \geq 0$; $CD = 0$ is the worst situation and the biggest CD is the optimum situation.

Suppose that sequenced data can be expressed in terms of row x and column y as well as time t; then

$$\mathbf{V}(t) = \begin{bmatrix} v(1,1,t) & v(1,2,t) & \cdots & v(1,Y,t) \\ v(2,1,t) & v(2,2,t) & \cdots & v(2,Y,t) \\ \vdots & \vdots & \vdots & \vdots \\ v(X,1,t) & v(X,2,t) & \cdots & v(X,Y,t) \end{bmatrix} = [v(x,y,t)]_{X \times Y}$$

where $\mathbf{V}(t)$ is the tth layer of the three-dimensional matrix ($t = 1, 2, \ldots, T$); X is the maximum row number; Y is the maximum column number, and T is the maximum value of the time variable.

Then, the curve-theorem-based approach (Yue et al., 2002) can be formulated as the following expressions in terms of the average in each column which consists of a nonlinear transformation $SAV(y, t)$, a nonlinear transformation $CAV(y, t)$, and a general index $CD(t)$:

$$SAV(y,t) = \frac{1}{X} \sum_{x=1}^{X} ABS\big(v(x+1,y,t) - v(x,y,t)\big) \tag{14.22}$$

$$CAV(y,t) = \frac{1}{X} \sum_{x=1}^{X} \frac{v(x+2,y,t) - 2v(x+1,y,t) + v(x,y,t)}{\left(1 + (v(x+1,y,t) - v(x,y,t))^2\right)^{3/2}} \tag{14.23}$$

$$CD(t) = \frac{1}{Y} \sum_{y=1}^{Y} \big(SAV^2(y,t) + CAV^2(y,t) + AV^2(y,t)\big)\big(1 + SAV^2(y,t)\big)^{1/2} \tag{14.24}$$

where $AV(y,t) = (1/X)\sum_{x=1}^{X} v(x,y,t)$ is an average value in terms of a column; $SAV(y, t)$ is an average slope in terms of a column; $ABS(number)$ represents the absolute value of the *number*; $CAV(y, t)$ is an average curvature in terms of a column.

14.3.2 Materials

14.3.2.1 Yellow River Delta

The Yellow River is the second longest river in China and has the highest silt-content of any river in the world. Since the Yellow River took the Daqing river as its course and flowed through Lijin county into the Bohai sea in 1855, the tail channel of the Yellow River has swung between north and south, and changed its course over 50 times, taking Ninghai of Kenli county as its central axle. This swinging and silting has created the Yellow River Delta. The delta is located at the mouth of Bohai Bay and Laizhou Bay (37°20′–38°10′N and 118°7′–119°10′E).

Rapid economic development and growth of the urban population on the Yellow River Delta began with the development of the petroleum industry. Since Dongying municipality was established in 1983, a new prospect for oil production has opened up, which includes two major urban districts; the Dongying district and the Hekou district. A petrochemical base for the Shandong Province is being developed on the delta. The development of oil fields and the petrochemical industry has promoted the development of enterprises in villages and towns, as well as the construction of infrastructures such as transportation systems, power stations, and communication systems.

14.3.2.2 NDVI

A number of vegetation indices have been developed to aid the interpretation of remotely sensed data (Bouman, 1992; Hurcom and Harrison, 1998).

The most popular vegetation index (Schowengerdt, 1997; Purevdorj, 1998) is the NDVI:

$$NDVI = \frac{\lambda_{NIR} - \lambda_{RED}}{\lambda_{NIR} + \lambda_{RED}} \tag{14.25}$$

where λ_{RED} is the spectral reflectance of the visible red band; λ_{NIR} is the spectral reflectance of the near-infrared band.

NDVI has been widely used in describing relationships between vegetation characteristics such as above-ground biomass, green biomass, and chlorophyll content (Tucker, 1985).

To make it easy to calculate the general index, we transform NDVI into

$$v(x,y,t) = \frac{1 + NDVI(x,y,t)}{2} \tag{14.26}$$

where $0 \leq v(x, y, t) \leq 1$; $t = 1$ and 2, respectively, correspond to December 1, 1976 and December 3, 1988; (x, y) represents the pixel location.

14.3.2.3 Comparability of Landsat MSS Data and Landsat TM Data

In the pattern decomposition method presented by Muramatsu et al. (2000), the spectral response patterns for each pixel in an image are decomposed into three components using three standard spectral patterns normalized to unity. The three standard spectral patterns are determined to correspond to water, vegetation, and soil. In spite of the differences in measured band number and wavelength, MSS data can be analyzed in the same parameter space as TM data. The difference in classification results between TM and MSS was 1.8%.

Our sampling analysis leads to a similar conclusion:

a. The area, $449 \leq x \leq 450$ and $1296 \leq y \leq 1300$, is covered by *imperata cylindrica* that is located in the core region of Yellow River Delta Natural Conservation. The linear regression equation is

$$NDVI_v^{TM} = 0.46NDVI_v^{MSS} - 0.08 \tag{14.27}$$

The correlation coefficient is 0.71.

b. The area, $2813 \leq x \leq 2814$ and $606 \leq y \leq 610$, is a residential quarter that remained the same between 1976 and 1988 according to our on-site investigation. The linear regression equation of the corresponding NDVI data is

$$NDVI_r^{TM} = 0.58NDVI_r^{MSS} + 0.02 \tag{14.28}$$

The correlation coefficient is 0.75.

c. The area, $2974 \leq x \leq 2975$ and $1888 \leq y \leq 1892$, is a sandy area that had almost no land-cover change during the 12 years because of the tides. The linear regression equation is

$$NDVI_s^{TM} = 0.43NDVI_s^{MSS} - 0.01 \tag{14.29}$$

The correlation coefficient is 0.80.

d. The area, $3136 \leq x \leq 3137$ and $2996 \leq y \leq 3000$, is located in the Bohai Sea, which is relatively far away from the changed continental river system. The linear regression equation is

$$NDVI_w^{TM} = 0.52NDVI_w^{MSS} + 0.09 \tag{14.30}$$

The correlation coefficient is 0.78.

14.3.3 Results

The inputs for the curve-theorem-based approach are two visualized algebraic matrixes (Figures 14.3 and 14.4). Each scene has 14057576 pixels that are divided into 4582 rows and 3068 columns. The size of each pixel is 30×30 m^2. The output includes the general index CD, the nonlinear transformation SAV, and the nonlinear transformation CAV.

According to the research results produced by Muramatsu et al. (2000) and our samples, if the land cover has no change between 1976 and 1988, the transformation curves should have the same progressive increase and convexity; if the transformation curves have different progressive increase or convexity between 1976 and 1988, the land cover must have undergone substantial changes. In other words, in addition to extracting land-cover changes in the whole area of the Yellow River Delta by comparing the general indexes, local land-cover changes can be found by comparing the progressive increase or the convexity between transformation curves.

Because the calculation results in terms of the average in each row could not provide us with distinct information on land-cover change, we can only report the results in terms of the average in each column.

14.3.3.1 General Index

Our results indicate that $0.1554 \leq AV(y,t) \leq 0.4799$, $0.0119 \leq SAV(y,t) \leq 0.0266$, and $-1.53E-05 \leq CAV(y,t) \leq 1.14E-05$. The general index in terms of the average in each column can be simplified as

$$CD(t) \approx \frac{1}{Y}\sum_{y=1}^{Y}\left(\left(AV(y,t)\right)^2 + \left(SAV(y,t)\right)^2\right)\left(1 + \left(SAV(y,t)\right)^2\right)^{1/2} \tag{14.31}$$

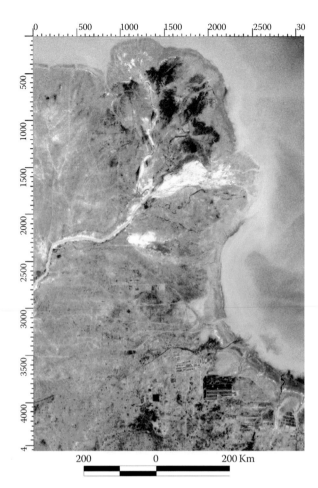

FIGURE 14.3
Visualization of the algebraic matrix transformed from the Landsat image acquired on December 1, 1976 over the Yellow River Delta.

By applying CD to the two images, we obtain $CD(1) = 0.1448$, $CD(2) = 0.2001$. Obviously, from 1976 to 1988, NDVI increased in the Yellow River Delta. We should note that if vegetation cover has no change, $CD(1)$ should be greater than $CD(2)$ according to the sample statistics presented earlier.

14.3.3.2 Nonlinear Transformation SAV

$SAV(y,t)$ provides us a rich amount of information on changes along rivers as seen in Figures 14.5 and 14.6:

- When $184 \leq y \leq 481$, which corresponds to the Zhanli river, $SAV(y,1)$ decreases progressively and $SAV(y,2)$ increases progressively; when

FIGURE 14.4
Visualization of the algebraic matrix transformed from the Landsat image acquired on December 3, 1988 over the Yellow River Delta.

$1156 \leq y \leq 1435$, which corresponds to the old course of the Yellow River, $SAV(y,1)$ is concave and $SAV(y,2)$ is close to a straight line.

- When $2088 \leq y \leq 2330$, which corresponds to the coastal boundary area where the west side of the Yellow River Delta meets with the Bohai sea at the outset, $SAV(y,1)$ is convex and $SAV(y,2)$ is concave.

- When $2463 \leq y \leq 2950$, in which $y = 2463$ is the mouth of the Yellow River flowing into the Bohai sea in 1976 and $y = 2950$ is at the mouth of the Yellow River flowing into the Bohai sea in 1988, $SAV(y,1)$ increases progressively and $SAV(y,2)$ decreases progressively.

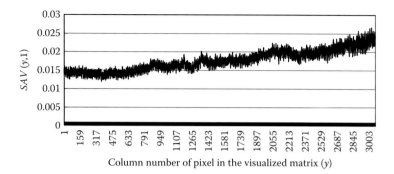

FIGURE 14.5
$SAV(y,1)$, a nonlinear transformation of NDVI in 1976 (in terms of average of each column).

In other words, $SAV(y,1)$ and $SAV(y,2)$ have a greater difference of progressive increase and convexity in these areas. This indicates that land cover of the areas along rivers has changed a great deal since 1976. According to our field visit, boats were able to sail in the Zhanli river in 1976, because there was water from the Yellow River flowing into the Bohai sea through the Zhanli river; however, this was not the case in 1988, by which time there was no water flowing from the Yellow River into the Zhanli river. The Zhanli river became a drainage canal, because the connection between the Zhanli river and the Yellow River had been artificially cut off. There was some water from the Yellow River flowing into the Bohai sea through the old course of the Yellow River in 1976; however, part of the old course of the Yellow River had become farmland and *Phragmites australis* (Cav.) Trin. ex Steud. land. There was a strong red tide in the Bohai sea near Yellow River Delta in 1988, while there was almost no tide in 1976.

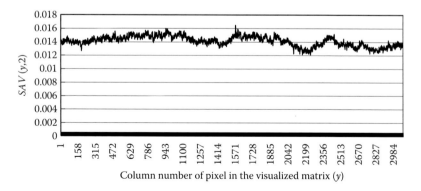

FIGURE 14.6
$SAV(y,2)$, a nonlinear transformation of NDVI in 1988 (in terms of average of each column).

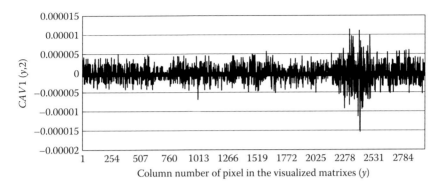

FIGURE 14.7
$CAV(y, 2)$, a nonlinear transformation of NDVI in 1988 (in terms of average of each column).

14.3.3.3 Nonlinear Transformation CAV

From $CAV(y, 2)$ as seen in Figure 14.7, we find that

- When $628 \leq y \leq 784$, where the two biggest urban districts of Dongying municipality, Dongying district and Hekou district, as well as many bigger towns of Dongying municipality such as Luozheng, Niuzhuang, Jixian, Kenli, Gaogai, Chenguan, Daying, Daozhuang, and Xiying, are located, the range of CAV is $-1.4993E - 06 \leq CAV1(y,2) \leq 1.8278E - 06$. In other words, the values in this range are very small and have an even distribution.

- When $2167 \leq y \leq 2454$, where the biggest oil extraction area of the Gudao oilfield is located, the range of CAV is $-1.531E - 05 \leq CAV1(y,2) \leq 1.141E - 05$. In other words, the biggest and smallest values are distributed in this range and a sharp change takes place.

14.4 Discussion and Conclusions

Many studies focused on the importance of habitat connectivity (Bowne et al., 1999; Tischendorf and Fahrig, 2000; Soendgerath and Schroeder, 2002). Schumaker (1996) thought that if indices of landscape pattern do indeed estimate habitat connectivity, then these indices should correlate well with predictions of dispersal success. However, he found that the nine common indices of landscape pattern were weakly correlated with the results from the dispersal model. Thus, he introduced a patch cohesion index to estimate the habitat connectivity, $PC(t) = \Sigma\left(P(t)(\sqrt{A(t)} - s)\right) / \Sigma\left(P(t)\sqrt{A(t)}\right) \cdot \sqrt{N(t)}/\sqrt{N(t)} - 1,$ where $A(t)$ and $P(t)$ are, respectively, the area and the perimeter of a patch, s is the

TABLE 14.3

Comparison of Patch Connectivity and Patch Cohesion

Period	1984	1991	1996
Patch connectivity $CO(t)$	0.3258	0.1959	0.0113
Patch cohesion $PC(t)$	0.9961	0.99596	0.9874
Total number of pixels $N(t)$	812796	952339	952866
Length of a pixel edge s	30 m	30 m	30 m

length of a pixel edge, $N(t)$ is the total number of pixels. Schumaker's (1996) results showed that patch cohesion index correlated remarkably well with estimates of dispersal success. The results (Table 14.3) from applying both the patch connectivity index and the patch cohesion index to the newly created wetland in the Yellow River Delta show that patch connectivity and patch cohesion have a very close linear relation. This is formulated as

$$PC(t) = 0.0289 \cdot CO(t) + 0.988 \tag{14.32}$$

where the correlation coefficient is 0.9175.

Landscape change detection in the newly created wetland of the Yellow River Delta supports the conclusions of Roberts et al. (2000), Malanson and Gramer, (1999) and Gustafson and Gardner (1996). Human impact does increase fragmentation. Broader habitat loss often increases both the distance between desirable habitat patches and the hostility of interpatch habitation for certain species. Thus, the probability of influx to the remnant patches of pioneer species is decreased. Increased fragmentation leads to an impairment of patch connectivity.

The case-study of the curve-theorem-based method for land-cover change detection in the Yellow River Delta demonstrates that the general index *CD* gives us an overall view of vegetation cover change, *SAV* provides us with useful information on environmental changes in rivers, and *CAV* is capable of highlighting the location of industrial and urban areas.

References

Adams, J.B., Sabol, D.E., Kapos, V., Filho, R.A., Roberts, D.A., Smith, M.O., and Gillepie, A.R. 1995. Classification of multispectral images based on fractions of end members: Application to land-cover change in the Brazilian Amazon. *Remote Sensing of Environment* 52: 137–154.

Anderson, J.R. 1977. Land-use change analysis using sequential aerial photography and computer technique. *Photogrammetric Engineering and Remote Sensing* 46: 1447–1464.

Baudry, J. and Merriam, H.G. 1988. Connectivity and connectedness: functional versus structural patterns in landscapes. In *Proceedings of the 2nd International Seminar of the International Association for Landscape Ecology,* August 1987, Muenster, 23–28. Muenster: Muenstersche Geographische Arbeiten 29.

Bauer, M.E., Burk, T.E., Ek, A.R., Coppen, P.R., Lime, A.D., Walsh, T.A., Walters, D.K., Befort, W., and Heinzen, D.F. 1994. Satellite inventory of Minnesota forest resources. *Photogrammetric Engineering and Remote Sensing* 60: 287–298.

Bouman, B.A. 1992. Accuracy of estimating the Leaf Area Index from vegetation indices derived from crop reflectance characteristics, a simulation study. *International Journal of Remote Sensing* 13: 3609–3084.

Bouziani, M., Goïta, K., and He, D.C. 2010. Automatic change detection of buildings in urban environment from very high spatial resolution images using existing geo-database and prior knowledge. *ISPRS Journal of Photogrammetry and Remote Sensing* 65(1): 143–153.

Bowne, D.R., Peles, J.D., and Barrett, G.W. 1999. Effects of landscape spatial structure on movement patterns of the hispid cotton rat *Sigmodon hispidus. Landscape Ecology* 14: 53–66.

Bryant, E.S., Birnie, R.W., and Kimball, K.D. 1993. A practical method of mapping forest change over time using Landsat MSS data: A case study from central Maine. In *Proceedings of the 25th International Symposium, Remote Sensing and Global Environmental Change,* Graze, Austria, April 4–8, 1993, 469–480. Ann Arbor: ERIM.

Chen, X.X., Vierling, L., and Deering, D. 2005. A simple and effective radiometric correction method to improve landscape change detection across sensors and across time. *Remote Sensing of Environment* 98(1): 63–79.

Coxeter, H.S.M. 1961. *Introduction to Geometry.* New York, NY: Wiley.

Ebdon, D. 1978. *Statistics in Geography.* Oxford: Basil Blackwell.

Engvall, J.L., Tubbs, J.D., and Holmes, Q.A. 1977. Pattern recognition of Landsat data based upon temporal trend analysis. *Remote Sensing of Environment* 6: 303–314.

Fang, C.Y., Chen, S.W., and Fuh, C.S. 2003. Automatic change detection of driving environments in a vision-based driver assistance system. *IEEE Transactions on Neural Networks* 14(3): 646–657.

Forman, R.T.T. 1995. *Land Mosaics: The Ecology of Landscapes and Regions.* Cambridge, Great Britain: Cambridge University Press.

Forman, R.T.T. and Godron, M. 1986. *Landscape Ecology.* London: John Wiley and Sons.

Fuller, R.M., Groom, G.B., and Jone, A.R. 1994. The land-cover map of Great Britain: An automated classification of Landsat Thematic Mapper data. *Photogrammetric Engineering and Remote Sensing* 60: 553–562.

Fung, T. and Siu, W. 2000. Environmental quality and its changes, an analysis using NDVI. *International Journal of Remote Sensing* 21: 1011–1024.

Gao, Z.Q., Liu, J.Y., and Zhuang, D.F. 1998. A study of Chinese farmland change based on Chinese resources and environment database. *Journal of Natural Resources* 13(1): 92–96 (in Chinese).

Garrison, W. 1960. Connectivity of the interstate highway system. *Papers of the Regional Science Association* 6: 121–137.

Gong, P. 1993. Change detection using principal component analysis and fuzzy set theory. *Canadian Journal of Remote Sensing* 19: 22–29.

Gustafson, E.G. and Gardner, R.H. 1996. The effect of landscape heterogeneity on the probability of patch colonization. *Ecology* 77: 94–107.

Haber, W. 1984. Über Landschaftspflege. *Landschaft + Stadt* 16: 193–199.

Haber, W. 1986. Über die menschenliche Nutzung von Ökosystemen: unter besonderer Berücksichtigung von Agrarokosystemen. *Verh. Gesellsch. für Ökologie* 14: 13–24.

Haber, W. 1987. Zur Umsetzung ökologischer Forschungsergebnisse in politisches Handeln. *Verhandlungen der Gesellschaft für Ökologie* 15: 61–69.

Haber, W. 1988. Über den Umweltzustand der Bundersrepublik Deutchland am Ende der 1980er Jahre. *Korrespondenz Abwasser* 35: 1084–1089.

Haber, W. 1990. Using landscape ecology in planning and management. In *Changing Landscapes: An Ecological Perspective*, eds., I.S. Zonneveld and R.T.T. Forman, 217–231. New York, NY: Springer-Verlag.

Haber, W. and Burkhardt, I. 1986. Protecting the environment by means of landscape ecology. *Universitas* (English edition) 28: 233–238.

Haber, W. and Burkhardt, I. 1988. Landschaftsökologie in der Landespflege in Weihenstephan. *Berichte zur deutschen Landeskunde* 62: 155–173.

Haggett, P. and Chorley, R.J. 1969. *Network Analysis in Geography*. New York, NY: St. Martin's Press.

Hall, F.G., Botkin, D.B., Strebel, D.E., Woods, K.D., and Goetz, S.J. 1991. Large-scale patterns of forest succession as determined by remote sensing. *Ecology* 72: 628–640.

Hart, J.F. 1954. Central tendency in areal distributions. *Economic Geography* 30: 48–59.

Hurcom, S.J. and Harrison, A.R. 1998, The NDVI and spectral decomposition for semi-arid vegetation abundance estimation. *International Journal of Remote Sensing* 19: 3109–3125.

Janssens, P. and Gulinck, K.H. 1988. Connectivity, proximity and contiguity in the landscape interpretation of remote sensing data. In *Proceedings of the 2nd International Seminar of the International Association for Landscape Ecology*, August 1987, Muenster, 43–47. Muenster: Münstersche Geographische Arbeiten 29.

Jensen, J.R., Rutchey, K., Koch, M.S., and Narumalani, S. 1995. Inland wetland change detection in the Everglades Water Conservation Area 2A using a time series of normalized remotely sensed data. *Photogrammetric Engineering and Remote Sensing* 61: 199–209.

Kennedy, R.E., Townsend, P.A., Gross, J.E., Cohen, W.B., Bolstad, P., Wang, Y.Q., and Adams, P. 2009. Remote sensing change detection tools for natural resource managers: Understanding concepts and tradeoffs in the design of landscape monitoring projects. *Remote Sensing of Environment* 113(7): 1382–1396.

Lillesand, T.M. and Kiefer, R.W. 1979. *Remote Sensing and Image Interpretation*. New York, NY: John Wiley and Sons.

Liu, D.S. and Chun, Y.W. 2009. The effects of different classification models on error propagation in land-cover change detection. *International Journal of Remote Sensing* 30(20): 5345–5364.

Loveland, T.R., Merchant, J.W., Ohlen, D.O., and Brown, J.F. 1991. Development of a land-cover characteristics data base for the conterminous US *Photogrammetric Engineering and Remote Sensing* 57 (11): 1453–1463.

Lowe, J.C. and Moryadas, S. 1975. *The Geography of Movement*. Boston: Houghton Mifflin Company.

Lunetta, R.S. 1998. Applications, project formulation and analytical approach. In *Remote Sensing Change Detection*, eds., R.S. Lunetta and C.D. Elvidge, 1–19. Michigan: Ann Arbor Press.

Lunetta, R.S., Knight, J.F., Ediriwickrema, J., Lyon, J.G., and Worthy, L.D. 2006. Land-cover change detection using multi-temporal MODIS NDVI data. *Remote Sensing of Environment* 105(2): 142–154.

Lunetta, R.S., Lyon, J.G., Sturdevant, J.A., Dwyer, J.L., Elidge, C.D., Fenstermaker, L.K., Yuan, D., Hoffer, J.R., and Werrackoon, R. 1993. North American Landscape Characterization: Research Plan. Report No. 600/R-93/135, EPA, American United States.

Malanson, G.P. and Cramer, B.E. 1999. Landscape heterogeneity, connectivity, and critical landscapes for conservation. *Diversity and Distribution* 5: 27–39.

Maldonado, F.D., Santos, J.R., and Graça, P.M.L. 2007. Change detection technique based on the radiometric rotation controlled by no-change axis, applied on a semi-arid landscape. *International Journal of Remote Sensing* 28(8): 1789–1804.

Masry, S.E., Crawley, B.G., and Hilborn, W.H. 1975. Difference detection. *Photogrammatric Engineering and Remote Sensing* 41(9): 1145–1148.

McDonnell, M.J. and Pickett, S.T.A. 1988. Connectivity and the theory of landscape ecology. In *Proceedings of the 2nd International Seminar of the International Association for Landscape Ecology*, 1987, Muenster, 17–21. Muenster: Muenstersche Geographische Arbeiten 29.

Menger, K. 1927. Zur allgemeinen Kurventheorie. *Fundamentals in Mathematics* 10: 96–115.

Menzel, D.H. 1970. *Survey of the Universe*. Englewood Cliffs, NJ: Prentice-Hall.

Miller, A.B., Bryant and Birnie, R.W. 1998. An analysis of land-cover changes in the Northern Forest of New England using multitemporal Landsat MSS data. *International Journal of Remote Sensing* 19: 245–265.

Mladenoff, D.J., Niemi, G.J., and White, M.A. 1997. Effects of changing landscape pattern and USGS land-cover data variability on eco-region discrimination across a forest–agriculture gradient. *Landscape Ecology* 12: 379–396.

Muramatsu, K., Furumi, S., Fujiwara, N., Hayashi, A., Daigo, M., and Ochiai, F. 2000. Pattern decomposition method in the albedo space for Landsat TM and MSS data analysis. *International Journal of Remote Sensing* 21(1): 99–119.

Nelson, R.F. 1983. Detecting forest canopy change due to insect activity using Landsat MSS. *Photogrammetric Engineering and Remote Sensing* 49: 1303–1314.

Oetter, D.R., Cohen, W.B., Berterretche, M., Maiersperger, T.K., and Kennedy, R.E. 2001. Land-cover mapping in an agricultural setting using multiseasonal Thematic Mapper data. *Remote Sensing of Environment* 76: 139–155.

Onur, I., Maktav, D., Sari, M., and Sönmez, N.K. 2009. Change detection of land-cover and land-use, using remote sensing and GIS: A case study in Kemer, Turkey. *International Journal of Remote Sensing* 30(7): 1749–1757.

Purevdorj, T., Tateishi, R., Ishiyama, T., and Honda, Y. 1998, Relationships between percent vegetation cover and vegetation indices. *International Journal of Remote Sensing* 19: 3519–3535.

Risser, P.G., Karr, J.R., and Forman, R.T.T. 1984. *Landscape Ecology: Directions and Approaches*. Illinois: Illinois Natural History Survey Special Publication Number 2.

Roberts, S.A., Hall, G.B., and Calamai, P.H. 2000. Analyzing forest fragmentation using spatial autocorrelation, graphs and GIS. *International Journal of Geographical Information Science* 14(2): 185–204.

Schowengerdt, R.A. 1997. *Remote Sensing: Models and Methods for Image Processing*. London: Academic Press.

Schreiber, K.F. 1988. Connectivity in Landscape ecology. In *Proceedings of the 2nd International Seminar of the International Association for Landscape Ecology, 1987,* Muenster, 11–15. Muenster: Muenstersche Geographische Arbeiten 29.

Schumaker, N.H. 1996. Using landscape indices to predict habitat connectivity. *Ecology* 77(4): 1210–1225.

Sesnie, S.E., Gessler, P.E., Finegan, B., and Thessler, S. 2008. Integrating Landsat TM and SRTM-DEM derived variables with decision trees for habitat classification and change detection in complex neotropical environments. *Remote Sensing of Environment* 112(5): 2145–2159.

Shaw, G. and Wheeler, D. 1985. *Statistical Techniques in Geographical Analysis.* New York, NY: John Wiley & Sons.

Shimbel, A. 1953. Structural properties of communication networks. *Bulletin of Mathematical Biophysics* 15: 501–507.

Shimbel, A. and Katz, W. 1953. A new status index derived from sociometric analysis. *Psychometrika* 18: 39–43.

Singh, A. 1989. Digital change detection techniques using remotely-sensed data. *International Journal of Remote Sensing* 10: 989–1003.

Skole, D. and Tucker, C. 1993. Tropical deforestation and habitat fragmentation in the Amazon: Satellite data from 1978 to 1988. *Science* 260: 1905–1910.

Sobrino, J.A. and Raissouni, N. 2000. Toward remote sensing methods for land cover dynamic monitoring: Application to Morocco. *International Journal of Remote Sensing* 21: 353–366.

Soendgerath, D. and Schroeder, B., 2002. Population dynamics and habitat connectivity affecting the spatial spread of populations—A simulation study. *Landscape Ecology* 17: 57–70.

Spivak, M. 1979. *A Comprehensive Introduction to Differential Geometry.* Houston, TX: Publish or Perish, Inc.

Stone, T.A., Schlesinger, P., Houghton, R.A., and Woodell, G.M. 1994. A map of the vegetation of South America based on satellite imagery. *Photogrammetric Engineering and Remote Sensing* 60: 541–551.

Swart, J. and Lawes, M.J. 1996. The effect of habitat patch connectivity on samango monky (*Cercopithecus mitis*) metapopulation persistence. *Ecological Modelling* 93: 57–74.

Taaffe, E.L. and Jr. Gauthier, H.L. 1973. *Geography of Transportation.* Englewood Cliffs, NJ: Prentice-Hall.

Taylor, A.E. 1958. *Introduction to Functional Analysis.* New York, NY: John Wiley & Sons, Inc.

Taylor, P.D., Fahrig, L., Henein, K., and Merriam, G. 1993. Connectivity is a vital element of landscape structure. *Oikos* 68: 571–573.

Tian, J.Y., Jia, W.Z., Dou, H.Y., Jiao, Y.M., Gao, K.J., and Cai, X.J. 1999. *Study on Biodiversity in the Yellow River Delta.* Qingdao: Qingdao Press (in Chinese).

Tischendorf, L. and Fahrig, L. 2000. On the usage and measurement of landscape connectivity. *Oikos* 90: 7–19.

Tucker, C.J. 1985. Satellite remote sensing of total herbaceous biomass production in the Senegalese Sahel: 1980–1984. *Remote Sensing of Environment* 17: 233–249.

Wang, L., Chen, J., Gong, P., Shimazaki, H., and Tamura, M. 2009. Land-cover change detection with a cross-correlogram spectral matching algorithm. *International Journal of Remote Sensing* 30(12): 3259–3273.

Warntz, W. and Neft, D. 1960. Contributions to a statistical methodology for areal distributions. *Journal of Regional Science* 2: 47–66.

Weismiller, R.A., Kristof, S.J., Scholz, D.K., Anuta, P.E., and Momen, S.A. 1977. Change detection in coastal zone environments. *Photogrammetric Engineering and Remote Sensing* 43: 1533–1539.

With, K.A., Gardner, R.H., and Turner, M.G. 1997. Landscape connectivity and population distributions in heterogeneous environments. *Oikos* 78: 151–169.

Wulder, M.A., Butson, C.R., and White, J.C. 2008. Cross-sensor change detection over a forested landscape: Options to enable continuity of medium spatial resolution measures. *Remote Sensing of Environment* 112(3): 796–809.

Xian, G., Homer, C., and Fry, J. 2009. Updating the 2001 National land-cover database, land-cover classification to 2006 by using Landsat imagery change detection methods. *Remote Sensing of Environment* 113(6): 1133–1147.

Yue, T.X. 1994. *Systems Models for Land Management and Real Estate Evaluation.* Beijing: China Society Press (in Chinese).

Yue, T.X. and Ai, N.S. 1990. A morphological mathematical model for cirques. *Glaciology and Cryopedology* 12(3): 227–234 (in Chinese).

Yue, T.X., Chen, S.P., Xu, B., Liu, Q.S., Li, H.G., Liu, G.H., and Ye, Q.H. 2002. Extracting environmental changes in Yellow River Delta from remotely sensed data using a curve-theorem based approach. *International Journal of Remote Sensing* 23(11): 2283–2292.

Yue, T.X., Haber, W., Grossmann, W.D., and Kasperidus, H.D. 1999. A method for strategic management of land. In *Environmental Indices: Systems Analysis Approaches,* eds. Y.A. Pykh, D.E. Hyatt and R.J.M.B. Lenz, 181–201. London: EOLSS Publishers Co Ltd.

Yue, T.X., Haber, W., Herzog, F., Cheng, T., Zhang, H.Q., and Wu, Q.H. 1998. Models for DLU Strategy and Their applications. *Ekologia,* 17 (Suppl. 1): 118–128.

Yue, T.X., Liu, J.Y., Jørgensen, S.E., and Ye, Q.H. 2003. Landscape change detection of the newly created wetland in Yellow River Delta. *Ecological Modelling* 164: 21–31.

Yue, T.X., Xu, B., and Liu, J.Y. 2004. A patch connectivity index and its change on a newly born wetland at the Yellow River Delta. *International Journal of Remote Sensing* 25(21): 4617–4628.

Zhao, Y.M. 1997. *Forestry Development and Natural Conservation of Yellow River Delta.* Beijing: China Forestry Press (in Chinese).

15

Simulation of Horizontal Wind Velocity[*]

15.1 Introduction

Wind plays a very important role in ecosystem change. Wind is the dominant disturbance on patterns of vegetation recovery (Schumacher et al., 2004). Many herbaceous species rely on wind as their most important dispersal vector (Schippers and Jongejans, 2005). Wind is also one of the major dynamic factors that cause extensive damage to ecosystems (Beinhauer and Kruse, 1994; Gardiner et al., 2000; Blennow and Sallnaes, 2004). Changes in the magnitude and spatial gradient of wind stress have significant effects on coastal pelagic ecosystems (Rykaczewski and Checkley, 2008). Ecological diversity in mountain ecosystems is considerably impacted by wind disturbance (Santos et al., 2010). This chapter focuses on the simulation of horizontal wind velocity above a nonuniform underlying surface.

Under thermally neutral conditions, steady-state flows over horizontally bare soil can be described by the well-known logarithmic law (Sutton, 1953; Mihailovic et al., 1999; Baldauf and Fiedler, 2003)

$$u(z) = \frac{u_*}{k} \ln \frac{z}{z_0} \tag{15.1}$$

where $u(z)$ is the horizontal velocity at height z; u_* is the friction velocity for the bare soil, which physically represents the shear stress $\tau = \rho u_*^2$ where ρ is the air density; k is the von Karman's constant taken to be 0.41; and z_0 is the roughness length of the bare soil.

The popular saying (Wieringa, 1993) "z_0 is the height at which wind speed becomes zero" is therefore true in a purely algebraic sense only if it is according to Equation 15.1. Research results from Blackadar and Tennekes (1968) showed that the logarithmic wind profile is not only a feature of the lower few meters over homogeneous terrain, but also rather a consistent description of any surface layer wind field up to heights of 30–100 m. Consequently, the roughness length z_0 is the optimal parameter for specifying terrain effects on wind (Wieringa, 1981).

---

[*] Dr. Wei Wang is a major collaborator of this chapter.

For vegetative surfaces (Sutton, 1953),

$$u(z) = \frac{u_*}{k} \ln \frac{z-d}{z_0} \tag{15.2}$$

where u_* is the friction velocity over the vegetation surface; d is a zero-plane displacement that is the mean height of the vegetation on which the bulk aerodynamic drag acts; and z_0 is the roughness length. According to this expression, the wind speed is zero at height $d + z_0$, but the logarithmic profile cannot be extrapolated that far downwards. When the quantities d and z_0 are known, the whole profile above a vegetative surface as well as the ratio (u_* / k) can be obtained if the wind at a single level is known.

Apparently, the transfer of momentum between short grass and the atmosphere does not differ so much from the corresponding exchange when bare soil is the underlying surface (Mihailovic et al., 1999). Over tall grass, the transfer of momentum into the atmosphere is more intensive since u_* becomes greater than u_{*g}. Difference in these velocity scales physically is from an effect of displacement and an increase in z_0.

Equation 15.2 is not valid when the height z is between the height of the vegetation h and some height z^* representing the lower limit of inertial sublayer. Its order of magnitude may vary between $z^* = d + 10z_0$ and $z^* = d + 20z_0$ (De Bruin and Moore, 1985). Since z_0 is around 10% of the canopy height, the thickness of the roughness sublayer may vary between one and two canopy heights. In models of biosphere–atmosphere exchange, when the underlying vegetative surface consists of patches of bare soil and plant communities with different morphological parameters, the level of inhomogeneity in the cover has to be taken into account in addition to a spatially varying displacement height (Mihailovic et al., 1999).

Experimental evidence indicates that the estimates of momentum transfer coefficient, K_m, above a vegetative surface were 1.5–2.0 times larger than a simple application of Equation 15.2 would indicate. Thus, Equation 15.2 can be modified as

$$u(z) = \frac{u_*}{\alpha_G k} \ln \frac{z-d}{z_0} \tag{15.3}$$

where α_G is a dimensionless constant estimated to be between 1.5 and 2.0 (Raupach and Thom, 1981; Massman, 1987). Equation 15.3 can only be valid for the lower part of the roughness sublayer (Mihailovic et al., 1999).

Above a nonuniform underlying surface, the nonuniformity is expressed with the surface vegetation fractional cover σ, which takes values from 0 when the ground surface is bare soil to 1 when the ground surface is totally covered by plants. Suppose that the underlying surface is a combination of only two homogenous portions characterized as σ and $1 - \sigma$, the wind profile can be formulated as (Mihailovic et al., 1999)

$$u(z) = \frac{u_*}{k}\left[\sigma(\alpha - 1) + 1\right]^{-1} \ln \frac{\left[\sigma(\alpha - 1) + 1\right]z - \sigma\alpha d}{\alpha^2 z_0} \tag{15.4}$$

where $u(z)$ is the horizontal velocity at height z; z_0 is the roughness length; u_* is a friction velocity above the nonhomogeneous surface; k is the von Karman's constant taken to be 0.41; d is the displacement height that is the mean height of the vegetation; and α is a dimensionless constant representing a correction to the mixing length in the roughness sublayer.

In terms of Equation 15.4, roughness length z_0, zero-plane displacement height d, dimensionless constant α, and vegetation fractional cover σ are the most important parameters of the horizontal velocity $u(z)$.

15.2 Estimation of the Parameters

15.2.1 Dimensionless Constant α

Comparing model simulations with observations, Laric (1997) has found that for short grass

$$\alpha^2 = \left(6.4 \, LAI\right)^{1/10} \tag{15.5}$$

for tall grass

$$\alpha^2 = \left(6.4 \, LAI\right)^{1/5} \tag{15.6}$$

and for forest

$$\alpha^2 = \left(3.2 \, LAI\right)^{1/2} \tag{15.7}$$

where LAI is the leaf area index.

For wheat, as an extension of the dimensionless constants of short grass and tall grass, α is generally expressed as

$$\alpha = \left(6.4 \, LAI\right)^{b} \tag{15.8}$$

where b is the wheat parameter of the dimensionless constant to be simulated.

LAI is the one-side foliage area per ground area (m^2/m^2) (White et al., 2000). The simplest and most practical way to estimate *LAI* is to investigate the relationships between *LAI* and values of various vegetation indexes by means of regression models (Asrar et al., 1985; Price and Bausch, 1995; Wulder, 1998; Brown et al., 2000; Qi et al., 2000; Vaesen et al., 2001; Chen et al., 2002). These vegetation indexes include the simple ratio vegetation index (Nemani et al., 1993), the reduced simple ratio vegetation index (Nemani et al., 1993), the perpendicular vegetation index (Wiegand and Richardson, 1987), the weighted difference vegetation index (Clevers, 1989), the NDVI (Goward et al., 1985; Yue et al., 2002), the soil-adjusted vegetation index (Huete, 1988), the atmospherically resistant vegetation index (Kaufman and Tanre, 1992), the soil and atmospherically resistant vegetation index (Kaufman and Tanre, 1992), the modified NDVI (Liu and Huete, 1995), and the feedback-based vegetation index (Huete et al., 1997). These relationships between *LAI* and the vegetation indices have the following generalized formulations:

$$LAI = a \cdot VI^3 + b \cdot VI^2 + c \cdot VI + d \tag{15.9}$$

$$LAI = a \cdot VI^b + c \tag{15.10}$$

$$LAI = a \cdot \ln(b \cdot VI + c) \tag{15.11}$$

where *VI* is a kind of vegetation index; *a*, *b*, *c*, and *d* are empirical parameters and vary with vegetation types.

In addition to the relationship between *LAI* and vegetation indexes, the LAI-2000 instrument was used in different vegetation types to derive *LAI* indirectly (Gower and Norman, 1991; Colombo et al., 2003), that is,

$$LAI = -2 \int_0^{\pi/2} \ln P(\theta) \cdot \cos\theta \cdot \sin\theta \cdot d\theta \tag{15.12}$$

where $P(\theta)$ is the gap fraction in five zenith angle θ ranges with midpoints of 7°, 23°, 38°, 53°, and 67°.

15.2.2 Vegetation Fractional Cover σ

For fractional vegetation cover, the mean vertically projected canopy area per unit ground area is formulated as (Kellerer, 1983; Jasinski and Crago, 1999)

$$\sigma = 1 - \exp\left\{ -\frac{n \cdot A_t}{A_p} \right\} \tag{15.13}$$

where n is the number of roughness elements; A_t is the mean vertically projected canopy area of a single roughness element, A_p is the unit area such as the pixel area.

Fractional vegetation cover, as an important element of climate models, was first introduced by Deardorff (1978). However, its specification from field-observations has been problematic (Zeng et al., 2000). It is a relatively simple parameter to obtain by means of satellite remote sensing, which is mostly formulated as (Baret et al., 1995; Wittich and Hansing, 1995; Wittich, 1997; Gutman and Ignatov, 1998; Zeng et al., 2000)

$$\sigma(i,j) = 1 - \left(\frac{N_v(i,j) - N(i,j)}{N_v(i,j) - N_s(i,j)} \right)^b \tag{15.14}$$

or

$$\sigma(i,j) = \frac{N(i,j) - N_s(i,j)}{N_v(i,j) - N_s(i,j) + \left(1 - d(i,j)\right)\left(N(i,j) - N_v(i,j)\right)} \tag{15.15}$$

where $N(i,j) = (\lambda_{NIR}(i,j) - \lambda_{RED}(i,j)/\lambda_{NIR}(i,j) + \lambda_{RED}(i,j))$, λ_{RED} (i,j) is the spectral reflectance of visible red band; λ_{NIR} (i,j) is the spectral reflectance of near-infrared band; $d(i,j) = (\lambda_{NIR}(i,j) + \lambda_{RED}(i,j))_v/(\lambda_{NIR}(i,j) + \lambda_{RED}(i,j))_s$; the subscripts v and s, respectively, denote values of over 100% vegetation cover and bare soil; b is the parameter to be simulated.

Zeng et al. (2000) modified the formulation of fractional vegetation cover so that it became independent of season and representative of the annual maximum green vegetation fraction for a given pixel. The modified formulation was expressed as

$$\sigma(i,j) = \frac{N_{\max}(i,j) - N_s(i,j)}{N_v(i,j) - N_s(i,j)} \tag{15.16}$$

where $N_{\max}$ (i,j) is the annual maximum value of the normalized difference vegetation index $N(i,j)$.

Many other studies (Defries et al., 1999, 2000a,b; Scanlon et al., 2002) classified fractional vegetation cover into three broad categories: σ_b (i,j), a portion of the land surface that always remains as bare soil, σ_w (i,j), a portion of the fractional cover that is woody vegetation, and σ_g/b (i,j), a remaining portion that consists of bare soil and herbaceous vegetation cover. Their relationship is formulated as (Scanlon et al., 2002)

$$\langle \sigma_b(i,j) \rangle + \langle \sigma_w(i,j) \rangle + \langle \sigma_{g/b}(i,j) \rangle = 1 \tag{15.17}$$

where $\langle \rangle$ operator represents spatial averaging; the subscripts b, w, and g/b, respectively represent the portion of the land surface that always remains as bare soil, the portion of the fractional cover that is woody vegetation, and the remaining portion that consists of bare soil and herbaceous vegetation cover.

The temporal mean of the observed NDVI at each pixel is equal to the sum of the NDVI weighted by the fractional cover types:

$$\overline{N_b(i,j)} \cdot \langle \sigma_b(i,j) \rangle + \overline{N_w(i,j)} \cdot \langle \sigma_w(i,j) \rangle + \overline{N_{g/b}(i,j)} \cdot \langle \sigma_{g/b}(i,j) \rangle = \overline{N(i,j)} \qquad (15.18)$$

where $\overline{}$ operator represents temporal averaging; the subscripts b, w, and g/b, respectively, represent the portion of the land surface that always remains as bare soil, the portion of the fractional cover that is woody vegetation, and the remaining portion that consists of bare soil and herbaceous vegetation cover.

For homogeneous canopies, the relationship between fractional vegetation cover and the leaf area index was formulated as (Choudhury et al., 1994; Baret et al., 1995; Wittich, 1997)

$$\sigma(i,j) = 1 - e^{-cLAI(i,j)} \qquad (15.19)$$

where *LAI* is the leaf area index; *c* is a constant to be simulated.

15.2.3 Roughness Length z_0 and Zero-Plane Displacement Height *d*

The most common geometric approach is to use the mean height of roughness elements to estimate zero-plane displacement height *d* and roughness length z_0 (Grimmond and Oke, 1999). The relations between the mean height of roughness elements and *d* and z_0 are formulated as

$$d = A_d \cdot H \qquad (15.20)$$

$$z_0 = A_0 \cdot H \qquad (15.21)$$

where *H* is the mean height of roughness elements; A_d and A_0 are constants to be simulated.

Garratt's result (1992) showed that $A_d = 0.67$ and $A_0 = 0.10$ were good overall mean values for land surfaces. Raupach (1992) noted that for field crops and grass canopies $A_d = 0.64$ and $A_0 = 0.13$, for forests $A_d = 0.8$ and $A_0 = 0.06$. Jasinski and Crago (1999) compared the various roughness estimates for the Landes Forest (Parlange and Brutsaert, 1989; Gash et al., 1989; Raupach, 1994; Jasinski and Crago, 1999) and concluded that A_0 usually lies between 0.02 and 0.2 and A_d mostly between 0.6 and 0.9. In dispersion modeling over urban areas, Hanna and Chang (1992) suggested that $A_d = 0.5$ and $A_0 = 0.10$ were useful approximations. Although this parameterization ignores many aspects, it does capture the most important parameter influencing turbulence near the surface and provides a basis for comparisons with more sophisticated models (Yang and Friedl, 2003).

A key plant parameter is the frontal area index λ that is defined as the ratio of the frontal area of roughness elements, from the mean wind direction, per unit ground area. It can be formulated as

$$\lambda = \frac{n \cdot A_f}{A_p}$$

(15.22)

where n is the number of roughness elements; A_f is the mean frontal area of an individual roughness element; A_p is the unit area such as the pixel area.

For isotropically oriented elements, the relationship between the frontal area index λ and canopy area index Λ, for example, the total (single-sided) area of all canopy elements over the unit ground area, can be formulated as $\Lambda = 2\lambda$. The canopy area index includes all canopy elements, both transpiring (living leaves and stems) and nontranspiring (dead leaves and stems), while the leaf area index only includes transpiring surfaces. A_d and A_0 are respectively formulated as (Raupach, 1992, 1994)

$$A_d = 1 - \frac{1 - \exp\left(-\sqrt{c_{d1}\Lambda}\right)}{\sqrt{c_{d1}\Lambda}}$$

(15.23)

$$A_0 = (1 - A_d)\exp\left(-k\frac{u_h}{u_*} - \Psi_h\right)$$

(15.24)

where c_{d1} is a free parameter; k is the von Karman's constant; Ψ_h is the roughness sublayer influence function; u_* is the friction velocity; u_h is the mean velocity at height h. When $\lambda \geq \lambda_{max}$, $(u_h/u_*$ is nearly constant 0.3, where λ_{max} is the point at which adding further roughness elements to the surface does not affect the bulk drag because the additional elements merely shelter one another. When $\lambda < \lambda_{max}$, (u_h/u_*) is the solution of equation

$$\gamma = \frac{u_h}{u_*} = \left(\frac{\exp(\lambda \cdot \gamma / 4)}{(C_S + C_R \cdot \lambda)^{1/2}}\right)$$

(15.25)

where C_S is the drag coefficient of the substrate surface at height h in the absence of roughness elements (about 0.003); C_R is the drag coefficient of an isolated roughness element mounted on the surface.

In addition to the geometric approach and the frontal area index, there have been many well-documented experimental determinations of the roughness of various surfaces, ranging from mobile surfaces (sea, moving sand or snow) to vegetation and towns based on data of land-cover types.

It can be concluded that the mean height of roughness elements (*H*), fractional vegetation cover (σ) and leaf area index (*LAI*) are the most essential parameters of the vertical wind profile under neutral conditions. Currently, wind speed models require either field validation of simulated *LAI*, σ, and *H*, or remotely sensed estimates of *LAI*, σ, and *H* to initiate them (Running et al., 1999). *LAI*, σ, and *H* measurements are critical for improving the performance of such models over large areas and this has prompted investigations of the relationship between ground-measured *LAI*, σ, and *H* and spectral vegetation indexes derived from satellite-measured data (Colombo et al., 2003).

15.3 Simulation of Vertical Wind Profile

15.3.1 Retrieval of the Essential Wheat Parameters at Yucheng Integrated Agricultural Experiment Station of the Chinese Academy of Sciences in Shandong Province

In the north of China, the first 10-day period of March to the first 10-day period of April is the optimum time for identifying wheat, among other crops, because wheat is the sole crop that has become green during this period. According to available NDVI data and LANDSAT TM data kept in archives, LANDSAT-5 data of March 30, 2000 are selected to identify wheat land-use types, of which the projection is chosen as Albers Conical Equal Area. Observation data in the field and other auxiliary data include: (1) data observed everyday over a 10-day period detailing the mean height, leaf area index, and fractional vegetation cover of wheat at the Yucheng Integrated Agricultural Experiment Station of the Chinese Academy of Sciences in the Shandong province (N36°49′52″, E116°34′17″) in 2000 (Table 15.1); (2) NDVI data collected everyday over a 10-day period from the NOAA-14 meteorological satellite; (3) 1:10000 land-use data; and (4) 1:10000 topographical data.

Data preprocessing can be achieved using the following steps: (1) collect 41 control points on the 1:10000 topographical map to geometrically correct the TM image (Figure 15.1); (2) conduct resampling by means of the proximity-element method in order to easily classify the image and keep it relatively proportional to the gray gradations of the original image, (3) create 25 × 25 m corrected TM data; (4) conduct a projection transformation of the NDVI data collected over the 10-day period from Lambert homolographic projection to Albers Conical homolographic projection; (5) conduct a registration of the NDVI data collected over the 10-day period using the 1:10000 land-use map of Yucheng in the Shandong province; (6) fit the original image by means of quadratic polynomial.

TABLE 15.1

Observation Values of Relative Parameters in 2000

Observation Date	Stem Height (cm)	Leaf Area Index	Fractional Vegetation Cover (%)
March 5, 2000	13.5	2.18	20
March 15, 2000	21.6	2.58	25
March 20, 2000	22.6	3.30	33
March 25, 2000	24.2	3.58	36
March 30, 2000	29.2	4.58	39
April 4, 2000	35.4	4.61	44
April 9, 2000	45.4	4.88	50
April 14, 2000	55	4.97	62
April 19, 2000	73.6	4.98	70
April 24, 2000	84.5	5.04	90
April 29, 2000	85	4.98	95
May 4, 2000	93	3.76	95
May 9, 2000	93	3.29	95
May 14, 2000	93.2	2.93	90
May 19, 2000	93.2	2.38	88
May 24, 2000	93.2	1.55	80

Source: Adapted from Yucheng Integrated Agricultural Experiment Station of Chinese Academy of Sciences in Shandong province.

The simulation results show that

$$LAI = \begin{cases} -51.915 + 11.603 \cdot \ln(NDVI_s) \\ \text{when } NDVI_s \text{ increases (correlation coefficient is 0.889)} \\ \\ -51.751 + 11.077 \cdot \ln(NDVI_s) \\ \text{when } NDVI_s \text{ decreases (correlation coefficient is 0.860)} \end{cases} \quad (15.26)$$

$$H = \begin{cases} -238.753 + 2.251 \cdot NDVI_s & \begin{array}{l} \text{when } NDVI_s \text{ increases} \\ \text{(correlation coefficient is 0.868)} \end{array} \\ 93.2 \text{ cm} & \text{when } NDVI_s \text{ decreases} \end{cases} \quad (15.27)$$

$$\sigma = \begin{cases} e^{28.308 + 5.679 \ln(NDVI_s)} \\ \text{when } NDVI_s \text{ increases (correlation coefficient is 0.920)} \\ \\ e^{-3.712 + 0.733 \ln(NDVI_s)} \\ \text{when } NDVI_s \text{ decreases (correlation coefficient is 0.889)} \end{cases} \quad (15.28)$$

where $NDVI_s = 100 \cdot (NDVI + 1)$; LAI is the leaf area index; H is the mean height of wheat; and σ is the fractional vegetation cover.

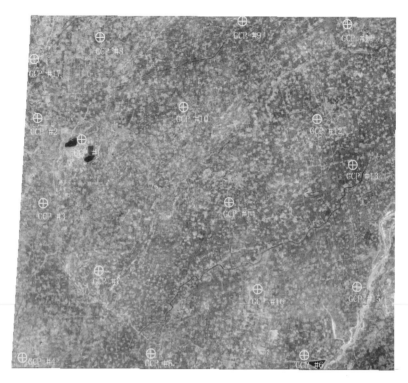

FIGURE 15.1
LANDSAT-5 TM image and control points for geometrical correction.

The scaled NDVI, $NDVI_s$, is derived from NOAA-14 data. $NDVI_s$ increases from early March to early May and decreases from early May to June. $NDVI_s$ ranges from 90 to 160 at the Yucheng Integrated Agricultural Experiment Station of the Chinese Academy of Sciences in the Shandong province.

15.3.2 Simulation of Horizontal Wind Velocity above the Wheat Surface at Yucheng Integrated Agricultural Experiment Station of the Chinese Academy of Sciences in the Shandong Province

According to Equations 15.4 and 15.8 as well as discussions on roughness length and displacement height, the horizontal wind velocity at height z and at time t is formulated as

$$u(z,t) = \frac{2.44 u_*(t)}{\sigma(t)((6.4\ LAI(t))^b - 1) + 1}$$

$$\times \ln \frac{\left[\sigma(t)((6.4\ LAI(t))^b - 1) + 1\right]z - 0.64 H(t)\sigma(t)(6.4\ LAI(t))^b}{0.13 H(t) \cdot (6.4\ LAI(t))^{2b}} \qquad (15.29)$$

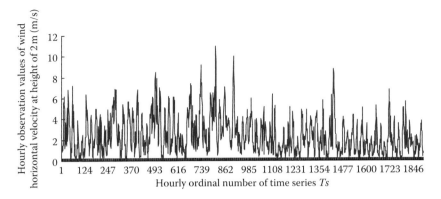

FIGURE 15.2
Hourly observation of horizontal wind velocity at a height of 2 m during the period from 9:05 pm on March 5, 2000 to 7:05 am on May 24, 2003 at the Yucheng Integrated Agricultural Experiment Station (when the observation time is 9:05 pm on March 5, 2000, $Ts = 1$; when the observation time is 7:05 am on May 24, 2000, $Ts = 1929$).

where $u_*(t)$ is the friction velocity over the wheat surface; $LAI(t)$ is the leaf area index; $H(t)$ is the mean height of wheat; $\sigma(t)$ is the fractional vegetation cover of wheat; b is the wheat parameter of the dimensionless constant to be simulated.

The simulation result most fits the observation of horizontal wind velocity at height 2 m when $b = 0.39$ (Figure 15.2). The horizontal wind velocity is expressed as (Yue et al., 2007)

$$u(z,t) = \frac{2.44u_*(t)}{2.06\sigma(t)(LAI(t))^{0.39} + (1 - \sigma(t))}$$

$$\times \ln \frac{\left[2.06\sigma(t)(LAI(t))^{0.39} + (1 - \sigma(t))\right]z - 1.32H(t)\sigma(t)(LAI(t))^{0.39}}{0.55H(t)(LAI(t))^{0.78}}$$

(15.30)

The hourly horizontal wind velocity at a height of 4 m during the period from 9:05 pm on March 5, 2000 to 7:05 am on May 24, 2000, which is simulated by means of formulation (Equation 15.30), is almost identical to the wind velocity in the observation (Figure 15.3). The correlation coefficient between the simulation results and the observation value is 0.998.

15.4 Discussion and Conclusions

It is found that the horizontal wind velocity above a nonuniform underlying surface is determined by roughness length, zero-plane displacement height,

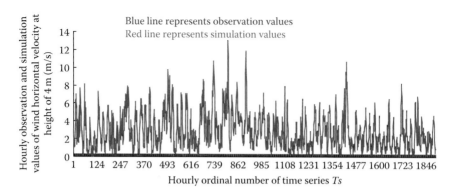

FIGURE 15.3
Comparison between hourly observation and the simulation of horizontal wind velocity at a height of 4 m during the period from 9:05 pm on March 5, 2000 to 7:05 am on May 24, 2003 at the Yucheng Integrated Agricultural Experiment Station (when the observation time is 9:05 pm on March 5, 2000, $Ts = 1$; when the observation time is 7:05 am on May 24, 2000, $Ts = 1929$).

the dimensionless constant, and vegetation fractional cover. The roughness length and zero-plane displacement height can be respectively expressed as a mathematical function of the mean height of roughness elements; the dimensionless constant can be formulated as a function of the *LAI*. Therefore, the mean height of roughness elements, fractional vegetation cover, and *LAI* are the most essential parameters in the formulation of horizontal wind velocity. The case study at the Yucheng Integrated Agricultural Experiment Station of the Chinese Academy of Sciences in the Shandong Province shows that the mean height of roughness elements, fractional vegetation cover, and *LAI* are closely related to the NDVI. A model of the vertical wind profile (MVWP), for example, the relationship between horizontal wind velocity and NDVI, is thus finally established. The simulated horizontal wind velocity is almost the same as the observed wind velocity at the Yucheng Integrated Agricultural Experiment Station, which means that MVWP is applicable to formulating horizontal wind velocity under thermally neutral conditions.

The spatial modeling of wind remains a significant issue in wind engineering studies. Wind fields were fitted by least-squares polynomial interpolation (Panofsky, 1949). Spline was used to spatially analyze wind fields (Fritsch, 1971). IDW was employed to interpolate wind velocity fields (Goodin et al., 1979). Statistical interpolation approaches, such as optimum interpolation (Dartt, 1972), diagnostic wind field numerical models (Porch and Rodriguez, 1987), and the Gauss–Markov theorem-based method (Feliks et al., 1996), were applied to time in a series of *in situ* observations to produce the required surface wind fields (Wang et al., 1998). Kriging and the artificial neural network (ANN) approach have been developed for spatial modeling and the prediction of wind velocity at any desired point, provided that there is a set of surrounding wind measurement sites (Oeztopal, 2006). All these approaches use black-box modeling with a set of input factors and output variables; the results of the input factors are then treated in a systematic manner. A combination of MVWP with HASM would create an alternative method for dynamically simulating wind velocity on different heights above various land covers.

References

Asrar, G., Kanemasu, E.T., and Yoshida, M. 1985. Estimates of leaf area index from spectral reflectance of wheat under different cultural practices and solar angle. *Remote Sensing of Environment* 17: 1–11.

Baldauf, M. and Fiedler, F. 2003. A parametrization of the effective roughness length over inhomogenous, flat terrain. *Boundary-Layer Meteorology* 106: 189–216.

Baret, F., Clevers, J.G.P.W., and Steven, M.D. 1995. The robustness of canopy gap fraction estimates from red and near-infrared reflectances: A comparison of approaches. *Remote Sensing of Environment* 54: 141–151.

Beinhauer, R. and Kruse, B. 1994. Soil erosivity by wind in moderate climates. *Ecological Modelling* 75/76: 279–287.

Blackadar, A.K. and Tennekes, H. 1968. Asymptotic similarity in neutral barotropic planetary boundary layers. *Journal of the Atmospheric Sciences* 25: 1015–1020.

Blennow, K. and Sallnaes, O. 2004. WINDA—A system of models for assessing the probability of wind damage to forest stands within a landscape. *Ecological Modelling* 175(1): 87–99.

Brown, L.J., Chen, J.M., Leblance, S.G., and Cihlar, J. 2000. A shortwave infrared modification to the simple ratio for LAI retrieval in Boreal forests: An image and model analysis. *Remote Sensing of Environment* 71: 16–25.

Chen, J.M., Pavlic, G., Brown, L., Cihlar, J., Leblance, S.G., White, H.P., Hall, R.J. et al., 2002. Derivation and validation of Canada-wide coarse-resolution leaf area index maps using high-resolution satellite imagery and ground measurements. *Remote Sensing of Environment* 80: 165–184.

Choudhury, B.J., Ahmed, N.U., Idso, S.B., Reginato, R.J., and Daughtry, C.S.T. 1994. Relations between evaporation coefficients and vegetation indexes studied by model simulation. *Remote Sensing of Environment* 50: 1–17.

Clevers, J.G.P.W. 1989. The application of a weighted infrared-red vegetation index for estimating leaf area index by correcting for soil moisture. *Remote Sensing of Environment* 29: 25–37.

Colombo, R., Bellingeri, D., Fasolini, H., and Marino, C.M. 2003. Retrieval of leaf area index in different vegetation types using high resolution satellite data. *Remote Sensing of Environment* 86: 120–131.

Dartt, D.G. 1972. Automated streamline analysis utilizing optimum interpolation. *Journal of Applied Meteorology* 11: 901–908.

Deardorff, J.W. 1978. Efficient prediction of ground surface temperature and moisture, with inclusion of a layer of vegetation. *Journal of Geophysical Research* 83: 1889–1903.

De Bruin, H.A.R. and Moore, C.J. 1985. Zero-plane displacement and roughness length for tall vegetation, derived from a simple mass conservation hypothesis. *Boundary-Layer Meteorology* 42: 53–62.

Defries, R.S., Hansen, M.C., and Townshend, J.R.G. 2000a. Global continuous fields of vegetation characteristics: A linear mixture applied to multi-year 8 km AVHRR data. *International Journal of Remote Sensing* 21: 1389–1414.

Defries, R.S., Hansen, M.C., Townshend, J.R.G., Janetos, A.C., and Loveland, T.R. 2000b. A new global 1-km dataset of percentage tree cover derived from remote sensing. *Global Change Biology* 6: 247–254.

Defries, R.S., Townshend, J.R.G., and Hansen, M.C. 1999. Continuous fields of vegetation characteristics at the global scale. *Journal of Geophysical Research* 104: 16911–16923.

Feliks, Y., Gavze, E., and Givati, R. 1996. Optimal vector interpolation of wind fields. *Journal of Applied Meteorology* 35: 1153–1165.

Fritsch, J.M. 1971. Objective analysis of a two-dimensional data field by the cubic spline technique. *Monthly Weather Review* 99: 379–386.

Gardiner, B., Peltola, H., and Kellomaeki, S. 2000. Comparison of two models for predicting the critical wind speeds required to damage coniferous trees. *Ecological Modelling* 129(1): 1–23.

Garratt, J.R. 1992. *The Atmospheric Boundary Layer*. London: Cambridge University Press.

Gash, J.H., Shuttleworth, W.J., and Lioyd, C.R. 1989. Micrometeorological measurements in Les Landes Forest during HAPEX-Mobility. *Agricultural and Forest Meteorology* 46: 131–147.

Goodin, W.R., McRae, G.J., and Seinfeld, J.H. 1979. A comparison of interpolation methods for sparse data: Application to wind and concentration fields. *Journal of Applied Meteorology* 18: 761–771.

Goward, S.N., Tucker, C.J., and Dye, D.G. 1985. North American vegetation patterns observed with NOAA-AVHRR. *Vegetation* 64: 3–14.

Gower, S.T. and Norman, J.M. 1991. Rapid estimation of leaf area index for forests using LI-COR LAI-2000. *Ecology* 72: 1896–1900.

Grimmond, C.S.B. and Oke, T.R. 1999. Aerodynamic properties of urban areas derived from analysis of surface form. *Journal of Applied Meteorology* 38: 1262–1292.

Gutman, G. and Ignatov, A. 1998. The derivation of the green vegetation fraction from NOAA/AVHRR data for use in numerical weather prediction models. *International Journal of Remote Sensing* 19: 1533–1543.

Hanna, S.R. and Chang, J.C. 1992. Boundary layer parameterizations for applied dispersion modelling over urban areas. *Boundary-Layer Meteorology* 58: 229–259.

Huete, A.R. 1988. A soil adjusted vegetation index (SAVI). *Remote Sensing of Environment* 25: 295–309.

Huete, A.R., Liu, H.Q., Batchily, K., and Van Leeuwen, W. 1997. A comparison of vegetation indices over a global set of TM images fro EOS-MODIS. *Remote Sensing of Environment* 59: 440–451.

Jasinski, M.F. and Crago, R.D. 1999. Estimation of vegetation aerodynamic roughness of natural regions using frontal area density determined from satellite imagery. *Agricultural and Forest Meteorology* 94: 65–77.

Kaufman, Y.J. and Tanre, D. 1992. Atmospherically resistant vegetation index (ARVI) for EOS-MODIS. *IEEE Transactions on Geoscience and Remote Sensing* 30: 261–270.

Kelllerer, A.M. 1983. On the number of clumps resulting from the overlap of randomly placed figures in a plane. *Journal of Applied Probability* 20: 126–135.

Laric, B. 1997. *Profile of Wind Speed in Transition Layer above the Vegetation.* Master Thesis in Serbian, University of Belgrade.

Liu, H.Q. and Heaute, A.R. 1995. A feedback based modification of the NDVI to minimize canopy background and atmospheric noise. *IEEE Transactions on Geoscience and Remote Sensing* 33: 457–465.

Massman, W. 1987. A comparative study of some mathematical models of the mean wind structure and aerodynamic drag of plant canopies. *Boundary-Layer Meteorology* 40: 179–197.

Mihailovic, D.T., Lalic, B., Rajkovic, B., and Arsenic, I. 1999. A roughness sub-layer wind profile above a nonuniform surface. *Boundary-Layer Meteorology* 93: 425–451.

Nemani, R., Pierce, L., Running, S.W., and Band, L. 1993. Forest ecosystem processes at the watershed scale: Sensitivity to remotely sensed leaf area index estimates. *International Journal of Remote Sensing* 14: 2519–2534.

Oeztopal, A. 2006. Artificial neural network approach to spatial estimation of wind velocity data. *Energy Conversion and Management* 47: 395–406.

Panofsky, H.A. 1949. Objective weather-map analysis. *Journal of Meteorology* 6: 386–392.

Parlange, M.R. and Brutsaert, W. 1989. Regional roughness of the Landes Forest and surface shear stress under neutral conditions. *Boundary-Layer Meteorology* 48: 69–76.

Porch, W. and Rodriguez, D. 1987. Spatial interpolation of meteorological data in complex terrain using temporal statistics. *Journal of Climate and Applied Meteorology* 26: 1696–1708.

Price, J.C. and Bausch, J.G. 1995. Leaf area index estimation from visible and near-infrared reflectance data. *Remote Sensing of Environment* 52: 55–65.

Qi, J., Kerr, Y.H., Moran, M.S., Weltz, M., Huete, A.R., Sorooshian, S., and Bryant, R. 2000. Leaf area index estimates using remotely sensed data and BRDF models in a semiarid region. *Remote Sensing of Environment* 73: 18–30.

Raupach, M.R. 1992. Drag and drag partition on rough surface. *Boundary-Layer Meteorology* 60: 375–395.

Raupach, M.R. 1994. Simplified expressions for vegetation roughness length and zero-plane displacement as functions of canopy height and area index. *Boundary-Layer Meteorology* 71: 211–216.

Raupach, M.R. and Thom, A.S. 1981. Turbulence in and above plant canopies. *Annual Review of Fluid Mechanics* 13: 97–129.

Running, S.W., Baldocchi, D.D., Turner, D.P., Gower, S.T., Bakwin, P.S., and Hibbard, K.A. 1999. A global terrestrial monitoring network integrating tower fluxes, flash sampling, ecosystem modeling and EOS satellite data. *Remote Sensing of Environment* 70: 108–127.

Rykaczewski, R.R. and Checkley, Jr. D.M. 2008. Influence of ocean winds on the pelagic ecosystem in upwelling regions. *Proceedings of the National Academy of Sciences* 105(6): 1965–1970.

Santos, M., Bastos, R., Travassos, P., Bessa, R., Repas, M., and Cabral, J.A. 2010. Predicting the trends of vertebrate species richness as a response to wind farms installation in mountain ecosystems of northwest Portugal. *Ecological Indicators* 10: 192–205.

Scanlon, T.M., Albertson, J.D., Caylor, K.K., and Williams, C.A. 2002. Determining land surface fractional cover from NDVI and rainfall time series for a savanna ecosystem. *Remote Sensing of Environment* 82: 376–388.

Schippers, P. and Jongejans, E. 2005. Release thresholds strongly determine the range of seed dispersal by wind. *Ecological Modelling* 185(1): 93–103.

Schumacher, S., Bugmann, H., and Mladenoff, D.J. 2004. Improving the formulation of tree growth and succession in a spatially explicit landscape model. *Ecological Modelling* 181(1): 175–194.

Sutton, O.G. 1953. *Micrometeorology*. New York, NY: McGraw-Hill Publishing Company Ltd.

Wang, W.S., Nowlin Jr., W.D., and Reid, R.O. 1998. Analyzed surface meteorological fields over the Northwestern Gulf of Mexico for 1992–94: Mean, seasonal, and monthly patterns. *Monthly Weather Review* 126: 2864–2883.

White, M.A., Asner, C.P., Nemani, R.R., Privette, J.L., and Running, S.W. 2000. Measuring fractional cover and leaf area index in arid ecosystems: Digital camera, radiation transmittance, and laser altimetry methods. *Remote Sensing of Environment* 74: 45–57.

Wiegand, C.L. and Richardson, A.J. 1987. Spectral components analysis: Rationale and results for three crops. *International Journal of Remote Sensing* 8: 1011–1032.

Wieringa, J. 1993. Representative roughness parameters for homogeneous terrain. *Boundary-Layer Meteorology* 63: 323–363.

Wieringa, J. 1981. Estimation of meso-scale and local-scale roughness for atmospheric transport modeling. In *Air Pollution Modeling and Its Application*, ed., C. de Wispelaere, 279–295. New York, NY: Plenum.

Wittich, K.P. 1997. Some simple relationships between land-surface emissivity, greenness and the plant cover fraction for use in satellite remote sensing. *International Journal of Biometeorology* 41: 58–64.

Wittich, K.P. and Hansing, O. 1995. Area-averaged vegetative cover fraction estimated from satellite data. *International Journal of Biometeorology* 38: 209–215.

Wulder, M.A. 1998. The prediction of leaf area index from forest polygons decomposed through the integration of remote sensing, GIS, UNIX, and C. *Computer and Geosciences* 24: 151–157.

Vaesen, K., Gilliams, S., Nackaerts, K., and Coppin, P. 2001. Ground-measured spectral signatures as indicators of ground cover and leaf area index: The case of paddy rice. *Field Crops Research* 69: 13–25.

Yang, R.Q. and Friedl, M.A. 2003. Determination of roughness length for heat and momentum over boreal forests. *Boundary-Layer Meteorology* 107: 581–603.

Yue, T.X., Chen, S.P., Xu, B., Liu, Q.S., Li, H.G., Liu, G.H., and Ye, Q.H. 2002. A curve-theorem based approach for change detection and its application to Yellow River delta. *International Journal of Remote Sensing* 23: 2283–2292.

Yue, T.X., Wang, W., Yu, Q., Zhu, Z.L., Zhang, S.H., Zhang, R.H., and Du, Z.P. 2007. A simulation of vertical wind profile under neutral condition. *International Journal of Remote Sensing* 28(10): 2207–2219.

Zeng, X., Dickinson, R.E., Walker, A., Shaikh, M., Defries, R.S., and Qi, J. 2000. Derivation and evaluation of global 1-km fractional vegetation cover data for land modeling. *Journal of Applied Meteorology* 39: 826–839.

16

Surface Modeling of Climatic Change and Scenarios*

16.1 Introduction

The simulations of ecosystems and their services require weather and climate data. There are two available sources for these data: (1) meteorological stations, and (2) general circulation models (GCMs). However, sparsely distributed meteorological stations are often unable to satisfy the data requirements of such studies. Most GCM simulations use coarse resolution with horizontal resolution being about 200–500 km (Xue et al., 2007). It is difficult to use GCMs to assess climate change impacts on various ecosystems on regional and local levels because of their coarse spatial resolution, although CGMs can provide a good overview of both current and future climates on a global level. Grotch and MacCracken (1991) found that the range of changes in temperature and precipitation predicted by different computer models is much broader at finer spatial scales; many shortcomings were apparent in the model simulations of the present climate. Von Storch et al. (1993) observed that simulations of GCMs were questionable on a regional level. Ciret and Sellers (1998) stated that the accuracy of regional climate predictions was linked to the spatial resolution at which GCMs operated; increasing the spatial resolution of GCMs would improve the simulation of climate, and hence, increase confidence in the use of GCM output for impact studies. Covey et al. (2003) argued that it was difficult to determine whether or not the models were good enough to be trusted when used to study climate either in the distant past or for future predictions. Raisanen (2007) concluded that many small-scale processes could not be simulated explicitly in current climate models. Prudhomme and Davies (2009) indicated that GCMs often resulted in different climate outputs from the same atmospheric and oceanic drivers, especially at regional scales.

One major problem is how to estimate values for locations where primary data is not available (Akinyemi and Adejuwon, 2008). Many dynamic and statistical downscaling approaches have been developed to improve the poor performance of GCMs at local and regional scales (Charles et al., 1999; Ashiq

---

* Drs. Ze-Meng Fan, Chuan-Fa Chen, and Xiao-Fang Sun are major collaborators of this chapter.

et al., 2010). For instances, a fine computational grid over a limited domain was nested within the coarse grid of a GCM in limited area models (LAMs) (Anthes, 1983; Giorgi, 1990; Walsh and McGregor, 1997). A numerical method for modeling climate on a regional scale was also developed, whereby large-scale weather systems were simulated with a global climate model (GCM), and the GCM output was used to provide boundary conditions needed for high-resolution mesoscale model simulations over the region of interest (Dickinson et al., 1989). A one-way nesting method was used to develop a 50 km regional-climate model driven by output from a GCM (Jones et al., 1995). A lateral nesting method was capable of improving/adding more climate information at different scales compared to GCM (Xue et al., 2007). A spectral nudging method was used to force a regional model to adopt prescribed large scales over an entire domain, not just at the lateral boundaries, while developing realistic detailed regional features consistent with the large scales (Radu et al., 2008).

The derivation of local-scale information from integrations of coarse-resolution GCMs with the help of statistical models fitted to present observations is generally referred to as the statistical downscaling approach (Zorita and von Storch, 1999). Statistical approaches have been widely used to bridge the gap between the large scale and the local scale. The basic idea behind statistical downscaling is to use the observed relationship between the large-scale and the local-scale climate variables for a projection of GCM results on a regional or local scale (Bergant and Kajfez-Bogataj, 2005). Weather-typing approaches (Brown and Katz, 1995) involved the grouping of local, meteorological variables in relation to different classes of atmospheric circulation. Transfer function was the most popular approach to downscaling, which derives from observational data using a statistical relationship (Hewitson and Crane, 1996; Murphy, 1999). Artificial neural nets were used in an empirical downscaling procedure to derive daily subgrid-scale precipitation from GCM geopotential height and specific humidity data (Crane and Hewitson, 1998), and to construct scenarios of daily temperature for a future decade (2090–2099) (Trigo and Palutikof, 1999). Using the analog method, a large-scale circulation simulated by a GCM was associated with local variables observed simultaneously with the most similar large-scale circulation pattern in a pool of historical observations (Zorita and von Storch, 1999). The multiway partial least-squares regression was introduced as an empirical downscaling tool in climate change studies (Bergant and Kajfez-Bogataj, 2005). A support vector machine (SVM) approach was proposed for the statistical downscaling of precipitation on a monthly time-scale (Tripathi et al., 2006). A statistical downscaling approach based on sparse Bayesian learning and relevance vector machine (RVM) was presented to model stream flow at river basin scale for the monsoon period (June, July, August, September) using GCM simulated climatic variables (Ghosh and Majumdar, 2008). An automated statistical downscaling (ASD) regression-based approach was presented and assessed in order to reconstruct the observed climate, which indicated that the modeling of temperature was one of ASD's strengths, whereas precipitation was a clear weakness (Hessami et al., 2008). A combined downscaling-disaggregation weather generator was developed for the multisite generation of hourly precipitation and temperature time series over complex terrain (Mezghani and Hingray, 2009).

GIS-based interpolation techniques have been widely used for the interpolation of observed or modeled point climatic variable data (Agnew and Palutikof, 2000; Yue et al., 2005, 2006, 2007a). Lloyd (2005) compared the performances of different interpolation methods, moving window regression (MWR), IDW, and kriging methods. The comparison results demonstrated that methods using elevation as secondary data performed better than others because of the relationships between climate, such as temperature, precipitation and evaporation, and elevation. A HASM-based downscaling method is developed by combining with statistical transfer functions on global, national, and provincial levels in this chapter.

16.2 Downscaling of HadCM3 Scenarios on a Global Level

DEM from the SRTM and observed data from meteorological stations were used to establish statistical transfer functions. HASM-PGS is driven by the transfer functions to downscale the outputs of HADCM3.

16.2.1 Data Sets

The spatial resolution of the SRTM DEM is 3 arc seconds, about 90 m (http://srtm.csi.cgiar.org). The DEM has an absolute vertical accuracy of ±16 m, a relative vertical accuracy of ±10 m for 90% of the data, and an absolute horizontal accuracy of ±20 m (Walker et al., 2007).

The climate change data set is taken from the China Meteorological Data Sharing Service System and Monthly Climatic Data for the World (MCDW). MCDW is a publication that summarizes climatic data from around the world on a monthly basis, prepared by the National Oceanic and Atmospheric Administration (NOAA), in cooperation with the World Meteorological Organization (WMO). Data are prepared by members of the WMO for selected stations and exchanged monthly via international telecommunications under the provisions of WMO Technical Regulations, WMO Publication No. 306 (Manual on Codes), and WMO Publication No. 386 (Manual on the Global Telecommunications System).

HadCM3, the third version of the Hadley Center coupled model, requires no flux adjustment and has a stable climate in the global mean (Collins et al., 2001). The HadCM3 climatic scenarios to be used in this chapter were developed by A1Fi, A2, and B2 experiments (Johns et al., 2003). The A1 scenario family describes a future world with very rapid economic growth, a global population that peaks mid-century and declines thereafter, and the rapid introduction of new and more efficient technologies. Its A1Fi group assumes that technological changes in the energy system will be fossil intensive (Watson et al., 2001). The A2 scenario family describes a very heterogeneous world with more rapid population growth but less rapid economic growth than A1, which is self-reliant and preserving of local identities. The B2 scenario family describes a world in which population increases at a lower rate than A2 and the emphasis is on finding local solutions to economic, social, and environmental sustainability. In the HadCM3A1Fi and HadCM3A2 scenario, greenhouse gas emissions eventually increase faster than in IS92a, whereas in B2, CO_2 emissions are lower than under IS92a. In HadCM3A1Fi and particularly in HadCM3A2 sulfur emissions increase but then decrease whereas, in HadCM3B2 sulfur emissions decrease throughout the twenty-first century. According to these three scenarios, in the next two decades the global-mean warming rate would be similar to that seen in the recent decades. However, the global-mean warming in HadCM3A1Fi and HadCM3A2 would be noticeably greater than that in HadCM3B2 by the middle of the twenty-first century, and would be greater than that in IS92a by the end of the century (Johns et al., 2003).

16.2.2 Methods

The DEM on spatial resolution 90×90 m is converted into that of spatial resolution $0.125° \times 0.125°$ by use of the resampling method (Efron, 1979). The relationships of precipitation and temperature with elevation and latitude are statistically analyzed on the basis of data from 2783 meteorological stations scattered across the Earth by using $0.125° \times 0.125°$ DEM as auxiliary data (Figure 16.1). The meteorological data include the temperature, precipitation, elevation, and geographical location of each meteorological station during the period 1949 to 2009. The statistical analysis indicates that mean annual precipitation (MAP) has no significant relationship with elevation and latitude while mean annual temperature (MAT) closely depends on elevation and latitude. Therefore, HASM-PGS, as seen in Chapter 3, is directly employed to downscale the precipitation scenarios of HadCM3 by using DEM and latitude data as auxiliary data.

A statistical analysis of temperature data from the 2783 meteorological stations indicates that temperature has a close relationship with latitude and elevation. The squared multiple correlations are respectively 0.962 and 0.914 in the northern hemisphere (*latitude* $\geq 0°$) and the southern hemisphere (*latitude* $< 0°$). The statistical transfer functions in the northern hemisphere and in the southern hemisphere can be respectively formulated as

$$Tnor = 39.147 - 0.725 \cdot Lnor - 0.004 \cdot Enor \tag{16.1}$$

$$Tsou = 31.514 + 0.414 \cdot Lsou - 0.004 \cdot Esou \tag{16.2}$$

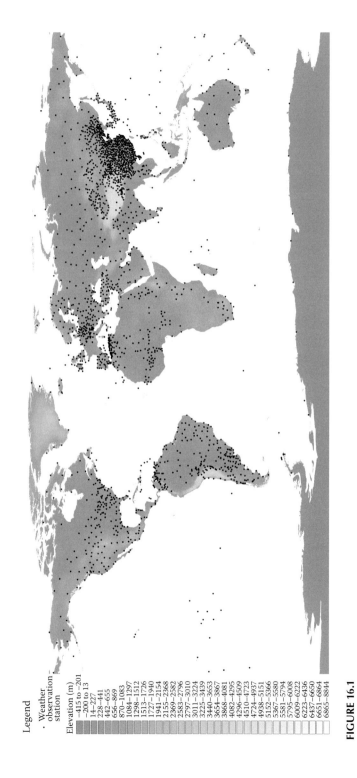

FIGURE 16.1

Global DEM and 2783 meteorological stations (black points).

where *Tnor, Lnor,* and *Enor* are respectively the simulated MAT, latitude, and elevation in the northern hemisphere; *Tsou, Lsou,* and *Esou* are respectively the simulated MAT, latitude, and elevation in the southern hemisphere.

The data of HadCM3A1Fi, A2, and B2 scenarios with a spatial horizontal grid point spacing of 3.75° longitude and 2.5° latitude have 7008 grid cells, of which the northern hemisphere and the southern hemisphere have 3504 grid cells each. The grid cells in the northern hemisphere are numbered 1, 2, ..., 3504 and the ones in southern hemisphere as 3505, 3506, ..., and 7008. The preprocessing intercepts of the statistical transfer functions of the MAT in the northern and southern hemispheres can be respectively expressed as

$$Inor_i(x,y,t) = Tnor_i(x,y,t) + 0.725 \cdot Lnor_i(x,y) + 0.004 \cdot Enor_i(x,y) \qquad (16.3)$$

$$Isou_j(x,y,t) = Tsou_j(x,y,t) - 0.414 \cdot Lsou_j(x,y) + 0.004 \cdot Esou_j(x,y) \qquad (16.4)$$

where $Inor_i$ (x, y, t), $Tnor_i$ (x, y, t), $Lnor_i$ (x, y), and $Enor_i$ (x, y) are respectively the processed intercept, MAT, latitude, and elevation of ith grid cell in the northern hemisphere according to HadCM3 data on spatial resolution $2.5° \times 3.75°$ where $i = 1, 2, ..., 3504$; $Isou_j$ (x, y, t), $Tsou_j$ (x, y, t), $Lsou_j$ (x, y), and $Esou_j$ (x, y) are respectively the processed intercept, MAT, latitude, and elevation of jth grid cell on the spatial resolution $2.5° \times 3.75°$ in the southern hemisphere where $j = 3505, 3506, ..., 7008$.

The processed intercepts $\{Inor_i$ $(x, y, t)\}$ and $\{Isou_j$ $(x, y, t)\}$ are respectively downscaled to the spatial resolution $0.125° \times 0.125°$ by combining HASM-MGS with Equations 16.3 and 16.4. The downscaled intercepts in the northern and southern hemispheres can be respectively expressed as $\{Inor_k$ $(x, y, t)\}$ and $\{Isou_l$ $(x, y, t)\}$, in which $k = 1, 2, ..., 2077201$ and $l = 1, 2, ..., 2074320$.

The improved transfer functions in the northern and southern hemispheres, based on the downscaled intercepts $\{Inor_k$ $(x, y, t)\}$ and $\{Isou_l$ $(x, y, t)\}$, are respectively formulated as

$$Tnor_k(x,y,t) = Inor_k(x,y,t) - 0.725 \cdot Lnor_k(x,y) - 0.004 \cdot Enor_k(x,y) \qquad (16.5)$$

$$Tsou_l(x,y,t) = Isou_l(x,y,t) + 0.414 \cdot Lsou_l(x,y) - 0.004 \cdot Esou_l(x,y) \qquad (16.6)$$

where $Inor_k$ (x, y, t), $Tnor_k$ (x, y, t), $Lnor_k$ (x, y), and $Enor_k$ (x, y) are respectively the downscaled intercept, MAT, latitude, and elevation of kth grid cell in the northern hemisphere on spatial resolution $0.125° \times 0.125°$ where $i = 1,2, ..., 2077201$; $Isou_l$ (x, y, t), $Tsou_l$ (x, y, t), $Lsou_l$ (x, y), and $Esou_l$ (x, y) are respectively the processed intercept, MAT, latitude, and elevation of lth grid cell on the spatial resolution $0.125° \times 0.125°$ in the southern hemisphere where $j = 1,2, ..., 2074320$.

Equations 16.5 and 16.6 are used to downscale temperature scenarios of HadCM3 on a global level.

16.2.3 Changes of Mean Annual Temperature and Precipitation

An analysis of the downscaled HadCM3 scenarios of A1Fi, A2a and B2a indicate that both the MAT and MAP would show an increasing trend on a global level during the periods T1 (1961–1990), T2 (2010–2039), T3 (2040–2069), and T4 (2070–2099) (Table 16.1). We can summarize Table 16.1 if we use MAT to represent the mean annual temperature (°C), MAP to represent the mean annual precipitation (mm), S to represent scenario, WMF to represent a weather main factor, WFV to represent the value of a weather main factor, WFVD1 to represent the difference by subtracting WFV in T1 from WFV in T2, WFVD2 to represent the difference by subtracting the climatic component value in T2 from its value in T3, and WFVD3 to represent the difference by subtracting the climatic component value in T3 from its value in T4.

16.2.3.1 Mean Annual Temperature and Precipitation Changes under Scenario A1Fi

The MAT on a global level would see an accelerated rising trend on an average during the period T1 to T4 (Figure 16.2). It would increase by 1.5246°C during the period T1 to T2, by 1.9676°C during the period T2 to T3, and by 2.3221°C during the period T3 to T4.

During the period T1 to T2, the temperature would in relative terms increase much less, from 0 to 1°C in northeast Asia, central Asia, the south of Australia, the northern edge of Antarctica, the edge around the African continent, the southern area and outer margins of South America, southwest Greenland, and the northwest outer margins of North America. During this

TABLE 16.1

Scenarios of Mean Annual Temperature and Precipitation on a Global Level on an Average

S	WMF	WFV in T1	WFVD1	WFV in T2	WFVD 2	WFV in T3	WFVD 3	WFV in T4
A1fi	MAT	3.2494	1.5246	4.7740	1.9676	6.7416	2.3221	9.0637
	MAP	558.1285	7.4418	565.5703	12.9426	578.5129	25.5084	604.0213
A2a	MAT	3.3297	1.3278	4.6575	1.4966	6.1541	1.9106	8.0647
	MAP	555.4442	7.6056	563.0498	10.7437	573.7935	19.9927	593.7862
B2a	MAT	3.3152	1.5470	4.8622	0.8922	5.7544	1.2367	6.9911
	MAP	555.4794	7.4070	562.8864	9.2784	572.1648	6.4048	578.5696

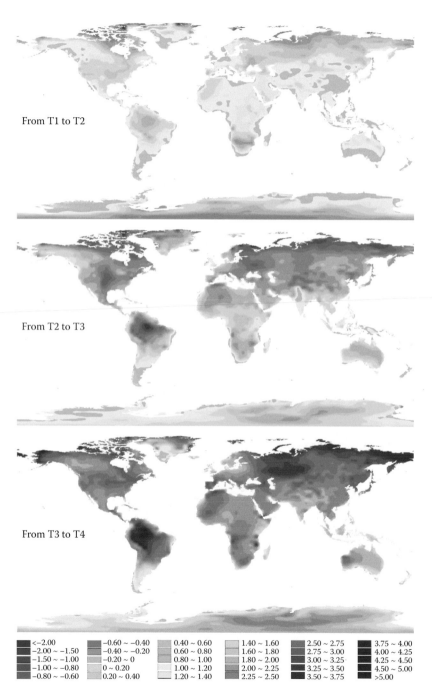

FIGURE 16.2
Changes of mean annual temperature in different periods under the A1Fi scenario.

same period, a warming rate of 1 ~ 2°C would be seen in Southwest Asia, north Australia, middle and northern Africa, middle South America, and northwest North America. The warming rate would be above 2°C in the northern Eurasian Continent, the northern area of North America, middle South America, northern Africa, and inland of Antarctica.

During the period T2 to T3, the warming rate would be above 2°C for most of the earth's surface, with the exception of southern Indonesia, the islands around the Island of New Guinea, the coastal area of the Island of New Guinea, northwest India, the edge of southern Africa, the southern corner and east border of South America, and the outer margins of Antarctica, which would see a temperature increase of less than 1°C in most of Antarctica, and southern India would see a temperature rise ranging between 1°C and 2°C.

During the period T3 to T4, most areas of the earth's surface would see an annual increase in the annual mean temperature of more than 2°C, with the exception of southwest Greenland, the outer margins of northwest Europe, the coastal area of southeast South America, the coastal area of south Australia, and the coastal area of southwest Africa, which would see warming rate of less than 1°C. The range of warming rates would be between 1°C and 2°C in the outer margins of Antarctica, the west and middle of Greenland, the southeast of South America, and northwest Europe.

On an average, the MAP would see an accelerating increase trend during the period T1 to T4 under the A1Fi scenario. The MAP would increase by 7.4418 mm during the period T1 to T2, by 12.9426 mm from T2 to T3, and by 25.5084 mm from T3 to T4 (Figure 16.3).

During the period T1 to T2, the MAP of most of the earth's surface would increase, particularly in eastern Indonesia, where it would reach more than 400 mm. But the MAP would decrease by more than 100 mm in the northeast area of South America, Latin America, southern Africa, the west, and middle areas around the Mediterranean Sea, and most of Australia and southwest Asia. The decrease range in western Greenland and the middle of Antarctica inland would be between 25 and 50 mm.

During the period T2 to T3, most of the earth's surface would see increased precipitation. The MAP would grow by more than 400 mm in south Asia, the Island of New Guinea and the Solomon Islands. But the MAP may decrease by more than 100 mm in the northeast of South America, southern Africa, the area around the Mediterranean, central Asia, and almost half of Australia.

From T3 to T4, most of the earth's surface would see more precipitation. The MAP would increase by more than 400 mm in the middle area of south Asia and the northern area of the Island of New Guinea. But MAP may decrease by more than 100 mm in northeast South America, western Africa, Latin America, western Australia, and the western areas around the Mediterranean.

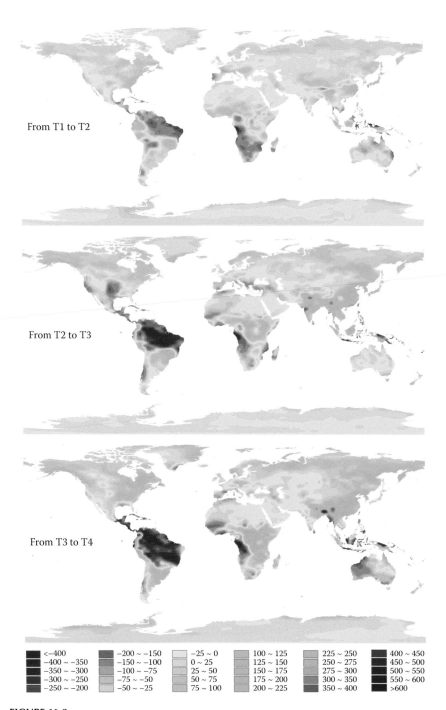

FIGURE 16.3
Changes of mean annual precipitation in different periods under the A1Fi scenario.

16.2.3.2 Mean Annual Temperature and Precipitation Changes under Scenario A2a

The MAT would have an accelerating increase trend under scenario A2a during the period T1–T4 (Figure 16.4). The temperature during the period T1–T2, T2–T3, and T3–T4 would respectively increase by 1.3278°C, 1.4966°C, and 1.9106°C.

During the period T1 to T2, areas where the temperature would increase by more than 2°C include northern Asia, northern Europe, northern North America, northern South America, the southern and northern inland of Africa, and the central area of Antarctica. The temperature may rise less than 1°C in southern Asia, southern Australia and the coastal areas of Australia, the southern area and edge of South America, the outer margin of the African continent, southern Greenland, western Canada, and west Europe.

During the period T2 to T3, the areas, where the temperature increase might not exceed 1°C, include the islands in southern Asia and around Australia, the outer margins of the African continent, west Europe, the southern corner of South America, the northeast and northern coastal areas of South America, the northwestern coastal area of North America and the western boundary area between Canada and the United States. The MAT would increase by more than 2°C in northeastern Asia, middle and Eastern Europe, northern Canada, Greenland, and the northern inland of South America.

From T3 to T4, the temperature increase would exceed 2°C for most terrestrial surfaces. Several relatively small areas might see an increase of less than 1°C, including southwest Greenland, the islands in the middle of Indonesia, the coastal areas of southern and western South America, the southern coastal area of Australia, New Zealand, the southwest coastal area of the African continent, and the outer margin of Antarctica.

The MAP would increase continuously on an average during the period T1 to T4 under the scenario A2a. The precipitation would increase by 7.6056, 10.7437, and 19.9927 mm, respectively during the periods T1 to T2, T2 to T3, and T3 to T4 (Figure 16.5).

During the period T1 to T2, most terrestrial surfaces would see greater precipitation. MAP might increase by more than 400 mm in eastern Indonesia and northeastern Papua New Guinea. But precipitation would decrease by more than 100 mm in Latin America, northern South America, southern Africa, western and eastern areas of Australia, Middle East Asia, and Southeast Asia.

From T2 to T3, the MAP would generally become higher. In the southeast Tibet Plateau and southwest coastal area of South America, precipitation might increase by more than 400 mm. But in southern areas of North America, Latin America, northeast South America, southwest Africa, and Northeast Australia, precipitation might decrease by more than 100 mm.

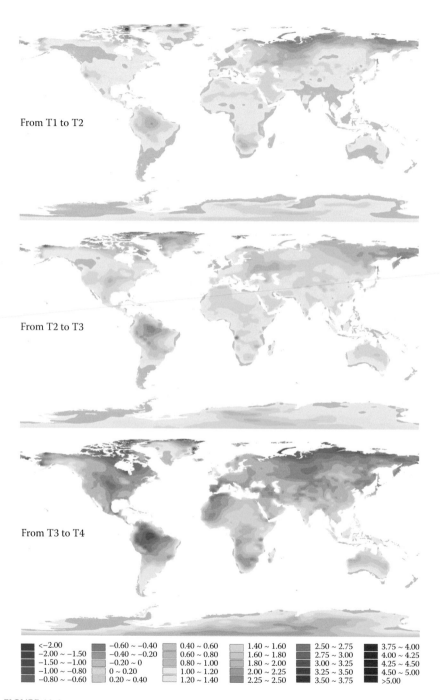

From T1 to T2

From T2 to T3

From T3 to T4

<−2.00	−0.60 ~ −0.40	0.40 ~ 0.60	1.40 ~ 1.60	2.50 ~ 2.75	3.75 ~ 4.00
−2.00 ~ −1.50	−0.40 ~ −0.20	0.60 ~ 0.80	1.60 ~ 1.80	2.75 ~ 3.00	4.00 ~ 4.25
−1.50 ~ −1.00	−0.20 ~ 0	0.80 ~ 1.00	1.80 ~ 2.00	3.00 ~ 3.25	4.25 ~ 4.50
−1.00 ~ −0.80	0 ~ 0.20	1.00 ~ 1.20	2.00 ~ 2.25	3.25 ~ 3.50	4.50 ~ 5.00
−0.80 ~ −0.60	0.20 ~ 0.40	1.20 ~ 1.40	2.25 ~ 2.50	3.50 ~ 3.75	>5.00

FIGURE 16.4
Changes of mean annual temperature in different periods under the A2a scenario.

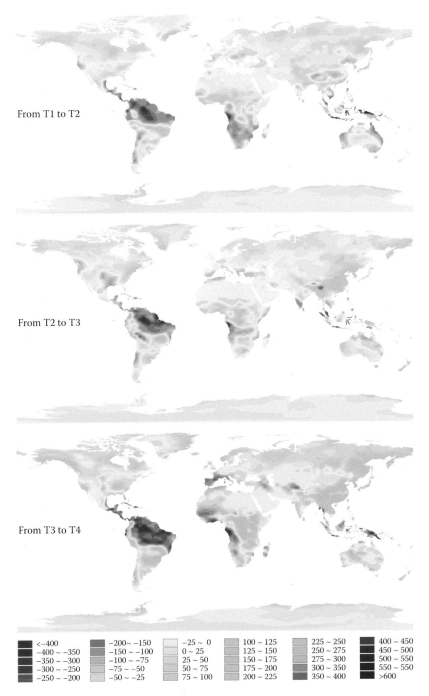

FIGURE 16.5
Changes of mean annual precipitation in different periods under the A2a scenario.

From T3 to T4, MAP would generally see a rising trend, increasing by more than 400 mm in southern Burma, western Indonesia, and Papua New Guinea. However, precipitation would decrease by more than 100 mm for a considerable part of the Earth's surface, including areas such as the northeast area of South America, Western Africa, the area around the Mediterranean Sea, and western Asia.

16.2.3.3 Mean Annual Temperature and Precipitation Changes under Scenario B2a

MAT would rise continuously during the period T1 to T4 under scenario B2a. It might increase by 1.5470°C, 0.8922°C and 1.2367°C, respectively in the periods T1–T2, T2–T3, and T3–T4 (Figure 16.6).

From T1 to T2, the MAT would increase by more than 2°C for most terrestrial surfaces, but there would be cooling in southern Saudi Arabia. The increase would be less than 1°C in southern Central Asia, the northern inland of Africa, the United Kingdom, Norway, the coastal area of west Europe, Iceland and southeast Greenland, the southern corner area of South America, northern Antarctica, and the outer margins of south Australia.

From T2 to T3, almost all terrestrial surfaces would be warmer, with the exception of western Iceland, the coastal areas of Antarctica along the Bellingshausen Sea and the Amundsen Sea, which would see decreases in temperature. The areas where temperature rise would be between 1°C and 2°C include the eastern inland of North America, northern Greenland, the northern inland of South America, southwest Africa, the inland of Portugal and Spain, northern Asia, southern Central Asia and western China, and some areas in western and eastern Australia.

From T3 to T4, the temperature of terrestrial surfaces would become higher, with the exception of Tasmania Island and the Abbott Ice Shelf of Antarctica, which would see decreases in temperature. The MAT would increase by more than 2°C in northern South America, northern North America, northern Asia, and the central area of Antarctica.

The MAP would increase by 7.4070, 9.2784 and 6.4048 mm, respectively during the periods T1–T2, T2–T3, and T3–T4 under scenario B2a on an average (Figure 16.7). From T1 to T2, most terrestrial surfaces would see increased precipitation. The MAP might increase by more than 400 mm in the eastern areas of Indonesia and Papua New Guinea. But there would be a decrease of more than 100 mm in northeast South America, northern Chile, Latin America, most of southern North America, southern Africa, most of the area around the Mediterranean Sea, the southeast area of Southeast Asia and eastern Australia. From T2 to T3, the MAP for most of the earth's surface would see an increase trend, but considerably large areas might see a decrease in precipitation. Precipitation might decrease by more than 100 mm in northeast South America. From T3 to T4, the precipitation on most areas would

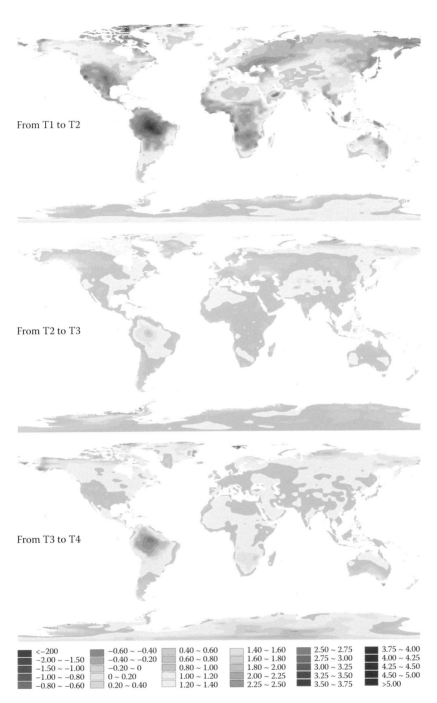

FIGURE 16.6
Changes of mean annual temperature in different periods under the B2a scenario.

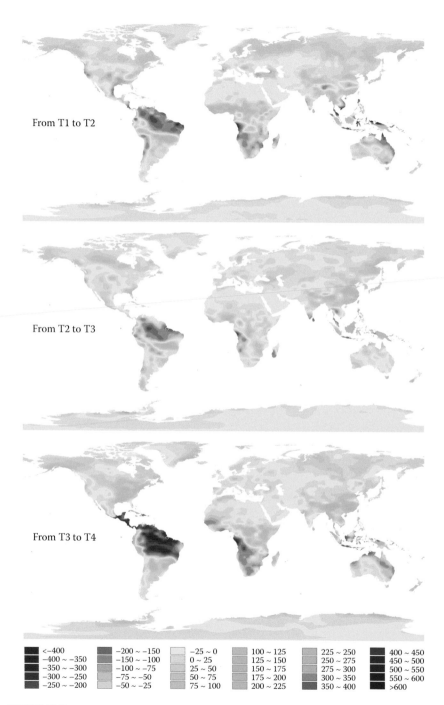

FIGURE 16.7
Changes of mean annual precipitation in different periods under the B2a scenario.

increase, but the MAP might decrease by more than 100 mm in Latin America, northeast South America, southern and northwest Africa, and northeastern and western Australia.

16.2.3.4 Brief Summary

In terms of the three scenarios of HadCM3 A1Fi, A2a, and B2a, both the MAT and MAP would see an increasing trend in the twenty-first century. The three scenarios indicate that those areas where the MAT would be lower than 0°C in Greenland and lower than −35°C in Antarctica would shrink. The area, in which the MAT would be higher than 40°C between north and south tropics, would expand. Scenario A2a has the highest starting temperature, while Scenario A1Fi has the lowest starting temperature, but the highest starting precipitation in period T1. Scenario A1Fi gives the fastest increase rates of both the MAT and MAP in the period T1 to T4, while B2a produces the slowest increased rates, meaning it outputs the lowest temperature and precipitation in the period T4.

16.3 Simulation of Climatic Surfaces on a National Level

16.3.1 Method

Owing to the small 11-year cyclicity of climate change in China (Yue et al., 2005), the MAT and MAP are calculated for every 11-year period according to 10-day observation data provided by 752 meteorological stations scattered across the whole of China during the period 1964–2007 (Figure 16.8). The period 1964–2007 is divided into four subperiods: C1 (from 1964 to 1974), C2 (from 1975 to 1985), C3 (from 1986 to 1996), and C4 (from 1997 to 2007). The MAT and MAP at each meteorological station are calculated in terms of these subperiods. These property feature data are transferred into spatial feature data in terms of coordinate information supplied by every meteorological station. SRTM DEM (http://srtm.csi.cgiar.org) of China on a spatial resolution of 90×90 m is transferred into DEM on a spatial resolution of 1×1 km using the resampling method. The DEM is used as auxiliary data to develop the statistical transfer functions of MAT and MAP (Figure 16.9).

16.3.1.1 Statistical Transfer Function of Temperature

A statistical analysis of the time-series data of temperature at the 752 meteorological stations during the period 1964–2007 indicates that the mean annual temperature has a close relationship with latitude and elevation. The

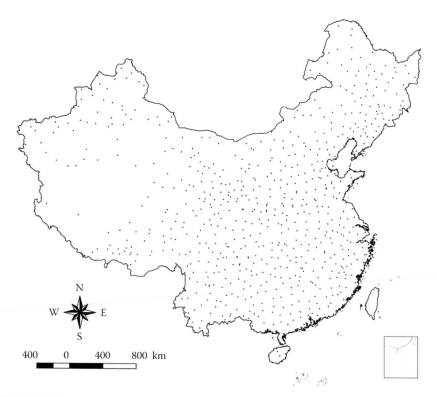

FIGURE 16.8
Spatial distribution of the meteorological stations in China (China Meteorological Data Sharing Service System).

statistical transfer function of MAT, with a multiple correlation coefficient of 0.9631, can be formulated as

$$Ts = 38.7582 - 0.7102 \cdot La - 0.0034 \cdot Ele \tag{16.8}$$

where Ts represents the simulated MAT; La latitude and Ele elevation.

Then, the preprocessing intercept of the statistical transfer function of MAT at meteorological station k at time t can be expressed as

$$Int_k(x,y,t) = T_k(x,y,t) + 0.7102 \cdot La_k(x,y) + 0.0034 \cdot Ele_k(x,y) \tag{16.9}$$

where $Int_k(x, y, t)$ is the preprocessed intercept; $T_k(x, y, t)$ the MAT from observed data at the meteorological station k; $La_k(x, y, t)$ the latitude and $Ele_k(x, y, t)$ the elevation at the meteorological station k.

The preprocessed intercept values $\{Int_k(x, y, t)\}$ at all meteorological stations are employed to drive HASM-PGS to obtain the intercept value at every

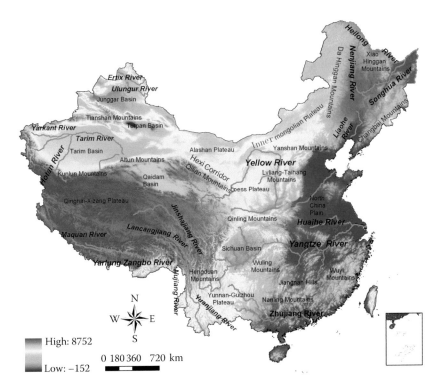

FIGURE 16.9
Digital elevation model of China.

grid cell *i*. The grid cell size is 1×1 km. 5000×4320 grid cells are involved. Finally, the simulated temperature at every grid cell *i*, in which $i = 1, 2, ..., 19606916$, can be formulated as

$$Ts_i(x, y, t) = IntT_i(x, y, t) - 0.7102 \cdot La_i(x, y) - 0.0034 \cdot Ele_i(x, y) \qquad (16.10)$$

where $Ts_i(x, y, t)$ represents the simulated temperature at time *t*; $IntT_i(x, y, t)$ the interpolated intercept value at grid cell *i*; (x, y) the coordinate of grid cell *i*; $La_i(x, y)$ the latitude of grid cell *i*; and $Ele_i(x, y)$ the elevation of grid cell *i*.

16.3.1.2 Statistical Transfer Function of Precipitation

Most of the atmospheric moisture is derived by the evaporation of ocean water and controlled by the transportation of air masses from the tropics to the polar areas. During this transportation, the air masses cool down, leading to continuous condensation and rain-out from low to high latitudes on both hemispheres (van der Veer et al., 2009). Latitude and longitude can be used to reflect the influence of general circulation and continentality on precipitation. Spatial variability in precipitation in the complex terrain is caused by the dependence of precipitation on altitude and windward effects (Franke et al., 2008). A statistical analysis of precipitation data from the 752 meteorological stations demonstrates that precipitation has a close relationship with the aspect index, the sky view factor, latitude, longitude, and elevation. The statistical transfer function of MAP with a multiple correlation coefficient of 0.8883 can be formulated as

$$
\begin{aligned}
Pre = {} & 2659.73384 + 0.00018 \cdot Lo - 0.00048 \cdot La \\
& - 0.08469 \cdot Ele - 40.47906 \cdot ICA - 107.59924 \cdot SVF
\end{aligned}
\tag{16.11}
$$

where *Pre* is the simulated MAP; *La* represents latitude; *Lo* refers to longitude; *Ele* is elevation; *ICA* is the impact coefficient of aspect on precipitation, and *SVF* the sky view factor.

If a digital elevation model is regarded as a bivariate function $z = f(x, y)$, then slope angle can be defined by

$$
Slope = \arctan\left(\sqrt{(f_x)^2 + (f_y)^2}\right) \cdot 57.29578
\tag{16.12}
$$

where *Slope* is the slope angle; $f_y = (\partial f/\partial y)$ and $f_x = (\partial f/\partial x)$.

The sky view factor can be formulated as

$$
SVF = \frac{1 + \cos(\pi \cdot (Slope/180^\circ))}{2}
\tag{16.13}
$$

Aspect is the direction that a slope faces. The value of each grid cell in an aspect grid indicates the direction in which the cell's slope faces. Aspect is measured counterclockwise in degrees. 0° represents due north, 90° due east, 180° due south, and 270° due west. The aspect due south has the biggest precipitation so that the value of 1 is given to the ICA because of the considerable effects of the Southwest Monsoon and Southeast Monsoon, that is, the effect of warm and wet airflow from the Pacific Ocean and Indian Ocean. A value of –1 is given to ICA due north. A value of 0 is given in flat areas. The ICA can be expressed as

$$
Asp = \begin{cases} -\cos\left(\pi \cdot \left(\dfrac{Aspect}{180^\circ}\right)\right) & \text{slope area} \\[2mm] 0 & \text{flat area} \end{cases}
\tag{16.14}
$$

The preprocessed intercept of the transfer function of MAP at meteorological station k at time t can be expressed as

$$IntP_k(x,y,t) = Pre_k(x,y,t) - 0.00018 \cdot Lo_k(x,y) + 0.00048 \cdot La_k(x,y)$$
$$+ 0.08469 \cdot Ele_k(x,y) + 40.47906 \cdot ICA_k(x,y)$$
$$+ 107.59924 \cdot SVF_k(x,y) \tag{16.15}$$

HASM-PGS is driven to obtain the intercept at every grid cell i on a spatial resolution of 1×1 km by the preprocessed intercepts $\{IntP_k(x, y, t)\}$ at all meteorological stations. The simulated precipitation at every grid cell i, $i = 1, 2, \ldots, 21600000$, can be formulated as

$$Pre_i(x,y,t) = IntP_i(x,y,t) + 0.00018 \cdot Lo_i(x,y) - 0.00048 \cdot La_i(x,y)$$
$$- 0.08469 \cdot Ele_i(x,y) - 40.47906 \cdot ICA_i(x,y)$$
$$- 107.59924 \cdot SVF_i(x,y) \tag{16.16}$$

where $Pre_i(x, y, t)$ is the simulated MAP at grid cell i; $IntP_i(x, y, t)$ is the interpolated intercept by HASM-MGS at grid cell i; $La_i(x, y)$ is the latitude of grid cell i; $Lo_i(x, y)$ is the longitude of grid cell i, $Ele_i(x, y)$ is the elevation of grid cell i, $ICA_i(x, y)$ is the impact coefficient of aspect on the precipitation at grid cell i; and $SVF_i(x, y)$ is the sky view factor of grid cell i.

16.3.2 Changing Trends of Mean Annual Temperature and Precipitation

The simulation results in terms of Equations 16.10 and 16.16 indicate that the MAT has shown an accelerating increase trend since 1964 (Figures 16.10 and 16.11). MAT rose 0.1631°C during the period C1 to C2, 0.3835°C from C2 to C3 and 0.7007°C from C3 to C4. The MAP increased by 4.7069 mm during the period C1 to C2, decreased by 2.1335 mm from C2 to C3 and continuously decreased by 2.3735 mm from C3 to C4 (Figures 16.12 and 16.13). The changes of MAT and MAP are summarized in Table 16.2. If ΔC21 represents the result of subtracting the value of a WMF in C1 from the value of a WMF in C2, ΔC32 represents the result of subtracting the value of a WMF in C2 from the one in C3, and ΔC43 of subtracting the one in C3 from the one in C4.

16.3.2.1 *Mean Annual Temperature*

During the period C1 to C2, most of China showed a cooling trend, with the exception of Northeast China, Inner Mongolia, northwestern Xinjiang, and

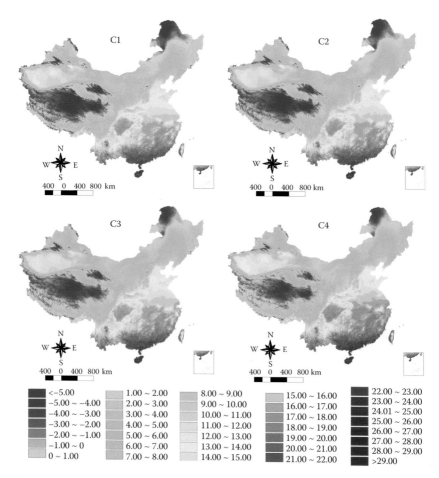

FIGURE 16.10
Mean annual temperature in different periods.

several local areas in the Tibet plateau. The range of temperature decrease was between 0.25°C and 0.50°C in almost all the cooling areas. Owing to the more than 2°C rise in temperature in western Inner Mongolia and local areas in the southwestern Tibet plateau, as well as a heating scope of between 0.5°C and 1.0°C in northeast Helongjiang and northwest Xinjiang, MAT across the whole of China increased by 0.1631°C on an average in the period C2 compared with the period C1.

During the period C2 to C3, Northeast China, the northern area of North China, and most of western China saw a warming trend. The temperature increased by more than 1°C in Northeast China, Inner Mongolia, northeastern Xinjiang, and the northern Tibet plateau. However, MAT showed a decreasing trend in Southwest China, east China, the southern margins of

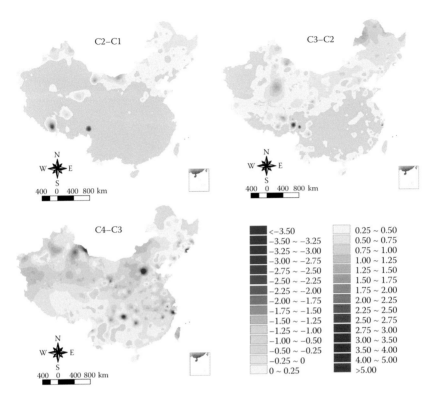

FIGURE 16.11
Changes of mean annual temperature in China.

Tibet plateau, and western Xinjiang. The decreasing scope was between 0.25°C and 0.50°C.

MAT increase sped up in most of China during the period C3 to C4. The rise in temperature was more than 2°C in northeastern Xinjiang, eastern Zhejiang, the area where Hebei and Shanxi meet, and the area on the border between Chongqing and Guizhou. However, there was a cooling trend in the eastern and western corner of Helongjiang, local areas of the Tianshan mountain, several local areas in the eastern Tibet plateau, several local areas in southwest China, and several small local areas in eastern China.

16.3.2.2 Mean Annual Precipitation

During the period C1 to C2, MAP had an increasing trend in most of China. This was particularly evident in the southeastern area of China and some parts of middle China, which saw increases of more than 200 mm. But there was a decreasing trend in the mountainous areas of Southwest China, the Shandong province, the northern Jiangsu province, and the eastern area of

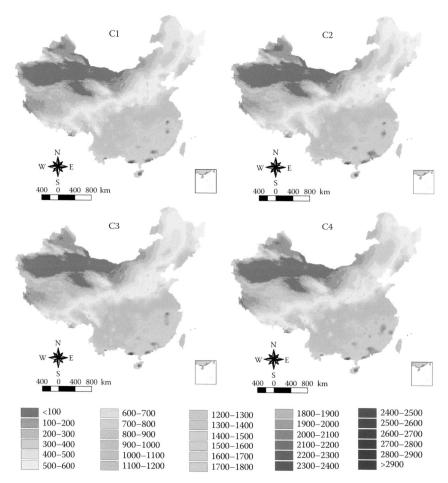

FIGURE 16.12
Mean annual precipitation in different periods.

Northeast China. The decrease amplitude was more than 100 mm (Figure 16.13).

From C2 to C3, MAP had a reducing trend on an average in the whole of China. The reduction was more than 250 mm in eastern Tibet, the boundary area between Chongqing and Guizhou, Fujian, southern Jiangxi, Guangdong, southern Henan, and the border between Shaanxi and Sichuan. Areas where MAP increased were mainly distributed in Northeast China, the Xinjiang Uygur autonomous region, the North China Plain, the lower reaches of the Yangtze River, and Taiwan.

During the period C3 to C4, MAP had a decreasing trend on average. Areas where MAP decreased were mainly distributed in Northeast China, eastern Inner Mongolia, the lower reaches of the Yangtze River and South

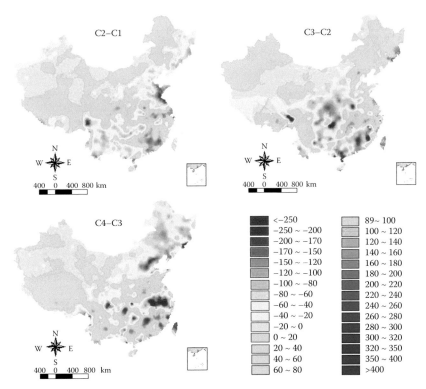

FIGURE 16.13
Changes of mean annual precipitation in China.

China. MAP decreased by more than 250 mm in Jilin, Beijing, Hebei, Anhui, Hubei, Zhejiang, Sichuan and Hubei. There was an increased MAP in Gansu, Qinghai, Tibet, Yunnan, Guizhou, Shaanxi, Henan, Shandong, and Taiwan.

In short, MAT had a rapid increase trend during the period 1964 to 2007. The fastest increase happened in the subperiod C3 to C4. MAP showed a considerable increase from C1 to C2, but decreased from C2 to C4. The changes of MAT and MAP had great spatial variety. Climate change was from warm-and-humid to warm-and-dry in the investigated period.

TABLE 16.2

Changing Trends of Mean Annual Temperature and Precipitation in China

	C1	ΔC21	C2	ΔC32	C3	ΔC43	C4
MAT (°C)	6.9495	0.1631	7.1126	0.3835	7.4960	0.7007	8.1967
MAP (mm)	612.7988	4.7069	617.5057	−2.1335	615.3722	−2.3735	612.9987

16.3.3 Scenarios of Mean Annual Temperature and Precipitation

16.3.3.1 Downscaling of HadCM3 Climatic Scenarios

Data sets on a spatial resolution of $3.75° \times 2.50°$ are downloaded from the IPCC Data Distribution Center (http://ipcc-ddc.cru.uea.ac.uk), which include A1Fi, A2 and B2 scenarios of HadCM3 of monthly mean temperature and precipitation. There are 7008 grid cells. The time periods are from 1961 to 1990, 2010 to 2039, 2040 to 2069, and 2070 to 2099; these are respectively termed as T1, T2, T3, and T4.

The China section (Figure 16.14) of the downloaded data sets are downscaled to a spatial resolution of 1×1 km by using transfer functions developed in Section 16.3.1 to drive HASM-PGS. The downscaled results show that temperature and precipitation should increase on an average in the next 100 years in China (Figures 16.15 through 16.26, Table 16.3). In Table 16.3, WMF is the abbreviation of weather main factor. $\Delta T21$ stands for subtracting WMF in T1 from WMF in T2, $\Delta T32$ stands for subtracting WMF

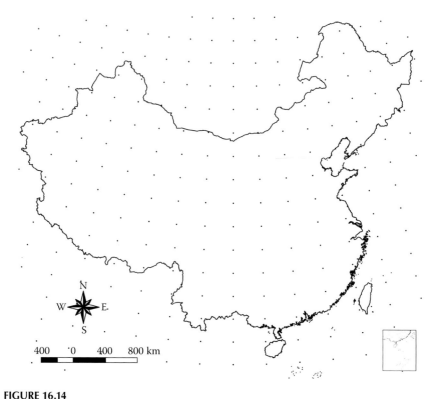

FIGURE 16.14
Spatial distribution of grid cells in China in terms of data sets of HadCM3 climatic scenarios (http://ipcc-ddc.cru.uea.ac.uk).

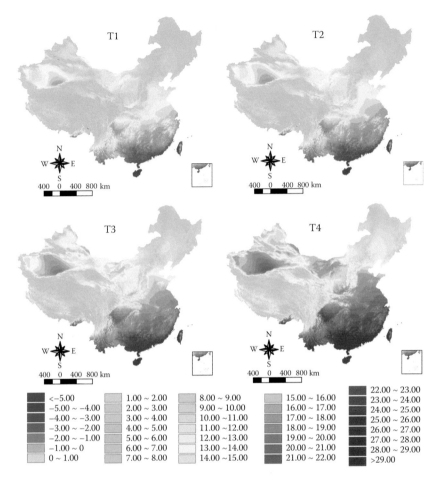

FIGURE 16.15
Mean annual temperature in China in different periods under scenario HadCM3 A1Fi.

in T2 from WMF in T3, and ΔT43 stands for subtracting WMF in T3 from WMF in T4.

16.3.3.2 Climatic Scenarios under A1Fi

MAT would have an accelerating warming trend under scenario A1Fi during the period T1 to T4 (Figures 16.15 and 16.16). It would become 0.8121°C warmer in the period T1 to T2, 1.4475°C warmer from T2 to T3, and 1.8778°C warmer from T3 to T4.

From T1 to T2, most of China would be warmer than 0.50°C, with the exception of most of the Tibet Plateau, which would have a warming range of between 0.25°C–0.50°C. In middle Xinjiang, the southern areas of Yunnan

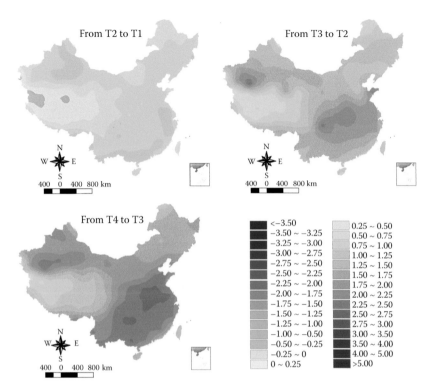

FIGURE 16.16
Temperature change under scenario HadCM3 A1Fi.

and Guangxi, and southwest Guangdong as well as several local parts in Fujian, Zhejiang, Jiangsu, and Anhui, temperature increase may reach above 1°C. It is noted that there would be a cooling trend in two local areas in the western Tibet Plateau.

From T2 to T3, MAT would increase at every grid cell of the terrestrial surfaces of China. In Tarim Basin, Sichuan Basin, and the Hubei province temperatures would increase by more than 2°C. Most of the Tibet plateau would see an increase range between 0.50°C and 0.75°C. The warming amplitude would be between 1°C and 2°C in the remaining areas of China.

From T3 to T4, MAT would be warmer more than 1°C in almost all areas of China with the exception of the western Tibet plateau, which would see an increase range of between 0.50°C and 0.75°C. The Tarim Basin, Hubei province, Henan province, and some local area in the Anhui province would all be warmer more than 2.5°C.

Under scenario of A1Fi, MAP would have an accelerating increase trend (Figures 16.17 and 16.18). It would increase by 36.4485 mm in the period from T1 to T2, by 72.5968 mm from T2 to T3, and by 93.6988 mm from T3 to T4.

During the period T1 to T2, most of the terrestrial surface of China would become wetter. MAP would increase by more than 200 mm in Jiangxi, Anhui, and Zhejiang. But it would become dryer in several local areas of the south Tibet plateau, northern Sichuan, eastern Gansu, southern Shaanxi, Shanxi, Hebei, Beijing, northern Shandong, Liaoning, the Zhujiang River Delta, middle Yunnan, and eastern Taiwan.

From T2 to T3, MAP would increase in China on an average. The areas where MAP would increase by more than 250 mm include the North China Plain, western Yunnan, Guangdong, and Taiwan. But MAP would show a decrease range between 40 and 60 mm in northeastern Tibet, southern Qinghai, and southern Xinjiang.

From T3 to T4, MAP would increase in almost all terrestrial surfaces of China with the exception of western Xinjiang, the Tarim Basin, and northern

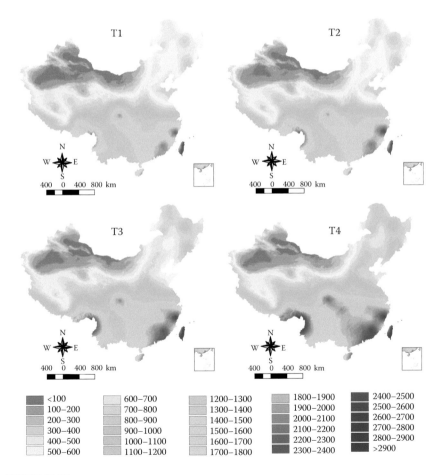

FIGURE 16.17
Mean annual precipitation in China in different periods under scenario HadCM3 A1Fi.

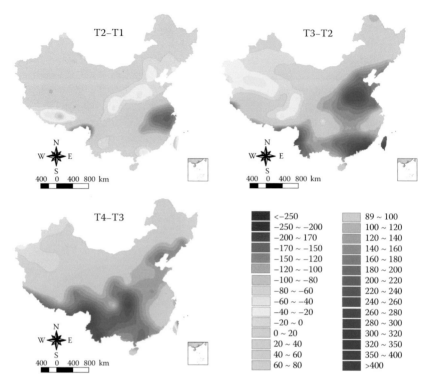

FIGURE 16.18
Precipitation change under scenario HadCM3 A1Fi.

Taiwan. In southern Tibet, western Yunnnan, Guizhou, Chongqing, and the border area between Hunan and Hubei, the increase would be more than 250 mm.

16.3.3.3 Climatic Scenarios under A2a

The rise in MAT would accelerate during the period T1 to T4 under scenario A2a of HadCM3 (Figures 16.19 and 16.20). MAT would rise to 0.8524°C from T1 to T2, 0.9283°C from T2 to T3, and 1.4966°C from T3 to T4.

From T1 to T2, MAT would rise considerably in most of China with the exception of the northwest of Tibet, which would see a cooling period. The rise would be more than 1°C in middle Xinjiang, Beijing, most of Hebei, the boundary area between Inner Mongolia and Liaoning, Henan, Anhui, Hubei, Hunan, Jiangxi, Chongqing, and Sichuan.

From T2 to T3, almost all of the terrestrial surface of China would become warmer with the exception of a small local area in western Tibet, which would experience cooling. The areas where MAT would increase by more than 1.25°C include Tarim Basin, Guizhou, Guangxi, Guangdong, Hunan, Jiangxi, Fujian, Anhui, Zhejiang, Jiangsu, and northeastern Shandong.

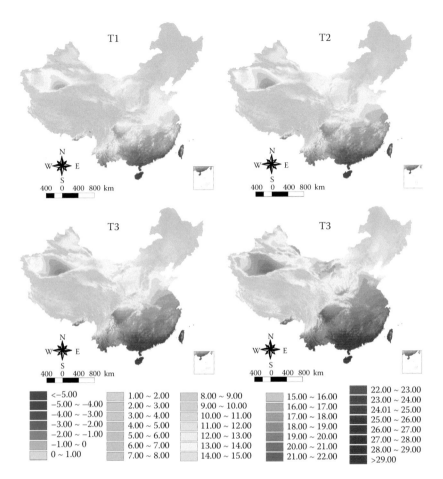

FIGURE 16.19
Mean annual temperature in China in different periods under scenario HadCM3 A2a.

From T3 to T4, all of the terrestrial surface of China would be much warmer. Most of China would be warmer more than 1°C, with the exception of the middle western Tibet Plateau in which the warming scope would be between 0.50°C and 0.75°C. MAT in the inland of the Tarim Basin would increase by more than 2.0°C.

MAP would show an accelerating increase trend during the period T1 to T4 under scenario A2a. MAP would increase by 9.2368 mm from T1 to T2, 71.5279 mm from T2 to T3, and 72.1649 mm from T3 to T4.

From T1 to T2, China would be wetter on an average. MAP would increase by more than 200 mm in the coastal area of southeast China but there would be a drying trend in the southeast Tibet Plateau, northwestern Yunnan, the Sichuan Basin, northeastern Guizhou, northeastern Hunan, Hubei,

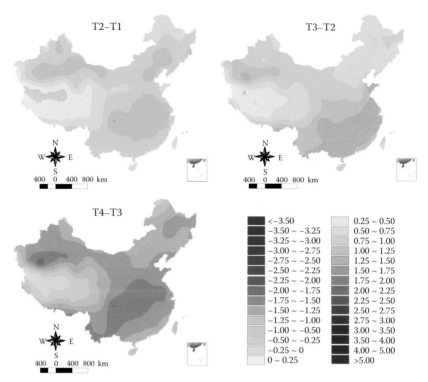

FIGURE 16.20
Temperature change under scenario HadCM3 A2a.

Northwestern Jiangxi, Anhui, southern Shaanxi, Henan, western Shandong, Hebei, Beijing, Tianjin, Liaoning, Jilin, and eastern Inner Mongolia.

From T2 to T3, MAP would increase in almost the entire terrestrial surface of China with the exception of several local areas in the middle of Tibet, western Xinjiang, central Qinghai, and southern Taiwan. MAP could increase by more than 250 mm in the southwestern margins of the Tibet Plateau, Chongqing, Guizhou, and the boundary area between Hunan and Hubei.

From T3 to T4, with the exception of middle and western Xinjiang and the boundary area between Inner Mongolia, Helongjiang and Jilin, almost all other areas of China would be much wetter. MAP would increase by more than 250 mm in southeast Tibet, western Yunnan, Sichuan, the border area between Gansu and Shaanxi, Henan, Anhui, the boundary area between Shandong and Jiangsu, Guangdong, and southern Taiwan.

16.3.3.4 Climatic Scenarios under B2a

MAT would continuously rise on an average under scenario HadCM3 B2a during the period T1 to T4 (Figures 16.23 and 16.24). It would increase

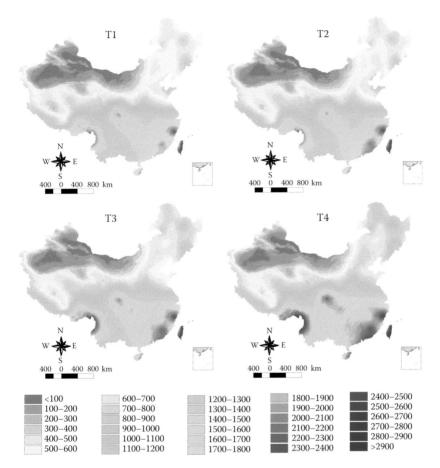

FIGURE 16.21
Mean annual precipitation in China in different periods under scenario HadCM3 A2a.

by 1.0069°C from T1 to T2, by 0.6669°C from T2 to T3, and by 0.7491°C from T3 to T4.

From T1 to T2, almost all of the terrestrial surface of China would have an increasing MAT. The warmest areas would include Northeast China, North China, Southwest China, and middle and eastern China, in which the range of MAT increase would be between 0.50°C and 2.0°C.

From T2 to T3, most of China would become warmer with the exception of the middle and western Tibet Plateau, in which the cooling range would be between 0.25°C and 0.50°C. The temperature would increase more than 1°C in the Tarim Basin, the border area between Sichuan and Chongqing, and the boundary area between Zhejiang and Fujian.

From T3 to T4, with the exception of southwest Tibet, all other terrestrial areas of China would become warmer. MAT would increase by more than

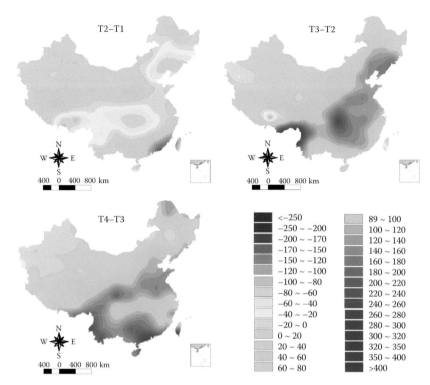

FIGURE 16.22
Precipitation change under scenario HadCM3 A2a.

1°C in the southwest corner of the Tarim Basin, eastern Yunnan, western Guangxi, Guizhou, southern Chongqing, northwestern Hunan, and middle Hubei.

MAP would show a considerable average increase under the scenario of HadCM3 B2a during the period T1 to T4 (Figures 16.25 and 16.26). MAP would increase by 25.7393 mm from T1 to T2, 39.6598 mm from T2 to T3, and 35.5541 mm from T3 to T4.

From T1 to T2, most of the terrestrial surface of China would be wetter, especially southern Guangdong, which would have a MAP increase of more than 200 mm. But MAP would decrease in the eastern Tibet Plateau, the area along the lower and middle reaches of Yangtze River, Taiwan, and Hainan.

From T2 to T3, MAP would increase in most of China. In the area along the boundaries of Bhutan, India and Myanmar, Henan, Anhui, Zhejiang, and Fujian, it would increase by more than 100 mm, but it would be dryer in the vast areas around Lasa, southern Xinjiang, and Jilin.

From T3 to T4, most of the terrestrial surface of China would become wetter. MAP would increase by more than 100 mm, particularly in the southwest boundary area of Tibet and northwest Yunan. But it would be

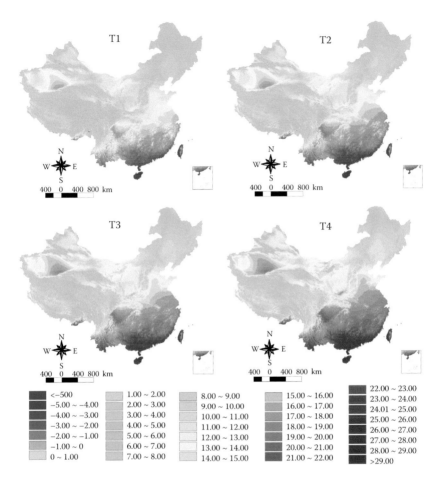

FIGURE 16.23
Mean annual temperature in China in different periods under scenario HadCM3 B2a.

dryer in western Xinjiang, northeastern Inner Mongolia, northern Hebei, western Liaoning, the boundary area between Helongjiang, Jilin and Inner Mongolia, and Jiangsu.

16.3.3.5 Brief Summary

The simulation results based on the three scenarios of HadCM3 A1Fi, A2a, and B2a indicate that both MAT and MAP would increase in twenty-first century. Scenario A1Fi has the highest initial MAT, but lowest initial MAP. A2a has the highest initial MAP, but the lowest initial MAT. A1Fi gives the fastest warming rate and precipitation increase. B2a presents the slowest increase rates of MAT and MAP. B2a shows the lowest MAT and MAP in the subperiod T4 compared with A1Fi and A2a.

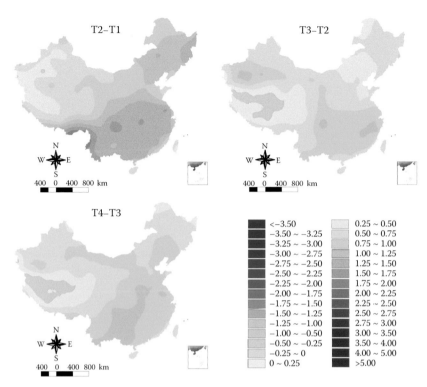

FIGURE 16.24
Temperature change under scenario HadCM3 B2a.

16.4 Simulation of Climatic Surfaces on a Provincial Level

16.4.1 Materials and Methods

Daily temperature and precipitation data from 100 meteorological stations scattered over and around the Jiangxi province (Figure 5.6 as seen in Chapter 5) are selected to develop statistical transfer functions of WMFs in the Jiangxi province during the period 1960 to 1999. DEM data set of the Jiangxi province on a spatial resolution of 90×90 m is downloaded from the USGS DEM data files (http://data.geocomm.com/dem). The MAT and MAP at every meteorological station are calculated in terms of the daily temperature and precipitation data during the period 1960–1999.

The transfer functions are developed by correlation analysis of the MAT and MAP with elevation. The statistical analysis shows that the MAT and MAP significantly correlate with elevation when the altitude is higher than

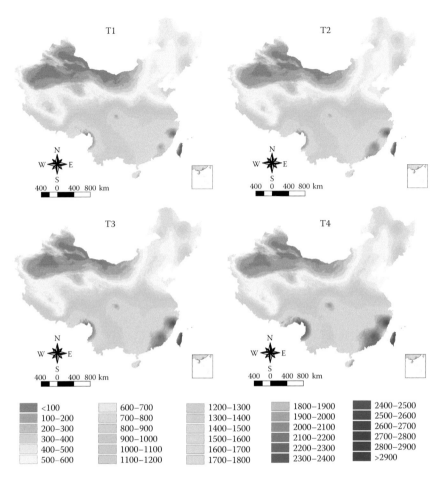

FIGURE 16.25
Mean annual precipitation in China in different periods under scenario HadCM3 B2a.

200 m. According to the climate data from the meteorological stations, of which altitudes are higher than 200 m, the correlation coefficient between MAT and elevation (R_t) is −0.9693 and the correlation coefficient between MAP and elevation (R_p) is 0.7389. When the analysis is limited to stations above 250 m in altitude, $R_t = -0.9608$ and $R_p = 0.9296$. R_t becomes −0.9959 and R_p increases to 0.9612 when data from stations higher than 300 m in altitude are analyzed statistically. Therefore, when the altitude is lower than 300 m, HASM is used directly to simulate the surfaces of MAT and MAP. The transfer functions as seen in Chapter 5, $Ts = -0.0065 \cdot Ele + 19.564$ and $Ps = 0.4021 \cdot Ele + 1558$ are combined with HASM when the altitude is higher than 300 m, in which Ts represents MAT, Ps is MAP, and Ele is elevation (higher than 300 m).

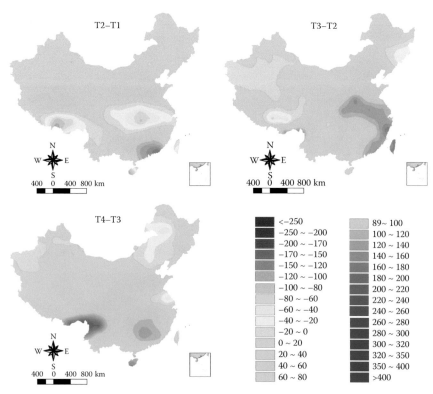

FIGURE 16.26
Precipitation change under scenario HadCM3 B2a.

16.4.2 Spatial Pattern of Mean Annual Temperature

MATs in the 1960s, 1970s, 1980s, and 1990s were 17.34°C, 17.07°C, 17.03°C, and 17.42°C in the Jiangxi province on an average (Table 16.4). In the period from the 1960s to the 1970s, MAT had a declining trend. The decadal decrease is 0.27°C on an average. But from the 1980s to the 1990s, MAT saw a rapid increase.

TABLE 16.3

Scenarios of Mean Annual Temperature and Precipitation

Scenarios	WMF	T1	ΔT21	T2	ΔT32	T3	ΔT43	T4
A1fi	MAT (°C)	10.5010	0.8121	11.3131	1.4475	12.7605	1.8778	14.6383
	MAP (mm)	799.2063	36.4485	835.6549	72.5968	908.2516	93.6988	1001.9505
A2a	MAT (°C)	10.4452	0.8524	11.2976	0.9283	12.2259	1.4966	13.7225
	MAP (mm)	808.7257	9.2368	817.9625	71.5279	889.4904	72.1649	961.6553
B2a	MAT (°C)	10.4551	1.0069	11.4620	0.6669	12.1290	0.7491	12.8781
	MAP (mm)	807.8286	25.7393	833.5679	39.6598	873.2277	35.5541	908.7818

TABLE 16.4

Changes of Mean Temperature and Mean Precipitation Every Decade since 1960s

| Period | Temperature | | Precipitation | |
	Mean Annual Temperature (°C)	Decadal Increase	Mean Annual Precipitation (mm)	Decadal Increase
1960s	17.34		1529.72	
1970s	17.07	−0.27	1662.83	133.11
1980s	17.03	−0.04	1667.85	5.02
1990s	17.42	0.39	1838.48	170.57

In 1970s, MAT decreased in the whole surface of Jiangxi province comparing with the one in the 1960s spatially (Figure 16.27). There was a particularly significant decrease in the area where Ying-Tan, Nan-Chang, Jin-De-Zhen, and Shang-Rao meet, in western Xin-Yu, in northeastern Yi-Chun, and in northern Ji-An. In the 1970s, isotherms of 17°C and 18°C as well as the northern isotherm of 19°C shifted 20 km to the south compared to their position in the 1960s; while the southern 19°C isotherm moved 19 km to the north (Figure 16.28). The area circled by 18°C in the east of Poyang lake in the 1960s disappeared in the 1970s.

In the 1980s, most of the Jiangxi province had a decreased MAT, with the exception of middle Jiu-Jiang, northeastern Jiu-Jiang, northern Jing-De-Zhen, northwestern Shang-Rao, the northern area where Shang-Rao and Ying-Tan meet, south Xin-Yu, southwestern Lin-Chuan and the western and southern boundary area of Gan-Zhou, compared with MAT in the 1970s (Figure 16.27). MAT decreased by 0.04°C on an average. Isotherms saw no great change (Figure 16.28).

In the 1990s, the MAT increased rapidly in the whole Jiangxi province compared to temperatures in the 1980s, especially in Jing-De-Zhen, northern Jiu-Jiang, and southern Shang-Rao. The temperature increased by 0.39°C on an average. All isotherms shifted about 25 km to the north on an average. The 17°C isotherm moved about 80 km to the north, compared to its position in the 1980s.

16.4.3 Spatial Pattern of Mean Annual Precipitation

Precipitation in the Jiangxi province generally saw a continuous increase trend during the period 1960s to the 1990s. MAP in the 1960s, 1970s, 1980s, and 1990s were respectively 1529.72, 1662.83, 1667.85, and 1838.48 mm. In the 1970s, precipitation increased for most of the Jiangxi province, with the exception of northern Ping-Xiang, southern Ping-Xiang, southern Yi-Chun, northwestern Ji-An, and central Lin-Chuan, compared to precipitation in the

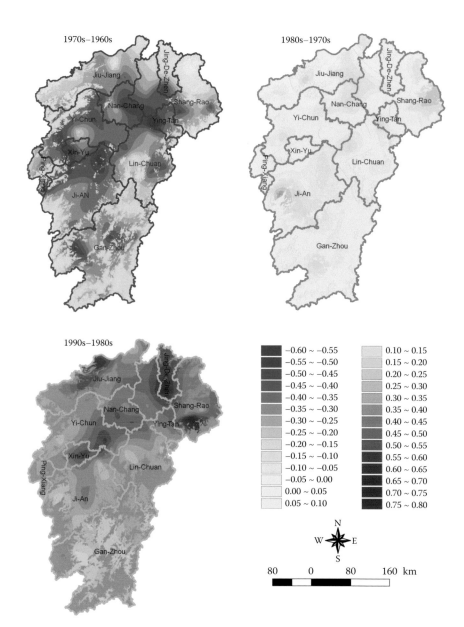

FIGURE 16.27
Mean annual temperature change.

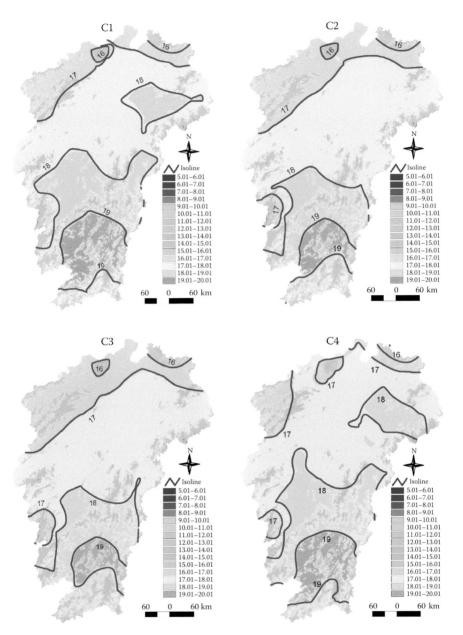

FIGURE 16.28
Spatial patterns of mean annual temperature during the period from the 1960s to the 1990s: (C1) in the 1960s, (C2) in the 1970s, (C3) in the 1980s, and (C4) in the 1990s.

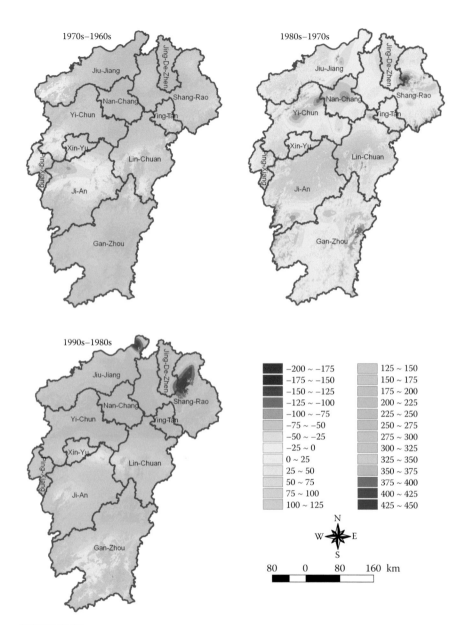

FIGURE 16.29
Precipitation change.

1960s (Figure 16.29). In the 1960s, the isoline of the lowest MAP was 1200 mm and the isoline of the highest was 1800 mm; the isolines on an average were between 1400 and 1600 mm (Figure 16.30). In the 1970s, the isoline with the lowest MAP was 1400 mm and the isoline with the highest was 1800 mm; the isolines in general were between 1500 and 1700 mm.

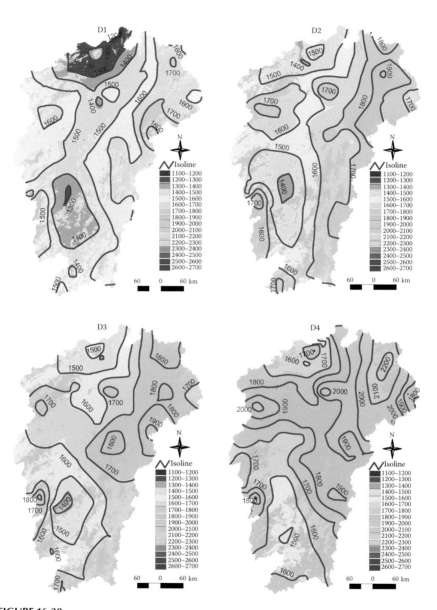

FIGURE 16.30
Spatial patterns of MAP during the period from the 1960s to the 1990s: (D1) in the 1960s, (D2) in the 1970s, (D3) in the 1980s, and (D4) in the 1990s.

In the 1980s, the MAP saw a slight increase, that is, 5.02 mm on an average, compared to precipitation in the 1970s. However, there was a great increase in Xin-Yu, northern Ji-An, northwestern Lin-Chuan, and the western boundary area of Jing-De-Zhen. The isoline with the lowest MAP was 1400 mm and the isoline with the highest was 1900 mm. The precipitation had great spatial variability.

In the 1990s, the MAP greatly increased in almost the entire Jiangxi province, with the exception of some areas of southern Gan-Zhou. Precipitation increased by 400 mm in the northern corner of Jiu-Jiang and middle of Shang-Rao on an average compared to precipitation in the 1980s. The isoline with the lowest MAP was 1500 mm and the isoline with the highest mean was 2200 mm. Isolines were between 1600 mm and 2000 mm on an average.

In short, in relative terms it was warmer and drier in the 1960s, cooler and wetter in the 1970s and 1980s, and warmer and wetter in the 1990s. During the period 1960 to 1999, MAT increased by 0.02°C and MAP increased by 77.19 mm per decade on an average.

16.5 Discussion and Conclusions

HASM has been successfully applied to simulating climate surfaces on provincial, national, and global levels by interpolating observed data from the sparsely distributed meteorological stations and downscaling outputs from the simulation of GCMs.

On a global level, statistical transfer functions are developed by establishing the regression relation of MAT with latitude and elevation in terms of digital elevation models and climate data from meteorological stations. The transfer functions are used to drive the preconditioned Gauss–Seidel method of HASM (HASM-PGS) to downscale the outputs of HadCM3 GCM from $3.75° \times 2.5°$ to $0.125° \times 0.125°$ of spatial resolution.

On the national level of China, the transfer function of MAT is formulated as a linear regression equation of latitude and elevation, of which the multiple correlation coefficient is 0.9631; the transfer function of MAP is expressed as a regression equation of the impact coefficient of aspect, the sky view factor, latitude, longitude, and elevation, of which the multiple correlation coefficient is 0.8883. These two transfer functions are employed to drive HASM-PGS for simulating surfaces of the MAT and MAP on a spatial resolution of 1×1 km in China.

On a provincial level, the relationship between temperature and elevation with a correlation coefficient of $R_t = -0.9959$ and that between precipitation and elevation with a correlation coefficient of $R_p = 0.9612$ are used as transfer functions to drive HASM-PGS to simulate the surfaces of MAT and MAP on a spatial resolution of 25×25 m in the Jiangxi province of China.

In short, the integration of HASM (Yue et al., 2007b) with statistical transfer functions (HASM-STF) is an alternative method for simulating the surfaces of weather factors. The statistical transfer functions are not difficult develop in terms of digital terrain model (DTM) and data from meteorological stations. HASM-STF is also useful for downscaling GCM outputs on coarser spatial resolution into finer spatial resolution to improve the poor performance of GCMs on local and regional levels.

References

Akinyemi, F.O. and Adejuwon, J.O. 2008. A GIS-based procedure for downscaling climate data for West Africa. *Transactions in GIS* 12(5): 613–631.

Anthes, R.A. 1983. Regional models of the atmosphere in middle latitudes. *Monthly Weather Review* 111: 1306–1335.

Ashiq, M.W., Zhao, C.Y., Ni, J., and Akhtar, M. 2010. GIS-based high-resolution spatial interpolation of precipitation in mountain–plain areas of Upper Pakistan for regional climate change impact studies. *Theoretical and Applied Climatology* 99: 239–253.

Bergant, K. and Kajfez-Bogataj, L. 2005. N–PLS regression as empirical downscaling tool in climate change studies. *Theoretical and Applied Climatology* 81: 11–23.

Brown, B.G. and Katz, R.W. 1995. Regional analysis of temperature extremes: Spatial analog for climate change? *Journal of Climate* 8: 108–119.

Charles, S.P., Bates, B.C., Whetton, P.H., and Hughes, J.P. 1999. Validation of downscaling models for changed climate conditions: Case study of southwestern Australia. *Climate Research* 12: 1–14.

Ciret, C. and Sellers, A.H. 1998. Sensitivity of ecosystem models to the spatial resolution of the NCAR Community Climate Model CCM2. *Climate Dynamics* 14: 409–429.

Collins, M., Tett, S.F.B., and Cooper, C. 2001. The internal climate variability of HadCM3, a version of the Hadley Centre coupled model without flux adjustments. *Climate Dynamics* 17: 61–81.

Covey, C., AchutaRao, K.M., Cubasch, U., Jones, P., Lambert, S.J., Mann, M.E., Phillips, T.J., and Taylor, K.E. 2003. An overview of results from the coupled model intercomparison project. *Global and Planetary Change* 37: 103–133.

Crane, R.G. and Hewitson, B.C. 1998. Doubled CO_2 precipitation changes for the Susquehanna Basin: Down-scaling from the genesis general circulation model. *International Journal of Climatology* 18: 65–76.

Dickinson, R.E., Errico, R.M., Giorgi, F., and Bates, G.T. 1989. A regional climate model for the western U.S. *Climatic Change* 15: 383–422.

Efron, B. 1979. Bootstrap methods: Another look at the jackknife. *The Annals of Statistics* 7(1): 1–26.

Franke, J., Haentzschel, J., Goldberg, V., and Bernhofer, C. 2008. Application of a trigonometric approach to the regionalization of precipitation for a complex small-scale terrain in a GIS environment. *Meteorological Applications* 15(2008): 483–490.

Ghosh, S. and Mujumdar, P.P. 2008. Statistical downscaling of GCM simulations to stream flow using relevance vector machine. *Advances in Water Resources* 31: 132–146.

Giorgi, F. 1990. Simulation of regional climate using a limited area model nested in a general circulation model. *Journal of Climate* 3: 941–963.

Grotch, S.L. and MacCracken, M.C. 1991. The use of general circulation models to predict regional climate change. *Journal of Climate* 4: 286–303.

Hessami, M., Gachon, P., Ouarda, T.B.M.J., and St-Hilaire, A. 2008. Automated regression-based statistical downscaling tool. *Environmental Modelling and Software* 23: 813–834.

Hewitson, B.C. and Crane, R.G. 1996. Climate downscaling: Techniques and application. *Climate Research* 7: 85–95.

Jones, P.D., Murphy, J.M., and Noguer, M. 1995. Simulation of climate change over Europe using a nested regional-climate model, I: Assessment of control climate, including sensitivity to location of lateral boundaries. *Quarterly Journal of the Royal Meteorological Society* 121: 1413–1449.

Johns, T.C., Gregory, J.M., Ingram, W.J., Johnson, C.E., Jones, A., Lowe, J.A., Mitchell, J.F.B. et al., 2003. Anthropogenic climate change for 1860 to 2100 simulated with the HadCM3 model under updated emissions scenarios. *Climate Dynamics* 20: 583–612.

Lloyd, C.D. 2005. Assessing the effect of integrating elevation data into the estimation of monthly precipitation in Great Britain. *Journal of Hydrology* 308: 128–150.

Mezghani, A. and Hingray, B. 2009. A combined downscaling-disaggregation weather generator for stochastic generation of multisite hourly weather variables over complex terrain: Development and multi-scale validation for the Upper Rhone River basin. *Journal of Hydrology* 377: 245–260.

Murphy, J.M. 1999. An evaluation of statistical and dynamical techniques for downscaling local climate. *Journal of Climate* 12: 2256–2284.

Prudhomme, C. and Davies, H. 2009. Assessing uncertainties in climate change impact analyses on the river flow regimes in the UK, Part 1: Baseline climate. *Climatic Change* 93: 177–195.

Radu, R., Deque, M., and Somot, S. 2008. Spectral nudging in a spectral regional climate model. *Tellus* 60A: 898–910.

Raisanen, J. 2007. How reliable are climate models? *Tellus A* 59: 2–29.

Trigo, R.M. and Palutikof, J.P. 1999. Simulation of daily temperatures for climate change scenarios over Portugal: A neural network model approach. *Climate Research* 13: 45–59.

Tripathi, S., Srinivas, V.V., and Nanjundiah, R.S. 2006. Downscaling of precipitation for climate change scenarios: A support vector machine approach. *Journal of Hydrology* 330: 621–640.

van der Veer, G., Voerkelius, S., Lorentz, G., Heiss, G., and Hoogewerff, J.A. 2009. Spatial interpolation of the deuterium and oxygen-18 composition of global precipitation using temperature as ancillary variable. *Journal of Geochemical Exploration* 101: 175–184.

von Storch, H., Zorita, E., and Cubash, U. 1993. Downscaling of global climate change estimates to regional scales: An application to Iberian rainfall in wintertime. *Journal of Climate* 6: 1161–1671.

Walker, W.S., Kellndorfer, J.M., and Pierce, L.E., 2007. Quality assessment of SRTM C- and X-band interferometric data: Implications for the retrieval of vegetation canopy height. *Remote Sensing of Environment* 106: 428–448.

Walsh, K.J.E. and McGregor, J.L. 1997. An assessment of simulations of climate variability over Australia with a limited area model. *International Journal of Climatology* 17: 201–223.

Watson, R. T., 34 coauthors and many IPCC authors and reviewers. 2001. *Climate Change 2001: Synthetic Report*, approved in detail at IPCC Plenary XVII, Wembley, UK, September 24–29, 2001.

Xue, Y.K., Vasic, R., Janjic, Z., Mesinger, F., and Mitchell, K.E. 2007. Assessment of dynamic downscaling of the continental U.S. regional climate using the Eta/SSiB regional climate model. *Journal of Climate* 20: 4172–4193.

Yue, T.X., Fan, Z.M., and Liu, J.Y. 2005. Changes of major terrestrial ecosystems in China Since 1960. *Global and Planetary Change* 48: 287–302.

Yue, T.X., Fan, Z.M., and Liu, J.Y. 2007a. Scenarios of land-cover in China. *Global and Planetary Change* 55(4): 317–342.

Yue, T.X., Du, Z.P., Song, D.J., and Gong, Y. 2007b. A new method of surface modeling and its application to DEM construction. *Geomorphology* 91(1–2): 161–172.

Yue, T.X., Fan, Z.M., Liu, J.Y., and Wei, B.X. 2006. Scenarios of major terrestrial ecosystems in China. *Ecological Modelling* 199: 363–376.

Zorita, E. and von Storch, H. 1999. The analog method as a simple statistical downscaling technique: Comparison with more complicated methods. *Journal of Climate* 12: 2474–2489.

17

Surface Modeling of Terrestrial Ecosystems*

17.1 Introduction

Since the early 1880s, there have been continuing efforts made in different earth surface sciences to classify the world's environment (Humboldt, 1807; Schouw, 1823; Merriam, 1892; Clements, 1916; Koeppen and Geiger, 1930; Thornthwaite, 1931). Since the categorization of previous systems designed to do this had been coarse and largely inapplicable, Holdridge (1947) devised a new classification system (HLZ). HLZ classification relates to the distribution of major terrestrial ecosystems (termed life zones) to bioclimatic variables. It is a scheme that uses the three bioclimatic variables derived from standard meteorological data to formulate the relation of climate patterns to broad-scale vegetation distribution. It has been widely used to project the impact of climate change on vegetation distribution (Post et al., 1982; Smith et al., 1992; Belotelov et al., 1996; Metternicht and Zinck, 1998; Dixon et al., 1999; Kirilenko et al., 2000; Peng, 2000; Powell et al., 2000; Xu and Yan, 2001; Yue et al., 2001; Yang et al., 2002; Chen et al., 2003, 2005; Chinea and Helmer, 2003; Kerr et al., 2003). The HLZ system has been employed to simulate evolutionary dynamic processes and vegetation patterns (Yue et al., 2005, 2006, 2007a; Feng and Chai, 2008).

Since the early 1980s, surface modeling of terrestrial ecosystems has been the subject of a great deal of interest and has achieved various results. For instance, it was found that climate was one of the main determinants of ecosystem ranges; distributions of many ecosystems were known to have changed with changing climates in the past (Huntley and Birks, 1983; Lavorel, 1999; McGlone et al., 2001; Mayle et al., 2004; Millennium Ecosystem Assessment, 2005; Biermann, 2007). Bioclimatic models, based on physiological constraints to plant growth and regeneration were developed to simulate the potential distribution of forest trees (Sykes et al., 1996). It was reported that rising temperatures and a redistribution of precipitation had a significant impact on plant physiology and led to major changes in potential vegetation structures, which could in turn provide feedback on climate through its effect on meteorological conditions at the land surface (Pielke et al., 1998; Betts et al., 2000). It was recognized that the IPCC climate scenarios for the twenty-first century represented major changes in environmental boundary conditions for the

---

* Dr. Ze-Meng Fan is a major collaborator of this chapter.

Earth's vegetation, but that these could not presently be identified locally because the regional pattern of a modified climate remained unpredictable (Cramer and Whittaker, 1999). The changes in regional and seasonal climate patterns could strongly influence the diversity and distribution of species and thus affect both ecosystems and biodiversity. The simulated result indicated that the climate in large areas would no longer be suitable for many plant species and many individuals of these species would disappear from the areas concerned (Bakkenes et al., 2002). Extinction risks for sample regions that cover around 20% of the Earth's terrestrial surface were assessed using projections of species' distribution for future climate scenarios (Thomas et al., 2004). For the sample regions used, it was found that about 35% of species and taxa would be committed to extinction under maximum climate change scenarios. Different aspects of risk arising from future climate change were assessed by quantifying changes in the spatial distribution of future climatic conditions compared to the conditions in the recent past, on the basis of species distribution models and extinction rates (Ohlemüller et al., 2006). With rapid advances in methods of climate surface modeling, it is now possible to simulate the distributions of ecosystems using their major climate drivers.

Great efforts have been made to set up the necessary data sets for surface modeling. Global vegetation and land-use databases on a spatial resolution of $1° \times 1°$ were compiled in digital form drawing on approximately 100 published sources complemented by a large collection of satellite images (Matthews, 1983), which made a considerable contribution to simulating the radiation balance of the earth and numerous biogeochemical cycles related to climate maintenance and climate change. After it was found that the absence of a global biodiversity map with a sufficient biogeographical resolution made many distinctive biotas unrecognizable, a data set of the terrestrial ecoregions of the world was developed to accurately reflect the complex distribution of the Earth's natural communities, which consisted of 14 biomes and eight biogeographic realms containing 867 ecoregions (Olson et al., 2001). The principal sources of climate data used for surface modeling were climate surfaces; these were generated by interpolating observed climate data sampled at varying intensities from across a region (Olwoch et al., 2003). A global distribution map of tropical dry forest was derived from the recently developed MODIS product of vegetation continuous fields, which depicted the percentage tree cover at a resolution of 500 m, combined with previously defined maps of biomes; the extent of tropical dry forest currently protected was estimated by overlaying the forest map with a global data set of the distribution of protected areas (Miles et al., 2006). A physical approach for generating leaf area index (LAI) was developed to produce a long time series of LAI products from MODIS and AVHRR data (Ganguly et al., 2008). Implementation of this approach indicated good consistency in LAI values retrieved from the NDVI (AVHRR mode) and from the spectral bidirectional reflectance factor (MODIS mode).

Mountain areas around the globe could explain why global ecological diversity patterns did not always follow a latitudinal gradient of decreasing ecological diversity from the tropics to the poles (Körner, 2000). The contribution of mountain areas to local, regional, and global ecological diversity is undisputed, but the processes generating the observed patterns are poorly understood (Burke, 2003). Projections of future species distributions and derived community descriptors cannot be reliably discussed unless model accuracy is quantified explicitly. A method was proposed to incorporate methodological uncertainty into the modeling process and derive robust estimates of species turnover across a range of climate scenarios (Thuiller, 2004). Understanding past, present, and future of ecosystem distribution strongly depends on the accuracy of DEMs, because ecological patterns and terrain are closely related (Tarolli et al., 2009).

17.2 Methods

17.2.1 HLZ Classification

HLZ classification divides the world into over 100 life zones in terms of mean annual biotemperature in degrees centigrade (MAB), average total annual precipitation in millimeters (TAP), and potential evapotranspiration ratio (PER) logarithmically. Biotemperature is defined as the mean of unit-period temperatures with a substitution of zero for all unit-period values below 0°C and above 30°C (Holdridge et al., 1971). Evapotranspiration is the total amount of water that is returned directly to the atmosphere in the form of vapor through the combined processes of evaporation and transpiration. Potential evapotranspiration is the amount of water that would be transpired under constantly optimal conditions of soil moisture and plant cover. The potential evapotranspiration ratio is the ratio of mean annual potential evapotranspiration to average total annual precipitation, which provides an index of biological humidity conditions. In other words, MAB, MAP, and PER at site (x,y) and in the tth year have the following formulation:

$$MAB(x,y,t) = \frac{1}{365} \sum_{j=1}^{365} TEM(j,x,y,t) \tag{17.1}$$

$$MAP(x,y,t) = \sum_{j=1}^{365} P(j,x,y,t) \tag{17.2}$$

$$PER(x,y,t) = \frac{58.93 MAB(x,y,t)}{MAP(x,y,t)} \tag{17.3}$$

where $TEM(j, x, y, t)$ is the value summing the hourly temperature above 0°C and below 30°C on the jth day and dividing by 24; $P(j, x, y, t)$ is the mean of precipitation on the jth day.

Suppose

$$M(x,y,t) = \ln MAB(x,y,t) \tag{17.4}$$

$$T(x,y,t) = \ln MAP(x,y,t) \tag{17.5}$$

$$P(x,y,t) = \ln PER(x,y,t) \tag{17.6}$$

$$d_i(x,y,t) = \sqrt{(M(x,y,t) - M_{i0})^2 + (T(x,y,t) - T_{i0})^2 + (P(x,y,t) - P_{i0})^2} \tag{17.7}$$

where M_{i0}, T_{i0}, and P_{i0} are standards of MAB logarithm, TAP logarithm, and PER logarithm at the central point of the ith life zone in the hexagonal system of HLZs. When $d_k(x,y,t) = \min\{d_i(x,y,t)\}$, the site (x, y) is classified into the kth life zone (Figure 17.1). The Equation 17.7 is valid for all HLZ ecosystems with the exception of desert and nival area. The criterion for desert is MAB > 3°C, TAP < 125mm, and PER > 2.0; the one for nival area is MAB < 1.5°C, TAP < 500 mm, and PER < 1.0.

17.2.2 Models Related to Spatial Patterns

On the basis of HLZ classification, the mean center model, the ecological diversity index and the patch connectivity index as seen in Chapter 14 are employed to analyze changing trends and scenarios of ecosystem spatial patterns in China. The mean center model is formulated as

$$x_j(t) = \sum_{i=1}^{I_j} \frac{s_{ij}(t) \cdot X_{ij}(t)}{S_j(t)} \tag{17.8}$$

$$y_j(t) = \sum_{i=1}^{I_j} \frac{s_{ij}(t) \cdot Y_{ij}(t)}{S_j(t)} \tag{17.9}$$

where t is the variable of time; I_j (t) is the patch number of HLZ type j; s_{ij} (t) is the area of the ith patch of HLZ type j; S_j (t) is the total area of HLZ type j; $(X_{ij}(t), Y_{ij}(t))$ is the longitude and latitude coordinate of the geometric center of the ith patch of HLZ type j; $(x_j(t), y_j(t))$ is the mean center of HLZ type j.

Shift distance and direction of HLZ type j in the period t to $t+1$ are respectively formulated as

$$D_j = \sqrt{\left(x_j(t+1) - x_j(t)\right)^2 + \left(y_j(t+1) - y_j(t)\right)^2} \tag{17.10}$$

$$\theta_j = arctg\left(\frac{y_j(t+1) - y_j(t)}{x_j(t+1) - x_j(t)}\right) \tag{17.11}$$

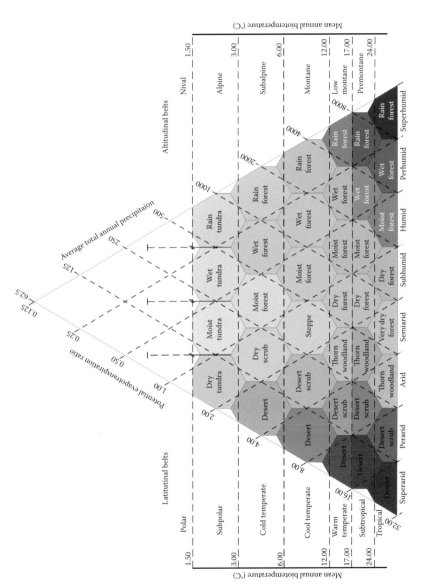

FIGURE 17.1
A modified conceptual model of Holdridge life zones.

where D_j is the shift distance of HLZ type j in the period t to $t+1$; θ_j is the shift direction of HLZ type j, which due east is $0°$, due north is $90°$, due west is $180°$ and due south is $270°$; $(x_j(t), y_j(t))$ and $(x_j(t+1), y_j(t+1))$ are respectively the coordinate of the mean center of HLZ type j in the years t and $t+1$. When $0° < \theta_j < 90°$, it is stated that HLZ type j shifts to the northeast during the period t to $t+1$; when $90° < \theta_j < 180°$, HLZ type j shifts to the northwest; when $180° < \theta < 270°$, HLZ type j shifts to the southwest; when $270° < \theta < 360°$ HLZ type j shifts to the southeast.

The ecological diversity index is expressed as (Yue et al., 1998, 2001, 2003, 2005, and 2007b)

$$d(t) = -\frac{\ln\left(\sum_{i=1}^{m(\varepsilon)}(p_i(t))^{1/2}\right)^2}{\ln(\varepsilon)} \qquad (17.12)$$

where t is the variable of time; $p_i(t)$ is the probability of the ith ecotope; $m(\varepsilon)$ is the total number of investigation ecotopes; $\varepsilon = (1/e + a)$, a is the area of studied region measured by hectares, and e equals 2.71828.

The patch connectivity index, $CO(t)$, is formulated as (Yue et al., 2003, 2004)

$$CO(t) = \sum_{i=1}^{m(t)}\sum_{j=1}^{n_i(t)}p_{i,j}(t) \cdot S_{i,j}(t) \qquad (17.13)$$

$$S_{i,j}(t) = \frac{8\sqrt{3} \cdot A_{i,j}(t)}{(\mathrm{Pr}_{i,j}(t))^2} \qquad (17.14)$$

where $S_{i,j}(t)$ is an index of movement efficiency of migrants or propagules in the patch (i, j); $A_{i,j}(t)$ and $\mathrm{Pr}_{i,j}$ are the area and the perimeter of the jth patch in the ith land-cover type, respectively; t is the variable of time; the coefficient $8\sqrt{3}$ is the ratio of the square of perimeter to the area of a hexagon; $p_{i,j}(t)$ is the proportion of the area of the jth patch in the ith type to the total area under investigation; $0 \le C(t) \le 1.1$ and when all patches have the shape of hexagon (6-gon), $C(t) = 1.0$.

17.3 Scenarios of Terrestrial Ecosystems on a Global Level

17.3.1 Downscaling of Weather Main Factors

The statistical transfer functions of weather main factors (WMFs) are developed using data set from the 2783 weather observation stations scattered across the earth, since 1949 (as seen in Figure 16.1) and DEM on spatial resolution $0.125° \times 0.125°$ on a global level. In the north hemisphere, the regression relationship among MAB (*MABnor*), latitude (*Lnor*), and elevation (*Enor*) can be formulated as, with squared multiple correlation of 0.912

$$MABnor = 29.743 - 0.383 \cdot Lnor - 0.003 \cdot Enor \tag{17.15}$$

In the southern hemisphere,

$$MABsou = 29.724 + 0.398 \cdot Lsou - 0.003 \cdot Esou \tag{17.16}$$

where *MABsou*, *Lsou*, and *Esou* represent MAB, latitude and elevation, respectively; the squared multiple correlation is 0.918.

The preprocessing intercepts of the statistical transfer functions of MAB in the northern and southern hemispheres can be respectively formulated as

$$IBnor_i(t) = MABnor_i(t) + 0.383 \cdot Lnor_i + 0.003 \cdot Enor_i \tag{17.17}$$

$$IBsou_j(t) = MABsou_j(t) - 0.398 \cdot Lsou_j + 0.003 \cdot Esou_j \tag{17.18}$$

where $IBnor_i(t)$, $MABnor_i(t)$, $Lnor_i$, and $Enor_i$ respectively represent the preprocessing intercept, MAB, the latitude, and elevation at grid cell i in the north hemisphere ($i = 1, 2, 3, \ldots, 3504$); $IBsou_j(t)$, $MABsou_j(t)$, $Lsou_j$, and $Esou_j$ are, respectively, the preprocessing intercept, MAB, the latitude, and elevation at grid cell j in the south hemisphere ($j = 3505, 3506, \ldots, 7008$); t is time variable; $Lnor_i \geq 0$ and $Lsou_j < 0$.

Then HASM-PGS is operated on the preprocessed data $IBnor_i(t)$ and $IBsou_j(t)$ by using $0.125° \times 0.125°$ DEM as the auxiliary data. The results on a spatial resolution of $0.125° \times 0.125°$, produced by HASM-PGS, include 2077201 grid cells in the northern hemisphere and 2074320 grid cells in the southern hemisphere. The data produced by HASM-PGS are presented as $IBnor_l(x, y, t)$ ($l = 1, 2, \ldots, 2077201$) in the northern hemisphere and $IBsou_k(x, y, t)$($k = 1, 2, \ldots,$ 2074320) in the southern hemisphere, in which (x, y) are the geographical coordinates of the corresponding grid cells in the northern and southern hemispheres. In terms of $IBnor_l(x, y, t)$ and $IBsou_k(x, y, t)$, the MAB surfaces can be formulated as

$$MABnor_l(x,y,t) = IBnor_l(x,y,t) - 0.383 \cdot Lnor_l(x,y)$$
$$- 0.003 \cdot Enor_l(x,y) \tag{17.19}$$

$$MABsou_k(x,y,t) = IBsou_k(x,y,t) + 0.398 \cdot Lsou_k(x,y)$$
$$- 0.003 \cdot Enor_k(x,y) \tag{17.20}$$

where $MABnor_l$ (x, y, t), $Lnor_l$ (x, y), and $Enor_l$ (x, y) are respectively the MAB, latitude, and elevation at the lth grid cell in the northern hemisphere; $MABsou_k$ (x, y, t), $Lsou_k$ (x, y), and $Esou_k$ (x, y) are respectively MAB, latitude and elevation at the kth grid cell in the southern hemisphere.

Surfaces of MAP on a spatial resolution of $0.125° \times 0.125°$, $MAP(x, y, t)$, is developed by operating HASM-PGS on precipitation data from HadCM3AlFi, A2 and B2 scenarios with a spatial resolution of $3.75° \times 2.5°$ by taking $0.125° \times 0.125°$ DEM as the auxiliary data. Then, the PER on a spatial resolution of $0.125° \times 0.125°$, $PER(x, y, t)$, can be calculated in terms of the MAB and the MAP produced by HASM-PGS.

17.3.2 Terrestrial Ecosystem Simulation

The downscaled surfaces of MAB, MAP, and PER are used to generate surfaces of HLZ terrestrial ecosystems by operating the HLZ model that consists of formulations from Equations 17.4 through 17.7. The simulation results show that the 0.0131 million hectares of warm temperate rain forest would only appear in the period T4 of scenario A1Fi, while the other 33 ecosystem types would appear in all periods under all scenarios (Tables 17.1 through 17.3). The nival area would account for the biggest proportion of terrestrial surface on an average in terms of the three scenarios, with a proportion of about 12%. The first three smallest ecosystem types would be subtropical rain forest, at 0.0745 million hectares, dry tundra, at 0.3859 million hectares, and tropical rain forest, at 0.771 million hectares on an average.

Under scenario A1Fi (Table 17.1), comparing ecosystems in period T_4 with ecosystems in period T_1, it is clear that the areas of subtropical moist forest, cold temperate wet forest, and nival area would see the greatest decrease, with a drop of 221 million hectares, 153 million hectares, and 148 million hectares on an average, respectively. Areas of tropical dry forest, cool temperate moist forest, and tropical very dry forest would see the biggest increase, of 242 million hectares, 189 million hectares, and 175 million hectares, respectively. Tropical rain forest and subpolar/alpine moist tundra would decrease respectively by 97% and 95%; cool temperate rain forest and tropical thorn woodland would increase by more than 123%; these would be the most sensitive ecosystems.

Under scenario A2a (Table 17.2, Figure 17.2), the areas which would see the greatest reduction would be subtropical moist forest, nival area, and subtropical dry forest, which would see a loss of 222, 132, and 130 million hectares, respectively, on an average. Areas of tropical dry forest, cool temperate moist forest, and tropical very dry forest would expand by 207, 201, and 177 million hectares, respectively. In T_4 period, 92% of moist tundra would disappear compared with its status in the T_1 period; tropical thorn woodland and tropical very dry forest would increase respectively by 108% and 97%.

TABLE 17.1

Area of Every Ecosystem Type Under HadCM3 AlFi (Unit: km²)

HLZ Ecosystem Type	T1	T2	T3	T4	Average	Proportion (%)
Polar/nival area	19,827,497	18,264,935	16,563,396	15,393,169	17,512,249	11.93490
Subpolar/Alpine dry tundra	5771	4627	4307	732	3859	0.00263
Subpolar/Alpine moist tundra	767,282	405,452	152,626	34,677	340,009	0.23172
Subpolar/Alpine wet tundra	4,246,990	3,582,775	2,282,096	1,152,526	2,816,097	1.91922
Subpolar/Alpine rain tundra	4,360,273	3,505,485	2,557,220	1,618,076	3,010,264	2.05155
Cold temperate dry scrub	234,286	152,286	82,627	47,622	129,205	0.08806
Cold temperate moist forest	4,476,318	4,510,837	4,046,413	3,274,076	4,076,911	2.77849
Cold temperate wet forest	12,388,477	11,567,468	10,002,475	7,798,559	10,439,244	7.11452
Cold temperate rain forest	2,041,052	2,190,934	2,179,071	1,814,101	2,056,289	1.40140
Cool temperate scrub	1,642,669	1,418,682	956,732	429,067	1,111,788	0.75770
Cool temperate steppe	4,052,525	4,280,052	4,055,191	2,789,424	3,794,298	2.58588
Cool temperate moist forest	9,539,050	11,664,606	14,361,821	15,220,008	12,696,372	8.65280
Cool temperate wet forest	3,367,327	3,098,278	2,853,411	3,510,628	3,207,411	2.18591
Cool temperate rain forest	141,988	164,875	207,374	316,352	207,647	0.14152
Warm temperate desert scrub	1,571,003	1,674,103	1,399,611	1,240,360	1,471,269	1.00270
Warm temperate thorn steppe	3,136,398	2,932,590	3,153,775	3,930,327	3,288,272	2.24101
Warm temperate dry forest	3,627,621	3,624,774	3,934,419	5,555,888	4,185,675	2.85261
Warm temperate moist forest	4,438,651	4,118,104	3,831,383	3,940,998	4,082,284	2.78215
Warm temperate wet forest	399,057	402,316	364,841	328,061	373,569	0.25459
Warm temperate rain forest				131		
Subtropical desert scrub	2,216,727	2,492,107	2,574,849	3,780,610	2,766,073	1.88513
Subtropical thorn woodland	4,883,440	5,493,943	5,725,090	4,025,833	5,032,076	3.42945
Subtropical dry forest	9,321,950	8,107,672	6,596,327	5,105,360	7,282,827	4.96337

continued

TABLE 17.1 (continued)

Area of Every Ecosystem Type Under HadCM3 AIFi (Unit: km²)

HLZ Ecosystem Type	T1	T2	T3	T4	Average	Proportion (%)
Subtropical moist forest	12,956,941	10,262,495	7,695,889	63,157,32	9,307,764	6.34340
Subtropical wet forest	1,153,828	782,120	765,078	713,201	853,557	0.58171
Subtropical rain forest	777	646	647	910	745	0.00051
Tropical desert scrub	1,903,121	2,249,367	2,666,780	3,750,324	2,642,398	1.80084
Tropical thorn woodland	2,656,098	3,517,057	4,809,334	6,149,521	4,283,003	2.91894
Tropical very dry forest	5,866,534	7,258,008	9,674,220	11,113,356	8,478,029	5.77792
Tropical dry forest	6,908,637	9,364,252	11,731,729	14,165,007	10,542,406	7.18483
Tropical moist forest	6,626,347	7,915,044	8,636,978	8,755,165	7,983,383	5.44081
Tropical wet forest	610,722	647,887	740,104	669,133	666,962	0.45455
Tropical rain forest	10,297	12,254	7966	324	7710	0.00525
Desert	11,371,830	11,081,023	12,132,162	13,807,738	12,098,188	8.2452

TABLE 17.2

Area of Every Ecosystem Type Under HadCM3 A2a (Unit: km²)

HLZ Ecosystem Type	T1	T2	T3	T4	Average	Proportion (%)
Polar/nival area	19,815,466	18,318,821	17,047,229	15,848,241	17,757,439	12.1020
Subpolar/Alpine dry tundra	5772	5956	4164	2323	4554	0.0031
Subpolar/Alpine moist tundra	712,473	422,571	226,257	57,769	354,767	0.2418
Subpolar/Alpine wet tundra	4,342,344	3,471,264	2,553,198	1,549,132	2,978,985	2.0302
Subpolar/Alpine rain tundra	4,269,133	3,654,703	2,773,226	2,078,371	3,193,858	2.1767
Cold temperate dry scrub	233,599	153,814	98,971	54,052	135,109	0.0921
Cold temperate moist forest	4,785,239	4,607,820	4,430,809	3,447,496	4,317,841	2.9427
Cold temperate wet forest	11,894,621	11,397,073	10,469,938	8,925,155	10,671,697	7.2730
Cold temperate rain forest	2,154,012	2,156,348	2,224,906	1,837,838	2,093,276	1.4266
Cool temperate scrub	1,676,017	1,397,970	1,109,732	736,538	1,230,064	0.8383
Cool temperate steppe	4,087,001	4,408,070	4,298,397	3,258,684	4,013,038	2.7350
Cool temperate moist forest	9,597,946	11,591,957	13,547,685	15,618,888	12,589,119	8.5797
Cool temperate wet forest	3,474,548	3,069,259	2,994,441	3,180,962	3,179,802	2.1671
Cool temperate rain forest	169,196	157,763	210,660	291,810	207,357	0.1413
Warm temperate desert scrub	1,489,279	1,689,695	1,433,625	1,394,966	1,501,891	1.0236
Warm temperate thorn steppe	3,128,854	2,971,114	2,866,108	3,431,340	3,099,354	2.1123
Warm temperate dry forest	3,475,981	3,563,127	3,717,566	4,876,644	3,908,330	2.6636
Warm temperate moist forest	4,296,230	4,219,637	3,904,582	3,588,386	4,002,208	2.7276
Warm temperate wet forest	439,801	343,980	377,730	378,017	384,882	0.2623
Subtropical desert scrub	2,302,531	2,733,901	2,589,250	2,905,309	2,632,748	1.7943
Subtropical thorn woodland	5,053,150	5,227,609	5,813,644	5,427,614	5,380,504	3.6669
Subtropical dry forest	9,676,523	8,420,912	7,149,281	5,786,619	7,758,334	5.2874
Subtropical moist forest	13,319,940	10,878,423	8,008,036	6,655,899	971,5574	6.6213

continued

TABLE 17.2 (continued)

Area of Every Ecosystem Type Under HadCM3 A2a (Unit: km²)

HLZ Ecosystem Type	T1	T2	T3	T4	Average	Proportion (%)
Subtropical wet forest	1,093,894	753,874	604,222	653,309	776,324	0.5291
Subtropical rain forest	777	516	384	778	614	0.0004
Tropical desert scrub	1,781,356	2,225,222	2,857,941	3,202,133	2,516,663	1.7152
Tropical thorn woodland	2,534,959	3,386,393	4,429,746	5,281,170	3,908,067	2.6634
Tropical very dry forest	5,443,611	7,128,922	8,779,740	10,735,396	8,021,917	5.4671
Tropical dry forest	6,724,082	9,192,037	11,591,285	12,926,509	10,108,478	6.8891
Tropical moist forest	6,360,093	7,468,062	8,661,488	8,897,841	7,846,871	5.3478
Tropical wet forest	621,368	640,072	620,643	803,873	671,489	0.4576
Tropical rain forest	6166	9810	2587	7966	6632	0.0045
Desert	11,779,591	11,083,645	11,344,200	12,907,944	11,778,845	8.0276

TABLE 17.3

Area of Every Ecosystem Type Under HadCM3 B2a (Unit: km²)

HLZ Ecosystem Type	T1	T2	T3	T4	Average	Proportion (%)
Polar/nival area	19,828,232	18,033,841	17,170,148	16,402,827	17,858,762	12.1711
Subpolar/Alpine dry tundra	5810	5632	5290	4333	5266	0.0036
Subpolar/Alpine moist tundra	708,450	454,150	262,973	143,148	392,180	0.2673
Subpolar/Alpine wet tundra	4,339,430	3,172,702	2,796,989	2,099,062	3,102,046	2.1141
Subpolar/Alpine rain tundra	4,259,324	3,428,466	2,981,591	2,554,326	3,305,927	2.2531
Cold temperate dry scrub	236,669	173,388	125,377	75,422	152,714	0.1041
Cold temperate moist forest	4,784,480	4,584,499	4,344,641	3,915,187	4,407,201	3.0036
Cold temperate wet forest	11,891,597	11,040,057	10,446,593	10,017,391	10,848,909	7.3937
Cold temperate rain forest	2,141,024	2,324,965	2,237,810	2,087,321	2,197,780	1.4978
Cool temperate scrub	1,692,320	1,543,464	1,246,426	1,106,648	1,397,214	0.9522
Cool temperate steppe	4,037,436	4,293,419	4,284,853	3,983,391	4,149,775	2.8282
Cool temperate moist forest	9,564,671	12,143,168	13,592,820	14,536,819	12,459,370	8.4913
Cool temperate wet forest	3,542,723	2,935,378	2,809,993	2,949,733	3,059,457	2.0851
Cool temperate rain forest	168,305	169,495	227,977	288,709	213,621	0.1456
Warm temperate desert scrub	1,489,912	1,445,090	1,535,665	1,541,989	1,503,164	1.0244
Warm temperate thorn steppe	3,120,387	2,928,309	2,673,630	2,784,780	2,876,776	1.9606
Warm temperate dry forest	3,456,466	3,329,720	3,563,538	3,954,664	3,576,097	2.4372
Warm temperate moist forest	4,305,697	4,037,392	3,860,702	3,870,607	4,018,600	2.7388
Warm temperate wet forest	440,600	327,181	316,923	310,530	348,808	0.2377
Subtropical desert scrub	2,291,814	2,224,246	2,592,634	2,802,750	2,477,861	1.6887
Subtropical thorn woodland	5,102,177	5,692,257	5,534,499	5,543,413	5,468,087	3.7266
Subtropical dry forest	9,676,155	7,763,489	7,115,532	5,783,660	7,584,709	5.1691
Subtropical moist forest	13,319,879	8,772,405	7,499,480	6,541,520	9,033,321	6.1564

continued

TABLE 17.3

Area of Every Ecosystem Type Under HadCM3 B2a (Unit: km^2)

HLZ Ecosystem Type	T1	T2	T3	T4	Average	Proportion (%)
Subtropical wet forest	1,083,842	544,277	545,776	532,588	676,621	0.4611
Subtropical rain forest	777	646	647	647	679	0.0005
Tropical desert scrub	1,779,918	2,217,299	2,496,655	2,750,044	2,310,979	1.5750
Tropical thorn woodland	2,540,624	3,303,558	3,901,417	5,016,411	3,690,503	2.5151
Tropical very dry forest	5,440,589	7,898,220	8,981,154	10,481,507	8,200,368	5.5887
Tropical dry forest	6,721,554	10,665,543	12,301,607	12,938,220	10,656,731	7.2628
Tropical moist forest	6,351,272	8,729,651	9,030,356	8,644,844	8,189,031	5.5810
Tropical wet forest	614,471	666,447	806,748	714,604	700,568	0.4775
Tropical rain forest	6055	7742	7032	4365	6299	0.0043
Desert	11,807,370	11,894,653	11,447,569	12,364,780	11,878,593	8.0956

Scenario B2a (Table 17.3) indicates that subtropical moist forest, subtropical dry forest, and nival area would have the largest reduction, of 226, 130, and 114 million hectares, respectively. Tropical dry forest, tropical very dry forest, and cool temperate moist forest would see the fastest expansion, with areas increasing by 207, 168, and 166 million hectares. Moist tundra, tropical thorn woodland, tropical very dry forest, and tropical dry forest would be the most sensitive ecosystems. Moist tundra would shrink 80%, while the three tropical ecosystems would increase by more than 92%.

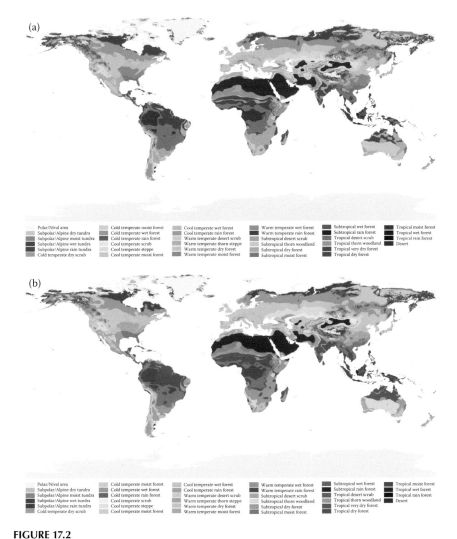

FIGURE 17.2
HadCM3 A2a scenarios of HLZ ecosystem distribution on average: (a) during the periods 1961–1990 (T1), (b) 2010–2039 (T2), (c) 2040–2069 (T3), and (d) 2070–2099 (T4).

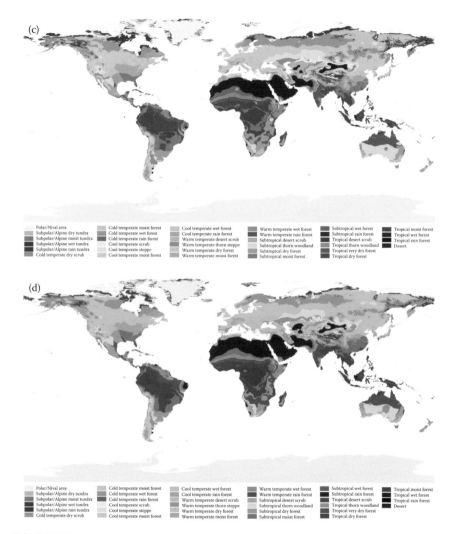

FIGURE 17.2
Continued

Ecological diversity would have a continuously decreasing trend under all the three scenarios (Table 17.4). During the period T_2 to T_4, the decadal decreasing rate would be 0.133% under scenario AlFi, under scenario A2a it would be 0.094%, and under B2a it would be 0.0651%. However, patch connectivity would have a fluctuating change. Scenario AlFi shows that patch connectivity would increase from T_1 to T_2 and then decrease. Scenario A2a indicates that patch connectivity would increase from T_1 to T_2, decrease from T_2 to T_3 and then increase again. Patch connectivity would have a

TABLE 17.4

Ecological Diversity and Patch Connectivity of HLZ Ecosystems

Scenarios	Period	Connectivity	Diversity
AlFi	T_1	0.1462	0.1697
	T_2	0.1491	0.1695
	T_3	0.1518	0.1683
	T_4	0.1512	0.1665
	Decadal increasing rate from T_2 to T_4 (%)		−0.1330
A2a	T_1	0.1472	0.1694
	T_2	0.1475	0.1693
	T_3	0.1416	0.1685
	T_4	0.1497	0.1672
	Decadal increasing rate from T_2 to T_4 (%)		−0.0940
B2a	T_1	0.1474	0.1694
	T_2	0.1446	0.1689
	T_3	0.1527	0.1685
	T_4	0.1513	0.1679
	Decadal increasing rate from T_2 to T_4 (%)		−0.0651

contrastive change under scenario B2a compared with patch connectivity in scenario A2a.

17.3.3 Brief Summary

The statistical transfer functions of biotemperature and precipitation are developed by operating HASM on data from the 2783 weather observation stations scattered across the earth, in which DEM on a spatial resolution of $0.125° \times 0.125°$ is employed as the auxiliary data. HadCM3AlFi, A2a, and B2a scenarios with a spatial resolution of $3.75° \times 2.5°$ are downscaled to a spatial resolution of $0.125° \times 0.125°$ in terms of the simulated climatic surfaces. Finally, three scenarios of HLZ ecosystems are produced. The results indicate that all polar/nival, subpolar/alpine and cold ecosystem types would see a continuous decreasing trend, including nival area, moist tundra, wet tundra, rain tundra, cold temperate desert, cold temperate dry scrub, cold temperate moist forest, cold temperate wet forest, and cold temperate rain forest as well as cool temperate scrub. With the exception of tropical rain forest, all other tropical ecosystem types would see an increase, such as tropical wet forest, tropical moist forest, tropical dry forest, tropical very dry forest, tropical thorn woodland, and tropical desert scrub as well as subtropical rain forest. The most sensitive ecosystems would be subpolar/alpine and tropical ecosystem types; notably 80% of alpine moist tundra would disappear in the period T_4 compared with the period T_1. Tropical thorn woodland would increase by more than 97%. In general, climate change would cause a continuous decrease of ecological diversity.

17.4 Terrestrial Ecosystem Simulation on a National Level

17.4.1 Methods

The statistical analysis of WMF data from 752 meteorological stations of China indicates that multiple correlations among MAB, latitude and altitude is 0.9643 on a national level. The linear regression equation can be formulated as

$$MAB = 32.6429 - 0.5015 \cdot La - 0.003 \cdot Ele \qquad (17.21)$$

where *La* and *Ele* are respectively the latitude and altitude.

As Section 16.3 of Chapter 16 demonstrates, the preprocessing intercept of the statistical transfer function of MAB at meteorological station k at time t can be expressed as

$$IB_k(x,y,t) = MAB_k(x,y,t) + 0.5015 \cdot La_k(x,y) + 0.003 \cdot Ele_k(x,y) \qquad (17.22)$$

where $IB_k(x,y,t)$ is the preprocessed intercept; $MAB_k(x,y,t)$ is the MAB from observed data at meteorological station k; $La_k(x,y)$ is the latitude and $Ele_k(x,y)$ is the elevation at meteorological station k.

The preprocessed intercept values $\{IB_k(x,y,t)\}$ at all meteorological stations are employed to drive HASM-PGS to obtain the intercept value at every grid cell on a spatial resolution of 1×1 km, in which 5000×4320 grid cells are involved. Finally, the simulated MAB at every grid cell i, in which $i = 1, 2, \ldots, 19606916$, can be formulated as

$$MAB_i(x,y,t) = IB_i(x,y,t) - 0.5015 \cdot La_i(x,y) - 0.003 \cdot Ele_i(x,y) \qquad (17.23)$$

where $MAB_i(x, y, t)$ is the simulated MAB at time t; $IB_i(x, y, t)$ the interpolated intercept value at grid cell i; (x, y) the coordinate of grid cell i; $La_i(x, y)$ the latitude of grid cell i and $Ele_i(x, y)$ the altitude of grid cell i.

The simulation process of the MAP surface is exactly the same as in Chapter 16. PER can be easily calculated in terms of Equation 17.3.

Owing to the 11-year-old small cyclicity of climate change in China, the period 1964 to 2007 is divided into four subperiods: C1 (from 1964 to 1974),

C2 (from 1975 to 1985), C3 (from 1986 to 1996), and C4 (from 1997 to 2007). Terrestrial ecosystems in the four subperiods are simulated by operating the HLZ model on the surfaces of MAB, MAP and PER on a spatial resolution of 1×1 km. The simulation process includes five steps: (1) converting the 1×1 km grid data into ASCII data by ArcToolbox module and AML program of ARC/INFO; (2) calculating the potential evapotranspiration in ASCII data by modularized programming of VC++ and then transforming the results into Grid data; (3) simulating the surfaces of biotemperature, precipitation, and evapotranspiration by means of HASM-PGS; (4) operating the HLZ model on the ASCII data by VC++; (5) transforming the ASCII data from the results of the HLZ model into 1×1 km grid data; and (6) converting the 1×1 km grid data into polygon coverage.

17.4.2 Changing Trends of Terrestrial Ecosystems in China

17.4.2.1 Spatial Distribution of HLZ Ecosystems

The simulation results show that 27 HLZ ecosystem types appeared in China during the period 1964 to 2007 (Figure 17.3 and Table 17.5). Nival area, alpine dry tundra, and alpine moist tundra were mainly distributed in the middle and western Tibet Plateau. Alpine wet tundra and alpine rain tundra were distributed in eastern the Tibet Plateau, the southern Qilian mountains, the Tianshan mountains, and the Altay Mountains. Cold temperate dry scrub mainly appeared in the low mountain areas of the southern Altay Mountains, around the Bogda Mountains and northern Kunlun mountains as well as the western Himalayan Mountains. Desert mainly distributes in Tarim Basin, Turpan Basin, Qaidam Basin, Alashan plateau, and Junggar Basin.

Warm temperate desert scrub was mainly distributed in peripheral areas of eastern the Badain Jaran desert and the Tengger desert, most parts of the Junggar Basin and the northwestern corner of the Gansu province. Cool temperate scrub mainly appeared in most of the areas between the Tianshan Mountains and Altay Mountains, the middle of Inner Mongolia, and north of the Qaidam Basin. Cool temperate steppe was mainly distributed in the northern Inner Mongolian Plateau, the Loess Plateau, and the western Northeast Plain. Warm temperate thorn steppe was mainly distributed in the low mountain areas of the southeastern Himalaya Mountains and the foothills of the northern Kunlun Mountain as well as in the Ili river basin and the Ebinur lake area of Xinjiang.

Cold temperate moist forest and cold temperate wet forest were mainly distributed in the eastern and southern Tibet Plateau and the Da-Hinggan-Ling mountains. Cold temperate rain forest, cool temperate moist forest, and cool temperate wet forest were mainly in the northeastern Tibet Plateau, the high mountain areas of the Qin-Ling Mountains, the southern Tai-Hang Mountains, the eastern Chang-Bai mountains, and the Xiao-Hinggan-Ling

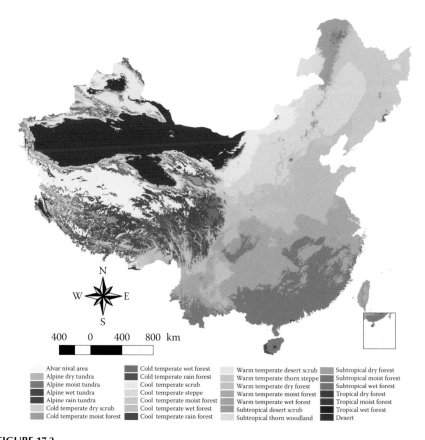

FIGURE 17.3
The spatial distribution of HLZ ecosystem on average during the period 1964 to 2007 in China.

mountains. Warm temperate dry forest was mainly in the high mountain areas of the Yun-Gui Plateau, the low mountain area of the Qin-Ling Mountains and most of the North China Plain.

Warm temperate moist forest and warm temperate wet forest were mainly found in the Yangtze River basin and the high mountain areas of Taiwan. Subtropical dry forest was mainly scattered in the Karst Region of the Yun-Gui plateau and western Taiwan. Subtropical moist forest and subtropical wet forest were mainly found in the southeastern Tian-Mu Mountains, the Wuyi Mountains, the Nanling Mountains, and the Jiulian mountains as well as the southern Yunnan province, Hainan province, and the low mountain areas of Taiwan. There were also a few tropical moist forests and tropical wet forests scattered in the Red river basin of the lower reaches of the Yuan-Jiang River, the estuary of the Zhu-Jiang River, and the central area of Hainan province.

TABLE 17.5

Classification Standards of MAB, TAP, and PER at the Central Point of the *i*th Life Zone in the Hexagonal System of HLZs

HLZ	Standard of MAB (°C)	Standard of TAP (mm)	Standard of PER
Alvar nival area	<1.5	<500	<1.0
Alpine dry tundra	2.1210	88.3880	1.4140
Alpine moist tundra	2.1210	177.7770	0.7070
Alpine wet tundra	2.1210	353.5520	0.3540
Alpine rain tundra	2.1210	707.1070	0.1770
Cold temperate dry scrub	4.2430	177.7770	1.4140
Cold temperate moist forest	4.2430	353.5520	0.7070
Cold temperate wet forest	4.2430	707.1770	0.3540
Cold temperate rain forest	4.2430	1414.2130	0.1770
Cool temperate scrub	8.4850	177.7770	2.8280
Cool temperate steppe	8.4850	353.5520	1.4140
Cool temperate moist forest	8.4850	707.1070	0.7070
Cool temperate wet forest	8.4850	1414.2130	0.3540
Cool temperate rain forest	8.4850	2828.4270	0.1770
Warm temperate desert scrub	14.2700	177.7770	5.6750
Warm temperate thorn steppe	14.2700	353.5520	2.8280
Warm temperate dry forest	14.2700	707.1070	1.4140
Warm temperate moist forest	14.2700	1414.2130	0.7070
Warm temperate wet forest	14.2700	2828.4270	0.3540
Warm temperate rain forest	14.2700	5656.8540	0.1770
Subtropical desert scrub	20.1810	177.7770	5.6750
Subtropical thorn woodland	20.1810	353.5520	2.8280
Subtropical dry forest	20.1810	707.1070	1.4140
Subtropical moist forest	20.1810	1414.2130	0.7070
Subtropical wet forest	20.1810	2828.4270	0.3540
Subtropical rain forest	20.1810	5656.8540	0.1770
Tropical desert scrub	33.9410	177.7770	11.3140
Tropical thorn woodland	33.9410	353.5520	5.6750
Tropical very dry forest	33.9410	707.1070	2.8280
Tropical dry forest	33.9410	1414.2130	1.4140
Tropical moist forest	33.9410	2828.4270	0.7070
Tropical wet forest	33.9410	5656.8540	0.3540
Tropical rain forest	33.9410	11,313.7100	0.1770
Desert	>3.0	<125	>2.0

17.4.2.2 Area Changing Trends of HLZ Ecosystems

An area analyses of the HLZ ecosystems in the periods of C1, C2, C3, and C4 (Figure 17.4, Tables 17.6 and 17.7) indicate that the 27 HLZ ecosystem types appeared in all the four periods, with the exception of cool temperate rain forest which only appeared in the periods C1 and C2. In terms of the area size, the 27 HLZ ecosystem types could be ordered from bigger to smaller as follows: desert (14.1415%) warm temperate moist forest (11.6954%), cool temperate steppe (11.0248%), subtropical moist forest (9.9470%), cool temperate moist forest (9.6541%), alvar nival area (9.1676%), warm temperate dry forest (7.8623%), cool temperate scrub (5.4823%), cold temperate moist forest (5.0814%), alpine wet tundra (4.0446%), cold temperate wet forest (3.2209%), alpine rain tundra (2.4392%), cold temperate dry scrub (1.8731%), alpine moist

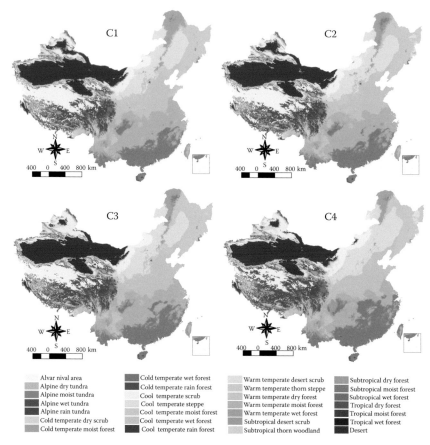

FIGURE 17.4
Spatial distributions of HLZ ecosystems in C1, C2, C3, and C4.

TABLE 17.6

Average Area and its Proportion of Each HLZ Ecosystem in the Four Periods in China (Units: km^2)

HLZ Type	C1	C2	C3	C4	Average	Proportion (%)
Alvar nival area	984,985	970,966	880,612	644,489	870,263	9.1676
Alpine dry tundra	10,438	5082	5796	431	5437	0.0573
Alpine moist tundra	170,020	151,327	154,845	127,681	150,968	1.5903
Alpine wet tundra	341,104	345,880	377,579	471,219	383,946	4.0446
Alpine rain tundra	216,542	241,386	224,371	243,883	231,545	2.4392
Cold temperate dry scrub	180,057	183,293	171,331	176,547	177,807	1.8731
Cold temperate moist forest	524,311	477,893	471,084	456,177	482,366	5.0814
Cold temperate wet forest	278,403	28.9489	31.0337	34.4775	30.5751	3.2209
Cold temperate rain forest	11,409	14,479	15,992	11,347	13,307	0.1402
Cool temperate scrub	480,095	487,256	533,846	580,487	520,421	5.4823
Cool temperate steppe	957,056	1,021,505	1,003,403	1,204,268	1,046,558	11.0248
Cool temperate moist forest	985,470	935,638	1,012,761	731,909	916,444	9.6541
Cool temperate wet forest	66,152	64,073	63,661	40,603	58,622	0.6175
Cool temperate rain forest	27	63	172	66	66	0.0007
Warm temperate desert scrub	30,650	30,027	53,133	112,400	56,552	0.5957
Warm temperate thorn steppe	11,131	9639	13,715	70,842	26,332	0.2774
Warm temperate dry forest	645,734	741,174	798,740	799,770	746,355	7.8623
Warm temperate moist forest	1,261,860	1,204,295	1,100,289	860,773	1,106,804	11.6594

continued

TABLE 17.6   (continued)

Average Area and its Proportion of Each HLZ Ecosystem in the Four Periods in China (Units: km²)

HLZ Type	C1	C2	C3	C4	Average	Proportion (%)
Warm temperate wet forest	10,881	14,063	12,858	6625	11,106	0.1170
Subtropical desert scrub	134	188	125	214	165	0.0017
Subtropical thorn woodland	1289	694	563	1442	997	0.0105
Subtropical dry forest	67,034	69,784	80,543	105,997	80,840	0.8516
Subtropical moist forest	845,222	836,136	900,933	1,194,712	944,251	9.9470
Subtropical wet forest	7165	5444	5067	4279	5489	0.0578
Tropical dry forest	4972	5346	5207	6170	5424	0.0571
Tropical moist forest	2730	2466	2577	2480	2563	0.0270
Desert	1,398,331	1,385,198	1,293,261	1,293,211	1,342,500	14.1415

TABLE 17.7

Area Change of HLZs in the Four Periods in China (Units: km²)

HLZ Type	From C1 to C2		From C2 to C3		From C3 to C4	
	Increased Area	Increasing Rate (%)	Increased Area	Increasing Rate (%)	Increased Area	Increasing Rate (%)
Alvar nival area	-14,019	-1.4232	-90,355	-9.3057	-236,122	-26.8134
Alpine dry tundra	-5356	-51.3141	714	14.0512	-5365	-92.5626
Alpine moist tundra	-18,693	-10.9945	3518	2.3246	-27,164	-17.5425
Alpine wet tundra	4776	1.4001	31,699	9.1649	93,640	24.8000
Alpine rain tundra	24,844	11.4731	-17,015	-7.0491	19,512	8.6964
Cold temperate dry scrub	3236	1.7972	-11,962	-6.5263	5216	3.0445
Cold temperate moist forest	-46,418	-8.8531	-6810	-1.4249	-14,906	-3.1643
Cold temperate wet forest	11,086	3.9821	20,848	7.2016	34,438	11.0970
Cold temperate rain forest	3070	26.9080	1513	10.4518	-4645	-29.0457
Cool temperate scrub	7161	1.4916	46,590	9.5618	46,640	8.7366
Cool temperate steppe	64,449	6.7341	-18,102	-1.7721	200,865	20.0184
Cool temperate moist forest	-49,832	-5.0567	77,123	8.2428	-280,851	-27.7313
Cool temperate wet forest	-2079	-3.1425	-412	-0.6431	-23,058	-36.2204
Cool temperate rain forest	36	134.0800	109	172.5312	-172	-100.0000
Warm temperate desert scrub	-623	-2.0339	23,106	76.9514	59,267	111.5466
Warm temperate thorn steppe	-1492	-13.4050	4076	42.2886	57,127	416.5253
Warm temperate dry forest	95,440	14.7801	57,566	7.7669	1029	0.1289
Warm temperate moist forest	-57,565	-4.5619	-104,006	-8.6363	-239,516	-21.7685
Warm temperate wet forest	3182	29.2404	-1205	-8.5675	-6233	-48.4788
Subtropical desert scrub	54	40.4838	-63	-33.3414	88	70.2475

continued

TABLE 17.7   (continued)

Area Change of HLZs in the Four Periods in China (Units: km²)

HLZ Type	From C1 to C2		From C2 to C3		From C3 to C4	
	Increased Area	Increasing Rate (%)	Increased Area	Increasing Rate (%)	Increased Area	Increasing Rate (%)
Subtropical thorn woodland	−595	−46.1852	−131	−18.8335	879	156.1139
Subtropical dry forest	2750	4.1031	10,758	15.4162	25,455	31.6041
Subtropical moist forest	−9086	−1.0750	64,798	7.7496	293,778	32.6082
Subtropical wet forest	−1721	−24.0250	−376	−6.9135	−788	−15.5462
Tropical dry forest	374	7.5280	−140	−2.6113	963	18.4932
Tropical moist forest	−264	−9.6805	111	4.4963	−96	−3.7341
Desert	−13,133	−0.9392	−91,938	−6.372	−48	−0.0037

tundra (1.5903%), subtropical dry forest (0.8516%), cool temperate wet forest (0.6175%), warm temperate desert scrub (0.5957%), warm temperate thorn steppe (0.2774%), cold temperate rain forest (0.1402%), warm temperate wet forest (0.1170%), subtropical wet forest (0.0578%), alpine dry tundra (0.0573%), tropical dry forest (0.0571%), tropical moist forest (0.0270%), subtropical thorn woodland (0.0105%), subtropical desert scrub (0.0017%), and cool temperate rain forest (0.0007%). The first 10 ecosystem types account for 88.0650% of the total land surface of China but the last several ecosystem types are very sensitive to climate change and human impacts.

During the period C1 to C4, five ecosystem types, continuously expanded. These were: alpine wet tundra, cold temperate wet forest, cool temperate scrub, warm temperate dry forest, and subtropical dry forest. Alpine wet tundra expanded 4776 km^2 from C1 to C2, 31,699 km^2 from C2 to C3, and 93,640 km^2 from C3 to C4. Cold temperate wet forest expanded by 1,1086 km^2, 20,848 km^2, and 34,438 km^2 respectively from C1 to C2, from C2 to C3 and from C3 to C4. Cool temperate scrub expanded 7176 km^2 from C1 to C2, 46,590 km^2 from C2 to C3, and 46,640 km^2 from C3 to C4. Warm temperate dry forest expanded by 95,440, 5.7566, and 1029 km^2 from C1 to C2, from C2 to C3 and from C3 to C4, respectively. Subtropical dry forest expanded by 2750, 10,758, and 25,455 km^2, respectively from C1 to C2, from C2 to C3, and from C3 to C4.

The six ecosystem types continuously shrank during these periods. They were: nival area, cold temperate moist forest, desert, cool temperate wet forest, warm temperate moist forest, and subtropical wet forest. Areas of the nival area decreased by 7.9463% every 11-year subperiod, on an average, reducing 14,019 km^2 from C1 to C2, 90,355 km^2 from C2 to C3, and 2,36,122 km^2 from C3 to C4. Areas of cold temperate moist forest decreased by 3.2487 on an average every subperiod, reducing 46,418 km^2 from C1 to C2, 681 km^2 from C2 to C3, and 14,906 km^2 from C3 to C4. Desert decreased by 1.8974% on an average every subperiod, reducing 19,225 km^2 from C1 to C2, 74,551 km^2 from C2 to C3, and 44,835 km^2 from C3 to C4. Cool temperate wet forest decreased by 9.6555% on an average every subperiod, reducing 2079 km^2 from C1 to C2, 412 km^2 from C2 to C3, and 23,058 km^2 from C3 to C4. Warm temperate moist forest decreased by 7.9463% on an average every subperiod, reducing 57,565 km^2 from C1 to C2, 1,04,006 km^2 from C2 to C3, and 2,39,516 km^2 from C3 to C4. Areas of subtropical wet forest decreased by 10.0680% on an average every subperiod, reducing 1721 km^2 from C1 to C2, 376 km^2 from C2 to C3, and 788 km^2 from C3 to C4.

17.4.2.3 Ecotope Diversity and Patch Connectivity

Both HLZ diversity and patch connectivity saw an increasing trend (Table 17.8). Their 11-year increase rates were respectively 0.2754% and 1.5499%. HLZ diversity and patch connectivity have the following regression equations respectively:

TABLE 17.8

Change Trends of HLZ Diversity and Patch Connectivity during the Four Periods

Periods	C1	C2	C3	C4	11-Year Increasing Rate (%)
HLZ diversity	0.1712	0.1711	0.1722	0.1731	0.2754
Patch connectivity	0.0370	0.0374	0.0391	0.0392	1.5499

$$DI(t) = 0.0007 \cdot t + 0.1702 \qquad (17.24)$$

$$CO(t) = 0.0009 \cdot t + 0.036 \qquad (17.25)$$

where $DI(t)$ is HLZ diversity; $CO(t)$ is HLZ patch connectivity; t is the time variable; and $t = 1, 2, 3, 4$, respectively correspond to C1, C2, C3, and C4. The correlation coefficients of both regression equations 17.24 and 17.25 are 0.94.

17.4.2.4 Shift Trends of HLZ Mean Centers

The mean centers of the HLZ ecosystems had different shift trends (Figure 17.5 and Table 17.9). Nival area shifted 7.7879 km to the southeast from C1 to C2, 38.4471 km to the northwest from C2 to C3 and 11.4966 km to the south from C3 to C4. Alpine dry tundra shifted 260.5925, 715.818, and 163.8877 km, respectively, to the northeast, southwest, and continuously southwest from C1 to C2, from C2 to C3, and from C3 to C4. Alpine moist tundra moved 80.4171, 359.2612, and 58.6707 km, respectively, to the northwest, southeast, and northwest from C1 to C2, C2 to C3, and C3 to C4. Alpine wet tundra shifted 58.6626 km to the west from C1 to C2, 59.1065 km to the northeast from C2 to C3, and 99.0687 km to the northwest from C3 to C4. Alpine rain tundra shifted 19.2602 km to the northwest from C1 to C2, 28.0638 km to the west from C2 to C3, and 135.5852 km to the west from C3 to C4.

The desert moved 21.0352, 48.7177, and 9.0569 km, respectively, toward the northwest, southeast, and northwest from C1 to C2, from C2 to C3, and from C3 to C4. The cold temperate dry scrub shifted 55.7868 km to the northwest from C1 to C2, 124.1248 km to the south from C2 to C3, and 109.7830 km to the southwest from C3 to C4. The cold temperate moist forest shifted 281.6452 km to the southwest from C1 to C2, 235.4414 km to the southwest continuously from C2 to C3 and 133.4956 km to the west from C3 to C4. Cold temperate wet forest shifted 182.4321, 38.7310, and 283.7616 km, respectively, to the northeast, southwest, and continuously southwest from C1 to C2, from C2 to C3 and from C3 to C4. The cold temperate rain forest shifted 33.6803 km to the west from C1 to C2, 63.2406 km to the northeast from C2 to C3, and 153.0424 km to the southwest from C3 to C4.

The cool temperate scrub shifted 7.7485, 90.1884, and 185.9929 km, respectively, to the southwest, northwest, and east from C1 to C2, from C2 to C3,

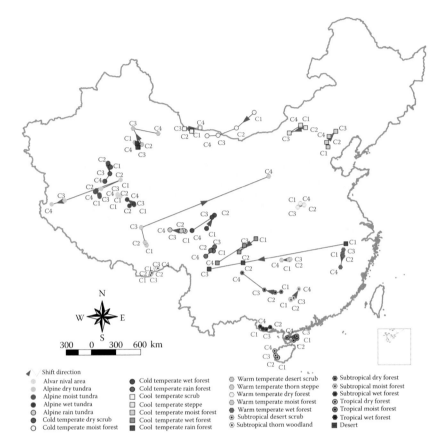

FIGURE 17.5
Shift trend of the mean center of each HLZ in the four periods in China.

and from C3 to C4. The cool temperate steppe shifted 62.2631, 206.1511, and 164.3963 km, respectively, to the southeast, southwest, and northeast from C1 to C2, from C2 to C3, and from C3 to C4. The cool temperate moist forest shifted 56.7606, 168.884, and 100.5148 km, respectively, to the northwest, northeast, and southwest from C1 to C2, from C2 to C3 and from C3 to C4. The cool temperate wet forest shifted 255.4687, 98.4038, and 384.2707 km to the southwest, northeast, and southwest. The cool temperate rain forest shifted 1252.9711 km to the west from C1 to C2 and 415.6455 km continuously to the west, but disappeared in C4.

The warm temperate desert scrub shifted 89.6367, 216.7490, and 309.8927 km, respectively, to the southeast, northwest, and east from C1 to C2, from C2 to C3, and from C3 to C4. The warm temperate thorn steppe shifted 39.1507, 179.2972, and 1644.9307 km, respectively, to the northwest, continuously northwest, and northeast from C1 to C2, from C2 to C3, and from C3 to C4. The

TABLE 17.9

Shift Trend of Each HLZ Type in the Four Periods in China (Units: km)

HLZ Type	From C1 to C2		From C2 to C3		From C3 to C4	
	Shift Distance	Direction	Shift Distance	Direction	Shift Distance	Direction
Alvar nival area	7.7879	Southeast	38.4471	Northwest	11.4966	South
Alpine dry tundra	260.5925	Northeast	715.8180	Southwest	163.8877	Southwest
Alpine moist tundra	80.4171	Northwest	59.2612	Southeast	58.6707	Northwest
Alpine wet tundra	58.6626	West	59.1065	Northeast	99.0687	Northwest
Alpine rain tundra	19.2602	Northwest	28.0638	West	135.5852	West
Cold temperate dry scrub	55.7868	Northwest	124.1248	South	109.7830	Southwest
Cold temperate moist forest	281.6452	Southwest	235.4414	Southwest	133.4956	West
Cold temperate wet forest	182.4321	Northeast	38.7310	Southwest	283.7616	Southwest
Cold temperate rain forest	33.6803	West	63.2406	Northeast	153.0424	Southwest
Cool temperate scrub	7.7485	Southwest	90.1884	Northwest	185.9929	East
Cool temperate steppe	62.2631	Southeast	206.1511	Southwest	164.3963	Northeast
Cool temperate moist forest	56.7606	Northwest	168.8840	Northeast	100.5148	Southwest
Cool temperate wet forest	255.4687	Southwest	98.4038	Northeast	384.2707	Southwest
Cool temperate rain forest	1252.9711	West	415.6455	West		

Warm temperate desert scrub	89.6367	Southeast	216.7490	Northwest	309.8927	East
Warm temperate thorn steppe	39.1507	Northwest	179.2972	North	1644.9307	Northeast
Warm temperate dry forest	49.5548	East	79.9478	Southwest	52.9940	Northeast
Warm temperate moist forest	6.0564	Southwest	20.9408	Northeast	105.6254	West
Warm temperate wet forest	36.7036	South	75.4987	North	208.6402	South
Subtropical desert scrub	2.5855	Southeast	32.2950	Southwest	29.0863	Northeast
Subtropical thorn woodland	14.8540	Northwest	18.5283	Southeast	17.5208	Northeast
Subtropical dry forest	72.1314	West	133.3046	Northwest	336.2482	Northwest
Subtropical moist forest	9.8195	Southeast	39.9031	Northeast	121.7762	Northeast
Subtropical wet forest	90.7260	East	56.4018	Southwest	78.4328	Northwest
Tropical dry forest	3.9603	West	32.1138	Northwest	56.0808	North
Tropical moist forest	46.5807	Southwest	63.3580	Northeast	120.1246	West
Tropical wet forest						
Desert	21.0352	Northwest	48.7177	Southeast	9.0569	Northwest

warm temperate dry forest shifted 49.5548, 79.9478, and 52.994 km, respectively, to the east, southwest, and northwest. The warm temperate moist forest shifted 6.0564, 20.9408, and 105.6254 km, respectively to the southwest, northeast, and west. The warm temperate wet forest shifted 36.7036, 75.4987, and 208.6402 km, respectively, to the south, north, and south.

The subtropical desert scrub shifted 2.5855, 32.295, and 29.0863 km to the southeast, southwest, and northeast from C1 to C2, from C2 to C3, and from C3 to C4. The subtropical thorn woodland shifted 14.8540, 18.5283, and 17.5208 km, respectively, to the northwest, southeast, and northeast. The subtropical dry forest shifted 72.1314, 133.3046, and 336.2482 km, respectively, to the west, northwest, and continuously northwest. The subtropical moist forest shifted 9.8195, 39.9031, and 121.7762 km, respectively, to the southeast, northeast, and continuously northeast. The subtropical wet forest shifted 90.726, 56.4018, and 78.4328 km, respectively, to the east, southwest, and northwest.

The tropical dry forest moved 3.9603, 32.1138, and 56.0808 km, respectively, to the west, northwest, and north from C1 to C2, from C2 to C3, and from C3 to C4. The tropical moist forest shifted 46.5807, 63.358, and 120.1246 km, respectively, to the southwest, northeast, and west.

In short, the HLZ ecosystem types which had the longest shift distances, for example, shift distances that were longer than 100 km in every subperiod, included the nival area, cold temperate moist forest, cold temperate wet forest, cool temperate wet forest, cool temperate rain forest, cool temperate steppe, cool temperate moist forest, warm temperate desert scrub, warm temperate thorn steppe, warm temperate wet forest, and subtropical dry forest. These ecosystem types are more sensitive to climate change in China. During the period 1964 to 2007, the nival areas had a trend shifting toward the northwest. The subtropical dry forest, subtropical moist forest, and tropical dry forest moved toward the north.

17.4.3 Scenarios of HLZ Ecosystems on a National Level

17.4.3.1 Area Change of HLZ Ecosystems in China

17.4.3.1.1 Area Change under HadCM3 AIFi

T1, T2, T3, and T4, respectively, represent the periods from 1960 to 1999, 2010 to 2039, 2040 to 2069, and 2070 to 2099. In all, 27 ecosystem types could be found in China (Figure 17.6). Of these 27 ecosystems, subtropical desert scrub, subtropical thorn woodland, and tropical wet forest would not appear in T1 and T2 (Table 17.10).

The ecosystem types found in all four periods can be placed in the following order in terms of their area sizes: cool temperate moist forest accounts for 16.0444%, subtropical moist forest for 12.9018%, warm temperate moist forest 10.1535%, nival area 9.1316%, alpine rain tundra 6.3936%, cold temperate wet forest 6.3704%, desert 6.3438%, cool temperate steppe 5.6072%, warm temperate

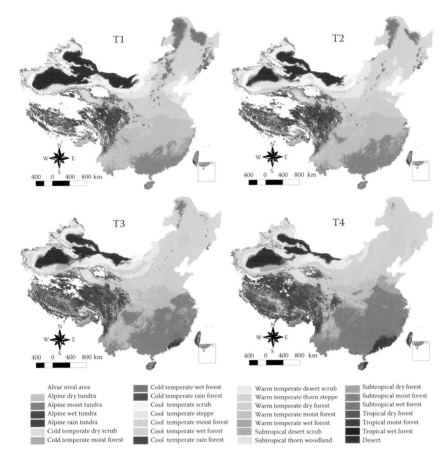

FIGURE 17.6
HLZs in T1, T2, T3, and T4 in terms of HadCM3 AIFi.

dry forest 4.5161%, cool temperate wet forest 3.7967%, cold temperate rain forest 3.4474%, warm temperate desert scrub 3.4099%, cool temperate scrub 3.1501%, warm temperate thorn steppe 2.0699%, cold temperate moist forest 1.9366%, alpine wet tundra 1.0025%, warm temperate wet forest 0.8599%, tropical moist forest 0.7270%, subtropical wet forest 0.5678%, cold temperate dry scrub 0.4541%, subtropical dry forest 0.3913%, subtropical desert scrub 0.3182%, cool temperate rain forest 0.2251%, tropical dry forest 0.0918%, alpine moist tundra 0.0590%, subtropical thorn woodland 0.0148%, and tropical wet forest 0.0001%. The first 10 ecosystem types would have a proportion of 81.2744% to the total land surface of China.

The ecosystems with continuously decreasing areas would include the nival area, alpine moist tundra, alpine wet tundra, cold temperate dry scrub, cold temperate moist forest, and warm temperate moist forest (Table 17.11). The decadal decreasing rate of nival area would be 5.5692%; the reduction

TABLE 17.10

Area of Each HLZ Types in Terms of HadCM3 AlFi (Unit: km²)

HLZ Type	T1	T2	T3	T4	Average	Proportion (%)
Alvar nival area	1,353,230	1,110,749	691,413	298,125	863,379	9.1316
Alpine moist tundra	13,618	5257	3071	366	5578	0.0590
Alpine wet tundra	137,260	114,231	83,699	43,967	94,789	1.0025
Alpine rain tundra	571,934	638,448	659,349	548,262	604,498	6.3936
Cold temperate dry scrub	96,485	48,729	19,722	6801	42,934	0.4541
Cold temperate moist forest	217,007	200,286	179,329	135,798	183,105	1.9366
Cold temperate wet forest	739,831	574,636	532,376	562,397	602,310	6.3704
Cold temperate rain forest	232,188	266,719	342,180	462,699	325,946	3.4474
Cool temperate scrub	405,190	416,211	259,266	110,690	297,839	3.1501
Cool temperate steppe	620,448	700,136	455,701	344,329	530,153	5.6072
Cool temperate moist forest	1,347,165	1,528,661	1,643,510	1,548,546	1,516,970	16.0444
Cool temperate wet forest	389,585	346,169	345,079	360,814	360,412	3.8119
Cool temperate rain forest	4161	8437	23,985	48,532	21,279	0.2251
Warm temperate desert scrub	119,431	282,665	391,508	495,975	322,395	3.4099
Warm temperate thorn steppe	44,802	180,452	199,358	358,211	195,706	2.0699
Warm temperate dry forest	251,999	269,168	525,831	660,945	426,986	4.5161
Warm temperate moist forest	1,374,571	1,206,781	808,622	450,004	959,994	10.1535
Warm temperate wet forest	50,912	65,271	78,722	130,302	81,302	0.8599
Subtropical desert scrub			39,306	81,049	30,089	0.3182
Subtropical thorn woodland			1414	4181	1399	0.0148
Subtropical dry forest	3801	4037	9217	130,913	36,992	0.3913
Subtropical moist forest	640,669	895,150	1,442,427	1,901,106	1,219,838	12.9018
Subtropical wet forest	15,159	25,450	82,951	91,198	53,689	0.5678
Tropical dry forest	2974	4534	8356	18,860	8681	0.0918
Tropical moist forest	11,733	16,105	73,341	173,775	68,738	0.7270
Tropical wet forest			19	6	6	0.0001
Desert	848,638	584,502	584,823	381,210	599,793	6.3438

TABLE 17.11

Area Change of Each HLZ Types in Terms of HadCM3 AlFi (Unit: km²)

HLZ Type	From T1 to T2		From T2 to T3		From T3 to T4	
	Increased Area	Increasing Rate (%)	Increased Area	Increasing Rate (%)	Increased Area	Increasing Rate (%)
Alvar nival area	−242,481	−17.9187	−419,336	−37.7526	−393,287	−56.8817
Alpine moist tundra	−8361	−61.3966	−2186	−41.5882	−2704	−88.0745
Alpine wet tundra	−23,029	−16.7776	−30,532	−26.7285	−39,732	−47.4702
Alpine rain tundra	66,514	11.6297	20,901	3.2737	−111,087	−16.8480
Cold temperate dry scrub	−47,755	−49.4954	−29,007	−59.5279	−12,921	−65.5140
Cold temperate moist forest	−16,721	−7.7054	−20,957	−10.4633	−43,531	−24.2742
Cold temperate wet forest	−165,195	−22.3288	−42,260	−7.3543	30,021	5.6390
Cold temperate rain forest	34,531	14.8721	75,462	28.2926	120,518	35.2207
Cool temperate scrub	11,021	2.7199	−156,944	−37.7079	−148,576	−57.3064
Cool temperate steppe	79,688	12.8437	−244,435	−34.9126	−111,372	−24.4397
Cool temperate moist forest	181,496	13.4724	114,849	7.5130	−94,963	−5.7781
Cool temperate wet forest	−43,416	−11.1142	−1090	−0.3148	15,735	4.5597
Cool temperate rain forest	4277	102.7860	15,548	184.2761	24,547	102.3421
Warm temperate desert scrub	163,234	136.6770	108,843	38.5058	104,467	26.6833
Warm temperate thorn steppe	135,650	302.7770	18,906	10.4770	158,853	79.6824
Warm temperate dry forest	17,169	6.8130	256,663	95.3541	135,114	25.6954
Warm temperate moist forest	−167,790	−12.2067	−398,159	−32.9935	−358,618	−44.3493
Warm temperate wet forest	14,359	28.2035	13,451	20.6076	51,580	65.5214
Subtropical desert scrub			39,306		41,742	106.1964

continued

TABLE 17.11 (continued)

Area Change of Each HLZ Types in Terms of HadCM3 AlFi (Unit: km^2)

HLZ Type	From T1 to T2		From T2 to T3		From T3 to T4	
	Increased Area	Increasing Rate (%)	Increased Area	Increasing Rate (%)	Increased Area	Increasing Rate (%)
Subtropical thorn woodland			1414		2767	195.6389
Subtropical dry forest	236	6.2224	5180	128.3022	121,696	1320.4038
Subtropical moist forest	254,481	39.7212	547,276	61.1379	458,679	31.7991
Subtropical wet forest	10,291	67.8853	57,502	225.9443	8247	9.9414
Tropical dry forest	1560	52.4597	3823	84.3161	10,503	125.6966
Tropical moist forest	4373	37.2689	57,236	355.3888	100,434	136.9417
Tropical wet forest			19		-0013	-69.3164
Desert	-264,136	-31.1247	321	0.0549	-203,613	-34.8162

from T1 to T2, from T2 to T3, and from T3 to T4 would be 242,481, 419,336, and 393,287 km^2 respectively. Alpine moist tundra would have a decadal shrinking rate of 6.9508%; the shrunk area would be 8361 km^2 from T1 to T2, 2186 km^2 from T2 to T3, and 2,704 km^2 from T3 to T4. 23,029 km^2 of alpine wet tundra would be reduced from T1 to T2, 30,532 km^2 from T2 to T3, and 39,732 km^2 from T3 to T4. The decadal decreasing rate of cold temperate dry scrub would be 6.6394% and its area would decrease by 47,755 km^2 from T1 to T2, by 29,007 km^2 from T2 to T3, and by 12,921 km^2 from T3 to T4. The decadal shrinking rate of cold temperate moist forest would be 2.6730%; it would shrink 16,721 km^2 from T1 to T2, 20,957 km^2 from T2 to T3, and 43,531 km^2 from T3 to T4. In a decade, 4.8044% of warm temperate moist forest would be lost on an average, decreasing by 16,779 km^2 from T1 to T2, 398,159 km^2 from T2 to T3, and 358,618 km^2 from T3 to T4. The fastest decrease of all these ecosystem types would happen in the period T3 to T4.

The ecosystems with continuous area expansion would include the cold temperate rain forest, cool temperate rain forest, warm temperate desert scrub, warm temperate thorn steppe, warm temperate dry forest, subtropical moist forest, subtropical wet forest, tropical dry forest, and tropical moist forest. Their decadal increase rates would be 7.0913%, 76.1747%, 22.5202%, 49.9672%, 11.5915%, 11.1382%, 238.8695%, 14.0527%, 35.8295%, 38.1545%, and 98.6485%, respectively.

17.4.3.1.2 *Area Change under HadCM3 A2a*

The 27 ecosystem types would appear under scenario A2a (Figure 17.7). Of these, tropical moist forest would only appear in T4, and subtropical desert scrub and subtropical thorn woodland would only appear in T3 and T4 (Table 17.12).

The 24 ecosystem types found in all four periods can be placed in the following order in terms of their area sizes: cool temperate moist forest accounts for 15.9969%, subtropical moist forest for 11.7907%, warm temperate moist forest for 11.3079%, nival area for 9.9797%, desert 6.6915%, alpine rain tundra for 6.5907%, cold temperate wet forest for 6.1793%, cool temperate steppe for 6.0799%, warm temperate dry forest for 4.0493%, cool temperate wet forest for 3.8301%, cool temperate scrub for 3.4062%, warm temperate desert scrub for 3.3774%, cold temperate rain forest for 3.1895%, cold temperate moist forest for 2.0752%, warm temperate thorn steppe for 1.6436%, alpine wet tundra for 1.073%, warm temperate wet forest for 0.8156%, tropical moist forest for 0.4536%, cold temperate dry scrub for 0.4375%, subtropical wet forest for 0.3774%, subtropical desert scrub for 0.2216%, cool temperate rain forest for 0.209%, subtropical dry forest for 0.0862%, tropical dry forest for 0.0618%, alpine moist tundra for 0.0594%, subtropical thorn woodland for 0.0168%, and tropical wet forest for 0.0001%. The first 10 biggest ecosystem types would account for 82.4960% of the total land surface of China.

The continuously shrinking ecosystems would include the nival area, alpine moist tundra, alpine wet tundra, cold temperate dry scrub, cold temperate

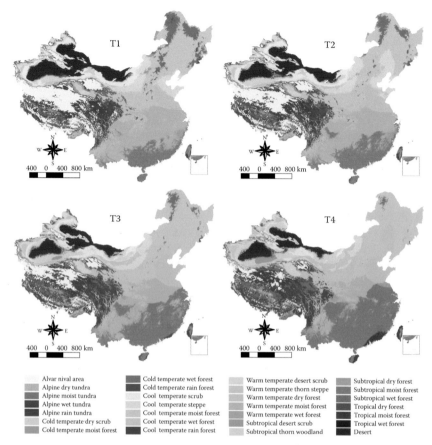

FIGURE 17.7
HLZs in T1, T2, T3, and T4 in terms of HadCM3 A2a.

moist forest, cold temperate wet forest, desert, cool temperate scrub and warm temperate moist forest (Table 17.13). Their decadal shrinking rates would be 4.7108%, 6.4377%, 4.1261%, 6.3468%, 1.9173%, 2.0768%, 2.8168%, 4.1980%, and 4.2153%, respectively.

The ecosystems with continuously expanded areas would include the cold temperate rain forest, warm temperate desert scrub, warm temperate thorn steppe, warm temperate dry forest, subtropical moist forest, subtropical wet forest, and tropical moist forest. Their decadal expanding rates would be 5.4762%, 19.3109%, 59.1297%, 14.5221%, 13.7019%, 35.3670%, and 62.1474%, respectively.

17.4.3.1.3 Area Change under HadCM3 B2a

There would be 27 ecosystem types under scenario B2a. Of these 27 ecosystems, subtropical desert scrub would appear only in T3 and T4, tropical wet

TABLE 17.12

Area of Each HLZ Types in Terms of HadCM3 A2a (Unit: km²)

HLZ Type	T1	T2	T3	T4	Average	Proportion (%)
Alvar nival area	1,351,638	1,126,623	850,946	460,210	947,354	9.9797
Alpine moist tundra	12,662	5250	3411	1250	5643	0.0594
Alpine wet tundra	141,398	111,507	94,802	59,718	101,856	1.0730
Alpine rain tundra	574,099	636,747	659,675	632,059	625,645	6.5907
Cold temperate dry scrub	84,984	51,532	20,129	9471	41,529	0.4375
Cold temperate moist forest	238,012	195,853	179,994	174,124	196,996	2.0752
Cold temperate wet forest	736,374	563,697	524,025	522,273	586,592	6.1793
Cold temperate rain forest	233,369	247,439	318,009	412,286	302,776	3.1895
Cool temperate scrub	419,217	391,044	310,279	172,838	323,345	3.4062
Cool temperate steppe	610,126	807,998	490,448	400,030	577,150	6.0799
Cool temperate moist forest	1,343,747	1,439,212	1,717,051	1,574,197	1,518,552	15.9969
Cool temperate wet forest	423,904	333,740	345,046	351,646	363,584	3.8301
Cool temperate rain forest	7486	4978	23,788	43,101	19,838	0.2090
Warm temperate desert scrub	134,191	258,775	392,504	496,979	320,612	3.3774
Warm temperate thorn steppe	33,071	112,552	171,630	306,836	156,022	1.6436
Warm temperate dry forest	209,200	278,403	415,455	634,522	384,395	4.0493
Warm temperate moist forest	1,437,677	1,298,000	968,820	589,238	1,073,434	11.3079
Warm temperate wet forest	54,201	52,837	82,487	120,155	77,420	0.8156
Subtropical desert scrub			16,227	67,923	21,038	0.2216
Subtropical thorn woodland			1469	4907	1594	0.0168
Subtropical dry forest	4026	1475	15	27,222	8184	0.0862
Subtropical moist forest	598,524	848,427	1,283,466	1,746,657	1,119,268	11.7907
Subtropical wet forest	12,141	21,916	36,984	72,256	35,825	0.3774
Tropical dry forest	2131	4947	9355	7024	5864	0.0618
Tropical moist forest	11,304	16,771	34,506	109,660	43,060	0.4536
Tropical wet forest				32	8	0.0001
Desert	819,303	683,077	542,284	496,206	635,218	6.6915

TABLE 17.13

Area Change of Each HLZ Types in Terms of HadCM3 A2a (Unit: km²)

HLZ Type	From T1 to T2		From T2 to T3		From T3 to T4	
	Increased Area	Increasing Rate (%)	Increased Area	Increasing Rate (%)	Increased Area	Increasing Rate (%)
Alvar nival area	−225,015	−16.6476	−275,676	−24.4693	−390,736	−45.9179
Alpine moist tundra	−7412	−58.5385	−1839	−35.0299	−2161	−63.3584
Alpine wet tundra	−29,891	−21.1393	−16,705	−14.9812	−35,084	−37.0078
Alpine rain tundra	62,648	10.9124	22,929	3.6009	−27,616	−4.1863
Cold temperate dry scrub	−33,452	−39.3628	−31,403	−60.9392	−10,657	−52.9459
Cold temperate moist forest	−42,159	−17.7130	−15,859	−8.0974	−5870	−3.2614
Cold temperate wet forest	−172,677	−23.4496	−39,672	−7.0378	−1752	−0.3344
Cold temperate rain forest	14,069	6.0288	70,570	28.5203	94,276	29.6458
Cool temperate scrub	−28,173	−6.7204	−80,765	−20.6537	−137,441	−44.2961
Cool temperate steppe	197,872	32.4314	−317,549	−39.3008	−90,418	−18.4359
Cool temperate moist forest	95,465	7.1044	277,839	19.3049	−142,854	−8.3197
Cool temperate wet forest	−90,165	−21.2700	11,306	3.3878	6600	1.9128
Cool temperate rain forest	−2508	−33.5041	18,810	377.8788	19,313	81.1882
Warm temperate desert scrub	124,584	92.8409	133,730	51.6781	104,474	26.6174

Warm temperate thorn steppe	79,482	240.3380	59,077	52.4889	135,206	78.7778
Warm temperate dry forest	69,203	33.0797	137,052	49.2280	219,067	52.7295
Warm temperate moist forest	-139,677	-9.7155	-329,181	-25.3606	-379,581	-39.1798
Warm temperate wet forest	-1364	-2.5159	29,650	56.1148	37,667	45.6647
Subtropical desert scrub			16,227		51,696	318.5851
Subtropical thorn woodland			1469		3438	233.9979
Subtropical dry forest	-2551	-63.3664	-1460	-98.9835	27,207	181,483.0790
Subtropical moist forest	249,903	41.7532	435,039	51.2759	463,191	36.0891
Subtropical wet forest	9775	80.5132	15,068	68.7519	35,272	95.3711
Tropical dry forest	2815	132.0983	4408	89.1172	-2331	-24.9164
Tropical moist forest	5467	48.3596	17,735	105.7475	75,154	217.7973
Tropical wet forest					32	
Desert	-136,226	-16.6271	-140,793	-20.6116	-46,078	-8.4970

forest would appear only in T4, and subtropical dry forest would disappear in T3 and T4 (Figure 17.8, Table 17.14). The other 24 ecosystem types would exist in all the four periods. They can be placed in the following order in terms of their area sizes: cool temperate moist forest accounts for 16.4176%, warm temperate moist forest for 11.6266%, subtropical moist forest for 11.4432%, nival area for 10.0989%, desert 6.7663%, alpine rain tundra for 6.751%, cool temperate steppe for 6.2111%, cold temperate wet forest for 5.9844%, cool temperate wet forest for 3.8706%, cool temperate scrub for 3.8553%, warm temperate dry forest for 3.589%, cold temperate rain forest for 3.2533%, warm temperate desert scrub for 2.98%, cold temperate moist forest for 2.0911%, warm temperate thorn steppe for 1.3479%, alpine wet tundra for 1.1283%, warm temperate wet forest for 0.8618%, cold temperate dry scrub for 0.5017%, subtropical wet forest for 0.4072%, tropical moist forest for 0.3927%, cool temperate rain forest for 0.2153%, subtropical desert scrub for 0.0689%, alpine moist tundra for 0.0687%,

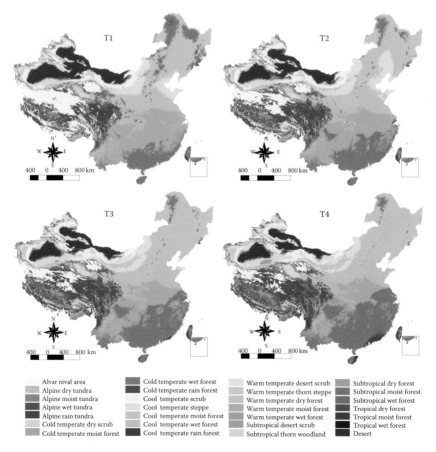

FIGURE 17.8
HLZs in T1, T2, T3, and T4 in terms of HadCM3 B2a.

TABLE 17.14

Area of Every HLZ Type in Terms of HadCM3 B2a (Unit: km²)

HLZ Type	T1	T2	T3	T4	Average	Proportion (%)
Alvar nival area	1,348,838	1,014,065	825,308	646,449	958,665	10.0989
Alpine moist tundra	12,531	6152	4111	3300	6524	0.0687
Alpine wet tundra	141,358	112,481	98,365	76,229	107,108	1.1283
Alpine rain tundra	573,639	660,640	668,737	660,411	640,857	6.7510
Cold temperate dry scrub	85,102	56,144	30,929	18,328	47,626	0.5017
Cold temperate moist forest	237,791	198,286	187,118	170,815	198,502	2.0911
Cold temperate wet forest	731,119	539,287	506,809	495,127	568,086	5.9844
Cold temperate rain forest	233,778	264,809	327,578	409,165	308,833	3.2533
Cool temperate scrub	418,927	408,193	348,798	287,996	365,979	3.8553
Cool temperate steppe	609,372	810,933	493,429	444,687	589,605	6.2111
Cool temperate moist forest	1,347,434	1,488,573	1,750,115	1,647,823	1,558,486	16.4176
Cool temperate wet forest	424,310	320,778	347,897	376,722	367,427	3.8706
Cool temperate rain forest	7463	5110	24,029	45,159	20,440	0.2153
Warm temperate desert scrub	134,966	226,053	347,535	422,999	282,888	2.9800
Warm temperate thorn steppe	34,011	115,129	151,520	211,149	127,952	1.3479
Warm temperate dry forest	212,660	294,039	397,503	458,584	340,697	3.5890
Warm temperate moist forest	1,439,511	1,229,965	997,109	748,163	1,103,687	11.6266
Warm temperate wet forest	54,307	48,776	85,435	138,716	81,809	0.8618
Subtropical desert scrub			4420	21,755	6544	0.0689
Subtropical thorn woodland			567	1613	545	0.0057
Subtropical dry forest	4042	2401			1611	0.0170
Subtropical moist forest	596,792	956,058	1,265,892	1,526,363	1,086,277	11.4432
Subtropical wet forest	12,103	22,491	37,915	82,091	38,650	0.4072
Tropical dry forest	2128	5072	7344	3065	4402	0.0464
Tropical moist forest	11,319	16,674	30,597	90,534	37,281	0.3927
Tropical wet forest				19	5	0.0001
Desert	819,283	690,682	553,726	505,540	642,308	6.7663

tropical dry forest for 0.0464%, subtropical dry forest for 0.017%, subtropical thorn woodland for 0.0057%, and tropical wet forest for 0.0001%. The first 10 ecosystem types would account for 83.025% of the total land surface of China.

The ecosystems whose areas would continually shrink include the nival area, alpine moist tundra, alpine wet tundra, desert, cold temperate dry scrub, cold temperate moist forest, cold temperate wet forest, cool temperate scrub, warm temperate moist forest, and subtropical dry forest (Table 17.15). The decadal decease rates of nival area, alpine moist tundra, alpine wet tundra, desert, cold temperate dry scrub, cold temperate moist forest, cold temperate wet forest, cool temperate scrub, and warm temperate moist forest would be 3.7195%, 5.2615%, 3.2910%, 2.7353%, 2.0119%, 2.3056%, 4.4319%, 2.2324%, and 3.4305%, respectively. The subtropical dry forest would disappear in T3.

The ecosystems with continuously expanding areas would include the cold temperate rain forest, warm temperate desert scrub, warm temperate thorn steppe, warm temperate dry forest, subtropical moist forest, subtropical wet forest, and tropical moist forest. Their decadal increase rates would be 5.3588%, 15.2437%, 37.2018%, 8.2601%, 11.1258%, 41.3050%, and 49.9886%, respectively.

17.4.3.1.4 Brief Summary

The three scenarios indicate that the biggest ecosystem types would include cool temperate moist forest, subtropical moist forest, nival area, and warm temperate moist forest. These would account for more than 48% of the whole land surface of China. The total average area of these four ecosystem types put together would be 4,560,181, 4,658,608, and 4,707,115 km², respectively, under scenarios AlFi, A2a, and B2a. The proportions of these four ecosystem types to the total land area of China would be 48.2313%, 49.0752%, and 49.5863%, respectively.

Nival areas would see the biggest loss. Their decadal decrease would be 75,365, 63,373, and 50,171 km², respectively, under scenarios AlFi, A2a, and B2a. Their decadal decrease rate would be 5.5692%, 4.7108%, and 3.7195%, respectively, under the three scenarios. The subtropical moist forest would see the biggest expansion in terms of the three scenarios. It would increase 90,031, 82,009, and 66,398 km² per decade on an average under scenarios AlFi, A2a, and B2a, respectively. Its decadal increase rate would be 14.0527%, 13.7019%, and 11.1258%, respectively, under the three scenarios.

If $t = 1$, $t = 2$, $t = 3$, and $t = 4$ represent T1, T2, T3, and T4, respectively, the squared correlation coefficients between HLZ diversity and the time variable t would be 0.9880, 0.9813, and 0.9920, respectively, under scenarios AlFi, A2a, and B2a. Their linear regression equations could be respectively formulated as

$$D_{A1Fi}(t) = 0.003 \cdot t + 0.1575 \tag{17.26}$$

TABLE 17.15

Area Change of Each HLZ Types in Terms of HadCM3 B2a (Unit: km²)

HLZ Type	From T1 to T2		From T2 to T3		From T3 to T4	
	Increased Area	Increasing Rate (%)	Increased Area	Increasing Rate (%)	Increased Area	Increasing Rate (%)
Alvar nival area	−334,772	−24.8193	−188,758	−18.6140	−178,858	−21.6717
Alpine moist tundra	−6378	−50.9009	−2042	−33.1829	−810	−19.7144
Alpine wet tundra	−28,877	−20.4284	−14,116	−12.5495	−22,136	−22.5039
Alpine rain tundra	87,001	15.1665	8096	1.2255	−8325	−1.2450
Cold temperate dry scrub	−28,958	−34.0274	−25,215	−44.9118	−12,601	−40.7425
Cold temperate moist forest	−39,505	−16.6133	−11,168	−5.6324	−16,303	−8.7128
Cold temperate wet forest	−191,832	−26.2381	−32,478	−6.0225	−11,681	−2.3049
Cold temperate rain forest	31,031	13.2737	62,769	23.7034	81,587	24.9060
Cool temperate scrub	−10,734	−2.5623	−59,394	−14.5506	−60,802	−17.4319
Cool temperate steppe	201,561	33.0769	−317,505	−39.1530	−48,741	−9.8781
Cool temperate moist forest	141,140	10.4747	261,541	17.5699	−102,292	−5.8449
Cool temperate wet forest	−103,532	−24.4002	27,119	8.4542	28,825	8.2855
Cool temperate rain forest	−2353	−31.5282	18,919	370.2248	21,130	87.9337
Warm temperate desert scrub	91,086	67.4883	121,483	53.7408	75,464	21.7139

continued

TABLE 17.15 (continued)

Area Change of Each HLZ Types in Terms of HadCM3 B2a (Unit: km^2)

HLZ Type	From T1 to T2		From T2 to T3		From T3 to T4	
	Increased Area	Increasing Rate (%)	Increased Area	Increasing Rate (%)	Increased Area	Increasing Rate (%)
Warm temperate thorn steppe	81,118	238.5062	36,391	31.6086	59,629	39.3538
Warm temperate dry forest	81,379	38.2672	103,464	35.1872	61,081	15.3663
Warm temperate moist forest	−209,546	−14.5568	−232,857	−18.9320	−248,946	−24.9668
Warm temperate wet forest	−5531	−10.1842	36,659	75.1565	53,281	62.3641
Subtropical desert scrub			4420		17,335	392.1700
Subtropical thorn woodland			567		1047	184.6062
Subtropical dry forest	−1641	−40.6030	−2401	−100.0000		
Subtropical moist forest	359,266	60.1995	309,834	32.4074	260,471	20.5761
Subtropical wet forest	10,388	85.8338	15,424	68.5770	44,176	116.5120
Tropical dry forest	2945	138.4075	2272	44.7960	−4279	−58.2604
Tropical moist forest	5355	47.3099	13,923	83.5013	59,938	195.8959
Tropical wet forest					19	
Desert	−128,602	−15.6969	−136,956	−19.8291	−48,186	−8.7021

$$D_{A2a}(t) = 0.0027 \cdot t + 0.1473 \qquad (17.27)$$

$$D_{B2a}(t) = 0.0022 \cdot t + 0.9841 \qquad (17.28)$$

where $D_{A1Fi}(t)$, $D_{A2a}(t)$, and $D_{B2a}(t)$ are, respectively HLZ diversity in terms of scenarios A1Fi, A2a, and A2b.

The decadal increase rate of HLZ diversity would be 1.2391%, 1.1657%, and 0.9453%, respectively under A1Fi, A2a, and B2a. Patch connectivity would increase during the period T1 to T2 and then continuously decrease in terms of all three scenarios (Table 17.16).

17.4.3.2 Mean Center Shifts

In terms of the three scenarios of mean center shift (Figures 17.9 through 17.11, Tables 17.17 through 17.19), the nival area, alpine rain tundra and cold temperate rain forest would have a shifting trend toward the northwest. The alpine moist tundra and alpine wet tundra would rapidly shrink in the Tibet plateau. The shrunk parts of the nival area, alpine moist tundra and alpine wet tundra in the Tibet plateau and Tian-Shan Mountains might be gradually replaced by alpine rain tundra and cold temperate rain forest with warming climate and increasing precipitation.

During the period 1960 to 2099, the ecosystem types that would see a shifting distance of more than 50 km include the cold temperate moist forest, cold temperate wet forest, cool temperate steppe, warm temperate thorn steppe, warm temperate dry forest, warm temperate wet forest, and subtropical dry forest. These seven ecosystem types would be more sensitive to climate change than other ecosystem types.

Climate warming could lead to the fast reduction of cold temperate moist forest and cold temperate wet forest in the Da-Hinggan-Ling mountains, the Xiao-Hinggan-Ling mountains, and the Chang-Bai mountains, which could make the mean centers of cold temperate moist forest and cold temperate wet

TABLE 17.16

Scenarios of HLZ Ecosystem Diversity and Patch Connectivity

Scenario	Index	T1	T2	T3	T4	Decadal Increase Rate (%)
A1Fi	HLZ ecosystem diversity	0.1726	0.1750	0.1793	0.1811	1.2391
	Patch connectivity	0.0397	0.0421	0.0405	0.0388	−0.5787
A2a	HLZ ecosystem diversity	0.1723	0.1736	0.1767	0.1803	1.1657
	Patch connectivity	0.0397	0.0410	0.0412	0.0401	0.2443
B2a	HLZ ecosystem diversity	0.1723	0.1739	0.1760	0.1789	0.9453
	Patch connectivity	0.0397	0.0409	0.0415	0.0414	1.1129

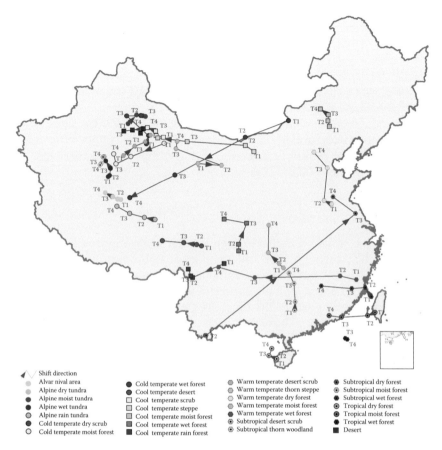

FIGURE 17.9
HLZ mean center shift in terms of HadCM3 AlFi scenario (T1, T2, T3, and T4 present the periods 1961 to 1990, 2010 to 2039, 2040 to 2069, and 2070 to 2099, respectively).

forest generally, move toward the southwest. Climate change could also drive the shift of cool temperate moist forest, cool temperate wet forest, warm temperate dry forest, warm temperate moist forest, and subtropical moist forest toward the north.

17.5 Ecosystem Changing Trends on a Provincial Level

17.5.1 Spatial Pattern of Ecosystems

The results obtained by operating the HLZ model on data set from the 100 meteorological stations scattered over and around the Jiangxi province (as

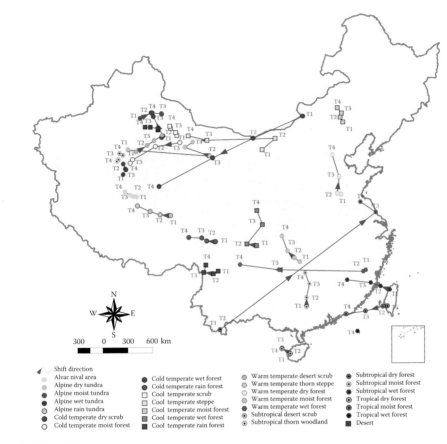

FIGURE 17.10
HLZ mean center shift in terms of HadCM3 A2a scenario (T1, T2, T3, and T4 present the periods from 1961 to 1990, 2010 to 2039, 2040 to 2069, and 2070 to 2099, respectively).

seen in Figure 5.6) indicate that during the period 1960 to 1999 there were six ecosystem types, including cool temperate wet forest, cool temperate rain forest, warm temperate moist forest, warm temperate wet forest, subtropical moist forest, and subtropical wet forest (Figure 17.12, Table 17.20). Cool temperate wet forests of about 1408 km² existed in the 1960s, growing to about 233 km² in the 1980s. These were mainly distributed in the upper areas of the Mu-Yi Mountains, the Jiu-Lin Mountains, and the Luo-Xiao mountains (Figure 5.6). With the exception of a decline from the 1970s to the 1980s, the cool temperate rain forest generally saw great expansion. In the 1960s and 1970s, cool temperate rain forest was mainly distributed in the lower areas of the Mu-Yi mountains, the Jiu-Lin mountains, and the Luo-Xiao mountains, while in the 1970s and 1990s it was mainly distributed in the upper areas of the mountains mentioned above. The warm temperate moist forest had an

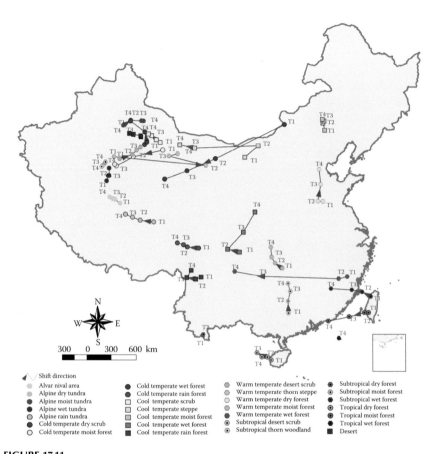

FIGURE 17.11
HLZ mean center shift in terms of HadCM3 B2a scenario (T1, T2, T3, and T4 present the periods 1961 to 1990, 2010 to 2039, 2040 to 2069, and 2070 to 2099, respectively).

expansion trend during the period from the 1960s to the 1980s and was mainly distributed in mountainous areas in northwestern Jiangxi and middle-southern Jiangxi; although most warm temperate moist forest was transformed into warm temperate wet forest by the 1990s. Areas of warm temperate wet forest saw a great increase on an average during the period from the 1960s to the 1990s, particularly between the 1980s and 1990s, in which the decadal increase was 11,353 km², mainly distributed in the middle and high mountainous areas. The subtropical moist forest was the most widely distributed ecosystem type in the Jiangxi province, covering almost all its plains and low mountainous areas. Subtropical wet forest appeared mainly in Jing-De-Zhen, Shang-Rao, and Ying-Tan in the 1990s and its area was about 8595 km².

TABLE 17.17

HLZ Mean Center Shift in Terms of HadCM3 AIFi (Units: km)

HLZ Type	From T1 toT2		From T2 to T3		From T3 to T4	
	Shift Distance	Direction	Shift Distance	Direction	Shift Distance	Direction
Alvar nival area	37.2194	West	60.6098	Northwest	99.6501	Northwest
Alpine moist tundra	141.0729	Northeast	78.3171	East	30.4775	Southeast
Alpine wet tundra	21.8417	Northeast	81.2397	North	101.5349	Northwest
Alpine rain tundra	129.9343	West	179.9046	West	167.9019	Northwest
Cold temperate dry scrub	98.4220	North	343.4599	Northwest	166.9453	East
Cold temperate moist forest	422.4274	Southwest	193.0952	Southwest	116.4851	North
Cold temperate wet forest	540.5310	Southwest	931.1582	Southwest	593.6569	Southwest
Cold temperate rain forest	58.7822	Northwest	153.0260	West	253.3328	West
Cool temperate scrub	87.6030	West	51.9332	North	104.8240	Northwest
Cool temperate steppe	122.6129	Northwest	679.0816	West	347.6060	West
Cool temperate moist forest	66.3066	North	67.9678	North	113.5406	Northwest
Cool temperate wet forest	75.7874	North	260.1738	Northeast	266.9697	West
Cool temperate rain forest	401.3900	Southwest	26.7518	Northwest	85.7699	Northwest
Warm temperate desert scrub	172.3590	Northeast	134.7576	Northeast	101.6592	Northeast
Warm temperate thorn steppe	274.1663	East	564.6226	Northwest	85.5257	North
Warm temperate dry forest	96.9735	Northwest	378.5829	North	239.1537	Northwest

continued

TABLE 17.17   **(continued)**

HLZ Mean Center Shift in Terms of HadCM3 AlFi (Units: km)

HLZ Type	From T1 to T2		From T2 to T3		From T3 to T4	
	Shift Distance	Direction	Shift Distance	Direction	Shift Distance	Direction
Warm temperate moist forest	72.8066	Northwest	176.5961	Northwest	316.9810	North
Warm temperate wet forest	195.7823	West	999.8136	West	439.4663	Northwest
Subtropical desert scrub					17.3054	West
Subtropical thorn woodland					24.5570	Southwest
Subtropical dry forest	9.3037	East	2245.0686	Northeast	328.7293	Northwest
Subtropical moist forest	100.6789	North	211.7613	North	129.6742	Northwest
Subtropical wet forest	129.0019	Northwest	160.3973	West	359.0190	West
Tropical dry forest	15.3132	Northeast	99.0169	Northwest	87.9915	Northeast
Tropical moist forest	97.3433	Southwest	309.4629	West	145.2467	West
Tropical wet forest					25.4139	Southeast
Desert	64.0197	Northeast	133.7741	West	101.2072	West

TABLE 17.18

HLZ Mean Center Shift in Terms of HadCM3 A2a (Units: km)

HLZ Type	From T1 to T2		From T2 to T3		From T3 to T4	
	Shift Distance	Direction	Shift Distance	Direction	Shift Distance	Direction
Alvar nival area	35.2513	West	44.4149	Northwest	69.7833	Northwest
Alpine moist tundra	145.0283	Northeast	126.5858	East	9.0322	Southeast
Alpine wet tundra	68.9656	Northeast	13.1552	Southeast	74.3745	Northwest
Alpine rain tundra	129.3539	West	125.9363	Northwest	166.8751	Northwest
Cold temperate dry scrub	23.3848	Northeast	253.5297	Northwest	43.5062	Northwest
Cold temperate moist forest	300.8838	West	310.7990	Southwest	50.3000	Southwest
Cold temperate wet forest	641.4670	Southwest	530.5808	Southwest	725.0678	Southwest
Cold temperate rain forest	105.2661	West	69.1078	Northwest	155.4221	West
Cool temperate scrub	37.7052	Northwest	77.0694	Northwest	27.4662	Northwest
Cool temperate steppe	201.6137	Northeast	809.1574	West	232.0738	West
Cool temperate moist forest	65.9999	Northwest	50.6007	Northeast	79.6391	Northwest
Cool temperate wet forest	117.7265	West	289.3060	Northeast	176.0143	Northwest
Cool temperate rain forest	64.7409	Southwest	114.8759	West	87.4074	Northeast
Warm temperate desert scrub	151.7895	Northeast	98.3391	Northeast	116.7359	Northeast
Warm temperate thorn steppe	985.0514	East	314.4679	Northwest	110.5791	Northeast
Warm temperate dry forest	48.6522	West	196.3621	North	267.2956	Northwest
Warm temperate moist forest	79.8392	Northwest	115.2804	Northwest	185.8942	Northwest
Warm temperate wet forest	33.0024	Southwest	999.4259	West	466.2325	West
Subtropical desert scrub					46.5526	Northwest
Subtropical thorn woodland					28.9257	Southwest
Subtropical dry forest	6.5363	Northeast	2295.9557	Northeast	220.8044	Northwest
Subtropical moist forest	108.2348	North	160.9662	North	137.2133	Northwest
Subtropical wet forest	116.1314	Northwest	126.8190	Northwest	283.1569	West
Tropical dry forest	20.5618	Northeast	115.1750	Northwest	4.6331	Southwest
Tropical moist forest	53.1696	West	218.0660	West	234.7223	West
Desert	13.1868	North	93.3342	West	62.0373	West

TABLE 17.19

HLZ Mean Center Shift in Terms of HadCM3 B2a (Units: km)

HLZ Type	From T1 to T2		From T2 to T3		From T3 to T4	
	Shift Distance	Direction	Shift Distance	Direction	Shift Distance	Direction
Alvar nival area	75.5579	Northwest	27.7668	Northwest	42.9994	Northwest
Alpine moist tundra	108.1856	Northeast	130.5999	East	26.3666	East
Alpine wet tundra	63.5463	Northeast	12.4665	East	88.5478	North
Alpine rain tundra	166.4244	West	99.9990	Northwest	91.4609	Northwest
Cold temperate dry scrub	38.6359	Northeast	121.8549	North	219.0454	Northwest
Cold temperate moist forest	365.9395	West	229.9398	Southwest	26.8438	Northwest
Cold temperate wet forest	919.4139	Southwest	368.1529	Southwest	283.5834	Southwest
Cold temperate rain forest	135.8805	West	53.5425	Northwest	76.0682	Northwest
Cool temperate scrub	56.3766	Northwest	71.9718	Northwest	43.0520	Northwest
Cool temperate steppe	209.7029	Northeast	759.6094	West	182.1454	West
Cool temperate moist forest	85.0069	Northwest	30.5507	Northeast	28.0870	Northwest
Cool temperate wet forest	145.2420	West	273.4521	Northeast	276.8163	Northeast
Cool temperate rain forest	68.9449	Southwest	108.5113	West	99.9375	Northeast
Warm temperate desert scrub	159.3361	Northeast	72.2974	Northeast	61.3474	Northeast
Warm temperate thorn steppe	1108.1778	East	454.1713	West	108.9489	Northeast
Warm temperate dry forest	61.0955	West	198.8580	North	180.7560	North
Warm temperate moist forest	104.6603	Northwest	82.8780	Northwest	118.9558	North
Warm temperate wet forest	105.3193	West	914.8551	West	336.4942	West
Subtropical desert scrub					18.2940	Northeast
Subtropical thorn woodland					15.7180	Southwest
Subtropical dry forest	12.0757	Northeast				
Subtropical moist forest	141.6918	North	116.4096	North	96.9020	Northwest
Subtropical wet forest	119.3966	Northwest	129.5004	West	250.5218	West
Tropical dry forest	56.6797	Northwest	135.3257	Northwest	76.8831	Southeast
Tropical moist forest	51.0334	West	192.6599	West	255.9286	West
Desert	13.3850	Northwest	91.8621	West	48.7417	West

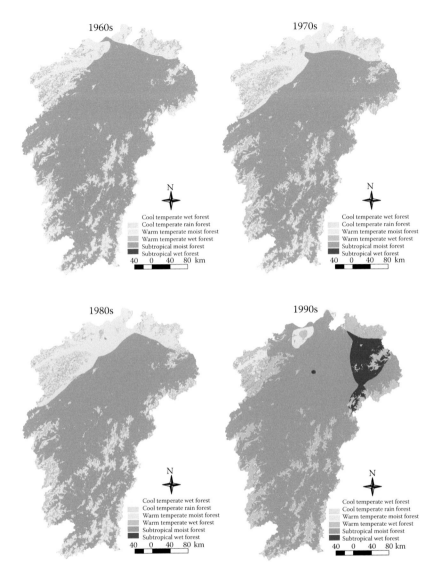

FIGURE 17.12
Spatial distributions of HLZ ecosystems during the period from the 1960s to the 1990s.

In terms of scaling diversity index $D(\varepsilon,r,t) = -\ln\left(\sum_{i=1}^{m(\varepsilon,r,t)}\left(p_i(\varepsilon,r,t)\right)^{1/2}\right)^2 /$ $\ln\varepsilon$ (Yue et al., 2007c), HLZ diversity in the Jiangxi province was 0.0533, 0.0574, 0.0584, and 0.0694 in the 1960s, 1970s, 1980s, and 1990s, respectively (Table 17.21). It had a decadal increase of 0.004. If t = 1, 2, 3, and 4 represent

TABLE 17.20

Area Change of HLZs in Four Decades in Jiangxi Province (Units: km²)

HLZ Type	1960s		1970s		1980s		1990s	
	Area	Proportion (%)	Area	Proportion (%)	Area	Proportion (%)	Area	Proportion (%)
Cool temperate wet forest	1408	0.84	0	0	233	0.14	0	0
Cool temperate moist forest	70	0.04	241	0.14	162	0.10	244	0.15
Warm temperate moist forest	44,572	26.70	50,390	30.18	55,373	33.17	19,581	11.73
Warm temperate wet forest	1099	0.66	8033	4.81	5787	3.47	17,139	10.27
Subtropical moist forest	119,811	71.76	108,295	64.86	105,404	63.13	121,401	72.71
Subtropical wet forest	0	0	0	0	0	0	8,595	5.15

TABLE 17.21

Average Changes of the HLZ Terrestrial Ecosystem Diversity
and Patch Connectivity from the 1960s to 1990s

Periods	Patch Connectivity	Diversity
1960s	0.0237	0.0533
From the 1960s to the 1970s	0.0006	0.0045
1970s	0.0243	0.0574
From the 1970s to the 1980s	−0.0007	0.0011
1980s	0.0236	0.0584
From the 1960s to the 1970s	0.0086	0.0101
1990s	0.0322	0.0694
Changes rate every 10 years	0.0021	0.0040

the 1960s, 1970s, 1980s, and 1990s, respectively, the equation of linear regression, with a correlation coefficient of $R^2 = 0.8835$, can be formulated as

$$D(\varepsilon, r.t) = 0.0048t + 0.0472 \tag{17.29}$$

In the scaling diversity index, p_i (ε, r, t) is the proportion of area of the ith ecotope to the area of the whole investigation region or the individual number of the ith species to the total individual number; $m(\varepsilon, r, t)$ is the total number of species or ecotopes under investigation; t represents time; $\varepsilon = (e+A)^{-1}$, A is the area of the investigation region measured by hectare; r is the spatial resolution of the used data set; and e equals 2.71828.

17.5.2 Shift Trends of HLZ Mean Centers

The cool temperate wet forest only appeared in the 1960s and 1980s (Table 17.22), and so their mean centers are only marked in Figure 17.13; however, they show no shift trend. Cool temperate rain forest shifted toward the northwest in general during the period from the 1960s to the 1990s. Its mean center in the 1970s shifted 34.72 km toward the northwest starting from its mean center in the 1960s. It shifted 9.76 km to the east in the 1980s compared to its mean center location in the 1970s. In the 1990s, its mean center shifted 9.81 km toward the west, taking its mean center in the 1980s as its start.

The mean center of warm temperate moist forest had a shift trend to the southwest during the period 1960 to 1999 on an average. Specifically, the mean center shifted 25.14 km to the north in the 1970s compared to its position in the 1960s. In the 1980s, the mean center shifted 2.01 km toward northeast compared to its position in the 1970s. In the 1990s, the mean center shifted 70.24 km toward the southwest taking the mean center in the 1980s as a starting point.

TABLE 17.22

Shift Trend of HLZ Mean Center from the 1960s to 1990s

HLZ Type	From the 1960s to the 1970s		From the 1960s to the 1970s		From the 1960s to the 1970s	
	Distance	Direction	Distance	Direction	Shift Distance	Direction
Cool temperate wet forest						
Cool temperate rain forest	34.72	Northwest	9.76	East	9.81	West
Warm temperate moist forest	25.14	North	2.01	Northwest	70.24	Southwest
Warm temperate wet forest	11.18	Southeast	6.16	North	8.05	Southeast
Subtropical moist forest	9.81	Southeast	4.19	South	18.23	Northwest
Subtropical wet forest						

The warm temperate wet forest had a shift trend toward the southeast during the last decades of last century. In the 1970s, the mean center shifted 11.8 km toward the south taking the mean centre in 1960s as a starting point. The mean center shifted 6.01 km toward the north in the 1980s compared with its position in the 1970s. The mean center shifted 8.05 km toward southeast in the 1990s, taking the mean center in the 1980s as a starting point.

The mean center of subtropical moist forest had a shift trend toward the west. The mean center shifted 9.81 km toward the southeast in the 1970s compared with its position in the 1960s. The mean center shifted 4.19 km toward the south in the 1980s using the mean center in the 1970s as a starting point. It shifted 18.23 km toward the northwest in the 1990s taking the mean center in the 1980s as a starting point.

In short, warm temperate moist forest and cool temperate rain forest had the biggest shift distances in the four decades and their mean centers became much closer in the 1990s than ever before. The subtropical moist forest moved toward the west and warm temperate wet forest toward the southeast. These ecosystems are very sensitive to climate change.

17.6 Discussion and Conclusions

In many cases, quantitative spatial analysis yielded different implications with different locations, which led to the development of methods to integrate

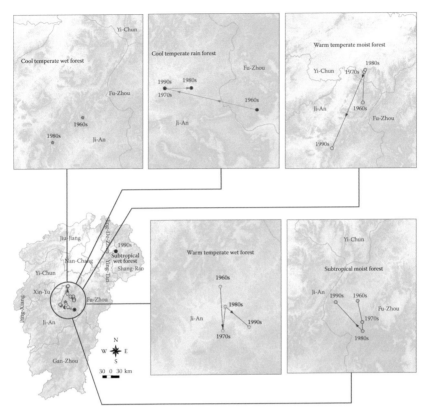

FIGURE 17.13
Shift trend of HLZ ecosystem mean center from the 1960s to the 1990s.

global and local controls. Ecosystem dynamics could not be recovered from dealing with global controls alone while ignoring local complications, or by treating local case studies in isolation from global factors. The key to a sustainable environment is to think globally and act locally; the key to understanding spatial systems is to think both globally and locally at once (Phillips, 2002). However, this idea is not so easy to quantify. It must deal with multiscale, across-scale, and multiresolution problems.

An HASM-based method for simulating terrestrial ecosystems (HASM-STE) is developed by constructing statistical transfer functions, which are used to downscale data on coarser spatial resolutions and to estimate values for locations where primary data is not available. HASM-STE is successfully applied to simulate terrestrial ecosystems on provincial, national, and global levels, in which the data on various spatial resolutions are assimilated. HASM-STE is an alternative method for analyzing the contribution of landforms and climate change to local, regional, and global ecological diversity by assimilating data from different sources on different scales.

References

Bakkenes, M., Alkemade, J.R.M., Ihle, F., Leemans, R., and Latour, J.B. 2002. Assessing effects of forecasted climate change on the diversity and distribution of European higher plants for 2050. *Global Change Biology* 8: 390–407.

Belotelov, N.V., Bogatyrev, B.G., Kirilenko, A.P., and Venevsky, S.V. 1996. Modelling of time-dependent biomes shifts under global climate changes. *Ecological Modelling* 87: 29–40.

Betts, R.A., Cox, P.M., and Woodward, F.I. 2000. Simulated responses of potential vegetation to doubled-CO_2 climate change and feedbacks on near-surface temperature. *Global Ecology and Biogeography* 9: 171–180.

Biermann, F. 2007. Earth system governance as a crosscutting theme of global change research. *Global Environmental Change* 17(3): 326–337.

Burke, A., 2003. Inselbergs in a changing world—Global trends. *Diversity and Distributions* 9: 375–383.

Chen, X.W., Zhang, X.S., and Li, B.L. 2003. The possible response of life zones in China under global climate change. *Global and Planetary Change* 38: 327–337.

Chen, X.W., Zhang, X.S., and Li, B.L. 2005. Influence of Tibetan Plateau on vegetation distributions in East Asia: A modeling perspective. *Ecological Modelling* 181: 79–86.

Chinea, J.D. and Helmer, E.H. 2003. Diversity and composition of tropical secondary forests recovering from large-scale clearing: Results from the 1990 inventory in Puerto Rico. *Forest Ecology and Management* 180: 227–240.

Clements, E., 1916. Climax formations of North America. *Plant Succession: An Analysis of the Development of Vegetation, Publication 242.* Washington, DC: Carnegie Institute of Washington.

Cramer, W. and Whittaker, R.J. 1999. Changing the surface of our planet—Results from studies of the global ecosystem. *Global Ecology and Biogeography* 8: 363–365.

Dixon, R.K., Smith, J.B., Brown, S., Merara, O., Mata, L.J., and Buksha, I. 1999. Simulations of forest system response and feedbacks to global change: Experiences and results from the U.S. Country studies Program. *Ecological Modelling* 122: 289–305.

Feng, Q.Y. and Chai, L.H. 2008. A new statistical dynamic analysis on vegetation patterns in land ecosystems. *Physica A* 387(14): 3583–3593.

Ganguly, S., Schull, M.A., Samanta, A., Shabanov, N.V., Milesi, C., Nemani, R.R., Knyazikhin, Y., and Myneni, R.B. 2008. Generating vegetation leaf area index earth system data record from multiple sensors. *Remote Sensing of Environment* 112: 4318–4343.

Holdridge, L.R. 1947. Determination of world plant formations from simple climate data. *Science* 105(2727): 367–368.

Holdridge, L.R., Grenke, W.C., Hatheway, W.H., Liang, T., and Tosi, J.A. 1971. *Forest Environments in Tropical Life Zones.* Oxford: Pergamon Press.

Humboldt, A.V. 1807. Ideen zu einer Geographie der Pflanzen nebst einem naturgemaelde der Tropenlaender. Tuebingen.

Huntley, B. and Birks, H.J.B. 1983. *An Atlas of Past and Present Pollen Maps for Europe 0-13,000 years ago.* Cambridge: Cambridge University Press.

Kerr, S., Liu, S.G., Pfaff, A.S.P., and Hughes, R.F. 2003. Carbon dynamics and land use choices: Building a regional-scale multidisciplinary model. *Journal of Environmental Management* 69: 25–37.

Kirilenko, A.P., Belotelov, N.V., and Bogatyrev, B.G. 2000. Global model of vegetation migration: Incorporation of climatic variability. *Ecological Modelling* 132: 125–133.

Koeppen, W. and Geiger, R. 1930. *Handbuch der Climatologie*, Teil I D. Berlin: Borntraeger.

Körner, C. 2000. Why are there global gradients in species richness? Mountains might hold the answer. *Trends in Ecology and Evolution* 15: 513–514.

Lavorel, S. 1999. Global change effects on landscape and regional pattens of plant diversity. *Diversity and Distributions* 5: 239–240.

Matthews, E. 1983. Global vegetation and land use: New high-resolution databases for climate studies. *Journal of Climate and Applied Meteorology* 22: 474–487.

Mayle, F.E., Beerling, D.J., Gosling, W.D., and Bush, M.B. 2004. Responses of Amazonian ecosystems to climatic and atmospheric carbon dioxide changes since the last glacial maximum. *Philosophical Transactions of the Royal Society of London Series B* 359: 499–514.

McGlone, M.S., Duncan, R.P., and Heenan, P.B. 2001. Endemism, species selection and the origin and distribution of the vascular plant flora of New Zealand. *Journal of Biogeography* 28: 199–216.

Merriam, C.H. 1892. The geographic distribution of life in North America. *Proceedings of the Biological Society of Washington* 7: 1–74.

Metternicht, G.I. and Zinck, J.A. 1998. Evaluating the information content of JERS-1 SAR and Landsat TM data for discrimination of soil erosion features. *ISPRS Journal of Photogrammetry and Remote Sensing* 53: 143–153.

Miles, L., Newton, A.C., DeFries, R.S., Ravilious, C., May, I., Blyth, S., Kapos, V., and Gordon, J.E. 2006. A global overview of the conservation status of tropical dry forests. *Journal of Biogeography* 33: 491–505.

Millennium Ecosystem Assessment. 2005. *Ecosystems and Human Well-being: Synthesis.* Washington, DC: Island Press.

Ohlemüller, R., Gritti, E.S., Sykes, M.T., and Thomas, C.D. 2006. Towards European climate risk surfaces: The extent and distribution of analogous and non-analogous climates 1931–2100. *Global Ecology and Biogeography* 15: 395–405.

Olson, D.M., Dinerstein, E., Wikramanayake, E.D., Burgess, N.D., Powell, G.V.N., Underwood, E.C., D'amico, J.A. et al., 2001. Terrestrial ecoregions of the world: A new map of life on Earth. *BioScience* 51: 933–938.

Olwoch, J.M., Rautenbach, C.J. de W., Erasmus, B.F. N., Engelbrecht, F.A., and van Jaarsveld, A. S. 2003. Simulating tick distributions over sub-Saharan Africa: The use of observed and simulated climate surfaces. *Journal of Biogeography* 30: 1221–1232.

Peng, C.H. 2000. From static biogeographical model to dynamic global vegetation model: A global perspective on modelling vegetation dynamics. *Ecological Modelling* 135: 33–54.

Phillips, J.D. 2002. Global and local factors in earth surface systems. *Ecological Modelling* 149: 257–272.

Pielke, R.A., Avissar, R., Raupach, M., Dolman, A.J., Zeng, X., and Denning, A.S. 1998. Interactions between the atmosphere and terrestrial ecosystems: Influence on weather and climate. *Global Change Biology* 4: 461–475.

Post, W.M., Emanuel, W.R., Zinke, P.J., and Stangenberger, A.G. 1982. Soil carbon pools and world life zones. *Nature* 298: 156–159.

Powell, G.V.N., Barborak, J., and Rodriguez, M. 2000. Assessing representativeness of the protected natural areas in Costa Rica for conserving biodiversity: A preliminary gap analysis. *Biological Conservation* 93: 35–41.

Schouw, J.F. 1823. *Grundzuege einer allgemeinen Pflanzengeographie*. Berlin: G. Reimer.

Smith, T.M., Shugart, H.H., Bonan, G.B., and Smith, J.B. 1992. Modeling the potential response of vegetation to global climate change. *Advances in Ecological Research* 22: 93–116.

Sykes, M., Prentice, I.C., and Cramer, W. 1996. A bioclimatic model for the potential distributions of north European tree species under present and future climates. *Journal of Biogeography* 23: 203–233.

Tarolli, P., Arrowsmith, J.R., and Vivoni, E.R. 2009. Understanding earth surface processes from remotely sensed digital terrain models. *Geomorphology* 113: 1–3.

Thomas, C.D., Cameron, A., Green, R., Bakkenes, M., Beaumont, L.J., Collingham, Y.C., Erasmus, B.F.N., et al., 2004. Extinction risk from climate change. *Nature* 427: 145–148.

Thornthwaite, C.W. 1931. The climates of North America according to a new classification. *Geographical Review* 21(4): 633–655.

Thuiller, W. 2004. Patterns and uncertainties of species' range shifts under climate change. *Global Change Biology* 10: 2020–2027.

Xu, D.Y. and Yan, H. 2001. A study of the impacts of climate change on the geographic distribution of *Pinus koraiensis* in China. *Environment International* 27: 210–205.

Yang, X., Wang, M.X., Huang, Y., and Wang, Y.S. 2002. One-compartment model to study soil carbon composition rate at equilibrium situation. *Ecological Modelling* 151: 63–73.

Yue, T.X., Fan, Z.M. and Liu, J.Y. 2007a. Scenarios of land-cover in China. *Global and Planetary Change* 55(4): 317–342.

Yue, T.X., Du, Z.P., Song, D.J. and Gong Y. 2007b. A new method of surface modeling and its application to DEM construction. *Geomorphology* 91(1–2): 161–172.

Yue, T.X., Fan, Z.M. and Liu, J.Y. 2005. Changes of major terrestrial ecosystems in China since 1960. *Global and Planetary Change* 48: 287–302.

Yue, T.X., Fan, Z.M., Liu, J.Y., and Wei, B.X. 2006. Scenarios of major terrestrial ecosystems in China. *Ecological Modelling* 199(3): 363–376.

Yue, T.X., Haber, W., Grossmann, W.D., and Kasperidus, H.D. 1998. Towards the satisfying model for biological diversity. *Ekologia* 17 (Suppl. 1): 129–141.

Yue, T.X., Liu, J.Y., Jørgensen, S.E., Gao, Z.Q., Zhang, S.H., and Deng, X.Z. 2001. Changes of HLZ diversity in all of China over half a century. *Ecological Modelling* 144: 153–162.

Yue, T.X., Liu, J.Y., Jørgensen, S.E., and Ye, Q.H. 2003. Landscape change detection of the newly created wetland in Yellow River Delta. *Ecological Modelling* 164: 21–31.

Yue, T.X., Ma, S.N., Wu, S.X., and Zhan, J.Y. 2007c. Comparative analyses of the scaling diversity index and its applicability. *International Journal of Remote Sensing* 28(7): 1611–1623.

Yue, T.X., Xu, B., and Liu, J.Y. 2004. A patch connectivity index and its change on a newly born wetland at the Yellow River Delta. *International Journal of Remote Sensing* 25(21): 4617–4628.

18

Surface Modeling of Land Cover*

18.1 Introduction

How climatic changes might affect land cover is one of the overarching questions of the land-use and land-cover change (LUCC) project—a core project of the international geosphere–biosphere programme (IGBP) and the international human dimensions programme on global environmental change (IHDP). Land-cover dynamics and its diagnostic models are one of LUCC's main focuses (http://www.geo.ucl.ac.be/LUCC/lucc.html). Various models for simulating land-cover scenarios have been developed. For instance, the published Special Report on Emissions Scenarios (SRES), land-cover scenario assumes that in any one of the four major SRES regions in the world, changes occur everywhere at the same rate; this includes changes in cropland, grassland, forest, and energy biomass crop production. Land-cover changes under A1, B1, and B2 marker scenarios are highly uncertain. The A2 marker scenario does not include land-cover change, and so changes under the A1 scenario are assumed to also apply to A2. The SRES land-cover scenarios do not include the effect of climate change on future land cover (Arnell et al., 2004). SRES A2 and B2 scenarios of IPCC were downscaled in a probabilistic cellular automata model (PCAM) to define the narrative scenario conditions of future urban land-use change. The results of the modeling experiments illustrate the spectrum of possible land-cover scenarios of the New York Metropolitan Region for the years 2020 and 2050 (Solecki and Oliveri, 2004).

A model for conversion of land-use and its effects (CLUE) was developed under the assumption that there is a dynamic equilibrium between the total population and agricultural production, and that land-cover changes occur only when biophysical and human demands can no longer be met through existing land use (Veldkamp and Fresco, 1996; Verburg and Veldkamp, 2001; Veldkamp and Lambin, 2001).

The input–output (I–O) model was used to develop land-cover scenarios in China, which were not spatially explicit and did not consider the possible impacts of climate change (Hubacek and Sun, 2001). Socio-economic changes are linked to different types of land via an explicit representation of land requirement coefficients associated with specific economic activities. The

---

* Dr. Ze-Meng Fan is a major collaborator of this chapter.

strong biophysical linkages are mainly manifested in the derivation of regional differences in the land requirement coefficients and typical I–O technical coefficients by means of an Agro-Ecological Zone (FAO and IIASA, 1993).

A spatially explicit stochastic methodology (SESM) was developed for simulating land-use changes at watershed level without the need to describe the complex relationships between biophysical, economic, and human factors (Luijten, 2003). Its transition probabilities were based on observed frequencies of actual conversions between forest, pasture, and scrub in the period 1946–1970. Land-use changes were simulated on a grid cell basis, in which each grid cell acts independently. Three explorative scenarios for the year 2025 were developed under the assumptions of Business as Usual (BU), Ecological Watershed (EW), and Corporate Farming (CF).

A land-use change modeling kit (LUCK) was developed for scenario generation in a grid-based discretization mode on catchment scale, which represents the spatial distribution of land-cover types in the landscape (Niehoff et al., 2002). The potential conversion of land-cover types in each grid is based on an evaluation of the characteristics of each grid as well as on its neighborhood relationships. Because land-cover changes happen successively, LUCK tries to simulate the dynamics by an iterative procedure.

The above typical models can be broken down into four categories: statistical, stochastic, dynamic, and integrated models. All these have their drawbacks. The published SRES land-cover scenarios have an assumption problem. PCAM puts more emphasis on urban ecosystems. CLUE emphasizes agricultural ecosystems. I–O can only address the question of what rates changes are likely to progress (quantity of change), but ignores the question of where land-cover changes are likely to take place (location of change). SESM limits its application to a watershed scale. This chapter tries to find a solution to these problems.

18.2 Land-Cover Classification and Transition Probability Matrixes

Land-cover classification in China was conducted using a combination of remote sensed data from AVHRR and geophysical data sets (Liu et al., 2003). The geophysical data sets include annual mean temperature, annual precipitation and elevation. China was first divided into nine bio-climatic regions using the long-term mean climatic data. For each of these nine regions, AVHRR data, the AVHRR-derived normalized difference vegetation index, and the geophysical data were analyzed to generate a land-cover map. The nine land-cover maps for individual regions were then brought together for the whole of China. An existing land-cover data set derived from Landsat Thematic Mapper (TM) images was used to assess the accuracy of the classification based on AVHRR and geophysical data. The accuracy of individual regions varied from 73% to 89%, with an overall accuracy of 81% for the whole of China. The land-cover types include cultivated land, woodland, grassland, built-up land, water area, wetland, nival area, desert, bare rock, and desertification land (Figure 18.1).

For establishing the transition probability matrix between HLZ types in T_1 on an average and land-cover types in 2000 (Table 18.1), in which T_1 represents the period from the year 1971 to 2000, it is essential to develop a grid-oriented code matrix. The code of grid (i, j) is formulated as (Yue et al., 2007)

$$C_{i,j}^{2000} = 1000A_{i,j}^{2000} + A_{i,j}^{T_1}$$ (18.1)

where $C_{i,j}^{2000}$ is the code of element (i, j) of the grid-oriented code matrix for the transition probability matrix; $A_{i,j}^{2000}$ is the type code of land-cover at grid (i, j) in 2000; and $A_{i,j}^{T_1}$ is the type code of the HLZ ecosystem at grid (i, j) in T_1.

To build land-cover scenarios for the year 2039, a new grid-oriented code matrix for the transition probability matrix from HLZ ecosystem types in T_1

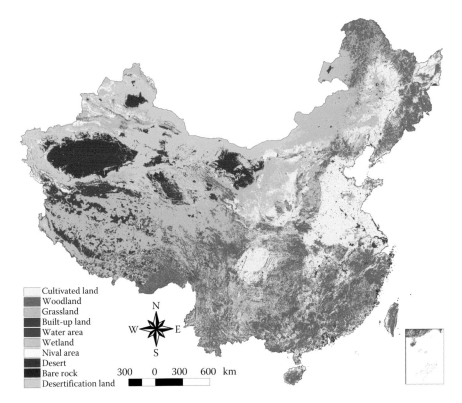

FIGURE 18.1
Land-cover map of China in 2000.

TABLE 18.1

Transition Probability Matrix from HLZ Ecosystem Types in T_1 on an Average to Land-Cover Types in 2000

HLZ type	Cultivated Land	Woodland	Grassland	Built-up Land	Water Area	Wetland	Nival Area	Desert	Bare Rock	Desertification Land
Nival	0.0004	0.0009	0.4717	0	0.0155	0.0018	0.0488	0.0006	0.2638	0.1089
Alpine dry tundra	0.0003	0.0231	0.6034	0	0.0049	0	0.0053	0.0163	0.0461	0.3007
Alpine moist tundra	0.0008	0.0767	0.7295	0	0.0304	0	0.0110	0.0047	0.0244	0.1226
Alpine wet tundra	0.0041	0.1694	0.6816	0	0.0339	0.0001	0.0077	0	0.0790	0.0243
Alpine rain tundra	0.0099	0.4694	0.4258	0	0.0084	0.0061	0.0057	0	0.0510	0.0238
Cold temperate dry scrub	0.0457	0.0671	0.6012	0.0001	0.0127	0	0.0026	0.0136	0.0096	0.2474
Cold temperate moist forest	0.0431	0.4317	0.4526	0.0006	0.0333	0.0095	0.0020	0.0003	0.0161	0.0109
Cold temperate wet forest	0.0402	0.6631	0.2490	0.0001	0.0051	0.0030	0.0026	0	0.0188	0.0182
Cold temperate rain forest	0.1388	0.5341	0.3222	0	0	0.0022	0	0	0.0010	0.0017
Cool temperate scrub	0.1097	0.0047	0.3718	0.0015	0.0041	0.0001	0	0.0786	0.0051	0.4245
Cool temperate steppe	0.3285	0.1013	0.4954	0.0030	0.0090	0.0178	0.0001	0	0.0007	0.0443
Cool temperate moist forest	0.3240	0.5083	0.1301	0.0047	0.0060	0.0239	0.0001	0	0.0008	0.0021
Cool temperate wet forest	0.1740	0.6697	0.1546	0.0001	0	0	0	0	0	0.0015

Cool temperate rain forest	0.2698	0.5930	0.0804	0.0091	0.0059	0.0005	0	0	0.0001	0.0412
Warm temperate desert scrub	0.0878	0.0004	0.1710	0.0012	0.0082	0	0	0.1442	0	0.5873
Warm temperate thorn steppe	0.2264	0.4831	0.1876	0.0018	0.0022	0	0	0	0	0.0988
Warm temperate dry forest	0.7009	0.1864	0.0751	0.0139	0.0114	0.0114	0	0	0	0.0009
Warm temperate moist forest	0.4152	0.4971	0.0569	0.0055	0.0252	0.0001	0	0	0	0.0001
Warm temperate wet forest	0.1169	0.8516	0.0290	0.0005	0.0020	0	0	0	0	0
Subtropical thorn woodland	0.0045	0.9676	0	0	0.0020	0	0	0	0	0.0259
Subtropical dry forest	0.3905	0.5177	0.0630	0.0136	0.0086	0	0	0	0	0.0065
Subtropical moist forest	0.3446	0.6009	0.0287	0.0137	0.0121	0	0	0	0	
Subtropical wet forest	0.2012	0.7888	0	0.0101	0	0	0	0	0	
Tropical dry forest	0.4083	0.1429	0.4365	0.0079	0.0044	0	0	0	0	
Tropical moist forest	0.4894	0.3384	0	0.0725	0.0997	0	0	0	0	
Desert	0	0.0091	0.0813	0	0.0041	0	0.0027	0.8752	0.0169	0.0107

and T_2 to land-cover types in 2039 is established, in which T_2 represents the period from the year 2010 to 2039. The code of grid (i, j) is formulated as

$$C_{i,j}^{2039} = 1000 A_{i,j}^{T_1} + A_{i,j}^{T_2} \tag{18.2}$$

where $C_{i,j}^{2039}$ is the code of element (i, j) of the grid-oriented code matrix for land-cover scenarios in 2039; $A_{i,j}^{T_1}$ is the type code of the HLZ ecosystem at grid (i, j) in T_1; and $A_{i,j}^{T_2}$ is the type code of the HLZ ecosystem at grid (i, j) in T_2.

In the process of building land-cover scenarios for 2039, if the HLZ type at grid (i, j) sees no change during the period T_1 to T_2, the land-cover type at grid (i, j) in 2039 would be assigned as the same as that in 2000. If the HLZ type at grid (i, j) converts from type K to type L, the land-cover type at grid (i, j) in 2039 would be assigned as the one that has a maximum transition probability to the HLZ type L in 2000.

Similarly, land-cover scenarios for the years 2069 and 2099 are built by establishing grid-oriented code matrixes for transition probability matrixes from HLZ types in T_2 and T_3 to land-cover types in 2069 and from T_3 and T_4 to 2099, in which T_3 represents the period 2040 to 2069 and T_4 the period 2070 to 2099. The codes of grid (i, j) in the years 2069 and 2099 are respectively formulated as

$$C_{i,j}^{2069} = 1000 A_{i,j}^{T_2} + A_{i,j}^{T_3} \tag{18.3}$$

and

$$C_{i,j}^{2099} = 1000 A_{i,j}^{T_3} + A_{i,j}^{T_4} \tag{18.4}$$

where $C_{i,j}^{2069}$ and $C_{i,j}^{2099}$ are respectively the codes of element (i, j) of the grid-oriented code matrix for land-cover scenarios in 2069 and 2099; $A_{i,j}^{T_k}$ is the type code of the HLZ ecosystem at grid (i, j) in T_k, $k = 2, 3$, and 4.

18.3 Spatial Distribution of Land-Cover Types

In terms of land-cover scenarios based on HadCM3A1Fi, HadCM3A2, and HadCM3B2 that are termed as scenario I, scenario II, and scenario III, respectively, the spatial distribution of land-cover types would have a very apparent regional variety in China (Figures 18.2 through 18.4). This is because of the regional difference of spatial hydrothermal distribution, the influence of the continental monsoon climate and human activities.

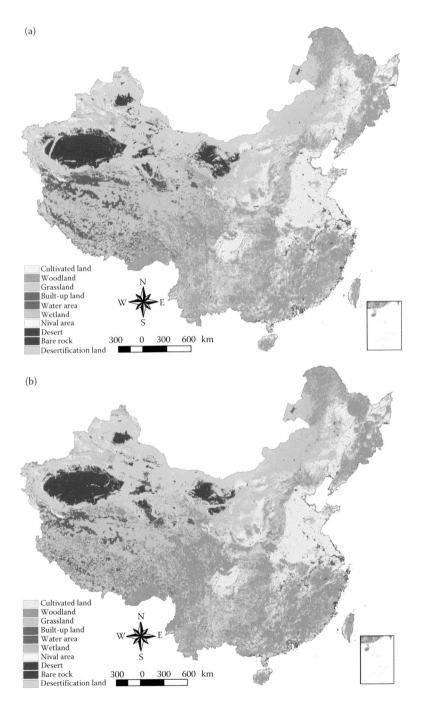

FIGURE 18.2
Land-cover scenarios I: (a) land cover in 2039; (b) land cover in 2069; and (c) land cover in 2099.

(c)

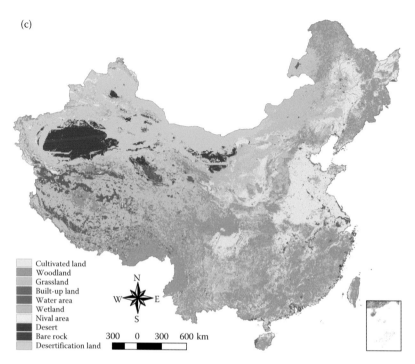

Cultivated land
Woodland
Grassland
Built-up land
Water area
Wetland
Nival area
Desert
Bare rock
Desertification land

N
W   E
S

300   0   300   600 km

FIGURE 18.2
Continued.

   In the next 100 years, cultivated land in China could in principle be divided
into two production areas; agricultural regions and livestock farming regions.
These would take the curve linking; the Da-Xiao Hinggan Mountains, Yulin,
Lanzhou, the east Qinghai-Xizang Plateau, and its southeast edge as their
border (Figure 18.5). Cultivated land would be centrally distributed in the
northeast plain, the north China plain, the middle and lower reaches of the
Yangtze River, the Sichuan Basin, the central Shaanxi plain, the Hexi corridor
in the Gansu province, and alluvial fan areas on the north and south of the
Tianshan Mountains. In addition, a great quantity of cultivated land would
be scattered on hilly areas in the south of China.
   The complicated terrain characteristics and heterogeneous climate of China
lead to great differences of spatiality in terms of woodland distribution. The
woodland in northeast China would mainly be distributed in the Da-Xiao
Hinggan Mountains, the Changbai Mountains, and the East Liaoning Basin;
in southwestern China, the woodland would mainly be distributed in the
area of the Himalaya Mountains and Hengduan Mountains in the east and
south of the Yalu Tsangpo River in Tibet, the mountainous range around the
Sichuan Basin, the Yunnan-Guizhou Plateau, and most hilly areas in Guangxi.
In southeastern China, the woodland would mainly be distributed in low

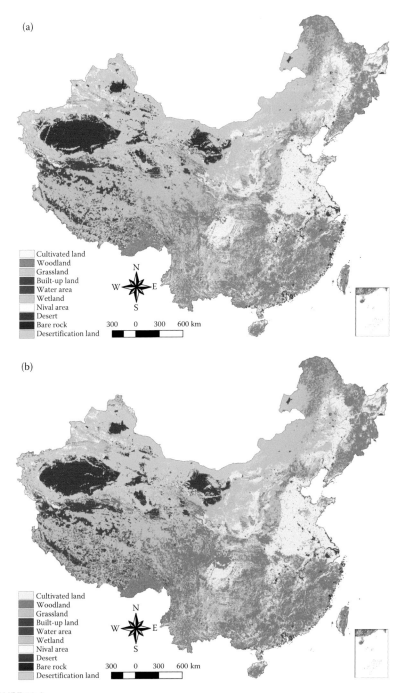

FIGURE 18.3
Land-cover scenarios II: (a) in 2039, (b) in 2069, and (c) in 2099.

(c)

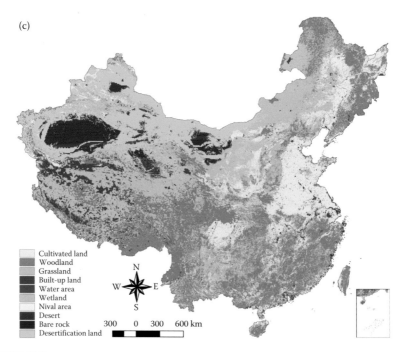

FIGURE 18.3
Continued.

mountainous and hilly areas such as the Wuyi Mountains, the Nanling ridges, and the Taiwan Mountains. In northwestern China, the woodland would mainly be distributed in the mountainous areas of the Tianshan Mountains, the Altai Mountains, the Qilian Mountains, the Ziwu Mountains, the Helan Mountains, the Liupan Mountains, and the Yinshan Mountains. In brief, the woodland in China would mainly be distributed in mountainous and hilly areas.

The grassland would mainly be distributed in western China and comparatively less distributed in eastern China. Geographically, the grassland would mainly be distributed in the Qinghai-Xizang Plateau, the Inner Mongolia Plateau, the Loess Plateau, the Tianshan Mountains, the Altai Mountains, and areas around the Tarim Basin. Some grassland would also be scattered on the hilly areas of Hunan, Hubei, Anhui, Fujian, Yunnan, Guizhou, Sichuan, Guangdong, Guangxi, and Taiwan, which would mix with the woodland.

The water areas include rivers and lakes (Yue et al., 2005). Rivers can be divided into oceanic systems that discharge into oceans and inland ones that start in mountainous areas and disappear in conoplain or flow into inland lakes. The oceanic system can be subdivided into the Pacific, Indian, and Arctic drainage basins, accounting for 64% of the total land area of China.

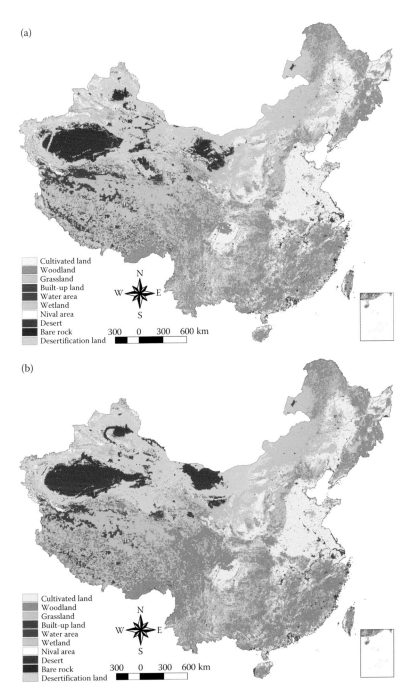

FIGURE 18.4
Land-cover scenarios III: (a) in 2039, (b) in 2069, and (c) in 2099.

(c)

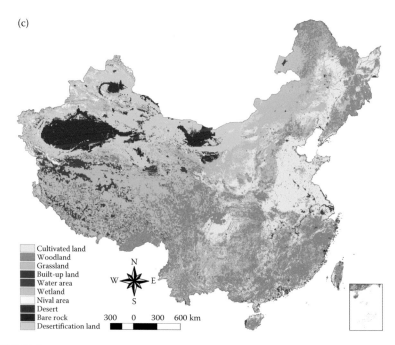

Cultivated land
Woodland
Grassland
Built-up land
Water area
Wetland
Nival area
Desert
Bare rock
Desertification land

N
W ⊕ E
S

300   0   300   600 km

FIGURE 18.4
Continued.

The inland system includes a few perennial rivers and has large tracts with no runoff, accounting for 36% of the total land area of China (Zhao, 1986). Five major lake regions can be identified: the northeast lake region, the northwest lake region, the Qinghai-Xizang lake region, the eastern lake region, and the southwest lake region. They include approximately 2800 natural lakes, each with an area greater than 1 km^2 and many reservoirs that are artificial lakes.

Built-up land would mainly be distributed in the areas near rivers or valleys with abundant water resources, convenient traffic, fertile soil, and plentiful products, which are determined by the characteristics of urban development. Because of the joint affection of various factors such as geographical features and socio-economic conditions in eastern China, the built-up land would mainly be distributed in the Northeast Plateau, the North China Plain, the Plain in the middle and lower reaches of the Yangtze River, the Yangtze River Delta and the Pearl River Delta in the next 100 years. In western China, the built-up land would mainly be distributed in the Sichuan Basin, the central Shaanxi plain, the Hetao Plain in Ningxia, the Hexi Corridor in Gansu and Xinjiang Oasis. In mountainous, hilly and plateau areas, built-up land would mainly be distributed in provincial and prefectural capitals, and other administrative areas.

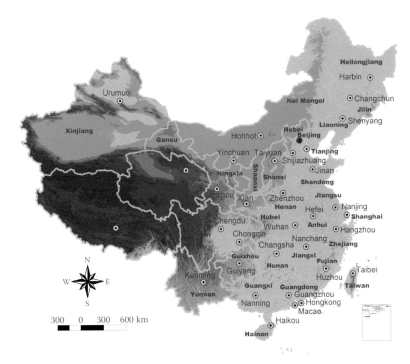

FIGURE 18.5
Provinces and provincial capitals in China.

Nival areas would mainly occur in the mountainous ridges of the Tianshan Mountains, the Qilian Mountains, the Kunlun Mountains, the Himalaya Mountains, and the Hengduan Mountains. Deserts would mainly occur in the Junggar Basin, the Tarim Basin, the Qaidam Basin, and the Inner Mongolia Plateau. Bare rocks would mainly be distributed in the Qinghai-Xizang Plateau, the east of the Turpan Basin, and the Karst area in the southwest China. Desertification land would mainly be distributed in areas outside deserts and areas inside meadows in northwestern China.

18.4 Area Changes of Land-Cover Types

Land-cover types which would increase in size in all the periods 2000 to 2039, 2039 to 2069, and 2069 to 2099 under all three scenarios include built-up land and desertification land. Woodland-type areas would increase in all three periods under all three scenarios, with the exception of a decrease in the period 2069 to 2099 under scenario I. Land-cover types which would decrease in size in all three periods include cultivated land, grassland, and nival areas. The water-type areas would decrease in all three periods under all three scenarios, with the exception of an increase in the period 2069 to 2099 under scenario I. During the period 2000 to 2039, wetland-type areas would decrease under all three scenarios; from 2039 to 2069, wetland areas would increase under scenarios I and II, but decrease under scenario III; from 2069 to 2099, wetland areas would increase under scenarios I and III, but decrease under scenario II. Under scenarios I and II, desert areas would increase during all three periods; under scenario III, desert areas would decrease during the period 2000 to 2039 but increase during the other two periods. Under scenario III, bare rock areas would decrease during all three periods; under scenario II, bare rock areas would decrease during the periods 2000 to 2039 and 2069 to 2099, but increase during the period 2039 to 2069 (Tables 18.2 through 18.4). Woodland would have the greatest increase rate, growing at 2.34% per decade, while bare rock would have the biggest decrease rate, shrinking 2.38% per decade.

In terms of the three scenarios of land-cover, most land-cover types would see similar change trends during the period 2000 to 2099. Woodland areas would increase by 42.4984 million hectares under scenario I, 50.9585 million hectares under scenario II, and 79.219 million hectares under scenario III. The desertification land would increase by 27.1433, 14.9857, and 4.2709 million hectares under scenarios I, II, and III, respectively. The built-up land would increase by 0.2679, 0.3559, and 0.5474 million hectares under scenarios I, II, and III, respectively. Grassland, cultivated land, and nival areas, which would continuously shrink in all three periods 2000 to 2039, 2039 to 2069, and 2069 to 2099, would decrease by 30.1825, 13.3035, and 1.8434 million hectares

TABLE 18.2

Area Change of Land-Cover Types under Scenario I (Units: km²)

Land-Cover Type	2000–2039		2039–2069		2069–2099	
	Area	Decadal Change Rate (%)	Area	Decadal Change Rate (%)	Area	Decadal Change Rate (%)
Cultivated land	−104000	−1.25	−1000	−0.02	−28000	−0.47
Woodland	303300	3.08	143500	1.73	−21700	−0.25
Grassland	−151200	−1.58	−25500	−0.38	−125100	−1.88
Built-up land	600	0.38	900	0.82	1200	1.02
Water area	−13800	−2.76	−11400	−3.44	13100	4.40
Wetland	−7900	−3.50	2400	1.67	3700	2.41
nival area	−4800	−1.51	−12800	−5.71	−900	−0.47
Desert	−15700	−0.72	−20600	−1.29	−100900	−6.57
Bare rock	−46700	−2.40	−117000	−8.88	69000	7.14
Desertification land	40300	0.82	41500	1.09	189700	4.81

TABLE 18.3

Area Change of Land-Cover Types under Scenario II (Units: km²)

	2000–2039		2039–2069		2069–2099	
Land-Cover Type	Area	Decadal Change Rate (%)	Area	Decadal Change Rate (%)	Area	Decadal Change Rate (%)
Cultivated land	−91071	−1.0958	−6482	−0.1088	−9979	−0.1680
Woodland	181233	1.8412	191106	2.4110	137246	1.6147
Grassland	−64003	−0.6678	−189823	−2.7132	−127870	−1.9897
Built-up land	329	0.2218	256	0.2281	2974	2.6314
Water area	−8419	−1.6907	−1885	−0.5413	−816	−0.2382
Wetland	−7863	−3.4703	4126	2.8193	−830	−0.5229
Nival area	−3925	−1.2372	−2819	−1.2465	−6065	−2.7859
Desert	−22136	−1.0094	−14169	−0.8977	−47304	−3.0800
Bare rock	−40956	−2.1079	4787	0.3587	−25499	−1.8906
Desertification land	56811	1.1536	14903	0.3857	78143	1.9991

under scenario I, 38.1696, 10.7532, and 1.2809 million hectares under scenario II, and 42.3194, 20.0045, and 1.8873 million hectares under scenario III, respectively (Tables 18.5 through 18.7).

The land-cover types that would see the greatest transformation include cultivated land, woodland, grassland, and desertification land. Cultivated land would mainly be converted into woodland, and the conversion rate would gradually decrease during the period 2000 through 2099. Woodland would mainly be converted into grassland. Grassland would mainly be

TABLE 18.4

Area Change of Land-Cover Types under Scenario III (Units: km²)

	2000–2039		2039–2069		2069–2099	
Land-Cover Type	Area	Decadal Change Rate (%)	Area	Decadal Change Rate (%)	Area	Decadal Change Rate (%)
Cultivated land	−125179	−1.5062	−44783	−0.7645	−30083	−0.5256
Woodland	406009	4.1246	180856	2.1028	205325	2.2456
Grassland	−173510	−1.8104	−112941	−1.6939	−136743	−2.1606
Built-up land	1708	1.1513	901	0.7741	2865	2.4057
Water area	−18839	−3.7833	−8936	−2.8194	−6227	−2.1462
Wetland	−12442	−5.4912	−4840	−3.6498	94	0.0796
Nival area	−5556	−1.7513	−6185	−2.7953	−7132	−3.5183
Desert	−4983	−0.2272	17800	1.0921	29914	1.7772
Bare rock	−79193	−4.0759	−50308	−4.1248	−60301	−5.6423
Desertification land	11985	0.2434	28436	0.7624	2288	0.0600

TABLE 18.5

Land-Cover Scenario I (Units: km²)

Land-Cover Type	2000	2039	2069	2099	Area Change	Change Rate per Decade (%)
Cultivated land	2077798	1973766	1972751	1944763	−133035	−0.64
Woodland	246088	276415	29076	2885864	424984	1.73
Grassland	2396064	2244831	2219328	2094239	−301825	−1.26
Built-up land	37088	37658	38587	39767	2679	0.72
Water area	124488	110736	99293	112399	−12089	−0.97
Wetland	56645	48706	51148	54843	−1802	−0.32
Nival area	79311	74511	61755	60877	−18434	−2.32
Desert	548258	532572	511949	411048	−13721	−2.50
Bare rock	485742	439075	322088	391041	−94701	−1.95
Desertification land	123123	1271499	1313005	1502663	271433	2.20

converted into woodland, and a considerable part of the transformed grass-land would be converted into desertification land. Desertification land would mainly be converted into woodland and grassland (Tables 18.8 through 18.16).

18.5 Ecotope Diversity and Patch Connectivity of Land Cover

The ecotope diversity of land-cover would have a decreasing trend, while patch connectivity would generally increase. Scenario I shows that on an average, the decease rate of ecotope diversity would be 0.1604% per decade

TABLE 18.6

Land-Cover Scenario II (Units: km²)

Land-Cover Type	2000	2039	2069	2099	Area Change	Change Rate (%)
Cultivated land	2077798	1986727	1980245	1970266	−107532	−0.52
Woodland	246088	2642113	2833219	2970465	509585	2.07
Grassland	2396064	2332061	2142238	2014368	−381696	−1.59
Built-up land	37088	37417	37673	40647	3559	0.96
Water area	124488	116069	114184	113368	−1112	−0.89
Wetland	56645	48782	52908	52078	−4567	−0.81
Nival area	79311	75386	72567	66502	−12809	−1.62
Desert	548258	526122	511953	464649	−83609	−1.52
Bare rock	485742	444786	449573	424074	−61668	−1.27
Desertification land	123123	1288041	1302944	1381087	149857	1.22

TABLE 18.7

Land-Cover Scenario III (Units: km²)

Land-Cover Type	2000	2039	2069	2099	Area Change	Change Rate (%)
Cultivated land	2077798	1952619	1907836	1877753	−200045	−0.96
Woodland	246088	2866889	3047745	325307	79219	3.22
Grassland	2396064	2222554	2109613	197287	−423194	−1.77
Built-up land	37088	38796	39697	42562	5474	1.48
Water area	124488	105649	96713	90486	−34002	−2.73
Wetland	56645	44203	39363	39457	−17188	−3.03
Nival area	79311	73755	6757	60438	−18873	−2.38
Desert	548258	543275	561075	590989	42731	0.78
Bare rock	485742	406549	356241	29594	−189802	−3.91
Desertification land	123123	1243215	1271651	1273939	42709	0.35

and the increase rate of patch connectivity would be 2.1591% per decade. Under scenario II, ecotope diversity would monotonically decrease and patch connectivity would monotonically increase; the decrease rate of ecotope diversity would be 0.1176% per decade and the increase rate of patch connectivity would be 2.1402% per decade. Under scenario III, the monotonically decreasing rate of ecotope diversity would be 0.246% per decade and the monotonically increasing rate of patch connectivity would be 3.8258% per decade (Table 18.17). In short, climate change would generally cause less ecotope diversity and more patch connectivity.

18.6 Mean Center Shifts of Land-Cover Types

The mean center of cultivated land would pace up and down in Nanyang of Henan province in the southwest of the North China Plain (Figure 18.6). Under scenario I (Table 18.18), during the period 2000 to 2039, the mean center of cultivated land would shift about 38 km to the southeast; during the periods 2039 to 2069 and 2069 to 2099, the mean center would shift to the northeast distances of about 44 and 36 km, respectively. Under scenario II (Table 18.19), during the periods 2000 to 2039 and 2069 to 2099, the mean center would shift to the southeast distances of about 47 and 17 km, respectively; from 2039 to 2069 the mean center would shift to the northeast, a distance of about 59 km. Under scenario III (Table 18.20), the mean center would shift to the southeast in all three periods, distances of about 40, 26, and 18 km during the periods 2000 to 2039, 2039 to 2069, and 2069 to 2099, respectively.

The mean center of woodland would pace up and down between Yichang of the Hubei province and Wanxian County in Chongqing. Under scenario I,

TABLE 18.8

Transformation of Land-Cover Types under Scenario I during the Period 2000 to 2039 (Units: km^2)

2000 \ 2039	Cultivated Land	Woodland	Grassland	Built-up Land	Water Area	Wetland	Nival Area	Desert	Bare Rock	Desertification Land	Total in 2000
Cultivated land	1973766	77686	11778	1098				943	10	1.517	2077798
Woodland		2412882	41087	18					809	6084	2460880
Grassland		228763	2120237					1731	1877	43456	2396064
Built-up land		537	18	36486				6		41	37088
Water area		8645	3965	56	110736			181		905	124488
Wetland		7470	466			48706				3	56645
Nival area		928	1883				74511		5	1984	7.311
Desert		435	1014					522364		24445	548258
Bare rock		9001	33259					206	436081	7195	485742
Desertification land		17803	31124					7141	293	1174869	1231230
Total in 2039	1973766	2764150	2244831	37658	110736	48706	74511	532572	439075	1271499	9497504

TABLE 18.9

Transformation of Land-Cover Types under Scenario I during the Period 2039 to 2069 (Units: km^2)

2069 / 2039	Cultivated Land	Woodland	Grassland	Built-up Land	Water Area	Wetland	Nival Area	Desert	Bare Rock	Desertification Land	Total in 2039
Cultivated land	1912007	43679	3580	212						14288	1973766
Woodland	49042	2512823	172971	1313	4258	5383	2	185	923	17250	2764150
Grassland	3256	285985	1881030	247	10	9	1		1602	72691	2244831
Built-up land	1098	246	9	35987	56					262	37658
Water area		7424	7843	4	94151					1314	110736
Wetland		2251	692			45756				7	48706
Nival area		1728	7308		181		60870		242	4363	74511
Desert	943	113	3034	806				490273	206	37016	532572
Bare rock		24597	92782						313995	7701	439075
Desertification land	6405	28754	50079	18	637		882	21491	5120	1158113	1271499
Total in 2069 period	1972751	2907600	2219328	38587	99293	51148	61755	511949	322088	1311005	9497504

TABLE 18.10

Transformation of Land-Cover Types under Scenario I during the Period 2069 to 2099 (Units: km²)

2069＼2099	Cultivated Land	Woodland	Grassland	Built-up Land	Water Area	Wetland	Nival Area	Desert	Bare Rock	Desertification Land	Total in 2069
Cultivated land	1900675	43396	22	1240				696		26722	1972751
Woodland	30463	2513929	275030	1374	10030	3888	2364	185	30283	40054	2907600
Grassland	4337	314917	1737341	727	6975	7	3178	666	57598	93582	2219328
Built-up land	33	1110	204	35932				500		808	38587
Water area		797	420	2	94694			25		3355	99293
Wetland		0181				50946				21	51148
Nival area		231	2369				53771		103	5281	61755
Desert		7	0982	99				393522		117339	511949
Bare rock		2610	11960	6					295792	11720	322088
Desertification land	9255	8686	65911	387	700	2	1564	15454	7265	1203781	1313005
Total in 2099 period	1944763	2885864	2094239	39767	112399	54843	60877	411048	391041	1502663	9497504

TABLE 18.11

Transformation of Land-Cover Types under Scenario II during the Period 2000 to 2039 (Units: km²)

2000 \ 2039	Cultivated Land	Woodland	Grassland	Built-up Land	Water Area	Wetland	Nival Area	Desert	Bare Rock	Desertification Land	Total in 2000
Cultivated land	1986727	66827	12492	311				364		11077	2077798
Woodland		2417957	34795	584					752	6792	2460880
Grassland		140931	2208783					728	1515	44107	2396064
Built-up land		494	12	36422						160	37088
Water area		4550	3314		116069			8		547	124488
Wetland		7407	451			48782				5	56645
Nival area		6	1871				75386		47	2001	79311
Desert		158	1457	100				516670		29873	548258
Bare rock		403	32140					559	442100	10540	485742
Desertification land		3380	36746					7793	372	1182939	1231230
Total in 2039	1986727	2642113	2332061	37417	116069	48782	75386	526122	444786	1288041	9497504

TABLE 18.12

Transformation of Land-Cover Types under Scenario II during the Period 2039 to 2069 (Units: km²)

2039 \ 2069	Cultivated Land	Woodland	Grassland	Built-up Land	Water Area	Wetland	Nival Area	Desert	Bare Rock	Desertification Land	Total in 2039
Cultivated land	1919417	55480	4562	357		6547	5	158	504	6911	1986727
Woodland	52461	2431319	139488	563	4364	386	1375	418	28611	6704	2642113
Grassland	3336	276005	1961269	183	2835					57643	2332061
Built-up land	8	838	18	36475						78	37417
Water area		8223	230	24	106714					878	116069
Wetland		2801	8			45973					48782
Nival area		1734	774				70360		55	2463	75386
Desert	364	361	1705		8			492876	559	30249	526122
Bare rock		20678	1585						418237	4286	444786
Desertification land	4659	35780	32599	71	263	2	827	18501	1607	1193732	1288041
Total in 2069 period	1980245	2833219	2142238	37673	114184	52908	72567	511953	449573	1302944	9497504

TABLE 18.13

Transformation of Land-Cover Types under Scenario II during the Period 2069 to 2099 (Units: km^2)

2069 \ 2099	Cultivated Land	Woodland	Grassland	Built-up Land	Water Area	Wetland	Nival Area	Desert	Bare Rock	Desertification Land	Total in 2069
Cultivated land	1932173	29073	60	315				2		18622	1980245
Woodland	24257	2628130	129198	2207	6085	525	999	315	12613	28890	2833219
Grassland	5813	234715	1825477	135	63	8	60	514	1588	73865	2142238
Built-up land	239	489	178	36358	20					389	37673
Water area		5058	211	7	106440					2468	114184
Wetland		1338				51545				25	52908
Nival area		2225	2993				63818		200	3331	7.2567
Desert		466	831	1099				444663		64894	511953
Bare rock		39175	7169						397368	5861	449573
Desertification land	7784	29796	48251	526	760		1625	19155	12305	1182742	1302944
Total in 2099 period	1970266	2970465	2014368	40647	113368	52078	66502	464649	424074	1381087	9497504

TABLE 18.14

Transformation of Land-Cover Types under Scenario III during the Period 2000 to 2039 (Units: km²)

2000 \ 2039	Cultivated Land	Woodland	Grassland	Built-up Land	Water Area	Wetland	Nival Area	Desert	Bare Rock	Desertification Land	Total in 2000
Cultivated land	1952619	100545	12927	581				2145		8981	2077798
Woodland		2393115	60446	1756					357	5206	2460880
Grassland		308602	2050071	5				3855	805	32726	2396064
Built-up land		611	26	36373				9		69	37088
Water area		11600	6478	2	105649			88		671	124488
Wetland		11773	667			44203				2	56645
Nival area		1260	3317				73755			974	79311
Desert		440	1105					522936		23777	548258
Bare rock		13876	54665					634	405254	11313	485742
Desertification land		25067	32852	79				13608	128	1159496	1231230
Total in 2039	1952619	2866889	2222554	38796	105649	44203	73755	543275	406549	1243215	9497504

TABLE 18.15

Transformation of Land-Cover Types under Scenario III during the Period 2039 to 2069 (Units: Million Hectares)

2069 / 2039	Cultivated Land	Woodland	Grassland	Built-up Land	Water Area	Wetland	Nival Area	Desert	Bare Rock	Desertification Land	Total in 2039
Cultivated land	1890787	24989	9295	2010	80			3506		21952	1952619
Woodland	6178	2827075	25833	861	201			46	394	6301	2866889
Grassland	4660	158541	2002371	506	17		1	2999	618	52841	2222554
Built-up land	521	1630	19	36305	2			49		270	38796
Water area	373	3725	3527	5	96336			360		1323	105649
Wetland	972	3453	404			39361				13	44203
Nival area		743	2738				67553	3		2718	73755
Desert	2224	431	4009					474018	338	62255	543275
Bare rock		9769	37287					743	354464	4286	406549
Desertification land	2121	17389	24130	10	77	2	16	79351	427	1119692	1243215
Total in 2069	1907836	3047745	2109613	39697	96713	39363	67570	561075	356241	1271651	9497504

TABLE 18.16

Transformation of Land-Cover Types under Scenario III during the Period 2069 to 2099 (Units: km^2)

2069 \ 2099	Cultivated Land	Woodland	Grassland	Built-up Land	Water Area	Wetland	Nival Area	Desert	Bare Rock	Desertification Land	Total in 2069
Cultivated land	1871116	19350	2992	1040	373	972		1737		10256	1907836
Woodland	517	3015362	21513	2113	13			15	287	7925	3047745
Grassland	5419	187455	1876167	14	11		1	1090	1138	38818	2109613
Built-up land		92	6	39387				16		196	39697
Water area		3099	2672		90089			66		787	96713
Wetland		677	195			38485				6	39363
Nival area		815	3887				60437			2310	67570
Desert	338	207	1690					533213	121	25483	561075
Bare rock		13192	44218					1349	293962	3520	356241
Desertification land	363	12821	19530	8				53503	288	1185138	1271651
Total in 2099 period	1877753	3253070	1972870	42562	90486	39457	60438	590989	295940	1273939	9497504

TABLE 18.17

Ecotope Diversity and Patch Connectivity

Scenarios	Period	Diversity	Connectivity
HadCM3A1FI	2000	0.0935	0.0528
	2039	0.0928	0.0687
	2069	0.0917	0.0701
	2099	0.092	0.0642
	Decadal increased ratio	−0.1604	2.1591
HadCM3A2	2000	0.0935	0.0528
	2039	0.0929	0.0624
	2069	0.0928	0.0661
	2099	0.0924	0.0641
	Decadal increased ratio	−0.1176	2.1402
HadCM3B2	2000	0.0935	0.0528
	2039	0.0924	0.0697
	2069	0.0918	0.071
	2099	0.0912	0.073
	Decadal increased ratio	−0.2460	3.8258

the mean center would shift about 20 km to the northwest during the period 2000 to 2039; about 256 km to the southwest from 2039 to 2069, and about 60 km to the northeast from 2069 to 2099. Under scenario II, the mean center would shift about 78 km to the northeast during the period 2000 to 2039, about 306 km to the southwest from 2039 to 2069, and about 69 km to the

TABLE 18.18

Shift Trend of the Mean Center under Scenario I

Land-Cover Type	2000–2039 Distance (km)	Direction	2039–2069 Distance (km)	Direction	2069–2099 Distance (km)	Direction
Cultivated land	38.3827	287.81°	43.7309	44.40°	36.3724	5.47°
Woodland	20.4859	123.48°	256.2602	227.36°	60.2040	86.84°
Grassland	54.7277	225.72°	124.7824	61.32°	80.0148	343.80°
Built-up land	25.6433	212.17°	58.9676	175.43°	177.9427	266.92°
Water area	128.7035	357.19°	195.5907	6.72°	197.0785	194.49°
Wetland	22.8663	289.37°	111.5089	58.65°	37.9343	229.43°
Nival area	11.3650	234.50°	25.7471	222.66°	116.5167	270.66°
Desert	20.3248	42.61°	37.9435	156.50°	145.7545	247.51°
Bare rock	15.0415	149.93°	28.1945	184.93°	53.9005	273.03°
Desertification land	110.9973	287.53°	93.2249	281.28°	157.8782	92.71°

TABLE 18.19

Shift Trend of the Mean Center under Scenario II

Land-Cover Type	2000–2039		2039–2069		2069–2099	
	Distance (km)	Direction	Distance (km)	Direction	Distance (km)	Direction
Cultivated land	46.5404	274.34°	59.3802	54.54°	16.5264	340.79°
Woodland	78.1206	25.12°	305.8616	228.24°	68.7200	106.78°
Grassland	112.1583	212.25°	219.4164	41.51°	23.2976	336.89°
Built-up land	60.3967	274.76°	7.1556	155.81°	225.3487	169.04°
Water area	65.7434	351.57°	56.0458	18.99°	18.7554	265.56°
Wetland	40.9152	263.08°	77.7505	66.25°	32.1454	48.15°
Nival area	18.4164	263.97°	20.0544	261.20°	56.9311	263.35°
Desert	32.7392	107.86°	3.1286	323.86°	110.0835	269.38°
Bare rock	12.5083	183.61°	12.2299	127.59°	9.1083	39.87°
Desertification land	125.2159	285.99°	29.8726	228.32°	141.0112	85.25°

northwest from 2069 to 2099. Under scenario III, the mean center would shift 45, 60, and 73 km to the southwest during the periods 2000 through 2039, 2039 through 2069, and 2069 through 2099, respectively.

The mean center of grassland would pace up and down in the area around Qinghai Lake. In terms of all three scenarios, the mean center would almost always shift to the southwest during the period 2000 through 2039, to the northeast, with the exception of scenario III, where it would shift to the

TABLE 18.20

Shift Trend of the Mean Center under Scenario III

Land-Cover Type	2000–2039		2039–2069		2069–2099	
	Distance (km)	Direction	Distance (km)	Direction	Distance (km)	Direction
Cultivated land	40.4861	288.14°	24.5506	326.00°	17.6988	342.30°
Woodland	45.1047	202.81°	59.8017	206.83°	73.0859	194.89°
Grassland	68.7917	199.08°	42.5383	152.29°	9.8802	330.50°
Built-up land	138.7902	294.62°	122.9710	223.38°	53.6349	202.16°
Water area	169.9664	354.13°	105.9959	356.35°	111.4750	358.56°
Wetland	43.8508	293.82°	40.2609	36.60°	10.4989	352.90°
Nival area	25.1487	133.53°	17.8007	241.86°	23.9020	229.60°
Desert	24.4768	35.55°	75.9650	321.27°	21.3177	306.87°
Bare rock	31.4136	147.31°	23.3175	141.45°	48.4996	216.60°
Desertification land	140.8968	280.97°	14.4511	125.04°	93.5899	79.66°

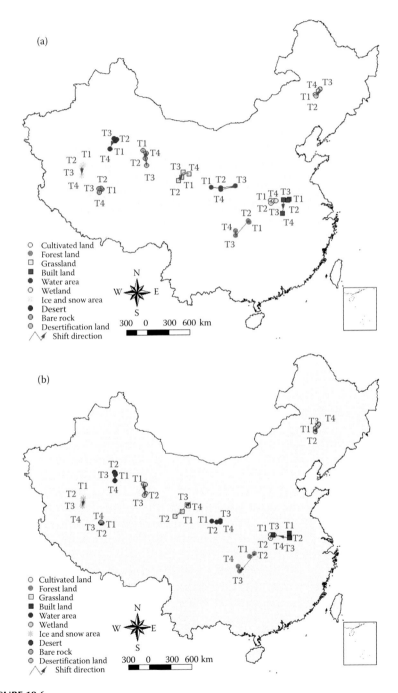

FIGURE 18.6
Shift trend of mean centers of land-cover types in China: (a) scenario I, (b) scenario II, and (c) scenario III.

(c)

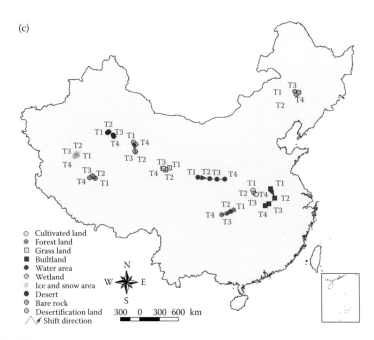

FIGURE 18.6
Continued.

northwest, during the period 2039 through 2069, and to the southeast from 2069 to 2099.

The mean center of built-up land would oscillate between the juncture of Anhui, Henan, and Hubei. In terms of scenario I, the mean center would shift 26 and 178 km, respectively to the southwest during the periods 2000 to 2039 and from 2069 to 2099, and 59 km to the northwest from 2039 to 2069. In terms of scenario II, the mean center would shift 60 km to the southeast during the period 2000 through 2039, 7 and 225 km to the northwest from 2039 to 2069, and from 2069 to 2099. In terms of scenario III, the mean center would shift 139 km to the southeast, 123 and 54 km to the southwest during the periods 2000 through 2039, 2039 through 2069, and 2069 through 2099, respectively.

The mean center of water area would oscillate between the upper reaches of the Weihe River in the south of Gansu. In terms of scenarios I and II, the mean center would shift to the southeast, northeast and southwest respectively during the periods 2000 to 2039, 2039 to 2069, and 2069 to 2099. In terms of scenario III, the mean center would shift 170, 106 and 111 km respectively to the southeast during the periods 2000 to 2039, 2039 to 2069, and 2069 to 2099.

The mean center of wetland would pace up and down in Tongliao in Northeast Plain. Under scenario I, the mean center would shift 23 km to the southeast from 2000 through 2039, 112 km to the northeast from 2039 through

2069, and 38 km to the southwest from 2069 through 2099. Under scenario II, the mean center would shift 41 km to the southwest from 2000 to 2039, 78 km and 32 km to the northeast from 2039 to 2069, and from 2069 to 2099, respectively. Under scenario III, the mean center would shift 44 km to the southeast from 2000 to 2039, 40 km to the northeast from 2039 to 2069, and 10 km to the southeast from 2069 to 2099.

The mean center of nival areas would oscillate between the juncture of the Kunlun Mountains and the Arjin Mountains. Under all three scenarios during all three periods, the mean center would almost always shift toward the southwest, with the exception of the period 2000 to 2039 under scenario III, where it would move toward the northwest.

The mean center of desert would pace up and down in the eastern area of Tarim Basin. Under scenario I, the mean center would shift 20 km to the northeast from 2000 to 2039, 38 km to the northeast from 2039 to 2069, and 146 km to the southwest from 2069 to 2099. Under scenario II, the mean center would shift 34 km to the northwest from 2000 to 2039, 3 km to the southwest from 2039 to 2069, and 110 km to the southwest from 2069 to 2099. Under scenario III, the mean center would shift 25 km to the northeast from 2000 to 2039, and 76 km and 21 km to the southeast from 2039 to 2069, and from 2069 to 2099, respectively.

The mean center of bare rock would pace up and down in the area around Kekexili Mountain in Qinghai-Xizang Plateau. Under scenario I, the mean center would shift 15 km to the northwest from 2000 to 2039, 28 km to the southwest from 2039 to 2069, and 54 km to the southeast from 2069 to 2099. Under scenario II, the mean center would shift 12 km to the southwest from 2000 to 2039, 12 km to the northwest from 2039 to 2069, and 9 km to the northeast from 2069 to 2099. Under scenario III, the mean center would shift 31 and 23 km respectively to the northwest from 2000 to 2039, and 2039 to 2069, and 48 km to the southwest from 2069 to 2099.

The mean center of desertification land would pace up and down in the northwest of Qaidam Basin. During the period 2000 to 2039, the mean center would shift 111, 125, and 141 km, respectively to the southeast under scenarios I, II, and III. From 2039 to 2069, the mean center would shift 93 km to the southeast under scenario I, 30 km to the southwest under scenario II, and 14 km to the northwest under scenario III. From 2069 to 2099, the mean center would shift 158 km to the northwest under scenario I, and 141 and 94 km, respectively to the northeast under scenarios II and III.

18.7 Discussion and Conclusions

Over the next 100 years, cultivated land would gradually decrease, especially, the land distributed in the north and south of the Tianshan Mountains, the Hexi Corridor of Gansu, the Loess Plateau, and south of the Inner Mongolia Plateau. The mean center of cultivated land would shift to the east in general. Woodland areas would greatly increase as a result of temperature and precipitation in China. Woodland and grassland in hilly areas would increase with the decrease of cultivated land. These results simulated in terms of climate scenarios coincide with the expectations of China's Grain-for-Green policy (Feng et al., 2005). In other words, climate change would favor the implementation of the Grain-for-Green policy in China.

With temperatures, precipitation, and evapotranspiration increasing in considerable areas of China, the nival areas would shrink, ecotope diversity would decrease, and desertification areas would expand at a comparatively slow rate. Desertification land in peripheral areas of the Tarim Basin and Junggar Basin would spread out in an irregular circular shape. Desertification in the Inner Mongolia Plateau would extend to the east and southeast. Desertification in the Loess Plateau would become much more serious.

In short, ecosystems in China would significantly benefit if efficient ecological conservation and restoration measures were implemented. However, if human activities exceed the regulation capacity of the ecosystems themselves, the ecosystems in China may deteriorate more seriously. Ecological conservation and restoration is a long-term and complex project. It involves undeveloped local governments and millions of farmers living in ecologically vulnerable areas. Government investment and financial subsidy are not enough. It is necessary to establish a set of policies to ensure ecological conservation, and restoration is carried out. Various economic activities must be subject to strict judicial controls, particularly in desertification areas.

The method for surface modeling of land-cover (SMLC) developed in this chapter is a stochastic model of land-cover change, dealing with both the location and quantity issues in an integrated way. It has no need to integrate with other simulation modules. It is comparatively simple to construct transition probability matrixes between HLZ types and land-cover types, which avoids the discontinuity problem between the past and future that appeared in the SRES land-cover scenarios. On the basis of the combination of the current spatial pattern of land-cover types with climatic change trends and scenarios, SMLC is able to generate reliable projections of the future and, in backward mode, the past, which is the ultimate objective of the LUCC project. SMLC could be conveniently used on any spatial scale because of the availability of rich LUCC data and IPCC data (http://ipcc-ddc.cru.uea.ac.uk).

The global-scale ecological unit of the biome has been extended to include human influence on ecosystems, which incorporates human population density, land use and land cover to describe anthropogenic effects on earth (Alessa and Chapin III, 2008). Climate and geology have shaped ecosystems and evolution in the past; however, human forces may now outweigh these, across most of the earth's land surface, as most of "nature" is now embedded within anthropogenic mosaics of land use and land cover (Ellis and Ramankutty, 2008). However, SMLC severely simplifies the land-cover system and focuses primarily on climatic attributes. Consequently, further SMLC research work will focus on a wide range of social, economic, and political factors.

References

Alessa, L. and Chapin III, F.S. 2008. Anthropogenic biomes: A key contribution to earth-system science. *Trends in Ecology and Evolution* 23(10): 529–531.

Arnell, N.W., Livermore, M.J.L., Kovats, S., Levy, P.E., Nicholls, R., Parry, M.L., and Gaffin, S.R. 2004. Climate and socio-economic scenarios for global-scale climate change impacts assessments: Characterizing the SRES storylines. *Global Environmental Change* 14: 3–20.

Ellis, E.C. and Ramankutty, N. 2008. Putting people on the map: Anthropogenic biomes of the world. *Frontiers in Ecology and the Environment* 6(8): 439–447.

FAO and IIASA. 1993. *Agro-Ecological Assessment for National Planning: The Example of Kenya*. Rome, Italy: Food and Agriculture Organization of the United Nations.

Feng, Z.M., Yang, Y.Z., Zhang, Y.Q., Zhang, P.T., and Li, Y.Q. 2005. Grain-for-green policy and its impacts on grain supply in West China. *Land Use Policy* 22(4): 301–312.

Hubacek, K. and Sun, L.X. 2001. A scenario analysis of China's land-use and land-cover change: Incorporating biophysical information into input–output modeling. *Structural Change and Economic Dynamics* 12: 367–397.

Liu, J.Y., Zhuang, D.F., Luo, D., and Xiao, X.M. 2003. Land-cover classification of China: Analysis of integrated AVHRR imagery and geophysical data. *International Journal of Remote Sensing* 24(12): 2485–2500.

Luijten, J.C. 2003. A systematic method for generating land-use patterns using stochastic rules and basic landscape characteristics: Results for a Colombian hillside watershed. *Agriculture, Ecosystems and Environment* 95: 427–441.

Niehoff, D., Fritsch, U., and Bronstert, A. 2002. Land-use impacts on storm-runoff generation: Scenarios of land-use change and simulation of hydrological response in a meso-scale catchment in SW-Germany. *Journal of Hydrology* 267: 80–93.

Solecki, W.D. and Oliveri, C. 2004. Downscaling climate change scenarios in an urban land-use change model. *Journal of Environmental Management* 72: 105–115.

Veldkamp, A. and Fresco, L.O. 1996. CLUE-CR: An integrated multi-scale model to simulate land-use change scenarios in Costa Rica. *Ecological Modelling* 91: 231–248.

Veldkamp, A. and Lambin, E.F. 2001. Predicting land-use change. *Agriculture, Ecosystems and Environment* 85: 1–6.

Verburg, P.H. and Veldkamp, A. 2001. The role of spatially explicit models in land-use change research: A case study for cropping patterns in China. *Agriculture, Ecosystems and Environment* 85: 177–190.

Yue, T.X., Fan, Z.M., and Liu, J.Y. 2007. Scenarios of land-cover in China. *Global and Planetary Change* 55(4): 317–342.

Yue, T.X., Wang, Y.A., Liu, J.Y., Chen, S.P., Qiu, D.S., Deng, X.Z., Liu, M.L., Tian,Y.Z., and Su, B.P. 2005. Surface modelling of human population distribution in China. *Ecological Modelling* 181(4): 461–478.

Zhao, S.Q. 1986. *Physical Geography of China*. New York, NY: John Wiley & Sons.

19

Surface Modeling of Soil Properties*

19.1 Introduction

Surface modeling of soil properties has become an important topic in soil science research. Soil surface modeling has been enhanced by the advancement of technology that enables the collection of remotely sensed imagery and DEMs as ancillary data for use in digital soil mapping. Surface modeling with ancillary data has attracted research in pedometrics, because soil surveying and the analysis of soil properties is still expensive and time-consuming (Minasny and McBratney, 2007).

Over the last few decades, geostatistics has received considerable attention from the soil industry as it provides a set of tools and methods for analyzing data distributed in space. Geostatistical methods, such as universal kriging (UK) (Burgess and Webster, 1980), kriging with an external drift (KED) (Goovaerts, 1997) and regression kriging (RK) (Odeh et al., 1994) as well as the analysis of covariance (Turner and Thayer, 2001), have become familiar to pedomatricians (Lesch and Corwin, 2008). Surface modeling of soil property by kriging is equivalent to the best linear unbiased estimation (Cressie, 1993). UK is the best linear unbiased estimation with a fixed-effect model that has some linear function of spatial coordinates. KED is the best linear unbiased estimation with a linear function of other secondary predictor variables.

The geostatistical methods for the surface modeling of soil properties were criticized by many scientists (Isaaks and Srivastava, 1989; Zhao et al., 2005; Magnussen et al., 2007). Goovaerts (1999) believed that the estimation by kriging was the best in the least-squares sense, because the local error variance is minimum, but the drawback of the least-squares criterion was that the local variation was smoothed. Lark et al. (2006) pointed out that a problem in UK and KED was finding a spatial variance model for the random variation, since empirical variograms estimated from the data by method-of-moments would be affected by both the random variation and the variation represented by the fixed effects. Emery and Ortiz (2007) concluded that most of the geostatistic models relied on a variogram model that quantified the spatial variability of the soil property under study, but the variogram model is generally based on the assumption that the data have a multivariate Gaussian distribution and may yield inaccurate results if this assumption is violated.

---

* Dr. Wen-Jiao Shi is a major collaborator of this chapter.

Robinson and Metternicht (2006) compared the accuracy of OK, lognormal ordinary kriging (LOK), IDW, and spline for interpolating seasonally stable soil properties such as pH, electric conductivity, and organic matter. They found that there was no single interpolator that could produce chief results for the generation of continuous soil property maps at all times. OK performed best for pH in topsoil and LOK outperformed both IDW and spline for interpolating electrical conductivity in topsoil. Spline surpassed kriging and IDW for interpolating organic matter.

When measurements are sparse or poorly correlated in space, the estimation of the primary attribute of interest is generally improved by accounting for secondary information originating from other related continuous or categorical attributes (Goovaerts, 1999). Some studies have indicated that important relationships exist between soil properties with several continuous variables, including DEM (Odeh et al., 1994, 1995; Bishop and McBratney, 2001; Florinsky et al., 2002; Mueller and Pierce, 2003), remote sensing data (Chen et al., 2000; Bishop and McBratney, 2001; Takata et al., 2007), and other soil attributes related to target variables (Knotters et al., 1995; Triantafilis et al., 2001). More specifically, significant improvements in soil map accuracy can be achieved using categorical information such as ecosystem information (D'Acqui et al., 2007), soil map-delineation (Liu et al., 2006a), soil map units (Wu et al., 2008), and rock types (Goovaerts, 1999). There are also some hybrid techniques of kriging which include categorical information for prediction, such as stratified kriging (Stein et al., 1988) and mosaic kriging (Castoldi et al., 2009).

HASM, which has been developed in recent years (Yue et al., 2007, 2008; Yue and Song, 2008), performs much better than the classical methods according to numerical tests, real-world studies of DEMs (Yue et al., 2010a, 2010c; Yue and Wang, 2010), and climate simulation (Yue et al., 2010b, 2010d). Moreover, HASM also performs better than kriging, IDW, and spline for interpolating soil pH (Shi et al., 2009). However, soil property data may vary severely within a short horizontal distance because of different soil environments, and thus it is difficult to interpolate accurately using HASM when such obvious environmental features are ignored. Therefore, HASM may attempt to incorporate secondary variables to improve interpolation efficiency.

Land-use types can be considered as the combination of soil types, vegetation cover, and anthropogenic effects on land surface. Land-use information, as a major factor of the spatial patterns of soil properties, is related to the spatial pattern of soil properties (Hu et al., 2007; Basaran et al., 2008). The spatial distribution pattern of soil organic carbon is approximately consistent with land-use types (Liu et al., 2006b). Hu et al. (2007) reported that land use was one of the main factors affecting soil organic matter levels. Basaran et al. (2008) studied the dynamic relationships between soil properties and land-use types. In this chapter, a method for high-accuracy surface modeling of soil properties (HASM-SP) is developed on the basis of HASM using land-use information as auxiliary data.

19.2 Materials and Methods

19.2.1 Investigation Region and Data Acquisition

The investigation region ($26°26'55'' \sim 27°37'35''$N, $114°17'49'' \sim 115°30'13''$E) is located in the middle part of the Jiangxi Province, China. It includes the Ji'an municipal district, Ji'an county, and Taihe county, covering 6156.92 km². Their annual mean precipitations are 1458, 1438, and 1381 mm and their annual mean temperatures are 18.1°C, 18.4°C, and 19.0°C, respectively. They have a typical subtropical monsoon climate. The elevation decreases from the periphery toward the center with altitudes ranging from 1204.5 to 42.0 m. The investigation region is a typical red soil hilly region of South China. According to the 1/1,000,000 scale soil maps reported by the National Soil Census Office, the soils are classified into seven groups: red soil, paddy soil, purple soil, fluvo-aquic soil, yellow soil, alluvial soil, and limestone soil (Figure 19.1). Red soils are the main type in the investigation region, predominantly derived from Quaternary red earth, Tertiary red sandstone, and granite, whose areas are 3945.22 km². Paddy soil, purple soil, fluvo-aquic soil, yellow soil, alluvial soil, and limestone soil are respectively 1645.49, 123.32, 71.49, 15.11, 7.17, and 0.01 km².

Samples of 150 topsoil (0–20 cm) (Figure 19.1) were collected in October 2007. Each one was air-dried and passed through a 2 mm sieve. The investigated soil properties include soil pH, alkali-hydrolyzable N (AN), total carbon (TC), total nitrogen (TN), total kalium (TK), aluminum (Al), calcium (Ca), magnesium (Mg), and zinc (Zn).

The data set of land use was built through the visual interpretation of Landsat TM images for 2004/2005 with a spatial resolution of 30×30 m². First, a series of TM images were processed, such as single band extraction and false color composition. Second, these images were geo-referenced and ortho-rectified through field-collected ground control points and high-resolution digital elevation models with the mean location errors less than 1.5 pixels (i.e., 45 m). The databases used to improve the interpretation include

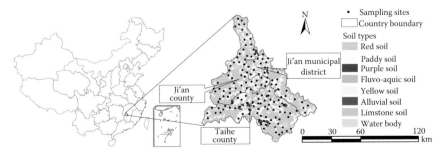

FIGURE 19.1
Soil types and soil samples in the investigation region.

time-series data for three time periods: (1) the late 1980s, including Landsat-TM scenes for 1987–1990; (2) the mid-1990s, including Landsat-TM scenes for 1995/1996; and (3) the late 1990s, including Landsat-ETM scenes for 1999/2000 (Liu et al., 2005). Based on these databases, the land-use types in 2007 were classified into six groups: woodland, cropland, grassland, built-up areas, water bodies, and unused lands.

The interpretation of the images and land-use classifications were validated through extensive field surveys in October 2007. We conducted ground truth checking in the investigation region and more than 700 field photos were taken using cameras equipped with global position systems (GPS). The average interpretation accuracy is 95.3% for land-use type classification. The land-use types in 2007 include woodland, cropland, grassland, built-up areas, water bodies, and unused lands (Figure 19.2). Woodland is the main area type with a total area of 3302 km². Cropland and grassland are 1957 and 397 km², respectively.

19.2.2 Methods

HASM-PCG, as seen in Chapter 3, uses data of soil sampling points to globally fit a surface through several iterative simulation steps. This surface is then used to interpolate soil property values at unknown points. The iterative simulation steps are summarized as follows (Yue et al., 2009): (1) conduct interpolation on the computational domain in terms of sampling data, which provides interpolated approximate values; (2) calculate the first fundamental coefficients and the second fundamental coefficients as well as the Christoffel symbols of the second kind; (3) solve HASM equations using an iterative process of preconditioned conjugate gradient; (4) repeat the iterative process until simulation accuracy is satisfied.

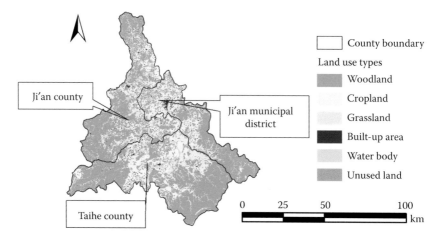

FIGURE 19.2
Land-use types in the investigation region in 2007.

19.2.2.1 Integration of HASM with Information on Land Use

Each observation $f(x_{i,k}, y_{j,k})$ of one specific soil property at grid (i,j) in the kth land-use type can be expressed as

$$f(x_{i,k}, y_{j,k}) = m(lu_k) + r(x_{i,k}, y_{j,k}) \tag{19.1}$$

where $m(lu_k)$ is the mean value of $f(x_{i,k}, y_{j,k})$ in each land-use type k and $r(x_{i,k}, y_{j,k})$ is the residual computed by subtracting the average soil property of the relative land-use type from the original value of soil properties.

It is assumed that $m(lu_k)$ and $r(x_{i,k}, y_{j,k})$ are mutually independent and that the variation of $r(x_{i,k}, y_{j,k})$ is homogeneous over the investigation region. The residuals are then used to interpolate the surface of residuals by means of HASM. The residuals interpolated by HASM are finally added to the soil property averages of the relevant land-use types as the final soil property values with land-use information (Shi, 2009).

19.2.2.2 Integration of Ordinary Kriging with Information on Land Use

The geostatistical mixed linear model (GMLM) can be formulated as (Haskard et al., 2007)

$$\mathbf{f}(x_i, y_j) = \mathbf{A} \cdot \mathbf{s}(x_i, y_j) + \boldsymbol{\eta}(x_i, y_j) + \boldsymbol{\varepsilon}(x_i, y_j) \tag{19.2}$$

where $\mathbf{f}(x_i, y_j)$ represents an $n \times 1$ vector of observed soil property data, (x_i, y_j) represents the corresponding survey location coordinates, $\mathbf{A}$ represents an $n \times p$ fixed data matrix that includes observed functions of sensor readings and possibly also the survey location coordinates, $\mathbf{s}(x_i, y_j)$ represents a $p \times 1$ vector of unknown parameter estimates, $\boldsymbol{\eta}(x_i, y_j)$ represents a 0-mean–second-order stationary spatial Gaussian error process, and $\boldsymbol{\varepsilon}(x_i, y_j)$ represents a vector of jointly independent normal $(0, \sigma_n^2)$ random variables.

OK, UK, KED, and RK models can all be derived as special cases of GMLM (Schabenberger and Gotway, 2005). UK, KED, and RK are all suitable for working with quantitative variables but not categorical information. OK is able to incorporate categorical information so that a geostatistic method for simulating soil properties is developed by integrating OK with land-use information (OK-LU).

19.2.2.3 Stratified Kriging

We assume that soil variables are isotropic and obey the intrinsic hypothesis within soil strata (Stein et al., 1988). Stratification of the region is based upon land-use types. Stratified Kriging (SK) was employed to compute a separate variogram for each land-use stratum and the sampled data within the land-use type was used only to develop the soil property surface within the land-use type (Stein et al., 1988). In this study, the total investigation region was partitioned into woodland, cropland, and grassland. In general, this method needs sufficient data to compute separate variograms for each land-use type (Voltz and Webster, 1990), but because only five sample points were collected in the grassland, and the grassland was generally near to woodland in this investigation region, we stratified the whole region into two strata. Woodland and grassland formed one stratum, and cropland formed another. Kriging was then performed within each stratum using a semivariogram model, and simulation of the entire region was carried out with respect to the stratified area (Stein et al., 1988).

19.2.2.4 Validation

Jackknife validation (Deutsch and Journel, 1998) is used for the validation of HASM-SP. In total, 120 training points are randomly selected from the 150 samples to simulate the soil property surfaces, and the remaining 30 samples are used as validation data sets. The mean error (ME), mean absolute error (MAE), and *RMSE* are employed to measure simulation accuracy. They are respectively formulated as

$$ME = \frac{1}{30}\sum_{i=1}^{30}(Sf_i - f_i) \tag{19.3}$$

$$MAE = \frac{1}{30}\sum_{i=1}^{30}|Sf_i - f_i| \tag{19.4}$$

$$RMSE = \sqrt{\frac{1}{30}\sum_{i=1}^{30}(Sf_i - f_i)^2} \tag{19.5}$$

where f_i is the sampled value and Sf_i is the simulated value of $f(x, y)$ at the ith sampled point.

The *ME* measures the bias of the simulated surface which should be close to zero for unbiased methods. The *MAE* and *RMSE* should be as small as possible for accurate interpolation.

19.2.2.5 Analysis of Variance

Analysis of variance (ANOVA) is a general statistical technique used to analyze deviation and identify other variation's sources in a measurement system (Kazerouni, 2009). Variance analysis includes between-groups and within-groups variance. The between-groups variance is called the effect variance and the within-groups variance is called the error variance (Turner and Thayer, 2001).

The variation σ_z^2 of $f(x_{i,k}, y_{j,k})$ can be shown as the sum of σ_m^2 between land-use types and σ_r^2 within land-use types:

$$\sigma_z^2 = \sigma_m^2 + \sigma_r^2 \tag{19.6}$$

where σ_m^2 represents the effect of the land-use type on the spatial variation of soil properties, and σ_r^2 shows the overall variation within land-use types. To compare the difference of soil properties among land-use types, the soil property data are grouped into three classes based on main land-use types. There are 61, 84, and 5 samples in woodland, cropland, and grassland, respectively. The variance of each soil property between and within land-use types are determined by using ANOVA with the software package SPSS 14.0 for windows. In ANOVA, Scheffee's test is conducted to *post hoc* multiple comparisons because of the different sample numbers in each land-use type.

19.3 Results and Analyses

19.3.1 ANOVA Analysis of Soil Properties among Different Land-Use Types

Table 19.1 shows an analysis of variances as a means of comparing the effects of different land-use types on tested soil properties, in which SS represents the sum of squares of deviation, MS represents the mean square deviation, df represents the degree of freedom, and Sig represents the significance level. We find that the variances σ_m^2 of most soil properties (except Zn), among land-use types are significant at the 0.01 or 0.05 level but Zn is significant at the 0.1 level. ANOVA analysis reveals that variances of tested soil properties among land-use types may play an important role in their spatial prediction in the investigation region.

19.3.2 Geostatistical Analysis

Analyses of the spatial correlation of residuals reflect good performance after removing the local mean within the same land-use type (Figure 19.3). Variograms of residuals for most of the soil variables tend to show a shorter range and a smaller sill, which indicates that the drift has indeed, been removed (Hengl et al., 2004). However, variograms of original values and residuals of soil Zn are similar, indicating that land-use type may not be the most important factor for the spatial distribution of soil Zn.

TABLE 19.1

ANOVA Analysis for Testing the Significance of Land-Use Type Effects on the Variances of Soil Properties

Soil Property	Source of Variance	df	SS	MS	F-Value	Sig
pH	Between land-use types	2	2.35	1.18	10.03	< 0.01[a]
	Within land-use types	147	17.26	0.12		
	Total	149	19.61			
AN, mg kg^{-1}	Between land-use types	2	66648.95	33324.47	14.16	< 0.01[a]
	Within land-use types	147	346016.39	2353.85		
	Total	149	412665.34			
TC, g kg^{-1}	Between land-use types	2	386.79	193.39	3.84	0.02[b]
	Within land-use types	147	7410.55	50.41		
	Total	149	7797.33			
TN, g kg^{-1}	Between land-use types	2	3.79	1.9	13.9	< 0.01[a]
	Within land-use types	147	20.06	0.14		
	Total	149	23.85			
TK, g kg^{-1}	Between land-use types	2	453.74	226.87	4.24	0.02[b]
	Within land-use types	147	7859.17	53.46		
	Total	149	8312.91			
Al, g kg^{-1}	Between land-use types	2	8928.5	4464.25	20.16	< 0.01[a]
	Within land-use types	147	32553.08	221.45		
	Total	149	41481.58			
Ca, g kg^{-1}	Between land-use types	2	466.23	233.12	64.01	< 0.01[a]
	Within land-use types	147	535.39	3.64		
	Total	149	1001.62			
Mg, g kg^{-1}	Between land-use types	2	140.5	70.25	16.72	< 0.01[a]
	Within land-use types	147	617.75	4.2		
	Total	149	758.24			
Zn, mg kg^{-1}	Between land-use types	2	6605.64	3302.82	2.59	0.08
	Within land-use types	147	187322.39	1274.3		
	Total	149	193928.03			

[a] Significant at the 0.01 level.
[b] Significant at the 0.05 level.

19.3.3 Comparison of the Performance of HASM-SP, OK-LU, and SK

A comparison of HASM-SP, OK-LU, and SK was conducted by assessing the feasibility of HASM-SP for simulating soil properties. To evaluate the performance of these three methods, the *MEs*, *MAEs*, and *RMSEs* were calculated by comparing the simulated values with the sampled values (Table 19.2). Most of the MEs of soil properties (except pH and Al) simulated by HASM-SP and OK-LU were much closer to 0 than SK. In general, most of the MAEs from HASM-SP (except N and Mg) were much lower than those of OK-LU

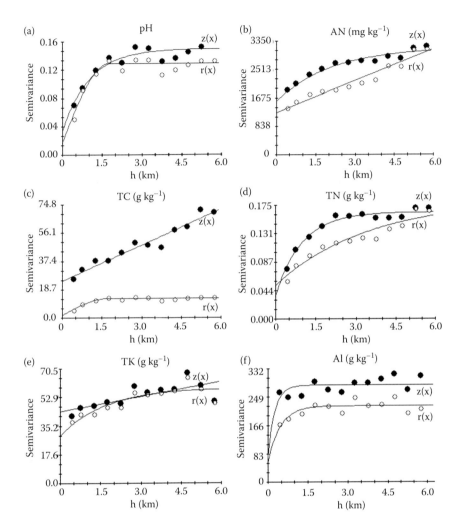

FIGURE 19.3
Semivariograms of original values and the residuals for soil properties: (a) pH, (b) AN, (c) TC, (d) TN, (e) TK, (f) Al, (g) Ca, (h) Mg, and (i) Zn. (● represents variograms of original data; ○ represents variograms of residual data).

and SK. HASM-SP always had the lowest *RMSEs* when compared with the other two methods. For example, *RMSEs* of AN simulated by HASM-SP were 1.6% and 41.5% smaller than *RMSEs* of AN simulated by OK-LU and SK, respectively. Rather higher *MAEs* and *RMSEs* were found in most cases for soil properties which were simulated by SK when compared to HASM-SP and OK-LU. This is because the number of data points available to estimate the variogram was not large enough to give reliable estimates (Stein et al., 1988). The results demonstrate that HASM-SP is an effective method for simulating soil properties.

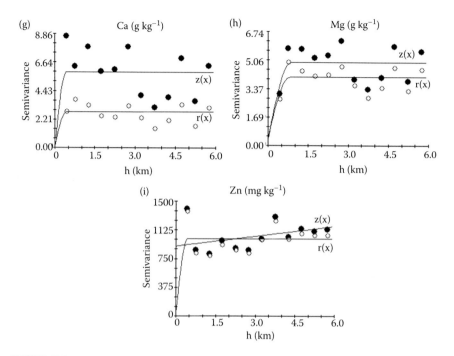

FIGURE 19.3
Continued

TABLE 19.2

MEs, MAEs, and *RMSEs* of HASM-SP, OK-LU, and SK

Method	Index	pH	AN (mg kg⁻¹)	TC	TN	TK	Al	Ca	Mg	Zn (mg kg⁻¹)
						(g kg⁻¹)				
HASM-SP	*ME*	−0.01	0.65	−0.2	0	−1.45	−5.38	−0.09	−0.58	−12.58
OK-LU		0.01	1.47	−0.15	0	−1.35	−5.32	−0.09	−0.59	−12.31
SK		−0.01	−4.72	−0.92	−0.04	−2.3	−3.49	−0.45	−0.69	51.96
HASM-SP	*MAE*	0.14	29.84	4.13	0.22	4.68	10.72	1	1.49	27.98
OK-LU		0.15	30.37	4.15	0.22	4.99	10.79	1.01	1.49	28.69
SK		0.19	40.7	5.28	0.29	6.23	11.59	1.34	1.72	53.36
HASM-SP	*RMSE*	0.19	36.69	5.07	0.31	5.94	13.71	2.6	2.23	36.87
OK-LU		0.2	37.28	5.1	0.32	6.31	13.77	2.61	2.23	37.14
SK		0.25	51.92	6.67	0.39	7.35	14.19	3.73	2.58	55.25

19.3.4 Comparison of Soil Property Maps Obtained by HASM-SP, OK-LU, and SK

In order to compare the map quality generated by HASM-SP, OK-LU, and SK, soil AN maps from the three methods are taken as examples. HASM-SP (Figure 19.4a), OK-LU (Figure 19.4b), and SK (Figure 19.4c) give many details

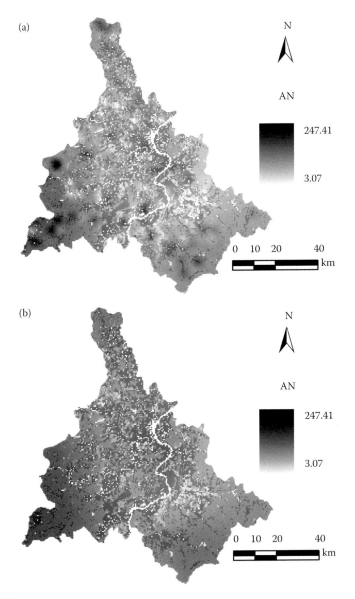

FIGURE 19.4
Soil AN (mg kg^{-1}) maps obtained by (a) HASM-SP, (b) OK-LU, and (c) SK.

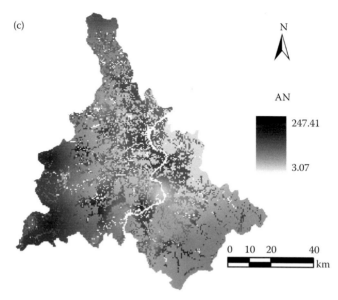

FIGURE 19.4
Continued

of AN in different land-use types for the whole map. Hence, the methods combined with land-use information can describe the local spatial variability more accurately.

To compare the performances of the three methods, the difference between OK-LU and HASM-SP is calculated by subtracting HASM-SP simulation values from OK-LU simulation values (Figure 19.5a). To compare SK and HASM-SP, the simulation values of HASM-SP are subtracted from the simulation values of SK (Figure 19.5b). To compare SK and OK-LU, the simulation values of OK-LU are subtracted from the simulation values of SK (Figure 19.5c). It is found that OK-LU has the most pattern similarities in common with HASM-SP but larger differences exist between them at some of the sampled points (Figure 19.1). Figure 19.5b and 19.5c show that the values of the soil AN, are estimated as lower in the mid-eastern woodland by SK than by the other two methods with land-use information. Table 19.2 shows that HASM-SP is an accurate method. Figures 19.4 and 19.5 suggest that HASM-SP could improve the simulation performance of soil properties by introducing information on the spatial changes of land-use types.

19.4 Discussion and Conclusions

HASM is a high-accuracy and high-speed method for DEM construction and climatic surface simulation. However, its performance may be not so good

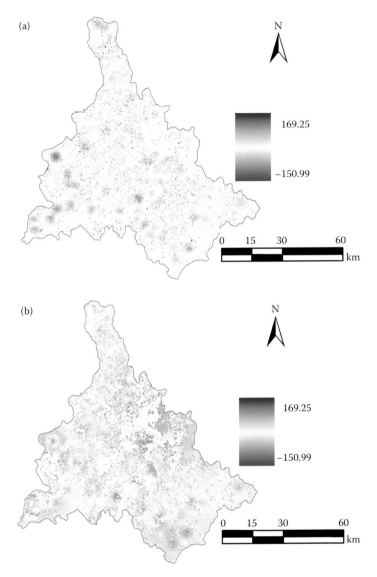

FIGURE 19.5
Performance differences between: (a) OK-LU and HASM-SP, (b) SK and HASM-SP, and (c) SK and OK-LU.

when it is directly used to simulate surfaces with abrupt boundaries. A derived method for HASM-SP is proposed by combining HASM with land-use information. When the performance of HASM-SP is compared to that of OK-LU and SK, the results indicate that HASM-SP is not only the most accurate method, but also the method that is able to provide the greatest amount of detail.

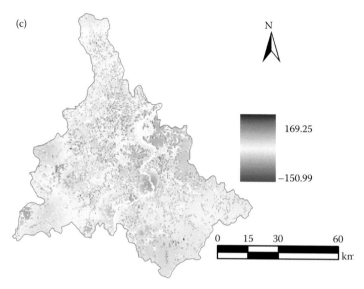

(c)

N

169.25

−150.99

0 15 30 60 km

FIGURE 19.5
Continued

Surface modeling of soil property faces two major limitations: the small number of available observations, and the nonlinearity of the relationship between environmental variables and soil properties (Ballabio, 2009). A possible approach to dealing with these limitations is to integrate categorical information, such as land-use information, into HASM. It is not difficult to combine other information, such as ecosystem types, vegetation cover, and lithology, into HASM to further heighten the simulation accuracy of soil properties. In addition, the DEM and climatic surfaces can be combined with HASM to improve the simulation quality of related soil attributes. HASM could be developed into a powerful tool with strong capacities for simulating surfaces of soil properties.

References

Ballabio, C. 2009. Spatial prediction of soil properties in temperate mountain regions using support vector regression. *Geoderma* 151: 338–350.

Basaran, M., Erpul, G., Tercan, A.E., and Canga, M.R. 2008. The effects of land-use changes on some soil properties in Indagi Mountain Pass—Cankiri, Turkey. *Environmental Monitoring and Assessment* 136(1–3): 101–119.

Bishop, T.F.A. and McBratney, A.B. 2001. A comparison of prediction methods for the creation of field-extent soil property maps. *Geoderma* 103(1–2): 149–160.

Burgess, T.M. and Webster, R. 1980. Optimal interpolation and isarithmic mapping of soil properties. III Changing drift and universal kriging. *Journal of Soil Science* 31: 505–524.

Castoldi, N., Stein, A., and Bechini, L. 2009. Agri-environmental assessment of extractable soil phosphorus at the regional scale. *NJAS—Wageningen Journal of Life Sciences* 56(4): 325–343.

Chen, F., Kissel, D.E., West, L.T., and Adkins, W. 2000. Field-scale mapping of surface soil organic carbon using remotely sensed imagery. *Soil Science Society of America Journal* 64(2): 746–753.

Cressie, N.A.C. 1993. *Statistics for Spatial Data*. New York, NY: John Wiley & Sons.

D'Acqui, L.P., Santi, C.A., and Maselli, F. 2007. Use of ecosystem information to improve soil organic carbon mapping of a Mediterranean Island. *Journal of Environmental Quality* 36(1): 262–271.

Deutsch, C.V. and Journel, A.G. 1998. *GSLIB: Geostatistical Software Library and User's Guide*. New York, NY: Oxford University Press.

Emery, X. and Ortiz, J.M. 2007. Weighted sample variograms as a tool to better assess the spatial variability of soil properties. *Geoderma* 140: 81–89.

Florinsky, I.V., Eilers, R.G., Manning, G.R., and Fuller, L.G. 2002. Prediction of soil properties by digital terrain modelling. *Environmental Modelling and Software* 17(3): 295–311.

Goovaerts, P. 1997. *Geostatistics for Natural Resources Evaluation*. New York, NY: Oxford University Press.

Goovaerts, P. 1999. Geostatistics in soil science: State-of-the-art and perspectives. *Geoderma* 89(1–2): 1–45.

Haskard, K.A., Cullis, B.R., and Verbyla, A.P. 2007. Anisotropic Matèrn correlation and spatial prediction using REML. *Journal of Agricultural, Biological, and Environmental Statistics* 12: 147–160.

Hengl, T., Heuvelink, G.B.M., and Stein, A. 2004. A generic framework for spatial prediction of soil variables based on regression-kriging. *Geoderma* 120(1–2): 75–93.

Hu, K.L., Li, H., Li, B.G., and Huang, Y.F. 2007. Spatial and temporal patterns of soil organic matter in the urban–rural transition zone of Beijing. *Geoderma* 141(3–4): 302–310.

Isaaks, E.H. and Srivastava, R.M. 1989. *Applied Geostatistics*. New York, NY: Oxford University Press.

Kazerouni, A.M. 2009. Design and analysis of gauge R&R studies: Making decisions based on ANOVA method. *Proceedings of World Academy of Science: Engineering and Technology* 40: 31–35.

Knotters, M., Brus, D.J., and Oude Voshaar, J.H. 1995. A comparison of kriging, co-kriging and kriging combined with regression for spatial interpolation of horizon depth with censored observations. *Geoderma* 67(3–4): 227–246.

Lark, R.M., Cullis, B.R., and Welham, S.J. 2006. On spatial prediction of soil properties in the presence of a spatial trend: The empirical best linear unbiased predictor (E-BLUP) with REML. *European Journal of Soil Science* 57: 787–799.

Lesch, S.M. and Corwin, D.L. 2008. Prediction of spatial soil property information from ancillary sensor data using ordinary linear regression: Model derivations, residual assumptions and model validation tests. *Geoderma* 148: 130–140

Liu, T.L., Juang, K.W., and Lee, D.Y. 2006a. Interpolating soil properties using kriging combined with categorical information of soil maps. *Soil Science Society of America Journal* 70(4): 1200–1209.

Liu, J., Tian, H., Liu, M., Zhuang, D., Melillo, J.M., and Zhang, Z. 2005. China's changing landscape during the 1990s: Large-scale land transformations estimated with satellite data. *Geophysical Research Letters* 32(2): 1–5.

Liu, D.W., Wang, Z.M., Zhang, B., Song, K.S., Li, X.Y., Li, J.P., Li, F., and Duan, H.T. 2006b. Spatial distribution of soil organic carbon and analysis of related factors in croplands of the black soil region, Northeast China. *Agriculture Ecosystems and Environment* 113(1–4): 73–81.

Magnussen, S., Næsset, E., and Wulder, M.A. 2007. Efficient multiresolution spatial predictions for large data arrays. *Remote Sensing of Environment* 109 (4): 451–463.

Minasny, B. and McBratney, A.B. 2007. Spatial prediction of soil properties using EBLUP with the Matérn covariance function. *Geoderma* 140: 324–336.

Mueller, T.G. and Pierce, F.J. 2003. Soil carbon maps: Enhancing spatial estimates with simple terrain attributes at multiple scales. *Soil Science Society of America Journal* 67(1): 258–267.

Odeh, I.O.A., McBratney, A.B., and Chittleborough, D.J. 1994. Spatial prediction of soil properties from landform attributes derived from a digital elevation model. *Geoderma* 63(3–4): 197–214.

Odeh, I.O.A., McBratney, A.B., and Chittleborough, D.J. 1995. Further results on prediction of soil properties from terrain attributes: Heterotopic cokriging and regression-kriging. *Geoderma* 67(3–4): 215–226.

Robinson, T.P. and Metternicht, G. 2006. Testing the performance of spatial interpolation techniques for mapping soil properties. *Computers and Electronics in Agriculture* 50: 97–108.

Schabenberger, O. and Gotway, C.A. 2005. *Statistical Methods for Spatial Data Analysis.* Boca Raton, FL: CRC Press.

Shi, W.J. 2009. *Spatial Analysis of Soil Environmental Elements Based on High Accuracy Surface Modelling—A Case Study of Red Soil Hilly Region in South China.* Doctoral Thesis, Graduate College of Chinese Academy of Sciences (in Chinese).

Shi, W.J., Liu, J.Y., Du, Z.P., Song, Y.J., Chen, C.F., and Yue, T.X. 2009. Surface modelling of soil pH. *Geoderma* 150(1–2): 113–119.

Stein, A., Hoogerwerf, M., and Bouma, J. 1988. Use of soil-map delineations to improve (Co-)kriging of point data on moisture deficits. *Geoderma* 43(2–3): 163–177.

Takata, Y., Funakawa, S., Akshalov, K., Ishida, N., and Kosaki, T. 2007. Spatial prediction of soil organic matter in northern Kazakhstan based on topographic and vegetation information. *Soil Science and Plant Nutrition* 53(3): 289–299.

Triantafilis, J., Huckel, A.I., and Odeh, I.O.A. 2001. Comparison of statistical prediction methods for estimating field-scale clay content using different combinations of ancillary variables. *Soil Science* 166(6): 415–427.

Turner, J.R. and Thayer, J.F. 2001. *Introduction to Analysis of Variance.* London: Sage Publications.

Voltz, M. and Webster, R. 1990. A comparison of kriging, cubic splines and classification for predicting soil properties from sample information. *European Journal of Soil Science* 41(3): 473–490.

Wu, C.F., Wu, J.P., Luo, Y.M., Zhang, H.B., and Teng, Y. 2008. Statistical and geostatistical characterization of heavy metal concentrations in a contaminated area taking into account soil map units. *Geoderma* 144(1–2): 171–179.

Yue, T.X., Chen, C.F., and Li, B.L. 2010a. An adaptive method of high accuracy surface modelling and its application to simulating elevation surface. *Transactions in GIS* (in press).

Yue, T.X., Du, Z.P., and Song, Y.J. 2008. Ecological models: Spatial models and Geographic Information Systems. In *Encyclopedia of Ecology*, eds., S.E. Jørgensen and B. Fath, 3315–3325. England: Elsevier Limited.

Yue, T.X., Du, Z.P., Song, D.J., and Gong, Y. 2007. A new method of surface modeling and its application to DEM construction. *Geomorphology* 91(1–2): 161–172.

Yue, T.X., Fan, Z.M., Chen, C.F., Wu, X.F., and Li, B.L. 2010b. Surface modelling of global terrestrial ecosystems under three climate change scenarios. *Ecological Modelling* (accepted).

Yue, T.X. and Song, Y.J. 2008. The YUE-HASM method. In *Accuracy in Geomatics, Proceedings of the 8th International Symposium on Spatial Accuracy Assessment in Natural Resources and Environmental Sciences*, Shanghai, June 25–27, 2008, eds., D.R. Li, Y. Ge, and G.M. Foody, 148–153. Liverpool: World Academic Union Ltd.

Yue, T.X., Song, D.J., Du, Z.P., and Wang, W. 2010c. High accuracy surface modeling and its application to DEM generation. *International Journal of Remote Sensing* 31(8): 2205–2226.

Yue, T.X., Song, Y.J., and Fan, Z.M. 2010d. The multigrid method of high accuracy surface modelling and its validation. *Computer and Geoscience* (accepted).

Yue, T.X. and Wang, S.H. 2010. Adjustment computation of HASM: A high accuracy and speed method. *International Journal of Geographical Information Sciences* (in press).

Zhao, C.Y., Nan, Z.G., and Cheng, G.D. 2005. Methods for modelling of temporal and spatial distribution of air temperature at landscape scale in the southern Qilian mountains, China. *Ecological Modelling* 189: 209–220.

20

Ecological Modeling and Earth Surface Simulation

20.1 Introduction

A model is an abstract of reality. It can be regarded either as a formal description of the essential elements of a problem, or as a description of the system-of-interest. Models can be classified into physical versus abstract, dynamic versus static, empirical versus mechanistic, deterministic versus stochastic, and simulational versus analytical. A system is an interlocking complex of processes characterized by many reciprocal cause–effect pathways (Watt, 1966; Swannack and Grant, 2008). In relation to physical science, a system is an organized collection of interrelated physical components characterized by a boundary and by functional unity (Grant et al., 1997). Ecological modeling on a global level is the process of simulating and analyzing the ecosystem-of-interest in terms of the principles of global ecology, which are absolutely essential in trying to understand the environmental problems affecting the world. For example, the simulation and analysis of global ecosystem services must calculate climate, soil, landform, hydrology, energy resources, and human impacts.

Global ecology is the study of ecological principles and problems on a worldwide basis (Southwick, 1996). It involves structure, process, and change (Botkin, 1982). Conventionally, global ecology focuses on earth system science and deals with biological, geographical, atmospheric, geological, and oceanographic issues. The most acute problem of global ecology is biosphere dynamics due to its anthropogenic impact on the biosphere and climate change (Kondratyev et al., 1992).

Earth system science is the study of how the earth works as a system of continents, oceans, atmosphere, ice, and life. It is based on our ability to measure key parameters and integrate that knowledge into earth system models. The earth system science concept fosters synthesis and the development of a holistic model, in which disciplinary process and action lead to synergistic interdisciplinary relevance (Johnson et al., 1997). Earth science researchers characterize the earth system and its interactions among its components with a network of satellite, airborne, and *in situ* sensors. Data from the network of sensors are used to describe land-cover change, ocean circulation, the cycling of water and carbon among land, atmosphere, and oceans. Earth system modelers use the resulting images, trend data, and flow depictions to understand the underlying processes and build computational models of the climate system, geophysical structure, biogeochemical processes, and the earth's response to incoming solar energy (King and Birk, 2004).

20.2 Earth Surface Modeling

Earth surface modeling has been through various phases of development, beginning in the middle of the nineteenth century (Table 20.1). Little progress in applying mathematics to earth surface issues was made until the latter half of the nineteenth century (Israel and Gasca, 2002). Beginning in the 1880s, several key achievements were made thereafter. In 1884, statistical analysis and elementary quantitative techniques started to be applied to the handling of biological information (Galton, 1884). In 1916, the probability theory was applied to study *a priori* pathometry (Ross, 1916). In 1920, the logistic curve was introduced to theoretical biology (Pearl and Reed, 1920). In 1925, statistical methods were developed for dealing with problems in population genetics (Fisher, 1925). In 1926, ordinary differential equations and integro-differential equations were applied to build the rational mechanics of biological association (Volterra, 1926). Earth surface modeling was schematized as follows: (1) characteristic properties of the natural phenomenon

TABLE 20.1

Major Steps in the Evolution of Ecological Modeling on a Global Level

Time Line	Event
Middle of the nineteeth century	First applications of quantitative methods for biological data
1916–1940	Development of statistical methods and fundamental biological models
Late 1940s	Application of game theory, systems theory, and information theory to biological systems and the development of other fundamental biological models
1950s	First earth's atmosphere model
1960s	Application of systems ecology and refinement of fundamental biological models
1970s	Active period of terrestrial ecosystem model development
1975	Publication of the first issue of Ecological Modelling
1980s and 1990s	Refinement of aid decision tools for model development
2000s	Revival of interest in differential equations

under study were identified; (2) a general description was set up in mathematical terms to obtain a certain number of differential equations, (3) qualitative methods were used to determine parameters, and (4) results were compared with empirical facts. During the late 1920s and the early 1930s, mathematical models were proposed for natural selection and evolution (Haldane, 1927; Wright, 1931). However, these specific studies focused on small-scaled systems and relatively few processes were modeled.

After a brief hiatus during World War II, studies on the earth system modeling of biospheres were eventually resumed in the late 1940s. However, it was not only the classical biometrical methods and differential equations that were in use; new mathematical tools such as game theory, system theory, and information theory were also employed. For instance, Leslie (1945, 1948) developed a population growth model, while Chapman (1950) proposed the idea of constructing a model of the earth's atmosphere and the space surrounding it on the basis of detecting the pull of the moon on the atmosphere by studying 6457 hourly records of air pressure made at Greenwich, England (Chapman, 1957). Sullivan (1961) and Skellam (1951) developed a model for spatial population distribution, while Beverton and Holt (1957) developed a dynamic relation model between fish population, which accounted not only for the reactions of particular fish population but also for the interactions between them.

The success of these early studies encouraged the development of a series of quantitative models. For instance, a model using the electrical analogue circuit was developed to analyze photosynthetic production, community metabolism, biomass, and species variety (Odum, 1960). Methods to increase the realism of Lotka–Volterra equations were also explored (Garfinkel, 1962, 1967a,b; Garfinkel and Sack, 1964), although Lotka–Volterra equations had previously been criticized (Smith, 1952). An animal population process was analyzed by developing a simple mathematical model (Holling, 1964). A one-dimensional model of radiative–convective equilibrium was developed (Manabe and Strickler, 1964). Plankton population dynamics were simulated by a linear differential equation set (Davidson and Clymer, 1966) and digital simulation models were developed for a salmon resource system, which were coded in DYNAMO (Paulik and Greenough, Jr., 1966).

In the 1970s, numerous computer-oriented mathematical models were developed for the analysis of entire ecosystems. For instance, grassland ecosystem models were developed, which involved about 40 state variables and several hundred parameters (Bledsoe et al., 1971; Anway et al., 1972; Patten, 1972). The first semimechanistic computer model of forest growth was developed to reproduce the population dynamics of trees in mixed-species forest type (Botkin et al., 1972). A generalized model for simulating lake ecosystems was developed in response to the growing need for models to enable humans to manage their environment (Park et al., 1974). A world model was developed for simulating the relationship between gross national product and human population growth (Jørgensen, 1975a). This period also saw the development of various model types to describe specific ecosystem properties; for instance, a linear differential model was developed to analyze productivity–stability relationships (Rosenzweig, 1971). Models which analyzed the diversity–stability relationship also became popular (Gardner and Ashby, 1970; May, 1972).

In 1975, the first peer-reviewed journal on ecosystem modeling, *Ecological Modelling*, was published. The journal attempted to combine three fields of science, namely mathematical modeling, system analysis, and computer techniques, along with ecology and environmental management (Jørgensen, 1975b). However, this generation of models suffered various problems, such as a lack of data and methods needed for obtaining measurements, inadequate modeling theory for macrosystem applications, and an inability to halt the propagation of errors, as well as uncertainties in model predictions (Patten, 1972; Shugart and O'Neill, 1979). In addition, these models assumed that processes occurring within the borders of a system were sufficiently uniform that one could adequately describe their dynamics by considering the total or mean properties of the system alone. The models emphasized structure and function in point-type systems that incorporated time but not space. Ecosystem models devoid of spatial consideration were ultimately found to be too abstract for use (Neuhold, 1975).

Since the middle of the 1980s, the translation of model results into geographical patterns has been rapidly developing (Jørgensen, 2002). For instance, a dynamic spatial simulation model composed of interacting cells was designed to simulate habitat changes as a function of marsh type, hydrology, subsidence, and sediment transport for a generalized coastal wetland area (Sklar et al., 1985; Costanza et al., 1990). GIS were used to spatially verify the close statistical relationship between the dependent variable and each of the independent variables selected by logistic regression modeling to better understand the dimensions of anthropogenic deforestation occurring throughout the tropical world (Ludeke et al., 1990). It was found that the storage and manipulation of large sets of data in GIS, which was facilitated by rapid advances in computing technology, presented opportunities for modeling spatial distribution and changing distribution on a regional scale (Elston and Buckland, 1993). The likely impact of climate change on the distribution and abundance of wildlife species was analyzed by means of spatial interpolation techniques (Aspinall and Matthews, 1994). Four predictive mathematical models were combined with GIS for population estimation in urban areas (Al-Garni, 1995). A spatially explicit forest community model was developed to generate estimates of potential natural vegetation for the entire potential forest area of Switzerland under today's climate as well as under altered climate regimes (Kienast et al., 1996). GIS and remote sensing were applied to aquatic botany in conjunction with simulation models (Caloz and Collet, 1997). The scale, pattern, and process relationships in fluvial and alpine environments were studied by means of optical and microwave remote sensing systems and GIS (Walsh et al., 1998).

GIS was employed to improve landscape classifications which could be useful for climate, socio-economic, or biodiversity modelers (Lioubimtseva and Defourny, 1999). A linkage between the GIS and water resources management model was developed to simulate river basin water allocation problem as a collection of spatial and thematic objects (McKinney and Cai, 2002). Forest ecosystems were spatially simulated and simulation uncertainty was analyzed (Larocque et al., 2006a, 2008). The revival of interest in differential equations in the early 2000s had a great impact in the field of earth surface modeling (Israel and Gasca, 2002), such as HASMs, which are based on the fundamental theorem of surfaces and formulated by PDEs (Yue et al., 2007, 2010; Yue and Song, 2008; Yue and Wang, 2010). A model, developed in combination with GIS, dealing with soil, water quality, superficial material, and aquifer properties was constructed to identify all areas of groundwater, vulnerable to nitrate pollution (Lake et al., 2003). Methods for surface modeling were used to simulate the change trends and scenarios of major terrestrial ecosystems (Yue et al., 2005, 2006). It was suggested that the development of spatially explicit disturbance historic information from historic accounts might enable future research to develop more precise quantitative disturbance history inputs at appropriate resolutions for ecosystem process models (McNeil et al., 2006). A data model, a statistical model and an ecological model were integrated into a GIS environment to support fine-scale vegetation community modeling (Accad and Neil, 2006). Spatially explicit models were used to estimate ecosystem services (Troy and Wilson, 2006; Yue et al., 2008). A simulation program that runs on GIS was developed to predict the multispecies size-structure dynamics of forest stands (Umeki et al., 2008). Several existing process-based models linked to economic valuation methods were integrated into a GIS platform to value ecosystem services (Gret-Regamey et al., 2008). Four different GIS-based habitat models were comparatively applied to analyze impacts of land-use changes, urbanization, and infrastructure developments on the fragmentation of natural habitats and biodiversity loss (Gontier et al., 2010). A spatial decision support system was developed to support land-use planning and local forestry by providing spatially explicit information on carbon sequestration and economic benefits under various scenarios of the carbon credit market (Wang et al., 2010).

20.3 Systems Ecology

In the 1960s, systems ecology appeared as a new and exciting sub-discipline of ecology (Olson, 1963). In systems ecology, a real world situation is abstracted into a mathematical model. The mathematical argument is applied to reach mathematical conclusions that are then interpreted into their physical counterparts (Van Dyne, 1966). Thermodynamics as a holistic view of ecosystems was employed to ecological modeling (Jørgensen, 1979, 1986). Models of energy, emergy, and environs were developed to provide holistic analyses of complex systems (Patten, 1982, 1985, 1992; Odum, 1983, 1988, 1994). Principles of energetics were applied to systems ecology at all scales (Odum, 1983, 1994). Systems ecology was defined as an approach to the study of the ecology of organisms which used the techniques and philosophy of systems analysis (Kitching, 1983). Systems ecology is characterized by the application of mathematical models to ecosystem dynamics. Systems ecology tends to differ from other ecological studies in the following attributes: (1) it considers ecological phenomena at large spatial, temporal, or organizational scales, (2) it introduces methodologies from other fields, (3) it emphasizes mathematical models, (4) it is orientated toward computer applications, and (5) it is willing to develop hypotheses about the nature of ecosystems (Shugart and O'Neill, 1979). Various approaches to applying systems ecology to earth systems have been suggested. One of the most common procedures of ecological modeling includes some or all of the following nine steps: (1) problem definition, (2) system identification, (3) decision on model type, (4) mathematical formulation, (5) decision on computing methods, (6) programming, (7) parameter estimation, (8) parameter validation, and (9) experimentation (Kitching, 1983). Four fundamental phases, that is, conceptual-model formulation, quantitative-model specification, model evaluation, and model use, were identified in the process of developing usable system models (Grant et al., 1997); these phases are now widely recognized. Phase 1 steps include stating the model objectives, bounding the system-of-interest, categorizing the components within the system-of-interest, identifying the relationships among the components of interest, representing the conceptual model, and describing the expected patterns of model behavior. Phase 2 steps include selecting the general quantitative structure for the model, choosing the basic time unit for the simulations, identifying the functional forms of the model equations, estimating the parameters of the model equations, coding the model equations for computer, executing the baseline simulation, and presenting the model equations. Phase 3 steps include assessing the reasonableness of the model structure and the interpretability of functional relationships within the model, evaluating the correspondence between model predictions and data from the real system and determining the sensitivity of model predictions to changes in the value of important parameters. Phase 4 steps include developing and executing the experimental design for the simulations, analyzing and interpreting the simulation results, examining additional types of management policy or environmental situations, and communicating the simulation results.

Recent progress in systems ecology can be found in the *Encyclopedia of Ecology* (Jørgensen and Fath, 2008) and the *Handbook of Ecological Modelling and Informatics* (Jørgensen et al., 2009).

20.4 Worldwide Collaborations

Since the middle of the nineteenth century, several international scientific societies have been promoted, which has greatly driven the worldwide collaboration (Table 20.2).

20.4.1 International Geophysical Year (IGY)

In 1851, Matthew Fontaine Maury, an American naval officer, proposed the first plan for coordinating worldwide weather observations at sea and ashore (Sullivan, 1961). This plan was discussed in an international conference held in Brussels in 1853. Representatives from Belgium, Britain, Denmark, France, the Netherlands, Norway, Portugal, Russia, Sweden, and the United States agreed that their warships would use standard procedures to record weather and oceanic phenomena wherever they went. A second conference was then proposed to map a program of international weather observations on land, which was not realized because of the Crimean War (1853–1856). Different organizations were formed thereafter: (1) in 1864, the International Association of Geodesy was established to test the theorem that fixed the equatorial bulge of the earth in terms of competition between the inward pull of

TABLE 20.2

Timeline of the Establishment of Worldwide Collaboration Organizations

Year	Event
1864	International Association of Geodesy
1875	International Bureau of Weights and Measures
1878	International Meteorological Organization (WMO)
1931	International Council of Scientific Union (ICSU)
1952	International Geophysical Year (IGY)
1963	Global Atmospheric Research Program (GARP)
1964	International Biological Program (IBP)
1970	Man and Biosphere Program (MAB)
1980	World Climate Research Program (WCRP)
1986	International Geosphere–Biosphere Program (IGBP)
1987	Intergovernmental Panel on Climate Change (IPCC)
1990	International Human Dimensions Program on Global Environmental Change (IHDP)
1991	International Program of Biodiversity Science (DIVERSITAS)
2001	Millennium Ecosystem Assessment (MA)
2001	Earth System Science Partnership (ESSP)
2005	Global Earth Observation System of Systems (GEOSS)

gravity and the outward tug of centrifugal force, (2) in 1875, the International Bureau of Weights and Measures was established to measure the earth's shape, and (3) in 1878, the International Meteorological Organization was established, a forerunner of today's World Meteorological Organization (WMO). In the late 1880s, weather charts were drawn for each day covering North America and Europe and extending deep into the Arctic. By the late 1930s, weather maps of the entire Northern Hemisphere were compiled, covering everyday from August 1932 to August 1933.

In the twentieth century, two key international organizations were founded: (1) in 1931, the ICSU was established, bringing together scientists from various international scientific unions with a common area of interest, and (2) in 1950, the WMO was established, which in 1951 became a specialized agency of meteorology, operational hydrology, and related geophysical sciences for the United Nations. In 1952, the ICSU General Assembly approved the proposal for the IGY. The IGY conference held in Uccle, Belgium, in April 1955 decided to establish three world data centers that housed a complete set of IGY data. World data centers A, B, and C was planned for the United States, the Soviet Union, Western Europe, Australia, and Japan, respectively. The IGY, jointly sponsored by WMO and ICSU, began on July 1, 1957 and ended on December 31, 1958 (Odishaw, 1958a,b). This period coincided with an expected peak of sunspot activity as well as several eclipses. The IGY investigations, for which about 60,000 scientists and technicians in 67 countries made observations and measurements at more than 4000 stations scattered across the world, were classified into three categories (Sullivan, 1961). The first category—the physics of the upper atmosphere, included solar activity, geomagnetism, aurora and airglow, ionosphere physics and cosmic rays, as well as rockets and satellites. The second category—the Earth's heat and water regime, included meteorology, glaciology, and oceanography. The third category—the Earth's structure and interior, included seismology, latitudes, and longitudes. IGY stimulated an interdisciplinary approach in which men trained in different fields worked together to resolve problems beyond the reach of any single specialist group.

20.4.2 International Biological Program and the Man and Biosphere Program

The IGY highlighted the advantages of international collaboration in geophysical science, which inspired biologists to think up similar international efforts (Worthington, 1975). The possibility of an IBP to sponsor the International Biological Year (IBY) was first considered by the Executive Board of ICSU in October 1959. The International Union of Biological Sciences (IUBS) presented a formal scheme for an International Program on Biology (IPB) to the executive committee of ICSU at its meeting in Lisbon in 1960. After a series of meetings, ICSU launched IBP at the first General Assembly of IUBS in Paris in July 1964. This 10-year IBP was limited to basic studies related to biological productivity and human welfare.

In September 1968, a conference on the "scientific basis for rational use and conservation of resources of the biosphere," held as a part of the intergovernmental conference convened by the United Nations Educational, Scientific and Cultural Organization (UNESCO) in Paris, led to the IBP process being transferred to the MAB. In contrast to the nongovernmental IBP, MAB was officially launched in 1970 as an intergovernmental program, which underwent an intensive phase of planning before it began after the first session held by its international coordinating council in 1971. It was likely to involve many scientists who had not been a part of IBP, although many of MAB's projects had their origins in IBP. The general objective of MAB was to develop a basis for the rational use of the natural resources of the biosphere, and to improve relationships between man and his environment on a global scale. It focused on the general study of the structure and functioning of the biosphere and its ecological divisions, systematic observation, changes brought about by man, the overall effects of these changes upon the human species, and providing education and information on the above issues (UNESCO, 1972). MAB was intended to develop the following specific projects: (1) identifying and assessing the changes in biosphere resulting from human activities, and assessing the effect of these changes on man, (2) studying and comparing the structure, functioning, and dynamics of natural, modified, and managed ecosystems, (3) studying and comparing the dynamic interrelationships between natural ecosystems and socio-economic processes, especially the impact of changes in human population, settlement patterns, and technology on the future viability of these systems, (4) developing ways and means to measure quantitative and qualitative changes in the environment in order to establish scientific criteria to serve as a basis for the rational management of natural resources, including the protection of nature, and the establishment of standards of environmental quality, (5) helping to bring about greater global coherence in environmental research, (6) promoting the development and application of simulation and other techniques for prediction as tools for environmental management, and (7) promoting environmental education in its broadest sense.

20.4.3 World Climate Research Program (WCRP)

On April 1, 1960, the United States launched its first meteorological satellite, TIROS 1, which allowed scientists to view the earth and its atmosphere from space. A very effective new tool for observing the weather had become available (Bolin, 2007). A WMO advisory committee on research was constituted to use this new tool operationally. In 1961, the United Nations agreed on a resolution to use satellites for observing the weather from space. An additional resolution of the United Nations in 1962 asked the WMO to develop its plan for an expanded program to strengthen methodological services and research. Responsibility for developing the expanded program of research was given to ICSU, specifically its International Union of Geodesy and Geophysics (IUGG). At the general assembly of IUGG held in the summer of 1963 in San Francisco, it was decided that IUGG should launch a truly international effort to prepare for the use of satellite technology in studies of the general circulation of the atmosphere to develop new methods for weather forecasting. In November 1963, an interunion committee on atmospheric science (CAS) was formed by IUGG that worked as the parent organization of the CAS.

The creation of CAS could be seen as the beginning of a series of global research programs in the field of environmental sciences. Its first meeting took place in 1965 at the WMO Headquarters in Geneva. Defined as an entirely research-oriented cooperative international meteorological and analytical program, its goal was to vastly improve our understanding of the general circulation of the global atmosphere. The program was called the GARP. In 1967, an agreement between ICSU and WMO on the formation of GARP was reached in Rome and a Joint Organizing Committee (JOC) of GARP was established. At the eighth session of JOC in 1973, a set of guiding principles was formulated, which concentrated on mathematical models, the mechanisms of climate fluctuations and the nature of climate change. In 1980, ICSU and WMO agreed that GARP would be transformed into a committee for international cooperation in climate research, and the WCRP was born. Since 1993, WCRP has also been sponsored by the Intergovernmental Oceanographic Commission (IOC) of UNESCO.

WCRP, sponsored by the ICSU, WMO, and IOC of UNESCO, is uniquely positioned to draw on the totality of climate-related systems, facilities and intellectual capabilities of more than 185 countries. The two overarching objectives of WCRP are, (1) to determine the predictability of climate, and (2) to determine the effect of human activities on climate. These two objectives underpin and directly address the requirements of the United Nations Framework Convention on Climate Change (UNFCCC), as well as contributing to many other international policy instruments. The WCRP is currently implementing a strategic framework titled Coordinated Observation and Prediction of the Earth System. Launched in 2005, this initiative aims to facilitate the analysis and prediction of earth system variability for use in an increasing range of practical applications of direct relevance, benefit, and value to society. In moving to provide a broader suite of products and services to a larger group of users, the WCRP is re-prioritizing its activities to optimize societal benefits. One of the primary WCRP pathways to application and end-user benefits will continue to be the integration of observations and models to generate new understanding, leading to enhanced benefits from climate predictions. The 2005–2015 WCRP strategy will help promote the creation of comprehensive, reliable, end-to-end global climate observations and models for the dual purpose of describing the structure and variability of the climate system and generating a fully consistent description of the state of the coupled climate systems for future prediction of climate (http://wcrp.wmo.int/wcrp-index.html).

20.4.4 Intergovernmental Panel on Climate Change

In 1980, ICSU, UNEP, and WMO jointly initiated the first international assessment of available knowledge on possible future human-induced climatic changes. A joint UNEP/WMO/ICSU international conference was organized in Villach, Austria, in October 1985 by the Austrian government after its completion of the report, "The Assessment of the Role of Carbon Dioxide and of Other Greenhouse Gases in Climate Variations and Associated Impacts." The UNEP, WMO, and ICSU agreed to form the advisory group on greenhouse gases (AGGG). The AGGG held its first meeting in July 1986 in Geneva. The scientific assessments in the late 1970s and 1980s, especially the report, "Our Common Future" (World Commission on Environment and Development, 1987), brought the climate change issue to the attention of the UN General Assembly in 1987. At this point, both the WMO congress and the UNEP governing council agreed that the executive heads of the two organizations should take steps to jointly organize an IPCC. Three working groups were formed at the first meeting of IPCC held in Geneva in November 1988, namely: (1) Assessment of Available Scientific Information on Climate Change, (2) Assessment of Environmental and Socio-Economic Impacts of Climate Change, and (3) Formulation of Response Strategies. The WMO took responsibility for the secretariat. The first assessment reports of the three working groups were titled: (1) Scientific Assessment (IPCC, 1990a), (2) Impact Assessment (IPCC, 1990b), and (3) Response Strategies (IPCC, 1990c). These assessments were completed in June 1990, and played a decisive role in the establishment of the UNFCCC. UNFCCC opened for signature at the Rio de Janeiro Summit of 1992 and came into force in 1994. The second assessment reports (IPCC, 1996a–c) were completed in 1995 after the reorganization of the IPCC, which provided key input for the negotiation of the Kyoto Protocol in 1997. The working groups I, II, and III, and the assessment reports, (a) Science of Climate Change, (b) Impacts, Adaptation and Mitigation, and (c) Economic and Social Dimensions of Climate Change, were approved in November, October, and July 1995, respectively. When the IPCC's second assessment reports were published, they faced severe criticism, but nonetheless gave impetus to subsequent work on the climate convention. In early 2001, the third assessment reports were completed (IPCC, 2001a,b). The contributions of working groups I, II, and III were adopted in January, February, and March 2001, respectively. In 2007, the fourth assessment was completed under the umbrella title "Climate Change 2007." The most comprehensive and up-to-date scientific assessment of past, present, and future climate change was obtained from working group I in February 2007. It was titled "The Physical Science Basis (IPCC, 2007a)." In April 2007, working group II gave an assessment of the impacts of climate change, the vulnerability of natural and human environments, and the potential for response through its report, "Impacts, Adaptation and Vulnerability." The report of working group III, "Mitigation of Climate Change," which was launched in May 2007 in Bangkok, provided an up-to-date overview of the characteristics of various sectors, mitigation measures that could be employed, costs, and specific barriers, and policy implementation issues. The IPCC continues to be a major source of information for negotiations under the UNFCCC.

20.4.5 International Geosphere–Biosphere Programme

In September 1986, the ICSU General Assembly began planning a new major transdisciplinary research programme, entitled the IGBP. The overall objectives of the IGBP were to describe and understand the interactive physical, chemical, and biological processes that regulate the total earth system, to understand the unique environment that it provides for life, to comprehend the changes that are currently occurring in the earth system, and to assess the manner in which these changes are influenced by human actions. A primary goal of the IGBP was to advance our capacity to predict change in the global environment. The development of this capacity would require the cooperation and complementary efforts of climate modelers, incorporating an appropriate level of understanding of relevant global biological, geological, and chemical processes into physical models of the earth system. The major scientific IGBP components and their interdependencies are terrestrial biosphere–atmospheric chemistry interactions, marine biosphere–atmosphere interactions, biospheric aspects of the hydrological cycle, and the effects of climate change on terrestrial ecosystems (Special Committee for the IGBP, 1988).

20.4.6 International Human Dimensions Programme on Global Environmental Change

In 1990, the IHDP was initially launched by the International Social Science Council (ISSC) as the human dimensions programme (HDP). In 1996, ICSU joined ISSC as cosponsors of the IHDP, and the secretariat was moved to Bonn, Germany. The IHDP is an international, interdisciplinary, nongovernmental science program dedicated to promoting and coordinating research. Its aims are to describe, analyze, and understand the human dimensions of global environmental change (Jaeger, 2003). The IHDP is designed around three main objectives of research, capacity building, and networking. IHDP has four core research projects: (1) LUCC, (2) global environmental change and human security (GECHS), (3) institutional dimensions of global environmental change (IDGEC), and (4) industrial transformation (IT). The LUCC's objectives are to obtain a better understanding of land-use and land-cover changes, and the physical and human driving forces behind them. Regarding the GECHS, its main goal is to advance interdisciplinary, international research and policy efforts in the area of human security and environmental change. The IDGEC analyzes the roles that social institutions play as determinants of the course of human/environment interactions. The IT research project focuses on the relationship between changes in industrial systems and changes in the environment. It also aims to discover ways of decreasing the environmental impact of industrial activities. In the meantime, three themes, carbon, water, and food systems have been selected for collaborative projects among the Global Environmental Change Research Programs of IGBP, WCRP, and the DIVERSITAS.

20.4.7 International Programme of Biodiversity Science DIVERSITAS

The International Programme of Biodiversity Science DIVERSITAS, was established in 1991, aimed at developing an international, nongovernmental umbrella program that would address the complex scientific questions posed by changes in global biodiversity (http://www.diversitas-international.org). Its founding sponsors, the UNESCO, the Scientific Committee on Problems of the Environment (SCOPE) and the IUBS, identified three projects: (1) the effects of biodiversity on ecosystem functioning (bioDISCOVERY), (2) origins, maintenance, and loss of biodiversity (ecoSERVICES), and (3) inventory and classification of biodiversity (bioSUSTAINABILITY). In 1996, ICSU and the International Union of Microbiological Societies (IUMS) joined DIVERSITAS as new sponsors, and seven core projects were added to this initial list. The seven core projects included, (1) the monitoring of biodiversity, (2) the conservation, restoration, and sustainable use of biodiversity, (3) soil and sediment biodiversity, (4) marine biodiversity, (5) microbial biodiversity, (6) inland water biodiversity, and (7) human dimensions of biodiversity. A partnership with the Convention on Biological Diversity was initiated in 1996. During the period 1991 to 1998, the DIVERSITAS Secretariat was hosted by MAB.

In 2001, the five sponsors of DIVERSITAS, ICSU, IUBS, IUMS, SCOPE, and UNESCO, decided to launch a second phase of the program. They opened a new Secretariat, hosted by ICSU in Paris, and called upon a task force of scientists to develop an international framework for biodiversity research. The resulting Science Plan, published at the end of 2002, clearly identified three interrelated areas for further development: (1) discovering biodiversity and predicting its changes, (2) assessing the impacts of biodiversity change on ecosystem functioning and services, and (3) developing the science of the conservation and sustainable use of biodiversity. In addition to the three thematic core projects, two integrated cross-cutting networks, the global invasive species program (GISP) and the global mountain biodiversity assessment (GMBA), were developed, which embraced issues addressed in all three core projects. In 2004, in addition to its three core projects and its two cross-cutting networks, DIVERSITAS established two new cross-cutting networks on agro-biodiversity and freshwater biodiversity. Since 2005, the convention for biological diversity (CBD) has been an ex-officio member of the DIVERSITAS scientific committee. In 2006, the DIVERSITAS Scientific Committee recognized the need to create a new core project aiming at providing an evolutionary framework for biodiversity science, as the scope of bioDISCOVERY was too large to efficiently manage the major issues related to evolution and systematics. This new core project, titled bioGENESIS, and bioDISCOVERY will focus on issues related to the monitoring of biodiversity, understanding the drivers of biodiversity changes and predicting those changes.

20.4.8 Millennium Ecosystem Assessment

In 2001, the millennium ecosystem assessment (MA) was initiated to respond to the needs described in the Millennium Report to the UN General Assembly in April 2000. In this Millennium Report, Kofi Annan stated that effective environmental policy must be based on sound scientific information, but there had never been a comprehensive global assessment of the world's major ecosystems (MA, 2003). The MA was designed to meet some of the assessment needs of the convention on biological diversity, the convention to combat desertification, and the wetlands convention, as well as the needs of other users in the private sector and civil society. The MA was carried out through four working groups on conditions and trends, scenarios, responses, and subglobal assessments. The millennium assessment focused on how humans had altered ecosystems, how changes in ecosystem services had affected human well-being, how ecosystem changes might affect people in future decades, and what types of responses could be adopted at local, national, or global scales to improve ecosystem management and, thereby, contribute to human well-being and poverty alleviation. Ecosystem services were defined as the benefits people obtain from ecosystems. These include the provisioning of services such as food and water; the regulation of services such as flood prevention, drought prevention, the prevention of land degradation, and the prevention of disease; the provision of supporting services such as soil formation and nutrient cycling; and the provision of cultural services such as recreational, spiritual, religious, and other nonmaterial benefits. The findings, contained in five technical volumes and six synthesis reports, provided a state-of-the-art scientific appraisal of the conditions and trends of the world's ecosystems, the services they provide and the options available to restore, conserve, or enhance the sustainable use of ecosystems.

On October 22, 2008, the ICSU along with the UNESCO and the United Nations University (UNU) launched a new program, "Ecosystem Change and Human Well-being," to understand the human impact on earth's life-support systems. It is a 10-year global research initiative and will be a part of the second MA which will build upon and strengthen existing global change research programs such as DIVERSITAS, IHDP, and IGBP. The initiative seeks to answer the most fundamental and policy-relevant questions concerning factors driving changes in ecosystem services, the impacts of those changes on human well-being, and opportunities to better manage human use and its impact on ecosystems (ICSU-UNESCO-UNU, 2008).

20.4.9 Earth System Science Partnership

At the first Global Environmental Change Open Science Conference in Amsterdam in 2001, the 1400 participants signed the Amsterdam Declaration on Global Environmental Change. The declaration called for the strengthening of cooperation amongst the global environmental change research programs. In response to the declaration, the four international global environmental change research programs, DIVERSITAS, IGBP, IHDP, and WCRP joined together to form the ESSP. The ESSP brings together researchers from diverse fields from across the globe, to undertake an integrated study of the earth system (www.essp.org). The declaration defines the earth system as a single, self-regulating system comprised of physical, chemical, biological, and human components (Loevbrand et al., 2009). Earth system science views the earth as a synergistic physical system of interrelated phenomena, processes, and cycles (Johnson et al., 2000). The goal of earth system science is to obtain a scientific understanding of the entire earth system on a global scale by describing how its component parts and their interactions have evolved, how they function and how they may be expected to continue to evolve on all timescales. Its challenge is to develop our capability to predict changes that will occur anytime between the next decade and the next century, both naturally and in response to human activities. Its fundamental purpose, which builds upon the traditional disciplines, is to emphasize relevant interactions of physical and dynamical properties that could extend over a spatial scale of anything from millimeters to the circumference of the earth, and over a timescale ranging from seconds to billions of years.

20.5 Global Models

20.5.1 General Circulation Models

The development of general circulation models (GCMs) was divided into three phases (Arakawa, 2000): the early numerical weather prediction model (NWP), the early general circulation model (EGCM), and the recent coupled atmosphere–ocean GCM. In 1904, Bjerknes first proposed the idea of weather prediction. In 1922, Richardson presented a weather prediction model with a basic structure not dissimilar to the current model; however, its forecast of the surface pressure at two points over Europe failed. In 1939, Rossy recognized the relevance of absolute vorticity advection to large-scale wave motions. In the late 1940s, the theories of baroclinic and barotropic instability (Charney, 1947), and the concept of equivalent-barotropy (Charney and Eliassen, 1949) represented major developments.

In 1950, Charney et al. successfully predicted 24-hr weather by using the adiabatic quasi-geostrophic equivalent-barotropic model, which indicated the appearance of NWP. In 1953, an adiabatic three-level quasi-geotrophic model was successfully used to predict the rapid development of a storm observed over the United States in November 1950 (Charney and Phillips, 1953). Although natural forecasts were not always successful, NWP became operational, first in the United States and then in many other countries. In 1956, Philips' numerical general circulation experiment recognized the close relationship between the dynamics of cyclones and that of general circulation. In 1959, Phillips demonstrated that nonlinear computational instability may occur in solutions of the nondivergent barotropic vorticity equation, which was the simplest nonlinear dynamical equation applicable to the real atmosphere.

Development of EGCM was distinguished into five generations. In 1968, Mintz described generation I of EGCM, which later became known as the Mintz–Arakawa model. This model included seasonal changes of solar radiation and long-wave cooling for each layer given as a function of the temperature at the lower level of the EGCM. Generation II of EGCM had the same two-level vertical structure as that of generation I; the horizontal domain covered the entire globe with uniform grid intervals in both longitude and latitude (Arakawa, 1969). Generation III of EGCM was the first multilevel model, which included a change of horizontal grid structure, the implementation of a bulk model for the planetary boundary layer (PBL), the inclusion of the Arakawa–Schubert cumulus parameterization, and the prediction of ozone mixing ratio with interactive photochemistry (Schlesinger and Mintz, 1979). In generation IV of EGCM, the vertical discretization followed Arakawa and Suarez (1983) for the troposphere above the PBL. The horizontal differencing of the momentum equation was based on the energy and potential enstrophy conserving scheme for the shallow water equations (Arakawa and Lamb, 1981). Generation V of EGCM was characterized by a change in the radiation scheme (Harshvardan et al., 1987), inclusion of an orographic gravity wave drag parameterization (Kim and Arakawa, 1995), inclusion of convective downdraft effects in the cumulus parameterization (Cheng and Arakawa, 1997), the revision of PBL moist processes (Li et al., 1999) and the inclusion of explicit predictions on ice and liquid clouds (Koehler et al., 1997).

For most applications of EGCMs, the geographical distribution of sea surface temperature (SST) was prescribed as an external condition, meaning that calculations of the surface heat flux and the vertical distribution of radiative heating did not have to be very accurate. NWPs and EGCMs instead concentrated on synthesizing hydrological processes, cloud processes, boundary-layer processes, radiation processes, and land processes. GCMs focused on interactions among ocean and land processes, radiation, and chemical processes, boundary-layer processes, cloud processes, and hydrological processes (Arakawa, 2000).

20.5.2 World Model for the Limits to Growth

In the early 1970s, a world model of limits to growth (WM) was developed and used as the prototype model for investigating five major trends of global concern: accelerating industrialization, rapid population growth, widespread malnutrition, the depletion of nonrenewable resources, and a deteriorating environment (Forrester, 1971; Meadow et al., 1972). The first version of the WM called World3, was written in the computer simulation language, DYNAMO. In 1992, on the basis of studying global developments between 1970 and 1990, the world model was updated and the results published in a book, Beyond the Limits (Meadows et al., 1992). World3 was updated into World3-91 for Beyond the Limits and the computer language was converted from DYNAMO to STELLA. In 2004, the 30-year update of limits to growth indicated that humanity was already in unsustainable territory but the human ecological footprint was still increasing (Meadows et al., 2004). The ecological footprint measures the amount of biologically productive land and water area required to support the demands of a population or productive activity, which is then compared to the carrying capacity. Productive land and sea areas support human demands for food, fiber, timber, and energy as well as providing space for infrastructure. These areas also absorb waste products from the human economy. The ecological footprint measures the sum of these areas that are often weighted according to their relative productivity and expressed in global hectares (Kitzes and Wackernagel, 2009). World3-91 was slightly updated and turned into World3-03, which showed the behavior of the human welfare index and the human ecological footprint for the period between 1900 and 2100. It concluded that it would take a long time to obtain political support for reversing current trends, bringing the ecological footprint back below the long-term carrying capacity.

20.5.3 Ecopath with Ecosim

A mathematical ecosystem model, ECOPATH, was constructed for an entire coral reef ecosystem and was used to estimate its standing stock and production budget (Polovina, 1984). ECOPATH was based on assuming mass balance in trophic flows and the idea that if some biomasses and/or flows are known, then others can be calculated under the mass-balance assumption. ECOPATH partitions the ecosystem into species groups and produces estimates of mean annual biomass, annual biomass production, and annual biomass consumption for each of the species groups, because each species group is an aggregation of species having common physical habitats, similar diets, and similar life history characteristics. However, ECOPATH only provides a static picture of ecosystem trophic structure by means of a linear equation system. Thus, dynamic ecosystem models, termed ECOSIMs, were derived from ECOPATH results by transferring the linear equation system to a system of differential equations. ECOSIM is a modeling tool for representing the spatially aggregated dynamics of whole ecosystems by combining relatively simple differential equations for biomass dynamics of some ecosystem components with delay-difference, age-size-structured equations for some key populations that have complex trophic ontogenies and use selective harvesting for older animals. The incorporation of ECOSIM into ECOPATH enabled a wide range of potential users to conduct fisheries policy analyzes that explicitly accounted for ecosystem trophic interactions. But ECOSIM had many limitations, such as overestimating the potential productivity of low-fecundity species, producing overly optimistic assessments of maximum fishing mortality rates for intermediate trophic levels, producing indeterminate outcomes in complex food webs, and producing misleading parameter estimates due to ECOPATH's equilibrium assumptions (Walters et al., 1997). ECOSPACE represents biomass dynamics over two-dimensional space as well as time. The Lagrangian approach and Eulerian approach are two possible ways to discretize dynamic relationships for the practical simulation of space–time patterns. The Lagrangian approach is to divide biomasses into large numbers of parcels assumed homogeneous, and to move these parcels about in space, while the Eulerian approach treats movement as flows of organisms among fixed spatial reference points or cells, without retaining information about the history of the organisms present at any point in any given moment. ECOSPACE takes the Eulerian approach. ECOSPACE is a spatially explicit model for policy evaluation that allows consideration of the impact of marine protected areas in an ecosystem context, relying on the ECOPATH mass-balance approach for most of its parameterization. The modeling approach, Ecopath with Ecosim (EwE), combines models for ecosystem trophic mass balance analysis (ECOPATH) with a dynamic modeling capability (ECOSIM), as a means of exploring the past and future impacts of fishing and environmental disturbances as well as optimal fishing policies. Ecosim models can be replicated over a spatial map grid (ECOSPACE) to allow the exploration of policies such as marine protected areas, while accounting for spatial dispersal effects (Christensen and Walters, 2004).

20.5.4 Integrated Model to Assess the Global Environment

According to Kram and Stehfest (2006), development of the integrated model to assess the global environment (IMAGE) can be distinguished into IMAGE 1.0, IMAGE 2.0, IMAGE 2.1, IMAGE 2.2, IMAGE 2.3, IMAGE 2.4, and IMAGE 3.0. In 1984, the National Institute for Public Health and the Environment (RIVM) of the Netherlands initiated a project, Reference Function for Global Air Pollution/CO_2, within which global modeling was defined as a quintessential activity (Rotmans, 1990). In 1986, IMAGE 1.0 was developed on the basis of prototypes constructed in the previous two years. IMAGE 1.0 was the first version of IMAGE. The primary objective of IMAGE 1.0 was to create a comprehensive picture of global climate change by integrating separate components relating to the world economy, atmospheric chemistry, marine and terrestrial biogeochemistry, ecology, climatology, hydrology, and glaciology into a synthetic framework. IMAGE 1.0 was based on the hypothesis of the greenhouse effect, and so its underlying principles would be greatly tarnished if the greenhouse hypothesis was to be rejected in future.

IMAGE 1.0 is a policy-oriented model that tries to capture as much as possible of the cause–effect relationship with respect to climate change. It was developed for the calculation of historical and future emissions of greenhouse gases on global temperatures and sea level rise, taking into account ecological and socio-economic interests in specific regions. IMAGE 1.0 consisted of interlinked modules that included emission modules, concentration modules, a climate module, a sea level rise module, and a socio-economic impact module. Each module described a specific element of climate change and the output of one module served as the input to the next.

In the emission modules, the trace gases included CO_2, CH_4, CO, N_2O, CFC-11, and CFC-12. Annual estimates of historical emissions were incorporated for the period 1900 to 1985. Four sets of scenarios were chosen for the period 1985 to 2100. The underlying scenario assumptions were based on different sources of trace gas emissions grouped as nature, energy, agriculture and industry. The world was divided into nine regions, the United States, OECD west, OECD Asia, Centrally Planned Europe, Centrally Planned Asia, the Middle East, Africa, Latin America, and South and East Asia. The main parameters included population, economic growth, end use and conversion efficiency, cost development for nonfossil fuels, environmental costs and carbon taxes. The input of the concentration module was taken from the output of the emission modules. The concentration module was linked with an ocean module and a deforestation module. The calculated trace gas concentrations provided input for the climate module. The resulting global mean equilibrium surface temperature rise was calculated by a climate feedback factor, in which the water vapor factor was explicitly taken into account. The resulting temperature changes provided input for the sea level rise module. The effect of global warming on potential sea levels was determined by five processes: the thermal expansion of ocean water, the melting of alpine glaciers, the ablation of the Greenland ice caps, the accumulation of the Antarctic, and natural trends. The socio-economic impact module was developed for describing the consequences of an accelerated sea level rise for the coastal defense systems of dikes, dunes, and intertidal areas.

IMAGE 1.0 was performed with four sets of scenarios. Scenario A assumed that economic growth would continue without being limited by environmental constraints. Scenario B assumed that environmental measures presently being considered to control environmental problems would be implemented. Scenario C assumed the enforcement of stricter environmental controls. Scenario D calculated the possibility of maximum efforts toward global sustainable development. In all scenarios, the world population was assumed to approach 10.8 billion in 2100, and it was assumed not to be influenced by greenhouse policies.

IMAGE's focus was then shifted to a regional base, and became known as version 2.0 (Alcamo, 1994). In this version, land-cover and land-use modeling were done on a resolution of $0.5 \times 0.5°$, drawing on experience with geographically explicit global models. IMAGE 2.0 was the first published global integrated model to have a geographic resolution. IMAGE 2.0 consisted of three linked clusters of modules: the Energy-Industry System (EIS), the Terrestrial Environment System (TES), and the Atmosphere–Ocean System (AOS).

Guided by the recommendations of international review meetings, further refinements and extensions were implemented in IMAGE 2.1 (Alcamo et al., 1998). Major steps included an improved computation of future regional energy use in EIS. Since the development of IMAGE 2.1, future fuel prices have influenced the selection of fuels in the model, depending on resource depletion on the supply side and price-dependent energy conservation on the demand side.

The third session of the IMAGE Advisory Board in 1999 resulted in a list of recommendations and suggestions for further development work on IMAGE (Tinker, 2000). One of the major changes of IMAGE 2.2 concerned the recommended two-track strategy for climate modeling. The earlier zonal-mean climate-ocean model in IMAGE was replaced by a combination of the simple MAGICC climate model and the Bern ocean model. For the new approach, the resulting global average temperature and precipitation changes were scaled using temperature and precipitation patterns generated by complex, coupled Global Circulation Models (GCMs). The widely accepted method of Schlesinger et al. (2000) for scaling patterns of aerosol-induced climate change was also adopted. A specific advantage of this method is that patterns from different GCMs can be used to explore uncertainties in the behavior of the global climate system. This IMAGE 2.2 version was used to contribute to work on the SRES of the IPCC (Nakicenovic et al., 2000); in particular, the B1 scenario (De Vries et al., 2000).

Participation in the IPCC-SRES process has greatly enhanced the capacity of both the model and IMAGE teams to explore scenarios and obtain results geared to the requirements of various international assessment processes. Experiences gained from the SRES process have, however, reinforced the desire to seriously reconsider the future of the IMAGE model. A set of model enhancements was identified and later initiated; these enhancements taken jointly will constitute the next generation model, IMAGE 3. However, parallel to this, a small set of model changes, internally referred to as version 2.3, and mainly concerned with the integration of energy crops and carbon plantations, was implemented for the analysis of mitigation options (Van Vuuren et al., 2007).

In IMAGE 2.4, grid-based population dynamics have been improved by introducing a new downscaling algorithm. The population within a grid cell is calculated by a proportional method using available country-specific data combined with world regional trends. Aggregated economic indicators like GDP, household consumption and value added in industry, services, and agriculture are used to estimate the demand for energy services. Energy supply chains with substantial technological detail are then selected on the basis of relative costs to meet the resulting final energy demand after autonomous and price-induced energy savings. Demand for and production of agricultural products on the basis of population changes and economic developments are simulated through a linkage to the global trade analysis project model. One of the most striking elements of IMAGE 2.4 is its geographically explicit land-use modeling, which considers both cropping and livestock systems on the basis of agricultural demand and demand for energy crops. IMAGE 2.4 uses the AOS model developed for IMAGE 2.2 (Eickhout et al., 2004). However, important nonlinear interactions between the land, atmosphere, and ocean cannot be studied with IMAGE 2.4 due to limitations of the current climate model and the natural vegetation module.

20.5.5 CENTURY Model

In 1987, a rangeland and cropland version of the CENTURY model was developed to simulate the effects of macro-environmental gradients as a first step toward simulating the effects of climatic change. The model incorporated multiple compartments of soil organic matter (SOM), simulated decomposition rates that vary as a function of monthly soil temperature and precipitation, and included both C and N flows (Parton et al., 1987). The model simulated primary productivity, soil nutrient dynamics, and soil water but focused mainly on changes in SOM. The first objective of the modeling exercise was to simulate the effects of climatic gradients on productivity and SOM over large areas; the second objective was to identify key soil properties that would allow the difference between soils within a single climatic zone to be simulated. The CENTURY model consisted of a soil and decomposition submodel, a plant submodel, and a nitrogen submodel. The soil and decomposition submodel contained three fractions of SOM: (1) an active fraction of soil C and N consisting of live microbes and microbial products; (2) a pool of C and N that is physically protected and/ or in chemical forms with more biological resistance to decomposition; (3) a fraction that is chemically recalcitrant and can be physically protected. The plant submodel simulated the monthly dynamics of C and N in live and dead aboveground plant material, live roots, structural and metabolic surfaces, and soil residual pools. The nitrogen submodel assumed that most N was bonded to C and that the C/N ratio of structural, active, slow, and passive fractions remained fixed.

In 1993, a grassland version of the CENTURY model was tested using observed data from 11 temperate and tropical grasslands around the world (Parton et al., 1993). In this version, biophysical submodels were added to calculate monthly evaporation and transpiration water loss, water content of soil layers, snow water content, and the saturated flow of water between soil layers. This improved version of the CENTURY model could be used to simulate the dynamics of carbon (C), nitrogen (N), phosphorus (P), and sulfur (S) for different plant–soil systems. Different plant production submodels for grassland, forests, and crops were linked to a common SOM submodel. The SOM submodel simulated the flow of selected elements through different inorganic and organic pools in the soil.

In 1994, a cropland version of the CENTURY model was developed to simulate long-term (10–1000 year) patterns in the dynamics of SOM (0–30 cm depth), plant production, and nutrient cycling (N, P, and S) using data collected on a monthly basis. The driving variables included monthly average maximum air temperature (°C at 2 m), monthly precipitation (cm), soil texture (sand, silt, and clay content), dead plant material nutrient and lignin content, and atmospheric and soil inputs of N (Parton and Rasmussen, 1994).

In 2002, the SOM submodel was modified to provide a better description of the litter and soil C and N dynamics of forests. The modifications included: (1) incorporation of additional woody litter pools, (2) allowing the N content of the SOM pool to vary, (3) constraining N mineralization to the active SOM pool, (4) incorporation of a small flux of mineral N to the resistant SOM pool, (5) allowance of mycorrhizal uptake of N, and (6) reformulation of temperature and moist effects on decomposition. This modified version of the CENTURY model was able to simulate a range of complex observations related to the C and N dynamics reported in a number of published studies (Kirschbaum and Paul, 2002).

In 2003, the point-based CENTURY model was integrated with spatially distributed data. Spatially explicit estimates of historical, current, and future soil C were produced for a semi-arid area in the Sudan using readily available data for temperature, precipitation, soil texture, and land cover (Ardoe and Olsson, 2003). The integration of the CENTURY model with spatially explicit data provided a flexible and powerful way to assess how different scenarios for land use, management, and climatic change can affect carbon dynamics on a regional scale.

20.5.6 Terrestrial Ecosystem Model

The terrestrial ecosystem model (TEM) is a process-based biogeochemical model that uses spatially referenced information on climate, elevation, soils, and vegetation to make monthly estimates of important carbon and nitrogen fluxes and pool sizes for the terrestrial biosphere. The first version of TEM (Raich et al., 1991) was used to describe and analyze the spatial and temporal patterns of terrestrial NPP in South America. The second TEM version (McGuire et al., 1992) was used to estimate NPP for potential vegetation in North America. The third version examined the response of NPP to temperature and carbon dioxide for temperate forests (McGuire et al., 1993), and its response to climate change predicted by the GCM for global terrestrial biosphere (Melillo et al., 1993). Version 4.0 of TEM was improved so that it was able to evaluate global and regional responses of NPP and carbon storage to elevated carbon dioxide (McGuire et al., 1997). TEM 4.1 was applied to investigating the dynamics of terrestrial carbon fluxes and storage in potential vegetation of the conterminous United States during the period 1900 to 1994 (Tian et al., 1999). TEM 4.2, in which the GPP equation of TEM 4.1 had been improved, was used to simulate the combined effects of climate variability, increased atmospheric CO_2 concentration, cropland establishment, and abandonment of CO_2 between the atmosphere and monsoon in Asian ecosystems (Tian et al., 2003). It simulated changes in carbon storage during three stages of disturbance: conversion from natural vegetation to cultivation, production, and harvest on cultivated land, and abandonment of cultivated land.

The net carbon exchange between the terrestrial biosphere and the atmosphere in TEM is represented by net ecosystem production (NEP), which is calculated as the difference between NPP and heterotrophic respiration (RH). NPP is calculated as the difference between gross primary production (GPP) and plant respiration (RA). Gross primary production represents the uptake of atmospheric CO_2 during photosynthesis and is influenced by light availability, atmospheric CO_2 concentration, temperature, and the availability of water and nitrogen (Tian et al., 2000). Plant respiration includes both maintenance and construction respiration, and is calculated as a function of temperature and vegetation carbon. The flux RH represents the microbially mediated decomposition of organic matter in an ecosystem and is influenced by the amount of reactive soil organic carbon, temperature, and soil moisture. The annual NEP of an ecosystem is equivalent to its net carbon storage for the year. For regional or global extrapolation with TEM, spatially explicit input datasets includes vegetation, elevation, soil texture, mean monthly temperature, monthly precipitation and mean monthly solar radiation. The input data sets are gridded at a resolution of 0.5° latitude and 0.5° longitude. In addition to the input datasets, TEM requires soil- and vegetation-specific parameters appropriate to a grid cell. Although many of the parameters in the model are defined from published information, some of the vegetation-specific parameters are determined by calibrating the model to the fluxes and pool sizes of an intensively studied field site.

20.5.7 Dynamic Integrated Climate–Economy Model

Nordhaus (1993) developed a dynamic integrated climate–economy (DICE) model to simulate the long-term interactions between global warming damages and world economic growth. Because the DICE model does not involve energy flow systems, Mori (1995) developed a DICE + e model, which imposed energy flows upon the DICE model to estimate the value of exhaustive resources, nuclear technology, and other options. The DICE + e was then extended to the multiregional approach for resource and industry allocation (MARIA) model to generate the international market prices of resources as well as the tradable carbon emission permits under certain constraints. MARIA has been applied to assess the potential contribution of fossil, biomass, nuclear, and other energy technologies and land-use changes to future greenhouse gas (GHG) emissions, global and regional GHG mitigation policies, and future GHG reduction options (Mori and Saito, 2004). Owing to the DICE model omitting induced innovation and overstating the costs of complying with carbon change policies, an ENTICE model (for Endogenous Technological change) was developed. It was used to compare the overall global welfare costs of carbon policy to the results of models in which technological change is exogenous (Popp, 2004). ENTICE makes use of empirical results on technological change in the energy industry to incorporate endogenous technological change into the DICE model of climate change, and simulates technological change that enhances energy efficiency. The ENTICE framework provides a detailed sensitivity analysis of different assumptions on features such as imperfect research markets and the potential crowding out effects of energy research and development (R&D). ENTICE model limitations include the fact that it dramatically simplifies policy by modeling the world as a single region, that it only designs innovations to improve energy efficiency, and that does not consider the uncertainty factor. A DICE94 + THC model was developed by amending the DICE Model to include an indicator of an emissions-dependent collapse of North Atlantic Thermohaline Circulation (THC) (Keller et al., 2004; Lempert et al., 2006). DICE94 + THC were run by repeatedly invoking the Stochastic Ranking and Evolution Strategy (SRES) algorithm developed by Runarsson and Yao (2000).

20.5.8 Model for Evaluating Regional and Global Effects of GHG Reduction Policies

Manne et al. (1995) proposed a model for evaluating the regional and global effects of GHG reduction policies (MERGE). The model was designed to be sufficiently flexible when used to explore alternative views on a wide range of contentious issues such as costs, damages, valuation, and discounting. It consisted of a series of linked modules representing the major processes of interest, which included the costs of reducing emissions of radiatively important gases, natural system disposition, reactions to the emission of these gases, and the reaction of human and natural systems to changes in the atmospheric/climatic system. MERGE had three major submodels: Global 2200, the climate submodel, and the damage assessment submodel. Global 2200 divides the world into five major geopolitical regions, the USA, other OECD nations (Western Europe, Japan, Canada, Australia, and New Zealand), FSU (the former USSR), China and the ROW (rest of world); Global 2200 is used to assess the economy-wide costs of alternative emission constraints at a regional and global level. The climate submodel describes the relationship between manmade emissions and atmospheric concentrations and their resulting impact on temperature. The damage assessment submodel divides damages into two categories, market and nonmarket, to quantify the impacts of climate change. Kypreos (2007) introduced two stylized backstop systems with endogenous technological learning (ETL) in the MERGE, one for electric markets and the other for nonelectric markets, to analyze the impacts of ETL on carbon-mitigation policy. He contrasted the results of the ETL backdrop system with a situation in which ETL was not used.

20.5.9 Asian Pacific Integrated Model

Matsuoka et al. (1995) developed the Asian Pacific Integrated Model (AIM). AIM was developed mainly to examine global warming response measures in the Asia-Pacific region, but was later linked to a world model enabling global estimates to be made. AIM comprises three main models, the GHG emission model (AIM/emission), the global climate change model (AIM/climate), and the climate change impact model (AIM/impact). The AIM/emission model consists of country-level, bottom-up type energy models and global-level, top-down type energy and land-use models. Emissions of SO_2, NO_x, and SPM are calculated within the AIM/emission model, and then used as input for the AIM/climate model. The absorption of CO_2 and heat by the oceans is calculated using an upwelling-diffusion model, which is a part of the AIM/climate model, with the oceans divided into a surface mixed layer and an intermediate layer that extends down to about 1000 m. Global mean temperature changes are calculated with an energy balance/upwelling diffusion ocean model, and used as input for the regional models. Data from the GCM experiments are used in order to estimate the regional distribution of climate parameters. They are coupled with the global mean temperature change calculated in the AIM/climate model. The interpolated climate distributions are used in the AIM/impact model, which calculates global and regional climatic impacts. The AIM/impact model mainly treats the impact on primary production industries, such as water supply, agriculture, forest products, and human health. It is necessary to consider many uncertainties in human activities such as population growth, economic development, and technological innovation as well as uncertainties in natural processes to estimate future CO_2 emissions and to make plans for climate stabilization. A variety of global and regional assumptions about population growth, economic growth, technological improvement, and government policies are entered into the models to provide estimates of energy consumption and land-use changes, which ultimately provide predictions of GHG emissions. The main goal of AIM is to assess policy options for stabilizing the global climate, particularly in the Asia-Pacific region; from the two perspectives of reducing GHG emissions and avoiding the impacts of climate change (Kainuma et al., 2004).

20.5.10 Model for Water–Global Assessment and Prognosis

The comprehensive assessment of the freshwater resources of the world was the first global-scale assessment of water resources (Kjellén and Mcgranahan, 1997). However, the assessment had many limitations. For instance, it lacked a global modeling approach for water availability and use; the smallest spatial units for which water stress indicators could be computed were whole countries; the important impact of climate variability on water stress could not be taken into account, and the impact of climate change on future water availability and irrigation requirements could not be assessed. To overcome these limitations and to achieve an improved assessment of the present and future water resources situation, a model for Water–Global Assessment and Prognosis (WaterGAP) was developed (Doell et al., 1999). It simulated the impact of demographic, socio-economic, and technological change on water use as well as the impact of climate change and variability on water availability and irrigation water requirements on a spatial resolution of $0.5° \times 0.5°$. It consists of the Global Water Use Model and the Global Hydrology Model. The Global Water Use Model comprises submodels for each of the water use sectors such as irrigation, livestock, households, and industry. Irrigation water requirements are modeled as a function of the cell-specific irrigated area, crop and climate, and livestock water use is calculated by multiplying livestock numbers by livestock-specific water use.

Household and industrial water use in grid cells are computed by downscaling published country values based on population density, urban population, and access to safe drinking water (Doell et al., 2003). The Global Hydrology Model is used to compute time series of monthly runoff and river discharge on a spatial resolution of $0.5° \times 0.5°$. Runoff is determined by calculating the daily water balances of soil and canopies as well as lakes, wetlands, and large reservoirs. The storage capacity of groundwater, lakes, wetlands, rivers, and routes rivers discharge through basins are taken into account according to a global drainage direction map. WaterGAP can be used in global assessments of water security, food security, and freshwater ecosystems. Time-variable gravity data of the Gravity Recovery and Climate Experiment (GRACE) satellite mission were incorporated directly into the tuning process of the Global Hydrological Model to improve simulations of the continental water cycle. By mapping time variations of the earth's gravitational field with GRACE since its launch in 2002, a global data set of mass variations close to the earth's surface has become available. After the removal of mass variations due to tidal and nontidal atmospheric and oceanic transport processes, the time-variable gravity data mainly represents water mass variations in continental hydrology, that is, the total water storage change on the continents (Werth et al., 2009).

20.5.11 Global Unified Metamodel of the Biosphere

A global model is not meant to predict every possible aspect of the world, and does not support either a purely optimistic or pessimistic view of the future. It represents a mathematical expression of the interrelationships between global concerns such as population, industrial output, natural resources, and pollution. Global modelers investigate what might happen if policies continue along present lines, or if specific changes are instituted. With this assumption in mind, Boumans et al. (2002) described the framework of a Global Unified Metamodel of the Biosphere (GUMBO). A metamodel is interpreted as a regression model of an actual simulation model (Kleijnen, 1987). GUMBO consists of five distinct modules: atmosphere, lithosphere, hydrosphere, biosphere, and anthroposphere. These are further divided into 11 biomes: Open Ocean, coastal ocean, forest, grassland, wetland, lakes/rivers, desert, tundra, ice/rock, cropland, and urban areas. The module of atmosphere facilitates exchanges of carbon, water, and nutrients across biomes and accounts for global energy balances; atmospheric dynamics are calibrated against atmospheric carbon and global temperatures. The module of lithosphere represents the solid uppermost shell of the earth, which includes soils and deposited sediments; lithosphere stocks are represented by silicate rocks, carbon reserves, and ore and fossil fuel deposits in rock and soil. The module of hydrosphere accounts for biome-specific stocks of water, carbon, and nutrients in surface and subsurface water bodies; surface water exchanges between continental and oceanic biomes are calculated to compensate for uneven distributions among biome-specific evapotranspiration. The module of biosphere defines biosphere as a self-regulating system sustained by large-scale cycles of energy and materials such as carbon, oxygen, nitrogen, certain minerals, and water; many of the ecosystem services provided by the biosphere are associated with the rate of photosynthesis; important factors for achieving optimum productivity are the nutrient availability in the lithosphere, temperature, light levels, carbon pressure in the atmosphere, soil moisture in the hydrosphere, and waste levels generated within the anthroposphere. The module of anthroposphere represents human social and economic systems; the anthroposphere harvests large amounts of material and energy from the larger system and discards waste at each phase along a production chain. The enormous number of variables in the GUMBO model makes it impractical to use analytical techniques to solve an optimum. In addition, with so many variables, it is virtually certain that there are many local optima for the model, and seeking a global optimum would probably be both futile and pointless. Therefore, GUMBO calculates the shadow prices for each of these optima and presents prices as a range.

20.5.12 Grid-Enabled Integrated Earth System Modeling

The long-term response of the carbon cycle has previously been assessed with simple box models, but these had lacked spatial dimensions, a hydrological cycle, and other climate variables. To address this problem, a Grid Enabled Integrated Earth system model, termed GENIE, was developed (Lenton et al., 2006, 2007). GENIE was composed of a 3-D frictional geostrophic ocean model, a model for energy moisture balance atmosphere, a dynamic and thermo-dynamic sea-ice model, a new land surface physics and terrestrial carbon model, and an ocean biogeochemical model. GENIE was designed to enable paleoclimate simulations, long-term future projections, large ensemble sensitivity studies, and coupling in an integrated assessment system.

The frictional geostrophic ocean model was 3-D, had eight depth levels, and shared a latitude–longitude horizontal grid of 36×36 equal area grid cells (about $5° \times 10°$) with other components of the coupled model. It included realistic bathymetry, an isoneutral and eddy-induced mixing scheme and spatially varying drag.

The atmospheric energy moisture balance model was based on an atmospheric model incorporating energy and moisture balance equations (Fanning and Weaver, 1996). Given the SST and specified surface wind field, the atmospheric model calculated the surface fields of air temperature, specific humidity, and heat and freshwater fluxes. The main changes of the model incorporated were the inclusion of seasonality, the separation of surface and atmosphere/cloud albedo, the inclusion of a relaxation time for precipitation and revision of the wind fields.

The new land surface physics and terrestrial carbon model was designed as a minimal representation of land surface physics, hydrology, and carbon cycling for use in a spatially resolved context. It had single pools of vegetation carbon, soil carbon, and soil water at each land grid point.

The ocean biogeochemical model was based on a single nutrient (PO_4) representation of ocean biogeochemical cycling. Phosphate concentrations in each surface ocean cell were restored with a characteristic time constant toward a prescribed target concentration. The excess PO_4 removed was converted into organic matter, while any deficit of model PO_4 compared to the target was left unchanged. A double exponential formulation was used to partition down through the water column the solute release arising from the remineralization of particulate matter. A double exponential formulation was also used to remineralize $CaCO_3$ in the water column. A uniform air–sea CO_2 transfer coefficient was employed to exchange CO_2 with a well-mixed atmosphere (Cameron et al., 2005).

20.6 Simulation Systems

20.6.1 Earth Simulator

In 1997, an Earth simulator was developed to simulate global warming. The primary function was to create a huge-scale supercomputer, which would achieve at least 5 teraflop/s for an atmospheric GCM. It first conducted the simulations of global atmospheric circulation and oceanic circulation on a horizontal resolution of 10 km. According to Descartes' paradigm, nature loves equilibrium and stability; natural system conditions seldom move away from the equilibrium point or the point of stability. However, since nature is an open system through which information, energy, material, and flow are always passing in and out, it would appear that nature actually loves nonequilibrium, instability and therefore, nonlinearity. The earth simulator provides an important tool necessary for creating a new paradigm of nonequilibrium/nonlinearity/openness (Sato, 2004).

20.6.2 Digital Earth System

Digital Earth was defined as a system providing access to what is known about the planet and its inhabitants' activities—currently and for any period in history—via responses to queries and exploratory tools (Gore, 1998). During the period 1998 to 2001, the Digital Earth Initiative, coordinated by the NASA-led Interagency Digital Earth Working Group (IDEW), tried to realize the integration of existing data from multiple sources in order to improve the application of geospatial data for visualization, decision support, and analysis (IDEW, 2001). During this period, the results of the 3-year IDEW effort included the Web Mapping Service (WMS) standard, a Digital Earth Alpha Version prototype, and a Digital Earth Reference Model (DERM). The Geospatial Interoperability Reference Model was produced as an extension of the Digital Earth Initiative. Since 1998, many advances have been made. For instance, Earthviewer and GeoPlayer appeared in 2001, while World Wind and Google Earth were released in 2003 and 2005, respectively.

Grossner et al. (2008) proposed the concept of a digital earth system and defined it as a comprehensive, mass distributed geographic information and knowledge organization system. The digital earth system was distributed because: (1) there were necessarily multiple, geographically dispersed data stores providing content; and (2) the processing load of server-based query and analytical processes needed to be shared for performance reasons. Geographic information refers to very broad information about well-defined locations on the Earth's surface.

20.6.3 Planet Simulator

The Planet Simulator was a Model of Intermediate Complexity (MIC) that could be used to run climate and paleoclimate simulations for timescales up to ten thousand years or more in an acceptable real time (Fraedrich et al., 2005a, b). The Planet Simulator was built to support numerical experiments for understanding the dynamics of the climate of the earth and earth-like planets.

The model was composed of a dynamical core, parameterizations and subsystems, and a graphical user interface. The dynamical core of the Planet Simulator was based on the moist primitive equations representing the conservation of momentum, mass, and energy. The dimensionless set of equations consisted of the prognostic equations for the vertical component of vorticity and horizontal divergence, the first law of thermodynamics, the equation of state with hydrostatic approximation, the continuity equation, and the prognostic equation for water vapor. The parameterizations and subsystems included boundary layers and diffusions, radiation, moist processes, clouds and dry convection, land surface and soil, and ocean and sea ice.

The Planet Simulator was configured and set up by the module graphical user interface, Model Starter. Model Starter was the fastest way to get the model running, allowing access to the most important parameters of the model preset to the most frequently used values. The graphical user interface for running the Planet Simulator had two main purposes. The first was to display model arrays in suitable representations; the second was to facilitate the interaction part of the graphical user interface, allowing the user to change selected model variables during the model run.

20.6.4 Global Earth Observation System of Systems

The GEOSS is currently being built by the Group on Earth Observations (GEO) on the basis of a 10-year implementation plan running from 2005 to 2015. The main purpose of GEOSS is to improve the monitoring of the state of the earth, increase the understanding of the earth's processes, and enhance the prediction of the behavior of earth systems. The focus of GEOSS is to produce societal benefits through more coordinated observations and better data sharing. GEOSS will provide an important scientific basis for sound policy and decision making in energy, public health, agriculture, transportation, and numerous other areas (Lautenbacher, 2006). It will contribute to reducing loss of life and property from natural and human-induced disasters, increasing our understanding of environmental factors affecting human health and well-being, improving the management of energy resources, assessing and adapting to climate variability and climate change, improving water resource management through a better understanding of the water cycle, improving weather forecasting and weather warnings, improving the management and protection of ecosystems, supporting sustainable agriculture, combating desertification, and helping to monitor and conserve biodiversity (Christian, 2005; MacPhail, 2009).

20.7 Discussion and Conclusions

Ecology has expanded from its traditional focus on organisms to include studies of the earth as an integrated ecosystem (Schlesinger, 2006). Global ecology is recording the basic parameters of our planet, such as its NPP, biogeochemical cycling, and anthropogenic effects by means of satellite technologies and simulation models. It has been widely accepted that earth surface modeling is a powerful tool for analyzing long-term ecological issues (Forrester, 1982; Meadows, 1982; Iyer, 1988; Rotmans, 1990; Claudine and Alain, 2002; Sheffield et al., 2006; Larocque et al., 2006b; Solidoro et al., 2009). Owing partly to the increasing availability and speed of computers and to the rapidly expanding global database from various international cooperation programs (Boumans et al., 2002), many global models, starting in the 1970s, have been developed. Examples include the model for the NPP of the entire land surface of the earth (Whittaker, 1970), the world model of the limits to growth (Forrester, 1971; Meadow et al., 1972), the Ecopath and Ecosim models (Polovina, 1984), the TEM (Melillo et al., 1993), the DICE model (Nordhaus, 1993), the model for evaluating regional and global effects of GHG reduction policies (Manne et al., 1995), the Asia-Pacific integrated model (Matsuoka et al., 1995), the input–output network-based model of emerging ecosystems (Patten et al., 1997; Fath and Patten, 1998), the model for WaterGAP (Doell et al., 1999), the GCMs (Arakawa, 2000), the CENTURY model (Kirschbaum and Paul, 2002), the GUMBO (Boumans et al., 2002), the IMAGE (Kram and Stehfest, 2006), and the efficient numerical terrestrial scheme (Williamson et al., 2006).

However, various problems in earth surface modeling and simulation systems remain. For example, no global models have yet achieved a satisfactory level of dynamic integration between the biophysical earth system and the human socio-economic system. The global digital terrain model with high accuracy has not been completed or combined into related global models, and statistical transfer functions still need to be developed. Spatial relationships are discernible only on a particular scale of reference (Albrecht and Car, 1999). These scale-dependent spatial analyses thus become studies of scale. But current GIS does not allow spatial simulation on multiscales. Although there is an increasing need in earth surface modeling for 3D-GIS, current GISs are mainly designed for 2D data processing and management. An alternative means of solving these problems is to develop high-accuracy and high-speed simulation systems, which could deal with huge data and multiscale issues in three dimensions, under consideration of a ground- and satellite-based global observation system with an optimal data-sharing mechanism.

References

Accad, A. and Neil, D.T. 2006. Modelling pre-clearing vegetation distribution using GIS-integrated statistical, ecological and data models: A case study from the wet tropics of Northeastern Australia. *Ecological Modelling* 198: 85–100.

Albrecht, J. and Car, A. 1999. GIS analysis for scale-sensitive environmental modeling based on hierarchy theory. In *GIS for Earth Surface Systems*, eds., R. Dikau and H. Saurer, 1–24. Berlin: Gebrueder Borntraeger Verlagsbuchhandlung.

Alcamo, J. 1994. *IMAGE 2.0: Integrated Modeling of Global Change*. Dordrecht: Kluwer Academic Publishers.

Alcamo, J., Leemans, R., and Kreileman, E. 1998. *Global Change Scenarios of the 21st Century—Results from the Image 2.1 Model*. Oxford: Elsevier Science.

Al-Garni, A.M. 1995. Mathematical predictive models for population estimation in urban areas using space products and GIS technology. *Mathematical and Computer Modelling* 22(1): 95–107.

Anway, J.C., Brittain, E.G., Hunt, H.W., Innis, G.S., Parton, W.J., Rodell, C.F., and Sauer, R.H. 1972. *ELM: Version 1.0. U.S. IBP Grassland Biome Tech. Rep. No. 156*. Fort Collins: Colorado State Univesity.

Arakawa, A. 1969. Parameterization of cumulus clouds. In *Proceedings of the WMO/IUGG Symposium on Numerical Weather Predicti*on, Tokyo, 1968, pp. IV-8–1-IV-8–6.

Arakawa, A. 2000. A personal perspective on the early years of general circulation modeling at UCLA. In *General Circulation Model Development*, ed., D.A. Randall, 1–65. New York, NY: Academic Press.

Arakawa, A. and Lamb, V.R. 1981. A potential enstrophy and energy conserving scheme for the shallow water equations. *Monthly Weather Review* 109: 18–36.

Arakawa, A. and Suarez, M.J. 1983. Vertical differencing of the primitive equations in sigma-coordinates. *Monthly Weather Review* 111: 34–45.

Ardoe, J. and Olsson, L. 2003. Assessment of soil organic carbon in semi-arid Sudan using GIS and the CENTURY model. *Journal of Arid Environments* 54: 633–651.

Aspinall, R. and Matthews, K. 1994. Climate change impact on distribution and abundance of wildlife species: An analytical approach using GIS. *Environmental Pollution* 86: 217–223.

Beverton, R.J.H. and Holt, S.J. 1957. *On the Dynamics of Exploited Fish Populations*. London: Her Majesty's Stationery Office.

Bjerknes, V. 1904. Das Problem der Wettervorsage, betrachtet vom Standpunkte der Mechanik und der Physik. *Meteorogisch Zeitung* 21: 1–7.

Bledsoe, L.J., Francis, R.C., Swartzman, G.L., and Gustafson, J.D. 1971. *PWNEE: A Grassland Ecosystem Model. U.S. IBP Grassland Biome Tech. Rep. No. 64*. Fort Collins: Colorado State University.

Bolin, B. 2007. *A History of the Science and Politics of Climate Change*. New York, NY: Cambridge University Press.

Botkin, D.B. 1982. Can there be a theory of global ecology? *Journal of Theoretical Biology* 96: 95–98.

Botkin, D.B., Janak, J.F., and Wallis, J.R. 1972. Some ecological consequences of a computer model of forest growth. *Journal of Ecology* 60: 849–872.

Boumans, R., Costanza, R., Farley, J., Wilson, M.A., Portel, R., Rotmans, J., Villa, F., and Grasso, M. 2002. Modeling the dynamics of the integrated earth system and the value of global ecosystem services using the GUMBO model. *Ecological Economics* 41: 529–560.

Caloz, R. and Collet, C. 1997. Geographic information systems (GIS) remote sensing in aquatic botany: Methodological aspects. *Aquatic Botany* 58: 209–228.

Cameron, D.R., Lenton, T.M., Ridgwell, A.J., Shepherd, J.G., Marsh, R., and Yool, A. 2005. A factorial analysis of the marine carbon cycle and ocean circulation controls on atmospheric CO_2. *Global Biogeochemical Cycles* 19: GB4027.

Chapman, S. 1957. Annals of the International geophysical year, Volume 1: the histories of the International polar years and the inception and development of the International Geophysical Year. London: Pergamon Press.

Charney, J.G. 1947. The dynamics of long waves in a baroclinic westerly current. *Journal of Meteorology* 4: 135–162.

Charney, J.G. and Eliassen, A. 1949. A numerical method for predicting the perturbations of the middle latitude westerlies. *Tellus* 1: 38–54.

Charney, J.G., Fjørtoft, R., and von Neumann, J. 1950. Numerical integration of the barotropic vorticity equation. *Tellus* 2: 237–254.

Charney, J.G. and Phillips, N.A. 1953. Numerical integration of the quasi-geostrophic equations for barotropic and simple baroclinic flows. *Journal of Meteorology* 10: 71–99.

Cheng, M.D. and Arakawa, A. 1997. Inclusion of rainwater budget and convective downdrafts in the Arakawa–Schubert cumulus parameterization. *Journal of the Atmospheric Sciences* 54: 1359–1378.

Christian, E. 2005. Planning for the Global Earth Observation System of Systems (GEOSS). *Space Policy* 21(2): 105–109.

Christensen, V. and Walters, C.J. 2004. Ecopath with Ecosim: Methods, capabilities, and limitations. *Ecological Modelling* 172: 109–139.

Claudine, S.L. and Alain, P. 2002. Environment: Modeling and models to understand, to manage, and to decide in an interdisciplinary context. *Natures Sciences Societes* 10(2): 5–26 (in French).

Costanza, R., Sklar, F.H., and White, M.L. 1990. Modelling coastal landscape dynamics. *BioScience* 40(2): 91–107.

Davidson, R.S. and Clymer, A. B. 1966. The desirability and applicability of simulating ecosystems. *Annals of the New York Academy of Sciences* 128: 790–794.

De Vries, B., Bollen J., Bouwman L., Den Elzen M., Janssen M., and Kreileman E. 2000. Greenhouse gas emissions in an equity-environment and service-oriented world: An IMAGE-based scenario for the 21st century. *Technological Forecasting and Social Change* 63: 137–174.

Doell, P., Kaspar, F., and Alcamo, J. 1999. Computation of global water availability and water use at the scale of large drainage basins. *Mathematische Geologie* 4: 111–118.

Doell, P., Kaspar, F., and Lehner, B. 2003. A global hydrological model for deriving water availability indicators: Model tuning and validation. *Journal of Hydrology* 270: 105–134.

Eickhout, B., Den Elzen, M., and Kreileman, E. 2004. The atmosphere–ocean system in IMAGE 2.2. *A Global Model Approach for Atmospheric Concentrations, and Climate and Sea Level Projections.* Bilthoven: Report 481508017, National Institute for Public Health and the Environment, the Netherlands.

Elston, D.A. and Buckland, S.T. 1993. Statistical modelling of regional GIS data: An overview. *Ecological Modelling* 67: 81–102.

Fanning, A.G. and Weaver, A.J. 1996. An atmospheric energy-moisture model: Climatology, interpentadal climate change and coupling to an ocean general circulation model. *Journal of Geophysical Research* 101: 15111–15128.

Fath, B.D. and Patten, B.C. 1998. Network synergism: Emergence of positive relations in ecological systems. *Ecological Modelling* 107: 127–143.

Fisher, R.A. 1925. *Statisitcal Methods for Research Workers.* Edinburgh: Oliver and Boyd.

Forrester, J.W. 1971. *World Dynamics.* Cambridge MA: Wright-Allen Press Inc.

Forrester, J.W. 1982. Global modelling revisited. *Futures* 14(2): 95–110.

Fraedrich, K., Jansen, H., Kirk, E., Luksch, U., and Lunkeit, F. 2005a. The Planet Simulator: Towards a user friendly model. *Meteorologische Zeitschrift* 14(3): 299–304.

Fraedrich, K., Jansen, H., Kirk, E., and Lunkeit, F. 2005b. The Planet Simulator: Green planet and desert world. *Meteorologische Zeitschrift* 14(3): 305–314.

Galton, F. 1884. *Life History Album.* London: Macmillan and Co. Ltd.

Gardner, M.R. and Ashby, W.R. 1970. Connectance of large dynamic (cybernetic) systems: Critical values for stability. *Nature* 228: 784.

Garfinkel, D. 1962. Digital computer simulation of ecological systems. *Nature* 194: 856–857.

Garfinkel, D. 1967a. A simulation study of the effect on simple ecological systems of making rate of increase of population density-dependent. *Journal of Theoretical Biology* 14: 46–58.

Garfinkel, D. 1967b. Effect on stability of Lotka–Volterra ecological systems of imposing strict territorial limits on populations. *Journal of Theoretical Biology* 14: 325–327.

Garfinkel, D. and Sack, B. 1964. Digital computer simulation of an ecological system based on a modified mass action law. *Ecology* 45: 502–507.

Gontier, M., Moertberg, U. and Balfors, B. 2010. Comparing GIS-based habitat models for applications in EIA and SEA. *Environmental Impact Assessment Review* 30: 8–18.

Gore, A. 1998. *The Digital Earth: Understanding our Planet in the 21st Century.* WWW document, http://www.isde5.org/al_gore_speech.htm.

Grant, W.E., Pedersen, E.K., and Marín, S.L. 1997. *Ecology and Natural Resource Management: Systems Analysis and Simulation.* New York, NY: Wiley.

Gret-Regamey, A., Bebi, P., Bishop, I.D. and Schmid, W.A. 2008. Linking GIS-based models to value ecosystem services in an Alpine region. *Journal of Environmental Management* 89: 197–208.

Grossner, K.E., Goodchild, M.F., and Clarke, K.C. 2008. Defining a digital Earth system. *Transactions in GIS* 12(1): 145–160.

Haldane, J.B.S. 1927. A mathematical theory of natural and artificial selection, part V: Selection and mutation. *Mathematical Proceedings of the Cambridge Philosophical Society* 23: 838–844.

Harshvardan, R.D., Randall, D.A., and Corsetti, T.G. 1987. A fast radiation parameterization for atmospheric general circulation models. *Journal of Geophysical Research* 92: 1009–1016.

Holling, C.S. 1964. The analysis of complex population processes. *The Canadian Entomologist* 96: 335–347.

ICSU-UNESCO-UNU. 2008. *Ecosystem Change and Human Well-being: Research and Monitoring Priorities Based on the Millennium Ecosystem Assessment*. Paris: International Council for Science.

IDEW. 2001. *The Big Picture: Digital Earth and the Power of Applied Geography in the 21st Century*. WWW document, http://www.digitalearth.gov/BigPicture.doc.

IPCC. 1990a. *Climate Change—The IPCC Scientific Assessment*. Cambridge: Cambridge University Press.

IPCC. 1990b. *Climate Change—The IPCC Impact Assessment*. Canberra: Australian Government Publishing Service.

IPCC. 1990c. *Climate Change—The IPCC Response Strategies*. Washington, DC: US National Science Foundation.

IPCC. 1996a. *Climate Change 1995—The Science of Climate Change*. Cambridge: Cambridge University Press.

IPCC. 1996b. *Climate Change 1995—Impacts, Adaptations and Mitigation of Climate Change*. Cambridge: Cambridge University Press.

IPCC. 1996c. *Climate Change 1995—Economic and Social Dimensions of Climate Change*. Cambridge: Cambridge University Press.

IPCC. 2001a. *Climate Change 2001—Mitigation*. Cambridge, UK: Cambridge University Press.

IPCC. 2001b. *Climate Change 2001—The Scientific Basis*. Cambridge, UK: Cambridge University Press.

IPCC. 2007a. *Climate Change 2007—The Physical Science Basis*. Cambridge University Press, Cambridge.

Israel, G. and Gasca, A.M. 2002. *The Biology of Numbers*. Berlin: Birkhaeuser Verlag.

Iyer, R.K. 1988. Information and modeling resources for decision support in global environments. *Information and Management* 14(2): 67–73.

Jaeger, J. 2003. The International Human Dimensions Programme on Global Environmental Change (IHDP). *Global Environmental Change* 13: 69–73.

Johnson, D.R., Ruzek, M., and Kalb, M. 1997. What is Earth system science? In *Proceedings of the 1997 International Geoscience and Remote Sensing Symposium*, Singapore, 1997, pp. 688–691.

Johnson, D.R., Ruzek, M., and Kalb, M. 2000. Earth system science and the Internet. *Computers and Geosciences* 26: 669–676.

Jørgensen, S.E. 1975a. A world model of growth in production and population. *Ecological Modelling* 1: 199–203.

Jørgensen, S.E. 1975b. About "Ecological Modelling." *Ecological Modelling* 1: 1–2.

Jørgensen, S.E. 1986. *Fundamentals of Ecological Modelling*. Amsterdam: Elsevier.

Jørgensen, S.E. 2002. *Integration of Ecosystem Theories: A Pattern*. Dordrecht: Kluwer Academic Publishers.

Jørgensen, S.E., Chon, T.S., and Recknagel, F.A. 2009. *Handbook of Ecological Modelling and Informatics*. Southampton, UK: WIT Press.

Jørgensen, S.E. and Fath, B. 2008. *Encyclopedia of Ecology*. England: Elsevier Limited.

Jørgensen, S.E. and Mejer, H. 1979. A holistic approach to ecological modelling. *Ecological Modelling* 7: 169–189.

Kainuma, M., Matsuoka, Y., Morita, T., Masui, T., and Takahashi, K. 2004. Analysis of global warming stabilization scenarios: The Asian-Pacific Integrated Model. *Energy Economics* 26: 709–719.

Keller, K., Bolker, B.M., and Bradford, D.F. 2004. Uncertain climate thresholds and optimal economic growth. *Journal of Environmental Economics and Management* 48(1): 723–741.

Kienast, F., Brzeziecki, B., and Wildi, O. 1996. Long-term adaptation potential of Central European mountain forests to climate change: A GIS-assisted sensitivity assessment. *Forest Ecology and Management* 80: 133–153.

Kim, Y.J. and Arakawa, A. 1995. Improvement of orographic gravity wave parameterization using a mesoscale gravity wave model. *Journal of the Atmospheric Sciences* 52: 1875–1902.

King, R.L. and Birk, R.J. 2004. Developing Earth System Science knowledge to manage the earth's natural resources. *Computing in Science and Engineering* 6(1): 45–51.

Kirschbaum, M.U.F. and Paul, K.Y. 2002. Modelling C and N dynamics in forest soils with a modified version of the CENTURY model. *Soil Biology and Biochemistry* 34: 341–354.

Kitching, R.L. 1983. *Systems Ecology*. London: University of Queensland Press.

Kitzes, J. and Wackernagel, M. 2009. Answers to common questions in Ecological Footprint accounting. *Ecological Indicators* 9: 812–817.

Kjellén, M. and Mcgranahan, G. 1997. *Comprehensive Assessment of the Freshwater Resources of the World*. Stockholm: Stockholm Environment Institute.

Kleijnen, J.P.C. 1987. *Statistical Tools for Simulation Practitioners*. New York, NY: Marcel Dekker, Inc.

Koehler, M., Mechoso, C.R., and Arakawa, A. 1997. Ice cloud formulation in climate modelling. In *Proceedings of the 7th Conference on Climate Variations*, Long Beach, CA, 1997, 237–242. American Meteorological Society.

Kondratyev, K., Ortner, J., and Preining, O. 1992. Priorities for global ecology now and in the next century. *Space Policy* 8: 40–48.

Kram, T. and Stehfest, E. 2006. The IMAGE model: History, current status and prospects. In *Integrated Modelling of Global Environmental Change, An overview of IMAGE 2.4*, eds., A.F. Bouwman, T. Kram and K.K. Goldewijk, 7–24. Bilthoven: Netherlands Environmental Assessment Agency (MNP).

Kypreos, S. 2007. A MERGE model with endogenous technological change and the cost of carbon stabilization. *Energy Policy* 35: 5327–5336.

Lake, I.R., Lovett, A.A., Hiscock, K.M., Betson, M., Foley, A., Suennenberg, G., Evers, S., and Fletcher, S. 2003. Evaluating factors influencing groundwater vulnerability to nitrate pollution: Developing the potential of GIS. *Journal of Environmental Management* 68: 315–328.

Larocque, G.R., Archambault, L., and Delisle, C. 2006a. Modelling forest succession in two southeastern Canadian mixedwood ecosystem types using the ZELIG model. *Ecological Modelling* 199: 350–362.

Larocque, G.R., Mauriello, D.A., Park, R.A., and Rykiel Jr., E.J. 2006b. Ecological models as decision tools in the 21st century. *Proceedings of a Conference Organized by the International Society for Ecological Modelling (ISEM)*, Quebec, Canada, August 22–24, 2004. *Ecological Modelling* 199(3): 217–376.

Larocque, G.R., Bhatti, J.S., Boutin, R., and Chertov, O. 2008. Uncertainty analysis in carbon cycle models of forest ecosystems: Research needs and development of a theoretical framework to estimate error propagation. *Ecological Modelling* 219: 400–412.

Lautenbacher, C.C. 2006. The Global earth observation system of systems: Science serving society. *Space Policy* 22(1): 8–11.

Lempert, R.J., Sanstad, A.H. and Schlesinger, M.E. 2006. Multiple equilibrium in a stochastic implementation of DICE with abrupt climate change. *Energy Economics* 28: 677–689.

Lenton, T.M., Marsh, R., Price, A.R., Lunt, D.J., Aksenov, Y., Annan, J.D., Cooper-Chadwick T. et al., 2007. Effects of atmospheric dynamics and ocean resolution on bi-stability of the thermohaline circulation examined using the Grid Enabled Integrated Earth system modeling (GENIE) framework. *Climate Dynamics* 29: 591–613.

Lenton, T.M., Williamson, M.S., Edwards, N.R., Marsh, R., Price, A.R., Ridgwell, A.J., Shepherd, J.G., and Cox, S.J. 2006. Millennial timescale carbon cycle and climate change in an efficient Earth system model. *Climate Dynamics* 26: 687–711.

Leslie, P.H. 1945. On the use of matrices in certain population mathematics. *Biometrika* 33: 183–212.

Leslie, P.H. 1948. Some further notes on the use of matrices in population mathematics. *Biometrika* 35: 213–245.

Li, J.L.F., Mechoso, C.R., and Arakawa, A. 1999. Improved PBL moist processes with the UCLA GCM. In *Proceedings of the 10th Symposium on Global Change Studies*, Dallas, Texas, January 10–15, 1999, 423–426. American Meteorological Society.

Lioubimtseva, E. and Defourny, P. 1999. GIS-based landscape classification and mapping of European Russia. *Landscape and Urban Planning* 44: 63–75.

Loevbrand, E., Stripple, J., and Wiman, B. 2009. Earth system governmentality reflections on science in the Anthropocene. *Global Environmental Change* 19(1): 7–13.

Ludeke, A.K., Maggio, R.C., and Reid, L.M. 1990. An analysis of anthropogenic deforestation using logistic regression and GIS. *Journal of Environmental Management* 31: 247–259.

MacPhail, D. 2009. Increasing the use of earth observations in developing countries. *Space Policy* 25(1): 6–8.

Manabe, S. and Strickler, R.F. 1964. Thermal equilibrium of the atmosphere with a convective adjustment. *Journal of the Atmospheric Sciences* 21: 361–385.

Manne, A., Mendelsohn, R., and Richels, R. 1995. MERGE, a model for evaluating regional and global effects of GHG reduction policies. *Energy Policy* 23: 17–34.

Matsuoka, Y., Kainuma, M., and Morita, T. 1995. Scenario analysis of global warming using the Asian Pacific Integrated Model (AIM). *Energy Policy* 23(4/5): 357–371.

May, R.M. 1972. Will a large complex system be stable? *Nature* 238: 413–414.

McGuire, A.D., Joyce, L.A., Kicklighter, D.W., Melillo, J.M., Esser, G., and Vorosmarty, C.J. 1993. Productivity response of climax temperate forest to elevated temperature and carbon dioxide: A North American comparison between two global models. *Climate Change* 24: 287–310.

McGuire, A.D., Melillo, J.M., Joyce, L.A., Kicklighter, D.W., Grace, A.L., Moore III, B., and Vörösmarty, C.J. 1992. Interactions between carbon and nitrogen dynamics in estimating net primary productivity for potential vegetation in North America. *Global Biogeochemical Cycles* 6: 101–124.

McGuire, A.D., Melillo, J.M., Kicklighter, D.W., Pan, Y., Xiao, X., Helfrich, J., Moore III, B., Vörösmarty, C.J., and Schloss, A.L. 1997. Equilibrium response of global primary production and carbon storage to doubled atmospheric carbon dioxide: Sensitivity to changes in vegetation nitrogen concentration. *Global Biogeochemical Cycles* 11: 173–189.

McKinney, D.C. and Cai, X.M. 2002. Linking GIS and water resources management models: An object-oriented method. *Environmental Modelling and Software* 17: 413–425.

McNeil, B.E., Martell, R.E., and Read, J.M. 2006. GIS and biogeochemical models for examining the legacy of forest disturbance in the Adirondack Park, NY, USA. *Ecological Modelling* 195: 281–295.

Meadows, D.H. 1982. Lessons from global modelling and modelers. *Futures* 14(2): 111–121.

Meadows, D.H., Meadows, D.L., Randers, J., and Behrens III, W.W. 1972. *The Limits to Growth*. New York: Universe Books.

Meadows, D.H., Meadows, D.L., and Randers, J. 1992. *Beyond the Limits*. Post Mills, VT: Chelsea Green Publishing Company.

Meadows, D.H., Randers, J., and Meadows, D.L. 2004. *Limits to Growth: The 30-year Update*. White River Junction, VT: Chelsea Green Publishing Company.

Melillo, J.M., McGuire, A.D., Kicklighter, D.W., Moore III, B., Vörösmarty, C.J., and Schloss, A.L. 1993. Global climate change and terrestrial net primary production. *Nature* 363: 234–240.

Millennium Ecosystem Assessment (MA). 2003. *Ecosystems and Human Well-Being*. Washington, DC: Island Press.

Mori, S. 1995. A long term evaluation of nuclear power technology by extended DICE + e model simulations–multiregional approach for resource and industry allocation (MARIA) model. *Progress in Nuclear Energy* 29(Suppl.): 135–142.

Mori, S. and Saito, T. 2004. Potentials of hydrogen and nuclear towards global warming mitigation—Expansion of an integrated assessment model MARIA and simulations. *Energy Economics* 26: 565–578.

Nakicenovic, N., Alcamo, J., Davis, G., De Vries, B., Fenhann, J., Gaffin, S., Gregory, K. et al., 2000. *Special Report on Emissions Scenarios: IPCC Special Reports*. Cambridge: Cambridge University Press.

Neuhold, J.M. 1975. Introduction to modeling in the Biome. In *Systems Analysis and Simulation in Ecology*, ed., Patten, B.C., 7–12. New York, NY: Academic Press.

Nordhaus, W.D. 1993. Rolling the "DICE": An optimal transition path for controlling greenhouse gases. *Resource and Energy Economics* 15(1): 27–50.

Odishaw, H. 1958a. International Geophysical Year. *Science* 128: 1599–1609.

Odishaw, H. 1958b. International geophysical year: A report on the United States program. *Science* 127: 115–128.

Odum, H.T. 1960. Ecological potential and analogue circuits for the ecosystem. *American Sciences* 48: 1–8.

Odum, H.T. 1983. *Systems Ecology: An Introduction*. New Jersey: Wiley-Interscience.

Odum H.T. 1988. Self-organization, transformity and information. *Science* 242: 1132–1139.

Odum, H.T. 1994. *Ecological and General Systems: An Introduction to Systems Ecology*. Boulder, CO: University Press of Colorado.

Olson, J.S. 1963. Analog computer models for movement of nuclides through ecosystems. In *Proceedings of First National Symposium on Radioecology, Colorado State University*, September 10–15, 1961, eds., V. Schultz and Jr. A.W. Klement, 121–125. New York, NY: Reinhold Pub. Corp.

Park, R.A., Mankin, J.B., Hoopes, J.A., Peterson, J.L., O'Neill, R.V., Bloomfield, J.A., Shugart, H.H. Jr. et al., 1974. A generalized model for simulating lake ecosystems. *Simulation* 21: 33–50.

Parton, W.J. and Rasmussen, P.E. 1994. Long-term effects of crop management in wheat-fallow: II. CENTURY model simulations. *Soil Science Society of America Journal* 58: 530–536.

Parton, W.J., Schimel, D.S., Cole, C.V. and Ojima, D.S. 1987. Division S-3–soil microbiology and biochemistry. *Soil Science Society of America Journal* 51: 1173–1179.

Parton, W.J., Scurlock, J.M.O., Ojima, D.S., Gilmanov, T.G., Scholes, R.J., Schimel, D.S., Kirchner, T. et al., 1993. Observations and modeling of biomass and soil organic matter dynamics for the grassland biome worldwide. *Global Biogeochemical Cycles* 7(4): 785–809.

Patten, B.C. 1972. A simulation of the shortgrass prairie ecosystem. *Simulation* 19: 177–186.

Patten, B.C. 1982. Environs: Relativistic elementary particles for ecology. *The American Naturalist* 119: 179–219.

Patten, B.C. 1985. Energy cycling in the ecosystem. *Ecological Modelling* 28: 1–71.

Patten, B.C. 1992. Energy, emergy and environs. *Ecological Modelling* 62: 29–69.

Patten, B.C., Straskraba, M. and Jørgensen, S.E. 1997. Ecosystems emerging: 1. Conservation. *Ecological Modelling* 96: 221–284.

Paulik, G.J. and Greenough, Jr., J.W. 1966. Management analysis for a salmon resource system. In *Systems Analysis in Ecology*, ed., K.E.F. Watt, 215–252. New York, NY: Academic Press.

Pearl, R. and Reed, L.J. 1920. On the rate of growth of the population of the United States since 1790 and its mathematical representation. *Proceedings of the National Academy of Sciences* 6: 275–288.

Phillips, N.A. 1956. The general circulation of the atmosphere: A numerical experiment. *Quarterly Journal of the Royal Meteorological Society* 82: 123–164.

Phillips, N.A. 1959. An example of nonlinear computational instability. In *The Atmosphere and the Sea in Motion: Scientific Contributions to the Rossby Memorial Volume*, ed., B. Bolin, 501–504. New York, NY: Rockefeller Institute Press.

Polovina, J.J. 1984. Model of a coral reef ecosystems. I. The ECOPATH model and its application to French Frigate Shoals. *Coral Reefs* 3: 1–11.

Popp, D. 2004. ENTICE: Endogenous technological change in the DICE model of global warming. *Journal of Environmental Economics and Management* 48: 742–768.

Raich, J.W., Rastetter, E.B., Melillo, J.M., Kicklighter, D.W., Steudler, P.A., Peterson, B.J., Grace, A.L., Moore III, B., and Vörösmarty, C.J. 1991. Potential net primary productivity in South America: Application of a global model. *Ecological Applications* 1(4): 399–429.

Richardson, L.F. 1922. *Weather Prediction by Numerical Processes*. Cambridge, MA: Cambridge University Press.

Rosenzweig, M.L. 1971. Paradox of enrichment: Destabilization of exploitation ecosystems in ecological time. *Science* 171: 385–387.

Ross, R. 1916. An application of the theory of probabilities to the study of *a priori* pathometry. *Proceedings of the Royal Society of London* A 92: 204–230.

Rossy, C.G. 1939. Relation between the intensity of the zonal circulation of the atmosphere and displacement of the semipermanent centers of action. *Journal of Marine Research* 2: 38–55.

Rotmans, J. 1990. *IMAGE: An Integrated Model to Assess the Greenhouse Effect*. London: Kluwer Academic Publishers.

Runarsson, T.P. and Yao, X. 2000. Stochastic ranking for constrained evolutionary optimization. *IEEE Transactions on Evolutionary Computation* 4(3): 284–294.

Sato, T. 2004. The Earth Simulator: Roles and impacts. *Parallel Computing* 30: 1279–1286.

Schlesinger, W.H. 2006. Global change ecology. *Trends in Ecology and Evolution* 21(6): 348–350.

Schlesinger, M.E., Malyshev, S., Rozanov, E.V., Yang, F., Andronova, N.G., De Vries, B., Gruebler, A. et al., 2000. Geographical distributions of temperature change for scenarios of greenhouse gas and sulphur dioxide emissions. *Technological Forecasting and Social Change* 65: 167–193.

Schlesinger, M.E. and Mintz, Y. 1979. Numerical simulation of ozone production, transport and distribution with a global atmospheric general circulation model. *Journal of the Atmospheric Sciences* 36: 1325–1361.

Sheffield, J., Goteti, G., and Wood, E.F. 2006. Development of a 50-year high-resolution global dataset of meteorological forcings for land surface modeling. *Journal of Climate* 19: 3088–3111.

Shugart, H.H. and O'Neill, R.V. 1979. *Systems Ecology*. Stroudsgurg: Dowden, Hutchinson & Ross, Inc.

Skellam, J.G. 1951. Random dispersal in theoretical populations. *Biometrika* 38: 196–218.

Sklar, F.H., Costanza, R., and Day, J.W., Jr. 1985. Dynamic spatial simulation modeling of coastal wetland habitat succession. *Ecological Modelling* 29: 261–281.

Smith, F.E. 1952. Experimental methods in population dynamics: A critique. *Ecology* 33: 441–450.

Solidoro, C., Bandelj, V., Cossarini, G., Libralato, S., and Canu, D.M. 2009. Challenges for ecological modelling in a changing world: Global changes, sustainability and ecosystem based management. *Ecological Modelling* 220: 2825–2827.

Southwick, C.H. 1996. *Global Ecology in Human Perspective*. New York, NY: Oxford University Press.

Special Committee for the IGBP. 1988. *Report No. 4 of IGBP: A Study of Global Change (IGBP)*. Stockholm, Sweden: The International Geosphere–Biosphere Programme.

Sullivan, W. 1961. *Assault on the Unknown*. New York, NY: McGraw-Hill Book Company, Inc.

Swannack, T.M. and Grant, W.E. 2008. Systems ecology. In *Encyclopedia of Ecology*, eds., S.E. Jøgensen and B.D. Fath, 3477–3481. Oxford: Elsevier.

Tian, H., Melillo, J.M., Kicklighter, D.W., McGuire, A.D., and Helfrich, J. 1999. The sensitivity of terrestrial carbon storage to historical climate variability and atmospheric CO_2 in the United States. *Tellus* 51B: 414–452.

Tian, H., Melillo, J.M., Kicklighter, D.W., McGuire, A.D., Helfrich III, J., Moore III, B., and Vörösmarty, C.J. 2000. Climatic and biotic controls on annual carbon storage in Amazonian ecosystems. *Global Ecology and Biogeography* 9: 315–335.

Tian, H., Melillo, J.M., Kicklighter, D.W., Pan, S.F., Liu, J.Y., McGuire, A.D., and Moore III, B. 2003. Regional carbon dynamics in monsoon Asia and its implications for the global carbon cycle. *Global and Planetary Change* 37: 201–217.

Troy, A. and Wilson, M.A. 2006. Mapping ecosystem services: Practical challenges and opportunities in linking GIS and value transfer. *Ecological Economics* 60: 435–449.

UNESCO. 1972. *International Co-ordinating Council of the Programme on Man and the Biosphere (MAB)*. Paris: SC/MD/26.

Umeki, K., Lim, E.M., and Honjo, T. 2008. A GIS-based simulation program to predict multi-species size-structure dynamics for natural forests in Hokkaido, northern Japan. *Ecological Informatics* 3: 218–227.

Van Dyne, G.M. 1966. *Ecosytsems, Systems Ecology, and Systems Ecologists*. ORNL-3957, Oak Ridge, TN: Oak Ridge National Laboratory.

Van Vuuren, D.P., Den Elzen, M., Lucas, P., Eickhout, B., Strengers, B., Van Ruijven, B., Wonink, S., and Van Houdt, R. 2007. Stabilizing greenhouse gas concentrations at low levels: An assessment of reduction strategies and costs. *Climatic Change* 81(2): 119–159.

Volterra, V. 1926. Fluctuations in the abundance of a species considered mathematically. *Nature* 118: 558–560.

Walsh, S.J., Butler, D.R., and Malanson, G.P. 1998. An overview of scale, pattern, process relationships in geomorphology: A remote sensing and GIS perspective. *Geomorphology* 21: 183–205.

Walters, C., Christensen, V., and Pauly, D. 1997. Structuring dynamic models of exploited ecosystems from trophic mass-balance assessments. *Reviews in Fish Biology and Fisheries* 7: 139–172.

Wang, J., Chen, J.M., Ju, W.M., and Li, M.C. 2010. IA-SDSS: A GIS-based land-use decision support system with consideration of carbon sequestration. *Environmental Modelling and Software* 25: 539–553.

Watt, K.E.F., 1966. The nature of system analysis. In *Systems Analysis in Ecology*, ed., K.E.F. Watt, 1–14. New York, NY: Academic Press.

Werth, S., Güntner, A., Petrovic, S. and Schmidt, R. 2009. Integration of GRACE mass variations into a global hydrological model. *Earth and Planetary Science Letters* 277: 166–173.

Whittaker, R.H. 1970. *Communities and Ecosystems*. New York, NY: MacMillan.

Williamson, M.S., Lenton, T.M., Shepherd, J.G. and Edwards, N.R. 2006. An efficient numerical terrestrial scheme (ENTS) for Earth system modeling. *Ecological Modelling* 198: 362–374.

World Commission on Environment and Development. 1987. *Our Common Future.* Oxford: Oxford University Press.

Worthington, E.B. 1975. *The Evolution of IBP*. London: Cambridge University Press.

Wright, S. 1931. Evolution in Mendelian populations. *Genetics* 16: 97–159.

Yue, T.X., Du, Z.P., Song, D.J., and Gong, Y. 2007. A new method of surface modeling and its application to DEM construction. *Geomorphology* 91(1–2): 161–172.

Yue, T.X., Fan, Z.M., and Liu, J.Y. 2005. Changes of major terrestrial ecosystems in China since 1960. *Global and Planetary Change* 48: 287–302.

Yue, T.X., Fan, Z.M., Liu, J.Y., and Wei, B.X. 2006. Scenarios of major terrestrial ecosystems in China. *Ecological Modelling* 199: 363–376.

Yue, T.X. and Song, Y.J. 2008. The YUE-HASM method. In *Accuracy in Geomatics, Proceedings of the 8th International Symposium on Spatial Accuracy Assessment in Natural Resources and Environmental Sciences*, Shanghai, 2008, eds., D.R. Li, Y. Ge and G.M. Foody, 148–153. Liverpool: World Academic Union Ltd.

Yue, T.X., Song, D.J., Du, Z.P., and Wang, W. 2010. High-accuracy surface modeling and its application to DEM generation. *International Journal of Remote Sensing* 31(8): 2205–2226.

Yue, T.X., Tian, Y.Z., Liu, J.Y., and Fan, Z.M. 2008. Surface modeling of human carrying capacity of terrestrial ecosystems in China. *Ecological Modelling* 214: 168–180.

Yue, T.X. and Wang, S.H. 2010. Adjustment computation of HASM: A high-accuracy and speed method. *International Journal of Geographical Information Sciences* (in press).

Index